HANDBOOK OF ELECTROMECHANICAL PRODUCT DESIGN

PETER L. HURRICKS

Eur. Ing., M.Sc., C.Eng., M.I.M

Advisory Editor

Steven Dimond

Computer Peripherals Bristol Division
Hewlett-Packard Limited

Longman Scientific & Technical,
Longman Group UK Limited,
Longman House, Burnt Mill, Harlow,
Essex, CM20 2JE, England
and Associated Companies throughout the world.

*Copublished in the United States with
John Wiley & Sons, Inc., 605 Third Avenue, New York, NY 10158*

© Longman Group UK Limited 1994

All rights reserved; no part of this publication may be
reproduced, stored in a retrieval system, or transmitted in any
form or by any means electronic, mechanical, photocopying,
recording, or otherwise without either the prior written permission
of the Publishers or a licence permitting restricted copying in the
United Kingdom issued by the Copyright Licensing Agency Ltd,
90 Tottenham Court Road, London, W1P 9HE.

First published 1994

British Library Cataloguing in Publication Data
A catalogue record for this book is available from the British
Library

ISBN 0-582-04083-3

Library of Congress Cataloging-in-Publication Data
  Hurricks, Peter L. (Peter Leonard). 1944–
      Electromechanical product design / Peter
  L. Hurricks : advisory
    editor, Steven Dimond.
        p.   cm.
    Includes index.
    ISBN 0-470-22152-6
    1. Electromechanical devices — Design and construction.
    2. Design.
  Industrial. I. Dimond, Steven. II. Title.
  TJ163.H87   1993
  621—dc20   93–13169
                                              CIP

Set by 4 in Compugraphic Times 9½/11

Printed in Great Britain
by Bookcraft Ltd., Bath, Avon

CONTENTS

Preface vii

Chapter 1 The design process 1

1.1 Overview 1
1.2 The importance of the design function 2
1.3 The phases of design 2
1.4 The proposal phase 4
1.5 The feasibility phase 7
1.6 The acquisition phase 12
1.7 The maturity phase 17
1.8 Engineering change control 18
References 22

Chapter 2 Design functions 24

2.1 Introduction 24
2.2 The role of the product design engineer 24
2.3 The design team 26
2.4 Design innovation and creativity 27
2.5 Product safety 31
2.6 Packaging 35
2.7 Design protection 36
2.8 The Japanese style 37
2.9 Concurrent engineering 39
References 41

Chapter 3 Product reliability and test 42

3.1 Introduction 42
3.2 The reliability concept 42
3.3 Reliability aspects of the design cycle 50
3.4 Predicting reliability 59
3.5 Evaluating reliability 65
References 74

Chapter 4 Adhesives 75

4.1 Introduction 75
4.2 The contacting process 75
4.3 Joint design 80
4.4 The setting process 86
4.5 Adhesive types 87
4.6 Adhesive selection procedure 102
References 104

Chapter 5 Dry bearings 105

5.1 Introduction 105
5.2 Wear performance factors 106
5.3 Available types and materials 120
5.4 Application factors 124
5.5 The design procedure 127
References 133

Chapter 6 Porous metal bearings 135

6.1 Introduction 135
6.2 Manufacture and properties 136
6.3 Variables affecting performance 137
6.4 Application factors 145
6.5 The design procedure 147
References 150

Chapter 7 Rolling element bearings 151

7.1 Introduction 151
7.2 Bearing construction 151
7.3 Design factors 162
7.4 Lubrication 173
7.5 Application factors 177
7.6 The design selection procedure 179
7.7 Other rolling element products 181
References 186

Chapter 8 Drive motors 187

8.1 Introduction 187
8.2 Terminology and formulae 196
8.3 a.c. induction motors 200
8.4 Induction motor speed control 207
8.5 a.c. commutator motors 212
8.6 d.c. motors 214
8.7 Brushed d.c. motor speed control 220
8.8 Brushless d.c. motors 224
8.9 Motor selection criteria 237
References 239

Chapter 9 Drives, belts and couplings 240

9.1 Introduction 240
9.2 Belts 240
9.3 Chain drives 255
9.4 Flexible couplings 258
9.5 The drive shaft 264
References 267

Chapter 10 Gears 268

10.1 Introduction 268
10.2 Gear types 268
10.3 Gear materials 271
10.4 Gear terminology 271
10.5 Gear application factors 278
10.6 Gear rating 284
10.7 Plastic gears 293
10.8 Gear transmissions 297
10.9 Gear selection process 300
References 302

Chapter 11 Clutches, solenoids and brakes 303

11.1 Introduction 303
11.2 Clutches and brakes 304
11.3 Clutch and brake application factors 305
11.4 Types of clutch and brake 309
11.5 Clutch and brake design procedure 318
11.6 Solenoids 320
11.7 Limited angle torquers 327
References 328

Chapter 12 Springs 329

12.1 Introduction 329
12.2 Spring materials 329
12.3 Mechanical aspects 329
12.4 Simple flat springs 333
12.5 Helical springs 334
12.6 Torsion springs 340
12.7 Tensator springs 342
12.8 Elastomer springs 344
12.9 Plastic springs 347
12.10 Gas springs 348
12.11 Shape memory alloy springs 350
References 352

Chapter 13 Surface engineering 353

13.1 Introduction 353
13.2 Electrochemical and electroless treatments 354
13.3 Chemical conversion coatings 367
13.4 Vapour coating processes 369
13.5 Thermal and thermochemical treatments 372
13.6 Vitreous enamelling 377
13.7 Sprayed surface coatings 379
13.8 Organic finishes 382
13.9 Zinc coatings 391
References 392

Chapter 14 Fasteners 395

14.1 Introduction 395
14.2 Threaded fasteners 395
14.3 Nuts 407
14.4 Washers 413
14.5 Theory of nut tightening 419
14.6 Retaining rings 425
14.7 Inserted fasteners 428
14.8 Self-tapping screws 433
14.9 Pins 435
14.10 Quick release fasteners 438
References 440

Chapter 15 Thermal management 441

15.1 The nature of thermal design 441
15.2 Principles of heat flow 442
15.3 Forced air flow 449
15.4 Other heat transfer devices 458
15.5 Thermal management of electronics 463
References 473

Chapter 16 Noise reduction 474

16.1 Introduction 474
16.2 Fundamentals 474

16.3 Measurement of sound 478
16.4 The generation of noise 489
16.5 Noise reduction methods 498
16.6 Practical noise diagnosis and treatments 512
16.7 Noise labelling 513
References 517

Chapter 17 Design certification 518

17.1 Introduction 518
17.2 Approaches to certification 519
17.3 Conformity in the single market 523
17.4 Safety standards 527
17.5 EC Directives 539
References 546

Index 547

PREFACE

Most engineering and design textbooks tend to arise out of Academia or Consultancy organizations. I am told that it is somewhat unusual for someone like myself, both earning a living in Industry and involved with Mechanical and Electromechanical Product design on a daily basis, to write such a technical text as this. On that basis therefore, I hope this book will provide a different style of approach which would be of wide interest to both Design Engineers in Industry and students setting out on their Engineering careers.

When planning this book, the approach taken was 'as a design engineer in the 1990s, what does he/she broadly need to know in order to drive the Product Development Process toward the Marketplace'. This book therefore attempts to fulfil that need, at least in part, by covering the areas of design activities, common component guidelines, design specified processes in the manufacturing equation, reliability, test and certification issues. It is by and large a practical text, emphasizing the techniques and methods that can be used.

There are four levels of design information provided by the handbook. Firstly the design process, the core activity in any engineering organization, together with organizational and other factors of influence. Secondly, common component design guidelines and design calculations are considered, together with comments on advantages, disadvantages and limits of usefulness. Thirdly, design guidance on those manufacturing processes that are dictated by the design, e.g. use of adhesives, application of fasteners and type of surface finish. Finally, at the outer level, encompassing the whole design, are the performance testwork and certification guidelines covering noise, reliability, packaging, safety standards and the end-product regulatory framework. The order of the Chapters is only of significance in that I have started with design acquisition and ended with design certification.

The notation used may appear to lack complete consistency from Chapter to Chapter, but it was considered desirable to use the notation already established in the literature relevant to the various topics considered. The symbols used are identified in each of the Chapters concerned. In each Chapter I have tried to concentrate on technical principles and practical guidelines, introducing the basic mathematics as necessary to enable design quantification. The mathematics has therefore been kept to the useful and practical rather than the theoretically interesting and not much else. The overall motivation therefore has been to give understanding and application of subjects integral to the design of electromechanical products. However, for those wishing to manipulate numbers rather than understand the basics of their craft, there are a number of very good computer programmes on the market suitable for some of the component design applications outlined in this book.

In reading this text, you will notice the large number of illustrations that have been provided by Industry. I would like to thank the many organizations who sent me product literature and allowed reproduction of illustrations and half-tones. Space does not allow detailed acknowledgements, but a number of individuals did go to a lot of trouble to obtain the required figures for me, and to them I am especially grateful. Whilst every effort has been made to trace the owners of copyright material, in some cases this has proved impossible and we take this opportunity to apologize to any copyright holders whose rights we may have unwittingly infringed.

I would also like to thank my advisory editor, Steve Dimond, who read through each Chapter as it was prepared and gave me very useful advice, suggestions and recommendations on presentation and technical content. I feel that the end product is much improved as a result of his help. My thanks also to my colleague Derek Wilsher for re-drawing a number of sketches for me into much more presentable formats. And finally, my thanks to my wife Susan for not only typing all the original text and spending many hours helping me with proof reading, but also providing me with encouragement during the many months this book has taken to prepare.

Peter L. Hurricks
Chichester 1993.

CHAPTER 1
THE DESIGN PROCESS

1.1 Overview

Engineering is the application of science and technology to meet human needs. The design engineer is an innovator and creator, meeting the need for new products and processes in a practical and economic fashion.

Design can mean many things to different people, but in this book we refer to electromechanical based machines and equipment in the design context. Almost everyone is involved with design in one way or another either as customer or producers, but there are no correct answers in the design process, only good designs today which may turn out to be not so good tomorrow. So what is 'good' design? As we shall see later, it is essentially the quality of the product and its creation that quantifies a design. In design, the differences between good and not so good may be very slight yet none the less absolute. There is always a worse way of making a piece part, assembly or product and it is nearly always cheaper. A 'good' design will provide a correct solution to a need both economically and practically. In the practical sense these aims need to be modified to include the following features:[1]

1. Fulfilment of functional purpose;
2. Economic in resource utilization, both in producer and in customer terms;
3. Having the desired properties;
4. Having lasting aesthetic qualities.

Satisfying a design need is the creation of a physical reality, but a design is always subject to certain problem-solving constraints, hence the solution at any point in time is invariable a compromise.

In recent years it has become clear that reduced product cost is not necessarily the only key to competitiveness, the nature and quality of the product itself are major non-price contributors. Many assessments[2] emphasize the importance of design in adding value to products as perceived by the customer, in particular technical performance and overall quality in terms of reliability, safety, operability, maintainability, appearance, etc. all of which are directly design related. To quote Margaret Thatcher:[3]

> I am convinced that Industry will never compete if it forgets the importance of good design; not just appearance but all the engineering and industrial design that goes into the product from the concept stage to the production stage and which is so important in ensuring that it works, it is reliable, it is good value and looks right. It is good design that makes people buy products and gives products a good name.

However, designing quality products is not necessarily enough in itself to ensure competitive advantage, one must also exploit the innovative process and pay attention to cost. Change is an ever-present fact of life and changes in the market-place will dictate the need for new designs as products become obsolete or sales decline. The responding business activity in the modern economy will view design and development as a corporate function, not as a separate entity, thereby transforming the role and position of the designer and broadening the design spectrum quite significantly. In the same way that market forces create a requirement for product design, the act of designing gives rise to change in the company. Managing this rate of change at the correct level determines profitability.

Despite modern design aid technologies, design and development costs are steadily increasing. With products of increasing sophistication plus increasingly stringent market acceptability features, i.e. safety, statutory approvals, liability, customization, etc. there is no guarantee that the design and development process will be successful in producing a product that will give a profitable return on investment. In other words the stakes are getting higher and the outcome becoming less certain.[4] However, by correctly managing the design challenge, and there are many facets of this, the process of technological development can be a fascinating and rewarding experience.

1.2 The importance of the design function

Design is now seen, increasingly, as the key to industrial success in all markets. Design, quite simply is all about making the right products for the right markets at the right time, the right quality, the right price, to the right specification and delivering the right performance. Good design does not guarantee success, but it is a very powerful contributor to it. Critical concepts like value for money, safety, maintainability and operability are all part of the design mix alongside the aesthetic and functional necessities. Carefully designed products must also look the part.

There is a very close analogy between design and quality for the two are, in reality, aspects of each other. The customer perception of quality is also strongly influenced by the whole design process. For example, with equivalent competitive products in the market-place, desirability is crucial. Conveying the idea of modern technical excellence by the function, feel and looks of the product provides power in the market-place. The drive for quality encompasses all functional areas within a company as well as their suppliers, ensuring that all interacting departments, both internal and external, understand each other's needs. Many companies adopt the concept of total quality management, ensuring a strong commitment to regular co-operation across functional and organizational boundaries. By involving Marketing, Manufacturing and Field Service engineers from the inception of each product, and designing for effective manufacturing, fewer engineering changes are needed when the design is released and competitive standards of quality can be achieved. Thus with a strong design-for-production commitment, downstream delays and surprises can be avoided.

In the present-day competitive environment, it is usually necessary to achieve a rapid turn-round in product design and acquisition, an emphasis not surprisingly on getting it right first time. In this context of the design function, the design engineer will have the benefit of modern design tools, such as Computer-aided Design (CAD), but more important is the quality and experience of the engineer. Much is written about teaching the design process, but engineering design is a unique discipline where good design skills can only be acquired by experience. To be successful, the engineering designer must have a sympathetic understanding of all the processes and factors that go into the creation of an electromechanical product. The benefits from achieving a right-first-time goal can be significant. Manufacturing costs that reflect expenditure in rectifying the effects of poor quality might otherwise be reduced considerably, together with shorter lead times and production schedules. By inference therefore, the later that a design error is detected the greater the cost of correcting it.

Good design has a clear business purpose with a strong link between it and competitiveness, profits and products. With design power it is possible to make rapid inroads into any market, but poor design undoubtedly means poor profits. It is an expensive process to design and launch a new product, hence it is important to analyse and synthesize all aspects of the design commitment in getting it right.

1.3 The phases of design

For introduction to the market-place, a phased product development process is an established method of providing general direction for managing a new product. The process may be called different things in different organizations, but it is intended to be flexible, adaptive and easily understood, accommodating the considerable latitude differences of individual products. Emphasis should not be placed on specific boundaries or phases of the process, rather that it is the general intent that is important.

For the purposes of understanding the design and development process in the general context of this book, it is necessary to assume the existence of a minimum number of functional groups within the organization concerned. For example, the main departments of activity may be summarized as follows:

Engineering. Responsible for all design, development and product test work up to the point where volume production starts. Design authority for the product is, however, maintained throughout the product life.

Manufacturing. The production arm, which not only will assemble the product but also undertakes product costing, specialized tooling, vendor liaison, commodity evaluation, etc.

Quality. Responsible for the incoming inspection and quality checks on all bought-out items plus the outgoing checks on the quality of the final manufactured product. While the production line will adjust and test the product, the Quality Department will audit the product once it leaves the production line, feeding back to Engineering.

Field Service. The group responsible for maintaining the product in the field and is therefore the inter-face between the customer and Engineering. Will also handle spares, replacements, repairs and warranties.

Marketing. Ideally not just a sales department but the company gatekeepers, probing market information assessing needs, communicating trends and evaluating the competition.

The phased development process will be presented as four zones or phases of activity, i.e.

1. The proposal phase — is there a profitable business case? — is it technically feasible?
2. The feasibility phase — can the product be built economically? — what is the performance likely to be? — design for demonstration.
3. The acquisition phase (prototype and pre-production) — can it be produced commercially? — is it a marketable product? — design for market-place.
4. The maturity phase — design for variant opportunities.

The relationship of activities hardware and documentation milestones in these phases is given by Table 1.1. This represents the essence of the process. Although no time-scales have been allocated, typically the proposal phase may only be a few months while the acquisition phase possibly a year or so, depending on product complexity.

In the following descriptions of the development phases, and in Table 1.1, various documents, plans or procedures and their purpose are described. Depending on the nature of the design project and the number of people involved, these documents will vary in formality and length. Documentation not only forces the designer or product team to investigate and prepare projects and plans more thoroughly but is also the most effective means of communicating the same message to the whole organization, ensuring a common goal.

During the proposal phase, a concept document is prepared which should provide a business case for the

Table 1.1 *The product acquisition process*

	Activities	Hardware	Documentation milestones
Proposal phase	New product proposal New product team Concept formulation (specification) Product safety development	None	Product proposal (outline design sketches) Profit projection and return on investment (business case) Resource plan Investigation reports
Feasibility phase	Establish team Project definition Planning stage/design plan Product strategies and architectures Rigs, calculations, analytical models Verify technical viability (demonstration) Verify business viability Subsystem testing	Test rigs Technology rigs (pre-prototypes) Breadboards	Engineering feasibility study Financial forecasts Reliability engineering plan Project plan Concurred product specification
Acquisition phase	Technical process verified Initial design optimization Software development Establish design performance versus targets Pre-production line tryout Production line balance Field trials	Prototypes (A models) Pre-production (B models) (pilot run)	Manufacturing plan Field service plan Design reviews Update product specification Update project plan Assembly procedure Manufacturing release
Manufacturing phase	Product launch Line ramp-up Field assessment	Production (C models)	Technical manual Spares list Test specification Settings and adjustments specification Component specifications Firmware/software release

product, an intent to satisfy a real or latent market need. Basically this provides the 'what' of a programme. Simultaneously or subsequently in the feasibility phase the Engineering community would evaluate the programme 'how', i.e. in terms of technology, architecture, scoped costings, performance prediction, etc. The acquisition phase represents a formalized commitment to the product and is the point where serious detail design work starts. Actual hardware performance and reliability as demonstrated by early hardware models will normally fall short of specification levels. Over time, the differences between demonstrated performance and maturity requirements diminish until, at the maturity phase, there is a confident projection that actual performance equals specified performance.

1.4 The proposal phase

This phase represents an ongoing activity practised in some form by the majority of successful manufacturing companies, not only to identify market segments and opportunities but also to assess customer needs. Normally this is a Marketing-led information-gathering activity, based on market quality information channelled through customer-orientated marketing and service organizations. The result should be a product concept which answers the market need and satisfies the customer.

Once the concept is accepted in principle, the process normally moves on to what we will call a New Products Team, or more generally perhaps to a combination of Marketing and Engineering effort. In either case the result is a proposal document outlining the business case for producing such a product. This proposal document, once agreed, is the initiator for serious engineering activity to start, although there may well be some multidisciplinary input procedure in planning the proposal. The proposal document should basically answer a number of questions concerning the proposed product, e.g.:

1. The question of need;
2. Product features;
3. Presentation of product;
4. The resource requirement;
5. Return on investment;
6. Opportunities;
7. Risks.

1.4.1 The question of need

The need is twofold in essence, whether or not the company needs the product and whether or not the market-place needs the product. Although the product is the same, the needs of both are quite different. Some typical questions that should be addressed are as follows:

What evidence exists from market awareness surveys?
What is the likely market share?
Is this new business or just helping to maintain a market lead, e.g. a variant?
What is the current situation and future outlook on competitive products?
When does the market need the product?
Does the product fit the company long-term plan?
Does the product provide for a significant advance in the market-place?

1.4.2 Features

For any product aimed at a fairly well-defined market slot, there are clearly a number of features essential for competitiveness, but from that point on feature inclusions should be examined critically in terms of what the market will stand. It is often far better to offer a feature set which will satisfy 90 per cent of the market-place at a competitive price than to offer a feature set that will satisfy 99 per cent of the market, but which makes the product very expensive. Put simply, incorporating 'wouldn't it be nice to have' features is a law of diminishing returns. Of course, in the unlikely event of there being no competition, a totally different set of rules apply. However, the general approach to features may be summarized as follows:

1. Maximize within reason;
2. Concentrate on the musts;
3. Select from wish list only where product acquisition costs are little affected;
4. Consider feature enhancements as variants/ customizable options/upgrades plus the enabling design concept. Customer then pays for what is essential for his needs;
5. Keep features list stable during product development.

1.4.3 Presentation

There is no doubt that the appearance of a product can induce an impression of quality, durability or reliability to a prospective customer. Hence the aesthetics of a product are all important, usually the more it stands out among the competition the better. Although product desirability is not so obvious in the business sales sector as it is with consumer products, the need for the product to appeal is still paramount. the main factors regarding the presentation of the product can be summarized as follows:

What does it look like, does it make an impression?
Should be simple to operate.
Needs to be operator/user friendly.
Understandable operating instructions.
Ergonomic and anthropometric aspects considered.
Good technical publications available as back-up.
What environment will it be used in, is the appearance appropriate?

1.4.4 Resource

In evaluating the business case, clearly the resource estimates on space, equipment, facilities, manpower, etc. and how these affect existing product support, are very important. Provision of resource costs money, and depending on the nature of the project, implies a company investment decision on the basis of the product proposal document. In those cases where a specialization is required, and which cannot be supported on a full-time basis, consultancy organizations or university technology centres can usually provide assistance. The resource questions can be summarized as follows:

Can total company funding be arranged or are technology/offset grants available?
Is manpower available in house?
Are specialists needed; can we recruit?
Are space, facilities and equipment available?
Any likely impact on current product manufacturing?

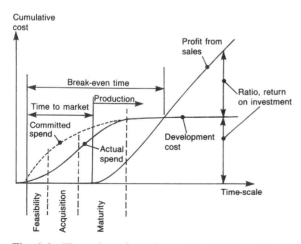

Fig. 1.1 Illustration of development cost and profit return

1.4.5 Return on investment

At the end of the day, to be a viable proposition the product needs to make a profit. The profit margin and return on investment calculations will depend on many factors, the cost of acquisition, the unit manufacturing cost, the product life cycle, likely volumes, etc. which, together with overheads, abandoned project costs, etc. determine the selling price. Thus, assuming the selling price is an acceptable one for the market-place to bear, development costs are defined and manufacturing cost targets fixed. A target ratio of return on investment should be set at a minimum at a given point in the maturity phase, (Fig. 1.1).

Additional margins may be created if the supply of consumables is also involved or the possibility exists for Field Service contracts to be negotiated. Spares also provide additional revenue, but hopefully good reliabillity would provide a better earner.

1.4.6 Opportunities

Apart from the basic decisions of whether the product fulfils the future needs of both the company and the market-place, there is the question of what additional opportunities may be created. For example:

Does the product allow the company to leapfrog the competition?
Is there potential for capturing a large share of competitors' market?
Do we have a unique patent situation from which a strong foothold may be established?

1.4.7 Risks

As a counterbalance to opportunities, there is the question of risk, especially where new technology or the development of an enabling technology is concerned. In this case it is wise to complete technology feasibility phase first to understand the design parameters involved before launching the total product acquisition cycle. Potential risks should be recognized early in the development and minimized if possible. On the lower level and in areas of industrial growth, one risk to any project is always the availability of suitable manpower. Finally, as the product starts to become mature, there is always the risk that it has missed its market slot by failing to achieve performance or cost targets, but where the project is well managed with an emphasis on design quality, this should not occur.

1.4.8 Development of the proposal

Having established an outline of the product proposal and the type of data format necessary for the business community, the information needs to be accumulated. Figure 1.2 outlines a flow diagram in which input is assembled from Sales/Marketing, Field Service, Competitive Products and Engineering organizations directed by a New Products Team. While this really reflects the procedure developed within a largish organization, such responsibility divisions may not exist within a smaller company, but the individual input items are, nevertheless, common.

(i) Sales/marketing input

Marketing departments often see themselves as only selling products, but in reality this is only part of their role. Having to deal directly with the customer, they alone are directly aware of what is in the market-place, how the competition performs, what the customer would really like to have and what he really thinks of your product. Unfortunately this type of investigative task does not return short-term financial benefit, but it is a very crucial activity as far as the company is concerned and should be encouraged.

(ii) Field service input

Service and repair engineers are well placed to feed back performance shortfalls that need to be addressed, hence they need to maintain reliability problem lists. Additionally, since service organizations work across the customer interface, it is essential that the maintainability and serviceability features are built in to the product at an early stage, since when on site the quicker a task can

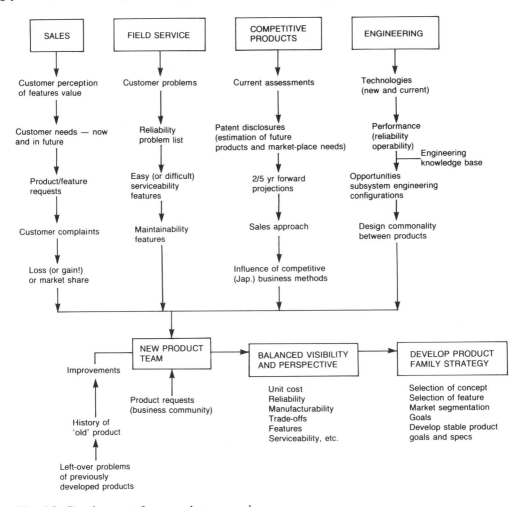

Fig. 1.2 Development of new product proposal

be completed the better. Not only does it save service engineer time, it also creates a better product quality perception in the eyes of the customer.

(iii) Competitive products

Not every organization has the luxury of a competitive products department, but the general desirability for such information still exists. It is usually not sufficient to base future products on the state of current market competition, rather to project forward to what the competition may be doing in a few years' time, e.g. by examining patent applications and disclosures, product trends, sales announcements, technology advances, etc.

(iv) Engineering input

The engineering input into the product proposal stage is not design input as such, rather an assessment of how various processes/functions were accomplished in previous products (not necessarily in-house), whether this is still 'state-of-the-art' or whether technology advances now enable opportunities that could reduce build costs, improve reliability, reduce noise, etc. Certain proposals will result from the 'if we were to do it again we wouldn't do it that way' past product history but likewise, if there is already some expensive hard tooling in place, there may be pressure to earn maximum capital from this and design for a degree of commonality. However, within any engineering organization, as personnel move on and product teams change, there is a tendency to invoke the 'not invented here' syndrome when new products are initiated. In terms of product cost and time-scale, penalties may accumulate if the engineering input is not effectively managed.

Given the inputs from various sources, the focal point for this information is the team or persons charged with the new product proposal preparation. some of the data assembled will conflict or will present a high risk, hence a balanced view needs to be taken on what is achievable within a realistic time-scale and budget, yet still fulfilling the market need. Some trade-offs on features/cost of acquisition/use of new technology, etc. will clearly arise, no product decision is clear cut, the final definition decision has to be based on a perceived balance of probabilities in relation to the quality of the material presented as the business case.

1.5 The feasibility phase

The feasibility phase is, as the name suggests, to establish general process and product feasibility. It is also often referred to as a definition or possibly a development phase. The purpose is to demonstrate a secure technology base, to establish the baseline architecture (schematic) of the product and to commit a product acquisition schedule. There are four basis activity areas:

1. Product introduction plans;
2. Technology tasks;
3. Product architecture;
4. Industrial design and human factors.

It is to be expected that a number of test rigs or breadboards would be built to understand any enabling technologies should the technology aspects of the concept document or design brief require it. For verification, and to aid performance understanding, analytical models may also be developed. The feasibility phase is also a planning phase when the product acquisition strategy, in terms of time-scales, resource requirements and costs, are mapped out. Marketing may also update their market forecasts, return on investment calculations and pricing plans.

Although this is not a product design phase *per se*, first cut product architectures will evolve, influenced somewhat by the technologies chosen and the industrial design concepts presented.

1.5.1 Product introduction plans

Based initially upon the product proposal, but more definitively later in the feasibility phase as the technology becomes clearer, an overall Engineering Project Plan may be produced that would identify the following key aspects of the project:

1. Time-scales and programme for the acquisition and manufacturing introduction phases.
2. Costs.
    (a) direct resource costs — labour, materials and consumables, equipment (as an annual equivalent), overheads, secretaries, rent, etc.;
    (b) cost of pre-production and initial production models;
    (c) unit manufacturing cost estimate;
    (d) projected tooling costs.
3. Reliability engineering programme.
4. Field trials and product support.
5. Quality assurance programme.
6. The product engineering specification.
7. Consideration of patentability.

Once the Engineering Project Plan has management concurrence and, by inference, a formal commitment to produce, a number of support group activities can commence, e.g. product marketing plan, manufacturing plan, customer service plan, cost of ownership analysis, etc.

A typical plan outlining the engineering activity in the acquisition and initial manufacturing phases is given in Fig. 1.3. However, depending on the complexity of the product, the cycle may be split into subsystems with subsystem and system designs of the prototype model being staggered. Conversely, if the design is an enhancement, a truncated acquisition stage may be more appropriate.

When a product concept is approved, clearly the cost of designing it, in terms of manpower resource, must be faced. There is probably an optimum here since if too little time is spent on design, there may be a high change rate and a prolonged line balance, while too much attention at the design stage may prolong the acquisition process unacceptably. However, given that the next model build is always more expensive than the previous one, the actual cost will to a large degree depend on design quality, especially when entering production, i.e. a quality design will result in lower production costs.

In addition to the product technical verification, one of the important exit documents from this phase is the product engineering specification, typically addressing the following critical aspects of the design requirement:

1. Safety;
2. Performance and reliability;
3. Features and functions;
4. Unit cost;
5. Manufacturability;
6. Serviceability and maintainability;
7. Ergonomics/industrial design;
8. Product architecture.

1.5.2 Technology tasks

In those product proposals where there is an enabling technology requirement, this will normally require both paper analyses and test-rig fixtures to understand performance latitudes and so establish process and product feasibility. Technical viability in this stage therefore relies heavily on system performance projection models.

Technology selection guidelines are as follows:

1. Select appropriate technologies;
2. Design test rigs/fixtures;
3. Understand key parameters and sensitivities that affect performance;
4. Establish the best performance window that can be expected from each technology;
5. Record what set-up parameters need to be controlled, and to what limits, in order to achieve the required performance latitude;
6. Understand the effects of all external variables on performance;
7. Match performance windows with product proposal goals.

Thus as far as the output from the above is concerned, the principal product is understanding. Hence by understanding the technology first, and specifying design latitude/tolerances, it often ensures a stable product design result and early market introduction. With early

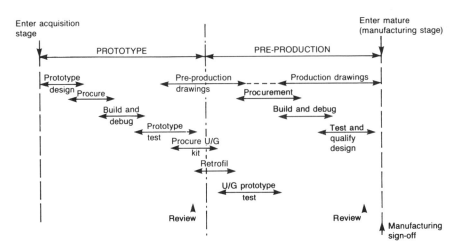

Fig. 1.3 Acquisition stage activities

technology definition it is possible to get product design right first time. However, it is often the case that with pressure applied to get a product into the field quickly, performance understanding is often done on the product itself and most design engineers have an experience of how disastrous that can be. Any competent designer can create a design to achieve a specific performance if the defined limits on controllable parameters are known and maintained.

A number of principles can be applied to technology tasks:

1. Mechanisms always fail in multiple ways; there is a need to take all failure modes into account in order to establish realistic operating windows.
2. Performance may be affected by a wide variety of external variables, e.g. temperature, humidity, etc. There is a need to discover all these and establish performance over a wider than expected range to ensure no operating performance cliff near the proposed product performance specification target.
3. As in the proposal phase, look at the history of old technologies to avoid wasting time and effort on repeating old mistakes.
4. Understand the engineering principles at work and so avoid surprises downstream.
5. The output is understanding, not just functional hardware. Avoid recommending a design when you do not know why it works.
6. Failure of a technology task to achieve the required target performance is a successful outcome if the reasons why are understood.
7. Material properties that can drift, will drift. Avoid these and latest 'new materials'.

1.5.3 Architecture

In product design cases where the configuration is less than straightforward, an outcome from the technology assessment will be some synthesis of ideas regarding the build concept of the machine or module, i.e. how the machine or module fits together. This cannot be a firm definition until the design process starts formally, but is mainly intended to produce an early viewpoint of the traditional long lead time items, especially where mould tools are likely to be involved, e.g. frames and covers.

Typical issues that will be addressed by the architectural concept are as follows:

How modular is the machine and does this enable simplified assembly and subsystem testing?
The ease of access for servicing, the removal of covers, removal of components, etc.
How individual modules, subsystems, components and PCBs can be packaged to give the most economical product volume.
The ease with which the machine can be moved as well as transported.

1.5.4 Industrial design and human factors

Industrial design is that part of the design function dealing with the aesthetic and functional impression aspects of product development. It is a task very often tendered to a professional consultancy, to improve the look and feel of an otherwise functionally adequate design.[5] Where a human interface is involved in the product usage, associated human factors will influence the industrial design. The relevance of the man−machine interface (MMI) underlines the importance of human factors in a design project, the human operator being seen as a flexible part of a closed control system loop with his/her mechanical environment. As such, an operator is never totally predictable and one learns always to expect the unexpected, hence a product designer should always avoid using himself as the end-user.[1] Specialized test facilities are now frequently used to observe the interactions of users with new products or products in the development cycle where reaction to various interface ideas would form part of the design process.

The industrial design and human factor aspects associated with the appearance and operability of the product can be summarized as follows, accepting that in reality there is a strong interaction of the two:

Industrial design — appearance and aesthetics, models and perspectives, colour schemes and logos.
Human factors — anthropometrics, ergonomics, physiological aspects, psychological aspects, biometrics, instructions and symbols.

These factors can now be considered in more detail.

(i) Aesthetics
This is in reality the application of a set of principles of good taste and appreciation recognition. Good design aesthetics in a product are not only pleasing to the eye and the mind but also convey and stimulate an outward impression of functional quality, efficiency and effectiveness. An old-fasioned design can give an impression of the product being out of date with current technology. Aspects of aesthetics are as follows:[5]

1. Symmetry — proposition and distribution about a product axis providing balance;
2. Continuity — a harmonious pattern in profile with the ordered arrangement of product elements;
3. Style — the visual quality of design which sets the product apart from the rest;
4. Impression — as to whether it fits into its environment and whether it looks fit for the purpose.

Often aesthetic considerations conflict with functional limitations, in which case there has to be an acceptable compromise. The industrial designer should always strive to minimize functional intrusions by choice of colours, texture, shapes, etc.

(ii) Models and perspectives
Both models (full size and scaled) together with perspective designs can make an invaluable contribution to the design process by developing the visual effect.[5] Also partly ergonomic, the full-scale model can assess the marketing potential of a new product in terms of appearance and operability. The use of three-dimensional CAD systems for visualization can speed up the assessment of designs, possibly before committing to an actual model.

(iii) Colour schemes
Generally the subject of much debate, but the correct choice of colours can have a very strong influence on product appeal and is therefore worthy of discussion. The use of colour contrast may be used to create an up-market impression, e.g. speed, technical superiority, etc. Again, good three-dimensional CAD systems with skilled use can give an early appreciation of the effects of colour choice.

(iv) Anthropometrics
This is the study of the physical size and dimensional variations of the human body. It is one of the inter-disciplinary aspects of human factor engineering, often involving principles and information outside the scope of design and development activities. When a human interface is involved, the difficulty that the design engineer faces is knowing what size of human body he is designing for. Normally, one would assess data based on broad headings of age, sex, nationality, etc. using the 90 per cent rule, i.e. dimensions based upon the 5th to 95th percentile of the population. An example of some typical European data is given in Fig. 1.4, in this case including an allowance for clothes.[5]

(v) Ergonomics
This is the study of people in relation to their working environment, the dynamic interaction of the operator and the machine. Ergonomic aspects are strongly related to industrial design of the product, e.g. the layout of control panels, the size and operation of control levers, etc. i.e. aspects which aim to maximize operator performance, efficiency and reliability, providing an easy learning curve and enhancing feelings of comfortableness.[6] Interactive tasks require movement and some force application on the part of the operator to complete successfully, e.g. loading machines, opening covers, etc. The analysis of these should be a standard feature of the design cycle since the problems associated with achieving a successful MMI can be greater than those associated with achieving machine performance. Single man—machine tasks may act efficiently in a individual sense, but combining two or more tasks may produce a synergistic or antagonistic result.

(vi) Physiological factors
Here the designer is concerned with the aspects of machine design that influence a response or reaction on the part of the machine user. In other words it is the cause of and the effect on body functions that is important in this context, rather than a medical understanding of how the senses operate. Various types of information is fed to the sensory system through receptors or sensors (sensation), the main types of interest in ergonomics are exteroceptor and proprioceptor sensors:

Exteroceptors: receive information about the state of the world outside the body, i.e. information transmitted from the environment to the operator, e.g. heat, noise, sight (visual).

Proprioceptors: tell the operator what his body is doing and his position relevant to the machine environment, e.g. touch, muscular (load sensing), neurological.

(vii) Psychological factors
In a general sense, the efficiency of an interactive task performed by a subject will depend upon the following:

1. His/her mental state — whether bored or well motivated by the machine and its operability.
2. Operator receptiveness — whether the machine is operator friendly or not and whether it functions reliably; in other words it does not create annoyance.
3. Sociological influence — country and class divisions that may influence the attitude of the user towards the machine or piece of equipment.

For successful machine operation, not only has there to be good user education (training, demonstrations, wall charts, etc.) but the operator has to demonstrate a desire to succeed. For example it is very easy to use technical equipment incorrectly and claim poor machine performance.

THE DESIGN PROCESS

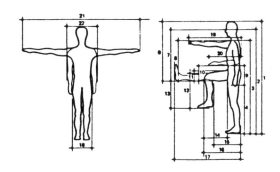

Key dimensions	Men aged 18 to 40 Percentiles			Women aged 18 to 40 Percentiles			Examples of applications to design problems
	5th	50th	95th	5th	50th	95th	
Standing							
1 Stature	1628	1737	1846	1538	1647	1758	95th: Minimum floor to roof clearance: allow for headgear, say 100 mm in appropriate situations.
2 Eye height	1524	1633	1742	1437	1546	1655	50th: Height of visual devices, transoms, notices etc.
3 Shoulder height	1328	1428	1428	1237	1333	1429	5th: Height for maximum forward reach.
4 Hand (knuckle) height	703	770	837	—	—	—	95th: Maximum height of grasp points for lifting.
5 Reach upwards	1972	2108	2284	—	—	—	5th: Maximum height of controls: subtract 40 mm to allow for full grasp.
Sitting							
6 Height above seat level	841	900	959	790	849	908	95th: Minimum seat to roof clearance; allow for headgear (men 75 mm, women 100 mm) in appropriate situations.
7 Eye height above seat level	726	785	844	676	735	794	50th: Height of visual devices above seat level.
8 Shoulder height above seat level	537	587	637	494	544	594	5th: Height above seat level for maximum forward reach.
9 Lumber height	—	254	—	—	—	—	50th: Height of table above seat.
10 Elbow above seat level	178	224	270	157	203	249	50th: Height above seat or armrests or desk tops.
11 Thigh clearance	124	149	174	121	146	171	95th: Space under tables.
12 Top of knees, height above floor	506	552	598	473	519	565	95th: Clearance under tables above floor or footrest.
13 Underside thigh, height above floor	402	435	468	385	418	561	50th: Height of seat above floor or footrest.
14 Front of abdomen to front of knees, distance	336	386	436	—	—	—	95th: Minimum forward clearance at thigh level from front of body or from obstruction, e.g. desk top.
15 Rear of buttocks to back of calf, distance	436	478	520	423	465	507	5th: Length of seat surface from backrest to front edge.
16 Rear of buttocks to front of knees, distance	568	614	660	542	584	626	95th: Minimum forward clearance from seat back at height for highest seating posture.
17 Extended leg length	998	1090	1182	—	—	—	5th (less than): Maximum distance of foot controls, footrest, etc., from seat back
18 Seat width	328	366	404	363	391	429	95th: Width of seats, minimum distance between armrests.
Sitting and standing							
19 Forward reach	773	848	923	600	675	750	5th: Maximum comfortable forward reach at shoulder level
20 Folded arm Forward reach	483	530	578	375	422	468	50th: Folded-arm reach for hand controls
21 Sideways reach	1634	1768	1902	1509	1643	1777	5th: Limits of lateral fingertip reach; subtract 130 mm to allow for full grasp.
22 Shoulder width	420	462	504	376	418	480	95th: Minimum lateral clearance in work space above waist.

Fig. 1.4 European anthropometric design details (adapted from Hawkes B. et al., *The engineering design process*, Pitman, 1984. Courtesy of Longman Group UK Ltd.). Dimensions in millimetres

(viii) Biometrics
This factor is concerned with what the body has, rather than size or function. typically these attributes are identification factors, i.e. fingerprints, blood type, speech form, handwriting characteristics, etc. which may present a user interface in the more technologically advanced product.

(ix) Instruction and symbols
The design and placement of instructions, illustrations and symbols is all important in conveying unambiguously to the machine or equipment user the intended sequence of events. Ideally, if the product is a multinational one, instruction is communicated via internationally recognized pictographic symbols, numeric sketches or icons, all without legends. A summary of the design guidelines for symbols is given by Osborne.[6]

A more expensive alternative, but one which can actively guide and assist a user through the machine operation, is the use of either a visual or an alpha-numeric display. In these cases it is the choice of wording in the message set that should be clear and helpful. Going one step further, graphics displays also enable symbols/sketches, hence potentially a very interactive feature. In those products provided with an operator manual, instructions are largely conveyed through the use of sequenced sketches. Such manuals (or flip pads) would be designed with the problems of the MMI and adequacy of customer education in mind.

Although the majority of symbols and instructions are intended to convey normal operating procedures, certain products or certain internal machine areas are dangerous or hazardous by their nature. However, benefits outweigh the risks provided that the latter are clearly made known to the end-user. With such products or situations, the provision of warnings are critical to product safety overall, see section 2.5.2.

1.6 The acquisition phase

The main outcome from the feasibility phase will be an Engineering Project Plan together with the concurred product specification. The product specification outlines service and maintenance concepts, performance requirements, the design appearance philosophy, the chosen technology set and architectural breakdown of the product. In other words it is the design brief. The Project Plan addresses schedules, costs and interdepartmental involvement requirements.

Entry to the acquisition phase may in fact be via other routes apart from new product development (Fig. 1.5), but in all cases it is a formal management commitment to go forward. Proven technology need not be an absolute requirement at this stage, but feasibility should have been assessed. This is also the point at which the rate of expenditure starts to increase[7] and for this reason, checkpoints may be built into this phase to monitor progress, but ideally not to apply major course corrections.

The purpose of the acquisition phase[1] is to design and build prototype machinery representing the complete product, which, may in fact be the first-time integration of all the feasibility technologies if no pre-prototypes were undertaken in the feasibility stage. From the configurational problems of this build cycle and the performance shortfalls identified on test, resolution is reflected in reworks and drawing updates. Subsequently, from a finalized bill of materials, Manufacturing will procure and build pre-production models prior to production start-up.

There are a number of key aspects to the acquisition phase and these are summarized in Table 1.2. These are addressed by the design team as an ongoing activity but more specifically as discussion items during design reviews. There are nominally two hardware stages characterizing the acquisition phase (Fig. 1.3) as given under the headings below.

(i) Prototypes (or 'A' models)
Designed and built within Engineering, the configuration is aimed at the final product intent, complete with all covers and controls confirming design feasibility and identifying/removing key risks. The models will be used for design evaluation (Table 1.2) and would be upgraded as far as possible in line with performance shortfalls and

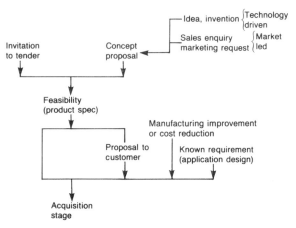

Fig. 1.5 Routes into the acquisition phase

THE DESIGN PROCESS

Table 1.2 *Key aspects of the product acquisition phase*

Key item	Associated factors
Product Schedule	The Design Plan
	Drawing Issue
	Prototype build and debug
	Pre-production build
	Line balance
Design Evaluation	Reliability
	Quality
	Manufacturability
	Serviceability
	Operability and safety
Cost	Parts cost and labour
	Service cost
	Tooling cost
	Jig and fixture costs
	Inventory costs
Other	Structured bill of materials
	Early vendor involvement
	Configuration control
	Spares lists
	Change rate
	Service procedures
	Problem management
	Definition of vendor base
	In-house manufacturing processes

allowed time-scales (Fig. 1.3). At completion of this stage, the design should be proven.

(ii) Pre-production (or 'B' models)

The object here is to produce a small number of machines using issued engineering drawings and documents (manufacturing intent) inclusive of changes identified from prototype evaluation work. This stage proves that the design is manufacturable. Configurational changes ideally should have been confirmed as prototype upgrades. The pre-production build is a limited line try-out designed to test the following:

1. The standard of issue documentation;
2. Assembly levels on the family tree;
3. Ease of assembly and assembly procedure;
4. Setting and adjustment procedures;
5. Parts made by the intended manufacturing process or vendor meeting specification.

To service these requirements the Manufacturing organization is likely to assemble a team involving Procurement, Industrial Engineering, Quality Assurance (QA) and Test Engineering. Manufacturing will report back to Design Engineering for corrective action by the design authority any documentation, functional or cost deviations found. The action and responsibility split for acquisition phase hardware is given in Table 1.3.

The elements of each model build can be presented in the following sequence of activities: (1) design, (2) build, (3) evaluate, (4) review.

If necessary, redesigns or fixes will go back through this cycle. These elements will now be further considered in detail.

1.6.1 Design

Either during the feasibility stage or on entering the acquisition stage, it is usual to assemble a 'design team'.

Table 1.3 *Generalized breakdown of action and responsibility levels within a typical manufacturing organization during prototype and pre-production hardware acquisition.*

Key activity	Prototypes		Pre-production	
	Action	Responsibility	Action	Responsibility
Build (inc. procure and debug)	Eng.	Eng.	Manuf.	Eng.
Performance test	Rely.	Eng.	Rely./QA	Eng.
Product cost	PCE	Eng.	PCE	Eng.
Serviceability (field trials)	FS	Eng.	FS	Eng.
Drawing issue	Eng.	Eng.	Eng.	Eng./FS/Manuf.
Approvals (test house)	—	—	Eng.	Eng.

Key: Eng. = engineering, Rely. = reliability, QA = quality assurance, Manuf. = manufacturing, FS = field service, PCE = product cost engineering.

This team can be regarded as an internal business unit, a cohesive operating entity with the commitment of engineers and product designers developing the product configuration and monitoring cost, schedule and performance. The design team develops the design plan, issues and releases drawings and is responsible for prototype and pre-production builds up to line balance. Design team members and their responsibilites are outlined in Chapter 2. Team size and span of control is generally determined by the complexity of the product or subsystem in terms of drawing count.

From the example acquisition schedule in Fig. 1.3, design stages overlap to some degree with test activities and machine upgrades. This is an intentional move away from the more rigid 'phase-gate' driven and overly long development cycles and to maintain a momentum of ongoing design improvement, then retesting as soon as information is available to enable further improvement.

As the design scheme is created, the following items are ongoing engineering considerations relating closely to the end-product, the actual drawing itself: specification factors, drawing definition, value engineering of design, tracking cost.

(i) Specification factors

The product specification issued as part of the feasibility phase represents the designer's brief to which he should seek compliance. The actual form of the specification is probably somewhat evolutionary within the organization concerned, but generally will cover design-related topics such as the follows:

1. Product description;
2. Application and scope;
3. Operating procedure required;
4. Control features;
5. Product architecture;
6. Mechanical and electrical descriptions;
7. Operating environment limitations;
8. Noise level;
9. Electrical noise immunity;
10. Packaging requirements;
11. Performance;
12. Test house approvals/safety requirement.

Additional design requirements will also originate from Field Service and Manufacturing plans.

(ii) Drawing definition

It is the general responsibility of the multidisciplinary design team to ensure appropriate design definition, or in other words ensure that the drawings themselves:

1. Conform to the required company and/or British Standards;
2. Are correctly toleranced with critical dimensions indicated;
3. Have appropriate data;
4. Have explanatory notes where necessary;
5. Convey the design intent unambiguously;
6. Have a fully completed title block, identifying material, finish, colour, etc.

In terms of the overall product, it is clear that the quality process starts here. Attention paid to drawing quality at this stage is time well spent in terms of keeping the drawing change rate down and achieving early manufacturing maturity.

(iii) Value engineering

There are a number of terms referring to the activities appropriate here, e.g. cost−benefit analysis, value analysis or value engineering, all of which aim to reduce cost of ownership and to provide Manufacturing with a more cost-effective engineering design in terms of reduced piece-part costs, easier parts procurement and lower levels of tooling, etc.

While the designer aims to produce a quality product and specification conformance, perfectionism, where this means specifying high accuracies for the sake of it, has no part in efficient engineering design. Cost consciousness is of greater importance, together with the realization that optimum design can only be achieved through patient reiteration. As the configuration moves downstream, however, the value of improvements to the baseline configuration should be evaluated by the design team on the overall probability of successful implementation, particularly so where scrap is involved or potential tooling changes.

It is easier to implement value engineering changes at the pre-production design stage where the prototype hardware is available for examination. Lessons learnt from the prototype build are also relevant. The general aim is one of the reducing product cost and some typical approaches here might be as follows:

1. Relax drawing tolerances where possible;
2. Use standard size stock for machining;
3. Allow only a restricted range of fasteners;
4. Avoid special materials;
5. Aim for parts commonality;
6. Amalgamate parts/reduce parts count;
7. Minimize special tooling;
8. Reduce need for precise settings and adjustment, etc.;
9. Assess compliance of manufacturing method to drawing requirements.

(iv) Cost tracking

Already referred to in section 1.5.1 in terms of piece part value engineering/cost reduction, the design team contract is a more wide-ranging requirement to monitor actual machine/product costing as an ongoing activity versus target costs. Since target costs represent the basis for the business case, the importance of the activity and its implications as the design becomes more mature, is fairly clear.

1.6.2 Build

The build cycle itself, whether mechanical or electrical hardware (e.g. PCBs), is shown schematically in Fig. 1.6. The concept is that any build and debug problems that result in drawing change requirements are fed back to the design authority. Since the pre-production models are built to confirm the accuracy of documentation and information held on product database, it is essential that problems at this build stage are carefully recorded.

The pre-production build cycle is also the point at which the procurement organization, materials control or purchasing department start to become involved with the product. At the prototype stage there is, in many cases, too fluid a configuration, so is generally best left within Engineering. However, the exception is the long lead item, invariably tooled items such as covers, complex mouldings, etc. Here the involvement starts when there is prototype build definition. Hence with long lead items, it is important that the responsible design engineer reaches an early outline consolidation of the part, although actual details within the part may be put on hold should there be a performance/design problem elsewhere undergoing resolution. Differences in the procurement cycle for the two model builds is shown in Table 1.4.

A further important activity, although already implied with long lead expensive items, is vendor involvement. To produce a quality product competitively today, no company can afford to ignore the experience and advice of vendors who, after all, are being used in a specialist role as suppliers if no production capability exists in house. It is not possible for a designer to appreciate all the technology and performance changes in outside manufacturing, hence issue drawing definition on high-profile parts needs to result out of discussion and compromise between all affected parties, i.e. design, manufacturing, procurement and vendor.

Table 1.4 *Acquisition phase procurement*

Prototype	Procurement cycle Pre-production
Some detailed drawings available	All drawings production intent
Sketches possible	Limited life tooling
No tooling undertaken	Part tooling
High cost of piece parts	Parts procured and inspected by Production
Limited build, 2–10 units	Changes must be documented and agreed to by Engineering
Parts procured and inspected by Engineering	
No formal change control system	
Long lead/early vendor involvement	

1.6.3 Evaluation

Design performance is assessed against a number of established criteria as outlined previously in this chapter. It is not just a question of reliability testwork, although this is clearly important, but also serviceability, operability, etc. as quantified by design requirements in the product specification.

In terms of Field Service or customer service requirements, the design compliance to specification can be evaluated in terms of a number of key issues as in Table 1.5.

The machine or product reliability clearly influences some of the servicing aspects. Both prototype and pre-production products will undergo reliability testing, some machines on a nearly continuous basis in order to detect and react to any design weaknesses as early in the acquisition phase as is possible. During the pre-production activity, since the product is, in a loose sense, a manufactured one, a Quality Assurance Team may well conduct independent testing to confirm approval or alternatively monitor all reliability test work. Reliability is discussed more fully in Chapter 3.

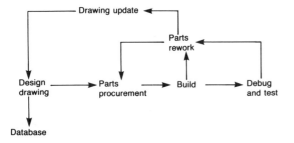

Fig. 1.6 The build process

Table 1.5 *Design evaluation factors*

Reliability	Establish the failure modes
	Evaluate failure rates
	Is reliability growth on target
	Operability
	Safety
Field Service	Ease of serviceability
	Mean time to repair (MTTR)
	Any requirement for special tools
	Key adjustments and settings required
	Need for a planned maintenance policy
	Spares list requirement
Manufacturing (pre-production)	General manufacturability
	Assembly levels as stock
	Special skill requirements
	Productivity estimates
	Correctness of drawings
	Accuracy of database
Quality	Conformance to design intent
	Measure consistency
	Maintenance of customer interests

Depending on the nature of the product, customer on-site testing could be undertaken to extend the test programme during the latter part of the acquisition phase. Before using live sites, the in-house performance would obviously be at or close to specification. Apart from confirmation of specification conformance, such field trials also evaluate user reaction and operability.

Thus while the purpose of the evaluation stage is to demonstrate the technology base and functional configuration conformance to the design requirement, it is nevertheless possible for some targets to be unrealistic. It is the purpose of the design team to recognize this and ultimately to negotiate performance trade-off if necessary.

Although the form of design evaluation described above takes the form of a systematic strategy, it is possible that it may take a simulation form, i.e. exercising computer or mathematical models. This is not so obvious in terms of appreciation, but as computer software becomes more sophisticated, it will become a potentially powerful procedure.

1.6.4 Review

In order that new products are developed and introduced into production in a controlled and fully documented manner, a procedure is required to review the design at critical points in the acquisition schedule. The design review is a systematic and independent evaluation by designers and specialists to provide assurance on design integrity and specification conformity.

A design review may take several forms, depending on the organization concerned and where the product is in the development cycle. As the product moves towards production, the formality of the review will increase as financial and resource commitment increase. Design reviews are used to enter the next development stage (Fig. 1.7), but generally in the sense of 'soft' gates, rather than 'hard' stops. The idea is not to have all the correct answers in place, rather to agree jointly that the probability of achieving the goals via the chosen technical approach is appropriate to the business objectives being sought.

The disadvantages in operating a rigid phase gate system are as follows:

1. Schedules become overly long as some problem areas struggle to keep pace.
2. Morale is affected if performance criteria have to be examined and a re-evaluation undertaken.
3. Production line start-up delays are inevitable and the product is late on to the market.
4. Development costs increase and return on investment decreases.

The make-up of the design review team should emphasize experience, it may be an 'internal' or an 'external' team, but it does not need to exercise authority, e.g. peer-group status. The review outcome is a recommendation based on technical assessment, it should not attempt to redesign or check design calculations, only reach a consensus on the performance, hence suitability of that design.[7] It is important that the review is rigorous enough but not overly strict. However, when projects are rolling and time is short, it is tempting to slip through a review without having all the answers. The designer's concern is often to avoid being the one who holds up progress, but it is usually the case that shortcomings become even more apparent later in the project.

The design review is a significant part of the BS 5750

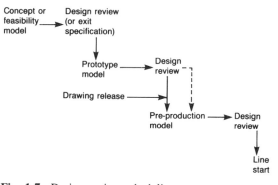

Fig. 1.7 Design review scheduling

quality assurance system.[8] This states that the supplier shall maintain the control of certain design functions, one of which is the establishment of design review or verification procedures. The object is to ensure that the design and development programme achieves its objectives by identifying problem areas before manufacture. The standard does not, however, specify how the actual review should take place although the holder of BS 5750 accreditation must clearly establish a design review procedure.

At the prototype stage, and possibly the feasibility stage, the purpose is to flush out potentially expensive problems early against design evaluation results referenced to the originating product specification. It will very much depend on particular situations as to how detailed the design review will become, i.e. some problems can be explored at the system level while others at the component level (Table 1.6). In general the review at this stage should address the following:

1. The Marketing requirement;
2. Functional performance and the design specification;
3. The technology set;
4. Costs;
5. Product schedule;
6. Manufacturability;
7. Maintainability;
8. Corrective actions;
9. Safety issues;
10. Packaging requirements;
11. Patents position.

The above will still apply at the pre-production review, but in this case Manufacturing will have built to formal documentation, hence the stability and accuracy of the database at the final design freeze will be an additional key issue. The following relevant criteria will feature in the pre-production review:

1. Design status of mechanical hardware, electrical hardware, software, configuration structure.
2. Reliability compliance with safety and approvals, functional performance targets.
3. Manufacturing status on critical components and materials, test equipment, special tooling, jigs and fixtures required, costs.
4. Quality assurance confidence in quality levels achievable, ability to test and inspect to documentation levels available.

The outcome of the pre-production review will ideally be a decision to launch the product. However, due to the required extent of the information base and the number of functional groups involved, it is rare to achieve a clean decision first time round.

Table 1.6 *Typical review levels of the design*

Configuration level	Key factors	
	Prototype review	Pre-production review
System	Reliability Manufacturability Quality Maintainability Operability Cost Schedule Risks Criticalities	Product specification Reliability Quality Manufacturability Serviceability Marketing plans Logistics support Packaging Cost and ROI Schedule Bill of materials
Subsystem (including software)	Functionality Modularity Problem list	Set-up adjustments Functionality Problem set
Lower level assembly	—	Assembly instructions Criticalities Inspection and test Tooling requirements
Piece part or component	Manufacturability Durability New materials	Vendor performance Cost and delivery Component specifications

1.7 The maturity phase

The maturity phase, also referred to variously as manufacturing introduction or production phase is the point at which total build control is vested in the Manufacturing organization with Engineering as a support function. The start of this phase is when the design and the design documentation is accurate and has been released by Engineering and concurred by Manufacturing and Field Service organizations. This will generally comprise the following:

1. All parts lists;
2. Assembly lists and the family tree;
3. Detail drawings and artworks;
4. Circuit diagrams and wiring schedules;
5. Settings and adjustments specification;
6. Test specifications;
7. Component specifications;
8. Firmware/software release specification;
9. Calibration procedures.

The purpose of the maturity phase is to obtain production capability, confirm manufacturing method (manufacturability), to identify any manufacturing process-related modification (in order to achieve the earliest possible optimization of both cost and production rate targets) and to launch the product. This is the point at which the 'customer perceived' performance should equate with the product performance specification, or at least converge to give a high degree of confidence that the desired requirement is achievable by the time of market introduction (Fig. 1.8). It is also the time at which the design starts to reach maturity and settle down, i.e. the drawing change rate decreases (Fig. 1.8).

The field trial is often a method whereby Management may obtain an initial performance observation of the product by placing at selected customer sites prior to actual launch. Following successful completion, the product would probably be transferred to current product status.

Design activities during the maturity or early maturity phase are related to documentation, problem-solving and drawing updates, plus the maintenance of the bill of materials. All configurational changes during this period are by necessity, formally controlled, these procedures being discussed below. Engineering documentation release at this stage is typically as follows:

1. Operator manual;
2. Installation instructions;
3. Set up procedures and specifications;
4. Technical manual;
5. Product brochures, etc.;
6. Approval certifications.

1.8 Engineering change control

The process of engineering change control or configuration management is a controlled method whereby design changes can be introduced into a manufactured product in a logical and properly planned manner. It is in the interests of the customer and also the quality of the product that changes should be managed properly. Most companies generate their own procedures which may vary from product to product, but more usually, with several products being manufactured at any one time, the change procedures are applied across the board.[9] The procedures outlined in this chapter are no more than typical, outlining a number of particular elements in the process.

The change process is disruptive to the normal flow of production work, but inevitable in competitive product situations where performance and product cost issues always arise. However, a company has to manage change effectively in order to meet the external market conditions, but having said that, too high a change rate results in chaos.

Some of the reasons for change are as follows:

1. Initial drawing release of new product;
2. Reduction in manufacturing costs;
3. Need to improve product performance/reliability;
4. Continuous product development;
5. Customized product/variants;
6. Obsolete components, source of supply dries up.
7. Serviceability improvement.

Rarely, if ever, does a product enter the market without some changes being required during initial introduction;

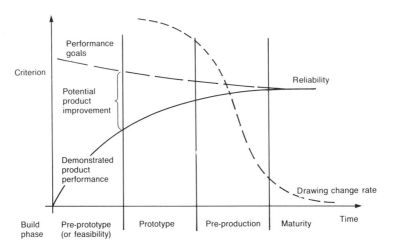

Fig. 1.8 Design life cycle

in general the more complex the product the more the changes. Changes will, if not carefully managed, cause delays and extend delivery dates, therefore there is a requirement for good control organization. However, no change, even if it is a modest one, should be introduced into the system without first a careful consideration of all the pertinent factors.

Of course there are always changes required for reasons of safety, reliability, performance, etc. that are mandatory, hence cost insensitive.

The types of engineering change control processes normally encountered within the manufacturing environment are as follows: pre-production change requests (PPCR); engineering change request (ECR); production concession/request; change notice (CN).

The relationship between change requests and other documentation activities in the development cycle is illustrated in Fig. 1.9.

With certain piece parts, e.g. mouldings, frames, covers, etc. long lead times are often involved. Consequently the Manufacturing engineer may wish to pull back some time-scale by requesting early drawing issue. Depending where this is in the design cycle, the level and confidence of the drawing information may vary (Fig. 1.9).

1.8.1 Pre-production change requests

This is a change document used to control drawing changes at the pre-released (pre-issue) drawing stage and is subject only to a modified or abridged control procedure. The pre-production stage is a small volume build, hence somewhat cost insensitive, in which the main focus is to get the drawings right and the product functional. Consequently the approval level on PPCRs is normally much reduced.

1.8.2 The engineering change request

Once a full release of essential documentation to commence parts manufacture and assembly has occurred, any changes required are subject to full modification control procedures using the initiating change request format. In many situations, especially where the overall drawing count is high, such change requests involve consideration of numerous issues, hence each one needs to be effectively managed. Such would be the task of a change control or modifications team, the leader of which needs to be a person of mature standing in the company familiar with the product line.

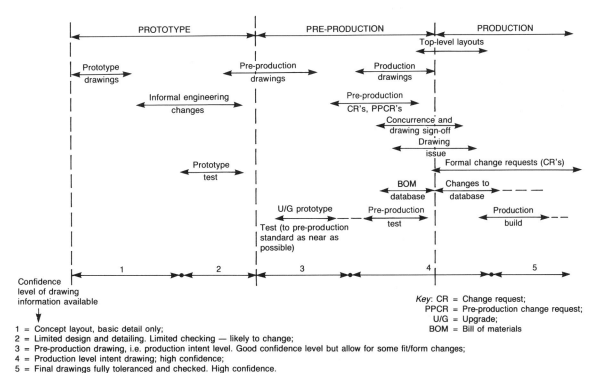

Fig. 1.9 Drawing documentation activities

When the ECR is submitted, the modifications team needs to assure itself that it has sufficient information available in order to make a decision, e.g.:

What is the cost effect, $+/-$?
What is the effect on product performance?
Are the rules on fit, form and functionality met?
What is the procurement lead time?
With current stock levels, scrap or rework?
Is the part common to another product?
Any tooling changes?
Are manuals, specifications, etc. affected?

A typical change request form is illustrated in Fig. 1.10. If insufficient information is available to enable a decision to be made, the request may well be referred for further investigation. An example of the procedural process is outlined in Fig. 1.11. Once accepted, the classification of the change is determined and whether the change is a modification or amendment to the bill of materials.

The classification system, typically a technical need action code, numerically 1–4, e.g. Table 1.7, indicates to both Manufacturing and Field Service how they should implement the change. Where an OEM (Original Equipment Manufacturer) customer is involved, this can be treated as a field situation.

Because of quality impact issues, QA concurrence should be required for all changes that affect fit, form, functionality or reliability. Thus QA approval and final ECR acceptance may well require prior test work and, if an OEM product, some form of customer approval. Once the change is released, Manufacturing and Procurement departments would agree a cut-in date and the effectivity date entered on the bill of materials listing.

1.8.3 Concessions

These are manufacturing requests to deviate from the controlled bill of materials on a temporary basis. Such requests are normally vetted by the Quality Department, the quality of the product is therefore maintained and the interests of the user safeguarded. As a configuration deviation document, it should be used with care and automatically superseded by issue or rejection of the associated ECR. A typical concession note is illustrated in Fig. 1.12. Concessions are of two types, temporary and change, as defined below.

(a) Temporary concession. This is a permit or temporary permission to use materials or components that differ from

Table 1.7 *Modification classication guide. The modifications classification defines the urgency with which a modification should be implemented. The classification is divided into two parts: In-house and Field/OEM. In both parts the modification is graded from 1 to 4, the implications of which are set out below.*

		In-house		
Grade	Brief description	Effect of material	Work in progress	Machines which have passed final inspection
1	Immediate and retrofit	Scrap or modify stores or orders immediately	Modify all work in progress immediately	Return to Production for immediate modification
2	As soon as possible	Modify immediately or scrap as soon as new parts are available	Modify all work in progress as soon as possible	Pass to Field
3	Programme at next convenient order	Use up stock up to machine serial number specified by Production	Programme with new material	Pass to Field
4	Documentation only	No change	No change	Pass to Field

Grade 1 implies that no unmodified machines leave the factory.

	Field/OEM
1	Modify at earliest possible moment
2	Modify at next service at which parts are available
3	Modify on failure, at customer request, or at Project Manager's discretion
4	Documentation change only

THE DESIGN PROCESS

ENGINEERING CHANGE REQUEST

ORIGINATOR		DATE		CHANGE No.	
MACHINE TYPE				DEPARTMENT MANAGER	
MODIFICATION CLASS		AFFECTED ITEMS	ISSUE	DESCRIPTION	
ASSOCIATED MODIFICATIONS		PART No.			
		ASSY No.			
CONCESSION No.					

DETAILS OF PROPOSED CHANGE	REASONS FOR PROPOSED CHANGE

COMMENTS		DECISION	ACCEPT/REJECT
		CHANGE NOTE ISSUE DATE	
		PRODUCTION CUT-IN DATE	

DOCUMENTS AFFECTED	YES	NO	FEATURES AFFECTED	YES	NO	ESTIMATED COSTS	£
Drawings			Safety			Documentation	
Engineering Database			Field Service			Design & Development	
Project Specification			Spares			Parts Scrap (Prodn.)	
Software Specification			Packaging			Parts Scrap (Field)	
Engineering Specs.			Approvals			Production Rework	
Component Specs.			Customer Approval			Tooling	
Manuals			Qualification Tests			Test Equipment	
Product Specification			Performance			Test House Certification	
Service Proceedures			Reliability				

+/- TO MACHINE UNIT COST	£	+/- LABOUR (ASSY) COSTS		UNIT COST FIELD MODIFICATION	

SIGNATURES (FOR ALL CHANGES)	DATE	SIGNATURES (AS REQUIRED)	DATE
ENGINEERING MANAGER		SALES & MARKETING	
QUALITY ASSURANCE		FIELD SERVICE	
PRODUCTION ENGINEERING		CUSTOMER	
MATERIAL CONTROL			
CHAIRMAN CHANGE COMMITTEE			

Fig. 1.10 Example of ECR

ELECTROMECHANICAL PRODUCT DESIGN

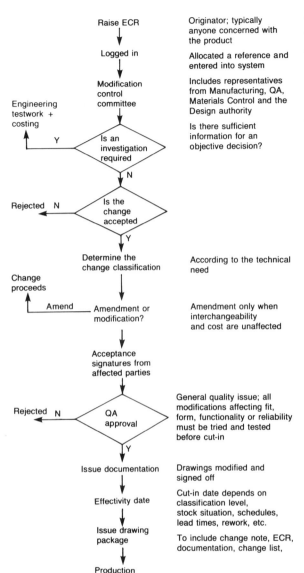

Fig. 1.11 The ECR process

the original drawing or specification. Typically this refers to discrepant parts identified on inspection that do not affect functionality.

(b) *Change concession.* This references a change request for immediate action by Manufacturing where a line stoppage situation exists, i.e. parts cannot be assembled, the product does not work, a temporary non-availability of stock, etc. In essence this is a Class 1 ECR authorized outside the normal change control procedures, but nevertheless still subject to the formal issue documentation process.

1.8.4 Change notice

An issue note, or as it is more usually referred to, a change notice (CN), is published to introduce any new or modified drawing, parts list or specification created as a result of change. In certain situations an issue note might be strictly referenced to a new drawing or documentation issue, however generated, with CNs referenced against alterations, i.e. raised issue level. In these cases, change references both sets of circumstance. Some company procedures, on the other hand, may choose to combine the ECR and CN into a single item to simplify documentation.

References

1. Ray M S *Elements of engineering design*. Prentice-Hall 1985
2. Roy R *Product design and technological innovation*. Open University Press 1986
3. *Engineering*, May 1982
4. Viscount Caldecote Investment in new product development. In *Product design and technological innovation*. Open University Press 1986
5. Hawkes, B, Abinett R *The engineering design process*. Pitman 1984
6. Osborne D J *Ergonomics at work*. Wiley 1982
7. Thompson G *The critical analysis of the design of production facilities*. I Mech. E, 1985
8. BS 5750
9. Leech D J, Turner B T *Engineering design for profit*. Ellis Horwood 1985

THE DESIGN PROCESS

CONCESSION REQUEST

ORIGINATOR	DATE	CONCESSION No.

PART A: ITEM IDENTIFICATION

SUPPLIER		GRN No.
PART No.	ISSUE	DESCRIPTION
ASSY No.	ISSUE	DESCRIPTION

No. OF ITEMS AFFECTED?   EXPECTED FINISH DATE

PART B: CONCESSION/PRODUCTION PERMIT REQUESTED

DESCRIPTION OF CONCESSION REQUIRED	REASON FOR REQUEST

FACTORS/FEATURES AFFECTED   YES/NO

Safety Approvals		Cost		Customer Approval	
Reliability		Spares		Delivery	
Function		Packaging			

PART C: CONCESSION AUTHORISATION

DEPARTMENT	COMMENT	SIGNATURE	DATE
Manufacturing			
Production Engineering			
Design/Development			
Quality Assurance			
On Behalf of Customer			

PART D: REASON FOR REJECTION OF CONCESSION REQUEST

Signed   Department

Fig. 1.12 Example of concession

CHAPTER 2
DESIGN FUNCTIONS

2.1 Introduction

In Chapter 1 we have outlined the mechanics of the design process itself, identifying elements in the process as clusters of design activity. These may run sequentially, but most likely overlapping to some degree. In practice there are any number of related wide-ranging functional activities that design engineers tend to carry out. Central to the design activity are awareness, organizational and managerial criteria that determine how ably the design engineer will function in his environment. Some of these criteria are considered in this chapter, from the organizational and co-ordinating role to the role of product champion, probing for innovative solutions to design problems and at the same time being aware of product legislation criteria. Finally, as an acknowledgement of effective market competition, some attention is given to the Japanese phenomenon and the Japanese use of the design function.

2.2 The role of the product design engineer

A designer is concerned with producing a product or solving a problem. The work has resulted from an actual or anticipated need within society and a company decision to proceed with the activity to address this need. However, the role of the design engineer goes beyond the designing feat itself, interacting with numerous support functions and activities. The design engineer needs to be involved in all stages of the design process and it is primarily his ability that determines the final product or solution form. It is the ability to design an efficient and effective product that sets apart the successful designer.

With small projects the design engineer may handle the whole product design process, whereas larger projects may be supervised by one or more project engineers with a number of designers involved. The project engineer is not necessarily an engineer *per se*, as he must carry a degree of organizational responsibility.

2.2.1 Types of design work

There are two main types of design work. Firstly there is development or adaptive design, where the start point is an existing design or product which is either being upgraded, updated or possibly used as a basis for an up-market or down-market variant (Fig. 2.1). Some producers deliberately avoid major product changes. Instead, by encouraging a process of continuous development, they achieve a machine of higher reliability at lowering cost over several years. Customization also comes into this category.

Secondly, the most demanding (and rewarding) work concerns the design of a new product that has no prior pedigree. It is sometimes referred to as creative design and represents the most challenging work. In practice, however, very few blank sheet designs are realized today, there is usually some aspect of the design which is historical.

2.2.2 The design engineering function

Design responsibility for a given product, subsystem or item involves more than just the physical design, it is for total integrity. In other words, negotiating development activity, ensuring performance goals are achieved, concurring manufacturability, etc. are all part of the job mix. Interfacing with numerous other job functions and departments, it has been said that the product design engineer must not only be a skilled negotiator but also well versed in the art of compromise.

DESIGN FUNCTIONS

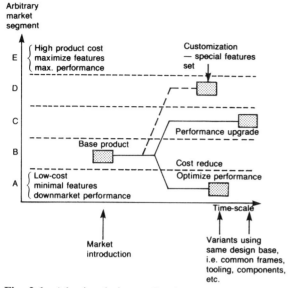

Fig. 2.1 Adaptive design application

The design engineering function is outlined schematically in Fig. 2.2. The main functional areas in which the design engineer operates are: configuration responsibility; development input; performance; drawing issue and database maintenance.

(i) Configuration responsibility

The advancement of a product design or system configuration is influenced by a range of design-related requirements from a number of affected parties. Often conflicts between different requirements result in trade-offs, usually dealt with as ongoing items. Aspects that the configuration design must address are as follows:

1. Cost;
2. Assembly;
3. Manufacturability;
4. Serviceability/maintainability;
5. Value engineering;
6. Operability;
7. Functionability;
8. Installability;
9. Product schedule;
10. Transportability;
11. Safety.

(ii) Development input

Typical development activities are as follows:

1. Establishing likely performance latitudes and sensitivities;
2. Response to performance problems;
3. Response to design problems (materials, mechanisms, etc.);

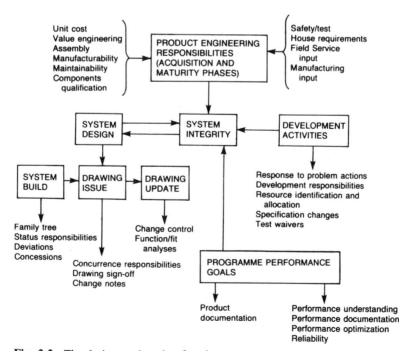

Fig. 2.2 The design engineering function

4. Specification changes;
5. Evaluation of design alternatives;
6. Component qualification/test.

Although certain development activities undertaken by the design team are essential for determining the design route, there are none the less some investigations best handled external to the team by specialist departments of consultants. Typically these are activities which can proceed in parallel with the mainline design.

(iii) Performance

Design performance is an ongoing evaluation of how well the design is doing in terms of reliability, cost, stability, mean time to repair (MTTR), etc. A major element of measuring performance (and evaluating the design) will be through reliability or stress-test work, especially in the earlier stages of design. Performance shortfalls will be reflected as an unstable design. As the design moves more towards Production, service cost and unit cost tend to be the performance targets. Tooling and spares cost estimates will reflect design quality.

Typical performance criteria are as follows:

1. Reliability growth, malfunction rates;
2. Service times;
3. Product documentation;
4. Service costs;
5. Machine cost;
6. Tooling costs;
7. Change rate.

(iv) Drawing issue and configuration control

As far as the design function is concerned, what is deliverable to the Manufacturing department is a drawing package, the confidence level of which can vary tremendously depending on the company, the product and the acquisition schedule. For acceptance by the Manufacturing department, the drawing package is recorded as a bill of materials on a database, outlining assembly levels (family tree). Past this point, the production line is balanced and any deviations and changes to assemblies and piece-parts should be subject to formal configuration control procedures. See section 1.8 for drawing-related functions.

2.3 The design team

Up to this point, the role of the product design engineer has been considered somewhat in isolation. In practice, a product design team may consist not only of design and development personnel but also representatives of Manufacturing, Field Services and the Quality Function. Whether the latter personnel are permanently located with the design team or only present at team meetings, depends on the nature of the product and company philosophy. Manufacturing will usually in turn coordinate and be responsible for inputs from Procurement, Industrial Engineering and Test Engineering.

An outline of member responsibilities is given in Table 2.1. The major responsibility of the team is to deliver

Table 2.1 *Design team member responsibilities*

Design	Manufacturing	Quality	Field service
Initial design concept	Tooling	Vendor training	Service procedures
Functional integrity	Vendor involvement	Supplier approval	Service documentation
Performance goals	Model build co-ordination	Quality audits	Service tools
Prototype build	Pre-production builds	Product verification	Spares list
Documentation issue	Assembly procedures	Warranty analysis	Service training
Test procedures	Test & set-up procedures	Field reliability	Customer training
Component qualification	Drawing concurrence	Quality circles	
Problem resolution	Procurement schedule	Total quality management	
Specifications	Line start		
Cost estimates	Co-ordination with:		
Materials	materials management		
Bill of materials	quality control		
Co-ordination with:	industrial engineering		
standards	product planning		
safety			
industrial design			
reliability			
model shop			
planning			

a functional design to a schedule that meets all the specification requirements of design, manufacturing and maintenance. The team would have developed design plans and be responsible for functional documentation issue, achieving integration goals, the build cycle and all relevant mileposts to line start-up.

The cost of ownership responsibilities of the team include the following:

1. Unit machine costs;
2. Service costs;
3. Tooling budget;
4. Inventory costs;
5. Special jigs/fixtures costs;
6. Premium payments;
7. Spares costs;
8. Change rate expenditure;
9. Standardization savings;
10. Value-engineered savings.

2.4 Design innovation and creativity

Much has been written about creativity and engineering innovation, what it is and what causes it to happen. The engineer can be classed as an innovator and creator of new products and processes that are aimed at solving problems or satisfying needs in a practical and economic fashion.[1] A doer if you like, or problem-solver, applying science and technology to solve problems and meet needs in society. So it is reasonable to ask what are the processes involved here and how can they be promoted. Firstly, however, some definitions. There is no complete agreement in the literature on the terminology used to describe the innovative spectrum, though the following outlines the generally accepted consensus:

1. *Scientific discovery.* Recognizes a natural object or discovery for the first time.
2. *Invention.* The creation of something technologically 'new' that has not previously existed and which is not obvious.
3. *Innovation.* Frequently used to describe the total process form concept through to the point of first commercial sale or social usage. To some extent it is awkward to refer to the whole process as innovation, but it is common practice in the literature.
4. *Process of technological innovation.* Broad interpretation to encompass the full range of activities from the initial design concept, through feasibility, assessment and into product acquisition, i.e. the technical, industrial and commercial steps to bring new products into the market.
5. *Creativity.* This is the initial step in the innovation process, whether the initial stimulus is a new form for an existing product or a novel principle and a technical opportunity, creativity is required. Both invention and design can be creative activities although the outcomes are different.

A design requirement or a design problem gives the design engineer the opportunity to use his inventive and creative skills. However, creativity and inventiveness are not the same thing as innovation. All engineers would like to produce solutions that display inventive ideas, but few in fact possess the skill. Likewise, many good inventors do not have the practical skills to successfully engineer and develop a total product.

Inventiveness involves new ideas and technologies, the ability to think of a new way of doing things, an applications opportunity to exploit, etc.

Creativity involves a blend of practical experience, innate ability and accumulated knowledge. To some extent it may be forced, but in general creativity is a resultant of open-ended and lateral thinking skills, flashes of thought, conceiving and evaluating alternative approaches. In other words, design engineers strive for creativity and inventiveness by searching for a 'best' solution.

2.4.1 Creativity

Creative individuals probably possess an ability for divergent thinking, are technically quite strong and have a positive belief in their ideas. Creativity in generating the inventive idea also implies a constructive desire for change. Some writers on creativity[2,3] emphasize the importance of chance in creative thought and generally speaking the 'act of illumination' or 'insight' does not usually occur without the backing of many years' preparation or experience. Throwing resources at a particular problem does not guarantee an outcome of creative ideas, it is far better to appoint competent designers in the first place.

Creative individuals are, however, still the most important source of original ideas but just having creative people on the payroll does not necessarily guarantee the development of new products, processes or procedures.[4] This assumption neglects all the other functions necessary for effective innovation. In analysing the creative process, a number of factors can be identified that either promote or hinder creativity (Table 2.2). Having recognized the

Table 2.2 *Factors in the development of the creative approach*

Positive	Negative
Adequate background preparation	Conditioned habits (conformity)
Review previous work	Fixed methods
Broad knowledge base	Repetitive behaviour
Assess all the possibilities	Prior training
Search for analogies	Fear of looking a fool
Get out of analytical gear (i.e. free wheel)	Inadequate knowledge
	One right answer
Encouraging environment — peers and management	Self-imposed barrier
	Failing to challenge the obvious
	Evaluating too quickly
	Inhibiting environment, e.g. restrictive management or over-critical peers

importance of creativity, a number of procedures exist that can be used specifically to promote creative analyses and stimulate or search for ideas. For example lateral thinking, brainstorming, synectics and morphological charts.

(i) Lateral thinking
This is the ability to make the brain move consciously away from a normal objective route to a solution and to explore divergent paths but, at the same time, ensuring that total diversification does not occur, i.e. continuously testing for association. Checklists, weighting systems and redefinitional aids may be useful in the open-ended thinking situation.

With lateral thinking one is allowed to be wrong on the way but right in the end,[5] i.e. one is not concerned about the nature of the information arrangement but instead where it might lead. It is not a matter of doing without judgement, but rather deferring it until later.

The main problem with this form of creativity is one's conditioning towards vertical thinking, the basis of our educational system where the expectation is about being right all the time, i.e. correct facts only as a method of ideas exclusion. Vertical thinking in itself, i.e. exclusive emphasis on being right all the time, is a major bar to new ideas, creativity and progress.

(ii) Brainstorming (forced creativity)
This is a widely applied technique to help groups of individuals create and recognize ideas and solutions.[6] The basic term has evolved into the principle of suspended judgement. As an extension of lateral thinking, deliberate alterations of thought processes are injected into a group of people so that they do not attempt imaginative and critical thought at the same time. It has been used with varying degrees of success, and often where it is unsuccessful the application may well not be appropriate. Without a successful outcome, people will be easily conditioned against the process. A poorly run session will likewise prove negative. The rules of brainstorming may be summarized as follows:

1. The group should not be overly large;
2. Elect a leader to record and provoke;
3. Choose people with various backgrounds;
4. The problem to be considered should be presented qualitatively rather than quantitatively;
5. Consider all the alternatives;
6. Seek quantity of ideas, not quality;
7. No criticism or judgement allowed during the session;
8. Encourage unusual and innovative ideas in addition to the more obvious ones;
9. Indicate possible combinations of ideas;
10. Objectively evaluate ideas after the session.

As a means of producing further stimulus, various analogous techniques can be used to support the main brainstorm session, e.g.:

Trigger session — where group members individually write down ideas before brainstorming them.
Wildest idea — within the context of the problem, an idea to break down thought boundaries and see what turns up.
Reverse brainstorming — turn the problem upside down and examine the paradox.

Table 2.3 is a summary of problems and subjects both suited to and unsuitable for brainstorming.

Table 2.3 *Suitability of subjects for brainstorming*

	Part A	
Type of problem	Suitability	Example
New product concepts	Lot of ideas needed to obtain balanced judgement	Expansion of product range Response to competition Market opportunities
Problem solving	Need rapid collection of views Identify maximum number of causes	Reliability problem Anticipating likely fail modes in advance
Management	Need a wide cross-section of viewpoints	Improving resource productivity Improving quality
Product improvements	Need a wide as possible collection of views	Reducing costs (value analysis) Improving reliability, serviceability, etc.

	Part B	
Types of problem	Unsuitability	Example
Diffuse and complex problems	Sessions only produce vague solutions which in essence just redefine the problem	How do we diversify the product range? What new technology should our products have?
Specific technical issues	Requires homogeneous group with experience	Specific mechanical or electrical design requirement Component selection
Problems with closed solutions	Only a small number of correct answers possible	What material should a piece-part be made from? Where can we get this piece-part made?

(iii) Synectics

Again a group behavioural activity for developing creative talent,[7] but unlike brainstorming, it depends on a leader to guide the process to a satisfactory conclusion. The main aim is to solve practical problems with a miscellaneous group of imagination and ability but of varied interest in the subject. A typical synectics session is conducted by having a group of about five to seven people and at least one of the group must be competent in the field of the posed problem. The typical synectics procedure is as follows:

1. A leader for the group is chosen, but is not allowed to contribute ideas of his own. However, he should be responsive to all contributions made by group members.
2. Firstly the leader will state the problem and the expert will answer all questions asked by the group and react to any ideas and comments advanced by them. This process removes any misconceptions and may produce some useful ideas.
3. In the next phase, each member of the group formulates a set of problems arising from the previous phase. The leader chooses one of these and declares a key concept.
4. The group then tries to think of as many analogies to the key concept as possible.
5. After a period of analogy searching, the leader starts the final phase by attempting to apply some of the unrelated concepts to the problem posed. If an idea looks useful and the group expert is in agreement, it is explored further.

Some elements of the synectics process, together with some positive and negative features, are outlined in Table 2.4.

(iv) Morphological charts

The aim here is to widen the area of search for a solution to a design problem by forcing divergent thinking. The method has been used to some success with engineering problems,[8] but it does require technical knowledge of

Table 2.4 *Elements of the synectics process*

Definitions	
Leader	Process not content oriented. Job is to enhance the operation of the group through his/her direction. Does not evaluate ideas or produce ideas
Client (the person with the problem)	Content not process oriented. Tries to be receptive to group stimuli from which new ideas and concepts might gel
Support group	Members try to provide fresh stimuli for the client. Not really concerned about learning technical details but try to throw in a lot of idea statements
Itemization	Systematic treatment of each idea or statement in a fair and constructive fashion. Careful records of each development should be kept during the meeting
Goal orientation	Attitude of mind that is constantly questioning the existing approach and seeking new ways of looking at a problem. Accept differences of opinion as positive attributes.

Positive aspects of process

Imply that the problem is an open-ended one
That there are different ways of approaching it
That the group can contribute useful ideas without having to become experts in the problem
Makes it easier to inject novelty and 'something different' from being outside the problem

Negative aspects of process

Might put the expert on the defensive
May hide a potentially good idea if replies are biased to reduce apparent value
The session could be directed towards previously covered ground
The support might end up with the expert's view of the problem thereby limiting their capacity to suggest new ideas
Risk that a person in the group might try to act on an individual basis, trying to win support for his/her ideas, instead of advancing the problem as a group activity

the problem environment itself, in other words the solution is bounded by known facts, the form of the design, which in itself may appear a more tangible and attractive process to engineers. The process outline is as follows:

1. Define the functions that any acceptable design must be able to perform. List these vertically on a chart. These should be reasonably independent of each other.
2. List horizontally, against each function, a wide range of sub-solutions or alternative ways of performing each function.
3. Select an acceptable set of sub-solutions, one from each function, to arrive at an alternative design approach. Scoring or weighting processes can be used to arrive at a rational choice between alternative sub-solutions.

2.4.2 Innovation

This represents the technical, industrial and commercial steps that bring new products on to the market or new processes into use. Technology innovation has been defined by the Organization for Economic Co-operation and Development (OECD)[9] as 'the transformation of an idea into a new or improved saleable product or operational process in industry or commerce'. It has, however, been suggested that the major innovative advances are more likely to result from a new technical capability (technology push), whereas smaller and more incremental improvements are likely to be need induced (market pull). Competition is exerted through the timing and cost of innovations and it is the close time and cost factor interplay that determines profitability. A good invention is not enough, the innovation cycle must be planned and be successful as a whole.[10]

Today, innovation is an organizational function although the most successful innovators in history have been inventor/entrepreneurs rather than merely producers of creative ideas. The entrepreneur has become the entrepreneurial function.[11] Hence,

Innovation = engineering team + entrepreneurial function

Successful innovations have few after-sales problems, have fewer technical bugs in production and they need less user modifications. There are a number of characteristics common to successfully innovative organizations and these can generally be summarized as follows:

1. Understanding the user requirements better;
2. Early and imaginative identification of a potential market;
3. Devote more effort to customer education;
4. Association with basic research, benefit from external technology;
5. A strong in-house R & D with the ability to fund it;
6. Use patents for protection;
7. Ready to take some risks;
8. Employing 'new-product' champions;
9. Encounter less internal opposition to innovation (i.e. change);
10. Good internal communications;
11. Co-ordination of all affected functions.

2.5 Product safety

2.5.1 Design liability

Product liability is not a new concept. It is the liability of designers, manufacturers, distributors and retailers for injury, loss or damage caused by *defects* in products which they design, produce or supply. There are three main causes of product defects, namely: a design defect, a manufacturing error, a failure to warn.

As alleged causes of liability suits, the breakdown of occurrence of the above is 39, 37 and 21 per cent respectively.[12] Hence not only is the quality of the design important but so is the quality of conformance to that design. Users expect a product to be safe, although statistically every product carries with it some risk. However, public awareness is changing as legislation is providing greater provision for compensation against defective products.

Liability for defective products can be dealt with in contract, tort and the Consumer Protection Act 1987. The relevant legislation on product liability is as follows:

EEC Directive on Liability for Defective products (85/374)
Consumer Protection Act 1987
Safety of Goods Act 1984
Provisions of — Trade Descriptions Act 1968
— Consumer Safety Act 1978 and 1986
— Health and Safety at Work Act 1974
— Sale of Goods Act 1979

With contractual liability, certain terms are implied into every contract for the supply or sale of goods, made in the course of a business. The goods must conform to the contract description, be reasonably fit for their purpose and be of merchantable quality.

Contractual liability is strict, if a product is defective, the seller acting in the course of a business will be liable, whether or not he was at fault. Tortious liability, on the other hand, is a breach of duty, other than under contract, with liability for damages. A manufacturer has a duty to the ultimate consumer to take reasonable care to ensure that, when his product is used in the manner intended, no physical damage is caused to the consumer, his property or his livelihood.

The 1987 Consumer Protection Act represents a milestone in bringing strict liability to the statute-book. In principle, anyone who supplies a defective (unreasonably unsafe) product will be liable for the damage it causes. Liability exists for 10 years after the sale.[13] Liability is for death, personal injury or loss of or damage to property caused by a defect in a product. 'Product' includes all goods and includes a product which contains another product. There is a defect in a product if the safety is not as one generally might expect. Additionally, the officers of a company (i.e. directors, secretary, etc.) may also be held guilty of an offence if they supply any consumer goods which do not comply with 'the general safety requirement' of the Act. (Under the definition of consumer goods, aircraft and motor vehicles are excluded.)

In determining the defectiveness of a product, the following are taken into account:

1. The way in which the product is marketed;
2. Any instructions provided;
3. Any warnings given;
4. Foreseeable application and intended use;
5. The time of sale.

Consequently the definition of a 'product' has to be revised to include the following:[12]

1. The actual product;
2. Packaging and labels;
3. Installation and use instructions;
4. Warranty documents;
5. Spares;
6. Sales literature, catalogues and advertising.

Thus what do these changes in strict liability mean to the designer? Undoubtedly many designers have regarded safety as an integral part of the design process and factors which affect fitness for use are expressed/implied as part of their design brief. Questions to be considered as part of design are 'Where is this product to be used, for what purpose and what is the likely abuse?' Familiarity with codes and standards is essential. Warnings are not a substitute for safer design, but correctly worded and placed labels have been favourably recognized by the courts.[12] Warning labels need very careful consideration, see section 2.5.2.

Thus a company must be reasonably careful when designing a product in that not only must the design be safe for the intended use but it should also allow for foreseeable misuse. Where injury results a simplified summary of the main points affecting design liability is as follows:[14]

1. If the risk of injury was not reasonably foreseeable then the designer cannot be liable;
2. If the risk of injury was reasonably foreseeable and obvious to the user, the designer cannot be totally liable;
3. If the risk of injury was foreseeable to the designer and reasonable design precautions were taken then the designer cannot be totally liable;

4. But if the risk of injury was reasonably foreseeable and was not obvious to the user but the designer took no reasonable steps to warn, then the designer is liable.

There is no simple answer how to reduce design-related risks, but each company must be aware of product design safety liability and establish a strategy to identify and manage any safety-critical parts as part of the design process. In some circumstances it may be necessary to design out certain product functions in order to preserve product safety. The design review, (section 1.6.4) is an appropriate vehicle for identifying and evaluating potential safety hazards and for checking safety features against appropriate criteria in the specification.

2.5.2 Hazards and risk management

The surest method of avoiding a product liability claim is via a product safety awareness programme which should ensure that the right measures are taken at every stage of the design, manufacture and marketing processes to identify and correct any defect that might present a danger to safety or health. Even if an accident then does occur, such a process can provide valuable evidence in court that a company did all it reasonably could to supply a safe product. The quantities considered should relate to the topics given under the headings below.

(i) General design
The design process should be carefully documented and records kept, since in the event of a product liability claim it might be necessary to show how decisions were reached. It is particularly important to record any contribution that customers or distributors may have made to the design process. New product specifications need to be reviewed early in the design cycle to identify safety features, as well as reviewing the safety aspects of the configuration during each design phase (section 1.6.4). The use of techniques such as Failure Mode and Effect Analysis (FMEA) (section 3.4.1) can be used to assess the risk potential in a design. Safety features should also be thoroughly tested (section 3.5). As a result of the Consumer Protection Act 1987, manufacturers and designers have added responsibility for carrying out research to discover potential risks which might arise from a product design.

The following factors which influence the design of a product would all come under scrutiny in the event of a product liability claim, some of which are also dealt with in section 1.5.4:

1. The level of customer education provided plus the experience and expertise of the persons expected to operate or maintain the product;
2. The environment in which the product is intended for use;
3. The intended use of the product;
4. Foreseeable misuse of the product;
5. Government laws and regulations or codes of practice and industrial standards (see Ch. 17);
6. The life of the product and its critical components.

(ii) Safety devices
The design should include as standard equipment appropriate devices to prevent injury from intended use or foreseeable misuse. There should be no difference in safety standards dependent on market destination. Plaintiff lawyers would take full advantage of any variation arguing that the safety devices should be on all products.

Normally no alteration to the basic design in respect of safety features should be agreed to for whatever reason. However, if such an arrangement is unavoidable, a disclaimer agreement must be obtained from the customer.

(iii) Labelling
Subject to language differences, labelling standards should be uniform, regardless of the market destination. Labels should use a mixture of pictographs and words and comply with six basic tests:

1. Indicate the level of the hazard. This can be equated to the seriousness of the hazard × the probability of the hazard occurring.
2. Adequately describe what the hazard is.
3. Show and tell. State what particular actions or uses of the product present the hazard.
4. Describe the consequences of use or misuse.
5. Instruct the user how to take care and avoid injury; avoid the hazard.
6. Locate labels where they can be read and present precise and definitive information.

The general hazard warning sign as applied to the majority of electromechanical products is a black triangle on a yellow background with a central '!' (Fig. 2.3).[15] Other signs may indicate the hazard within the triangle, e.g. fire, radiation, moving parts, etc. Three basic types of hazard warning labels are illustrated in Fig. 2.4 and in practice supplemented by the appropriate text, e.g. Fig. 2.5.[16]

When attaching safety labels to the product, the following factors should be taken into consideration:

DESIGN FUNCTIONS

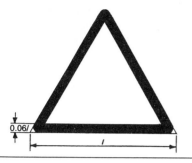

Warning signs
Background colour shall be yellow
Triangular band shall be black
The symbol or text shall be black and placed centrally on the background. Yellow shall cover at least 50% of the area of the safety sign

Safety colour	Meaning or purpose	Examples of use	Contrasting colour (if required)	Symbol colour
Yellow	Caution, risk of danger	Indication of hazards (fire, explosion, radiation, chemical, etc.) Warning signs Identification of thresholds, dangerous passages, obstacles	Black	Black

Fig. 2.3 Design layout of a warning sign

1. The print must be large enough to read, taking into account the likely viewing distance and lighting conditions.
2. The wording must stand out from the background and standard warning symbols must be used.
3. The label should be placed where it can be seen, but not where the user gets too close to the hazard before seeing it.
4. The warning label should be legible throughout the life of the product, i.e. the print should be rub and fade resistant as well as scratch and peel resistant.[17]
5. The warning label must be permanently attached so that it remains in place throughout the life of the product. If adhesive backed, the adhesive should be durable with age.

(iv) Instruction manuals

The product manual is referenced in sections 1.7 and 3.3.3 as a documentation activity during the maturity phase. However, there are clearly safety implications, not only in the way that instructions for use and maintenance are presented but also in their availability and accuracy. The following guidelines relate to manuals:

1. Keep it simple. It is important not to overestimate the capacity of the user. Simple language should be used and technical terms avoided.
2. It should include a reference to the importance of operator and maintenance personnel training.
3. Instructions must be complete.
4. Instructions must be technically and factually accurate. If products are modified, it is essential to amend the instructions. Updating of instructions should be considered as part of the maturity phase.
5. Instructions must be feasible and practical. The user must be able to carry out the actions required safely.
6. Instructions should apply to one product at a time if at all possible.
7. All warnings shown on the product should be repeated in the instruction manual. It is important to explain why the instructions should be followed and what will happen if they are not. The manual should contain warnings concerning:
   (a) the consequences of tampering with the product and removing safety devices;
   (b) the hazards of repair using unauthorized parts;
   (c) the hazards of repair by untrained personnel;
   (d) the consequences of foreseeable misuse and improper maintenance;
   (e) the inherent dangers in the product and how to avoid them, i.e. finger traps, rotating parts, etc.
8. The manual should be attached to the product. The design should incorporate an internal pocket in the cover, etc. for safe keeping. Users should be told to keep the manual in a safe place.
9. Send the manual with the product.

High standards must also apply to sales and promotional material, data sheets and product safety information, all of which is supplied to purchasers. All advertising claims

An immediate hazard with a likelihood of severe personal injury or death if instructions are not followed

The presence of hazards or unsafe practices which could result in severe personal injury or death if instructions are not followed

Possible hazards or unsafe practices which could result in minor injury, product or property damage if instructions are not followed

The guidelines provided will assist in the safe operation of the machine system

Fig. 2.4 Types of hazard warning

must be justified and wording such as 'foolproof' and 'fail-safe' should never be used. A product is likely to be judged defective if user information is incorrect or inadequate. This also applies to design risks that might emerge after the product is supplied.

2.5.3 Standards and codes

The design engineer will frequently refer to, and make substantial use of, a wide range of standards and codes of practice, particularly during the acquisition phase

DESIGN FUNCTIONS

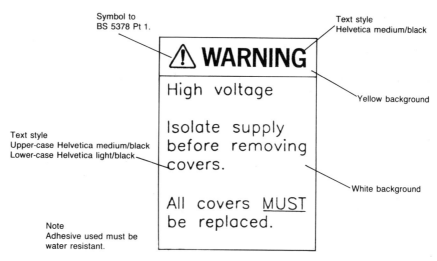

Fig. 2.5 Example of warning label

design activities. They provide a readily acceptable set of guidelines (Ch. 17). Many have no legal requirement enforcing their use, although in practice, conformity is invariably a contract requirement. Their use enables a common standard of good design practice and a safeguard against product hazards.

2.6 Packaging

Outside the factory, product packaging influences the customer's perception of the quality of that product, not only in terms of the artwork and appearance but also in respect of the physical protection afforded. In respect of the latter, it is important to ensure that the quality level of the product does not diminish during shipment to the customer. Product damage during shipment is expensive for everyone, hence the product design engineer should be sensitive to the handling procedures likely to be encountered, using laboratory test work to identify and correct weaknesses in the product as early in the design cycle as possible. The ideal situation is a sufficiently rugged product that requires a minimum amount of additional protection from the package. Packages should be designed to be as easy as possible to handle in order to reduce shipping damage. Very heavy packages should be on a pallet.

In general terms it is important to know or understand the fragility of the product and to be able to compare this to the expected hazards of the distribution environment. This may be expressed as follows:

Distribution hazards ≡ product durability + packaging protection

The essence of balancing this equation can be expressed in Fig. 2.6 where 'inherent ruggedness' is assessed from laboratory shock, vibration and environmental testwork. There is clearly some trade-off between product 'ruggedness' and the amount of packaging protection. It can be argued that only enhancements in ruggedness are of value to the customer since the packaging is thrown away.[18]

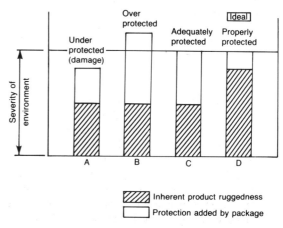

Fig. 2.6 Demonstration of hazards of the distribution environment (from Kent, G.C., *Industrial Engineering*, August 1989, p. 51. Courtesy of IBM)

2.7 Design protection

In the commercial world, industry needs to take care of its own interests by protecting its inventions, new technologies, processes and products in the market-place and prevent other people using designs or logos which make a particular business identifiable. Ideas, designs, techniques and know-how fall into an area known as intellectual and industrial property and are protected by laws on patents, copyrights, designs and trade marks.[19] Some laws offer protection without any effort from the design team, whereas others require registration, professional advice and an expense to process. A company will often leave the complexities of obtaining and protecting intellectual property to patent agents and legal specialists, simply because of the legal jargon and the unfamiliarity of the system. However, professional advice is expensive and is included in the cost of ownership. A company must balance the cost of protection against the likely commercial value of the innovation. Where technology is changing rapidly, there is clearly less value in protection.

In August 1989, the Copyright Designs and Patents Act 1988 came into force in the UK, implementing some significant changes in the protection of industrial designs. Whether functional or decorative, mundane or innovative, the options available to designers of 'industrial production' on how to protect their designs in light of the new legislation, are summarized in the following subsections.

(i) Patents

Patents are granted for the protection of inventions capable of commercial application in a product. In return the inventor receives a 20-year monopoly from the state. Although patents may be granted in virtually any field, the invention must fulfil certain criteria to be granted patent status:

1. *The idea must be new.* It should not be known anywhere in the world by the public in any form. Disclosing the product idea prematurely to a customer or potential supplier before the patent application is filed can destroy the novelty requirement. During the feasibility phase, when information regarding the enabling technology of a new product is often too sketchy to file a patent application, the contractual obligation of design team confidentiality ought to apply.
2. *The idea must be inventive.* The idea must not be routine or trivial. The test is whether someone 'skilled in the art' would regard the idea as being obvious.
3. *The idea must be capable of industrial application.* Whatever the industry may be, patents generally protect the tangible rather than the intangible, e.g. theories are excluded.

(ii) Copyright

This differs from patents in a number of respects. No registration procedure is necessary since copyright comes into existence as soon as the original work is created. Copyright can protect designs that have no inventive feature and that would therefore not qualify for patent protection. It can also be used to prevent someone devising a similar work independently. Protection for industrially produced products is now restricted to cases where the article itself is an 'artistic' or 'literary' work, e.g. design drawings, product literature, instruction manuals, etc. excluding functional articles and machine parts. Copyright will apply for 25 years.

(iii) Unregistered design right

This is a new creation to provide a regime for the protection of functional designs. The design right arises automatically and covers the internal or external shape of articles. It does not confer a monopoly on the design, merely a right to prevent copying, and lasts 10 years from the first marketing or 15 years from the design, whichever is the less. The right is subject to the 'must fit' and 'must match' exceptions which provide that the right will not apply to features of shape or construction that (a) enable the article to be connected to another so that they both function (must fit), or (b) are dependent on the appearance of another article of which the article is intended by the designer to form part (must match). An example of unregistered design right protection might apply to a filter casing except for the actual means of connection to another item.

(iv) Registered design

If an article has an appearance that appeals to the eye of a potential buyer, then providing the features do not depend on a method of construction or the form of the article, then it can be registered as a design. Such features can be shape, configuration, pattern, etc. which must be a result of an industrial process. The appearance has to be relevant to the buying decision. Registered design protection is comparable to that of patents in that the owner has an enforceable monopoly, even against an entirely independent designer. Protection is for 25 years. In common with patents, the registration can be invalidated by prior disclosure or use of the design.

2.8 The Japanese style

The story of the Japanese economy since the Second World War has been one of remarkable success to the point where Japan is the major industrial competitor to the Western world. Everyone is aware of the many product areas now dominated by Japanese products. Since they are clearly getting it right consistently it is worth while devoting a section of this book to the way design and product development is handled in Japan. This is not to say that we should follow Japanese methods in total — differences in European and Japanese cultures would not allow that. In essence, British and Japanese design management is very similar in terms of their objectives and technology base, and the design process outlined in Chapter 1 closely parallels that of the Japanese in all important aspects. Having read this section, many design engineers will recognize the advantages and the motivating aspects of the Japanese style. It is true to say that the vast majority of engineers in the UK want to do a good job, but they become stifled and frustrated by company systems. In Japan that cannot happen.

We shall consider the following aspects of Japanese engineering and the product acquisition process:

1. The Japanese business environment;
2. The Japanese work regime;
3. The design process;
4. Product initiation;
5. Product acquisition; and
6. The production process and vendor utilization.

Now there are clearly aspects of Japanese culture, i.e. hierarchy in society, social problems, work ethics, etc. that influence employee attitudes which are possibly inappropriate in a design engineering context, but which nevertheless are considered by some to be very relevant to overall business performance. Some of these have been discussed elsewhere.[20]

2.8.1 The Japanese business environment

There are continuous, confident and committed links underlying the commercial realism in Japan between industry and government investment banks and universities, all sectors of the economy recognizing that co-operation at national level is vital. Co-operation is also apparent between competitors, companies often working together to set new technology standards. There is very high investment in R & D, not only in design but also in manufacturing systems, which is obvious in the technical infrastructure for its products. Long-term company survival in Japan is more important than short-term profitability, and investment and training is continuous. In times of recession there is significant retraining.

The Japanese company will always take into account very carefully the changing needs of the market, it will extrapolate to the future and plan a response,[21] i.e. it is always planning the next products. Design is central to corporate policy, therefore it has direct management visibility, managers are more at home with the design emphasis and consequently a design engineer will have senior status in the system.

2.8.2 The Japanese work regime

This is designed to promote co-operation and team effort, it is the Japanese way, a natural outcome of their culture. Attributes of their work regime in engineering design areas may be summarized:

1. Individual achievement is not so important in Japan as it is in the UK. Individual expression is not encouraged, teamwork is the rule and work pressure is to conform. Japanese companies will reward those who help to make the team better.
2. The work ethic is for exclusive togetherness, encouraging consensus and compromise, so avoiding confrontation. The cohesion of the working group is paramount.
3. Japanese management tries not to be overly tough and impose decisions. Instead it would prefer to delay decisions and request further information if it feels it has an unpopular choice to make.
4. Communications are very good at all levels, everyone has a clear visibility of the goals.
5. Employees demonstrate clear mutual respect, there is none of the 'them' or 'us' or 'not invented here' syndromes. This, of course, makes the management task a lot easier.
6. Good education and motivation are taken for granted. Functional boundaries are simply not there, with employees expected to have more than one role. There is clearly an emphasis on organizational flexibility with effort concentrated on breaking down the barriers.
7. The work environment is not one of finger pointing, an engineer will monitor his own performance and ask himself how he can improve his own job effectivity.
8. The Japanese engineer's activity span is very broad. Although higher academic standards are often clearly

visible, he will prepare his own sketches, layouts, detailing work and checking, undertake model builds, debug and test as well as evaluate the technology. The Japanese engineer is thus clearly very much an all-rounder.

9. Peer pressure is a very strong Japanese motivator within groups for a design engineer to produce quality work. A commitment to high-quality standards is the stated key to success and all team members are so encouraged. A Japanese will lose face when he delivers late or poor-quality work.

2.8.3 The design process

The Japanese have a great respect for the design process, in new product development the emphasis is on speed and flexibility, discipline within the team is achieved by focusing on the objectives, decision-making is timely and vendor co-operation/involvement is essential.

The design schedule is a linear work flow from initial concept to pre-production machinery, no formal phase gates between model builds but exhaustive testing and near-constant meetings to discuss performance and assess progress. This is enabled by the project team, a multidisciplinary activity fusing design resource with sales, production, marketing, patents, etc. The project team in Japan is very much viewed as an entity, people joining and leaving as required, the make-up changing as the product develops into production, so providing the momentum and drive.

The design cycle is rapid and iterative, quality and schedule are most important, with cost and features managed in parallel but not to the detriment of schedule. Effective designs are productionized very rapidly, compared to the West, and this is due in part to an early and stable product specification and also to the 'hands-off' design approach, the rapid, multiple and empirical hardware design iterations, the multi-disciplinary team consensus process and the direct management involvement with design. Design integration is performed by the program manager, he is a hands-on generalist who maintains the overall layout and manages the system interfaces.

In Japan, quality has always been a way of life, being built into the manufacturing system at the design stage. Quality is something for the whole company, not just the production line. In fact the design process is managed to ensure that no final inspection is required.

2.8.4 Product initiation

New product development is achieved by consensus and team effort. The planning team consists of personnel from Marketing, Field Service, Engineering and Planning departments which aim to reach an early agreement. They refer to this consensus activity as the 'upstream control system', and trade-offs, performance, etc. will be agreed that both strengthens the product and proves beneficial to all.

Although the Japanese design to achieve timely customer needs in both the global as well as domestic markets, in Japan the prime market intelligence vehicle is the customer, whereas in the West we concentrate on market research and competitive product evaluations. Although Japan has four times as many engineers per capita as the UK,[20] a knowledge of general business skills and marketing is expected of design engineers.[21] Designers are responsible for collecting market information themselves, especially consumer life-style and social behaviour data, and are encouraged to travel overseas, to be close to the markets and be aware of changing trends.[20]

The engineering process is initiated with a set of product and product performance specifications. Defining the product objective is known as *Nemawashi* (the 'root-binding' procedure) which infers agreement from all functional departments. Although the Japanese may take a long while to come to a decision, once that decision is made there are very few difficulties in carrying it out.

2.8.5 Product acquisition

Upon completion of the feasibility stage the drawing process is initiated. In Japan, product development is an iterative process, a firm specification and rapid hardware cycles combine to produce the stable product objectives. Understanding results from doing; change is expected early in the development cycle and is a natural element in the process, early trial and error empirically results in product quality and a stable configuration at the point of launch. Many versions of pre-launch hardware are built and extensively tested, aiming to approach a mature design early in the acquisition cycle, so removing the need for expensive downstream test work. To produce the hardware-mature design at the end of acquisition, the early development of the pre-production intent configuration is important; hardware fit and form problems are sorted out early by instant model iteration, and critical problems are identified and worked with the help of vendor involvement. In the UK, the prototype, prototype upgrade and pre-production are the typical model-build iterations (outlined in Chapter 1). But in Japan there is an engineering model, an engineering model upgrade, a prototype, a post-prototype and pre-production build, all in very compressed time-scales.

To reflect design intent, it is usual to use a soft documentation system, i.e. often hand-drawn or modified drawings are utilized for configuration control. There is much less attention to formalized drawings in early builds.

2.8.6 Production and vendor utilization

Purchased materials in Japan can account for 70 per cent of the manufacturing costs of electromechanical products. Buyers, therefore, must rely on vendors as though they constitute another department, and in recognition of the critical role of purchased parts in their industry, the Japanese have evolved a different style as follows:

1. Subcontractors/vendors are utilized in the design of complex, long lead parts or subassemblies.
2. Vendors are selected early for critical parts, they are utilized on site and will have access to hardware.
3. The relationship between vendor and buyer is one of interdependence and co-operation rather than conflict.
4. The vendor will often verify function prior to shipment, hence no goods-in inspection required.
5. Vendor selection is not only based on samples and price but also on loyalty and dependability.
6. Because quality is important, buyers will reject substandard components even if this slows down the production schedule, thus manufacturers make strenuous efforts to co-operate with vendors to achieve high quality standards. However, surprisingly enough, preference for quality can be subordinated to the need for mutual understanding and trust with the supplier.[22]
7. The vendor participates in the preparation of drawings to ensure manufacturability and cost effectiveness.
8. Lead times are very much shorter in Japan.

2.9 Concurrent engineering

Concurrent or simultaneous engineering offers an integrated approach to new product introduction. Using product champions and multifunction teams, it ensures that design, production engineering, manufacturing, marketing and sales groups work in parallel from concept through to final production. The concept is not new, it has been practised for some time in one form or another by the more enlightened organizations. Concurrent engineering reflects the aims of this book and in particular those given in Chapters 1 and 2. The exact definition of concurrent engineering will differ in its application from one organization to the next, but an accepted explanation[23] proposes 'concurrent engineering as a systematic approach to the integrated concurrent design of products and their related processes, manufacture and support ... this approach intended to cause designers from the outset to consider all elements of the product life cycle from concept through to sales including quality, cost, schedule and user requirements'.

For many companies the term 'concurrent engineering' merely puts a label on activities that have been growing for years, the driving pressure being competition, the continual need for higher quality products and faster development cycles, even as products become more complex. The Japanese in particular have long recognized the value of a team approach to product creation, problem-solving and continuous improvement (section 2.8) to enhance product quality and timing using simultaneous engineering.

The ability to engineer new products concurrently provides for the critical competitive advantage in terms of response to market change, product quality and timing. The ability stems from four different aspects:

1. A clear vision of the market position and a clear means of achieving customer enthusiasm.
2. An organizational structure that makes it easy to communicate horizontally and encourages teamwork.
3. Information systems that ensure knowledge and data are freely available to everyone that needs it.
4. An entrepreneurial spark to encourage initiative and risk-taking.

Concurrent engineering does not merely consist of computer-aided design, engineering and manufacture (CAD/CAE/CAM); nor is it strictly management strategy — what makes concurrent engineering different is its unique blend of culture and technology. Teamwork plays a vital role in concurrent engineering, it is built around cross-function teams from Design, Manufacturing, Materials Control and Marketing, requiring extensive interdisciplinary co-operation and integration. Major decisions are borne of consensus, an agreement of the most logical solution. However, the situation must be properly managed, emphasizing the communications paths rather than organizational hierarchies. In the design process, CAD/CAM/CAE tools do play a vital role, providing 'soft' models and components to help all team members visualize what it is, how it can be assembled and how it will work. However, computer tools cannot do it alone, they do not replace the talent, innovation and experience of the development team.

Improved quality is also a fundamental goal of concurrent engineering. Most of this comes in the form of paying attention to detail, getting it right first time, catching mistakes when they are easy to fix, i.e. designing it in rather than adding it on. Alongside and incorporated into the concurrent engineering philosophy are a number of techniques, principles and concepts designed to integrate design and system manufacture, reduce lead times, increase product quality, reduce life-cycle cost and so on. These are summarized in Table 2.5.

Along with the concurrent engineering ground swell, there comes the inevitable hype, speculation and misconception. It might be a common-sense idea whose time

Table 2.5 *Summary of some techniques that may be associated with concurrent engineering*

Design for assembly (DFA)	Covers the use of systematic evaluation techniques which generally promote the importance of certain desirable characteristics in a product design such as reducing the number of parts, ensuring parts are easy to handle and easy to insert or fix. At a higher level of interpretation, DFA refers to rules, advice and DFM guidelines
Design for manufacture (DFM)	The objectives of the DFM approach are to identify product concepts that are inherently easy to manufacture, to focus on component design for ease of manufacture and assembly, and to integrate manufacturing process design and product design to ensure the best matching of needs and requirements. The process considers product form, function and fabrication as well as the procedures underlying the design process
Just in time (JIT)	A philosophy of production which concentrates on throughput or responsiveness of the whole system as key measures rather than utilization and return on investment of individual machines or processes. Applied to production, JIT leads to minimized work in progress to reduce money tied up in stock, expose quality or process control problems early and provide the ability to respond quickly to demand
Intelligent manufacturing systems (IMS)	In the context of concurrent engineering, IMS involves the further development of production technology through scientific systemization, applied to, e.g. human factors, new materials, technology standardization, information infrastructures, etc.
Quality engineering (QE)	The price of a product represents a cost to the customer, and so does poor quality. The QE goal is to reduce total loss to the customer. Applying the quality–cost trade-off means it is necessary to predict quality loss at product design and production phases and thus means any engineering that serves the customer's requirements for appropriate and reliable products
Knowledge-based engineering (KBE)	This is intelligent product definition, based on a product's function, backed by a complete set of rules which enable the designer to meet a functional specification. The key is that the KBE system captures both the geometric and non-geometric information that defines design intent
Design for logistics (DFL)	This is a phrase which is used to point out that the product design has an effect on the logistics of its manufacture; the aim is to make products JIT friendly. The process would typically identify long lead items, common assembly levels, sub-assembly modules, unique components, etc.
Poka-Yoke	This literally means 'mistake proofing' and is a concept promoted by Toyota. It advocates the use of low cost in-house process quality control mechanisms to reduce the need for costly inspection or reliance on statistical quality control. The assumption is that when defects occur, the system and not the operator is at fault. Principles are to control upstream as closely as possible to the source of the defect, establish controls in relation to the problem severity and do not delay improvement by over-analysing
Product delivery process (PDP)	Process used by Xerox for product development and production. PDP brings together the functions of development, manufacturing, marketing and customer operations in a joint managing partnership for product delivery. The phases of a product life cycle are identified as preconcept, concept, design, demonstration, production, launch and maintenance. The PDP process is regarded as an extension of a quality philosophy into the product cycle in a way consistent with the JIT philosophy in production
Taguchi	This is a strategy for off-line quality control conducted at the product and process design stages of the manufacturing cycle to improve product manufacturability and reliability and to reduce product development and lifetime costs. The method seeks to identify a combination of design parameter values by conducting a series of factorial experiments and statistical techniques

has come, but the message sometimes gets diluted. It is a product management philosophy that is personal to the organization practising it, there are no rights or wrongs, better or worse routes. The chosen path should evolve from a company's own culture and support technologies.[24] However, without exception, the bottom line is to maximize the product's profit potential by getting it into the market-place quicker than previous development practices may have allowed.

References

1. Duderstadt J, Knoll G, Springer G *Principles of engineering*. Wiley 1982
2. Schon D *Technology and change*. Pergamon 1967
3. Glegg G L The design of the designer. In *Product design and technical innovation*. Open University Press 1986
4. Roberts E B Generating effective corporate innovation. In *Product design and technical innovation*. Open University Press 1986
5. de Bono E *Lateral thinking*. Ward Lock 1970
6. Osborne A F *Applied imagination*. Scribner.
7. Gordon W J J *Synectics, the development of creative capacity*. Harper 1961
8. Jones J C *Design methods*. Wiley 1970
9. OECD *The measurement of scientific and technical activities*. 1981
10. Whelan R Knowing how to deal with know-how. *The Engineer* **264** (6826): 32 (1987)
11. Kelly P et al Introducing innovation. In *Product design and technological innovation*. Open University Press 1986
12. Roche J G Design implications of product liability. *Int. J. Quality and Rel. Management* **2**:7 (1988)
13. Reducing liability risks. *OEM Design* April 1988: 25
14. Abbot H *Safer by design*. The Design Council 1987
15. BS 5378: Part 1 *Safety signs and colours*
16. *Instructions for consumer products*. HMSO 1988
17. BS 4781 *Specification for PSA plastic labels for permanent use*
18. Sanders R J et al Proper packaging enhances productivity and quality. *Indust. Eng.*, Aug 1985: 51
19. Davies A If a secret's worth having. *Micro Decision* Feb 1988: 72
20. Evans W Japanese management, product design and corporate strategy. In *Product design and technical innovation*. Open University Press 1986
21. Japan Design Study Tour. The Design Council 1987
22. Banyard P Damaged by a Japanese myth, *Procurement Weekly* **16** (38): 12 (1988)
23. Mills R A et al The future of product development. *CAE* **10**(10): 38 (1991)
24. Bishop R A matching of culture speeds time to market. *Eureka* **11**(9): 102 (1991)

CHAPTER 3
PRODUCT RELIABILITY AND TEST

3.1 Introduction

Reliability should be considered as a system, or equipment, characteristic from the design concept through to the end of the product working life. Reliability experience in one generation of products should be used in designing the next (e.g. Ch. 1, Fig. 1.2). By 'reliability' is meant the expectation, or statistical probability, that a product will produce an expected level of performance under specified conditions of use over a specified period of time. In this respect, reliability is a quality characteristic and as such, a commercial and saleable commodity.

The reliability of a product is strongly influenced by decisions made during the design process. If the design foundation is unsound however, it may be possible to correct it, but often only at a high price.[1] Usually one would expect design problems to be resolved on an iterative basis during the design and development stages, but the final reliability value, generally speaking, will appear as a trade-off between time-scale and resource expenditure. Very high reliability has to be paid for both in the design and production processes and sales cost,[1] but on the other hand, the smaller number of failures in service will require less maintenance action hence necessitating fewer spares.

It is not the intention in this chapter to investigate reliability theory in great depth, as there are a number of standard works on the subject,[1–3] though some discussion is necessary for definition purposes. When evaluating the performance of other than simple electromechanical assemblies, standard theory is not appropriate since failure rates will be made up of a number of constituent failure mechanisms and the maintenance policies applied. Thus the perceived 'reliability' of a product is partly determined by the maintainability requirements of the equipment. Standard reliability theory and failure distributions are more appropriate to life testing of individual components.

3.2 The reliability concept

3.2.1 Reliability basics

Reliability (R) can be defined simply as the probability that an item will perform a required function under stated conditions for a specified period of time, i.e. it is a number lying somewhere between 0 and 1. Reliability is therefore probabilistic. Thus

$$R(t) = \frac{\text{number of surviving items at given time } t}{\text{number of items at start } (t = 0)}$$

$$= \text{probability of survival}$$

Hence if

Probability of failure $= F(t)$ (probability of item ceasing to perform its required function)

then $R(t) + F(t) = 1$.

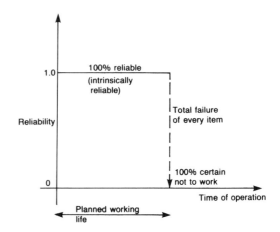

Fig. 3.1 Idealized reliability relationship

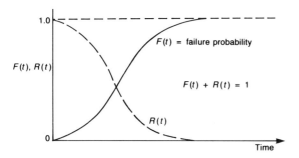

Fig. 3.2 Failure probability plotted as a cumulative distribution function

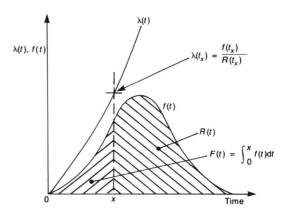

Fig. 3.3 Representation of $\lambda(t)$ and $f(t)$ functions

Ideally, it would make the design process easier if the relationship of reliability to time appeared as in Fig. 3.1, i.e. there being a well-defined period when all items would be certain to work. But in practice unreliability increases with time, a cumulative function (Fig. 3.2).

From Fig. 3.2, one can define an elementary failure rate, i.e. $dF(t)/dt$, where $F(t)$ refers to the original population and $dF(t)/dt$ to a proportion of the population existing at the instant to which the rate refers. In other words, the failure rate may be conveniently represented by a frequency distribution histogram, which when sample size $\to \infty$ and time interval $\to 0$, the curve may be described by a probability density function (pdf), represented by $f(t)$ (Fig. 3.3). The area under the curve is always unity. Hence

$$f(t) = \frac{dF(t)}{dt}$$

$$= \frac{\text{no. of items failing/unit time at time } t}{\text{no. of items at start}}$$

However, it is customary practice to express failure rate in relation to the current population of items, i.e. those items still at risk, often referred to as a statistical hazard concept in reliability terms, i.e.

$$\lambda(t) = \frac{\text{no. of items failing/unit time at time } t}{\text{no. of surviving items at time } t}$$

$$= \frac{f(t)}{R(t)}$$

from which

$f(t)$ = probability of surviving at time t (reliability)
× probability of failure at time t (failure rate)

Both $f(t)$ and $\lambda(t)$ are pictorially represented in Fig. 3.3. For a discussion of the above statistical terminology and other equivalent terms used in reliability see ref. 1.

It is often assumed that failures occur at a constant rate, in which a mean failure rate λ is used as a special case of $\lambda(t)$ where:

$$R(t) = e^{-\lambda t} = e^{-t/\theta} = \exp\left[-(t/MTBF)\right]$$

where t = time at which reliability is $R(t)$, θ = MTBF (mean time between failures), λ = a constant, (time)$^{-1}$.

In the general sense, it can be shown that[4]

$$\theta \int_0^\infty R(t)dt$$

$$= \int_0^\infty e^{-\lambda t} dt = \frac{1}{\lambda}$$

If the failure rate is constant, after one MTBF, the probability of survival $R(t) = e^{-t/\theta} = e^{-1} = 0.37$.

3.2.2 Patterns of failure

In any practical engineering situation the reliability of an item cannot be based on simple statistics, but rather as a joint probability function between the strength distribution of the component and the loading distribution experienced from the environment in which it is placed. Here the terms of 'strength' and 'load' are meant to infer a generic sense, inherent property and applied working parameters. For example, Fig. 3.4 illustrates the concept. Just because an item is inherently weak, it does not mean it will fail; it may never encounter a load greater than its strength. Figure 3.4 is simplified, engineering situations are more complex than this, but it does illustrate the role that the spread of both distributions play in determining reliability. Here it is assumed that the strength of the component is time-independent.

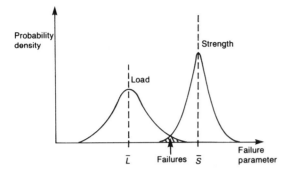

Fig. 3.4 Load–strength distribution (interfering example)

From this example it is possible to define two basic parameters, i.e.

$$\text{Safety margin} = \frac{\bar{S} - \bar{L}}{\sqrt{(\sigma_s^2 + \sigma_L^2)}}$$

$$\text{Loading roughness} = \frac{\sigma_L}{\sqrt{(\sigma_s^2 + \sigma_L^2)}}$$

where $\bar{S}$ = mean strength, $\bar{L}$ = mean load, σ_s = standard deviation of strength, σ_L = standard deviation of load.

In specifying a safety margin, clearly the use of a safety factor $\bar{S}/\bar{L}$ is totally inadequate since the result is sensitive to relative lateral position and insensitive to the spread of the distributions. The use of the safety margin above overcomes these drawbacks and allows it to be associated with a given reliability.

Loading roughness defines shapes, i.e. the relative spread of the load and stress distributions. For a constant safety margin, smooth and rough loading roughnesses are illustrated in Fig. 3.5. Since the distributions shown in Fig. 3.5 have no interacting tails, the situation is said to be intrinscially reliable. Mechanical equipment is mostly characterized by rough loading and electrical equipment by smooth loading. This approach can be used to estimate the probability of failure from experimental strength and stress distributions.

In the ideal case of normally distributed load and strengths, the relationship between the failed proportion and the safety margin is shown schematically in Fig. 3.6. There is an intermediate region in which failure is very sensitive to changes in loading roughness, while at low safety margins failure probability is high. Once the safety margin exceeds a value of 3–5, the design is intrinsically reliable. However, if the load and strength distributions are skewed giving significant interference, a safety margin of 8 may be necessary if the loading roughness is high.

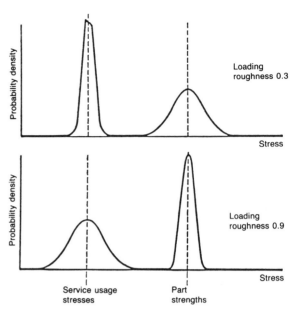

Fig. 3.5 Illustration of load–strength distributions at extremes of loading roughness and constant safety margin

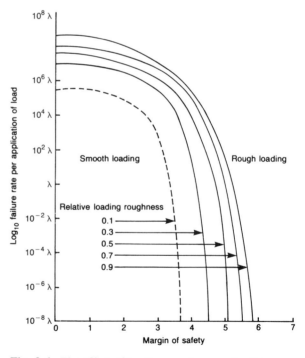

Fig. 3.6 The effect of loading roughness and safety margin on constant failure rates

Taking a simplistic view of reliability it would be useful to be able to refer to a complete system. In which case, where for reasons of economy, the reliability of mechanical and electromechanical systems are designed as series events: (a) failure of one component causes failure of the system; (b) failures are independent of each other, then either

$$R = \bar{R}^n$$

where $\bar{R}$ = mean component reliability, n = no. of components in series, if each component in the system is assumed, by the product rule, to have active reliability, which is more appropriate as the loading roughness decreases, or

$$R = \bar{R} \times (1.0)^{n-1} = \bar{R}$$

if it is assumed that in conditions of similar loading, one component can withstand the load as well as any others. This is more appropriate if the loading is rough (Fig. 3.5). An exact representation of any system is complicated, ideally rough or smooth is never achieved, hence neither rule is valid, the truth is somewhere in between.

There are three basic ways in which the pattern of failures can change with time, the failure rate increasing,

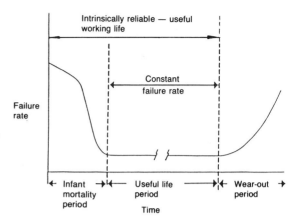

Fig. 3.7 Bathtub curve

decreasing or constant. These features can be illustrated by use of the well-known 'bathtub curve' (Fig. 3.7), which is frequently invoked as a conceptual rather than a mathematical explanation of the failure rate of a system and its entire life, failed items being replaced or repaired. The regions of this curve are summarized in Table 3.1. An illustration of component stressing in the constant failure rate portion of the curve is shown in Fig. 3.8.

3.2.3 Failure distribution estimation

A number of distributions are commonly used in reliability work for estimating failure and repair time distribution functions. Reliability data are empirical and often limited, so in engineering terms to achieve physical significance, one would examine distribution types for a best-fit judgement in the first instance. In this way the prediction of reliability will be enhanced, enabling the formatting of maintenance policies. The form of the distribution will also give a clue as to the nature of the failures and the dominant modes. For a detailed study of distribution functions, the reader is referred to books on statistics. Table 3.2 is a summary of some initial tests of distribution form and a brief explanation of the main types are given under the headings below.

(i) Normal (or Gaussian) distribution

$$f(t) = \frac{1}{\sigma\sqrt{(2\pi)}} \exp\left[-\frac{(t-\mu)^2}{2\sigma^2}\right]$$

$$F(t) = 1 - \int_0^\infty f(t)\,dt$$

$$\lambda(t) = f(t)/R(t)$$

Table 3.1 *Summary of the bathtub curve characteristics*

Phase	Descriptor	Explanation
1. Decreasing failure rate	Early failures Burn-in Infant mortality Learning period	Weak items Poor design Installation error Poor component quality (manufacturing faults) Operator error (inexpensive) Maintenance error Occurs especially if design is new, decreasing rapidly as design faults, component quality, incorrect settings, manufacturing errors, etc. are overcome and users become educated
2. Constant failure rate	Random failures Useful life CFR period	Random stress-related problems depending on safety margin and loading roughness Not due to any basic cause Constant rate if failures are independent.
3. Increasing failure rate	Wear-out failure Characteristic of non-maintained systems	Long-life items 'wearing out' Failure due to corrosion, fatigue, wear, insulation breakdown, etc.

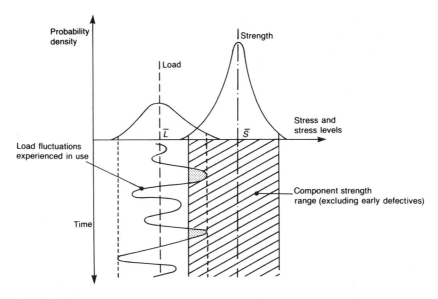

Fig. 3.8 Load—strength interaction in constant failure rate portion of the bathtub curve

Table 3.2 *Some initial tests of distribution*

Graph	Σ failures versus Σ time	— straight line ≡ constant failure rate
Graph	Σ % failures versus time on test	— straight line ≡ normal distribution
Graph	Σ % failures versus time on test (Weibull)	— straight line ≡ Weibull fn. with $\gamma = 0$ — if not straight, estimate γ, subtract and replot to get a straight line
Graph	$\lambda(t)$ versus t	— if increasing, either Weibull or normal — if increase then levels off or falls, either lognormal or γ Weibull possible — if decreasing, either hyperexponential or Weibull

where σ = standard deviation scale parameter, σ^2 = variance, μ = mean.

The normal distribution is the most important of statistical distributions because 80 per cent of naturally occurring variations are normally distributed. However, in reliability statistics it appears most applicable in describing the wear-out period failures. The normal distributions that best fit field data can be easily determined by plotting on standard probability paper. Figure 3.9 represents normalized life distributions for various industrial products and human beings. Interestingly, the standard S-shaped normal distribution curve is for oil-lubricated hypoid gears.

(ii) The log-normal distribution

This distribution has wide applicability in reliability and

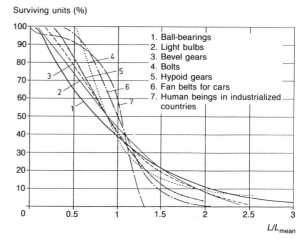

Fig. 3.9 Life distribution of various industrial products and humans. Courtesy of SKF (UK) Ltd.

maintainability, especially where material failure growth mechanisms are concerned. Much of the justification for this distribution is empirical, implying that the log (component life) is normally distributed as follows:

$$f(t) = \frac{1}{t\sigma(2\pi)^{1/2}} \exp\left[-\frac{(\ln t - \mu)^2}{2\sigma^2}\right]$$

where μ and σ are the mean and standard deviations of the $\log_e$ data.

Median = e^μ, $\mu = \log_e$ median, i.e. 50 per cent at cumulative frequency, mean = $e^{\sigma^2/2}$

(iii) The exponential distribution
As previously outlined, for a constant failure rate:

$$R(t) = e^{-\lambda t}$$

$$F(t) = 1 - e^{-\lambda t}$$

hence

$$f(t) = \lambda e^{-\lambda t} = \frac{1}{\theta}\exp\left(\frac{-t}{\theta}\right)$$

and mean life

$$\theta = \frac{1}{\lambda} = \text{MTBF}$$

This describes the useful life period of the bathtub curve and as such is the most widely used distribution in life testing, relating to failure as a result of random events rather than a systematic wear-out mechanism of single cause.

(iv) Weibull distribution
The Weibull is a flexible distribution that can be used to represent any form of rate/age curve. It is largely empirical in origin, but it is one of the most useful distributions for analysing failure data. In its modified form we can write

$$R(t) = \exp\left[-\frac{(t - \gamma)^\beta}{(\eta - \gamma)}\right]$$

where η = characteristic life, a scaling constant representing the time, from $t = 0$ by which 63.2 per cent of the population can be expected to fail whatever value is assigned to β; β = shape parameter, if $\beta = 1$ an exponential equation results, if $\beta = 3.4$, equation is approximately normal; γ = locator or minimum life parameter, shows where curve is to start, i.e. if $t = 0$, $\gamma = 0$ and if, as is often the case, $\gamma = 0$.

$$R(t) = \exp\left[-\left(\frac{-t}{\eta}\right)^\beta\right]$$

now

$$f(t) = -\frac{dR(t)}{dt} = \frac{\beta(t - \gamma)^{\beta-1}}{(\eta - \gamma)^\beta}\exp\left[\frac{-(t - \gamma)^\beta}{(\eta - \gamma)}\right]$$

and hence

$$\lambda(t) = \frac{\beta(t - \gamma)^{\beta-1}}{(\eta - \gamma)^\beta}$$

$$= \frac{\beta t^{\beta-1}}{\eta^\beta} \quad \text{when } \gamma = 0.$$

It is normal practice to estimate the constants of the Weibull distribution using Weibull probability paper.

If $\beta > 1$, this indicates an increasing $\lambda(t)$, probably symptomatic of wear-out. If $\beta = 1$, constant failure rate exponential; i.e. η = mean life θ. If $\beta < 1$, decreasing $\lambda(t)$. For an example where $\gamma = 0$, see Fig. 3.10.

If $\gamma \neq 0$, then the initial shape of the Weibull plot will be curved, i.e. convex if $\gamma > 0$, e.g. Fig. 3.11, or concave if $\gamma < 0$, i.e. if items under test had accumulated unrecorded operating time before the start. Transformation of data (Fig. 3.11) into a straight line for Weibull analysis is demonstrated in Fig. 3.12 by drawing equally spaced horizontal lines such that

$$\hat{\gamma} = t_2 - \frac{(t_3 - t_2)(t_2 - t_1)}{(t_3 - t_2) - (t_2 - t_1)}$$

One of the main applications of Weibull is in the life distribution of rolling element bearings since failure is dominated by a surface fatigue mechanism, and therefore statistically predictable. For design purposes, 'life' is taken as the time when 10 per cent only have failed at a given operating condition, termed the L_{10} or B_{10} life. For a given load rating,[5] distribution of bearing lives will be

$$F = 1 - e^{-0.105\left(\frac{L - L_0}{L_{10} - L_0}\right)^\beta}$$

where L = life (operating time), L_{10} = rated life, L_0 = minimum life exceeded by all bearings and $L_0 = 0.05 L_{10}$ in practice. Raw data for some grease lubricated deep groove ball-bearings is shown in Fig. 3.11 for which $L_0 \neq 0$. Adjusting for minimum life gives[6]

$$\frac{L}{L_{10}} = 0.95\left(\frac{\ln R}{\ln 0.9}\right)^{0.9} + 0.05$$

Since L is a quoted parameter in bearing manufacturers' catalogues, it is possible to calculate bearing life at any level of reliability.

(v) Pareto distribution
A Pareto analysis of failure events and their frequency

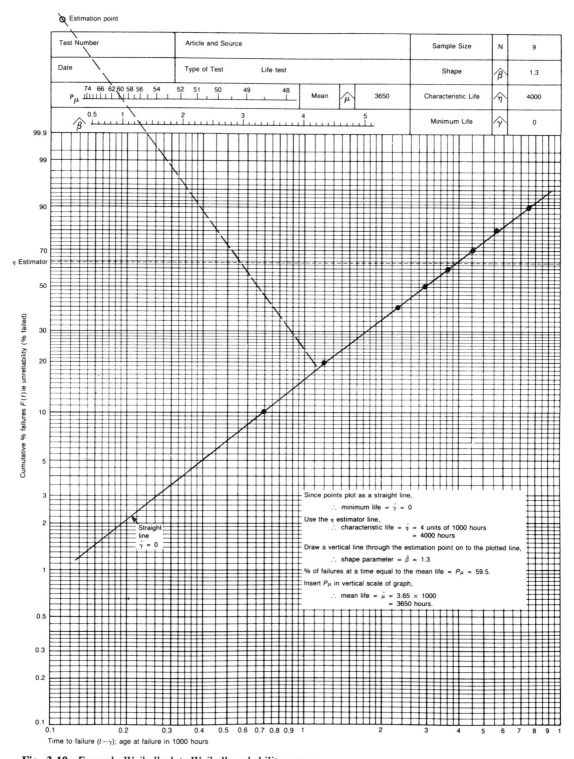

Fig. 3.10 Example Weibull plot; Weibull probability paper

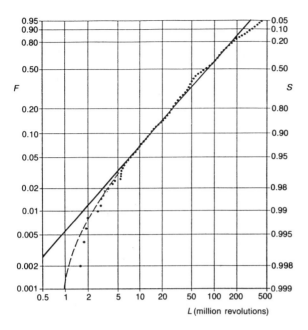

Fig. 3.11 Life distributions for grease lubricated deep groove ball bearings. Courtesy of SKF (UK) Ltd.

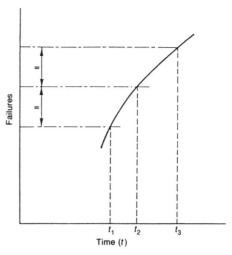

Fig. 3.12 Estimation of t

is a widely used and powerful method of focusing attention on major problems. A typical example of failure frequencies is shown as a histogram form in Fig. 3.13, the frequency of failure plotted against cause in decreasing

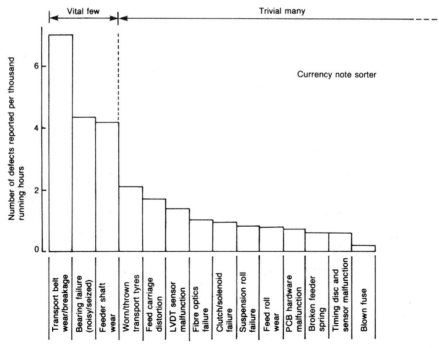

Fig. 3.13 Typical Pareto plot illustrating the segregation of the vital few from the trivial many for a currency note sorter

order, a very rapid drop in frequency over the first few items followed by a long tail of low-frequency items. The analysis can also include cost, in which case the failure distribution economics may well differ from the failure frequency results in cause ranking. Returning to Fig. 3.13, it can be seen that three items contribute some 60 per cent of all failures, i.e. Pareto's 'vital few' which are responsible for the main system behaviour, while the remainder are Pareto's 'trivial many', each of which plays less of a role in the overall machine behaviour.

In reliability improvement, one would naturally concentrate effort on the 'few', but depending on where the product itself is within the design cycle, so the major failure causes might be expected to change. Although the Pareto concept is fairly simple, a lot depends on the equivalent isolation of the failure mode. For example, if a failure is lumped within a subsystem, it may appear as one of the vital few, while as an individual failure it could equally well appear as part of the trivial many when compared to the rest of the system.

3.3 Reliability aspects of the design cycle

3.3.1 The reliability programme

From Chapter 1, we saw reliability emerge as one of the key activities in the design process, from the planning aspect in the feasibility stage and through testwork and performance evaluation in the acquisition phase to the design exit reviews. The ideal relationship of reliability to drawing change rate during the design process was outlined in Fig. 1.8, converging to a fixed level during the maturity phase. In practice, however, there is nearly always a temporary blip[1] when products are first introduced to the market-place (Fig. 3.14). As the design proceeds from concept to detailed definition, the reliability risks should be controlled by a standardized systematic approach, reviewing the design and charting the reliability growth at various stages, the process formalized into the reliability programme.

The assumption at this stage is that the ongoing reliability evaluation is in-house, rather than a vendor appraisal or an external conformity test. The reliability department acts in the role of an independent assessor, divorced from Engineering. For obvious reasons, a design engineer should not conduct his own reliability tests. The aim of the reliability programme is to ensure that adequate and effective effort is brought to bear on reliability as a principal quality measure during all phases of the product life cycle and that these activities are properly integrated into the design process.[7] Design deficiencies do in fact affect most products that are produced, and as most manufacturers will testify they are progressively more expensive to correct as the development proceeds. Also it is often not very practical to change the design once production has started, hence it is essential that disciplines are used to identify and correct the defect as early in the design cycle as possible.[8] Risks are low where the safety

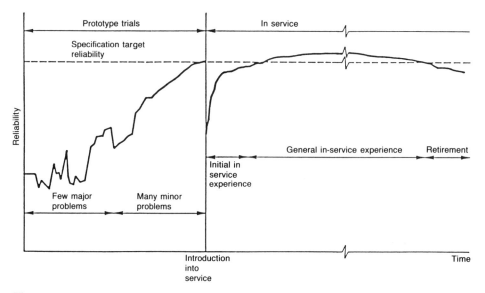

Fig. 3.14 Typical lifetime variation of reliability (from ref. 1). Courtesy of Macmillan Publishers (UK) Ltd

margins are very wide (Fig. 3.5), but the reliability programme and reliability disciplines are necessary where the risks and costs of failure are other than low.

With most industrial products and systems, reliability is synonymously inferred by tracking maintenance requirements. This could represent a number of machine stoppages/breakdowns or occasions of performance dropping below a predetermined level etc. represented on a time base, e.g. '*MTBF*', failures per 10 operations, breakdowns per 10 miles, etc. Such data accumulated during the reliability programme can then be projected forward to formulate Field Service policy and specify field maintenance intervals. To reach this stage, a number of reliability processes may be necessary. These are summarized in Table 3.3, activities largely based on testwork. Reliability testing, together with subsequent corrective action and design feedback carried out during the acquisition phase, is an effective means of increasing the reliability of complex electromechanical items and increasing confidence of adequate growth and maturity target attainment.

The collection of information and data during the development stage is therefore vital. The whole purpose of design and development iteration is to provide the information to progressively resolve the factors of uncertainty surrounding the design, its manufacture and use.[7]

Within some organizations there is the need for mandatory reliability design procedures encompassed within a Reliability Manual[8] covering company reliability practice, much the same way as quality practice. Subjects include the following:

1. Corporate policy for reliability.
2. The reliability organization.
3. Reliability design procedures, namely: (a) design

Table 3.3 *Key reliability and maintainability events in the design cycle*

	Feasibility phase	Acquisition phase		Maturity phase
		Prototype	Pre-production	
Requirements	Reliability specification Acceptance criteria definition Performance requirements	Determination of the failure modes Stress case analysis Redundancy analysis Maintainability analysis	Ongoing performance testing Environmental testing Accelerated testing Endurance testing Reliability demonstration	Maintainability demonstration Reliability growth
Activities	Determine reliability proposal requirements Reliability specification generation Reliability feasibility study	Assess operability and human factors Determine failure rates Stress and worst-case testing Variance analysis of controllable and uncontrollable parameters Establish the failure modes FMEA and FTA Safety analysis Maintainability analysis Design audits Sub-assembly and component testing Reliability growth demonstration Performance analysis and design feedback	Qualification testing Environmental loading tests Functional reliability tests Life and endurance tests Performance confirmation Reliability growth analysis	Screening and burn-in tests Durability testing
Output	Engineering specification including reliability and performance Reliability engineering plan	Reliability analysis/growth statement Problem list and failure rates Maintainability analysis	Field service programme Safety and approvals Actual reliability versus target Problem management, Pareto's few or many	Verification of reliability and quality

analysis techniques; (b) parts reliability rating policy; (c) preferred materials; (d) critical items listing; (e) design for reliability rules.
4. Standard test methods and procedures, both internal and company external specifications/standards.
5. Reliability procedures and results collection, namely: (a) data collection and methods of analysis; (b) failure analysis; (c) awareness and review procedures (d) corrective action requests; (e) warranty specification.

3.3.2 Designing for reliability

Initial design concepts, influenced by the overall reliability requirement, will depend upon the following:

1. The functional complexity of the item or product;
2. The state of development and technical understanding available with new components, materials or technologies considered for use.
3. The manner in which the product will be used, the overall equipment life requirement and the type of field service or maintainability that can be expected.
4. The environmental window expectation for usage conditions, both operational and non-operational, for packaging, storage and transportation.
5. The cost target.
6. The size envelope.
7. Any special requirements, i.e. safety.

There is no set of absolute rules on how to design a reliable item. Most of the key issues would be covered unconsciously by experienced design engineers, applying engineering principles and common sense coupled with attention to detail, looking for the likely failure modes. It is a systematic activity requiring creative intellect (Ch. 2). However, creative design of course is most effective when unconstrained, but an integrated reliability programme has to be disciplined if reliability and the quality effort are to be supported. Some design departments may use checklists[1] in a reliability design sense, the usefulness of such a process being greatly enhanced with constant feedback of field experience.

Aspects of the design process that can have a particular impact on reliability may be summarized as follows:[1]

1. Keep it simple, what is not there cannot go wrong.
2. Do not introduce value engineering improvements at the expense of reliability, although clearly there are trade-offs between parts cost and maintainability in certain situations.
3. Ideally, use known items with a known history.
4. With totally new design concepts, determine performance latitude and sensitivities with test rigs and fixtures first, i.e. aim to understand the technology early in the design process.
5. The purchase of an outside component must be carefully done and written into the design with agreed interface and performance specifications.
6. Facilitating straightforward manufacturability with minimal special jigs, fixtures or set-up adjustments will reduce build errors and possible field problems.
7. Confirm mechanical design choices mathematically wherever possible, e.g. bearings, load deflections, expansion, etc.
8. Analyse the failure modes (section 3.4).
9. Cater for rough handling, transportation and user abuse as likely events.
10. Allow for possible test house approval requirements in design of electrical, mechanical, safety and material selection aspects.
11. Understand the requirements and the rules. Can be expensive to correct downstream.
12. Take precautions to ensure radiated noise interference immunity. Software-driven electrical hardware is particularly at risk.

3.3.3 Designing for maintainability

We have already noted that the reliability of a product is not only controlled by the quality of the design but also by the maintainability of its constituent components. Maintainability of a system is defined as the probability that it will be restored to operational efficiency within a given time if specified maintenance is performed on it. Designing products that require maintenance always involves some compromise. In the make-up of the design team (section 2.3) there should ideally be a Field Service representative to ensure that the interests of the service engineer are represented. If manufacturing assembly is difficult, then maintenance in the field will be almost impossible.[4]

This section will refer only to the equipment design, i.e. the active elements involved in maintenance and the quality and effectiveness of such.

The general types of maintainability design features that should form part of the design consensus are outlined as follows:

1. *Access*. Clearly this is the most important feature inasmuch that ease of accessibility, ease of observation and ease of replacement all reduce maintenance times. Low reliability parts should be the most accessible and most easily removable with the minimum of disturbance or likelihood of damage to

the rest of the machine. Any dials, gauges or displays that need to be routinely checked should be easily visible.

2. *Adjustments*. The amount of adjustment required during normal operation can be minimized or eliminated by generous design tolerance or take-up features aimed at a low sensitivity to wear and drift. Such enabling design features will also aid manufacturability. It is desirable, wherever possible, to avoid difficult or critical adjustment requirement since the Field Service engineer is quite likely to choose the easier route and change the component regardless of cost. Where there is some risk of tools slipping and causing internal damage during initial installation and/or subsequent adjustment, guide holes and channels should be provided.

3. *Fasteners*. Ideally use captive screws and fasteners, since there then is no risk of losing loose screws/washers, etc. in the equipment itself. Where captive fasteners are not possible, some commonality of sizes should be insisted upon. With access to short-life components, the use of snap-in/quick-release fasteners are a good idea.

4. *Handling*. With replaceable modules, a design eye should be kept on their weight, size, shape and gravity centre. The module or item when removed must not only be mechanically stable when on its own and unsupported but should not have an unexpected mass concentration. Handles or finger grips should be provided wherever possible to reduce the temptation to use components for lifting purposes. Also some protection will be required for sharp edges (grommet strip) or drawing instructions to chamfer corners and/or deburr sharp edges. Tendencies to over-miniaturized or to overly compact components should be avoided since accidental damage is then less likely.

5. *Test features*. It is quite costly to provide meters and test displays on equipment purely for servicing requirements, although there are clearly some circumstances where built-in features are justifiable. However, since the majority of maintainable equipment today is microprocessor driven, it is usual to write into the software a self-test or diagnostic task which not only routinely checks the state of the machine on power-up but can also provide operator feedback. A more detailed diagnostics routine would be available to the Field Service engineer for rapid problem identification.

6. *Identification*. Test points, leads, terminals and connectors, etc. need to be clearly marked and identified/colour coded in a way that is simple and unambiguous. Depending on the machine end use and approval authorities, hazard warning symbols may also be a requirement. The design should make the interchange of non-identical items impossible, e.g. handed components, PCBs, etc. Part number identification may also be necessary on some items.

7. *Personnel*. One should assume, when designing for maintainability, that only a low-grade calibre is available. However, the skill level employed is not the whole story since it will also depend on the training given, initial as well as ongoing (say with field bulletins), motivation and product dedication. When designing for access, the consideration and comfort of maintenance personnel should be taken into account, e.g. the type of manipulative actions required, limits of reach, any awkward bending, undue strength requirement likely, etc.

8. *Manual*. Often stated as the most vital repair tool of all, and depending on how well it is written, depends how well the maintainability tasks are carried out. Writing a manual is difficult and time consuming. Not many people do this task well, quite often it is a hurried attempt to get something in place when the product is launched. A well-written manual is achieved by following a logical procedure. It should be accurate, complete and with information easily accessible. It should contain fault-finding algorithms and functional block diagrams to illustrate adjustment/checking/installation procedures, plus some emphasis on any safety hazards or any precautions necessary when working on the machine.

The design-influenced maintainability features outlined are not only enabling features, some also influence the equipment repair time (ERT) or individual repair time. The cost implications of reliability and maintainability are discussed in section 3.3.5. Together with the trend towards further reducing scheduled maintenance is the trend towards shorter repair times. Unlike maintainability, repair times are not influenced by failure rates, but rather determined by the practical features of the design. How well the design enables timely repair tasks to be carried out can be assessed using a scored checklist,[9] weighted in favour of ease of accomplishment. It is not intended to reproduce the whole process outlined in Ref. 9, rather to summarize the design-related aspects.

The checklist is in three groupings: (1) physical design factors; (2) facilities; (3) maintenance skills.

(i) Physical design factors

Access (external). Determines if the external access is adequate for visual inspection and manipulative actions. Scores would apply to the external packaging (frames/covers) as related to ease of maintenance, i.e. design for external visual and manipulative actions which would proceed internal maintenance actions. The maximum score would be for access uninhibited by obstructions, i.e. cables, panels, supports, etc.

Latches and fasteners (external). Determines if screws, clips, latches or fasteners outside the assembly require special tools, or if significant time was incurred in removing such items. Scoring would relate the external equipment packaging and hardware to maintainability design concepts. Time consumed with preliminary external disassembly will be proportional to the type of hardware and the range of tools needed to release them. The maximum score would be obtained for captive external latches/fasteners, no special tools required and requiring only a fraction of a turn for release, i.e. quick-release fasteners (Ch. 14).

Latches and fasteners (internal). As above, but for internal fastenings and preference for standardization throughout the equipment.

(ii) Facilities

External test equipment. Determines if, and how much, external equipment is needed to complete the maintenance action. The type of repair considered ideal from a maintenance point of view would be one which did not require any external test gear. It follows, therefore, that a maintenance task requiring test equipment would involve more task time for set-up and adjustment and should therefore receive a lower maintenance score.

Connectors. Determines if supplementary test equipment requires special fittings, special tools or adaptors to adequately perform tests on the electromechanical system or subsystems. During troubleshooting, the minimum need for test equipment adaptors or connectors indicates that a better maintainability condition exists. A high score therefore applies to regular test leads (probes, alligator clips) which can be directly plugged in or otherwise secured to test points.

Jigs and fixtures. Determines if supplementary materials such as unique tools, tackle, alignment jigs, etc. are required to complete the maintenance action. The use of such items during maintenance would indicate the expenditure of major maintenance resource (time and tools) and pin-point specific deficiencies in the design for maintainability. Maximum score is for no supplementary materials.

(iii) Maintenance skills

This checklist would evaluate the personnel requirements relating to the physical, intellectual and attitude characteristics required of the field service engineer in relation to the task he has to undertake. The scoring parameters that describe technician capability, relative to the task, can be grouped as follows:

1. Arm, leg and back strength;
2. Endurance and energy;
3. Manual dexterity, eye/hand co-ordination and neatness;
4. Visual sharpness;
5. Logical deduction and analysis;
6. Memory, things and events;
7. Planning ability and resourcefulness;
8. Alertness, cautiousness and accuracy;
9. Concentration, persistence and patience;
10. Initiative and effectiveness.

A maintenance action acquires a high score when there is minimum effort required on the part of the technician. This score will also relate to the environmental conditions under which the equipment is designed to work.

3.3.4 Product maintenance

This section will deal with both the terms and mathematics describing maintainability, the various types of maintenance action that define field reliability and the apportionment of responsibility within the engineering community for field problems.

(i) Definitions

There are a number of terms and parameters used quantitatively to describe the maintainability aspects of a product, but no hard and fast rules. Some parameters may be more appropriate than others in any given situation. Definitions are as follows:

$MTBF = \theta =$ observed mean time between consecutive failures during a certain period in the life of items that are repaired or replaced

$$= \frac{\text{total cumulative time between failures}}{\text{total number of failures}}$$

$MTTF =$ observed mean time to failure of items that are not repaired or replaced

$$= \frac{\text{total cumulative times to failure}}{\text{total number of failures}}$$

Here *MTTF* is more a measure of reliability, whereas *MTBF* is a mean up-time since by definition, it excludes down-time. For the negative exponential distribution only, it can be shown that

$MTBF = \dfrac{1}{\lambda} = MTTF$

MDT = mean down-time

= period during which equipment is in a failed state, often difficult to define since it should refer to normal conditions of use time rather than total time and should be expressed in percentile terms

$MTTR = \phi$ = mean time to repair, i.e. the average time spent on repairs

= $\dfrac{\text{total no. of maintenance action hours}}{\text{no. of maintenance actions}}$

often used in percentile terms, e.g. 'the 90th percentile repair time shall be 60 min'.

MAR = maintenance action rate or repair rate, i.e. the number of equipments restored to working order in unit time

= $1/\phi = \mu$ (compared with $1/\theta = \lambda$)

Availability = $\dfrac{\text{up-time}}{\text{total time}}$

= $\dfrac{\text{mean up-time}}{\text{mean up-time + mean down-time}}$

= $\dfrac{MTBF}{MTBF + MDT}$ = steady-state availability

Availability, a useful parameter describing the amount of time the equipment is available, is determined by both the reliability and maintainability of the item. For consistency, times should be operational hours only.

Utilization factor = $\dfrac{MTBF}{MTBF + MTTR + MTIBR}$

where $MTIBR$ = mean time in between repair.

If the durations of each maintenance action for a piece of equipment are plotted as a frequency distribution curve, it is often found that the result is approximated to by either a negative exponential or a log-normal curve (Fig. 3.15).

Negative exponential The pattern is less common than the log-normal, but it is typical of equipment systems requiring a number of small adjustments or tweaks and occasional lengthy repairs. A straight line results from plotting a cumulative curve on log (1 cycle)-normal graph paper.

If specific maintenance is performed on a piece of equipment, we can define maintainability as the prob-

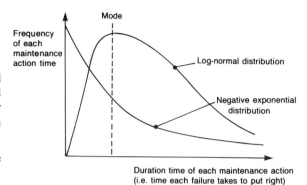

Fig. 3.15 Types of time to repair distributions

ability that a system will be restored to operational efficiency within a given time, t_c. Since θ and ϕ are mathematically equivalent in this case, then

Maintainability = $M = 1 - \exp[(-t_c)/\phi]$

Log-normal A straight line results from plotting a curve of log cumulative per cent maintenance times vs. log time. Observations suggest that fault location time, active repair time and total down-time are frequently log-normal. Often there is a minimum time for all maintenance jobs, i.e. the time it takes to get to and check out the failed equipment.

(ii) Maintenance categories

Most maintenance involves the replacement of a component or components that have failed, are about to fail or that can be restored to the 'as-new' condition by resetting or adjustment. Thus in any piece of electro-mechanical equipment, the age of the parent machine and the replaced part(s) are no longer the same. The quoting of an overall bulk *MTBF* is conveniently done for many products, but is often overly simplistic since the equipment may contain a wide range of components, hence a wide range of failure modes. Although components might need replacing, the equipment never does, it continues on even if very little of the original may eventually be left. However, there comes a point when this is no longer an economic proposition (section 3.3.5).

Maintenance is in two main categories, scheduled (preventative) and unscheduled (repair). The mix of policies to adopt in practice is an economic one and optimized for the particular failure pattern. A schematic breakdown of the maintenance spectrum is shown in Fig. 3.16. The unscheduled maintenance rate (UM rate) or *MTBF* deduced from equipment failures is usually quoted as the perceived reliability. The practical types of

ELECTROMECHANICAL PRODUCT DESIGN

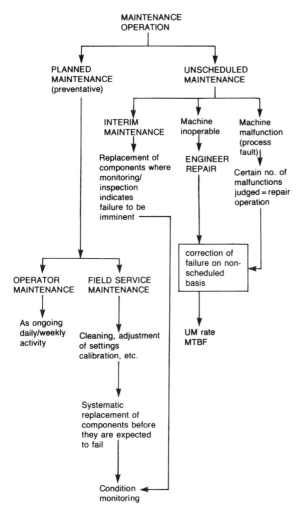

Fig. 3.16 The product maintenance spectrum

maintenance action encountered can be considered in more detail as follows:

Planned maintenance (PM). Also referred as interim, scheduled, inspection or preventative maintenance. This is generally the systematic adjustment of performance or replacement of components before they are projected to fail, i.e. in the wear-out period. It is only justified if the failure rate with it is lower than without it. Disturbances caused by regular machine 'repairs' may in themselves result in some unscheduled failures, hence some organizations may choose a policy of 'intelligent neglect'. However, planned maintenance can be done at convenient times to suit the user, and would normally be offered with a service contract.

On-condition maintenance. In certain cases it is possible with sensors, measurements, etc. to monitor item integrity continuously and thus be certain of projecting failure in the wear-out period.

Operator maintenance. As in ongoing daily or weekly activity it is possible with some forms of equipment to assign basic high-frequency maintenance tasks, i.e. cleaning, emptying, etc. to the operators themselves. Where possible, equipment design should allow for this.

Unscheduled maintenance (UM). This is the correction of a failure on a non-scheduled basis, it results in machine down-time and involves a Field Service engineer call-out. Some failures are predictable, but some occur randomly on a longevity basis as the loading environment may vary. Planned and/or inspection maintenance would logically be done at the same call.

Process maintenance. Apart from the obvious criterion where a component fault makes the machine inoperable, there is an additional UM category which would be defined as an x no. of process faults, causing the performance/output of the equipment to deteriorate to an unacceptable level, requiring unscheduled adjustment/maintenance operation.

During the design cycle and reliability test procedures, it will be possible to establish a failure pattern or to define the likely causes of unreliability (Pareto's vital few), outlook a mature reliability target and propose a maintenance plan. Depending on the product and company philosophy, the plan may or may not propose optimized scheduled maintenance intervals. Relying solely on service engineer call-out will result in a maximum UM rate, possibly a low perceived reliability and least effective use of his/her time. It is now very commonplace to adopt interim maintenance, a lot of unknowns are then swept up in the regular monitoring.

Even though the engineering design itself may be intrinsically reliable at the point of release to manufacturing, the actual field reliability observed will be compounded by manufacturing errors, servicing errors and levels of abuse from operators unfamiliar with the equipment. Thus machine failures/failure causes are often attributable, see Fig. 3.17. All of these causes are controllable within limits.

3.3.5 Reliability economics

A reliability and maintainability programme sets out to ensure products with acceptable reliability, and maintainability, at an acceptable cost, rather than high reliability

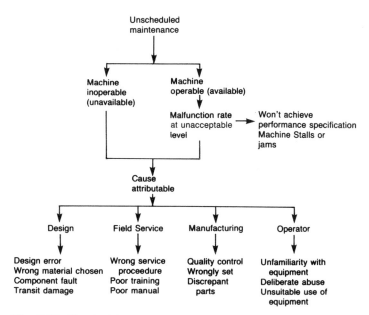

Fig. 3.17 Unscheduled product maintenance

and maintainability at whatever the cost.[1] The cost of providing reliability and maintainability, their interrelationship and the cost of maintenance form a complex equation. It is difficult to generalize, but the following points should be noted:

1. A greater investment in reliability should lead to reduced maintenance costs;
2. Money spend on maintainability reduces repair time (and cost);
3. Money spent on preventative maintenance enhances reliability;
4. It is not usually the case that too much money and effort is spent on reliability, all failure modes are worth discovering and correcting up to a point.

Reliability programmes can be expensive, but for any product there is optimum expenditure for optimum benefit (achieved realiability). This trade-off is usually presented schematically in terms of cost of ownership (Fig. 3.18). This shows that the cost of ownership (procurement cost + operating cost) will be reduced by enhanced reliability and maintainability costs, but increased by the activities costs necessary to achieve them. Figure 3.18 illustrates the philosophy of minimum or optimum cost, and while no one doubts the trend, in practice it is difficult to establish the optimum point. Clearly, however, the position of the optimum point will be influenced by the quality of the design team — the better the original design, the greater the level of achieved reliability at no extra cost.

Cost of ownership may also be variously referred to as total cost, whole life cost or life-cycle cost.

It is often very difficult to justify the expense of an inexact quantity such as reliability. It is tempting to trust design experience, but it is unlikely that even well-run development processes will pay off. Discovery of a potential weakness from reliability testing at an early stage of the product development cycle may cost several orders of magnitude less than if it is subsequently found in standard production hardware. Thus the earlier the reliability programme can be applied to the development process, the more cost effective are the reliability activities likely to be.

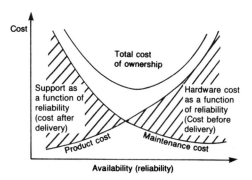

Fig. 3.18 The cost of ownership pattern of optimum reliability

The economic trade-off between procurement cost and operating cost is not the complete answer. The decisions must also be weighted by the seriousness of a failure, or other performance shortfalls. For example, in a piece of medical equipment or railway safety equipment, say, there would be a bias towards acceptably high reliability (higher initial cost).

(i) The reliability–maintainability trade-off

In a generalized sense it is possible to compensate for low reliability by high maintainability and vice versa, as we have already suggested. However, the situation is one which is balanced by the desire for high reliability and low maintenance costs, i.e. achieving a high level of availability.

If ΔC_m is the cost of unit improvement in maintainability and ΔC_r = cost of unit improvement in reliability. Hence if

$$\frac{\Delta C_m}{\Delta C_r} > 1$$

the cost of adding reliability can be less than the maintenance it is replacing and hence should proceed in that direction, i.e. add reliability. Now

$$\text{Availability} = \frac{MTBF}{MTBF + MTTR}$$

For a constant availability situation using MTBF as a measure of reliability and MTTR as a measure of maintainability, it can be shown that there are extremes where availability is largely independent of reliability or maintainability (Fig. 3.19), but there exists a middle ground where the unit cost ratio can be shown to influence the reliability/maintainability balance.

(ii) Durability

Maintained products, unlike components, never strictly wear out, but it is nevertheless possible for products to be uneconomic to maintain after a certain age. As equipment ages, due to the reliability spread on components, an increasing number will need replacing. However, very expensive repairs even during the normal life expectancy would not be worth while. It has been suggested that log cumulative maintenance cost is a linear function of the log age of the product,[10] i.e.

$$C_M = Bt^K$$

where B, K are constants, t = age, C_M = mean cumulative maintenance cost, and total cost

$$C_c = A + Bt^K$$

where A = procurement cost of equipment. Average total cost/unit operating time

$$\overline{C}_c = \frac{A}{t} + Bt^{K-1}$$

which decreases with time until a minimum is reached, thereafter increasing continuously with age. Minimum cost occurs when

$$\frac{d\overline{C}_c}{dt} = 0$$

hence

$$\frac{A}{t^e} - B(K-1)t^{K-2} = 0$$

or

$$t = \left(\frac{A}{B(K-1)}\right)^{1/K}$$

which is termed the 'durability' of the equipment.

This illustration is simplistic, but the bottom line is an economic limit for the period a piece of equipment may be operated and maintained. To judge whether an equipment is worth repairing, at time t its notional value is

$$C_c - \overline{C}_c t = A + Bt^K - \frac{AKt}{(K-1)}\left[\frac{B(K-1)}{A}\right]^{1/K}$$

3.3.6 Launch problems

As shown in Fig. 3.14, the introduction of a new product into service is typically accompanied by an initial drop in reliability before regaining the reliabililty growth expectation from the design cycle. To reduce this effect to a minimum, many companies will produce field trial units for initial machine observation at selected customer sites, basically at producer cost, prior to the initial product launch. If there are unexpected field reliability problems, that way at least Pareto's 'vital few' can be actioned early on. All launch problems come down to some form of human error or human unreliability, typically cause attributable as in Fig. 3.17, i.e.

1. *Design error*. Mostly due to inadequate design of the operator interface or insufficient strength to allow for transportation.
2. *Operator error*. Unfamiliarity with new equipment and/or inadequate instruction as to operability.
3. *Manufacturing error*. Unfamiliarity with new design in its manufacture, settings and adjustments, and its final inspection.
4. *Handling error*. Disregard of manufacturer's instruc-

tions on handling, stacking, transportation and storage.

Human error[11] is defined here as failure to perform a prescribed task (or the performance of a prohibited action) that could result in damage to equipment or disruption of scheduled operation. In real life most systems require some human participation irrespective of the degree of automation. Wherever people are involved, errors will be made. Thus the possibility of human error leading to unreliability should be considered throughout the life of the item or product. The main aspects of initial launch problems/unreliability are given under the heading below.

(i) The operator role

Consideration should always be given to the way in which the reliability of equipment or product may be increased by the reliability of the operator. Perhaps not enough time and effort tends to be put into operator quality control, but real gains are possible for those organizations investing in sound training programmes backed up by effective feedback. The following points can be made in respect of the operator environment.

1. Equipment abuse can clearly cause failure. Some hostile conditions cannot be avoided, but design limits can be specified. However, what is often abuse to the designer may be normal practice for the user.
2. With new equipment, operator unfamiliarity will initially result in unfriendliness, especially if the operator's perception of the equipment reliability does not match his/her expectation of what the reliability should be.
3. Unintentional abuse is a real cause of failure, the only remedy here is good operator training, documentation, instruction and communication.

(ii) Design assessment

If the product design is not produced in the true knowledge of the required operating conditions, then unreliability results. Obviously it is in the best interests of the manufacturer to ensure that his design meets customer expectations, including the provision of adequate instruction manuals. Depending on supplier guarantees, terms of contract, etc. a large user of equipment can, depending how he sees his role, feed back information to the design company on the performance of their product if he is interested in jointly improving reliability. The isolated purchaser, on the other hand (consumer products), cannot do a lot to influence design, it is generally left to approval bodies and consumer associations to assess performance and safety. The subject of equipment commissioning and test house approvals is dealt with in more detail in Chapter 18.

(iii) Packaging effectiveness

Depending on the nature of the product, the two problem areas reflecting the quality of the packaging are transit damage and storage problems.

Storage problems relate to the warehousing methods used, the stacking procedures, the number of units that will stack on a pallet and whether or not the original design has any vulnerable parts external to the main cover. If the equipment is large, it should be adequately protected when moved on a fork-lift truck. The most difficult situation for packaging to withstand is where side clamp lifts are used. Extremes of temperature and humidity may exist in storage (and transportation) facilities hence the packaging should not deteriorate.

Equipment transit will be another test of the packaging. Basically if the product can be dropped it will be, hence the need to evaluate packaging with drop, bump and vibration testing during the latter stage of the pre-production reliability testwork. Additionally this assesses the structural stability and integrity of the product. Although shock and vibration testwork in the laboratory will pin-point some areas of concern, it is unlikely to reproduce all the failure modes. Consequently, if the size of the product permits, it can be worthwhile to dispatch items by road/rail hauliers around the country on a packaging trials basis. The following points should be considered:

1. The tougher the product can be designed, the cheaper the packaging can be made. For high-volume, low-cost products this is especially true.
2. The packaging must be designed to suit the transport requirements. (A parallel with product design meeting its usage environment.)
3. Some transport environments are very demanding e.g. extremes of temperature, long drop heights, etc. making it very difficult to reproduce the condition of such packaging in the laboratory when trying to understand the history of a particular transport problem.

3.4 Predicting reliability

It is increasingly clear that engineering design attention to reliability throughout a development programme is necessary if the high reliability now being demanded of electromechanical products is to be met. Experience shows that any programme simply relying on demonstration tests for production/customer acceptance, do not

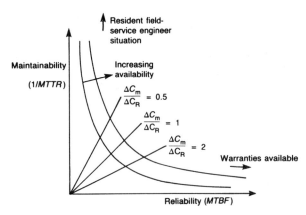

Fig. 3.19 The reliability–maintainability trade-off

usually meet reliability objectives. The problem is how to predict product reliability at various stages in the acquisition cycle, especially in the early phases where hardware is either limited, non-existent or immature with results hard to come by. Two analytical predictive failure techniques are presented which may be used in these circumstances. However, such models do nothing to determine failures or suggest cures, they are merely appraisal techniques which can be used to identify causes.

Once results and/or predictions are available, it is possible to monitor and predict a reliability growth pattern as well as estimating the effort required to achieve a mature reliability target within a given time-scale. No growth model can span a major change undertaken during the development cycle, one then has to return to original appraisal techniques.

3.4.1 Failure analysis

One of the reliability tasks identified during the acquisition phase was analysis of the design to determine possible modes of failure and their effect on system or equipment operation. The two common forms of reliability (unreliability) analysis used are failure mode and effect analysis (FMEA) and fault tree analysis (FTA). Other techniques such as Quirk's reliability index[12] etc. are also known, but we shall concentrate on FMEA and FTA. These two techniques differ inasmuch that FMEA starts with a failure mode and FTA has to find it. In other words when considering failure paths, FMEA is a bottoms-up analysis, working from effect to cause, whereas FTA is a tops-down analysis working from cause to effect. Each technique has its own value and area of applicability.

(i) FMEA

This is really a formalized approach to an experienced design engineering activity, investigating (and designing against) the repercussions and knock-on effects of a lower level piece part or component failure. It does require that the failure mode be known, or expected, beforehand, but the process may be selective, not every failure cause needs to be investigated. The basic level to start at must be chosen so that interactions are minimized. When the failure is related to a criticality index rating, the method is variously referred to as failure mode effect and criticality analysis (FMECA) or potential failure mode and effect analysis.

FMEA is a systematic method that considers the failure modes of various items and then determines impact on the system or equipment as a whole. This is achieved by breaking the system down into a hierarchy of decreasing levels of assembly and dividing each level into a number of functional items. The analysis starts by considering the failure modes of an item at the lowest level (Fig. 3.20), then determining their effects on the output of that level. These effects, in turn, become the failure modes of subsequent assemblies at successively higher levels until the overall system effect is established. The end result can be either a Pareto-type distribution, i.e. identifying the dominant failure modes, in which case accuracy is not necessary, or a quantitative judgement of the cumulative failure distribution (all modes).

FMEA is an established design analysis technique for assessing reliability, but the mechanics of the method are not so well defined. If is potentially a very beneficial and productive task within the reliability programme, although it can be costly in time and effort if done rigorously. However, the analysis[1] can provide a basis for the following:

1. Planning a development programme;
2. Planning a reliability and quality improvement programme;
3. Planning a maintainability programme and calculating MTTR for each failure mode;
4. Assessing the design process integrity when field performance data become available.

The approach in FMEA (Fig. 3.20) is as follows:

1. Identify the limits of the system and the level at which the analysis is to start.
2. Take each lowest level item in turn and list all failure modes. A good idea is to use a checklist (failure mode analysis).
3. Determine failure frequency (f) of modes listed in (2) during a given time interval. It may be easier to

PRODUCT RELIABILITY AND TEST

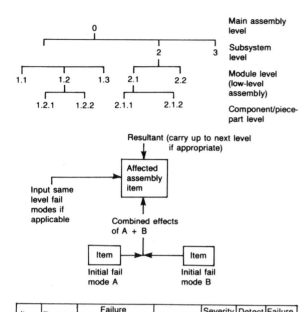

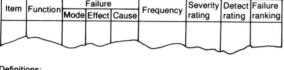

Definitions:

Level	identifies the assembly level
Failure mode	manner in which a unit can fail
Failure effect	consequences of a failure mode on operation of an assembly
End effect	a failure mode having no further effect at the current level
Final effect	a failure effect at the highest level (top-level assembly)
Criticality (or severity rating)	a relative measure of the severity of a failure and its frequency of occurrence
Detection rating	the probability of the cause of failure being detected by normal quality control procedures before reaching the customer

Fig. 3.20 Example of FMEA unit identification and worksheet format

use probability encoding to estimate L_5, L_{10}, L_{90} lives, etc. and to plot on Weibull paper. The failure frequency could also be assigned a rating level of 1–10 based on predetermined frequency bandwidths.

4. Examine effects of all failure modes at lower level on next level items. The resulting failure modes at that level are then examined for effects at the next higher level and so on (failure effect analysis).
5. Assign a rating level (α) to each effect in (4) recorded at the highest level. Use value of α between 0 and 10, i.e. 0 = no effect; 10 = major impact.
6. Derive a criticality, $f\alpha$, for each primary fail mode. A major problem occurring a few times is thus more critical than a small frequently occurring problem. If there are more than one effect resulting out of the failure mode then criticality = $f\Sigma\alpha$.
7. As a check on the manufacturing and quality process, a detection ranking index, n, may be assigned on the basis of a 1–10 rating, i.e. 1 = detection of a failure or quality departure before reaching the customer, and 10 = unlikely to be detected.
8. A risk priority number (RPN) = $fn\alpha$ can be calculated to more accurately predict what will happen in the field.

Now from the above, Pareto distributions may be produced for the following:

1. Frequency of modes of failure;
2. Relative repair costs from *MTTR* for each failure mode;
3. Criticality rating of each failure mode;
4. Risk priority rankings of failure modes.

Each of the above may produce different major reliability items to work on. The effect, and hence ranking, of the same failure mode may vary considerably depending on the level at which the analysis is completed. However, the reliability intent here of the process is to resolve the 'vital few' fail modes in the next hardware design iteration or upgrade. The use of FMEA analysis sheets are generally recommended.[13]

(ii) FTA

In those cases where a major system failure effect can be identified from the beginning, it may be helpful to use an FTA to identify possible causes. This consists of working an analysis down through subsystems, lower-level assemblies and components, identifying all possible causes. In other words, the top or final event is traced back to the primary failure mode cause — a 'chain' process or reversed FMEA. This requires a detailed logic diagram of the system to be prepared. Since the FTA is a logic flow process, standard logic and event symbols are used in the presentation, Fig. 3.21.

Probabilities can be allocated to all part failure modes, from which the system failure probabilities, hence part criticality ratings, may be obtained. The following assumptions may be made to develop the process:

1. The fault events (or component failures) are exponentially distributed in time;
2. The fault tree is redundancy free, i.e. no repeated events;
3. Component failures are random.

The fault tree OR/AND gate failure-rate expressions are derived from

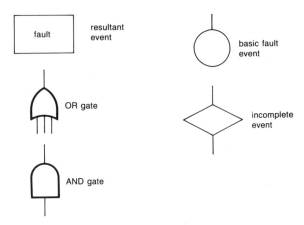

Fig. 3.21 Fault tree logic symbols

$$\lambda(t) = -\frac{dR(t)}{dt} \times \frac{1}{R(t)}$$

(a) OR gate Logically this gate corresponds to a series system, such that

$$\lambda s(t) = \sum_{i=1}^{n} \lambda_i = \text{'system' failure rate}$$

i.e. the OR gate output = the sum of its inputs, a series network reliability.

(b) AND gate This corresponds to a logically connected parallel configuration system. A parallel network reliability is then given by

$$\lambda p(t) = \frac{\sum_{i=1}^{n} \lambda_i (Z_i - 1)}{\sum_{i=1}^{n} Z_i - 1}$$

where

$$Z = \frac{1}{1 - e^{-\lambda t}}$$

Considering the failure-rate evaluation fault tree of Fig. 3.22, the event failure rates of gates 0, 1, 3 and 4 are

$\lambda T_0 = \lambda_1 + \lambda_2 + \lambda T_3 + \lambda T_2$
$\lambda T_1 = \lambda_1 + \lambda_2$
$\lambda T_3 = \lambda_5 + \lambda_6$
$\lambda T_4 = \lambda_3 + \lambda_4$

and

$$\lambda T_2 = \frac{\lambda_7 (Z_7 - 1) + \lambda T_4 (ZT_4 - 1)}{Z_7 ZT_4 - 1}$$

(iii) Reliability growth

Reliability growth has been defined as 'the systematic planning for reliability achievement as a function of time and other resources, controlling an ongoing rate of achievement by re-allocation of resources based on comparisons between planned and assessed reliability values'.[14]

Products in the early development stage tend to be less reliable than in later development phase, invariably have reliability and performance deficiencies and are the subject of development testing to find problem causes and to take corrective action. The improvements in reliability and performance which results will depend on the number and the effectiveness of the fixes that are incorporated into the system. This period of reliability growth, and including the product introduction phase, was first monitored by Duane.[15] He observed a linear log/log relationship between cumulative failure rate (*MTBF*) and cumulative time, namely:

$$\log\theta_c = \log\theta_0 + \alpha(\log T - \log T_0)$$

where θ_0 = cumulative *MTBF* at start of monitoring period, T_0 = start of monitoring period, θ_c = cumulative *MTBF*, and

$$\theta_c = \theta_0 \left(\frac{T}{T_0}\right)^\alpha$$

where α = rate of *MTBF* growth, i.e. the effectiveness of the reliability programme in correcting failure modes; the lower it is the longer it takes to achieve the target. The instantaneous *MTBF*, θ_i, can be obtained by differentiation of the above to give

$$\theta_i = \frac{\theta_c}{1 - \alpha}$$

The Duane model is thus very straightforward and easy to apply, either in terms of *MTBF* or failure rate, e.g. Fig. 3.23, the plot of θ_i is parallel to θ_c and displaced from it by $1 - \alpha$. The values of α can be interpreted as follows:

$\alpha = 0.4/0.6$: reliability is the main objective; programme dedicated to elimination of failure modes; use of stress testing to pin-point modes of failure; rapid corrective action.

$\alpha = 0.3/0.4$: normal reliability priorities; 0.4 is average value; well-managed analysis and corrective actions.

$\alpha = 0.2$: routine attention to reliability testwork; corrective actions for top failure modes only.

$\alpha = 0/0.2$: no reliability improvement priorities and no failure analysis conducted; concentration is on performance but not on reliability; low priorities for corrective actions.

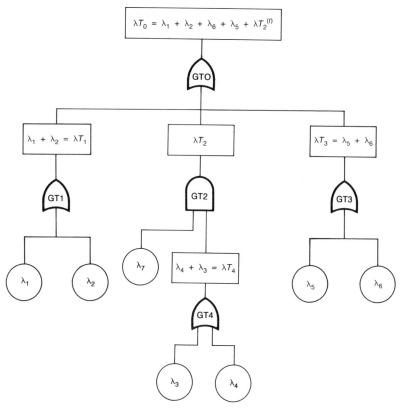

Fig. 3.22 A hypothetical failure rate evaluation fault tree (from Dhillon and Singh, ref. 3. Courtesy of B.S. Dhillon)

The method is applicable to a population with a number of failure modes (progressively corrected) and can be used to monitor MTBF growth and/or assess the test time required to attain a target rate if α is assumed. In Fig. 3.23 T_0 and θ_0 represent a known status. However, the analysis is sensitive to the starting assumptions. In the early stages of product development the reliability results are somewhat uncertain, hence some judgement needs to be exercised in choosing the starting assumptions in the embryonic stage.

The use of the Duane growth curve in practice is shown in Fig. 3.24[8] for the reliability growth of a photocopier. In this case the horizontal axis is taken as number of photocopies rather than time. It can be seen that there are noticeable reliability jumps at the early hardware iterations, typically where the vital few are being worked on. This is because once identified, some failure modes take a while to fix, often several modifications are installed at the same time and subsystem wear-outs may also occur. At the end of a test phase, it is usual to find two sets

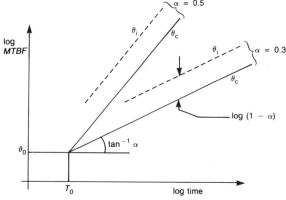

Fig. 3.23 Reliability growth — Duane model

of results: (1) a demonstrated value based on the raw trial data; (2) a projected reliability value based upon the problem fix effectiveness of the changes, i.e. the 'adjustment procedure' or delayed fix projection.

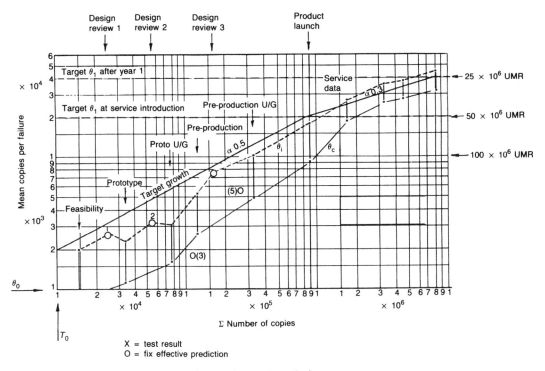

Fig. 3.24 Duane growth plot for a photocopier product design

Both (1) and (2) are plotted in Fig. 3.24. The projections are often optimistic in the early stages of development for the reasons, previously given. As the reliability programme becomes advanced, more attention is paid to the trivial many, the reliability growth is then much smoother and the growth models will then be more valid. It thus can be seen that by generalizing, reliability growth can be planned and managed.

In the simplest case, reliability growth is estimated by considering the failure data and the planned corrective action; e.g. if 20 per cent of failures are caused by a mode with an 80 per cent fix effectivity assigned, the reliability improvement is 16 per cent. However, the accuracy of projections based on the impact of fixes is likely to be biased, although it is a common engineering procedure. A recent model, which is an extension to the Duane model, structures the adjustment procedure and deals in particular with statistical bias.[14]

In the adjustment model, there are two types of failure rate λ_A and λ_B such that

λ_A = failure rate of 'A' components which because of size, cost, etc. is not cost effective to carry out a design change

λ_B = modes found during test that will be incorporated as delayed fixes

so system failure rate

$\lambda_A + \lambda_B = r(0)$ at $T = 0$

and at time T

$$r(T) = \frac{1}{T}\left[N_A + \sum_{i=1}^{m}(1 - d_i)N_{i(B)}\right]B(T)$$

hence

$r(0) - r(T)$ = reduction in system failure rate
$MTBF = r(T)^{-1}$

where N_A, N_B = no. of observed type A and B failures, d_i = fix effectiveness factor = percentage decrease in X_i after a corrective action has been made for the ith-type B mode and $B(T)$ = bias term of the Poisson distribution = $\mu d\, h(T)$

where

$$\mu d = \frac{1}{M(T)}\sum_{i=1}^{m(T)} d_i = \text{mean of effectiveness factors}$$

and

$h(t)$ = Poisson distribution intensity function (or rate)
= $\lambda \beta t^{\beta-1}$

where M = no. of types of failure mode, λ, B = constants estimated from a log/log plot of the field data, i.e. $\log N(T) = \log \lambda + \beta \log T$; it may also be shown that $(1 - \beta) \equiv$ the Duane α.

3.5 Evaluating reliability

In this section we shall consider some of the many possible ways that equipment reliability can be tested, evaluated or assessed. Testing itself is only one of the possible elements in a reliability programme, the reliability assurance of the design will also reflect service failure patterns, service environmental effects and role of the user. However, to quantify reliability, i.e. the performance and durability of the product, we have to test in some form or another. It should be remembered though that testing examines the functionality of the design, not the electromechanics.

3.5.1 Reliability testing

There are four major categories of reliability testwork listed as follows, the more well-known and specific forms of testing identified under each of the headings below.

(i) Performance testing
In general the emphasis here is on identifying the modes of failure rather than frequency, to explore the operational limitations of the design, e.g.: (a) equipment reliability test; (b) design and development testing; (c) feasibility testing; (d) stress testing; (e) environmental testing; (f) transportation, packaging and storage tests.

(ii) Determination testing
Testing undertaken to ascertain reliability, i.e. used to provide information, generally with the principle of upgrading reliability by upgrades and successive test analysis: (a) reliability determination testings; (b) milestone testing; (c) ongoing evaluation (test and fix); (d) reliability growth programme; (e) endurance tests; (f) environmental performance tests.

(iii) Compliance testing
Used to show whether reliability meets an agreed target and normally as a condition of acceptance either as a final product or possibly as an approval for continuing into the next phase of development, e.g.: (a) reliability compliance test; (b) qualification test; (c) acceptance testing; (d) type testing; (e) verification test; (f) demonstration test (g) sequential testing; (h) field test.

(iv) Production testing
As a means of ensuring optimum product reliability in the field, a number of testing procedures are often built into the production process, in most cases a form of compliance testing, e.g.:

(a) 'burn-in' testing;
(b) quality audit testwork (part of quality function);
(c) set-up evaluation and test;
(d) functional testing (module and system levels);
(e) automatic testing (ATE).

The applicability of the above test types within the design cycle is given in Table 3.3. However, there is no one authority on what comprises an individual test method type, it will depend on what is most suitable for a particular company's needs and where the emphasis is put. With the compliance testing in particular, it will be necessary to draft and agree a specification between parties prior to test. Only those tests which are necessary should be specified. Operational reliability requirements should be expressed in usage-orientated terms and reliability requirements should be specified in test-orientated terms.[16]

It is not the intent here to discuss all the above types of test, rather to concentrate on those forms of test most widely used. The technology taskwork referred to in the feasibility phase is early performance testing, aimed at establishing operating windows. This continues during the early part of the acquisition phase, but changes into determination testing as integrated hardware becomes available, the reliability (and maintainability) aspects of the design then most likely monitored via an ongoing growth programme (section 3.4.1). The main purpose of qualification testing is to ensure that products meet the targets laid down in the engineering product specification, an internal process, while the purpose of acceptance/type/ verification or demonstration testing is to verify that the product meets the requirements of, or the targets laid down by, a third party.

3.5.2 Stress testing (failure modes)

One of the main problems in demonstrating high reliability is achieving an MTBF with confidence, without involving an associated large amount of testwork. The response is usually to accelerate failure modes by increasing the loading/stress levels and thence fixing the problems by widening the operating window. There are two types of failure:

1. *Hard failure.* In essence this is part or component breakage, seizure, burn-out, etc., the result of which the equipment is inoperative until the failure is replaced.
2. *Soft failure.* This refers to the situation where equipment performance (output) is reduced below specification for some or all of the time. The equipment is still operative, but would not provide user satisfaction in the market-place.

In the case of hard failures, it is possible to accelerate failure modes by decreasing the safety margin (section 3.2.2), i.e. increasing the 'loading' (Fig. 3.4) if the stress pattern is known. It is customary to look for a relationship between failure rate and stress level, e.g. fatigue, although this is not always necessary, since it is the use of high stress levels to rapidly identify failure modes that is important in reliability improvement.[17] As an alternative to a stress or accelerated test, a step-stress endurance test is sometimes employed where the loading is continuously increased until failure occurs. In terms of physical durability, therefore, these types of failure are controllable via the design specification. Soft failures, on the other hand, represent the performance limitation of the design, i.e. an established latitude rather than a physical failure rate. Performance latitude determination is often the area where most stress testing is concentrated.

Usually the number of separate, but not independent variables involved in a test make it impossible for the effect of each to be assessed individually. To analyse one variable at a time with the rest constant would be illogical in terms of resource requirement. The following will consider the concept of multiparameter tests as applied to performance evaluation and stress testing.

The understanding process will start early in the design cycle, often in the feasibility phase, but with the task of improving performance latitude via failure mode evaluation. The concept is outlined in Fig. 3.25, where aspects defining the physical configuration of the design are initially considered in parallel with process and environmental aspects (required performance latitude) which are then jointly tested during early reliability and/or development activities to produce a problem set (the 'vital few'). This is then fed into the reliability growth and problem management programmes tracking hardware through the acquisition phase. The terminology referred to in Fig. 3.25 can be defined as follows:

1. *Controllable parameter.* Chosen by the designer in relation to the requirements of the product engineering specification. If enabling technology is being developed, however, the end requirement may be less well defined.
2. *Sensitivities.* The limits on controllable parameters which, when determined, result in a successful performance.
3. *Uncontrollable parameters.* These will vary depending on the operational environment and whether any processing or product media is involved. They also depend on the type of product.
4. *Latitude.* The limits of uncontrollable parameters within which no failure modes occur, hence successful performance.

Performance stress testing is about establishing the failure thresholds of the known failure modes against the widest possible range of uncontrolled parameters (market flexibility) and how these parameters are affected by the configuration, feeding back into the controllable parameter design process. For the simplified case of one input parameter, the idea of stress level hierarchy relative to a given market, is shown in Fig. 3.26, for both specification limits (a) and operating boundary (b). The specification points in Fig. 3.26(a) represent points of equal market 'stress', not necessarily uniformly distributed about a nominal performance. Figure 3.26(b) translates the market 'stress' requirement into an idealized trade-off between the set-up latitude available between the two parameters A and B. A flow diagram outlining the aspects of performance stress testing is given in Fig. 3.27. An example of an operating window is shown in Fig. 3.28. This demonstrates a particular type of paper feeder performance latitude in relation to the paper weight being fed and the net force on the feed head. Such a feeder will normally be found in photocopiers and it is the resultant use of this type of stress-induced window that determines how good the final design performance will actually be.

The following points apply to the operating window concept as related to a product design:

1. A small operating window results in large piece part and Field Service costs, i.e. it is difficult to manufacture within the window and difficult to keep it in operation.
2. A large window is preferred.
3. The window can almost certainly be enlarged by adding features at extra cost to control normally uncontrollable parameters. A trade-off analysis is often needed which is the most cost effective if done during the acquisition phase.
4. Focus testing on stress conditions rather than establishing failure rates. Stress testing will quickly identify improvements and lead to enlarging the operating window.

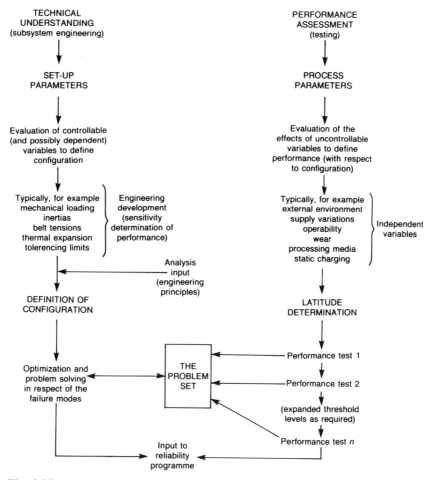

Fig. 3.25 The performance latitude improvement process

5. Uncontrolled parameters should always be probed beyond the expected range.

Now, critical input parameter combinations can also define stress levels by assuming a simple relationship between specification limits; however, stresses will be more severe and performance harder to achieve. For example, a three-dimensional stress case is defined schematically in Fig. 3.29. In this case, critical input parameters are normalized and stress at a point defined as

$$\bar{S} = W_1 S_1 + W_2 S_2 + W_3 S_3$$

where W = weighting factor between 0 and 1, Sn = stress value, single-parameter condition.

It can be seen that the stress levels at the specification corners are high by definition, and in practice the marketplace is unlikely to be sensitive to extremes. Hence by joining specification limits via a simple curve, this avoids excessive stress testing, but is still more severe than the one-dimensional case.

3.5.3 Environmental testing

Environmental testing is designed to demonstrate with some degree of assurance that an electromechanical product will survive and perform under specific environments, either by simulating or reproducing the effects of that environment. However, the conditions of operational use are not always well defined so for this reason, environmental tests are generally stress tests designed to accelerate the operative failure mode but not to introduce new ones. There are two basic types of environmental tests, but a very large range of tests[18] as follows:

1. *Climatic tests*. Basically the interaction of temperature and/or humidity effects on the equipment both

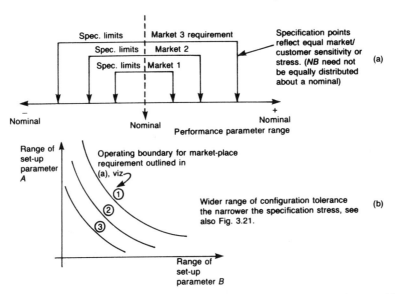

Fig. 3.26 Examples of a single-dimension stress specification and the general effect on operating boundary position

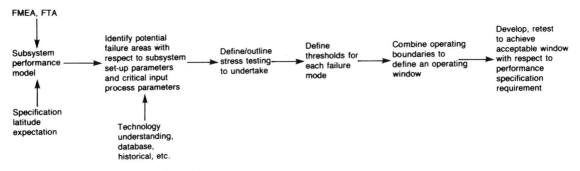

Fig. 3.27 Stress testing procedure

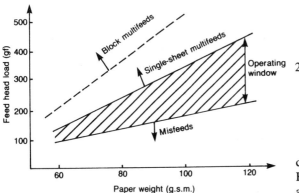

Fig. 3.28 Operating window for buckle type of paper feed mechanism

during operation and during storage, e.g.: (a) cold temperature; (b) dry heat; (c) damp heat, steady state and cyclic; (d) corrosive atmospheres; (e) mould growth; (f) air pressure; (g) dust and dirt.

2. *Transportation tests*. These tests nominally reproduce the conditions experienced during handling, evaluating both equipment and its packaging, e.g.: (a) impact, shock and bump (b) drop (free-fall); (c) vibration (shaking); (d) acceleration/deceleration.

There are clearly a large number of possible test combinations, both variable parameters and severity. However, it is important to agree a realistic specification and with sufficient consideration, these may be reduced by selecting a few standard groupings. Insistence on encompassing the extremes, Fig. 3.29, is expensive in

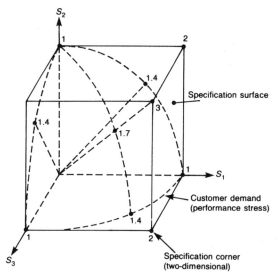

Fig. 3.29 Example of three-dimensional stress

resource terms. Tests are designed to provide information on the following:

1. The ability of the equipment to operate within specified limits of temperature/humidity/air pressure, etc. or any other environmental condition or combination.
2. The ability of the equipment to withstand storage and transportation, both with and without suitable packaging as appropriate.

Given that a number of environmental tests are usually undertaken, a testing sequence should be chosen so as to produce the best possible stress case, i.e.

1. A test with rapid temperature cycling should come first in the sequence as this may promote mechanical stressing that leaves the equipment more sensitive to later testwork.
2. For the same reason, testing for solder/connector integrity should come early in the test sequence.
3. Mechanical tests accentuate failure effects initiated by temperature cycling.
4. Damp heat (cyclic or soak) testing applied at the end of the sequence will not only determine long-term behaviour in a humid environment but also reveal the effects of preceding thermal and mechanical stresses.

(i) Climatic tests

Typically, one would expect the following conditions to apply to normally internally sited equipment:

Temperature. A minimum of 10 °C if operational, but probably -10 °C in transportation or storage. At elevated temperature, a probable 35 °C maximum. If the equipment is used outside the above limits, then degraded performance would be expected. For particular outside equipment specified at temperature extremes, an increased design cost is inevitable.

Humidity. Typically 40–70 per cent with degraded performance to 20 and 80 per cent. Storage conditions in warehouses would only normally exceed this range, non-condensing conditions assumed.

Within the temperature–humidity matrix there is usually specified a normal operating envelope in which normal performance is expected. Operating equipment outside of this range would be expected to produce a performance degradation (Fig. 3.30), the reliability in the outer zones modified by a performance factor. For example, if the probability of an event occurring in normal operation is 0.01, the probability is increased to 0.013 if the performance factor is 0.8. Alternatively, multiply the unscheduled maintenance rate in that environment by 0.8 to equate it to normal operation, assuming equal times spent at each environment.

A comprehensive summary of climatic effects in single environments is given in Table 3.4.

(ii) Transportation tests

There is little value in investing large sums in the acquisition of inherently reliable products if they are subsequently damaged by inadequate packaging and handling. Packaging needs to match the needs and weaknesses of the contents. A significant proportion of early launch problems (section 3.3.6) result from inadequate packaging in storage and during transportation. Many alternative types of packaging are available today, the choice dependent on cost, size and delivery.

Shock testing For small items, the shock from manual handling is the most common cause of damage, i.e. generating large 'G' forces for short times. The severity of the distribution environment in terms of drop height, increases as the weight of the package decreases (Fig. 3.31). Cushioning systems using plastic foams, etc. prevent damage by spreading the impact energy of the fall over a longer time interval, thus reducing the peak forces transmitted to the product. It is important to achieve the right amount of cushioning — too little is ineffective as also is too much packaging of too high a compression density, if it is too stiff for the product weight and product surface area in contact with it.

Vibration tests Unlike shock, which is a random

Table 3.4 *Single environmental effects*

Environment	Principal effects	Typical failures resulting

High temperature	Thermal ageing:
	    oxidation
	    cracking
	    chemical reaction
	Softening, melting, sublimation
	Viscosity reduction, evaporation
	Expansion
		Insulation failure, mechanical failure, increased mechanical stress, increased wear on moving parts due to expansion or loss of lubricating properties

Low temperature	Embrittlement Ice formation Increased viscosity and solidification Loss of mechanical strength Physical contraction	Insulation failure, cracking, mechanical failure, increased wear on moving parts due to contraction or loss of mechanical strength and to loss of lubricating properties, seal and gasket failure

High relative humidity	Moisture absorption or adsorption
	Swelling
	Loss of mechanical strength
	Chemical reaction:
	    corrosion
	    electrolysis
	Increased conductivity of insulators
		Physical breakdown, insulation failure, mechanical failure

Low relative humidity	Desiccation Embrittlement Loss of mechanical strength Shrinkage Increase of abrasion between moving contacts	Mechanical failure, cracking
High pressure	Compression, deformation	Mechanical failure, leaks (failures of sealing)
Low pressure	Expansion Reuced electric strength of air Corona and ozone formation Reduced cooling	Mechanical failure, leaks (failure of sealing) flashover, overheating
Solar radiation	Chemical, physical and photochemical reactions Surface deterioration Embrittlement Discoloration, ozone formation Heating Differential heating and mechanical stresses	Insulation failure See also 'High temperature'
Sand or dust	Abrasion and errosion Seizure Clogging Thermal insulation Electrostatic effects	Increased wear, electrical failure, mechanical failure, overheating

Corrosive atmospheres	Chemical reaction:
	    corrosion
	    electrolysis
	Surface deterioration
	Increased conductivity
	Increased contact resistance
		Increased wear, mechanical failure, electrical failure

Wind	Force application Fatigue Deposition of materials Clogging Erosion Induced vibration	Structural collapse, mechanical failure See also 'Sand or dust' and 'Corrosive atmospheres'
Rain	Water absorption Temperature shock Erosion Corrosion	Electrical failure, cracking, leaks, surface deterioration
Hail	Erosion Temperature shock Mechanical deformation	Structural collapse, surface damage
Snow or ice	Mechanical loading Water absorption Temperature shock	Structural collapse See also 'Rain'
Rapid change of temperature	Temperature shock Differential heating	Mechanical failure, cracking, seal damage, leaks
Ozone	Rapid oxidation Embrittlement (especially rubber) Reduced electric strength of air	Electrical failure, mechanical failure, crazing, cracking
Acceleration (steady state) Vibration Acoustic noise Bump or shock	Mechanical stress Fatigue Resonance	Mechanical failure, increased wear of moving parts, structural collapse

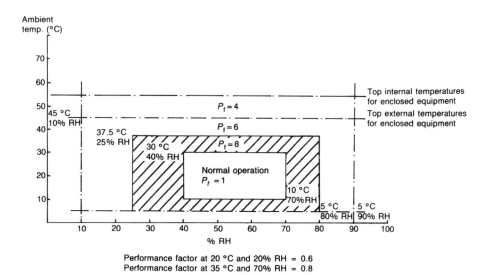

Fig. 3.30 Allowances for extreme temperature and humidity effect on performance

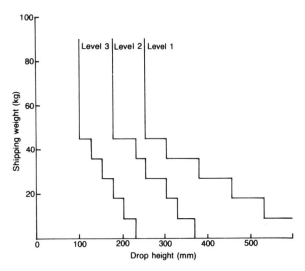

Notes: Intensity level: level of test intensity as based on the probability (ref. 22): of occurrence in a typical shipment cycle. This is a pre-established test level based on the product value, to acceptable level of product damage, to shipping environment, etc.

Level 1: A high level of test intensity and a low probability of occurrence. A severe environment

Level 2: Use this test regime unless conditions dictate otherwise

Level 3: Low level of test intensity, but with a high probability of occurrence.

Fig. 3.31 Manual handling test procedure at three levels of intensity; from ref. 20

occurrence, vibration affects the product at various levels throughout the entire journey to the customer, depending on the modes of transport chosen. It is therefore important during the design cycle to establish whether vibration in transit is likely to cause product damage. Vibration sweep testing in different planes will excite the natural frequency of various components. Damage can occur when a critical component resonates and fatigues over a period of time.

If a critical frequency corresponds to that expected of the transport means (lorries typically 3–20 Hz), then suitable cushioning or damping will have to be designed into the packaging. If there are dominant frequencies, endurance testing for a prescribed time is then undertaken at that frequency.[18] Alternatively, a specified time per axis over a number of sweep cycles could be specified as an endurance test.

Vibration severity is defined by a combination of three parameters: (1) frequency range; (2) vibration amplitude; (3) duration of test (sweep cycles or time).

A sweep cycle is a traverse of a specified frequency range once in each direction, e.g. $3 \rightarrow 2000 \rightarrow 3$ Hz with normally sinusoidal motion. Typically the sweep rate is 1 octave min^{-1} whence

No. of octaves/or sweep cycle = $6.65 \log\left(\dfrac{f_2}{f_1}\right)$

where f_2 = upper frequency limit of sweep and f_1 = lower frequency limit of sweep.

Bump test This test is designed to reproduce the effects of overload transportation, i.e. repeated jolting and bumping (repeated low-energy impacts), generally complex and random in nature and occurring over variable time periods.[18] Not therefore the easiest of tests to simulate, Table 3.5 gives some idea of typical tests used.

3.5.4 Sequential testing

This is one of a number of types of possible compliance tests but sequential testing is important since procedures are documented and standardized to a degree, whereas other compliant testing is often a unique agreement between supplier and customer. This type of test, also known as a fixed demonstration test, tends to achieve

Table 3.5 *Examples of typical bump test conditions*

Peak acceleration (g)	Severity (m s^{-1})	Pulse duration (ms)	Velocity change (m s^{-1})	No. of bumps	Typical applications
10	100	16	1.0	1K	General robustness test for transported and secured equipment
15	150	6	0.6	4K	Normal robustness test for transported general items
40	400	6	1.5	1K	Items carried loose in wheeled vehicles (road and rail)

PRODUCT RELIABILITY AND TEST

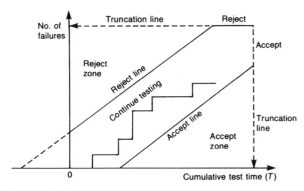

Fig. 3.32 Truncated sequential compliance test

faster results than fixed time testing and is based on sequential probability ratios.[19] Testing produces a familiar staircase plot, Fig. 3.32, until the failure plot crosses a decision line. Crossing the decision line denotes that the test criteria have been met. The decision lines are truncated to provide a reasonable maximum test time calculated as multiples of a specified *MTBF*. Here *MTBF* is used as the index of reliability.

To apply, it is assumed that the failure data are exponentially distributed, i.e. a constant failure rate. If the relationship between probability of acceptance and failure rate is given as a cumulative distribution, then

α = producer's risk (probability of rejecting a good item with high *MTBF* θ_0)
β = consumer's risk (probability of accepting an item with low *MTBF* θ_i)

and

θ_0 = design *MTBF*
θ_i = the specified, low limit or agreed MTBF to be demonstrated

where $d = \theta_0/\theta_i$ = design or discrimination ratio/index then in Fig. 3.32, the lines are constructed from the following equation:

$$y = \left(\frac{1}{\theta_i} - \frac{1}{\theta_0}\right)\frac{T + C}{\log d}$$

where T = cumulative test time, and

$$C = \frac{\log_e(1 - \beta)/\alpha}{\log d} \quad \text{for the reject line}$$

and

$$C = \log_e \frac{(1 - \alpha)/\beta}{\log d} \quad \text{for the accept line}$$

If the risks are reduced then the lines move further apart and the tests will take longer. If the design index is reduced bringing the two MTBFs closer together, then the lines will be less steep, making it harder to pass the test. The selection of a test plan to use (Table 3.6) will depend on the acceptable degree of risk and on the cost of testing.

3.5.5 Field feedback

Once launched, and by definition out of the direct control of the design team, it is important further to assess reliability in the field and to do this via formal reporting documentation. This is necessary to ensure that feedback is consistent and of adequate depth. Field information is valuable since it reflects maintenance activities conducted under real conditions. Any properly designed component should fail in wear-out mode, and since all deterioration can be controlled by maintenance, equipment reliability is then controlled by and dependent on maintenance. A maintenance progrmme is possible if the equipment is inherently reliable or predictably unreliable. Failures are

Table 3.6 *Fixed length test plans*

Test plan	Decision risks (%)		Design ratio (d)	Test duration ($\times \theta_0$)	Reject $\geq$ failures	Accept $\leq$ failures
	α	β				
IX	10	10	1.5	45.0	37	36
X	20	20	1.5	21.1	18	17
XI	10	10	2.0	18.8	14	13
XII	20	20	2.0	7.8	6	5
XIII	30	30	2.0	3.7	3	2
XIV	10	10	3.0	9.3	6	5
XV	20	20	3.0	4.3	3	2
XVI	30	30	3.0	1.1	1	0

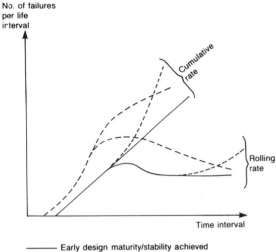

Fig. 3.33 Illustrations of field reliability (unscheduled maintenance)

Legend:
— Early design maturity/stability achieved
– – – Initial design immature; benefiting from ongoing necessary reliability improvements; unstable design with reliability improvements continuing into the field
- - - - Early reliability maturity achieved but deteriorates after a fixed period (designed obsolescence/poor design durability or deteriorating maintenance policy)

generally only rectified after they occur, i.e. maintenance repair policy (section 3.3.4), the level of unscheduled maintenance determined by scheduled maintenance intervals. Figure 3.33 illustrates three possible field scenarios depending on the basic reliability of the equipment and the planned maintenance policy adopted.

More maintenance is generally required with age as an increasing number of components in equipment pass into wear-out. However, 'hard' failures are the straightforward situation to report from the field, what is not so straightforward are the 'soft' failures that impair performance and the failures that are largely subjective and a function of user expectations. Consequently when analysing field data, a full definition of failure should be available in each case. To be able to compile statistical reliability data, field returns are essential and in particular they:

1. Indicate deficiencies and generate reliability growth programmes if necessary;
2. Provide reliability and product quality trends;
3. Provide vendor component performance ratings;
4. Contribute statistical reliability and maintainability data for future product development;
5. Enable the spares provisioning policy to be formulated;
6. Enable planned maintenance intervals to be revised;
7. Enable the product life cycle cost equation to be carried out.

References

1. Carter A D S *Mechanical reliability*. Macmillan 1986
2. Ireson W G *Reliability handbook*. McGraw-Hill 1966
3. Dhillon B S, Singh C *Engineering reliability*. J. Wiley 1981
4. Smith D J *Reliability and maintainability in perspective* Macmillan 1981
5. *Standards of the Anti-friction Bearing Manufacturers' Association*. New York 1972
6. Bergling G The operational reliability of rolling bearings. *Ball Bearing J.*, **188**: 1 (1976)
7. BS 5760, Part 1.
8. O'Connor F D T *Practical reliability engineering*. Heyden 1981
9. US MIL STD Handbook 472 *Reliability prediction* 1966
10. Clapham J C R Operational life of equipment. *Operational Research Quarterly* 8: 181 (1957)
11. Hagen E W Human reliability analysis. *Nuclear Safety* 17: 315 (1976)
12. Jones C *Design methods*. Wiley 1970
13. Leech D J, Turner B T *Engineering design for profit*. Ellis Horwood 1985
14. Crow L H *On Methods for reliability growth assessment during development*. In J K Skwirzynski (Ed) *Electronic systems effectivness and life cycle costing*, Springer-Verlag 1983
15. Duane J T Learning curve approach to reliability monitoring. *IEEE Transactions on Aerospace* **2**: 563 (1964)
16. IEC Publication 605 *Equipment reliability testing*
17. Nixon F Techniques of improvement in safety and failure of components. *Proceedings of the Institution of Mechanical Engineers* **184**: 3B (1969/70)
18. BS 2011 *Basic environmental testing procedures* (also IEC 68)
19. US MIL-STD-781 *Reliability qualification and production acceptance tests* 1967
20. ASTM Standard D4169 *Pre-shipment test procedures* 1986

CHAPTER 4
ADHESIVES

4.1 Introduction

It has been the practice of the aerospace industry for some time to regard adhesives as engineering materials and consequently it has seen the most important advances in structural adhesives technology. The same has not been true for the rest of manufacturing industry, although traditional fixing methods are gradually being superseded, especially so in automotive applications. It is estimated[1] that 30 per cent of all industrial joining is now done with adhesives. Threadlocking is a specific low-duty application area where adhesives are more widely used to supplement threaded fastener performance.

There is no such thing as a universal adhesive, the choice of adhesive depending not only on the joint performance required but also on the manufacturing specification. For good joint performance the design engineer must appreciate the importance of correct joint design, no adhesive will perform satisfactorily if the potential is thrown away on a poor joint configuration. There are many adhesive types available, but the adhesive manufacturers are not often keen to set out rules, consequently suitability has to be empirically assessed, requiring time and resources within the design cycle.

The following points can be made regarding adhesive bonding:

1. The adhesive system must be considered as an integral part of the design cycle fairly early on, and the commitment to it stressed in order to obtain the best results.
2. An adhesive bond obtained with a joint originally designed for a mechanical fixing can never be as good as the situation where the use of an adhesive has been designed in.
3. The laws of nature cannot be changed to suit the available production facilities. The Production Department must appreciate the need for correct surface preparation treatment, adhesive application techniques and curing time. Adequate quality control should be assured.
4. The designer must consider the advantages of adhesives, be aware what the materials will do, how the structure will perform and plan accordingly. There are stress concentrations in the glue line which should be appreciated.

In this chapter we explore the nature of the contacting process itself, what actually enables surfaces to be glued together and types of joint design to best exploit the magnitude of the bond. The various types of engineering adhesive available will also be discussed along with application suitability. Table 4.1 broadly outlines the advantages and disadvantages of adhesive bonding when compared to traditional mechanical fastening. Although a number of disadvantages are listed, most of these disappear as long as the application is purposely designed for an adhesive, in which case the benefits of such a joining technology can be significant.

4.2 The contacting process

Although a section on the adhesive contacting process may appear as mechanistic materials science, it is important for the design and development engineer to understand in general terms what is actually occurring at the glue line during the bonding process and what makes the joint stick. Zygology, the science of joining things, is often linked to adhesive bonding.

4.2.1 Surface energy and wetting

The adhesive process is a reduction in total surface area by the process of wetting. The forces achieving molecular

Table 4.1 *Advantages and disadvantages of adhesive bonding*

Advantages	Disadvantages
Almost any engineering material can be joined together, no compatibility problems	Joints are mostly permanent
Continuous joint with load-bearing capability, stresses tend to be lower than with mechanically fastened joints	Temperature range of operation tends to be limited to 150 °C, poor resistance to elevated temperatures
Joints do not invite stress concentrations	Durability of joint is difficult to predict, especially in an hostile environment
Bonded joints are stiff compared to other fastening methods, can use lighter alloys or thinner gauges	Effect of moisture likely to reduce strength of bond
Materials can be selectively incorporated in structure wherever required — hence capability of bonding composites	Durability is adversely affected by inadequate processing procedures
Access to materials easier, i.e. only required from one side	Polymers are weak compared to metals so for equivalent joint strength of metallic fasteners, must increase joint area
Process can be automated easily, assembly is simplified, no need for accurate hole alignment, etc.	Almost always use lap joints, butt joints are invariably not strong enough
Fewer post bonding operations required (e.g. no cleanup after welding)	Careful preparation of adhesive and adherends is often necessary, the procedures required are not normally used by the engineering industry
No fasteners visible on finished article	
Smooth contours possible in finished materials, joints neat and unobtrusive	Manufacturing cycle may be limited by application and cure times, depending to some extent on the particular adhesive choice
Generally a low-temperature process with little risk of distortion, damage, metallurgical change or residual stressing	Use of an adhesive may require conformity to the appropriate code of practice, emissions, handling precautions, etc.
Simultaneous sealing (insulation) can provide for a moisture and chemically resistant barrier, minimum contact corrosion between dissimilar materials, vibration damping and noise reduction, thermal insulation, electrical resistance and gap filling	If surface coatings, pre-treatments or platings are bonded, they may well cause inherent joint weakness
	NDT methods of joint evaluation are limited
	Adhesive joints, by definition, are permanent, hence the elements cannot be separated without destroying the joint
Cost reductions	More susceptible to failure when subject to cleavage and/or peel forces
Flexibility in production, the ability to add different components to standard piece-parts	Shortage of accurate/useful design data
IR and microwave radiation resistance	Total lack of design codes of practice and standard solutions
Strength/weight ratios and dimensional stability of anisotropic materials can be improved by cross bonding	Communications between manufacturer and user relatively poor and compounded by lack of mutual understanding
Generally lower capital costs incurred	

adhesion are short-range interactions across the adhesive-adherend interface, the whole basis of adhesion science being concerned with the relationship between solid and liquid surfaces.[1] To promote molecular adhesion, spreading out or wetting the surface with a liquid will result in intimate contact, provided that certain parameters are understood. Basic to the subject of wetting or wettability is the contact angle, θ, between a drop of liquid and a plain solid surface (Fig. 4.1) is independent of liquid volume. When $\theta > 0°$ the liquid is said to be non-spreading while if $\theta = 0°$, the liquid is said to completely wet the solid, spreading freely over the surface at a rate dependent on viscosity and surface roughness. However, every liquid wets or adheres to surface to some extent, i.e. $\theta \neq 180°$. In other words if we equate adhesion with wettability, there is always some adhesion of any liquid to any solid; $\cos \theta$ is a useful direct measure of wettability.

Thus with two flat surfaces separated by a thin layer of liquid having a low contact angle, strong adhesion will result. The effect arises from the liquid interfacial free energy and surface energy of the solid. For example, in Fig. 4.2:

γ_L = surface tension of drop of liquid L resting on surface S
γ_S = surface energy
γ_{SL} = interfacial free energy

so that for the drop of liquid in static equilibrium with

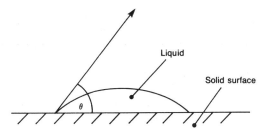

Fig. 4.1 Sessile drop of liquid on flat surface

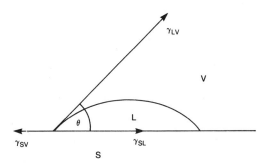

Fig. 4.2 Equilibrium of surface forces

its saturated vapour pressure

$$\gamma_{SV} = \gamma_{SL} + \gamma_{LV}\cos\theta$$

where V is the surrounding gaseous phase.

Although the above equation is deceptively simple and valid as an explanatory model, in practice it does present experimental difficulties. Every liquid with a low γ will always spread on a clean high γ surface at ordinary temperatures unless an adsorbed film, surface oxide or surface contaminant converts it to a low γ surface having a surface energy less than the surface tension of the liquid. Oily surfaces are a case in point here. Hard solids have surface energies increased substantially in some cases by the normal thin protective oxide layer on most metals. Plastics are low-energy surfaces and since adhesives are also polymers, it is generally more difficult to bond some plastics than metals.

Wettability may also be measured by the reversible work of adhesion W_A, where

$$W_A = \gamma_{SV} + \gamma_{LV} - \gamma_{SL}$$

expresses thermodynamically that the reversible work of separating a liquid and solid must be equal to a change in the free energy of the system, i.e. the sum of two energies minus the interfacial energy. However, one cannot neglect the adsorption of liquid vapour on the surface of a solid, such that

$$W_A = \gamma_{SV} + \gamma_{LV} + \pi - \gamma_{SL}$$

where $\pi = \gamma_S - \gamma_{SL}$ = the 'spreading pressure' of a monolayer of surface vapour, i.e. surface energy decrease due to adsorbed vapour. Therefore

$$W_A = \gamma_{LV}(1 + \cos\theta) + \pi$$

Now the energy required to produce rupture in the liquid (*NB* zero contact angle) W_C is equivalent to the work of cohesion in the liquid, i.e.

$$W_C = 2\gamma_{LV}$$

When perfect wetting occurs

$$W_A = 2\gamma_{LV} + \pi$$

hence

$$W_A > W_C$$

thus one would always expect cohesive failure in the liquid (or adhesive) and not at the interface. Where interface failure does occur, it is usually the result of a weak surface layer, grease film, surface contamination, etc.

Some values of surface energy (critical surface tension) are given for various surfaces in Table 4.2. A low γ liquid or adhesive will rapidly spread out and intimately contact a high γ surface, while a high γ liquid will bead up and not wet a low γ surface. Silicone low γ materials demonstrate both types of behaviour (a) when used as adhesives (sealants), and (b) when used as surface release agents for pressure-sensitive adhesives (PSA) or unintentionally present as a surface contaminant. If, for example in Table 4.2 surface energies are known, it should be possible to predict likely adhesive/adherend combinations.

When a liquid adhesive solidifies, the reversible work of adhesion, W_A, will still be close to the calculated values provided that the adhesive is able to deform during setting and release any stress concentrations. In bonded joints, poor wetting tends to produce greater stress concentrations at the free surface of the adhesive, sites where failure is likely to be initiated. If the contact angle is small, a prompt adhesive action is obtain by pressing the two adherends together until a thin liquid separation layer remains. Thereafter a strong and useful joint is obtained if the viscosity of the layer is increased as the adhesive sets. This analysis is simplistic, but it follows that there are three basic requirements for an adhesive,[2] i.e. that it should (1) wet the surfaces, (2) solidify, increasing cohesive strength and (3) be sufficiently deformable to dissipate elastic stresses during joint formation or set.

With plastics, a useful property for predicting compatibility between adhesive and adherend is the solubility

Table 4.2 *Surface energy and solubility parameter values*

Material	Surface energy γ (erg cm^{-2})[b]	Critical surface tension γ_c[a] (dyne cm^{-1})[b]	Solubility parameter δ (cal cm^{-1})
Aluminium	900	—	—
Al$_2$O$_3$	740	—	—
Copper	1120	—	—
Iron	1500	—	—
Tungsten	2300	—	—
FeO	580	—	—
Glass	310	—	—
PTFE	19.5	18.5	6.2
Silicone(polydimethyl)	—	24	7.2
Polythene	28.2	31	7.9
Natural rubber	—	36	8.1
Butyl rubber	—	27	7.7
Polyurethane	—	29	—
SBR	—	—	8.1
Polystyrene	43	33	9.1
Neoprene (chloroprene)	—	38	8.6
Nitrile rubber	—	—	8.7
PMMA	33	39	9.1
Polyvinyl acetate	—	37	9.4
PVC	29.1	39	9.5
Epoxy resin	—	—	10.3
Phenolic resin	—	—	11.5
Nylon 6.6	—	43	13.6
Water	—	72	23.4
Hexane	—	—	7.3
Carbon tetrachloride	—	—	8.6
Toluene	—	—	8.9
Benzene	—	—	9.2
Chloroform	—	—	9.3
Acetone	—	—	10.0
Isopropanol	—	—	11.5
Methanol	—	—	14.5
Ethylene glycol	—	—	14.0

[a] Critical surface tension of a surface is defined as the value of the surface tension of a liquid that will just spread on the surface.
[b] Value of surface energy γ generally given units of erg cm^{-2} = dyne cm^{-1}, i.e. a tensile force/unit length commonly referred to as the surface tension.

parameter δ (Table 4.2). This is a convenient measure of intermolecular attraction or ease of mixing. Polymers of closely approximating δ's are more likely to form homogeneous mixtures than those with widely differing parameters.[3] This fact can be used as an additional tool to surface energy in adhesive selection, since combination is most likely to take place when adhesive and adherend are most alike in solubility parameter. Additionally the joint is more environmentally stable, e.g. water resistant. However, actual surface energy measurements on an adherend will of course more accurately represent the surface condition for bonding. Solubility parameter may also be used in choosing a suitable solvent for making an inexpensive cement to directly bond polymers. For example, solvents and polymers with similar values of solubility parameter in Table 4.2 tend to be miscible, e.g. acetone and PVC. As a rule of thumb, if δ (polymer) = δ (solvent) $\pm$ 1.5, then the polymer will dissolve. The result is a solvent cement.

4.2.2 Surface preparation

Many suitable surfaces do not bond well unless properly prepared, the degree of adhesion and environmental durability depending on the degree of surface preparation. Most adhesives will not tolerate surface contamination although one type, based on acrylic or modified acrylic resins, is tolerant to low levels of oil contaminant,[4] displacing the oil from the surface without weakening its internal structure significantly. Some types of surface treatment will present bonding problems, e.g. coated, painted and plated surfaces that, compared to a steel substrate say, are weak and will easily break away under load when bonded. All engineering metals are covered with surface oxide, some such as zinc oxide on galvanized plate are weak and friable, whereas others, e.g. Al$_2$O$_3$ adhere well to the base metal and can be selectively anodized to improve the adhesive bond further. Some plastics are difficult to bond because of a low surface energy and some mouldings cause problems with plasticizers, mould release agents, etc.

Consistency and durability of bonds normally means surface preparation prior to bonding, followed by assembly as soon as possible after pre-treatment. Procedural details are normally available from adhesive manufacturers.[5] Treatment is required not only to clean the surfaces but also provide faying surfaces that are more readily wettable by the adhesive. Typical of the methods available are the following:

1. *Abrasion*. Improves adhesion through increasing the bonding area, also removing surface contaminant and any weak surface layers. Grit blasting is also an effective method of pre-treatment. Fresh surfaces are generally better than old ones.
2. *Solvent cleaning*. Suitably compatible solvents can be used to remove dirt, oil and in the case of plastics, any mould release agent. Certain solvents will dissolve plastics (Table 4.2). Detergent washing is an alternative. For metals, a combination of solvent degreasing and scrubbing abrasion is generally quite effective.

3. *Flame treating.* Often effective in increasing the surface energy of polythene, polypropylene, acetals and polytetrafluoroethylene (PTFE) by passing through the oxidizing portion of a gas flame. The incorporation of oxygen atoms in the surface structure then confers a more bondable surface. Use with caution.
4. *Primers.* Can be used to promote adhesion by stabilizing or strengthening a weak surface, or alternatively to protect a surface after its pre-treatment. Primers are now available for PTFE and other low-energy difficult surfaces.[6]
5. *Chemical modification.* Really as alternatives to abrasion, resulting in an alteration to the chemical and possibly mechanical structure of the surface. For example, acid etching can promote a stable oxidized surface with a rough contour.
6. *Corona discharge.* A process that works with all difficult polymers, increasing surface energy by incorporating surface oxygen, e.g. as (3). The process[7] involves spraying charge on to the surface with the advantage of no obvious surface change. High-gloss surfaces retain their lustre. The process is clean and straightforward, prior degreasing is not generally necessary.

4.2.3 Adhesive compatibility

Since adhesives and plastic adherends are both polymers, the possibility of chemical incompatibility between the two is feasible, resulting in stress cracking. This occurs in plastic mouldings containing residual internal stresses if the incorrect adhesive is applied. A similar phenomenon may occur with such mouldings exposed to aggressive environments or cleaning compounds. As a group, the amorphous thermoplastics, e.g. acrylonitrile–butadiene–styrene (ABS), acrylics, polycarbonate, polystyrene, polyphenylene oxide (PPO), etc. are susceptible to stress cracking and should be bonded with caution. Crystalline thermoplastics and thermosets are generally satisfactory, e.g. acetal, nylon, polyester, epoxy and phenolic. Stress cracking problems may be minimized by (a) bonding parts only in a low stress condition, annealing after moulding if necessary, and (b) using a minimum of adhesive to effect the cure, then removing the excess.

4.2.4 Glue line integrity

The frequently held belief that the thinner the glue line the better the joint is not necessarily so. There is an optimum thickness which will tend to be in the region of 0.10–0.15 mm for tensile applications, but the actual joint strength will depend on materials and type of adhesive. Adhesives operating below their glass transition temperature produce stronger joints with a thinner glue line, whereas rubbery adhesives, those operating above their glass transition, are much less sensitive to glue-line thickness. Where the adhesive is fairly thick, there is little effect on strength, the joint failure load remaining fairly constant. However, where the adhesive layer is very thin, approaching the surface roughness of the components or adhesive filler particle size, or even squeezed out leaving large areas of metallic contact, then there may not be sufficient adhesive material present to transmit loads between the components.

Voids can result from gas bubbles entrapped in the adhesive or air trapped as a result of the method of adhesive application and distribution. Occasionally, gases may also be evolved during the solidification process itself. However, as long as the adhesive coverage is sufficient to support the applied load then neither voids in the glue line nor starved edges will affect the joint strength.

Fully coating one adherend with adhesive then applying the second surface normally (Fig. 4.3) is not ideal since air is likely to be trapped in the joint. This will produce voids and also inhibit wetting. Air bleed-out can only be accomplished by adherend movement, a technique which

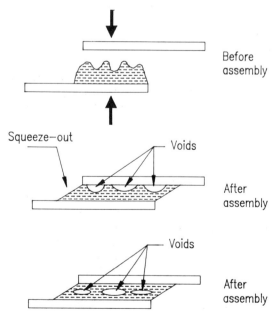

Fig. 4.3 Air entrapment in single plane joints

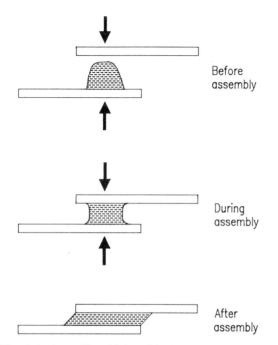

Fig. 4.4 Assembly of joint with single bead of adhesive

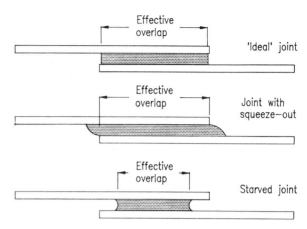

Fig. 4.6 Adhesive distribution laterally

should not be used if at all possible. Air entrapment can be avoided by ensuring mutual convexity, e.g. Fig. 4.4, a single adhesive bead spreading outward and wetting the bonding surfaces as the components are brought together. With more than one bead of adhesive, however, there may be voids in the joint.

The use of assembly pressure alone to fill a joint is unlikely to result in an even adhesive distribution. The most likely occurrence is some degree of squeeze out or a starved joint (Fig. 4.5). Squeeze-out may be considered as unsightly in the finished product and may need to be removed, but it has been found that excess adhesive squeeze-out can reduce stress concentrations at the joint edge,[8] hence remove only if essential. An alternative to squeeze-out is partial starvation (Fig. 4.6), requiring good control of the application process to ensure adequate joint area.

4.3 Joint design

When two materials are bonded, the resulting composite has at least five elements,[9] (Fig. 4.7) (1) adherend 1; (2) adhesive bond to adherend 1; (3) the adhesive; (4) adhesive bond to adherend 2; (5) adherend 2.

A joint is held together by the cohesive forces of both component and adhesive polymer plus the interfacial

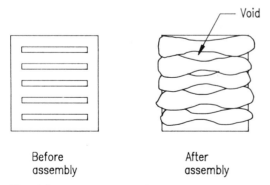

Fig. 4.5 Voids between adhesive beads

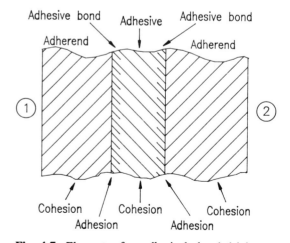

Fig. 4.7 Elements of an adhesively bonded joint

adhesive bond. Each element in the train is equally important to structural integrity. The strength of an adhesive joint will be the strength of its weakest member. With strong substrates, however, failure would be adhesive at the interface or cohesive within the glue, but failure will not occur at an interface if the adherend surface has been properly prepared and there is adequate wetting. In other words, adhesion between the adhesive and substrate should be greater than the cohesion within the glue line (or in some cases greater than the cohesive strength of the adherend itself).

For maximum effectiveness, adhesive bonds should be designed with the following general principles in mind:

1. Stress the adhesive in the direction of maximum strength;
2. Provide the maximum bond surface area;
3. Make the adhesive layer as uniform as possible;
4. Maintain a thin and continuous adhesive layer;
5. Avoid stress concentrations.

Adhesive joints are strongest in pure shear or when carrying both a shear and a compression load, but weakest in peel and cleavage. The designer should ensure that service stresses are orientated along the direction of greatest adhesive strength, i.e. maximize the adhesive in shear. Types of adhesive loading are shown schematically in Fig. 4.8. The vast majority of adhesive joints found in product manufacture are subject to tensile loads, e.g. Fig. 4.9.

1. *Butt joint.* Limited in its load-carrying capacity since the cohesive strength of the adhesive is likely to be much lower than the cohesive strength of the adherend, thus the joint is unlikely to support much load in tension.

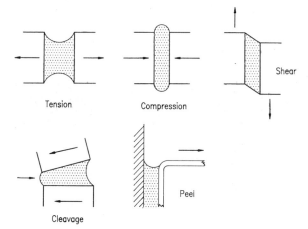

Fig. 4.8 Types of joint loading

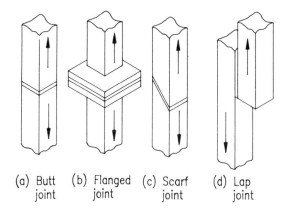

(a) Butt joint (b) Flanged joint (c) Scarf joint (d) Lap joint

Fig. 4.9 Simple tensile joints

2. *Flanged joint.* A means of reducing adhesive stress by distributing the load but is limited, and with metal components, insufficient.
3. *Scarf joint.* A scarf joint, or bevelled lap, is an effective way of eliminating stress concentrations at the ends of the joint and permitting a uniform stress over the whole of the bonded area. However, this does present a significant production problem.
4. *Lap joint.* This is the most effective means of increasing the area of the joint and fortunately the easiest to manufacture. The adhesive is stressed predominantly in shear, and with a large overlapping joint area possible, is the most durable of bond types.

The above joint types are sometimes referred to as structural and the adhesive as a structural adhesive, usually where there is some form of overlap. Other types of joint such as coaxial bonds are often loosely referred to as mechanical. Wherever possible, angular joints and butt joints can be redesigned to minimize tensile, peel and cleavage forces and become more effective against the stress direction. Figure 4.10 outlines some typical modifications to improve bond strength by arranging a more adequate bonding area. The principles involved in Fig. 4.10 should be self-explanatory.

Bonded structures are inherently stiffer than their spot-welded counterparts, for example in the case of a box section beam, the flexural stiffness and ultimate strength can be doubled by going from a welded to a bonded structure.[10]

Because the lap shear joint is the most suitable joint type applicable to adhesive bonding and hence the most frequently used, we shall consider the mechanical aspects in more detail. Any joint types specific to a class of adhesives are dealt with in section 4.5.

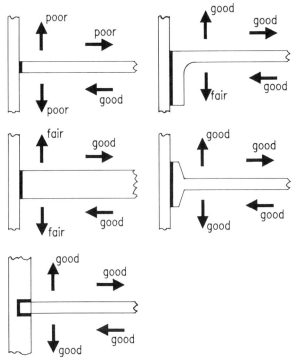

Fig. 4.10 Joint stress evaluation

4.3.1 The lap shear joint

(i) Adherend stresses

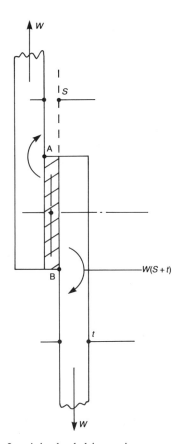

Fig. 4.11 Lap joint loaded in tension

Consideration of the joint behaviour in tension enables the performance of loaded joints in general to be demonstrated. Figure 4.11 illustrates a symmetrical joint with an applied tensile load. However, because of the thickness of the components, it is not possible to align the load direction precisely along the adhesive edges. Consequently there is always a resulting couple which produces an uneven stress concentration in the adhesive where the load is offset. By taking moments about the joint centre, it can be shown that

Clockwise moment = anticlockwise moment
$$= W(S+t) = M$$

where W = applied load/unit width of joint, S = adhesive thickness, t = material thickness, M = bending moment, conventional beam theory.

An illustration of the typical stress concentrations produced at points A and B are shown in Fig. 4.12, demonstrating that the edges of the lap carry a high proportion of the load. The load-carrying capacity of this type of joint is limited, more suitable for use with thin

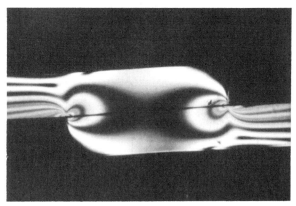

Fig. 4.12 Photoelastic model of stress concentrations in a loaded joint. Courtesy of Ciba-Geigy Polymers

sheet materials where the couple is then small. With thicker components and a given adhesive, there is a limit to the achievable joint strength.

The distribution of maximum tensile stress in the

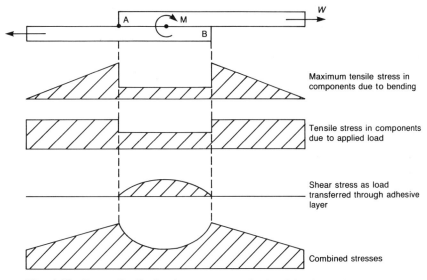

Fig. 4.13 Stress distribution in components for lap joint in tension

adherends due to bending is shown in Fig. 4.13. This is additional to the applied tensile load. Shear stresses are also present as the load is transferred in shear through the adhesive layer, the overall stress distribution in the joint being composed of three types of stress.

(ii) Adhesive stresses

The adhesive will be subjected to cleavage as well as shear, a twofold stress situation. The existence of tensile stresses in the adherend will give rise to shear in the adhesive. For maximum efficiency, the joint should be capable of carrying the same load as the components, hence

$$W = P_A Db = P_C tb \quad \text{and} \quad D = \frac{P_C t}{P_A}$$

where P_A = tensile strength of adhesive, P_C = yield stress of adherend, D = joint overlap, t = adherend thickness and b = width of joint. As a basic requirement in joint design, this gives a simple relationship between overlap and component thickness where stresses can be ideally considered as uniform. However, with lap joints in tension, there are stress concentrations at points A and B (Fig. 4.11) due to component bending, so affecting the stress in the adhesive.

Considering a lap joint with both plates of same metal and thickness, the adhesive subject to a shear strain γ (Fig. 4.14), then

$$\tau = G\gamma = \frac{CW \cosh Cx}{2b \sinh (CD/2)}$$

where $C = \sqrt{(2G/stE_C)}$, G = shear modulus of adhesive, τ = shear stress of adhesive and x = distance ± from midpoint of lap, from which it is possible to plot a shear stress distribution in the adhesive (Fig. 4.14). The maximum adhesive shear stress occurs at $x = \pm D/2$, i.e.

$$\tau \text{ (max)} = \frac{CW}{2b \tanh (CD/2)} = \frac{E_a}{E_c} (f_t + f_b)$$

$$= \frac{KW}{bD}$$

$$= K \times \text{average shear stress}$$

where

K = stress distribution factor

$$= \frac{CL}{2 \tanh (CL/2)}$$

$$= \frac{AC}{BC} \text{ (Fig. 4.14)}.$$

and E_a = modulus of elasticity of adhesive, E_c = modulus of elasticity of component, and f_t = tensile stress in component

f_b = bending stress

$$= \frac{6M}{bt^2} \text{ outside the joint}$$

$$= \frac{3M}{2bt^2} \text{ inside the overlap}$$

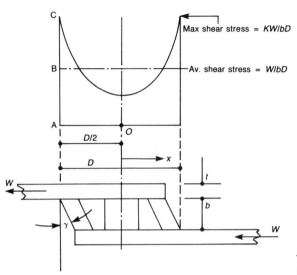

Fig. 4.14 Stress distribution in adhesive

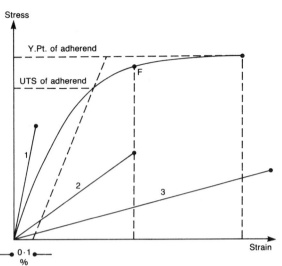

Key to adhesives: (1) Strong, brittle type of adhesive;
(2) Typical high strength epoxy;
(3) Low strength, high elongation adhesive.

Fig. 4.15 Behaviour of different adhesives in sheet metal lap joint

substantially reducing at the point of overlap. By reducing K, this reduces the stress concentration of the joint, i.e. by reducing the shear modulus of the adhesive, increasing glue-line thickness, etc. A stiff adhesive produces a higher degree of stress at the edge of a joint (stress concentration) than does a more ductile adhesive, the stress distribution of the latter tending to be more uniform.[8]

(iii) Joint strength

It is found in practice that bonded components start to yield at loads considerably less than loads required to yield the component material on its own. This is a result of the stress concentrations e.g. Fig. 4.12. Likewise, the load at which the joint fails is somewhat below the true strength of the adhesive.

Consider the stress–strain behaviour of three types of adhesive (Fig. 4.15) when compared to mild steel:

Adhesive 1. The strongest adhesive, but can only withstand limited strain, hence the bond will fail readily when the shear stress in the joint exceeds the ultimate shear strength of the adhesive material.

Adhesive 2. Typical of a high-strength epoxy adhesive. Adhesive failure cannot occur until the mild steel has yielded to point F, i.e. the adhesive stretches to absorb the deformation in the mild steel. However, yielding of the adherend at the outermost fibres will probably cause the adhesive to fail locally, reducing the cross-sectional area of the joint. If at the same time the applied load is increasing, eventual cohesive failure will occur when the remaining area is no longer able to support it.

Adhesive 3. The adhesive will not fail although it has the lowest tensile strength of all the adhesives. It is therefore possible to increase the load-carrying capacity of a single joint simply by raising the elasticity of the adhesive. The degree to which an adhesive will absorb the strain of a yielding component depends on the elongation of the component at failure.

From the above it can be seen that with a flexible adhesive, the mechanical properties of joint components can have a greater effect on the ultimate breaking load of a lap joint than the mechanical properties of the adhesive. For example, with identical tests on mild steel and aluminium, failure loads for aluminium is much less,[11] joint failure is thus not just failure of the adhesive.

Although joint failure has been inferred as the point at which cohesive failure of the adhesive occurs, there are other possible failure definitions, i.e.

1. Stress to cause adherend failure, e.g. proof stress or yield point;
2. The percentage elongation of adhesive at fracture;
3. The tensile strength of the adherend, although the stronger the adherend the more likely that failure will be in the adhesive.

The point at which the stress concentration sites A and B (Fig. 4.11) start to yield is often referred to as primary

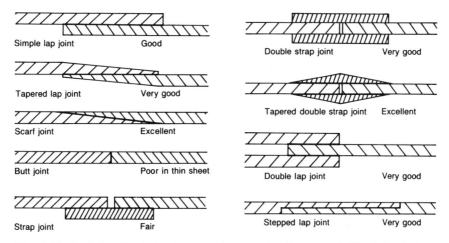

Fig. 4.16 Basic bonding joints between sheet metals. Courtesy of Ciba-Geigy Polymers

failure and as secondary failure when the joint breaks. In a large number of reported testwork in suppliers' catalogues, etc. secondary failure is reported, whereas primary failure would be more relevant for a design application.

(iv) Types of lap joint

A bonded joint element essentially should be load bearing and permanent. The simple lap joint may not always perform well in load-bearing structures, since even modest distortion can cause failure at apparent low loading levels. A number of lap joint variants can overcome these shortfalls, e.g. Fig. 4.16. However, some of these are often expensive or impractical from the manufacturing point of view. Aluminium extrusions can often employ optimum joint designs.[8] When joining plastics by means of a lap joint, supplementary fasteners may sometimes be necessary to reduce peel and cleavage forces.

In a simple rigid lap joint, the strain on the adhesive can be reduced if, for example, the thickness of component 1 (Fig. 4.17) is reduced to line BC, the removed chamfer being largely redundant material. By reducing material thickness at point B, B will extend with point A as a result of deformation in component 2, thereby reducing the stress concentration. The greater the degree of taper, the more the maximum adhesive shear stress approaches the average shear stress value for a given joint configuration.

(v) Physical parameters

Given a particular adhesive, there are a number of joint design parameters that can be expected to influence joint

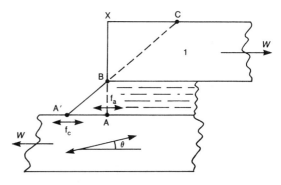

NB In practice, due to bending f_c will act at a small angle θ.

Fig. 4.17 Adhesive layer deformation — lap joint in tension

performance, e.g. overlap distance, joint width, material thickness and material yield point. The strength of a lap assembly in shear is directly proportional to the width of the bond. Strength is also improved by increasing the length of the overlap — up to a point; the unit strength does not increase at the same rate with successive increases in overlap, e.g. Fig. 4.18.

From section 4.3.1(ii),

$$\text{Joint strength} = W = \frac{2S \tanh CD/2}{C}$$

for unit width of joint based on working adhesive shear strength S. Now varying the overlap D, for small values of D

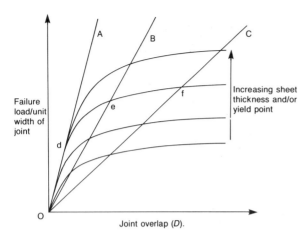

OA = theoretical relationship based on average adhesive shear strength; OB = metal yield point line, strictly tensile stress only; OC = points where increase in overlap gives no increase in breaking load.

Fig. 4.18 Single tensile lap tests for given material

$$\tanh \frac{CD}{2} \to \frac{CD}{2} \quad \text{hence} \quad W \propto D$$

for large values of D,

$$\tanh \frac{CD}{2} \to 1 \quad \text{hence} \quad W \to \text{constant}$$

This is consistent with the curve in Fig. 4.18. For values of $C \approx 1.5$, the joint will yield approximately 90 per cent of its theoretical strength, any further increase in overlap is of little value unless yielding redistributes the stress, i.e. above point f in Fig. 4.18. For a particular material curve, joint failure at or above point e will show some permanent set, evidence of yield or material necking. However, between points d and e is probably the area of most practical interest to designers. Below point d, the adhesive fails effectively in pure shear, i.e. the joint overlap area is very small. Figure 4.18 also illustrates that the strength of the component has a direct influence on the strength of the joint, although clearly the load asymptote approaches a value which is typical of the adhesive.

4.4 The setting process

To develop cohesive strength, an adhesive must set. The exception is a pressure-sensitive adhesive. Setting starts as soon as the application is complete and occurs as rapidly as possible consistent with ensuring as low as possible a level of defects, i.e. voidage, residual stresses, etc. Internal stresses and stress concentrations in the adhesive that result from solidification or cooling can result from (a) differences in thermal expansion of adherend/adhesive, (b) density increase or volume reduction.

In many applications the 'matching' process is neglected as not critical, but in general the strength of an adhesive joint is considerably decreased by development of internal stress concentrations. Poor wetting of the adherend tends to produce greater stress concentrations at the free surfaces of the adhesive, where failure is most likely to be initiated. As contact angles increase, i.e. the free adhesive surface concavity increases, the stress concentration increases (Fig. 4.19). Ideally, as an adhesive sets it should do so to a stress-free condition. This is possible with rubbery adhesives, but if setting to a rigid solid, too fast a setting will result in unequal shrinkage stresses. Thus either rely on the adhesive shrinking as little as possible or set the joint as slowly as possible. Reactive chemical cure adhesives shrink least while solvent evaporation types shrink most.

An adhesive must be applied to a substrate in fluid form to wet out the surface and leave no voids, hence it must be of sufficiently low viscosity. Structural adhesives (high-modulus types) set by a chemical reaction and include anaerobics, cyanoacrylates, toughened acrylics, epoxies, etc. Non-structural adhesives (i.e. low-modulus types) set by a physical change and include hot melts and contact adhesives. The transition of fluid to solid, i.e. viscosity increase, may be accomplished in several ways as detailed below.

4.4.1 Cooling of a thermoplastic

By convention, a 'hot-melt' material, often denotes a lower molecular weight, less viscous material than thermoplastics. Heating must result in a sufficiently high fluidity to achieve melting and the bond made fairly quickly after application as a result of heat transfer into

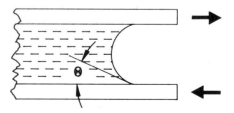

Fig. 4.19 Adhesive stress concentrations in a lap joint

the adherend. This can be a problem in bonding cold metals. Such adhesives generally have good wettability and can be used at low temperatures. Hot melts are compounded with waxes, tackifiers and plasticizers.

4.4.2 Release of solvent or carrier

Solvent base adhesives contain the adhesive as a suspension, solution or admixture with an organic solvent. These liquids lower the viscosity sufficiently to permit wetting of the substrate, but once this is accomplished the carrier has to be removed, either by absorption or evaporation before mating of the surfaces. Consequently solvent-based adhesives need a long open time, so slowing down production. The resulting strength is generally limited and so not often used for structural applications, but such adhesives generally have good wettability and can be used at low temperatures. Also there are the health and safety aspects of using solvent-based materials to consider.

Contact adhesives are based on synthetic elastomers such as neoprene, nitrile and urethane dissolved in organic solvents with additions of various tackifying resins. The principle of the contact adhesive is shown in Fig. 4.20. After the evaporation of solvent, the adhesive polymer is left in the amorphous state. Pressing together will diffuse rubber molecules through the surface zone, rapid diffusion causing the contact interface to disappear. After some time the polymer will haze as it starts to crystallize, increasing strength and rigidity.

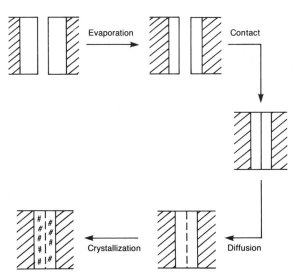

Fig. 4.20 Principle of the contact adhesive

In a solution, i.e. solvent cements, the concentration of polymer is < 30 per cent. Too high a 'solids' content will give high viscosities and the resultant dope showing poor wettability. Although Table 4.2 provides a means of producing solvent cements for many types of plastic, commercial material will be based on a mixture of solvents.

4.4.3 Polymerization in situ

This covers most of the adhesive types used in industry, e.g. epoxy, polyurethane, cyanoacrylate, acrylic, etc. The advantage is that strength can be developed in the glue line after the adherends have been brought together. The polymerization 'reaction' or setting process is usually exothermic. Where water may be a by-product of polymerization, e.g. with phenolics, the joint will need to be clamped under pressure while setting. Various methods are used to initiate the chemical reaction, e.g. mixing of catalyst, heat, absorption of atmospheric moisture, surface priming, etc.

4.4.4 Pressure-sensitive adhesives (PSAs)

These adhesives, unlike other classes do not undergo a progressive increase in viscosity. Instead they are permanently in the intermediate tacky stage. They are, however, becoming an important means of joining items in the manufacturing process. One merit is that as they wet out so inadequately, they can be removed from the adherend without marking or leaving a residue. While the deficiency in adhesive strength is deliberate, low cohesive strength prevents heavy-duty applications. It is inevitable that a permanently tacky material will be easily ruptured in load-bearing situations.

4.5 Adhesive types

Because adhesive application and usage is not an exact science, a majority of this chapter has concentrated on the mechanics of adhesion and the necessary requirements to achieve a good bond. This should give the design engineer a better appreciation of the application and type of adhesives to use. It is not possible to consider all the types of adhesive currently available, rather those adhesive families that are predominantly used in general manufacturing industry will be examined further. A general summary of the advantages and disadvantages of these adhesive types is presented in Table 4.3 together with some mechanical properties of the cured adhesive.

Table 4.3 *Comparison of the advantages and disadvantages of using different adhesives*

Adhesive type	Anaerobic	Cyanoacrylate	Reactive acrylic
Advantages	Elimination of many types of mechanical keys, lockwashers and pins Simplifies mechanical assembly Fit easily into high-volume production methods Cure with minimum shrinkage Ability to gap fill, can avoid scrap if press fits not dimensionally correct	Rapid set time Versatile; good shear and tensile bond strengths No heat curing required; bond at room temperature Can bond dissimilar materials Wide range of adherends Can be easily automated Widely accepted by industry Used as supplied No jigs or fixtures required Quite good chemical resistance	No mixing of components prior to application is necessary Rapid bonding at room temperature Tolerant to some surface contamination Fairly good impact, peel and shear strengths Minimal shrinkage on curing Gap filling Good environmental and chemical resistance Variable open time possible Activator/adhesive ratio does not need to be precise Bonds wide range of substrates Useful in a wide temperature range Activators can be pre-applied to items and stored in advance of bonding
Disadvantages	Cost Inability to deal with large gaps	Skin contact should be avoided Ingestion of vapours may produce irritation Expensive; but a little adhesive goes a long way, i.e. cost per application is competitive Not generally good gap fillers Adherends should mate well Surfaces need to be clean, dry and oil free Limited moisture resistance Limited peel and impact strength Risk of blooming	Limited pot life when pre-mixed Some activators are odorous Some products are flammable Some products contain skin irritants

Properties	Threadlock	UV curing	Structural	1500 cP	40 cP	
Tensile strength (N mm^{-2})		6–12	21	30	20	16
Z elongation at fracture			16			11
Y modulus (N mm^{-2})			435			615
Shear stress (N mm^{-2})	10–30	27				24
Shear modulus (N mm^{-2})						350

Hot melt	Epoxy	Organic solvent	Pressure sensitive
Fast bonding operation; applicable to production work	Tough adhesives; includes some of the strongest adhesives available	Good wetting properties (low θ) so good adhesion to low-energy surfaces, plastics etc.	Clean and quick
Solvent free			No special equipment required
Recovery of substandard assemblies possible	Good resistance to water, heat and chemicals	Rapid solvent removal	Items can be pre-prepared by suppliers ready for use
Rapid achievement of final bond strength	Open time is significantly longer than with other adhesives	High m.pt. polymers may be used; hence heat-resistant bonds possible	Cheap
Low-cost material	No flammable components used in these adhesives	Rigid or flexible adhesives possible	Simple concept to use
No mixing required			Consistent bond line thickness and consistent adhesive strength
Clamps and fixtures often unnecessary	No loss of volatiles during cure	Versatile; few limitations on adherends	
Application gives consistent product	Maintenance of dimensional stability during cure; negligible shrinkage		Can be tailored to production line requirements
	Low creep		No special tooling
Poor generic heat resistance in terms of bond strength	Low peel strength in the basic system	Provision for solvent escape necessary hence permeable adhesives only	Adhesive bond strength/unit area is fairly low
Creep and cold flow at room temperature if stressed; i.e. memory set possible	Not that tolerant to surface contamination	Joining delay as with contact types	Creep under sustained loading
Limited adhesion to smooth surfaces, inc. metals, glass plastics etc.	Heating required for single component types	Health and safety aspects of solvent loss	Often lack of adhesion and tack at low temperatures
Application equipment can be expensive and complex	Expensive	Control over solvent loss to avoid stress accumulation in rigid joints	
Control of coating thickness is limited	Curing can be slow		
Some adherends require a pre-heat	Two-part types need mixing and pot life can be low	High mol. wt polymers for strength means high viscosity or low concentration	
Limited assembly time	Some curing agents may be toxic		
Bonded area is limited by the open time of adhesive			
Application temperature is critical; too low gives a weak bond, too high could thermally degrade			

Low mol. wt polyamide	Med mol. wt polyamide	EVA	2-pt cold cure	1-pt cold cure
11.7	3.1	0.9–1.5	18–43	51–68
15	700	500	1–3	1–5
290	25	14–65	3200–4200	1250–7700
				43
				2000

Of major industrial importance is the acrylic polymer adhesive family comprising the anaerobics, cyanoacrylates, ultraviolet (UV) curing, toughened reactive, pressure-sensitive and aerobic adhesives. The suitability of a given type depends on the initiation and mechanism of polymerization. Other adhesive families are the hot melts and the epoxies, both of which represent traditional and well-established material groups. Solvent-based adhesives are not considered in detail, since by their very nature are not preferable in a manufacturing environment.

4.5.1 The 'toughened' adhesive

The so-called toughened adhesive was initially developed some 20 years ago which it was found that brittle epoxy resins could be toughened by incorporating small amounts of reactive liquid polymer. Subsequent analysis of fractured surfaces revealed a two-phase system with small rubber particles ($\sim 1 \mu$m diameter) embedded in the glassy epoxy matrix. The general concept has also been applied to acrylic and anaerobic adhesives, with an improvement in peel strength, impact resistance and tensile lap test results.[4] The generic class of adhesive has also been called 'reactive fluid', 'second generation' or 'toughened' by adhesive manufacturers.

A great deal of effort has been put into this class of adhesives in order to achieve better joint performance via a second phase dispersion, resulting in lower adhesive tensile strength but greater elongation. The situation is illustrated typically by Fig. 4.15, a brittle adhesive[1] modified to give a tensile performance as adhesive,[2] but it is misleading to imply an increase in toughness. The second phase rubber dispersion is said to act as a barrier inhibiting crack propagation, i.e. crack-stopping particles.[4] However, reference to Fig. 4.13 and section 4.3.1 can explain performance purely on the basis of elongation without recourse to crack growth inhibition. In any case, cohesive adhesive failure is likely to be sudden and catastrophic, localized particles not having much effect. On the other hand, fatigue resistance may well be improved.

4.5.2 Anaerobic adhesives

Anaerobics are one-component, solventless acrylic-based liquids and pastes (dimethacrylates) that cure into tough polymeric materials when deprived of oxygen, e.g. in the threads of a fastened nut or bolt or more generally between two major surfaces that exclude air. Materials in this group are known as adhesives despite the fact that they are poor adhesives. Otherwise known as locking or retaining fluids,[12] their main function is to jam or mechanically lock parts together. The toughened variants may be used as structural adhesives. Anaerobics are effective in use because of high wettability and high compressive strength. Although brittle, the stresses found in threadlock and retaining situations are, in the main, compressive. The advantages and disadvantages of anaerobics are given in Table 4.3.

For historical reasons, they are packaged in half-full, thin-walled polyethylene containers. Such packaging allows for oxygen diffusion plus an air reservoir above the liquid to prevent polymerization and cross-linking. However, modern formulations make these less critical requirements,[8] i.e. stabilizers are added to prevent gelation. Although oxygen exclusion results in polymerization, it is nevertheless at a slow rate. There is another key reaction catalysed by a transition metal surface, e.g. iron, copper, etc. that speeds up the polymerization process, but the exact nature of the surface mechanism has yet to be defined. There are three categories of cure speed, depending on the materials to be joined:

1. *Active surface (fast curing).* For example, with iron, copper, brass, phosphated or black oxide surfaces, 50 per cent of ultimate strength is obtained in 10 min–2 hours at room temperature, depending on viscosity.

2. *Inactive surface (slow cure).* On inactive metal surfaces e.g. anodized aluminium, zinc plate, stainless steel, etc. the concentration of transition metal ions is very low, hence cures can be very long, remaining uncured for long periods. However, modern modifier materials, i.e. primers or accelerators, can increase cure reliability although different surfaces will still have different rates of set. Figure 4.21 illustrates the relative cure time differences for active/inactive surfaces and a low-viscosity material.

3. *Passive (non-curing).* Such surfaces are typically plastics. For use with anaerobics, therefore, liquid accelerator sprays have to be used on the contacting surfaces. Anaerobic compositions will, however, attack solvent-sensitive plastics.

Anaerobic compounds were among the first materials used to supplement the strength and durability of mechanical fasteners, i.e. to improve the break-loose torque (section 14.2.6) and vibration resistance (section 14.3.3). The products cover three main areas of interest in mechanical design:

1. *Threadlocking.* It has been shown that single-component resins available to provide vibration

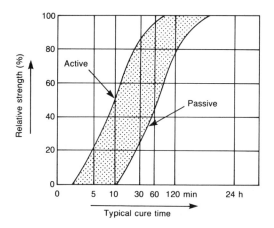

Fig. 4.21 Effect of surface on anaerobic cure rate. Courtesy of Loctite UK

resistance can out-perform mechanical locking devices.[13] They can be easily dispensed, cure rapidly and strength can be controlled to allow easy disassembly. Microencapsulated coating is an alternative to the liquid resin (section 14.2.6).

2. *Retaining*. Anaerobic adhesives are now widely used to overcome production difficulties associated with press-fit and interference-fit assemblies, e.g. the location of bearings, pulleys, gears, etc. on shafts, spreading the load-bearing area and eliminating the need for expensive tight tolerances. As with other anaerobic resins, retaining compounds are simply applied to the parts concerned and self-cure after the items are assembled.

3. *Bonding*. A combination of urethane and anaerobic chemistry has developed a series of high-strength, durable and convenient anaerobic structural adhesives, employing the primer/adhesive concept, i.e. adhesive applied to one surface and a surface activator to the other. This has the effect of simplifying the production bonding operation, plus providing a fast cure.

Some aspects that affect the setting time and ultimate performance of the joint are as follows:

1. *Surface condition*. In common with acrylics generally, anaerobics can cope with contaminated surfaces, the monomer being able to dissolve surface oil and contaminants and yet still harden. A lower strength and a longer cure time should be anticipated the greater the surface contamination.

2. *Clearance*. Anaerobics can generally cope with diametral clearances up to 0.025 mm, the thixotropic grades extending this to 0.4 mm. For optimum joint strength, the diametral clearance should not exceed 0.10 mm (Fig. 4.22). Interference fits only marginally improve joint strength. Robust performance follows the ability of anaerobics to make intimate surface contact with metal, avoid void formation and so evenly distribute the load over the hardened joint area. With a normal interference fit, only ~20 per cent of the available joint area is in contact. With an included anaerobic, it has been suggested that load spreading is much enhanced.[8]

3. *Length of engagement*. Ideally it should be twice the diameter of the cylindrical part concerned.

4. *Surface roughness*. The recommended surface roughness range is $2-10\,\mu m$ cla, e.g. Fig. 4.23. This should be considered in conjunction with clearance (Fig. 4.22) in order to achieve ideal bond performance.

5. *Strength*. Anaerobic adhesives are classified into broad strength and viscosity grades depending on the application, typically as in Fig. 4.24. Within any strength band, the actual bond strength achieved will

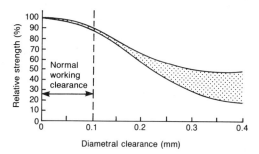

Fig. 4.22 Typical relationship between strength of anaerobic adhesive bond and diametral clearance. Courtesy PERA International

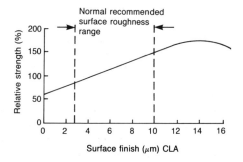

Fig. 4.23 Effect of surface roughness on anaerobic bond. Courtesy PERA International

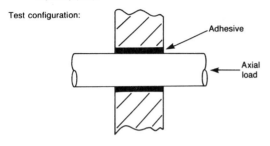

	Viscosity (mPa)		
	125	500	Thixotropic paste
Low strength	6	—	4.5
Med. strength	10	17.5	10
High strength	20[a] (0.1)	22[a] (0.15)	15 (0.4)

Notes: Shear values (Nmm^{-2}); mild steel surfaces; 0.05 mm diametral clearance fit; 140 mm^2 bond area; 20 °C reference temperature in air.
[a] Otherwise referred to as retaining adhesives. ( ) = gap fill capability (mm).

Test configuration:

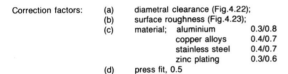

Correction factors:	(a)	diametral clearance (Fig.4.22);	
	(b)	surface roughness (Fig.4.23);	
	(c)	material; aluminium	0.3/0.8
		copper alloys	0.4/0.7
		stainless steel	0.4/0.7
		zinc plating	0.3/0.6
	(d)	press fit, 0.5	

Fig. 4.24 Static shear strength values (in tensile) for different types of anaerobic thread-locking compounds (coaxial joint)

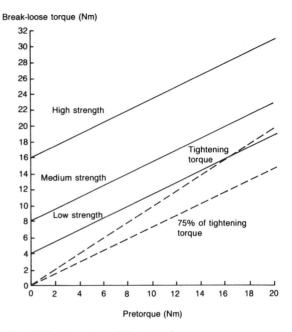

Fig. 4.25 Retention of threaded fasteners: torque resistance augmentation demonstrated by anaerobic adhesives of differing strengths on M8 bolt. Courtesy of Permabond Europe

depend on a number of factors, e.g. bond line thickness, surface roughness, joint area, surface condition, etc. The strength grade chosen should reflect the application, e.g. if future disassembly is necessary or whether a permanent union is required. By selecting the correct strength grade to augment a fastener application, it is possible to ensure that the fastener is not overloaded during disassembly (Fig. 4.25).[12]

6. *Temperature limitation.* Service temperature has an effect on the strength of a bonded joint and anaerobics are no different in this regard. At 120 °C the adhesive has 50 per cent of its room temperature strength. Operating continuously at an elevated temperature, the adhesive tends to heat age slowly, possibly reducing the service life expectation.

4.5.3 Anaerobic structural adhesive

Unlike the anaerobic threadlocking and sealing compounds, the structural anaerobics are true adhesives. They are based on resin systems that cure to flexible tough polymers, providing strength and ageing properties similar to other structural adhesives. These products have good toughness (when compared to early formulations), elongation and gap fill properties. Typical bond performance on a variety of substrates is shown in Table 4.4, but again with plastics (passive surfaces) bond strengths are not high. Benefits of the anaerobic cure system in structural adhesives including its greater applicability are as follows:

1. Adhesives can be cured without heat or mixing. Surface activators provide rapid cures and high bond strengths.
2. Cure rates are controlled by appropriate surface activators. Fixture times range from 15 sec to 30 min.
3. The single-component adhesives are stable during storage and use. No mixing required. Simplicity in use.

Table 4.4 *Summary of anaerobic structural adhesives properties*

Substrate*	Bond strength (Nmm^{-2}) (tensile shear test)
Sandblasted steel	24.2
Degreased steel	12.4
Oily steel	12.1
Zinc plate	9.7
PVC	7.3†
Nylon	3.5
Phenolic	3.2
ABS	0.7
Acrylic	0.4
Temperature range	−60 to 230 °C
Max. gap cure	1.5 mm
Max. joint strength	at < 0.5 mm gap
Elongation	> 50%
Oil insensitivity	fair
Thermoplastic bonding	fair

* Plastic substrates abraded.
† Substrate failure.

4. No pot-life limitations, adhesives do not gel in applicators.
5. Adhesive remains uncured outside of the bond lines, thus easy to clean up.
6. Low toxicity levels, no volatile solvents and classified only as mild skin sensitizers.
7. Some types may be cured with UV light, addressing many of the problems that exist in the bonding of glass. Clarity and refractive index are closely matched.

4.5.4 Cyanoacrylate adhesives

Cyanoacrylate adhesives were first introduced in the late 1950s and were first hailed as 'wonder adhesives' or 'superglues' because of their ability to bond a wide variety of materials (Table 4.5) with rapid setting times. Over the last 30 years they have come a long way as adhesives and have been used to solve numerous assembly problems on a wide variety of products. Their ability to bond dissimilar materials is a major advantage to the point where, in some applications, product design is dependent on the use of a cyanoacrylate adhesive. Cyanoacrylates are also well known as biological glues.[14] The advantages and disadvantages in using cyanoacrylate adhesives are given in Table 4.3.

As supplied, they are liquid acrylic monomers, the majority having a water-like appearance. When spread into a thin film between two surfaces to be bonded, the monomer will polymerize to a high-molecular-weight polymer producing a strong bond in a few seconds. Cyanoacrylates will bond most types of elastomer immediately, e.g. nitrile, ethylene-propylene rubber (EPDM), styrene-butadiene rubber (SBR), Neoprene, polyurethanes, natural rubber, etc. in most cases the bond strength surpassing the cohesive strength of the rubber. Averaged tensile lap shear strength values for plastics are given in Table 4.6.[15] Limited bond strengths are obtained with the low γ plastics such as polyethylene and polypropylene so some surface modification may be necessary (section 4.2.3).

Specific aspects of cyanoacrylates are discussed under the headings below.

(i) Chemical type
Virtually all are either methyl or ethyl 2-cyanoacrylate. For performance purposes there is little difference between the two types, although certain plastics, e.g. PVC, polycarbonate, etc. produce better bonds with the ethyl type. The adhesives will contain a thickener to vary viscosity, a plasticizer to improve impact strength and a stabilizer to control shelf life. A problem often noticed with cyanoacrylates is 'blooming' — this is a white stain on the bond line caused by the rapid condensation of cyanoacrylate vapours. This can be overcome by using a butyl ester grade, a larger monomer molecule, hence surface concentration is less, thus the adhesive is slower curing and relatively weaker but has a longer open time. The adhesives can be coloured by the addition of suitable dyes.

(ii) Viscosity
Basic cyanoacrylate monomers have a viscosity of 3cP, hence are very fluid, penetrate easily and are the fastest setting, but clearly the adhesive can easily seep out and so they are not ideal for production use. Increasing the viscosity will reduce the tendency for dripping and running and allow for easier adhesive control and distribution on the faying surfaces. Medium-viscosity products, 70–150 cP, i.e. slightly slower setting to allow more open time, probably offer a good general balance of properties for most uses. With even higher viscosity grades, (500–50 000 cP) a limitation of cyanoacrylates is overcome in that gap filling is possible, up to 0.5 mm or possibly more. Likewise the higher viscosity grades may be used on porous surfaces where the lower viscosity grades tend to be absorbed. Viscous adhesive formulations can also be prepared *in situ* by initiating polymerization with UV, light, thermal activation, etc.

Table 4.5 *Adhesive selection chart*

Adherend	Natural rubber	Neoprene	Nitrile	Polyurethane	Hot melt EVA	Hot melt polyamide	Anaerobic acrylic	Cyanoacrylate	Tough acrylic	Epoxy	Solvent adhesive (D)	Notes
Elastomers												
Natural	•	•	•					•	•			
Neoprene		•						•	•			
Nitrile			•					•	•			
Polyurethane			•	•				•	•	•		
Butyl		•						•	•			A
Polymers												
PVC (plasticized)			•	•				•		•		B
PVC (unplasticized)		•	•	•	•			•	•	•		
Polyester			•	•				•	•	•		B
Nylon			•	•				•	•	•		
Acrylic		•	•	•	•			•	•	•	•	B
Polystyrene		•						•	•	•		
Cellulosics	•	•	•	•	•			•	•		•	
Thermosets		•	•	•	•		•	•	•		•	A
Polyethylene					•							A
PTFE			•									A
ABS		•	•	•				•	•	•	•	B
Polycarbonate		•	•	•				•	•	•	•	B
Acetal			•					•	•			
Expanded rubber and plastics												
Natural		•						•	•			
Latex foam	•	•										
Neoprene		•										C
Polyurethane (rigid)	•	•	•	•	•			•	•	•		C
Polyurethane (flexible)	•	•	•	•	•			•	•	•		C
PVC (rigid)	•	•	•	•						•		
PVC (flexible)			•	•				•	•	•		C
Polystyrene					•					•		
Miscellaneous												
Aluminium and alloys	•	•	•	•	•	•	•	•	•	•		
Copper and alloys				•	•	•	•	•	•	•		
Steel	•	•	•	•	•	•	•	•	•	•		
Tinplate	•	•	•	•	•			•		•		
Galvanized	•	•	•	•	•			•	•	•		A
Glass	•	•		•			•	•	•	•		
Ceramics				•	•		•	•	•	•		

Notes A — substrates need special pre-treatment; B — risk of stress cracking; C — activator may cause discolouration; D — should be 'bodied' before bonding.

(iii) Setting mechanism

Fast-forming bonds are a result of an anionic polymerization initiated by a weakly basic species, i.e. rapid polymerization in the presence of a slightly alkaline surface having absorbed water on it. The presence of a slight amount of surface acidity, e.g. surface treatment residues, can delay bonds up to 3 min or more. An essential feature of the adhesive is an acidic stabilizer designed to prevent polymerization in the container during storage. Neutralization of the stabilizer will allow setting

Table 4.6 *Average bond strengths for cyanoacrylate adhesives*

Type of adherend	Average bond strength (Nmm^{-2})	
PTFE	—	(3.0)
Polyethylene	0.2	(4.5)
Polypropylene	0.8	(6.5)
Polyester	1.3	
Polyacetal	1.5	
Natural rubber	2.8	
Nylon	2.8	
Polystyrene	2.9[a]	
Acrylic	3.8[a]	
Polysulfone	4.1[a]	
Rigid vinyl	5.5[a]	
Glass-filled nylon	5.7	
ABS	5.8[a]	
Polycarbonate	8.6[a]	
Phenolic	9.0[a]	
Aluminium	14.5	
Mild steel	22.0	

Notes Failures are averaged lap shear strength; test temperature 20 °C; lap joints are like material on like; ( ) are results reported with surface primer.[23]

[a] Substantial risk of substrate failure.

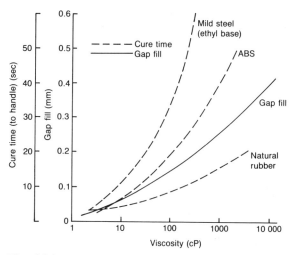

Fig. 4.26 Performance guide to cyanoacrylate adhesives as a function of viscosity

to commence, i.e. partly ionized water molecules, found on all normal engineering surfaces exposed to the atmosphere, will overcome the stabilizer. Thus the most important factors affecting setting time are the types of material being bonded and the conditions of the bonding surfaces. These parameters may be summarized as follows:

1. Joints between smooth, hard surfaces set more rapidly than between porous or rough surfaces.
2. Too low an ambient humidity (low moisture content) could delay curing.
3. Cure could be hindered if the bond gap is too wide for moisture alone to affect the join.
4. Surface deposits/contamination of an acidic nature will delay the cure.
5. Cure times are in the order metals > plastics > elastomers.

The relationship of gap fill and cure times to adhesive viscosity is shown in Fig. 4.26. Although the above description of cyanoacrylates is a general description of this class of adhesives, products are now available that deal rapidly with acidic or porous surfaces or are 'toughened' to raise both tensile shear and peel test strengths.

4.5.5 Reactive acrylic adhesives

This is a series of acrylic-based products evolving out of the anaerobics and cyanoacrylates. They are two-part adhesives combining an acrylic monomer with a rubber base and using a surface-applied activator to effect the cure. The base adhesive is a thickish syrup consisting of a colloidally disposed elastomer in a monomer or monomer/polymer solution. The resultant adhesive is toughened by the elastic domains resulting from the dispersed elastomer.[16]

Shrinkage of the base resin during cure is 20 per cent, reduced by the addition of other polymers. Since methyl methacrylate is a solvent for oil, this type of adhesive is tolerant to some surface contamination.

Reactive acrylic adhesives, or reactive bonding systems, cure fairly rapidly, achieving at best some 80 per cent of final bond strength in 8–10 min with an optimum joint at room temperature ambient, thus often making them a first choice for structural bonding requirements. In addition, their mechanical performance, temperature resistance and wide adherend compatibility range makes them one of the most versatile classes of adhesives. The comments regarding stress cracking in amorphous plastics should, however, be considered (section 4.2.3). Table 4.7 summarizes basic performance and adhesive properties.

As with the anaerobics, cure time is a function of the materials bonded and the bond gap (Figs 4.27, 4.28). Cure through a gap is a function of air inhibition, diffusion of accelerator and the type of monomer. The advantages

Table 4.7 *Summary of reactive acrylic adhesive properties*

Rate of cure	slow to fast
Adjustment time (sec)	5–30 depending on activator
Handling strength time (mins)	0.25–25 depending on viscosity
Gap fill (mm)	up to 1.25 mm, optimum 0.1 to 0.25; cure time depends on gap
Viscosity (cP)	11 000–90 000 depending on cure rate required
Shear strength (Nmm^{-2})	up to 22
Peel strength (Nmm)	up to 8
Substrate	Average bond strength in shear (Nmm^{-2}) (tensile shear test)
Steel (solvent wipe)	18.1
Steel (press oiled)	16.1
Aluminium (solvent wipe)	8.6
Aluminium (acid etched)	18.2
Brass	12.6
Copper	11.0
ABS	4.7
Nylon	4.5
Acrylic	3.4
Polycarbonate	9.6
Polypropylene (untreated)	0.8
PVC	8.1

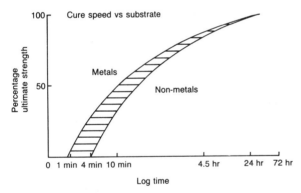

Fig. 4.27 Reactive acrylics; effect of substrate on cure time. Courtesy of Loctite UK

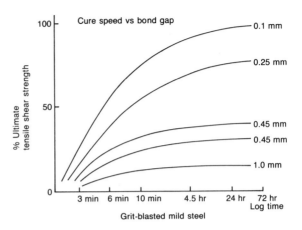

Fig. 4.28 Reactive acrylics; cure speed as a function of gap. Courtesy of Loctite UK

and disadvantages of this material class are given in Table 4.3. Most of these disadvantages have now been overcome with a new type of non-volatile reactive acrylic adhesive, termed 'aerobic'.[17] These adhesives maintain, and in some cases improve on, the performance advantages of second-generation acrylics without the drawbacks of odour and flammability. The term 'aerobic' is purely for convenience, it is used to infer a diminished sensitivity to air inhibition of thick layers, i.e. it will not cure simply with air exclusion, a pre-applied activator is required.

4.5.6 Hot melt adhesives

Hot melts are 100 per cent non-volatile thermoplastic materials, typically solid at room temperature, that must be melted to use and fluid to apply. They provide a most rapid method of securing adhesion as neither solvent loss nor chemical curing reactions are involved. The adhesive itself should have a low surface tension when molten so that it can wet a wide range of surfaces and the temperature coefficient of viscosity should be high, i.e. very fluid at the running temperature but rapidly sets as temperature falls. This, however, should not result in a brittle nature when the joint is closed and the adhesive hardens.

A polymer is the essential component, the backbone or strength of the hot melt, but on its own would probably lack the molten properties of tack, working range and wettability. The required properties have been achieved historically with a 'blend' of polymers rather than a single molecule. Most hot melts thus consist of the base polymer, a tackifying resin and a viscosity modifier. The following types of polymer are used in hot melts:

Ethylene vinyl acetate (EVA). Ethylene ethyl acrylate (EEA). Both copolymers of polyethylene. Have to be blended with resins and waxes in order to achieve sufficient tackiness and viscosity when molten as well as cohesional strength when solid. Because these polymers crystallize to achieve final strength, this may not be achieved for some time after reaching room temperature.

ADHESIVES

Polyamide. Usually a lower viscosity at the application temperature than EVA since as 'pure' polymers, they tend to be characterized by relatively sharp melting-points followed by a rapid decrease in molten viscosity. Some types are available offering good water resistance.

Polyester. Crystalline with a relatively sharp melting-point. The strongest of the hot melts. High application temperatures ~240°C, hence high heat resistance.

Pressure-sensitive hot melts. Bond well to most materials, good peel strength and tackiness, but little elevated temperature strength. Becoming an increasingly preferred mounting technique comparable with mechanical fasteners. Mainly based on thermoplastic rubbers.

Although well known in bookbinding and carton sealing, EVA and EEA adhesives are not extensively used in the engineering industry. A blend of EVA, resin and wax does not provide properties capable of competing with fasteners and spot welds, etc. (Table 4.8) although this may be perfectly adequate for unstressed securing applications, e.g. components on PCBs. However, to compete with conventional fastening methods, a high-strength polymer is essential. Polyesters and especially polyamides have taken hot melts into the area of high-performance product assembly. The advantages and disadvantages of hot melt adhesives are outlined in Table 4.3, the chief feature among these being the ability rapidly to achieve the final bond strength. The following factors predominate when considering a hot melt for industrial use:

1. *Physical strength.* Clearly a thermoplastic adhesive will be suspect if it is inherently weak. The higher the molecular weight of the polymer the greater the strength and mechanical properties, but also the greater the molten viscosity (Figs 4.29, 4.30). High

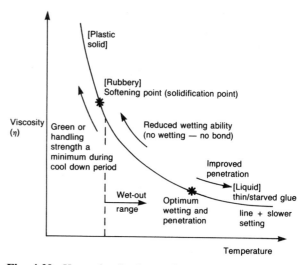

Fig. 4.29 Hot melt adhesive application characteristics

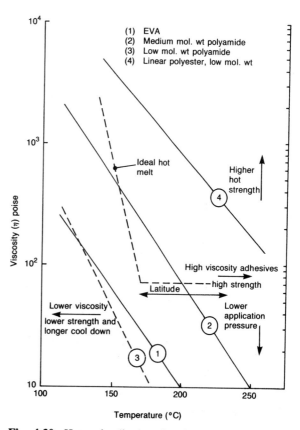

Fig. 4.30 Hot melt adhesive viscosity–temperature relationship

Table 4.8 *Mechanical properties of hot melt joints*

	Tensile lap shear strength at room temp. (Nmm^{-2})		Peel strength at room temp. (Nmm)	
	Steel	Aluminium	Steel	Aluminium
Polyester	40	19	11	5
Med. mol. wt polyamide	23	22	4	0.5
Low mol. wt polyamide	4.3	6.2	—	—
EVA	1.0[a]	0.9[a]	—	—
PSA	—	0.6[a]	—	—

[a] Failure is predominantly cohesive; i.e. joint strength is equivalent to shear strength of adhesive.

strength and low viscosity are opposing factors in hot melt technology so workable compromises have to be made.

2. *Specific adhesion*. Must possess a chemical affinity for the substrate since it can only flow over it when molten and is unlikely to take much advantage of mechanical roughness.
3. *Tack*. This is important for all adhesives, no less for hot melts, in promoting wettability.
4. *Thermal characteristics*. Need to achieve an ideal viscosity for the application and a working latitude of viscosity over the running temperature range.
5. *Flexibility*. The importance of this depends on the end use of the joint. A brittle hot melt though would have severe limitations. Generally blended with an elastomeric material to improve toughness.

In using hot melt adhesives it is important to understand the thermal balance between adhesive and adherend. If the adherend is relatively cool and at the same time has a high thermal conductivity, e.g. metal, it is possible for a poor joint to be formed if the adhesive is cooled before it has a chance to wet out the surfaces. A low-conductivity surface (plastic) or a preheated surface (metals) is desirable. However, too hot an adherend will produce a starved glue line. When the hot melt cools, viscosity is increasing rapidly and wettability is rapidly decreasing. The adhesive has little or no green handling strength during the cool-down until the plastic state is reached, hence pressure must be applied to the joint until the temperature has fallen sufficiently. Figures 4.29 and 4.30 outline the basic rheological aspects of hot melt bonding.

One interesting process which has emerged on to the market[18] is the idea of foaming the hot melt. Inert gas is mechanically introduced into the hot melt at the application stage and a number of advantages are claimed:

1. *Longer open time*. Gas bubbles act to retain heat within the adhesive, increasing open time which in turn infers increased wetting and penetration of the surface.
2. *Reduced running*. Foams are thixotropic and tend to remain more uniform when applied.
3. *Faster set time*. When the joint is made with the adhesive in compression, the foam collapses, releasing gas to produce a thinner glue line, more spreading hence a faster set. Voidage is assumed to be avoidable.

4.5.7 Epoxy adhesives

Epoxy adhesives have gained a very wide acceptance in structural applications because of excellent adhesion, toughness, chemical resistance and environmental performance. Of all the thermosetting adhesives, epoxies are probably the widest used, directly as a result of the properties attainable. There are available a very wide range of resin-hardener systems, formulations can produce flexible, rigid, tough, soft or hard adhesive bonds to almost any required property. Formulation can be designed for operation in the range -150 to $200\ °C$. For these reasons it is difficult to produce a simplified performance statement here. Some general points about this class of adhesives are as follows:

1. Resin-hardener systems are available that cure at room temperature as well as systems requiring extreme heat cures.
2. Surface preparation is important for epoxies, a clean and preferably abraded surface is necessary.
3. Epoxies bond most metals, ceramics and thermosets; adhesion is not always satisfactory with thermoplastics.
4. There are no simple relationships between the bulk mechanical properties of a cured epoxy resin and its properties as an adhesive.

Epoxies are infusible thermoset polymers formed by the *in situ* reaction between active chemical groups, the epoxy base material requiring a curing agent to convert to a thermoset. A great variety of chemical reagents can be used as curing agents or hardeners, most being an optimized combination of designed for particular properties or cure rate. Epoxy adhesives have existed for 50 years or so, the technology now having entered a fairly mature phase. Unmodified epoxy adhesives tend to have good shear strength but tend to be brittle (Fig. 4.15) and thus of limited usage for many applications. However, over the years much success has been achieved in modifying the composition with the addition of 'tougher' materials. Blends of epoxies with elastomers or thermoplastics were introduced in the mid 1960s to overcome these performance limitations, i.e. increasing impact resistance, peel strength, etc. in the cured resin making them acceptable for use in general engineering. Nitrile rubber is one blend with a high usage, the general features of toughened adhesives being described in section 4.5.1. Bond strengths are outlined in Table 4.9. Toughened epoxies are available in single (hot cure) or two-part (cold or warm cure) pastes or liquid systems. The advantages and disadvantages of epoxies are outlined in Table 4.3. Some of the advantages can be expanded further to underline the acceptability of this class of adhesives:

1. *Adhesion*. Epoxies have a high specific adhesion for metals, a variety of functional groups can also provide affinity for plastics. Although viscosities are

Table 4.9 *Toughened epoxy adhesive bond strengths*

Adherend	Average bond strength (Nmm^{-2})	
	2-part cold or warm cure epoxy	1-part hot cure epoxy
Mild steel	27	27
Aluminium	29	30
Stainless steel	26	27
Galvanized mild steel[a]	30	16
Copper	24	18
Brass	21	21

Notes: Failures are averaged lap shear strengths; adherends abraded and degreased.

[a] Not abraded.

generally high, low-viscosity types can be formulated to produce improved wetting. When properly cured the strength of the glue line is high, often resulting in failure of the adherend, rather than in the epoxy or at the interface.

2. *Shrinkage*. Epoxies are 100 per cent solids and react without the release of water vapour or other volatiles during curing. Essentially no shrinkage occurs compared, say, to the acrylics. These factors result in less residual stress in the glue line and the possibility of bonding at contact pressure, i.e. very suitable for the production line.

3. *Cure*. Most two-part systems cure at room temperature in a few hours, but the cure system can be accelerated with heat. Handling times can be upward of a few minutes. Figure 4.31 shows the effect of ambient cure temperature on open time, pot life, cure time, etc. of a typical two-part thixotropic epoxy mix. Single-component epoxy systems contain a latent heat-activated hardener and need to be heated to 100/250 °C to effect a cure, though this may be done in stages. Normally fixturing would be required throughout the cycle to ensure stability of the assembly over long cure times, to obtain good adhesive–adherend contact and to minimize glue-line thickness.

4. *Environmental*. Epoxies are insensitive to moisture and have an unmatched combination of chemical and high-temperature resistance. They have proved to be particularly durable over a wide range of demanding regimes.

4.5.8 PSAs

In 1935 the first pressure-sensitive label appeared; today pressure-sensitive adhesives, or pressure-sensitive tapes

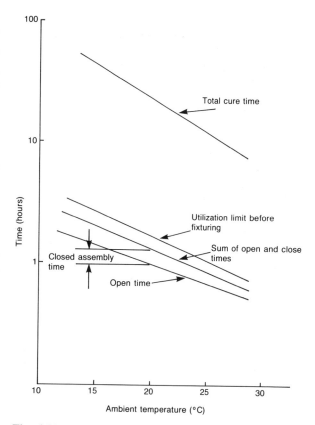

Fig. 4.31 Effect of external ambient on product parameters for an epoxy adhesive

as they are more commonly known in the wider context, are a major industry in their own right. The first thing that identifies a PSA is its tack or stickiness. A PSA is defined as an adhesive which in the dry state is permanently and aggressively tacky at room temperature, adhering firmly to a variety of dissimilar surfaces upon contact and requiring no activation from water, solvent or heat.[19] To be pressure sensitive, adhesives must be viscoelastic under conditions of use, i.e. a viscous liquid at every stage of its useful life. The strength of the bond depends on the application pressure, hence the justification for the term 'pressure sensitive'.

Tack is defined as the total energy required for the rapid separation of a film of the liquid, and by controlling the composition, a desired level of tack can be obtained. It is the property of a material which enables it to form a bond of measurable strength immediately on contact with another surface. Because of the viscoelastic nature (strength of adhesive being strain rate dependent), a high peel adhesion may be obtained at high strain rates, but creep performance of PSAs is poor, particularly at elevated temperature. If the adhesive is cross-linked, creep performance is more stable (Fig. 4.32). The behaviour of PSAs in relation to temperature, or alternatively strain rate, is shown in Fig. 4.33. The peel force and form of adhesive separation from the adherend show a strong dependence on pulling rate.[20] Types of PSA are summarized in Table 4.10, and the advantages/disadvantages in Table 4.3.

Pressure-sensitive adhesives are coated on to release

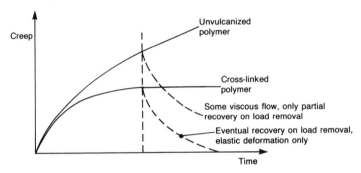

Fig. 4.32 Creep behaviour of PSAs under load

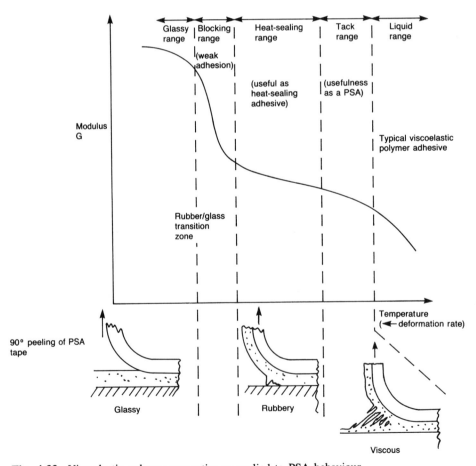

Fig. 4.33 Viscoelastic polymer properties as applied to PSA behaviour

Table 4.10 *Comparison of pressure sensitive adhesive types*

	Rubber based	Solvent based	Foamed hot melt
	Natural rubber is most common type, cheapest and versatile	Solvent-based rubbers and acrylic systems coated from solution	Acrylic-based foamed adhesive in tape form
	Rubber on its own has low tack and adhesion, thus is compounded to modify properties	Used typically for permanent applications	Strong adhesive bonds
		Resins can be cross-linked to give desired adhesive characteristics and improved creep resistance	Good ageing properties
	Blends of natural rubber and SBR produce excellent adhesives		Claimed to have peel strengths up to 6 times that of rubber-based PSAs
	Hot-melt materials based on SBRs produce adequate adhesive and cohesive strength	Acrylics have inherently good environmental resistance and ageing properties	Operating temperature range -30 to 150 °C
	General bond strengths are low although initial tack is very high	Good bond strength, initial tack may be low	Because adhesive itself is a 'closed cell foam', a separate foam carrier is not required in gap-fill, vibration situations etc.
Room temp. peel addhesion; N/25 mm	To 15	20/25	To 85
Room temp. tensile: Nmm^{-2}	Approx. 0.3	—	Approx. 0.8
Static shear; kg (170 hr at RT)	—	—	Approx. 1.5
Shear stress; Nmm^{-2} (RT)	—	—	To 0.62

paper, the adhesive (or antistick) properties require that the release paper has a surface energy much lower than that of the adhesive. Silicone-coated paper meets this requirement.

There are three basic tape constructions (Fig. 4.34) always supplied coated on to a substrate:

1. *Single coated*. In their simplest form as masking tapes, insulation tapes, etc. or with a release backing used for labelling, gaskets, seals, etc. Such adhesive uses could be paper, polymeric film, fabric, foam or metal film backed, the range being exceptionally wide.
2. *Transfer coated*. Simply for transferring a controlled thickness and width of adhesive on to a surface prior to bond formation. Able to gap fill.
3. *Double sided*. Comprises a carrier or support coated with adhesive on both sides and reeled up with a separate release coated interliner. Adhesives can be the same or different. Used to fix components *in situ* and replace mechanical fastening systems. Properties of carrier can also be utilized, e.g. foams to provide damping, rough surface conformity, sealing, etc.

In using PSAs, the following notes may be helpful in their application:

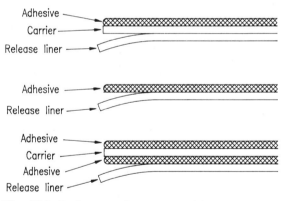

Fig. 4.34 Basic types of pressure-sensitive tape construction

1. The application of uniform pressure is required, e.g. roller.
2. Cleanliness of the substrate(s) is essential.
3. Temperature at application is important in achieving maximum bond strength faster. Being viscoelastic, bonding is a flow process by nature, the adhesive aiming to achieve conformity with the substrate. Temporarily reducing adhesive viscosity (Fig. 4.33), achieves a faster bond rate.

4.6 Adhesive selection procedure

As previously mentioned, there is no such thing as a universal adhesive, adhesives have to be selected on the basis of the application. Also within each adhesive type, there are several varieties to meet specific application requirements. A frequent problem becomes one of choosing from a range essentially based on the same adhesive, but with slight changes in adhesive chemistry and formulation. There is no single adequate classification system for adhesives, and all that any selection process can do is qualitatively to suggest one or two generic types that are worthy of further consideration.[21] Despite a considerable adhesive chemistry database and the use of adherends, in the majority of applications, that are well understood, very little has been written to enable a design engineer to design a bonded joint.[22] Although this chapter outlines the basis principles of load distribution, there is none the less a big gap between the adhesive manufacturer and the designer.

To design a joint, it is necessary to select materials with stress/strain properties that complement each other. Both adherend and adhesive are strained by the same amount since they are stuck together, therefore to prevent premature failure of the adhesive, its elongation to fracture must be greater than that of the adherend (Fig. 4.13). Although manufacturers' data sheets for a given adhesive may give different joint strengths for different materials, the adhesive must remain just as strong in each case if the failure is cohesive. However, adhesives are not used near the ultimate strength as measured by the overlap shear joint. Working stresses are at least 20 per cent of the lap shear strength and for this reason, there will not be much differentiation between the different types of structural adhesives, at least for metals (Tables 4.4, 4.5, 4.7, 4.9).

However, the general guidelines laid down in this chapter must be followed to achieve a durable end result, the bonded joint must be designed with the adhesive in mind and not arise from a substitution for some other method of joining. Bearing in mind the obvious limitations in attempting to generalize adhesive choice, Table 4.5 has been prepared as a preliminary selection guide. It is by no means definitive, but is intended as a starting-point. Table 4.11 lists the extent of the likely information requirement when putting an adhesive joint into production.

A general adhesive system selection procedure is considered in more detail as follows:

(i) Establish the performance requirement

A performance specification can serve a very useful

Table 4.11 *Factors affecting adhesive selection*

Process parameter	Influencing variables
1. Adhesive behaviour	Properties and mixing conditions Application temperature Vacuum degass prior to application Adhesive outgassing during solidification Any corrosive effect on adherend Storage life Coverage and viscosity
2. Bond quality	Surface finish Degree of surface contamination Wettability of adherend Component temperature Rate of cure Voids in glue line Residual solidification stresses
3. Bond strength	Degree of surface contamination Creep and flow under load Glue line thickness Bond area Accuracy of location and assembly fixtures Solidification time
4. Bond life	Type of loading Static or cyclic stress Max/min environmental temperature Relative humidity range Relative thermal expansion of joint Service time Surface corrosion/stress cracking Structural changes in adhesive

function provided that the requirements are clearly defined. Aspects of this are summarized as follows:

1. *Materials to be joined.* Normal engineering materials present little bonding difficulty. Certain surface finishes may require care (section 4.2.2). Some plastics have low-energy surfaces and may prove difficult.
2. *Joint design.* Decide on the local geometry of the joint. Initial design should maximize shear and compression and minimize cleavage and peel forces (Figs 4.10, 4.16). Avoid stress concentrations.
3. *Service loads.* Estimate shear forces and whether impact, cleavage or tensile loads also involved. Also likelihood of creep (cold flow) and/or fatigue.
4. *Operating environment.* Any unusual requirements?
5. *Service life.* Assessment of durability in operating service environment difficult. No procedure or testwork will accurately predict life.
6. *Suitability.* Decide if adhesive bonding is a suitable

bonding technique given the above initial evaluation of the requirement.

(ii) Establish the production requirement

Process requirements can be outlined in terms of a manufacturing specification, the underlying aim here is most likely to be a feasible, reliable and predictable process that can be carried out in normal engineering environments.

1. *Adhesive application*. Determine the optimum method of distributing adhesive on the adherend faces in respect of rheological properties and pot life. Automated processes will largely determine the viscosity to be used.
2. *Assembly time*. Specifically the production time allowable to coat the joint surfaces adhesively, plus the open time of the adhesive.
3. *Coverage*. Ensure uniform glue line (section 4.2.4). Follow the rules (Figs 4.3–4.6). Entrapped air pockets can result in adhesive being pushed out during assembly or cure. Avoid adhesive run-out by use of gap-fill types.
4. *Joint cure*. Whether heat or catalyst cure adhesive preferred. If joint needs to be fixtured, decide the allowable time-scale for achieving load-bearing strength. Best results demand the correct temperature specified for the adhesive. Room-temperature curing never achieves as high a lap shear stress as obtained with an elevated temperature cure.
5. *Joint preparation*. Preference for surface abrasion and degrease and/or use of surface primers (section 4.2.2). Primers involve a separate manufacturing stage. Some adhesives only function on correctly primed surfaces, some types tolerate surface contamination (Table 4.3). Increasing surface roughness increases wettability but also increases minimum glue-line thickness.
6. *Bonding pressure*. Choice is restricted if pressure cannot be applied to the joint.
7. *Post-bonding operations*. Whether joint durability is required during further production activities, e.g. spraying, stoving, washing, etc.

(iii) Selection of adhesive

Based on (i) and (ii) above, produce a short list of the most likely adhesives. Use Tables 4.3 and 4.5 as selection guides. Differentiation in terms of joint strength depends on the load situation prevailing.

1. *Strength requirement*. If cleavage and/or tension stresses are high, look for an adhesive with high peel strength. Be prepared to sacrifice some shear strength. toughened adhesives provide best impact resistance. Flexible adhesives are also very impact resistant, but mechanically poor elsewhere.
2. *Adhesive form*. In a large number of cases this can be suited to the convenience of the manufacturing process, i.e. whether filmic, liquid, paste, etc. Supported filmic adhesives are naturally one-component, but can be used with a primer. They have the advantage of easier working and uniform glue-line thickness. Low-viscosity liquids can cause production difficulties, use a paste or gel.
3. *Compatibility*. Ensure compatibility requirements of substrates and the adhesive system (section 4.2.3). Some adhesives such as epoxies are very adherend tolerant.
4. *Testwork*. Assesss the need to enable a more optimized choice.

(iv) Check manufacturer's data

This is the main area of uncertainty, in particular:

1. Good results are not guaranteed. Performance depends on empirical trials and correct application every time.
2. Manufacturer's information sheets can be misleading or irrelevant. Information is often limited to particular case-studies. Translating these generalities into specifics is totally inadequate for design purposes.
3. Amateurish descriptions impress no one, i.e. 'miracle in a tube', 'superglue', etc.
4. Data presented are always specific to a particular test apparatus, so making it impossible to use directly except in a comparative sense. An indication of the start of failure is really required. Mechanical properties of adhesives are rarely reported, if at all.

On the other hand, where the adhesives industry will provide substantial information is in the areas of surface preparation, the application process, cure requirements etc. Basically the practical aspects that enable the bond to be made.

(v) Optimize the design

This is an iterative process based on (i)–(iv) above, checking the selection against the requirements and optimizing the joint design and assembly methods as follows:

1. *Stress concentrations*. Usually will use some form of lap joint, but minimize the stress concentrations at the ends (Fig. 4.16). Also use chamfers, i.e. BC (Fig. 4.17). The more resilient the adhesive, the

more uniform the stress distribution (Figs 4.14, 4.15).

2. *Load-carrying capacity*. Essential that stresses are in direction of adhesives' greatest load capacity, i.e. in tension, compression or shear. Calculate bond area requirement from known load. Assume primary failure 20 per cent of ultimate. Adjust design of joint if required (Figs 4.14, 4.16, 4.17).

3. *Design principles*. Check effect of overlap distance and joint width on load capacity (Fig. 4.16 and section 4.3.1). Calculate if end stresses exceed the tensile strength of the adhesive (Table 4.3 and section 4.3.1), if so redesign.

4. *Setting*. Understand how the adhesive cures, together with any production implications. Effect of likely shrinkage stresses on joint performance (section 4.4).

(vi) Proving testwork

Unfortunately there is little option but to undertake some form of final selection proving of the adhesive, and each application must be evaluated. This gives the opportunity to design the testwork not only for feedback into design, but also to explore joint performance sensitivity in relation to production variables.

References

1. Allen K W Adhesion and adhesives. *Chem. Brit.* **22**(5): 451 (1986)
2. Zisman W A Influence of constitution on adhesion. In I Skeist (Ed) *Handbook of adhesives*. Van Nostrand 1977
3. Wake W C General properties of polymers used in adhesives and sealants. In B S Jackson (Ed) *Industrial adhesives & sealants*. Hutchinson Benham 1976
4. Less W A Adhesives for bonding metals. *Engineering* Jan 1981: 21
5. Ciba-Geigy Co *The users guide to adhesives* 1982
6. Superglues get to grips with the unstickables. *The Engineer* **268** (6945): 36 (1989)
7. Corona discharge cycle enhances bonding. *Eureka* **4**(9): 36 (1984)
8. Lees W A *The design of bonded joints*. Permabond Adhesives 1989
9. Skeist I et al Introduction to adhesives. In *Handbook of adhesives*. Van Nostrand 1977
10. Lees W A Designing with adhesives in mind. *Metals & Materials* **1**(7): 431 (1985)
11. Coover H W et al Cyanoacrylate adhesives. In *Handbook of adhesives*. Van Nostrand 1977
12. Lees W A *Adhesives in engineering design*. The Design Council 1984
13. Murray B D et al Anaerobic adhesives. In *Handbook of adhesives*. Van Nostrand 1977
14. Gross L et al Medical and biological adhesives. In I Skeist (Ed) *Handbook of adhesives*. Van Nostrand 1977
15. Blair R H Cyanoacrylate adhesives, the product assembly tool in a tube. In *Designing with today's adhesives*. ASC Spring Seminar 1979
16. Mellor S Acrylic engineering adhesives. *Suppl. Poly. Pain. Colour. J.*, March 1981: 10
17. Bachman A G Aerobic acrylic adhesives. In *Adhesive chemistry, development and trends*. Plenum Press 1984
18. Hughes F T Foamed hot melt adhesives. *Adhesives Age*. Sept 1982
19. Hodgson M E Pressure sensitive adhesives and their application. In K W Allen (Ed) *Adhesion 3*. Applied Science 1979
20. Aubrey D Viscoelastic basis of peel adhesion. In W Allen (Ed) *Adhesion 3*. Applied Science 1979
21. Lees W A Adhesive selection. In *Proceedings of the conference on structural adhesives in engineering*. Institution of Mechanical Engineers 1986
22. Adhesive bonding, a practical approach. *Engineering Materials and Design* Sept 1987: 3
23. Non-stick plastics. *Engineering Materials and Design*, **33**(6): 31 (1989)

CHAPTER 5
DRY BEARINGS

5.1 Introduction

The majority of non-metallic bearings are based on polymers and polymer composites. However, only a very small proportion of the very large number of commercially available materials are suitable for dry bearing applications. In general, cost is not the major deciding factor in the preference for dry bearings, it is the outstanding characteristic of plastics to operate under marginal lubrication or dry conditions without damage to, or wear of, the counterface, i.e. compatibility with metals in both frictional and wear terms. However, plastics do suffer from lack of strength and dimensional stability at temperatures acceptable to other bearings, hence the design of dry bearings does require special care. Design procedures, application guidelines and material properties are provided in this chapter to enable successful bearing design to be carried out.

This chapter will concentrate on the common polymeric dry bearing materials that are used in product design. Apart from polymers, other types of dry bearing clearly do exist, i.e. carbons and graphites, ceramics, surface-deposited solid lubricants, etc. but these do not find much application in the normal production environment. So what polymer characteristics determine its possibility as a dry bearing material? Two aspects of polymers determine physical characteristics. Firstly, the manner in which the polymers link dictates whether the material is a thermoplastic or a thermoset.[1] Secondly, the degree of branching within each polymer chain determines how closely the chains pack together, so influencing density and mechanical properties. This packing behaviour also creates two classes of thermoplastic, amorphous and crystalline. The vast majority of successful dry bearing polymers are crystalline. The characteristics of both amorphous and crystalline thermoplastics are compared in Table 5.1.

Non-metallic bearings are preferred in operating environments where there is no adequate lubrication present or where a combination of high load, low speed or intermittent motion makes lubrication difficult. This has to be a very attractive option for many types of equipment, fit for life concept, maintenance free, etc. Many potential dry bearing applications have, however, been unsuccessful due to the failure of the designer to appreciate the differences between metallic and non-metallic materials. Thermoplastics have poor creep strength, a low softening temperature, high thermal expansion coefficients and often will absorb liquids. These and other characteristics therefore have to be allowed for.

The dry bearing industry is a fairly stable one, there has not been any major breakthrough in the last 20 years, rather continued and improved use of the very well-established materials such as nylons, acetals and polytetrafluoroethylenes (PTFEs). The lowest wear rates are demonstrated by the reinforced, filled and woven PTFEs. Filled nylons, acetals and phenolics are not remarkable for their wear resistance, but they are low cost and used in non-critical situations. Recently liquid lubricant impregnated nylons/acetals have entered the market-place, giving very low wear rates and cost-effective solutions to a number of design problems.

A lot of effort has gone into optimizing the type and concentration of fillers used in bearing plastics. Multi-component composites are now common, e.g. polymer plus reinforcing fibre plus secondary solid lubricant. High-molecular-weight polyethylene (UHMWPE) has proved a useful low-temperature material and, like polytetrafluoroethylene (PTFE), capable of significant improvement with fillers or internal lubrication. A number of high-temperature polymers are also now available, e.g. polyimide, polyetherketone (PEEK), polyethersulphone (PES) and polyphenylenesulphide (PPS), but in general they are more expensive and

Table 5.1

Amorphous	Crystalline
Broad softening range — thermal agitation of the molecules breaks down the weak secondary bonds. The rate at which this occurs throughout the formless structure varies producing a broad temperature range for softening	Sharp melting-point — the regular close-packed structure results in most of the secondary bonds being destroyed at the same time
Usually transparent — the looser structure transmits light so the material appears transparent	Usually opaque — the difference in refractive indices between the two phases (amorphous and crystalline) causes interference so the material appears translucent or opaque
Low shrinkage — all thermoplastics are processed in the amorphous state. On solidification the random arrangement of polymers produces little volume change and hence low shrinkage	High shrinkage — as the material solidifies from the amorphous state the polymers take up a closely packed, highly aligned structure. This produces a significant volume change manifested as high shrinkage
Low chemical resistance — the more open random structure enables chemicals to penetrate deep into the material and to destroy many of the secondary bonds	High chemical resistance — the tightly packed structure prevents chemical attack deep within the material
Poor fatigue and wear resistance — the random structure contributes little to fatigue or wear properties	Good fatigue and wear resistance — the uniform structure is responsible for good fatigue and wear properties

Examples of amorphous and crystalline thermoplastics

Amorphous	Crystalline
Polyvinyl chloride (PVC)	Polyethylene (PE)
Polystyrene (PS)	Polypropylene (PP)
Polycarbonate (PC)	Polyamide (PA)
Acrylic (PMMA)	Acetal (POM)
Acrylonitrile–butadiene–styrene (ABS)	Polyester (PETP, PBTP)
Polyphenylene (PPO)	Fluorocarbons (PTFE, PFA, FEP and ETFE)

furthermore do not have the lowest wear rates. Particular materials will be dealt with in section 5.3.

5.2 Wear performance factors

5.2.1 Running-in

To produce an acceptable dry bearing application, one has to understand the wear process. In the initial stages of dry bearing wear, the wear rate is relatively high. Subsequently both the metal counterface and polymer bearing surfaces can be modified by abrasion or transfer effects to bring about a wear rate reduction, producing a stable steady-state wear condition (Fig. 5.1). If the counterface roughness is increased, the initial wear rate (running-in wear) increases significantly but actual steady-state wear rates, e.g. FC, EB and DA in Fig. 5.1, are not that different because of counterface modifications via the transfer and/or abrasion mechanism mentioned.[2] The appropriate choice of counterface material and finish will normally produce an insignificant level of running-in wear, i.e. 0.03 mm or less being typical. The risk in evaluating bearing performance from a single fixed time-period measurement is also illustrated in Fig. 5.1, i.e. OC, etc. producing results greatly in error. Factors that affect the running-in process are now considered in more detail.

(i) Counterface roughness

During initial sliding against a rough metal counterface, metal surface features (asperities) penetrate the softer polymer and wear occurs via permanent deformation, shear or microcutting. With smooth metal surfaces however, wear is more fatigue orientated. The topography of the counterface is the main factor affecting the magnitude of the wear rate, it applies not only to the initial stages of sliding with the surface 'as-prepared', but also to the latter stages where the topography is generated by the sliding process itself.[3]

In the initial stages of sliding

Wear rate (polymer) $\propto R_a^n$

where $n = 1.5-3$ depending on the polymer and R_a = surface roughness.

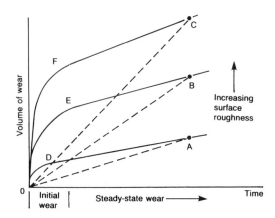

Fig. 5.1 Typical dry bearing wear behaviour based on journals of increasing roughness

The following general points can be made concerning counterface roughness:

1. With unfilled polymers, it is generally reckoned that a counterface roughness of $0.2-0.4$ μm R_a is useful in minimizing the initial wear rate.
2. There are no tribological advantages to be gained below 0.05 μm R_a, the cost will also rise significantly.
3. A smooth counterface is more important with non-abrasive fillers. With glass fibre-filled PTFE a similar wear result is obtained with a counterface roughness over the range $0.05-0.5$ μm R_a.
4. 0.6 μm R_a is considered as the coarsest shaft finish for good wear characteristics, and one of 0.4 μm R_a is preferred.
5. Increase of counterface hardness will offset the initial run-in smoothing effect. Hence attention should be paid to reducing the roughness of harder counterfaces.
6. The surface roughness R_a value is not a unique guide to the effect of the counterface on initial wear, a ground surface of 0.2 μm R_a will generally produce less wear than an abraded surface of 0.2 μm R_a.
7. The surface finish of the polymer itself is not critical. Bearings moulded with finishes $0.6-1.3$ μm R_a perform just as well as those with polished surfaces.

(ii) Transfer film formation

As with all mild wear situations[4] the two rubbing surfaces attempt an initial process of surface modification which, if successful, will produce an acceptable steady-state wear regime which remains operative as long as the sliding conditions remain fairly stable. This involves transfer of polymer to the metal counterface via a combination of mechanical cutting, adhesion and fatigue mechanisms,[5] a complex and not particularly well-understood process. The result is a smoother counterface (shaft, thrust face, etc.) and a partial polymer/polymer sliding situation.

With unfilled polymers, transfer can be an advantage or disadvantage to the wear process depending on the topography generated. With the high elongation crystalline polymers, e.g. PTFE, acetal, nylon, etc. transfer to the counterface produces a smoother surface. With the amorphous polymers, e.g. polymethylmethacrylate (PMMA), epoxies, polystyrene, etc. transfer may be in the form of irregular lumps, effectively increasing counterface roughness and hence the wear rate.

With fillers added to the polymer, it is possible for transfer to occur from either or from both.[5] Solid lubricant fillers, e.g. MoS_2 readily transfer to a metal counterface and helps explain why such fillers reduce friction. With glass fibre reinforced PTFE on the other hand, the transfer film is predominantly pure PTFE. In general, however:

1. Solid lubricant fillers such as PTFE, MoS_2 and graphite can contribute to the formation of fairly smooth transfer films and so effect wear reduction.
2. In some composites the stability of the transfer film is affected by high humidity (water condensation), this removes the transfer film and the wear rate increases. Thus for some dry bearing materials the worst environment is an alternating wet/dry one.
3. The transfer of lamellar solid lubricants is in fact not that efficient and contents > 10 per cent v/v (volume−volume basis) are required, thus reducing the mechanical strength of the polymer.
4. At lower concentrations, PTFE is more effective than a lamellar solid as an additive. Also, PTFE can transfer to metals in the presence of water.
5. The transfer film is very dependent on the sliding geometry, i.e. laboratory test rig performance behaviour may not duplicate the intended design application.

(iii) Abrasivity of fillers

It is not generally appreciated that fillers and reinforcing fibres in polymers are abrasive towards metals. Their abrasive nature is partly intrinsic and partly due to an impurity content in the fillers themselves. Mildly abrasive fillers, e.g. carbon, graphite and bronze can act to remove high spots on the counterface and hence act beneficially to assist bedding-in, but with abrasive fillers such as glass fibre, the counterface may be seriously damaged if of relatively low hardness. The relative abrasive effects of different fillers are shown in Fig. 5.2. The running-in effects of filler abrasion in practice may be summarized as follows:

1. A significant part of the variability detected in polymer wear properties can result from the effect of fillers on the counterface surface.
2. Clean surfaces of unfilled polymers do not mechanically damage metal surfaces during relative sliding. Counterface abrasion is therefore directly attributable to either the fillers or reinforcement present.
3. Fillers that roughen the counterface, clearly increase the wear rate of the composite, e.g. glass-filled plastics on a soft non-ferrous metal. Also metal particles may become embedded in the composite.
4. The relative hardness of filler and counterface is an important consideration in bearing application

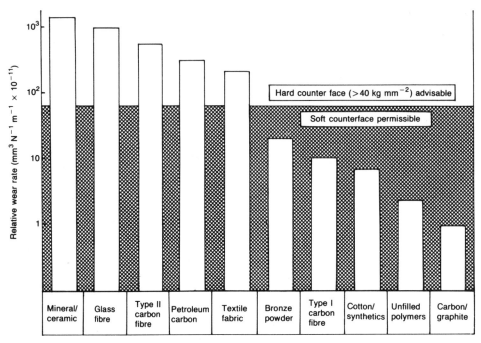

Fig. 5.2 Relative abrasiveness of dry bearing fillers and materials using a brass counterface, ref. 7. Courtesy of ESDU International

design, i.e. abrasive fillers require a harder counterface (Fig. 5.2).

5. Under certain circumstances there may be an advantage in deliberately adding small quanitities of abrasive to the polymer composite (e.g. 0.1 per cent of Al_2O_3 particles) to promote low steady-state wear by deliberately polishing the counterface during running-in.
6. With the more abrasive fillers (Fig. 5.2) and high bearing pressures, there is a risk that bearing wear debris will contribute significantly to high wear rates.

5.2.2 The PV factor

In technical literature, a *PV* value is generally quoted by manufacturers as a material performance guide. It is a useful concept, related both to wear rate and temperature rise at the sliding surface, but it has often been criticized as misleading and a source of confusion and unreliable performance prediction.[6] However, since it is such a widely quoted parameter, with no obvious well-documented equal, this section will deal with the *PV* concept in some depth, outlining the boundaries of applicability to dry bearing application design.

The normal equilibrium wear equation is generally represented as follows:

$$v = \frac{KWl}{P_m}$$

where v = wear volume, K = wear coefficient, W = normal load, l = sliding distance and P_m = mean yield pressure, but for polymers, the specific wear-rate concept is introduced such that in the steady-state wear regime (Fig. 5.1):

$$K = \text{specific wear rate} = \frac{v}{Wl}$$

where K is regarded as a material constant. However, with bearings, it is not so much the wear volume that is important, rather the depth of wear that leads to an increase in clearance. Now

$$v = Ah$$

where h = wear depth, A = apparent contact area W/P and P = applied pressure, i.e. specific loading/unit of projected bearing area. Thus

$$h = \frac{vP}{W} = KdP = KtPV$$

where V = sliding velocity and t = time. Hence

Wear rate $\dfrac{h}{t} \propto PV$

where PV is the widely used performance criterion or PV factor.

In the above treatment, pressure P is based on a projected bearing area (Fig. 5.3), where $A = Ld^*$. In US literature, K is often referred to as a 'wear-factor' and given units of in min ft^{-1} lb ft^{-1} hr^{-1} which is only numerically correct in the context of radial wear. Conversion factors between the various systems of units that might be encountered in dry bearing design are given in Table 5.2.

A quoted PV value by itself says nothing about actual bearing wear, but it can be used as a performance concept and will be quoted in two forms:

1. The 'limiting-PV' above which wear increases rapidly either as a consequence of thermal effects or overload. In other words it is not a particularly safe design limit.
2. The PV factor for continuous operation at some arbitrarily defined wear rate, e.g. 25 μm 100 hours^{-1}.

In neither case is the PV factor a unique criterion of performance, being valid only over a restricted range of P and V (Fig. 5.4). The following comments relate to the PV curve:

1. There is only inverse proportionality between P and V over a very restricted range of conditions when PV is actually constant. Also it is only in this limited range where the specific wear rate is constant.
2. The bearing temperature increases with PV as the softening point of the polymer is approached. Thus heat removal from the bearing is important with the low softening point materials.
3. Quoted PV values, and especially limited PVs are fairly meaningless without some knowledge of the bearing geometry and heat transfer characteristics/conditions to which they refer.
4. The maximum allowable bearing pressure, e.g. Fig. 5.4, i.e. the static load-carrying capacity is

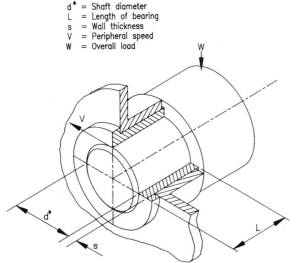

Fig. 5.3 Illustration of bearing area parameters. Courtesy Du Pont de Nemours Int. SA

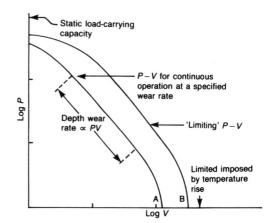

Fig. 5.4 The PV relationship for dry bearing materials

Table 5.2 *Conversion factors between various systems used in wear performance specification*

	Imperial units		Metric units		SI
Specific wear rate K in V/Wd	$\dfrac{10^{-12} \text{ in}^3}{\text{ft lbf}}$	≡	$\dfrac{1.2 \times 10^{-12} \text{ cm}^3}{\text{cm kgf}}$	≡	$\dfrac{1.2 \times 10^{-8} \text{ mm}^3}{\text{N m}}$
Wear factor K in radial wear context	$\dfrac{10^{-10} \text{ in}^3 \text{ min}}{\text{ft lbf hr}}$	≡	$\dfrac{1.2 \times 10^{-8} \text{ cm}^3 \text{ min}}{\text{m kgf hr}}$	≡	$\dfrac{1.9 \times 10^{-8} \text{ mm}^3}{\text{N m}}$
PV factor	$\dfrac{10^3 \text{ lbf ft}}{\text{in}^2 \text{ min}}$	≡	$\dfrac{20 \text{ kgf m}}{\text{cm}^2 \text{ min}}$	≡	$\dfrac{0.035 \text{ MN m}}{\text{m}^2 \text{ s}}$

approximately determined by one-third material compressive stress. However, some polymers are viscoelastic, hence prone to creep and permanent deformation under load.

5. Deformation is usually a function of the bearing material thickness, the thinner the material, the less the percentage deformation for a given load, especially if firmly bonded to a substrate.
6. A limiting PV value does not define a given wear rate as such. Simply stated, it defines the maximum load and velocity of the material and bypasses the wear consideration.
7. The operating PV of a dry bearing material will fall below the limiting PV by a distance depending on the allowable wear rate.
8. The concept of a unique PV factor characterizing a dry bearing is not particularly valid, a PV relationship is more informative.
9. Large bearings have lower PV limits than smaller bearings and the limit is lower for thrust rather than journal bearings. Lubrication can raise the limit.

Apart from using PV and limiting-PV values to classify dry bearing performance, it is often common practice to determine the K-value or specific wear rate. However, K does in fact vary with P, V, PV, time, counterface finish and also, importantly, bearing geometry. The latter applies since PV is an energy term and μPV represents the frictional energy generated at the interface. The greater PV the greater the heat generation and the greater the wear rate. Thus:

Heat flow $\propto$ projected bearing area A

Temperature rise $\propto \dfrac{\mu VW}{A} \propto \mu PV$.

The thermal aspects of dry bearing design is dealt with in greater detail in section 5.2.4.

5.2.3 Test methods

There is no general technical agreement on what represents a standard dry bearing test method. Variously 12–25 mm diameter bearing bores or thrust washers are used by manufacturers. Hence with a lack of relevant standards, the definition of a material's PV value is left to the supplier, consequently quoted results are far from uniform. Given the right layout and test equipment, low heat generation and good thermal conductivity, it is clearly possible to achieve high PVs. Often the PVs obtained via test specimens do not adequately duplicate the practical situation, the material likely to have a different PV value when operated as an actual bearing. Thus one should be careful in design data comparisons.

There are two methods commonly used in the determination of a PV curve:

1. *Limiting PV*. This is fairly straightforward for any given plastic and is established by load-stepping, e.g. Fig. 5.5. A torque arm is attached to the bearing and both frictional torque and bearing temperature are continuously monitored. At a particular velocity, the load is increased in stages with the temperature and friction allowed to stabilize at each load, until at a given high load friction and temperature will not stabilize so producing high wear. The load corresponding to the last region of stability is taken as the load limit. At each intermediate load, conditions are maintained for a half-hour or so in order to reach equilibrium. Repetition at different velocities enables the whole of the limiting-PV curve to be assembled.
2. *Normal PV*. This corresponds to a specified wear rate. It is a more time-consuming process and involves the measurement of rate of wear over a large number of loads and speeds.

Limiting PV relationships are easy to determine in laboratory tests, but are of little design value since the limit represents the point where the bearing is destroyed by melting or extrusion. Such values contain no factor of safety and should be applied with caution. On the other hand, PV values referenced to an acceptable wear rate are useful for design purposes, and although difficult and tedious to accumulate, data are available for a number of materials.[7]

In order that wear testwork is meaningful and allows a real evaluation of a bearing design, the following points should be observed:

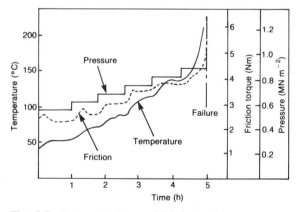

Fig. 5.5 Typical variation of friction and temperature during 'limiting PV' test determination

1. A description is given of the test equipment and the tests carried out.
2. The hardness and surface quality of the counterface is stated.
3. As far as possible, wear tests should simulate the expected running conditions.
4. Wear should be determined at regular intervals and curve shapes presented.

5.2.4 Bearing surface temperature

The disadvantage of plastics from the bearing point of view is their low thermal conductivity (compared to metals) which limits frictional heat dissipation and hence restricts PV. For example, bearing wear tests on polymers can vary, a bearing pressed into a metal housing will give a better result than a bearing not surrounded by metal as a result of the thermal conductivity path. The PV limit is really determined by the condition which develops when the rate of heat input into the bearing equals the rate of heat dissipation through the system. Temperature is the reason for the deviation of the PV relation from linear at high speeds (Fig. 5.4), i.e. via frictional heating.

The bearing surface temperature can be considered to have two components, the mean bearing temperature, basically the balance between rate of generation and rate of dissipation of heat, and secondly the so-called 'flash-temperature' confined to localized surface points of real contact. The latter is academically interesting but difficult to apply quantitatively, and really only relevant at sliding speeds $> 0.1\, V/V_{max}$, where V_{max} is typically as in Fig. 5.4. Wear rate corrections above this speed will be considered in the design procedure, section 5.5. Thus for most practical purposes the mean surface temperature (T_m) in the steady-state wear region may be considered as

$$T_m = T_a + \Delta T$$

where

$$\Delta T = C\mu PV$$

such that ΔT = mean temperature rise of the whole of the apparent contact area as a result of the equilibrium of frictional heat generated at the rubbing surface and its dissipation through the bearing assembly, C = thermal resistance coefficient, characterizing the overall resistance to heat dissipation from the rubbing surfaces for the overall bearing design configuration and μ = dynamic friction coefficient.

The above is a semi-empirical approximation derived for design purposes. A more exact relationship is sometimes quoted as

$$Q = \mu W V$$

where Q = amount of heat generated per unit time and W = applied load.

A detailed procedure for deriving C values is given in refs 7 and 8. Table 5.3 illustrates the potential range of C values for various design conditions. A good average value which describes the majority of applications is $1.2 \times 10^{-3}\,°C\,m\,s\,N^{-1}$. It also follows that for a journal bearing, the extent of frictional heating will rank in the following order:

Oscillating load > unidirectional load > rotating load

The effects of temperature on polymers are basically twofold:

1. *Mechanical stability.* As the polymer heats up, both load-carrying capacity and hardness decrease as the bonding between polymer chains weakens. The temperature defining the upper limit of the useful working range may be variously referred to as 'heat distortion temperature', 'maximum operating

Table 5.3 *Examples of thermal resistance coefficient, C (units of C in $10^{-3}°C\,m\,s\,N^{-1}$)*

Heat dissipation from mating surface (shaft)	Poor (multi-bearing shaft + hot ambient)			Average			Good (cooled)		
Housing heat dissipation	Poor (plastic)			Average (steel)			Good (cooled metal)		
Thermal conductivity of bearing material; W m^{-1} °C	0.2	1	10	0.2	1	10	0.2	1	10
ON/OFF ratio	cont	3	3	3	cont	3	3	3	cont
On time (secs)	—	20	5	20	—	6	5	20	—
C-values									
6 mm dia. journal bearing	7.7	5.5	2.7	0.5	0.9	0.2	0.1	0.1	0.4
10 mm dia. journal bearing	11	7.7	3.8	0.7	1.3	0.2	0.2	0.1	0.6
20 mm dia. journal bearing	22	15.4	7.6	1.4	2.6	0.4	0.4	0.2	1.2

temperature', 'softening temperature', 'continuous upper service temperature', etc.

2. *Oxidative stability.* Strength changes also occur with increasing temperature as a result of chemical reactions within the polymer itself or between the polymer and its environment, usually oxygen.

In the design of polymer bearings, it is usually assumed that the *PV* limit corresponds to the melting-point of the particular polymer, above which the increase in the frictional heating produces a dramatic increase in wear rate. In fact the point is reached where the bearing pressure approaches approximately one-third the compressive strength, so increasing the real contact area and the scale of the damage. However, in practice a continuous no-load service temperature is considered as the upper limit, e.g. Table 5.4, this being slightly in excess of the maximum operating temperature recommended by plastics manufacturers.

Thermoplastics are thermal insulators and the generation of frictional heat is a function of pressure, velocity and friction. Bearing surface temperature is believed to govern the wear rate of plastics, so the shortcomings of plastics have to be compensated for by a suitable design of bearing application. Above a certain speed level, depending on the heat flux of the friction couple, even a very light load variation may disturb the thermal equilibrium. In some proprietary materials, low thermal conductivity (and high thermal expansion) have been offset wholly or partly by the use of certain fillers (metal, graphite) or by using polymer as a thin surface layer on a metal backing.

The specific wear rate, or wear factor (section 5.2.2) is a useful starting-point in dry bearing design. The specific wear rate is in general a function of operating conditions, primarily pressure and temperature. Figure 5.6 shows in schematic form how the specific wear rate can vary with bearing pressure and temperature. At low pressures and at normal ambient (to 25 °C) the specific wear rate is independent of both P and T. The pressure and surface temperature limits (Table 5.4) for some common bearing materials are shown in Fig. 5.7, defining the general boundary of acceptable or modest wear behaviour. Performance bands are necessary in Fig. 5.7 because of the wide range of filled compositions available commercially. Bearing in mind that as operating temperature increases towards the performance boundaries in Fig. 5.7 for a given material, so the wear rate increases. The resulting wear rate multiplier is given in Fig. 5.8.

Using the relationship outlined in Fig. 5.7, this is capable of producing a more informative indication of a material's wear performance if it is expanded into a three-dimensional $K-P-T$ relationship.[6] This then emphasizes the importance of bearing surface temperature and pressure as fundamental factors which determine the wearing behaviour (Fig. 5.9). Unfortunately, although such plots provide a comprehensive picture of a material's wear performance, such plots do not widely exist for design purposes. From Fig. 5.9, the lowest and relatively constant wear factors are obtained in a limited pressure and temperature range, outside which the wear factor increases. Along the curve CB, the rapid rise corresponds

Table 5.4 *Permissible bearing working temperatures*

Material	Melting-point (ASTM D789) °C (*PV* limit)	Continuous no-load service temperature (°C)
Acetal	180	130
Polyamide	230	130
Polyethylene		80
Unfilled PTFE	320	250
PTFE + 25% glass	320	240
PTFE + 40% graphite	320	240
PTFE + 40% bronze	320	260
Thin layer PTFEs	320	270
Glass-reinforced nylon	255	130
Polyimide		290
Filled polyimide		290
Filled polysulphone		180
Filled PES		210
PEEK		305
Oil-filled nylon		145

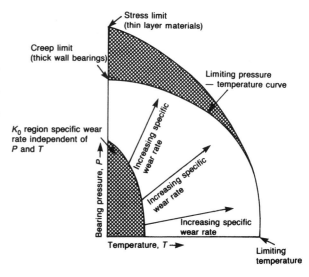

Fig. 5.6 Bearing temperature and pressure limits, ref. 7. Courtesy ESDU International

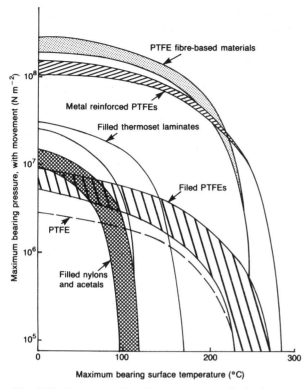

Fig. 5.7 Pressure and surface temperature limits for some dry bearing materials. From Crease A B, 'Design data for the wear performance of rubbing bearing surfaces'. *Tribology* **6**(1): 15 (1973). By permission of the publishers, Butterworth Heinemann Ltd ©

to the bearing pressure approaching the useful material wear resistance limit in terms of its strength. Along curve CA, the specific wear rate increases as the material softens. These result in pressure–temperature curves at constant wear rate, curve BA representing the limiting condition above which material performance is unstable. The pressure–temperature curves in Fig. 5.9 and Fig. 5.7 are directly comparable with the PV curve in Fig. 5.4, since the limiting V at pressure P is the sliding speed which results in the surface temperature reaching its limiting value. However, while the limiting PT curve is a material characteristic, limiting PV curves are not, because they depend not only on the material but also on the heat-dissipation characteristics of the bearing configuration and its ambient temperature.

As a general rule, the following design points facilitate heat removal from a bearing assembly:[7]

1. The provision of substantial shaft and housing areas. Fins can be used if necessary. Where bearings share a common shaft, the individual effective shaft areas may be reduced resulting in decreased heat dissipation from each bearing assembly.
2. The provision of air access to as much surface area as possible.
3. The provision of substantial heat-flow paths with all links as thermally conductive as possible. Factors that reduce the thermal conductivity of joints are orientation across the heat-flow direction, low fastening loads, a rough surface finish, gaskets or sealants.
4. The use of shaft, housing and bearing materials with high thermal conductivities or reduced bearing thickness if the bearing material has low conductivity.
5. Thermally insulate joints between the bearing housing and sources of heat.

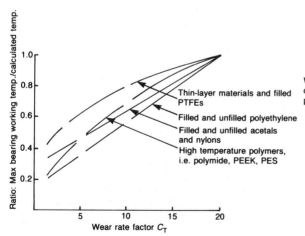

Fig. 5.8 Wear rate correction factor for bearing temperature

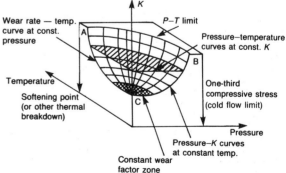

Fig. 5.9 Typical variation of wear factor K with bearing pressure and surface temperature

6. Forced air ducting or cooling, often conveniently achieved by means of a fan on the bearing shaft.
7. If the housing can dissipate heat more readily than the shaft, consider locating the thermoplastic bearing on the shaft in the form of a sleeve rather than bushed into a housing. This also has pressure advantages. The problem is deciding whether the housing or shaft is the best heat sink.

5.2.5 Fillers and additives

There are two basic groups of materials added to dry bearing polymers, primary lubricants and reinforcement fillers, both having the aim of reducing the bearing wear rate and the dynamic friction coefficient. The effect on counterface wear, however, may be different (section 5.2.1).

(i) Fillers
The incorporation of a filler has the following effects:

1. Wear resistance can be increased by three orders of magnitude.
2. Creep resistance under load can be increased by a factor of 10.
3. Depending on the filler used, thermal conductivity of the bearing material can be increased by a factor of up to 10.
4. Depending on the filler used, thermal expansion may be reduced by a factor of 5.

Fillers are effectively limited to about 40 per cent by volume if reasonable mechanical strengths are to be maintained. Combinations of filler can give rise to greater improvements in wear properties than for the individual filler types on their own, thus synergistic wear rate improvement seemingly is the only advantage in filler combination.

Although fillers and reinforcements increase the strength and stiffness of a polymer, the ductility of polymers is usually reduced correspondingly. Since during steady-state conditions of bearing wear, adhesion and fatigue effects predominate, reinforcements are extremely effective in wear reduction. Most added fillers are abrasive to a degree towards the counterface (section 5.2.1(iii)). The effect of fillers on the mechanical and thermal properties are interrelated with those on frictional and wear properties. For example, additions of glass fibre increase strength and therefore the load capacity of the bearing, but may also increase friction and therefore reduce the limiting speed. Conversely the addition of solid lubricants tends to reduce strength and load-carrying capacity, but also reduces friction and frictional heating thereby increasing the limiting speed.[5]

For optimizing the mechanical properties of a composite, the polymer and filler should create a homogeneous structure, equally sharing stress. However, in the sliding situation this may not be always correct.[5] Carbon and glass fibres are known to support the load preferentially, the tribological properties not greatly dependent on the polymer matrix type. Filler content may be quantified as follows:

1. *Low filler contents* (3–10 per cent v/v): where mechanical properties are more important in an application than maximum wear resistance, i.e. tensile strength, elongation and flexibility.
2. *Intermediate loadings* (10–20 per cent v/v): used where good wear resistance is of primary importance and when good physical properties are desirable at the same time.
3. *High filler contents* (20–35 per cent v/v): used for low wear bearing applications where tight tolerances need to be maintained over a period of time — above 35 per cent wear properties can tend to fall off sharply, hence a good optimum is ~ 30 per cent.

The wear resistance of a thermoplastic polymer is generally enhanced in the order bronze > carbon/graphite > ceramics (including glass fibre) > other metals (except bronze) > MoS_2. The functions of the more common types of filler are summarized in Table 5.5.

(ii) Internal lubricant additives
Internally lubricated thermoplastics offer advantages in classical dry bearing design as well as providing tribological opportunities for materials not usually classed as bearing polymers. In recent years a number of products have appeared where the polymer is impregnated with a lubricant, an additive rather than a filler. Although these additives do nothing to improve mechanical properties, they do appear to reduce wear rates by at least a factor of 2 over comparable types. Typical examples of additive effects in nylon and acetal are shown in Table 5.6. Internal lubricants considered are as follows:

1. *TFE addition*. Provides primarily self-lubrication in a thermoplastic base resin, greatly improving friction and wear behaviour. The actual TFE fluorocarbon powder added is a low-molecular-weight micropowder (0.1–0.5 μm) that can either be used as a dry lubricant or incorporated into the polymer. From Table 5.6 it can be seen that at low concentrations TFE is a more effective additive than lamellar solids. Wear resistance, friction and the *PV* factor all reach

Table 5.5 *Fillers used with dry bearing polymers*

Filler	Form	Effect
1. Glass	Usually milled fibres. A minimum L/D ratio of 10 : 1 is generally used. Glass beads are unsuitable	Wear resistance improvements directly related to bond effectiveness between glass and resin. Most widely used filler. Improves both long- and short-term mechanical properties of the polymer. Abrasive to opposing surface
2. Carbon and graphite	In the form of high-purity powdered coke or natural/synthetic graphite. Particle size usually <60 μm, shape irregular. Ability to lubricate depends on structure, purity and particle size	Reduces friction but not wear rate. Graphite requires presence of water vapour (condensing fluid), hence operates well in an aqueous environment. Do not use in dry gases or a vacuum. Graphite helps to reduce friction and stick–slip. Reducing the graphitic order (i.e. carbon) increases filler abrasivity. Graphite will transfer to counterface during run-in, carbons will not
3. Carbon fibre	Type I carbon fibre is treated at 2500 °C and shows graphitic behaviour. Type II carbon fibre is heat treated at 1500 °C and is wholly non-graphitic	Most systems $>15\%$ will dissipate static. Results in thermoplastic resins with highest strength, heat distortion temperature and fatigue resistance available commercially. Friction and wear rate reduction determined by the degree of carbon fibre graphitization. High-strength type II fibres are more abrasive than high-modulus type I fibres. As fibre content increases, wear rates of different composites becomes equivalent
4. Bronze	As irregular or spherical shaped particles, usually <60 μm dia.	Produces high wear resistance and high load capabilities, but limited rubbing speeds. Friction coefficient approx. 0.2. Additions of graphite or metal oxides improve temperature and speed limitations. Specifically used as a filler for PTFE
5. Metal oxides	Notably lead oxide in fine powder form, 5–10 μm in dia.	Raises temperature and speed limitations of polymers
6. Molybdenum disulphide	Dispersed particles of fine solid lubricant	Depends on the polymer. Improved mechanical properties and wear resistance when added to nylon materials. Will increase wear rate of some materials. A lamellar solid lubricant which may reduce friction and stick–slip marginally

an optimum value >15 per cent TFE, both for amorphous as well as crystalline thermoplastics.

2. *Silicone oil.* Silicone fluids are generally incompatible with common thermoplastics, hence it is straightforward to produce a well-dispersed system by mechanical compounding. The result is a continuous phase of unmodified plastic structure with silicone fluid droplets of a fairly narrow size distribution. High oil viscosities $\sim 12\,500$ cS provide the best wear rate reduction. Small concentrations of silicone can be as effective as higher loadings of TFE (Table 5.6). In contrast to organic lubricants, silicones do not reduce the heat deflection temperature of the polymer.

5.2.6 Friction

As far as polymers are concerned, there is no such thing as a unique value of the coefficient of friction, static or dynamic, hence the values quoted within this chapter should be regarded as approximate guides only. The basic laws of friction are very rarely obeyed for polymers, i.e.

1. Friction is dependent on the applied load;
2. Friction is dependent on the apparent area of contact;
3. Friction is speed dependent.

For a single contact area, elasticity theory gives

$$A_r \propto W^{-0.67} \quad \text{or} \quad \mu \propto W^{0.33}$$

Table 5.6 *Effect of internal lubrication on polymers*

	% TFE[b]	% SO[c]	% MoS	% Graphite	PV factor[a] (MN m² m s⁻¹) Normal	PV factor[a] (MN m² m s⁻¹) Limiting	Specific wear rate, $K \times 10^{-7}$ mm³ N m⁻¹ [d]	μk
ACETAL								
Base polymer					0.11	0.12	12.3	0.21
Polymer +	5				0.18	0.35	7.6	0.18
	10				0.25	0.44	5.7	0.17
	15				0.26	0.44	3.8	0.15
	20				0.26	0.44	3.2	0.15
		2				0.31	5.1	0.12
	18	2				0.52	1.7	0.11
			10		0.07	0.24	47.5	0.22
				10	0.07	0.18	13.0	0.22
NYLON								
Base polymer					0.07	0.07	38	0.26
Polymer +	15				0.25	0.96	2.9	0.19
	20					0.71	2.9	0.19
		2				0.14	9.5	0.12
	18	2				0.86	2.1	0.11
	[e]					0.59		0.12
			10			0.44	30	0.30
				5			11.4	0.19
	15		2.5		0.25	0.61	2.85	0.22

[a] At 0.5 m s⁻¹ sliding speed.
[b] Low molecular weight modified PTFE.
[c] High-viscosity polysiloxane.
[d] At limiting PV condition against steel.
[e] Cast nylon with integral lubricant.

hence

$F = aW^n$

where n is a value $>0.67 >1$, A_r = real area of contact, W = applied load, μ = frictional coefficient, F = frictional force and a = constant.

The above behaviour is to be expected in an elastic contact situation, although in practice n will be somewhat greater than 0.67 since deformation will not be entirely elastic. As sliding conditions become more severe, $n \to$ unity and the frictional force will then become contact area independent. However, polymers exhibit viscoelastic behaviour, so the deformation properties are rate dependent, and nowhere is this more in evidence than the variation of friction coefficient with speed (or temperature). Clearly, therefore, μ is not a materials constant. A typical variation of μ with bearing pressure and speed is shown in Fig. 5.10 for plain acetal. A decrease of μ with bearing pressure increase will apply to all polymers, from the above analysis, and μ will generally increase with increase of sliding speed, although there are exceptions.[7] Frictional values are quoted for other materials in Tables 5.6–5.8. It is impossible to be precise about the coefficient of friction, but the following observations can be made:

1. μ is no indication of wear rate.
2. The exceptionally low reported μ-value for PTFE of

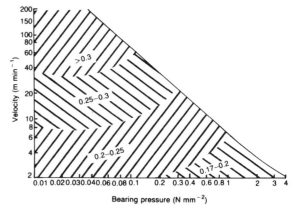

Fig. 5.10 Variation of friction coefficient with bearing pressure and speed. Acetal polymer versus hardened and ground steel. Courtesy of Du Pont de Nemours Int. SA

Table 5.7 *Performance summary of some dry bearing materials*

Material	Max. V (m s^{-1})	Max. P N m^{-2} × 10	PV @ 0.5 m s^{-1} MN m^{-2} m s^{-1}		Specific wear rate K_0 (light duty) 10^{-7} mm^3 N^{-1} m^{-1}	μ	K (W m^{-1} °C)	h max.
			Normal	Limiting				
Unfilled polymers								
Acetal	2	10	0.11	0.12	14	0.1/0.4	0.23	
Nylon	2.2	10	0.07	0.07	30	0.1/0.4	0.24	
Polyethylene (UHMW)	1.5	10		0.06	4.4		0.4	
PTFE	3			0.06	600	0.1	0.25	
High-temperature polymers								
Unfilled polyimide	8	50			19/76	0.1/0.3	0.33/0.38	
+15% graphite	8	50	0.1	10.5	6.5	0.1/0.3	0.68/1.0	
+50% graphite	8	50		10.5	4/5	0.1/0.3	1.5	
+15% graphite, 10% TFE	8	50		3.5	5	0.1/0.3	0.64/0.9	
+15% MoS$_2$	8	50			20	0.1/0.3	0.35/0.42	
+20% PTFE/MoS$_2$	8	50			1.8	0.1/0.3	0.22	
Unfilled PEEK		50			190	0.1/0.3	0.3	
Filled PEEK		50			6	0.1/0.3	0.3	
PES + carbon fibre		50			6	0.1/0.3	0.3	
Filled PTFEs								
+5% glass fibre		10				0.38		
+15% glass fibre		10		0.39	0.09	0.41		
+25% glass fibre		10		0.46	0.12	0.45		
+5% glass, 5% MoS$_2$		10				0.41		
+15% glass, 5% MoS$_2$		10				0.47		
+12.5% glass, 12.5% MoS$_2$		10		0.62	0.09	0.5		
+40% bronze		10				0.62		
+40% bronze, 5% MoS$_2$		10				0.62		
+55% bronze, 5% MoS$_2$		10		0.44	0.13	0.73		
+20% bronze, 20% graphite	4	10				1.0		
+40% graphite	4	10						
+15% graphite	4	10		0.6	0.12	0.48		
+ Mica		10		0.07		0.76		
Thin-layer materials								
PTFE, Pb impregnated; porous bronze on steel backing	12.5	130	1.75	3.5	0.3	0.05/0.2	40	0.05
Filled PTFE/Sn bronze mesh	6.5	120		0.35	0.5	0.1/0.2	7.6	
Glass-reinforced PTFE in phosphor bronze mesh	6.5	120			0.8	0.1/0.2	0.32	0.13
Woven PTFE/glass fibre + thermoset backing on metal	2	200	0.7		0.4	0.05	0.24	0.2
Woven PTFE/cotton fibre + thermoset backing on metal	2	200	0.7		0.6	0.03/0.2	0.24	0.5
PTFE flock (chopped fibre filled thermoset with polyimide cloth backing	2	200			2	0.03/0.2		0.3

Note: See Table 5.6 for filled acetals and nylons.

0.02–0.05 applies only to very low speeds and high loads. The vast majority of PTFE bearings operate at μ-values of 0.1–0.2.

3. As the polymer filler content increases, the μ-value departs away form the base polymer value to become more characteristic of the filler.

4. The μ-value will also depend on the counterface and the rubbing geometry.

Table 5.8 *Speed related friction of some polymers*

	$\mu_{max.}$	Sliding speed at max. μ (m s^{-1})
Polyethylene	0.7	3
Polystyrene	0.5	1
PMMA	0.6	2
Nylon 6.6	0.7	11
Acetal	0.5	11
Polypropylene	>0.8	>30
PTFE	0.2	>30

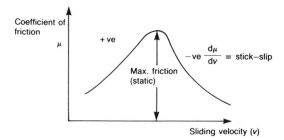

Fig. 5.11 Friction–velocity effect possible during sliding

5. Crystalline thermoplastics are fairly independent of the surrounding atmosphere (i.e. air, vacuum, inert gas, etc.) while amorphous thermoplastics and thermosets are, in frictional terms, more susceptible to the environment.
6. Crystalline thermoplastics show most frictional variation with speed while amorphous thermoplastics and thermosets show least.

One of the more noticeable features of some polymers in certain dry sliding situations is a tendency to audible squeaking. There is no evidence that this is destructive to the sliding situation, but nevertheless it has a high nuisance value. It is attributable to the variation of frictional coefficient at the sliding interface. This results in stick–slip, especially if the system is susceptible to vibration or if one surface has a certain degree of elastic freedom. It often arises at low speeds where the strength of mechanical junctions (adhesion theory of friction) are often greater than those occurring at high sliding speeds, i.e. where $\mu_s > \mu_k$. The motion between sliding surfaces depends on the friction-velocity characteristics of the surfaces involved as well as the mechanical properties of the system. If the friction coefficient is independent of velocity, or increases with velocity increase, the system is stable and the resulting motion is always smooth. If the coefficient of friction decreases with increased velocity, the motion is intermittent and the system may well respond to potential variations in frictional force. Even if the moving parts are extremely rigid, the surface irregularities may be capable of elastic deformation, sufficient to set up vibrations or stick–slip. Because polymers exhibit viscoelastic behaviour there is often a frictional peak in the μ-velocity relationship, a critical velocity condition being attained when the overall system damping changes from positive for smooth sliding to negative for oscillations (Fig. 5.11). The maximum value of the frictional coefficeint is often associated with the static coefficient, μ_s. However, this is not necessarily always the case, neither does a maximum μ-value always imply viscoelastic effects, e.g. a decreasing μ_k from temperature effects.

There are two basic types of stick–slip which can occur with dry bearings:

1. *Harmonic oscillations, velocity controlled.* Observed mainly at high sliding velocities; relative velocity between two sliding components never reaches zero; requires a friction–velocity curve with negative slope.
2. *Regular stick–slip, time controlled.* The classical form of stick–slip; occurs when $\mu_s > \mu_k$; assumes μ_s is a function of time of contact; increasing velocity reduces static contact time and stick–slip then has a smaller amplitude.

Generally the harder the material the lower the speed at which a maximum occurs. In Table 5.8 are some values for sliding dry against mild steel. Above the max. friction value $d\mu/dv$ is negative. It is by no means an isolated tribological phenomenon, for example stick–slip is the mechanism causing squeaking of door hinges, brake squeal and clutch judder. The resulting variations in frequency are due to the different materials, masses and stiffnesses involved. Unstable surface motion can be overcome in many situations to ensure that stable sliding conditions are maintained. The most obvious method of easily removing the squeaking is to lubricate. A thin film of a low shear strength solid is mostly sufficient to prevent stick–slip between a polymer and metal surface. Many polymer ingredients produce this effect, e.g. silicones to give the polymer self-lubricating properties, but frequently a trace of oil, grease or water is sufficient. In difficult cases it is possible that lubrication could reduce the overall friction coefficient but also accentuate the negative friction characteristics. In this case boundary lubricant additives would have to be used.

5.2.7 Lubrication

To some this may defeat the object of using a dry bearing as, after all, many design applications are undertaken to be lubricant free. However, having said that, the effect of limited lubrication can be to increase the service life of a bearing quite dramatically, to several times normal. Compared with conventional bearing metals, the effect is very significant (Table 5.9). Integral or structurally dispersed lubricant additives have already been considered in section 5.2.5. In this case we refer to externally applied oil or grease, the only requirement being that the lubricant should wet the polymer surface (Table 4.2). There are three degrees of lubrication:

1. *Initial.* A few drops of oil or a small amount of grease should be applied before the bearing is installed. This assists initial bedding-in, it prevents heat build-up and allows the surfaces to conform, especially with a soft metal counterface. This does not facilitate higher loads but does prolong service life. The bearing design should be dimensioned for dry running.
2. *Periodic.* This is recommended to increase both life and the load/speed range. However, if high temperatures are expected, lubricant selection should be carried out carefully to avoid any chemical reaction.
3. *Continuous.* Such situations are rare in practice. The lubricant not only cools the bearing but can satisfy a high-speed application as long as the oil film is maintained.

Polyacetals are particularly suitable for lubrication as they are impermeable to oils and greases and consequently do not swell. They have therefore found many applications in bearings and gears.

Where the lubricant is not circulated, the increase possible in the *PV* limit is proportional to the reciprocal of the frictional change, i.e.

Table 5.9 *The wear of 0.25 mm thick plastic lined bushes compared with conventional bearing metals. All bushes steel backed with a plain bore; 33.6 MN m^{-1} m s^{-1} PV, 24.8 N mm^{-1} stop–start tests, 9 min off 1 min on with a-drop-minute oil lubrication*

Bush lining material	Wear (μm)
Acetal copolymer	5.1
Acetal homopolymer	5.6
Nylon 66	12.7
Tin-based white metal	17
20% tin–aluminium	64.4
Lead bronze (25% Pb, 5% Sn)	Seized after 5–10 h

$$PV \text{ limit (lubricated)} = PV \text{ limit (dry)} \times \frac{\mu \text{ (dry)}}{\mu \text{ (lubricated)}}$$

which may apply to some materials also with integral lubrication, but in the more appropriate case where there is thin film (boundary) lubrication (mixture theory):

$$\frac{K \text{ (boundary)}}{K \text{ (dry)}} \approx \left[\frac{\mu \text{ (boundary)}}{\mu \text{ (dry)}} \right]^{3/2}$$

and

$$\mu \text{ (boundary)} = \mu \text{ (dry)} + (1-\beta)\mu \text{ (lubricated)}$$

where K = specific wear rate and β = contact area fraction where contact is unlubricated.

Because polymer elastic moduli are low, fluid film conditions can occur, and hence lead to reductions in wear, under conditions more severe than those which can be used with metals that would lie well within the boundary lubrication regime. With plastics under boundary conditions, however, lubricants may either increase or decrease wear.[9]

A novel lubrication concept is grease-filled, spiral-groove plastic bearings.[10] These provide the possibility of full fluid lubrication in small bearings, rotating in one direction only and moulded as part of the rotating shaft. The sealing action of the grooves prevents lubricant leakage and allows long periods of fluid film lubrication and as such are an attractive alternative to porous metal bearings (Ch. 6). The L_{10} life of spiral bearings is more than an order of magnitude better than porous bronze bearings under the same conditions (Fig. 5.12) spiral bearings showing a very low initial failure rate.

5.2.8 Counterface material

The wear rates of polymers and composites are greatly influenced by the properties of the counterface material against which they slide. The following guidelines can be used for design purposes:

1. The wear rate of sleeve bearings will reduce as the hardness of the steel shaft increases. For example, increasing the hardness of a steel counterface from 220 to 530 kg mm^{-2} reduces the wear rate of acetal by a factor of 0.7.
2. The abrasivity of certain fillers will dictate a certain level of counterface hardness (Fig. 5.2 and section 5.2.1).
3. The use of a hardened shaft will reduce the amount of initial break-in wear of the polymer.
4. Aluminium should never be used as a counterface material unless anodized to >350 kg mm^{-2}.

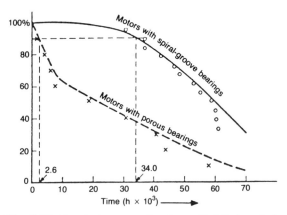

Fig. 5.12 Survival curves of motors with blind copical spiral-groove bearings and of motors with porous bearings found from life tests. Reprinted by permission of the Council of the Institution of Mechanical Engineers from Remmers G., Grease Filled Spiral Groove Bearings in Plastic, *Proc. Conf. Lubrication and Wear*, I. Mech. E., 1977

5. Other non-ferrous metals are not normally used for a counterface unless slow running and/or lightly loaded. However, initial lubrication on assembly may result in a reasonable solution with a soft shaft.
6. Slots in the counterface, i.e. an interruption to the bearing surface for debris collection, are believed to assist in promoting a mild wear regime.
7. For the same counterface roughness, a ground surface is preferable to a turned surface, but in either case a fine polishing operation is often beneficial.
8. Hardened steel or chromium-plated counterfaces are better than mild steel or stainless steel. Stainless steel often proves too soft, in many cases, although hard metals and stainless steels are preferred for forming transfer films.
9. Under certain conditions of low speed and light loads, it is possible to run plastic on plastic. With crystalline thermoplastic materials, dissimilar polymers should be used, whereas with amorphous polymers, similar composites are possible.
10. With plastics, wear is often greatest on the moving part.

5.3 Available types and materials

5.3.1 Commercial polymers

This section deals with polymeric bearing materials that are available commercially, the great majority based on one of six generic polymers, i.e. polyamide, polyacetal, fluorocarbon resin, polyimide, polyethylene and phenolic. It is the diversity of fillers and reinforcement structures available that give rise to such a wide range.

A number of manufacturers do produce solid bushings, and a guide of preferred sizes also exists,[11] but for commercial reasons there is often very little or no explanation of material. With thin-layer materials, on the other hand (section 5.3.2), the designer can establish material construction as an indicator of performance. However, the market for bearing materials is becoming more complex and the choice is widening, the trend towards the 'tailoring' of performance to meet the needs of specific end-use applications will have a future tribological impact.

(i) Fluorocarbons

The most commonly known is PTFE, but other types are also used for bearing applications. All except PTFE and PVF are melt processable, PTFE in bulk form being produced by compression moulding plus free sintering. Most filled PTFEs are consequently directional in their properties, i.e. mould direction versus cross-direction can vary by a factor of up to 2:1. Typical fillers are glass fibre (up to 30 per cent), carbon fibre (up to 30 per cent) or PTFE (up to 15 per cent). Operating temperatures can be up to 180 °C. The different types may be summarized as follows:

PTFE: polytetrafluoroethylene. Low friction coefficient decreasing with load increase. Low strength and stiffness. Expensive. Cannot be injection moulded. Process by ram extrusion or compression and sintering.
FEP: fluorinated ethylene-propylene. Melt processable. Expensive but very high impact strength. Similar properties to PTFE, i.e. dense, inert and thermally stable. Use as a coating.
ETFE: ethylenetetrafluoroethylene. High stiffness for a fluoropolymer. Can be moulded and easily processed.
ECTFE: ethylene-chlorotrifluoroethylene. Good tensile, creep and wear resistance. Melt processable but very expensive. Lowest moisture permeability of any polymer.
PFA: perfluoralkoxyethylene. Good toughness. Retains strength well at elevated temperatures.
PVDF: polyvinyledenedifluoride. Melt processable, high tensile strength and good wear resistance. Cannot be glass fibre reinforced.

(ii) Polyamides

Nylon 6 and 6/6 are the popular choices for engineering applications. The major difference between the types is the degree of water absorption. Although this results in some volume swelling the major drawback is the decrease in physical properties which occurs as a result of water absorption. Typical fillers are glass fibre (up to 30 per cent). Maximum operating temperatures would be 70–80 °C irrespective of filler content. The various nylon types may be summarized as follows:

Nylon 11. Expensive. Low impact strength and low water absorption. Used for gears and cams.
Nylon 12. Expensive, tough with low water absorption. Used for precision engineering components.
Nylon 6/6. Strongest and stiffest polyamide but low melt viscosity and shrinkage make it difficult to process. Used for bearings, gears and cams.
Nylon 6. Easy to process but high rate of water absorption. Less crystalline than 6/6 which reduces mould shrinkage. Used for bearings, gears and cams.
Nylon 6/10. Retains toughness at low temperatures. Used for precision items.
Nylon 6/12. Used where abrasion resistance is required.

(iii) Polyacetals

These are highly crystalline resins with good all-round properties, widely used in bearing applications. Typical filler is glass fibre. There are two basic forms, both having similar properties at room temperature:

Copolymer (POM). Crystalline, creep resistant, wear resistant and low moisture absorption. Can be immersed in water. Not flame retardant. Used for gears, bearings, etc. Long-term stability.
Homopolymer. Generally harder and more rigid with slightly higher tensile and flexural strength. Highest fatigue resistance of any unfilled commercial polymer.

(iv) Polyethylene

Polyethylenes are straightforward polymers available in several physical grades, increasing average molecular weight improving wear resistance, impact strength, etc. but at the expense of processability. Consequently UHMWPE offers an excellent combination of wear resistance and toughness. Maximum operating temperature, however, is 55 °C.

(v) High-temperature polymers

Considered as a group, PEEK, PPS, PES and polyimide materials have the advantage of high operating temperatures, but this does not mean that they will guarantee performance where an application has been underdesigned. Like all other polymers, their use in a given bearing application has to be designed. Normally these materials would not be considered for general bearing applications. However, detailed understanding of these materials is still at the development stage. Available fillers are glass fibre, carbon fibre, solid lubricants or PTFE. Operating temperatures may be up to 250 °C. The different types may be summarized as follows:

PEEK: polyetherketone. Good thermal stability and fatigue resistance. Expensive and difficult to process.
PPS: polyphenylenesulphide. Good temperature resistance and dimensional stability. Low friction and good wear resistance but notch sensitive. Mould processable but expensive.
Polyimide: thermoset polymer, no melting-point as such. Wide temperature range and high thermal fatigue resistance. High friction at normal ambients. Expensive.
PES: polyethersulphone. Long-term thermal ageing resistance. High residual stresses in mouldings. Expensive.

5.3.2 Thin-layer materials

Most are composite materials in the form of thin layers, 0.3–0.4 mm thick, bonded to a metallic substrate. The thickness of the bearing layer is deliberately kept small to maximize substrate load support, increase thermal conductivity and decrease the effect of thermal expansion. Such bearings can operate at high loads with low radial clearances. A metal backing allows location of the bearing by interface fit. All the layer materials are based on PTFE as a main ingredient, the various types of thin bearings differing in the form of the incorporated PTFE and in the way it is supported and effectively secured to the metal backing.

There are four basic forms of construction as given under the headings below.

(i) Interwoven PTFE and other fibres

In fibre form (spun extruded filament) PTFE is appreciably stronger and stiffer and less susceptible to creep under load, than the bulk solid; PTFE drawn into fibres induces a high degree of orientation, providing an increase in wear resistance. The contacting surface weave is predominantly PTFE, up to 80 per cent, with the rear surface some other type of fibre or yarn to facilitate adhesive bonding, e.g. glass, cotton or polyimide. The matrix resin material is usually the substrate adhesive as well (Fig. 5.13). With a metal backing to the bearing, sliding speeds up to 0.25 m s^{-1} and bearing temperatures up to 200 °C are possible.

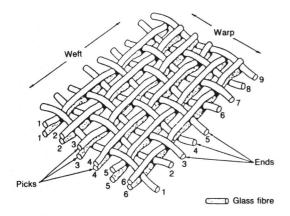

Fig. 5.13 Dry bearing liner of PTFE fibre interwoven with glass fibre. © Crown Copyright. Reproduced by permission of the Controller of the HMSO

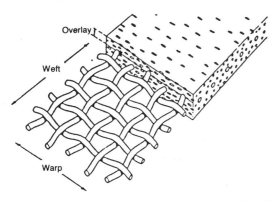

Fig. 5.14 Dry bearing liner of plain weave fabric with PTFE dispersed in resin. © Crown Copyright. Reproduced by permission of the Controller of the HMSO

(ii) Dispersed PTFE

As an alternative to a PTFE weave, PTFE can be dispersed as an irregular-shaped particulate throughout a fibre-reinforced resin matrix (Fig. 5.14). Since it is difficult to bond PTFE (section 4.2.3) a thermoset resin is used to lock it in by shape. Clearly there is great scope for modification in (i) and (ii) and a number of possibilities exist. Load capacities for (i) and (ii) for static or low sliding speeds are very high, $\sim 400 \text{ N mm}^{-2}$ maximum, but thermal considerations will limit speeds at high loads to less than 0.01 m s^{-1}.

(iii) Reinforced PTFE

This group of thin-layer bearings is basically a sintered PTFE layer with carbon fibre or metal mesh support, with or without steel backing. Where there is a steel backing, phenolic or polyimide resins are generally present.

(iv) Impregnated porous bronze

By far the most successful thin-layer dry bearing material is porous bronze sintered onto a steel backing and filled with a lead −PTFE mixture (Fig. 5.15). It has been known for some time that there is a synergistic effect between copper, lead and PTFE, improving significantly the tribological performance of PTFE. Without the lead addition, e.g. for food processing machinery, PV performance is reduced (Fig. 5.16). However, the maximum load capacity at 120 N mm^{-2} is less than the fabric composites because of the intrinsic weakness of the bronze particle/particle sinter bond in shear. Variations of this type of constriction include porous bronze impregnated with acetal, PPS, PVDF/PTFE and PEEK, all of these,

Fig. 5.15 Examples of polymer impregnated thin layer bearings with and without lubrication pockets. Courtesy of Glacier Vandervell Ltd

DRY BEARINGS

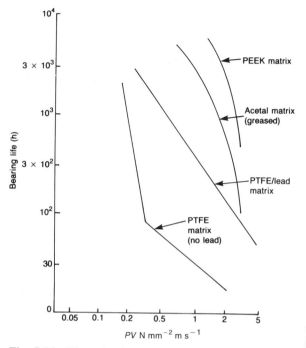

Fig. 5.16 Wear of polymer impregnated porous bronze thin layer bearings.

Fig. 5.17 Some examples of commercially available split bearings. Courtesy of Thomson Industries, Port Washington, NY

however, being intended for use under conditions of marginal lubrication rather than dry sliding. For this purpose, such bearings are provided with lubricant pockets, e.g. Fig. 5.15. Initial grease lubrication on assembly provides greatly increased life over oil lubrication. Mineral greases are preferred for normal running conditions (Fig. 5.16), but >80°C synthetic oil-based greases have to be used.

5.3.3 Split bearing liners

These are split bearings manufactured from tape or moulded to form. Various styles are available (Fig. 5.17). Referred to as bearing liners or split-curled bearings, all have a narrow axial slot, known as the 'compensation gap' not only to simplify installation but also to accommodate any bearing dimensional variations during use. These bearings are used mostly for slow speed and oscillatory applications, claiming some advantage over solid plastics and metals for the latter. The material thickness range is 0.4–1.5 mm, a typical rule of thumb being 0.4 mm for shafts up to 13 mm + 0.25 mm for each additional 13 mm of diameter.

Because of the slot, it is claimed that lower running clearances are possible, hence improved load distribution and a longer life from a more precise close-fit design application. Dimensional variations as a result of moisture absorption and frictional heating are accommodated by the gap, allowing the liner to expand and contract circumferentially around the inner periphery without altering the bore. Liner thickness is deliberately kept low to both allow rapid heat dissipation and also improve the maximum load-carrying ability without creep. In some applications the liner is also free to rotate, thereby avoiding wear concentrations. In such instances the compensation gap is a scarf cut.

1. *Canned liners.* Moulded types are typically either nylon 6.6 or acetal, e.g. Fig. 5.18. A radial lubrication slot is also possible with this type, the connecting compensation gap then also helping distribute lubricant along the bearing. The most economical form of canned liner, however, is to cut from tape. In this case the material is filled PTFE

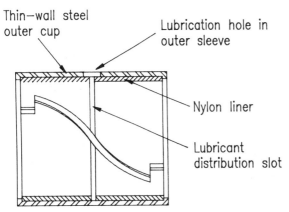

Fig. 5.18 Moulded can bearing liner. Courtesy of Thomson Industries, Port Washington, NY

for which the load-carrying capacity can be above $0.8\,\mathrm{MN\,m^{-2}\,m\,s^{-1}}$ with some applications[12] and a maximum pressure of $17.2\,\mathrm{MN\,m^{-2}}$. Chevron-type end cuts tend to retain the liner more positively if the bearing motion is reciprocating. A bronze or stainless mesh backed version is also available.[7]

2. *Plain liners*. As with canned bearings, these can either be moulded components (Fig. 5.17) or cut from PTFE-based tape. To provide proper retention, the undercut on the shaft or housing should be one-half to two-fifths of the liner thickness. Some retaining methods are shown in Fig. 5.19. One particular type of bearing foil has a glass-filled PTFE matrix supported by an $AlMg_3$ intermetallic compound making possible static loads up to $400\,\mathrm{MN\,m^{-2}}$.[13]

3. *Snap-fit*. These are thin lined bearings designed for installation in metal sideplates without the use of tools, e.g. Figs 5.17 and 5.19. A double flange provides positive location with the larger designed to take thrust.

5.3.4 Integrally lubricated polymers

Externally applied lubricants to polymers and polymer composites has been considered in sections 5.3.2 and 5.2.7. Encapsulated lubrication was considered as an additive in section 5.2.5. Commercial varieties are essentially based on silicone oil, and results would indicate that small concentrations can be as effective as high loadings of PTFE (Table 5.6). The addition of small quantities of silicone produces slight reductions in tensile strength but little effect on modulus and flex properties.

Silicone fluids are generally incompatible with common thermoplastics, hence attractive as a lubricant for acetals and nylons.[14] The most direct approach in obtaining these dispersions is mechanical blending or compounding prior to compression moulding.

Since it is possible, and necessary in some cases, to produce certain polymers by the sintering route, e.g. nylon, it is also possible to adjust the process to leave residual porosity and subsequently impregnate with lubricant. The result is the porous plastic bearing, an alternative to porous metal.

5.4 Application factors

5.4.1 Aspect ratio

In theory, the performance of a dry bearing should be independent of the length/diameter ratio. However, with large ratios, there is an increased probability of local surface heating due to out-of-roundness, misalignments and stress concentrations. Bearing lengths are normally recommended to be ≤ 1.5 diameter of the bearing, ideally an optimum ratio is unity. Above this, there is the possibility of increased friction and low life. Below unity, housing location problems may be introduced, but there is then less sensitivity to shaft deflection and misalignment and the advantage of easier escape for wear debris. Where a long bearing is indicated, it is advisable to consider the use of two separated bearings.

With thrust bearings it is normal practice to keep the

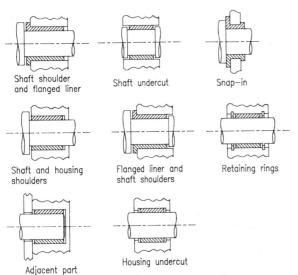

Fig. 5.19 Retaining methods for split liner bearings

outside/inside diameter ratio <2 to avoid trapping wear debris and to help maintain surface flatness.

5.4.2 Wall thickness

Often this is dictated by the overall assembly design, but if a choice is possible, a good starting-point is about one-tenth of shaft diameter. The following factors influence thickness:

1. Increase wall thickness of a bearing subject to impact or shock loads.
2. Decrease bearing thickness if normal pressure is causing bearing deformation/flow under load.
3. Reduced thickness facilitates the dissipation of frictional heat.
4. Reduced thickness minimizes the effect on running clearance arising from dimensional changes resulting from thermal effects, water absorption, etc.
5. Reduced wall thickness enables a higher PV.

The following equation results in a wall thickness that should be suitable for most applications:

$$s = 0.06d^* + 0.25$$

where s = wall thickness (mm) and d^* = shaft diameter (mm).

However, best bearing performance can be expected when the wall thickness is at a minimum since thinness should facilitate better heat transfer to a metal housing and surroundings. This should be commensurate with thickness being capable of holding an interference fit with the housing. Peripheral support is provided to thin bearings in the form of cans or by forming the bearing surface on to a steel backing.

In view of the low softening point of plastics, it is important to design maximum heat removal into the bearing from the start. This can be done simply by using thin bearings.

5.4.3 Clearance

The most important consideration when designing a bearing is the bearing/shaft running clearance. If excessive wear occurs under allowable operating conditions, it is nearly always due to insufficient clearance. This is probably responsible for more dry bearing failures than anything else. There is, however, an optimum clearance in practice, difficult to define precisely but nevertheless above and below which there is a wear rate increase. Experience has shown that the best results for solid polymers are obtained with a running clearance of $5\,\mu\text{m mm}^{-1}$ of bearing diameter with a minimum of $125\,\mu\text{m}$.

This clearance is much greater than that required for lubricated metallic types of bearing and is based both on dimensional instability during use, water absorption, frictional heating, etc. plus the requirement to form a transfer film on the shaft. These phenomena collectively result in 'close-in'. The recommended clearances for dry bearing materials at their operating temperature is given in Fig. 5.20, the actual clearance depending both on the type of bearing material and the bearing construction. Smaller clearances are possible at light loads and low speeds where running temperatures are low, also with metal backed thin layer bearings where there is less expansion. Thin walled split liners in a metal housing are intermediate (Fig. 5.20).

If the dry bearing bushings are of a standard manufacture, then internal bearing clearances need not be calculated since these will already have been determined by the manufacturer for the relevant thermoplastic. If the bushing is not standard, then the clearance should be determined. Factors affecting the bearing diameter are as follows:

1. Dimensional tolerance due to machining limits; shaft, housing and bearing.

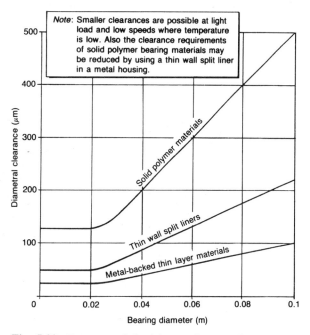

Fig. 5.20 Recommended minimum clearance for journal bearings (at operating temperature)

2. Effect of bore closure when press fitting into the housing. The degree of interference fit is material dependent.
3. Frictional heating of both shaft and bearing material under running conditions.
4. Swell effects due to adsorption of atmospheric moisture.

These effects are represented schematically by Fig. 5.21. Thus:

Bush OD: D_{min} = max. housing dia. + min. interference $D_{max} = D_{min}$ + manufacturing tolerance

Bush ID: d_{min} = material swell allowance + bore closure + thermal allowance + running clearance + max. shaft dia. (temp. corrected) − machining tolerance

Some explanation of the individual dimensional factors is as given under the headings below.

(i) Swell allowance

It is only necessary to allow for moisture adsorption if the bearing is required for permanent operation at high humidity. Most polymers are hygroscopic to a degree, but the effect is only appreciable with polyamide materials. The close-down resulting from moisture adsorption is proportional to wall thickness (Fig. 5.22). The increase/decrease of bearing bore is presented as a diametral clearance factor for the most common types of polyamide bearing materials. It is standard practice to quote mechanical properties for equilibrium conditions at 50 per cent RH (Fig. 5.22). It is possible to overcome water-absorption problems in nylon materials with a suitable heat treatment in an oil bath.[15] Fluorocarbons (e.g. PTFE) do not absorb moisture, but they do undergo a transition in the range 15/30 °C accompanied by a 1 per cent volume change.

(ii) Bore closure

Alternatively interference or press-fit allowance. Press fitting is the normal method of bearing retention in a housing and it is assumed that the actual interference fit on assembly is reflected 100 per cent into the bearing ID. Defined as the amount of oversize, the bush press-fit allowance is defined as a proportion of the nominal housing bore diameter as follows:

PTFE and UHMW polyethylene: 10–20 $\mu m\, mm^{-1}$
Nylons acetals, PEEK, etc. : 5–10 $\mu m\, mm^{-1}$
Reinforced thermosets : 2–3 $\mu m\, mm^{-1}$

Where significant fluctuating ambient/operating temperature variations are expected, stress relaxation (cold flow) of the interference fit may take place so that, in practice, supplementary mechanical retention may be required.

The maximum allowable interference fit, I, can be determined as follows, such that

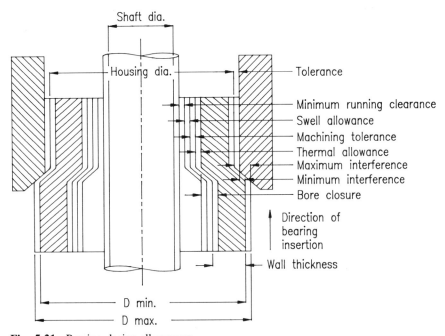

Fig. 5.21 Bearing design allowances

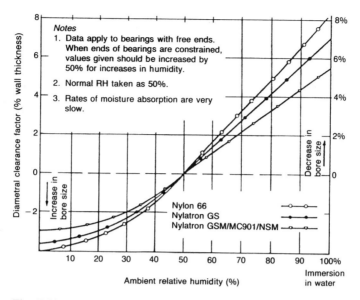

Fig. 5.22 Moisture allowance for Polypenco nylon bearings at varying relative humidities. Courtesy of Polypenco Ltd; Welwyn Garden City

$$I = \frac{DP}{E}\left[\frac{1 + (b/D)^2}{1 - (b/D)^2} - \nu\right]$$

where ν = Poisson's ratio, P = internal material pressure, D = housing diameter, E = Young's modulus of bearing material and b = required bearing inside diameter after fitting

and

$$P = \frac{\sigma}{S}\left[\frac{1 - (b/D)^2}{1 + (b/D)^2}\right]$$

where σ/S = design stress for long application use, σ = tensile strength at yield point and S = safety factor (typically 2 or 3).

(iii) Thermal allowance

This represents a reduction in bearing clearance, thermal close-up of the bore, as a result of the operational surface temperature of the bearing (on average a ΔT of 30/65 °C).

Assuming a rigid housing then

$$d_{T_2} = d_{T_1}(1 - \alpha_1 \Delta T)$$

where α_1 = coefficient of linear expansion, d_{T_2} = bore diameter at temperature T_2, d_{T_1} = bore diameter at temperature T_1 and ΔT = temperature rise. Or, more accurately, calculating the bore diameter using volume changes in bearing size, and assuming as before a rigid housing bore then

$$v_{T_2} = v_{T_1}(1 - 3\alpha_1 \Delta T)$$

and

$$v_T = \pi/4\,(D^2 - d^2)L$$

where L = length of bearing and v_T = volume at temperature T.

(iv) Running clearance

The working clearance is defined as the actual clearance between the shaft and the bearing bush at the operational temperature (Fig. 5.20).

One interesting commercial design to overcome the problems of clearance is shown in Fig. 5.23. Here a thin-walled plastic bearing is designed with a compensation cone to eliminate the effect of expansion of the bearing material by permitting the outer sleeve to expand inward and the inner sleeve to expand outward, thereby minimizing lateral dimensional changes. The thin bearing wall minimizes radial effects. The compensation cone also provides a self-aligning feature. However, there is no immediate heat sink available to the bearing.

5.5 The design procedure

Although a number of important design features have already been mentioned in this chapter, there are probably

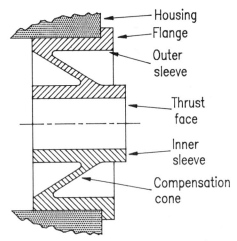

Fig. 5.23 Compensation cone principle to take up bearing tolerances

three important design goals for successful dry bearing application:

1. Design for minimum heat generation at the rubbing interface and design for maximum heat dissipation throughout the rest of the assembly.
2. Design to support the load without unacceptable deformation, distortion or cold flow of material at the operating temperature and pressure.
3. Design for an acceptably low rate of wear in order to provide maximum life and reliability.

However, dry bearing design is not an exact science, it is difficult and sometimes impossible to consider realistically all the factors involved, hence design analyses will only identify likely materials, actual performance behaviour has to be observed. The performance analysis and material selection procedure to be described here is intended to enable material selection for specified applications. Unsuitable materials can be identified at each stage of the procedure and by a process of elimination, those likely to provide the required performance ultimately emerge. Final selection can then be made on the basis of availability, cost and practical testwork.

The design process is outlined by the flow chart in Fig. 5.24. This is broken down into subactivities as follows:

(i) Specifying the application

Certain basic information regarding the intended application is initially required to start the selection procedure. This is as follows:

1. Required life t. The specified life determines the acceptable maximum wear rate of the bearing (section 5.2.2). Time in this context only refers to time the bearing is in motion.
2. Wear depth h. The maximum displacement of the moving component from its ideal location relative to the fixed component (section 5.2.2). Includes bedding-in wear.
3. Maximum allowable friction torque, M, where critical.
4. The bearing load, W. For initial selection purposes the bearing load is assumed to be steady. In journal bearings, it may be steady or rotate with the shaft.
5. Type of motion. Determine the frequency of rotation or oscillation, n, and in the case of oscillatory motion, the angle between the limits of rotation θ. The rubbing velocity V, can be calculated from Table 5.10. See also section 5.2.2.
6. Shaft diameter, $d*$.
7. Wall thickness, s (section 5.4.2).
8. Aspect ratio $L/d*$ (section 5.4.1) and Fig. 5.3, hence calculate projected bearing area A (section 5.2.2).
9. Bearing pressure P (Table 5.10 and section 5.2.2).

(ii) Initial material selection

Having specified the application and evaluated the main bearing parameters, a check should be made to ascertain whether the PV requirements are realistic (Fig. 5.25), in terms of dry rubbing bearings. Performance values for specific materials are given in Tables 5.6, 5.7 and Fig. 5.26.

(iii) Estimation of temperature rise

Once materials have been identified, friction coefficients are also available (Tables 5.6, 5.7). However, in order to estimate the temperature rise due to frictional heating, ΔT (section 5.2.4) it is also necessary to estimate

Table 5.10 *Summary of bearing parameters*

Parameter	Type of bearing	
	Journal	Thrust
A (apparent contact area)	dL	$\frac{\pi}{4}(D^2 - d^2)$
P (bearing pressure)	$\frac{W}{A}$	$\frac{W}{A}$
V (continuous rotation)	$\pi d n$	$\pi D_m n$
V (oscillatory motion)	$\pi d n \left(\dfrac{\theta}{180}\right)$	$\pi D_m n \left(\dfrac{\theta}{180}\right)$

Notes: n = frequency of rotation or oscillation; D_m = mean diameter of thrust face; θ = angle of oscillation; V = velocity.

DRY BEARINGS

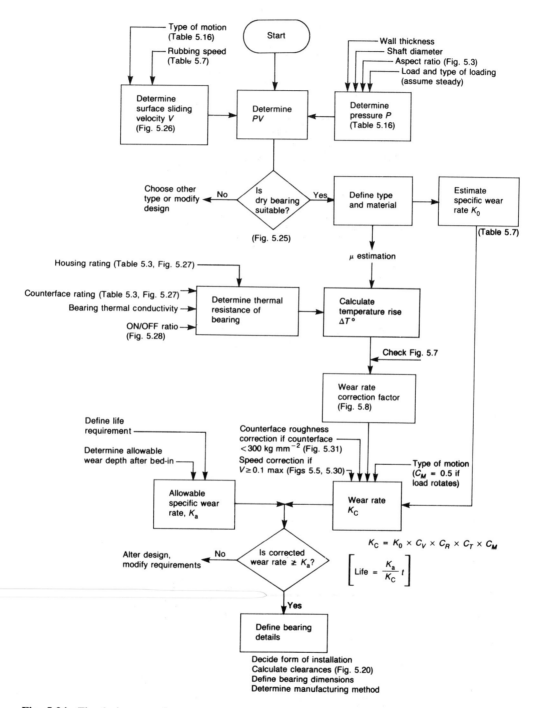

Fig. 5.24 The design procedure

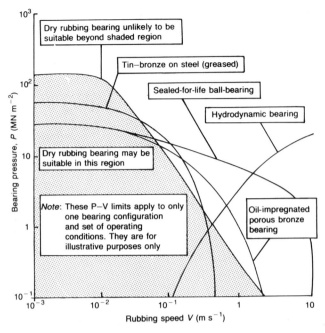

Fig. 5.25 Typical P–V limits for different bearing types, ref. 7. Courtesy of ESDU International

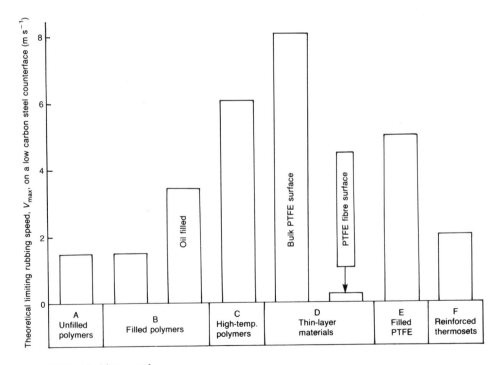

Fig. 5.26 Limiting speeds

the thermal resistance coefficient C of the bearing assembly,[7,8,16] as follows:

$$C = \frac{R_0 \, 0.1 d^*}{\alpha}$$

where R_o = uncorrected thermal resistance coefficient, α = intermittent operation correction factor and d^* = shaft diameter (mm).

The above will apply to a normal range of bearing sizes encountered in electromechanical products. Some examples of thermal resistance coefficients are given in Table 5.3. The value of R_0 is based upon a consideration of the heat-dissipation capability of housing, counterface and the bearing itself. Typical values of R_0 are outlined in Fig. 5.27, where K = thermal conductivity of bearing material in watts per metre °C. As a general rule:

1. If the housing is a poor thermal conductor, the bearing conductivity will have little effect on R_0.
2. If the shaft is a good thermal conductor, the housing itself has little effect over a wide range of bearing thermal conductivities.
3. If both shaft and housing are good thermal conductors, the value of the bearing conductivity has little effect on R_0.
4. The bearing conductivity has the most noticeable effect when the shaft is a poor thermal conductor.

If the bearing operation is an intermittent one then some allowance can be made for reduced bearing temperature by consideration of both the bearing ON/OFF ratio and also the on time duration. Values of α can be determined from Fig. 5.28.

Having determined C, then for known materials, values of ΔT can be calculated (section 5.2.4). From Table 5.4 and Fig. 5.7 the temperature rise can be checked against material type to ensure acceptable wear behaviour, remembering that the actual bearing temperature T is expressed as follows:

$$T = T_a + \Delta T \quad \text{(section 5.2.4)}$$

Unsuitable materials should be eliminated.

(iv) Estimation of wear rate

From the data collected in (i), estimate the allowed specific wear rate, K_a, as follows:

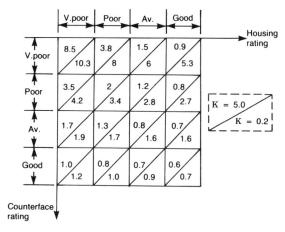

Fig. 5.27 Values of R_0

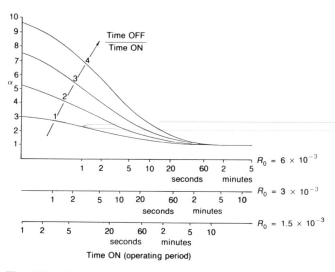

Fig. 5.28 Estimation of intermittent operation correction factor

$$K_a = \frac{h}{PVt} \text{ (section 5.2.2)}$$

Having identified suitable materials in (ii), modified by eliminating unsuitable materials in (iii), the uncorrected specific wear rate, K_0, can be determined from Table 5.7. Table 5.6 and Fig. 5.29 give the specific wear rates at PV limits for a number of materials; K_0 is pressure and velocity independent over a limited range and is thus a basic material property, e.g. Fig. 5.6. It relates to a wear rate (h/t) of 25 μm per 100 hours (unidirectional loaded journal bearing) and is used as a starting-point to determine the actual specific wear rate of the material. The corrected specific wear rate (K_c) is defined as

$$K_c = K_0 \times C_V \times C_R \times C_T \times C_M$$

where C_V = speed correction factor, C_R = counterface roughness factor, C_T = bearing surface temperature factor and C_M = loading factor.

Here K_c will normally exceed K_0 to a greater or lesser extent, depending on how the conditions of actual application differ from the light duty conditions on which the K_0 values are based.

1. Sliding speed correction factor C_V. Polymer materials with inadequate limiting rubbing speed will already have been eliminated in (ii). Determine V from Table 5.7 and Fig. 5.24. However, if the rubbing speed is above a critical ratio, wear rates will increase as in Fig. 5.30.
2. Counterface roughness factor C_R. The roughness effect for an unhardened steel shaft is shown in Fig. 5.31 which indicates the appropriate factor for various materials and counterface roughness.
3. Surface temperature factor C_T. The bearing surface temperature has been estimated in (ii). As a ratio of the maximum bearing working temperature (Fig. 5.7) the temperature wear rate factor may be estimated from Fig. 5.8. The temperature effect on wear rate is independent of the way in which the temperature rise is achieved.[17]
4. Loading factor C_M. In continuous rotation $C_M = 1$ if the load is fixed relative to the bearing material, but = 0.5 if the load rotates. For oscillating motion, C can be taken as 1 although lower values apply for some materials.

(v) Estimation of bearing life

If the value of K_c estimated in (iv) for the various materials considered in the analysis is greater than K_a, then these materials should be eliminated. If no material options remain at this stage then the design should be altered or requirements modified. The life of selected bearing materials may be estimated from

$$t_p = \frac{K_a t}{K_c}$$

where t_p = predicted life and t = required life.

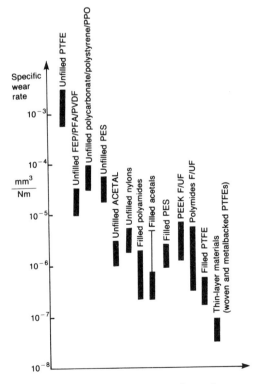

Fig. 5.29 Specific wear rate ranges for various polymers at their PV limits

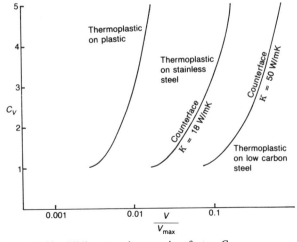

Fig. 5.30 Sliding speed correction factor C_V

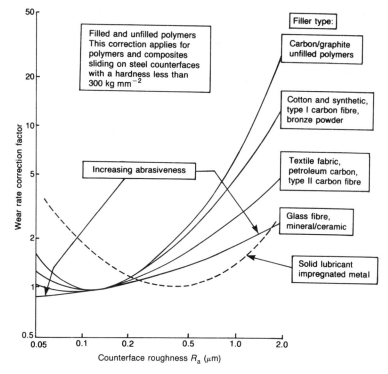

Fig. 5.31 Wear rate correction factor for counterface roughness, ref. 7. Courtesy of ESDU International

(vi) Determine bearing details

These have mainly been referred to in section 5.4 as follows:

1. *Form of installation.* If an interference fit, housing interference requirements are given in section 5.4.3. With other fixing techniques, sufficient support must be provided for the polymer and/or the bearing thickness reduced in order to minimize cold flow.
2. *Calculation of clearance.* This is dealt with in section 5.4.3. Because polymers have fairly high expansion coefficients and also swell with humidity, plastic bearings generally require generous clearances compared to other bearing types.
3. *Bearing dimensions.* Partly defined by the application (i) as well as (1) and (2) above.
4. *Manufacturing method.* In common with most plastics, bearing materials can exhibit anisotropy depending on the direction of force used in fabrication, i.e. whether rolled, extruded, etc. Injection moulded or compression moulded (sintered) components will generally demonstrate uniform tribological properties if the mould design is correct.

Other manufacturing operations that effect performance will be stress relief applied, how often applied plus the degree of crystallinity relative to thermal history.

References

1. George J Chemistry of polymers, *Plastics in Engineering* 39. Findlay Publications 1987
2. Lancaster J K Dry bearings. A survey of materials and factors affecting their performance. *Tribology* Dec 1973: 219
3. Hollander A E et al An application of topographical analysis to the wear of polymers. *Wear* 25: 155 (1973)
4. Hurricks P L Some metallurgical factors controlling the adhesive & abrasive wear resistance of steels — a review. *Wear* 26: 285 (1973)
5. Lancaster J K Polymer based bearing materials. *Tribology* Dec 1972: 249
6. Crease, A B The wear performance of rubbing bearings — improved data for design. *Proceedings of the 3rd Leeds–Lyon Symposium on Wear of Non-metallic Materials.* Institution of Mechanical Engineers 1976

7. ESDU Item 87007, *Design & materials selection for dry rubbing bearings*. Nov. 1987
8. ESDU Items 78026, 78027, 78028, 78029 and 79002, *Equilibrium temperatures in self-contained bearing assemblies, Pts 1 to 5*. Oct 1978 to Nov 1979
9. Lancaster J K Lubricated wear of polymers. *Proceedings of the 11th Leeds–Lyon Symposium on Tribology*. Butterworths 1985
10. Remmers G Grease filled spiral groove bearings in plastic. *Proceedings of the Conference on Lubrication & Wear*. Institution of Mechanical Engineers 1977
11. BS 4480: Pt. 2: 1974
12. Dineen J Designing with split-curled bearings of Teflon. *J. Teflon* **102** —D
13. *Bearing foil P30*. Ina Bearing Co. Product Manual 1986
14. Duncan Aetal Silicone lubricated thermoplastics. *Modern Plastics*, Nov. 1972: 114
15. Benedyk J C Plain bearings, an international survey. *SPE Journal* **26**(4): 78 (1970)
16. Wilson W H et al Heat flow in the dissipation from self-contained bearings. *Proceedings of the 6th Leeds–Lyon Symposium on Tribology*. Institution of Mechanical Engineers 1980
17. Anderson J C The wear and friction of commercial polymers and composites. In *Friction and wear of polymer composites*, Elsevier 1986

CHAPTER 6
POROUS METAL BEARINGS

6.1 Introduction

Oil-retaining porous metals have been in use now some 60 years and during this period the product has changed very little. It is sometimes described as a 'self-contained oil' type of journal bearing. The porous metal bearing generally finds application in one of two situations:

1. As a cheap, readily available bushing material which is fitted and lubricated by conventional means, and where the self-lubricating properties arising from the porosity are not required;
2. As a convenient, ready-made solution for a low-duty shaft/bearing situation where relubrication is undesirable, impossible, expensive or likely to be forgotten and where the only requirement is low cost and reliable operation throughout the life of the machine or component.

The latter situation is addressed by this chapter and is of most interest to designers of small machines, office equipment, rotating components, etc. The advantages and disadvantages of this type of bearing are given in Table 6.1. Design flexibility is allowed for by a wide selection of bearing sizes and shapes while variation in bearing composition and the sintering process allows for a wide choice of material properties. However, while the use of the porous metal bearing is widely accepted, it does have a major drawback in that the design technology associated with this product has not advanced over the past 30 years in any significant way. It is still not possible to quantify a porous metal bearing application from first principles and calculate the life expectancy. Consequently, designing with a porous metal bearing is not an exact science, and although some quantification is possible where there are continuously running hydrodynamic conditions, even here

Table 6.1 *Advantages and disadvantages of porous metal bearings*

Advantages	Disadvantages
Very wide choice of standard sizes and shapes	Machining will impair surface porosity, thus shaft, housing diameters and fitting tolerances must be specified to match standard sizes
Specials possible with structural requirements met	
No need to supply oil lines, the oil reservoir is a metallic sponge	Unsuitable for static and impact loads >48 N mm^{-2} of projected bearing area (porous bronze) unless porosity is reduced significantly but the working life may then be reduced to an unsatisfactory level
Very simple application of a lubricated journal bearing	
Oil which would normally be lost at the ends of a journal bearing is reabsorbed by the porous metal, filtered and used again	
	For higher loads use a low-porosity bronze or porous iron
	After fitting, bearing bore should not be reamed, broached or machined as any form of metal cutting action will reduce surface porosity — tapered drifts should be used if bore diameter has to be increased
	Temperature limited; roughly speaking a rise of 10°C halves the life of a given oil (in terms of oxidation)
	Cannot accurately determine life unless temperature of the bearing is measured. The boundary aspects of PMBs do not seem as well covered as hydrodynamic
	Not tolerant to shaft misalignment (i.e. rigid journal bearings). Drastic reductions in PV are possible

accurate life assessment will involve some experimentation and bearing temperature measurement. A powder metal bearing running under boundary conditions with no additional lubricant reservoir would be very difficult to assess for reliability. Thus although this section will outline the design parameters that affect bearing life, and how these interact, the approach with most powder metal bearing applications tends to be part empirical.

The 'life' of a sintered metal bearing is largely controlled by the lubricant, i.e. once the quality and the quantity of the lubricant drop to certain levels, the bearing, as with any other journal type lubricated bearings, starts to wear out. Most of the factors that influence lubricant performance are sensitive to temperature, so if the running temperature of the bearing is known, some estimate of lubricant deterioration, hence bearing life, is possible. The design application of a porous metal bearing should be for conditions that help promote optimum lubrication, i.e. hydrodynamic as opposed to boundary lubrication (Fig. 6.1). Under boundary lubricated conditions, bearing 'life' is a function of the boundary additives and the 'lubricity' of the oil together with the wear rate of the sintered bearing itself.

Porous metal bearings are rarely suitable for high load and speed applications; they are more appropriate for low-duty applications of low load and moderate speed. Seizure will always occur with a sintered bearing if the speed is high enough.[1] The types of porous metal bearings on the market are as follows, typically as in Table 6.2.

1. Plain cylindrical bushes for force fitting into cylindrical housings. The wall thickness should be >1/16 the length.
2. Flanged cylindrical bearings with thrust load capability. The wall thickness should be greater than one-tenth the length and the flange width greater than twice the wall thickness.
3. Self-aligning bearing where the (OD) is formed as part of a sphere for fitting into a self-aligning housing.
4. Thrust washers typically of thickness >3 mm.

6.2 Manufacture and properties

Porous metal bearings are produced by a partial compaction of fine metal powders to the desired shape (green compact) following by high-temperature sintering in a reducing atmosphere at about $0.8 \times$ the absolute melting-point of the metal. The sintered compact is then re-pressed (sized) to restore dimensional accuracy, provide a good surface finish and, by work hardening, increase the elastic limit in the surface layers. The result is a strong and rigid metal 'sponge', consisting of an interconnecting network of channels and reservoirs. The amount of porosity depends mainly on the degree of powder compaction, not changing significantly in fact during sintering and resizing operations, while the size of the pores depends on the particle size of the powder.

Table 6.2 *Standard specifications for porous metal bearings*

	Composition	Notes	Dry density ($g\,cm^{-3}$)	Porosity (%)	Tensile strength ($MN\,m^{-2}$)
a	89/10/1 Cu/Sn/graphite	General-purpose bronze (normally supplied unless otherwise specified). Reasonably tolerant to unhardened shafts	5.8 6.0 6.3	31 29 25	55 62 76
b	91/8/1 Cu/Sn/graphite	Lower tin bronze. Reduced cost. Softer	—	7–25	—
c	85/10/5 Cu/Sn/graphite	High-graphite bronze. Low loads. Increased tolerance towards oil starvation	—	18–27	—
d	86/10/3/1/ Cu/Sn/Pb/graphite	Leaded bronze. Softer. Increased tolerance towards misalignment.	—	18	—
e	>99% soft iron	Cheaper than bronze. Unsuitable for corrosive conditions. Hardened shafts preferred	5.2 5.5 5.8 6.0	34 30 26 23	76 85 97 105
f	97.5/2.5 Fe/graphite	Graphite improves marginal lubrication and increases tolerance towards unhardened shafts	—	27	—
g	2–5% Cu/Fe	Increasing copper content increases strength and cost. Series forms the most popular range of porous iron bearings	5.2 5.5 5.8 6.0	34 30 26 23	85 124 155 174
h	10–25% Cu/Fe	Hardened shafts preferred	5.8 6.1	27 23	170 214

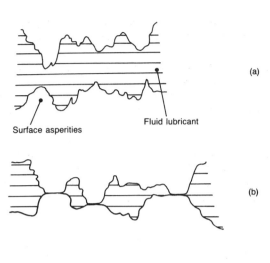

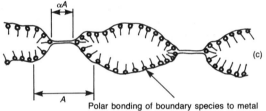

Load supported over area A. Metallic junction formed through boundary film over area αA.

Local friction = force to shear metallic junction + force to shear lubricant film

$\Delta F = (\alpha A\, Sm) + (1-\alpha) A S_L$

where Sm and S_L are shear strengths of metal junction and lubricant film

Fig. 6.1 Schematic representation of lubrication regimes. (a) fluid film lubrication (hydrodynamic); (b) mixed lubrication; (c) boundary lubrication

Porosity is typically in the range 20–40 per cent while pore size is usually 1–30 μm diameter. The actual bearing porosity for any given application is usually a compromise between material strength and oil content. Oil impregnation is by vacuum and heat until >90 per cent of the pores are lubricant filled. The usual impregnating oil used by most manufacturers is an SAE 20 or 30 refined, oxidation-inhibited oil.

Being a powder metallurgy production process, the composition range is fairly limitless with up to 5 per cent by weight of non-metallics possible, i.e. graphite, sulphides, but in practice it is usual to adhere to standard materials. Table 6.2(a–h) is a summary of standard specifications for porous metal bearings. The graphited tin bronze (a) is the general-purpose alloy giving a good balance of properties between strength, wear resistance

and conformability. Softer versions contain lead (d) or have a reduced tin content (b). Graphite increases the safety factor if oil replenishment is forgotten, (c), but at the expense of mechanical strength. Where rusting is not likely to be a problem, it may be possible to use the cheaper and stronger iron-based types (e), but these have a low safety factor against oil starvation especially when mild steel shafts are used. Both added graphite, (f) and copper (g,h) improve the situation.

As well as porous metal bearings, fully compacted bearings can be made by powder metallurgy techniques to incorporate a proportion of solid lubricant such as graphite. A common matrix is bronze or iron with 4–10 per cent of graphite. Under completely dry conditions at room temperature, the performance of this material group can be disappointing; there are better alternatives available. At bearing temperatures ~250 °C these materials apparently come into their own,[2] but this is an area outside the scope of this book.

6.3 Variables affecting performance

6.3.1 *Permeability, porosity and density*

The quantity of porosity present determines both the oil-retaining capacity and the mechanical strength of the bearing (Fig. 6.2), and depending on the powder composition, the bearing density. The higher the porosity, the lower the density and the lower the thermal conductivity. Porosity values generally quoted in product literature are average values, since the porosity at the ends of the bearing is less than in the centre. Most bearing properties are a function of porosity so the effect of a porosity gradient on performance may need to be considered in marginal applications. Axial variation of porosity is mainly due to die/powder wall friction resulting in a non-uniform pressure distribution during powder compaction. Any circumferential variation is due to non-uniform replenishment of the die.

Permeability is a measure of resistance to fluid flow in a porous medium, hence provides an indication of the ease with which oil can flow through the bearing wall. It varies with both porosity and pore size, hence low permeability can be achieved by reducing pore size without reducing porosity. Pore size in turn is a function of both the size distribution in the original powder and the sintering operation itself. The variation of permeability with density is given in Fig. 6.3. Sintering in the presence of a liquid phase gives a substantial increase in permeability.[3] With porous bronze for example, the particle

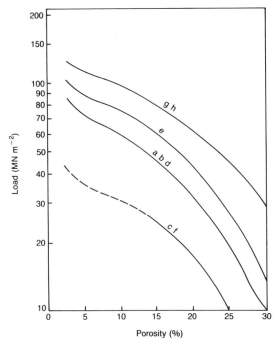

Fig. 6.2 A general guide to the maximum static load capacity (including impact loads) of a wide range of compositions and porosities. Curves are based on a length:diameter ratio of about 1 and assume a rigid housing. For interpretation of materials see Table 6.2

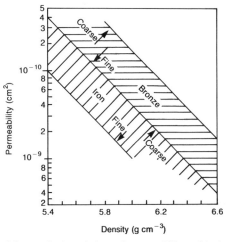

Fig. 6.3 Typical variation of permeability with density for porous metal bearings

For a given porosity, the addition of graphite to a porous bronze decreases permeability (Fig. 6.4). For a given premix, the sizing operation also results in a permeability decrease. Both a wide range of pore sizes and the level of interconnected porosity present are shown to be significant factors in the rate of oil flow allowed during operation of the porous bearing. However, the shape of the pores is far from geometrically uniform, hence as a measure of the space between sintered particles, porosimetry data all generally presented in size bandwidths to give a measure of the size of the channels. Table 6.3 is a summary of porosity data for a given premix, with and without graphite addition and before and after use. Also included in Table 6.3 is a measure of the interconnected porosity. Clearly a number of 'blind' pores exist. The amount of non-available porosity generally increases as porosity and permeability decrease. The ratio interconnected porosity:total porosity should be as close to unity as possible since values < 1 correspond to an increasing level of redundant porosity, not effective

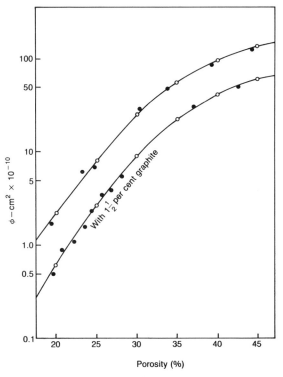

Fig. 6.4 Relationship between permeability ϕ and porosity for porous bronze, with and without graphite. Reprinted by permission of the Council of the Institute of Mechanical Engineers from *Proc. Conf. Lubrication and Wear*, 1957

size of the tin is adjusted to be finer than, but nevertheless compatible with, the size distribution of the copper to provide homogeneity yet avoid segregation during compaction.

Table 6.3 *Porosity summary of typical commercial porous bronze bearings*

	Total porosity (vol. %)	Interconnected porosity (vol. %)	Ratio Int. porosity / Tot. porosity	% of interconnected porosity in a given size range					
				$d>9$	$9>d>5$	$5>d>3$	$3>d>1$	$1>d>0.1$	$0.1>d>0.04$
				(d in μm)					
Basic bearing	27.2	24.4	0.89	1.8	58.2	10	20	6.7	3.3
1% graphite	27.7	21.7	0.78	1.9	27.7	18.5	33.3	16.7	1.9
Basic bearing after use	—	—	—	16.7	75	0	0	8.3	0
Bearing + 1% graphite after use	—	—	—	26.3	68.4	0	0	5.3	0

Note: Nominal composition Cu–10%Sn

in lubrication and reducing thermal conductivity. As sinter density increases, the amount of redundant porosity increases.

6.3.2 Lubrication

Assuming that the mechanical installation is correct and all other factors are equal, the porous metal bearing will perform well as long as the operating conditions are such to create fluid film lubrication. As far as the design engineer is concerned, this is basically the main problem with porous bearings, being able to design confidently for reliability. This section provides a view of the widely held lubrication model used in porous bearing design applications, but it is partly qualitative unless supporting experimentation is conducted in parallel with the design process.

There are three types of lubrication situation that can arise with porous metal bearings:

1. *Continuous lubrication*. This may be where the bearing runs continuously in an oil bath, pumped supply or has a wick/reservoir replenishment system. Here conditions may be more akin to a normal journal bearing situation. Simple wick systems are referred to in section 6.4.4.
2. *Impregnated oil only lubrication*. By far the major usage, the fit for life application where the retained oil provides the only lubricant. To achieve fluid film conditions with porous bearings, higher values of speed and oil viscosity are necessary. This is to be expected if a portion of the film leaks away through bearing pores, hence explains why porous bearings are widely used at high speed and low loads only.
3. *Stop–start lubrication*. At start-up, or indeed at any point where operating conditions fail to maintain a fluid film, metal-to-metal contact will then occur and hence there is the likelihood of bearing/shaft wear.

The extent of this will depend on the boundary characteristics of the oil and the wear coefficients of shaft and bearing. However, because boundary effects in normal applications are transient, it is very difficult to build in boundary lubrication effects into life estimates.

To obtain reasonable life, a fluid film situation should be maintained. However, to put this into perspective a small bearing will in fact only contain a few drops of oil from which several thousand hours of life are expected! A porous metal bearing OD 25 mm, inside diameter (ID) 19 × 12 mm length will hold 16 mm³ of oil. Hence the performance requirement of the oil itself is therefore very demanding, the life of the bearing being closely associated with the life of the oil and its running temperature.

(i) Hydrodynamic lubrication

The build-up of hydrodynamic pressure (Fig. 6.5)

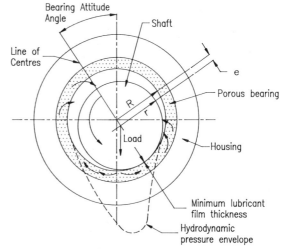

Fig. 6.5 Circulation of oil within the pores of a porous metal bearing

supports the load in a similar manner to a non-porous bearing,[4] but because of porosity, oil can be forced from the oil film in the loaded area into the bearing wall. This then encourages circulation of the oil through the pores in the wall of the bearing away from the load line towards the unloaded area. In normal operation the oil forced into the bearing wall in the pressure area is balanced by a similar quantity of oil drawn out of the pores in the unloaded area. In effect the leakage of a proportion of the oil film pressure through the bearing pores will result in a loss of load-carrying capacity of the film. The pattern of oil circulation within the bearing is depicted in Fig. 6.5. Oil escaping from the ends of the bearing is either reabsorbed by the porosity or drawn into the running clearance of the loaded zone. However, a quantity of oil is steadily lost from the bearing (Figs 6.6, 6.7).

The rate of oil loss is approximately proportional to log running time. As a safeguard therefore, it is recommended that porous bearings should be reimpregnated when the oil content has fallen to ~70 per cent. At 50 per cent residual content the bearing is likely to seize.

In terms of continuous lubrication conditions, porous metal bearing operation can be quantified by two dimensionless parameters, the Sommerfeld reciprocal $1/\Delta$ and the permeability factor ψ. These are defined as follows:

$$\psi = \frac{\phi t \times 1.9 \times 10^{-7}}{c^3}$$

where t = wall thickness of the porous metal (m), c =

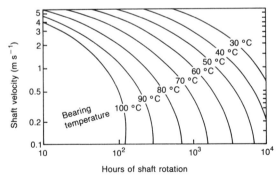

Fig. 6.6 Bearing temperature as a function of shaft velocity and time of running. The need to replenish the oil in the pores arises because of oil loss (which increases with shaft velocity) and oil deterioration (which increases with running temperature). Curves relate to the preferred standard bearings in Table 6.2. From Morgan V T, porous metal bearings. In *Tribology Handbook*, Butterworth–Heinemann, 1973

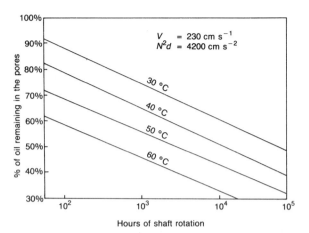

Fig. 6.7 Typical oil loss with time and temperature. Courtesy of Lincoln Publications

diametral clearance (m) and ϕ = permeability of the porous metal (units of m²).

More commonly ψ is also known as the oil leakage factor. The higher the value of ψ the faster the oil pressure will leak into the porosity. For a non-porous bearing $\psi = 1$. The unit of permeability of a porous substance is defined by 1 m³ of a fluid of dynamic viscosity 1 Pa s that will flow through a section 1 m thick and of section 1 m² in 1 s at a pressure differential of 1 Pa. In the SI system the unit of permeability has no special name, though in the cgs system it is referred to as a darcy (D). In which case

$$1 \text{ D} = 0.987 \times 10^{-12} \text{ m}^2$$

The load factor, or Sommerfeld reciprocal, $1/\Delta$ is defined as follows:[5]

$$\frac{1}{\Delta} = \frac{\eta N d^2 \times 10^{-3}}{\pi P c^2}$$

where η = viscosity (cP) depending on the grade of oil and the bearing operating temperature, N = shaft speed in revs/sec, d = shaft diameter, P = projected bearing pressure (N m⁻²), c = diametral clearance and c/d is the clearance ratio.

The above equations can be evaluated to determine whether it is theoretically possible to obtain hydrodynamic conditions for a given design. This may be inferred from Fig. 6.8 at various bearing length/diameter ratios based on an arbitrary eccentricity ratio of 0.8 (Fig. 6.5). The eccentricity ratio ϵ is defined as the distance between centres due to load line displacement, i.e.

$$\epsilon = \frac{\text{distance between centres}}{\text{radial bearing clearance}}$$

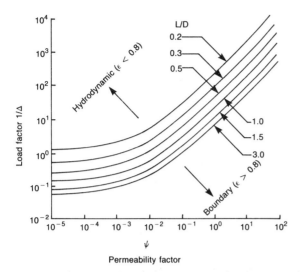

Fig. 6.8 Practical design limitation for ensuring hydrodynamic lubrication based on $e = 0.8$

$$= \frac{e}{c} = \frac{e}{R - r}$$

where e = distance between centres, R = bearing bore radius and r = shaft radius.

Here $1/\Delta$ increases with the permeability factor, the effect being more marked for a shorter bearing or a higher eccentricity ratio, i.e. the load-carrying capacity of a porous metal bearing decreases with increasing ψ. For a given c/d ratio, an increase in ψ implies a higher matrix permeability, i.e. a greater oil-carrying capacity for the bearing and hence a greater ability for sustained self-lubrication, but a compromise is necessary to determine a practical ψ for design purposes.

The reason for the choice of $\epsilon = 0.8$ is that when conditions are such that the attitude of the shaft in the bearing is 80 per cent of half the running clearance, metal-to-metal contact is likely to occur. To use a theoretical ratio of $\epsilon = 1$ (surface contact) as a base would not allow a sufficient safety margin to take care of shaft and bearing roughness. With oil-retaining porous metal bearings, the limit of hydrodynamic lubrication is reached when the rate of leakage back into the porosity is greater than the rate of film creation. At this critical $1/\Delta$ (Fig. 6.8), hydrodynamic lubrication is lost and friction may rise as boundary lubrication conditions are established. However, Fig. 6.8 is an ideally lubricated case, whereas in practice, restricted only to the oil in the pores, the critical $1/\Delta$ may well be an order of magnitude larger, so pushing the bearing out of the hydrodynamic regime.

Another factor identified above as affecting the creation of a lubricant film is viscosity, a temperature-sensitive property (Fig. 6.9). The bearing running temperature will depend both on the operating conditions as well as on the heat dissipation characteristics of the housing assembly. Since the oil is continuously reused, it does not provide any cooling function and will slowly oxidize if maintained at an elevated bearing temperature for long times.

(ii) Boundary lubrication

Strictly speaking, the boundary condition referred to in Fig. 6.8 represents both mixed film as well as boundary film lubrication. These are illustrated schematically in Fig. 6.1. Mixed film conditions are those where the oil film cannot completely separate surface asperities such that surface wear is manifest. True boundary conditions occur when no fluid film is developed. An illustration of the different lubrication regimes is possible if the coefficient of friction μ is plotted against $\eta N/P$ (Fig. 6.10). The point of minimum friction occurs at the point of minimum hydrodynamic oil film thickness.

Failure of an oil film to prevent metal/metal interactions

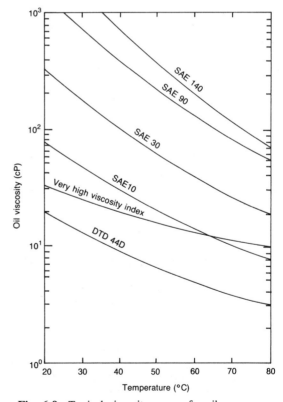

Fig. 6.9 Typical viscosity curves for oils

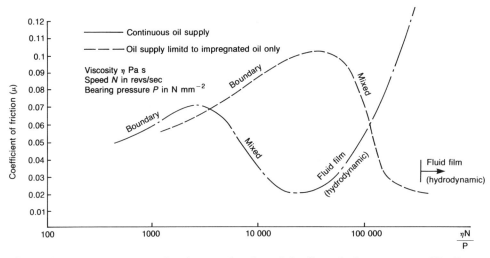

Fig. 6.10 Friction of a porous bearing as a function of the dimensionless parameter N/p. From Morgan, ref. 5

between bearing and shaft arises when the bearing load is too high or the relative speed too low. Such conditions also occur during start-up or shut-down. The degree of metallic contact and the proportion of the operating time over which it occurs, determines the severity of damage to both bearing and lubricant. However, even simple lubricants, such as the plain mineral oils used for impregnation, have some ability to reduce friction and wear under such conditions.[6] This is attributed to base liquid surface activity or some minor impurity or component. This property is generally called 'oiliness' or 'lubricity', but both terms are ill-defined and unsatisfactory. Many different types of additive are used in lubricating oils to improve boundary lubrication, e.g. vegetable or animal oils and fats and their derivatives.[6] However, to improve the oxidation resistance of the lubricant, polar boundary species tend to be refined out, thus the oil used in practice tends to be a compromise.

Porous bearings frequently become dry in certain regions as the shaft rotates, and in consequence air is drawn in, emulsifying the oil film. As a result, the load capacity of the oil film and the PV factor are reduced when compared to that achievable with perfect lubrication. The oil will also become charged with wear particles, but during recirculation it is believed that some debris filtering will occur, the porous metal acting as its own filter. However, despite the obvious difficulties with conditions of marginal lubrication, porous metal bearings do find reliable application if the operating conditions are low duty.

6.3.3 Running temperature

The factors that determine the useful life of a porous metal bearing are many, but they can be simplified in terms of one important parameter, the running temperature. Temperature controls the operating conditions of the bearing through the following:

1. Oil viscosity (Figs 6.9, 6.10);
2. The rate of oil deterioration with time;
3. The change in running clearance through differential thermal expansion.

The running temperature is proportional to bearing load, bearing speed, the coefficient of friction and the heat dissipation characteristics of the assembly, i.e.

Rate of heat generation/ unit area of bearing surface $\propto PV\mu$

An estimate of bearing temperature rise can be made by equating power input to heat loss, e.g.

$$PV\mu = \frac{\Delta T}{C}$$

where C = thermal resistance coefficient (see section 5.2.4), ΔT = temperature rise above ambient and V = shaft surface velocity.

Determination of the running temperature follows an iterative loop (Fig. 6.11). An outline procedure for resolving this is given in ref. 5. If the oil viscosity is known, it is possible to estimate μ; Fig. 6.10 represents

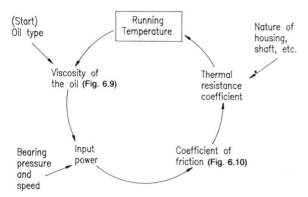

Fig. 6.11 Determination of bearing temperature — the iterative loop

the frictional behaviour of porous bronze bearings in terms of the non-dimensional parameter $\eta N/P$ (SI units), in all lubrication regimes. Figure 6.10 is often referred to as a Stribeck curve[7] and can be used to provide an indication of the lubrication state.

When the porous metal bearing operates with only the original impregnated oil, the angular extent of the oil film is reduced (less oil), the peak oil pressure increases, $1/\Delta$ increases and the lubrication curve moves to the right (Fig. 6.10), i.e. it is more difficult to promote hydrodynamic conditions. For hydrodynamic conditions and normal ambients, the design is probably quite safe up to a temperature rise of 40 °C. The oil film under boundary conditions is thin (Fig. 6.1), with the coefficient of friction approximately 0.1. The mixed lubrication situation is dealt with in section 5.2.7.

The thermal resistance coefficient can be estimated from section 5.2.4, but because friction levels are lower for porous metal bearings than for most plastics, most C values will be in the range $0.3-2 \, °C \, m \, s \, N^{-1}$.

6.3.4 Bearing clearance

Modest changes to the running clearance produce only small changes to the operating conditions of a porous metal bearing because changes to $1/\Delta$ and ψ tend to cancel out, but there is a second-order effect. With an increase in clearance there is a small friction and running temperature reduction, but the bearing tends to be less hydrodynamic.

The correct choice of running clearance is a compromise between achieving a low running temperature (large clearance) and the need for precise radial location of the shaft. Figure 6.12 gives some assistance on the choice of running clearance ratio c/d (= $R - r/R$)

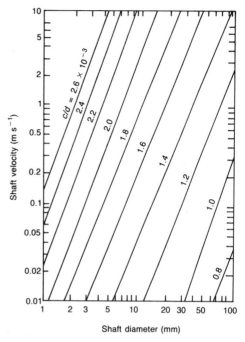

Fig. 6.12 Guidance to the choice of clearance ratio

according to shaft diameter and velocity. It is often quoted as a rule of thumb that the design clearance should be 0.1 per cent. However, for small bearings this is inadequate, resulting in increased bearing friction and running temperature (Fig. 6.13)[8] in which case a clearance ratio of 0.15−0.2 per cent is probably optimum. Too large a clearance ratio will result in uneven wear, excess shaft deflection and early failure.

Where the porous metal bearing is an interference fit in a housing, the running clearance will change with variation of temperature. In a loosely held, keyed or spherical bearing, the clearance change is straightforward to calculate. In the case of a force-fitted bearing, the change of clearance, Δc, with temperature can be deduced as follows:[5]

$$\Delta c = \theta[2R(B-S) - FD(B-H)]$$

where Δc = change in clearance, θ = temperature change (°C), B = coefficient of linear expansion of bearing, S = coefficient of linear expansion of shaft, H = coefficient of linear expansion of housing, D = housing bore diameter (fitting bearing OD), F = stiffness factor = m/n, m = reduction in bore diameter of bearing when fitted and n = bearing/housing interference.

The relative stiffness factor of housing and bearing, F, is less than unity in the majority of applications, typically

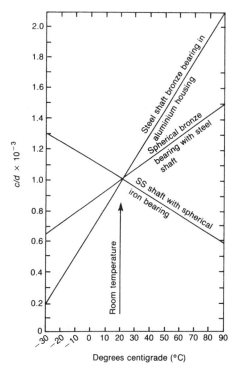

Fig. 6.13 Change in clearance with temperature

0.6–0.9. The higher values are for thin housing walls and vice versa. Some typical case illustrations are given in Fig. 6.13.

6.3.5 The running-in process

As with any running-in process, some degree of surface contact will occur, producing a slight amount of wear debris until the shaft and bearing settle down to a conforming topography and the coefficient of friction eventually stabilizes. Initial metal/metal contact results in surface smearing and plastic deformation, i.e. partially closing some pores in the load line, locally reducing permeability (decreasing ψ) and tending to make the bearing more hydrodynamic (Fig. 6.8). Thus there is a degree of 'automatic compensation' in the running-in process inasmuch that if a bearing is run at a $1/\Delta$ value below lift-off, the process goes on reducing ψ until a critical $1/\Delta$ is reached, in effect increasing the load-carrying capacity and PV value. On the other hand, if the process starts too far below the critical $1/\Delta$, then instead of smearing, the surfaces seize. In this context, a cautious reference may be made to deliberate under-design,[5]

relying on an initial boundary/mixed situation to produce hydrodynamic conditions. The action is of course irreversible.

The same effect will occur from factors that increase $1/\Delta$, e.g. accidental overloading, a reduction in η due to overheating, stopping and starting, all of which cause the load, or a proportion of the load, to be carried in mixed or boundary lubrication. The following points have been made regarding the bearing surface behaviour during the running-in process[9,10] (see also Table 6.3):

1. Most of the interconnected small surface pores are closed during run-in.
2. The extent of surface deformation and pore closure caused by smearing is higher the lower the bearing clearance.
3. In a porous bearing having a uniform distribution of large irregular pores, these resist closure so resulting in a lower running temperature.
4. If the initial bearing porosity is too fine there is a danger that the running-in process might result in subsequent problems during adverse conditions.
5. The oil will quickly darken due to the generation of extremely fine wear particles.
6. If temperature build-up occurs faster than permeability decreases, catastrophic wear can occur.

When there is no additional or supplementary lubrication available, the running-in process can be accelerated by using periodic stop-starts rather than by continuous running, e.g. Fig. 6.14. This shows the friction curve stabilizing after some 960 hours of continuous running with a particular bearing configuration while the same result is achieved after only 24 hours of intermittent running. Intermittent running allows the dissipation of the additional frictional heat produced from the partial boundary conditions existing during run-in. An additional technique to assist running-in is to add a few drops of lubricating oil to the shaft/bearing during assembly, e.g. Fig. 6.15.[8]

In the majority of porous bearing applications, however, we are dealing with extremely thin lubricating films and there is no question of full film lubrication in the conventional sense. The process is fragile inasmuch as there is often only a narrow tolerance to changes in operating conditions, i.e. in the event of a load increase or change of load direction, the friction increases sharply and the running-in process starts over again. If the shaft is cyclically loaded, which is quite likely in practice, shaft/bearing conformity will be difficult to achieve so that mixed lubrication may be permanent.[10] The running-in effect explains why statically loaded porous metal bearings behave so well on life tests but not in the

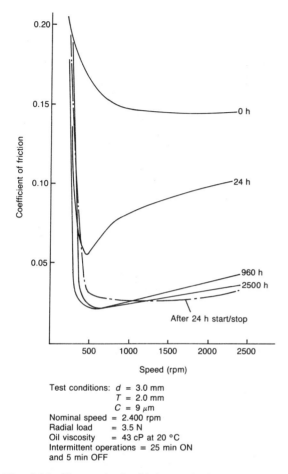

Fig. 6.14 Changes in the friction–velocity curve with time (running-in) for small porous bronze bearings, ref. 10. Courtesy of A L Braun

Test conditions:
- d = 3.0 mm
- T = 2.0 mm
- C = 9 μm
- Nominal speed = 2.400 rpm
- Radial load = 3.5 N
- Oil viscosity = 43 cP at 20 °C
- Intermittent operations = 25 min ON and 5 min OFF

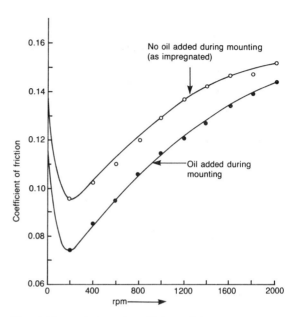

Fig. 6.15 Variation of coefficient of friction with speed for ISO VG46 oil, ref. 8. Courtesy of Professor R Raman, Indian Institute of Technology, Madras

field. Caution should clearly be exercised when using those results in designing for variable load conditions. The events in the running-in process are summarized in Fig. 6.16.

6.4 Application factors

6.4.1 Shaft roughness

With porous metal bearings it is usual to specify a shaft roughness of 0.4–0.8 μm Ra. However, larger-diameter bearings are more tolerant to greater shaft roughness. The lower the roughness, the better the bearing performance and less debris generated during running-in.

In low-duty applications, friction decreases with decrease in shaft roughness, the reduction being very significant in the 0.1–0.2 μm Ra region (Fig. 6.17). Although this is an expensive finishing area, temperature rise as a function of sliding speed is also much reduced. With increased loading, boundary conditions arise earlier with higher roughness shafts, whereas low roughness shafts help to support hydrodynamic conditions to greater loadings.[8] It has been estimated that each improvement of 0.1 μm Ra below 0.4 μm Ra produces a service life increase of 25 per cent.

6.4.2 Shaft material

Should be 0.4 per cent carbon steel as a minimum standard, i.e. low carbon or free machining steels should not be used. High-carbon or heat-treated steel is preferable for heavier duty applications. If corrosion resistance is required, the shaft should either be martensitic stainless steel or either hard electroless nickel or flash chrome plated. Austenitic stainless steel should not be used for shafting except for very light duty. As a general rule, stainless steel shafts will provide lubrication difficulties under boundary or mixed lubrication conditions, so should be avoided if at all possible.

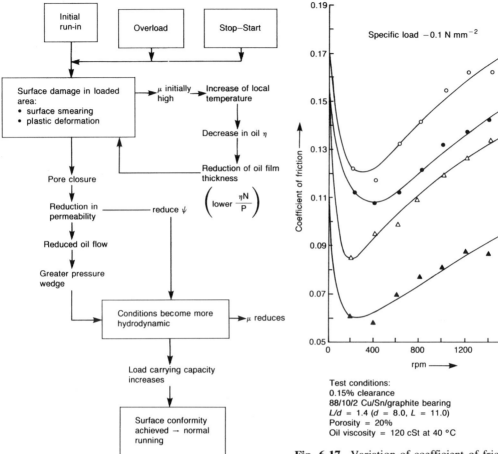

Fig. 6.16 Diagrammatic representation of the running-in process

Fig. 6.17 Variation of coefficient of friction against speed for various roughness of journals

6.4.3 Press-fit installation

The interference between the bearing OD and the housing bore should ideally be

$$0.025 + 0.0075 \sqrt{D} \text{ mm}$$

where D = OD of bearing. The interference should not fall below $0.5 \times$ or rise above $2 \times$ that given by the above equation when based on the max/min tolerance stack-ups of both housing and bearing. Too tight a bearing fit can damage both components, whereas with too low an interference, the bearing could work loose. An adequate chamfer should be provided on the housing.

Actual press-fit values should, however, be reduced both for long bearings and thin-walled bearings. A long bearing requires less of an interference fit because it has high surface area compensation. With a thin-walled bearing there is the risk of distortion if the fit is too tight. The amount of bore closure depends largely on the wall thickness and can be estimated as a percentage of the interference (Fig. 6.18). A steady squeeze action is recommended for fitting the bearing; impact loads will almost certainly cause damage. If it is subsequently necessary to increase the bore diameter, this must be done with a suitable burnishing tool.[11] This should never be attempted with any form of cutting or grinding tool since the surface pores will smear, reduce oil permeability and produce a spiral surface texture resulting in an Auger effect to transport oil out of the system.

6.4.4 Supplementary lubrication

To increase the quantity of oil available in a porous metal bearing assembly, an oil-soaked pad is often fitted in

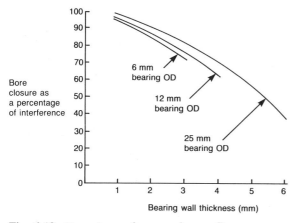

Fig. 6.18 Bore closure for normal press-fit porous bronze bearings

contact with the outside surface of the bearing. As the oil content of the porous metal is reduced, the bearing is able to recharge itself from this pad. The positioning of the pad relative to the pressure wedge extent needs to be considered in the exchange of oil between bearing and pad.[13]

6.5 The design procedure

Despite their overall simplicity, the design analysis of a porous metal bearing presents a number of obvious difficulties as previously outlined. However, having assessed the basic load-carrying capacity, the application suitability and the design life will really depend on three aspects, the lubrication regime itself, the quantity of the lubricating oil and its quality. For example:

1. *Type of lubrication*. Bearings have to work over a wide range of conditions of lubrication, hydrodynamic to boundary and 'mixed friction' regimes. While they may operate hydrodynamically when full of oil, the performance changes to 'mixed friction' as the oil content within the bearing decreases from leakage and evaporation.
2. *Quantity of oil*. To operate effectively the bearing must contain a sufficient volume of oil to cope with the load, speed and contact area and to circulate the oil within the pores at the required rate. In either 'mixed friction' or boundary lubrication regimes a bearing can go on working for a long time as long as a quantity of oil is present in the bearing and steps are taken to minimize the oil loss or to provide ancillary reservoirs.
3. *Quality of oil*. How effective the oil will be in sustaining conditions of adequate lubrication basically depends on the running temperature and the stability of the oil at that temperature, together with the amount of wear debris and air entrapment carried around the load area.

Although it is not possible to take into account all aspects of lubricant behaviour in the long term, the design analysis can explore the major factors that influence short-term lubricant behaviour and try to ensure that the bearing design is initially as hydrodynamic as possible. The process is outlined by the algorithm in Fig. 6.19, and broken down into sub-activities as follows.

(i) Specify the application

Certain basic information should be available to the design engineer at the start of the bearing selection procedure. This is as follows:

1. The life requirement. Many porous bearing applications are 'fit and forget', hence there has to be an economic life requirement as far as the bearing user is concerned. This is compared later to the calculated life as to whether a supplementary lubrication policy is required.
2. Define bearing dimensions. It is normal practice initially to consider a length/diameter ratio (l/d) of 1, although other ratios are available, the preferred range being ~0.8–1.6.[12]
3. Define the maximum load and speed.
4. It is common practice to refer to bearing capacity in terms of PV, e.g. typically $1.75 \, \text{MN m}^{-2} \, \text{ms}^{-1}$ for bronze, but this is very over-simplified and takes no account of the lubrication conditions under which the bearing operates. This parameter therefore cannot be used in a safe design sense, but a pressure–velocity diagram does give a useful guide to the heat dissipated (Fig. 6.20).
5. Choose the running clearance (Fig. 6.12) used later.

If the application is clearly non-preferred or even borderline (Fig. 6.20), readjust the design parameters or consider the additional requirements. Porous metal bearings are not suitable for use where stringent requirements of the bearing are likely, e.g. high reliability, precise shaft movement.

(ii) Material parameters

In order to progress further with the design analysis, it is necessary to quantify a few bearing material parameters:

1. Determine the type of bearing material and estimate

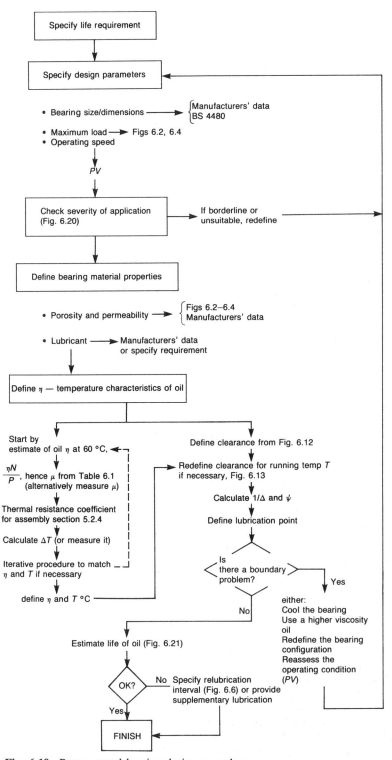

Fig. 6.19 Porous metal bearing design procedure

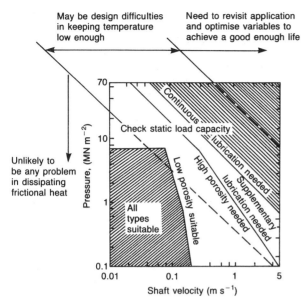

Fig. 6.20 A general guide to the severity of the duty. Based on an length:diameter ratio of approximately unity

porosity (Table 6.2). Check radial crushing strength (static load) acceptable for the application (Table 6.2 and Fig. 6.2).
2. Obtain bearing permeability from manufacturers' data sheets or alternatively use Fig. 6.4 as typical of 90/10 bronze.
3. Ascertain the type and grade of lubricant, hence viscosity/temperature characteristics, e.g. Fig. 6.9.

(iii) Estimation of running temperature

Because oil viscosity is a major contributor in determining the lubrication mechanism, either an estimate is required of the running temperature or an experimental confirmation (Fig. 6.19). See also section 6.3.3.

If the calculated temperature rise does not match the original viscosity estimate, it will be necessary to repeat steps (1)–(3) above in an iterative process matching end temperature to the same viscosity/temperature point on the oil rheology curve (Fig. 6.9).

(iv) Lubrication regime

The initial choice of bearing clearance reflected the guidelines in Fig. 6.12. This should now be reassessed in relation to the estimated running conditions and the bearing mounting configuration (section 6.3.4). For a press-fit assembly, interferences and bore closures are dealt with in section 6.4.3.

From the clearance, calculate the parameters $1/\Delta$ and ψ. By reference to Fig. 6.8, the lubrication state for a given L/D ratio can be established. This curve strictly only applies where there is full lubrication, i.e. when the bearing is new or where there is supplementary lubrication. For those bearings operating with the original oil film only, the operating point should be hydrodynamic or closely hydrodynamic. If the application is too boundary (Fig. 6.8), either of the following should be applied:

1. Cool the bearing or redefine the housing as a heat sink in order to increase η (clearance then also varies);
2. Redefine the operating conditions by reducing PV;
3. Redefine the bearing configuration;
4. Choose a higher-viscosity oil (Figs 6.9, 6.21).

(v) Bearing life

Assuming the oxidation characteristics of the oil reflect the bearing life requirement (Fig. 6.21), mechanical loss of oil from the bearing by creep or centrifugal effects, or straightforward evaporative loss, need to be estimated (Figs 6.6, 6.7). Rate of oil loss $\propto$ log running time (Figs 6.6, 6.7). The recommended replenishment interval, at 70 per cent level, is given in Figs 6.6 and 6.7 or 1000 hours — whichever is the less.

If the relubrication interval is not acceptable, consider supplementary lubrication (section 6.4.4) or move the lubrication operating point as above.

(vi) General points

There are a number of tips or useful advice that go with the application of porous bearings, some of which have already been referred to in the text. They are worth summarizing as follows:

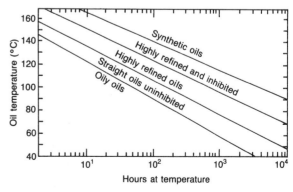

Fig. 6.21 Relative oxidation life of oils as a function of temperature, ref. 5. Courtesy of the Institute of Materials

1. Do not select oils that are not compatible with common mineral oils unless replenishment or initial fitting by the user with the wrong oil can be safeguarded.
2. Do not use grease.
3. Avoid lubricants with dispersed solid lubricants unless testwork has proven the application.
4. The design is always a compromise. Although the thickest wall and the greatest porosity provide the greatest oil volume, they do not necessarily give the longest life and optimum running conditions.
5. Misalignment of shaft and bearing is often the reason for premature failure. It reduces clearance and the *PV* factor considerably.
6. Long-term bearing performance can only be predicted from short-term tests on the basis that surface damage occurs in the initial stages and running conditions remain stable.

References

1. Eudier M et al Practical working conditions for sintered bearings. *Powder Met.* **12**(24): 417 (1969)
2. Pratt G C A review of sintered metal bearings. *Powder Met.* **12**(24): 356 (1969)
3. Morgan V T et al Mechanism of lubrication of porous metal bearings. *Proceedings of the Conference on Lubrication and Wear*. Institution of Mechanical Engineers 1957
4. Cameron A *Basic lubrication theory*. Longmans 1971
5. Morgan V T Porous metal bearings. *Powder Met.* **12**(24): 426 (1969)
6. Hurricks P L et al Interaction of lubricants and materials. *Trans. I. Mar. E.* **85**(A7): 157 (1973)
7. DeGee A W J Selection of materials for lubricated journal bearings. *Wear* **36**: 33 (1976)
8. Raman R et al Effect of roughness, clearance, oil and material composition on the frictional characteristics of sintered bearings. In *National Conference on Industrial Tribology*. India 1984
9. Cheng J A et al Structure, property and performance relations in self-lubricating bronze bearings. *Int. J. Powder Met.* **22**(3): 149 (1986)
10. Braun A L Porous bearings. *Tribology International* 1982: 235
11. Morgan V T The application of porous metal bearings. *Ind. Lub. & Tribology* **24**(3): 129 (1972)
12. BS 4480: 1974
13. Morgan V T Porous metal bearings *Tribology Handbook* section A8. Butterworths 1973

CHAPTER 7
ROLLING ELEMENT BEARINGS

7.1 Introduction

Rolling element bearings play a vital role in engineering. The prime users of these bearings are the transportation, electric motor, domestic appliance, hand-held power tool and business machine industries. These industries tend to set the trends and establish bearing standards by virtue of their volume demands. For the cost, precision and material quality of the modern bearing, it has to be seen as very good engineering value. Advantages and disadvantages of rolling element bearings are summarized in Table 7.1.

Topics will be covered in this chapter only to the extent of defining suitable bearing performance based on the design and selection process. Many bearing manufacturers' catalogues refer in detail to aspects of fitting practice, methods of lubrication and types and sizes of bearings, hence only the necessary basic aspects need to be included here. Discussion on Hertzian contact stress and bearing kinematics can be found in more theoretical texts.[1,2]

Rolling element bearings generally consist of two rings with a set of elements running in the tracks. Standard shapes of rolling element include the ball, cylindrical roller, needle roller and tapered roller. The elements maintain the shaft and supporting structure (housing) in a spaced radial relationship. The rolling elements themselves are usually held in a circumferentially spaced relationship by a separator or cage. The shape of the rolling element divides bearings into two main classes, ball- and roller-bearings. The main direction of bearing load creates a further distinction between radial and thrust bearings.

The main failure mode of rolling bearings is metal fatigue. Wear is mainly associated with the introduction of contaminants. The sealing arrangements, therefore, must be selected with regard to the hostility of the environment. Notwithstanding the fact that the responsibility for the basic design of ball- and roller-bearings rests with the bearing manufacturer, the design engineer must form a correct appreciation of the duty to be performed by the bearing and not only be concerned with correct bearing selection but also with correct conditions of installation.[1]

Also considered with this chapter are some commonly used product variants borne out of rolling element technology. These are the ball nut, the linear bushing, the one-way clutch and plastic bearings.

7.2 Bearing construction

7.2.1 Types of rolling element bearing

(i) Deep groove ball-bearings

This is the most popular, versatile and cheapest type of rolling element bearing (Fig. 7.1). This type has the ability to sustain medium radial and light axial loads, operate at low levels of frictional torque and is suitable for high speeds. The bearings can be fitted with shields or seals for lubricant retention. When grease lubricated with two shields/seals, the bearings are referred to as 'lubricated for life'. Allowed misalignment is 0.3 per cent.

These may be manufactured with or without ball filling slots. With filling slots, a larger number of balls can be accommodated in a unit, hence the radial load capacity increases but at the expense of the axial load capacity, i.e. only approximately 20 per cent of the continuous race types.

With continuous raceways, the deep groove ball-bearing supports a thrust load capacity up to approximately 70 per cent of the radial load capacity, not only due to the high degree of conformity (radii ratio

Table 7.1 *Advantages and disadvantages of rolling element bearings*

Advantages
1. Precision manufacture. Available off the shelf with precise shaft and housing dimensions, catalogued to ISO standards
2. Good alignment of rotating machine components is maintained and, with the exception of straight rollers, capable of supporting both radial and thrust loads
3. Catalogue data are fairly comprehensive; enables designer to identify max. load, max. speed, the required alignment and service life
4. A low and fairly constant torque from 0 to max. speed is coupled with high load capacity, more or less speed independent
5. Service life can be determined from simple calculations
6. Generally requires minimum lubrication. Expensive lubrication systems can be obviated by the requirement for only a small quantity of lubricant
7. Early warning of impending failure signalled by increasing noisiness at the same speed of rotation
8. Can be used for mounting a shaft in any position in space
9. Short axial length
10. Within reasonable limits, changes in load, speed and operating temperature have only a slight effect on the satisfactory performance of rolling element bearings

Disadvantages
1. May be some infantile mortality failures which will not show up in the fatigue data
2. Less capacity to withstand shock when compared to a journal bearing
3. Noisier in normal operation than a journal bearing
4. Dirt, etc., can limit life. The lubrication environment must be clean
5. Initial cost is always higher than a dry or porous metal bearing
6. Requirement for tight mounting tolerances

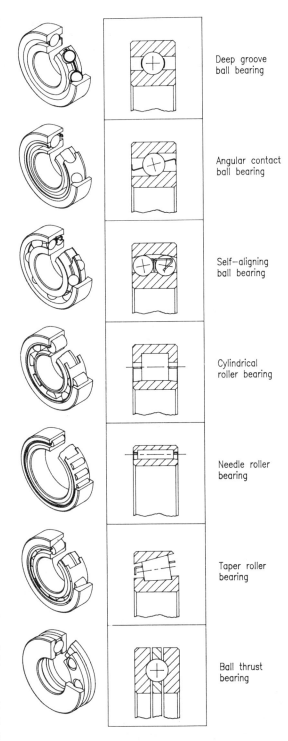

Fig. 7.1 Selection of common bearing types

track/ball) but because only a few balls carry the radial load while all balls carry the thrust load. Types with uninterrupted shoulders may also be referred to as Conran bearings.

(ii) Angular contact bearings

In single-row types, the raceways are so arranged such that forces are transmitted from one raceway to another through a pre-defined contact angle (Figs 7.1, 7.2). These are recommended for use where the magmitude of the thrust component is high enough to preclude the use of radial-type bearings. Preferred contact angles are 15°, 20° and 25°; 30° and 40° are available but non-preferred. Above 45° contact angles, bearings are classed as thrust types. For the smaller contact angles, the outer race has one heavy shoulder and the other counterbored to within a small distance of the bottom of the raceway. For the higher contact angles, both races are counterbored (Fig.

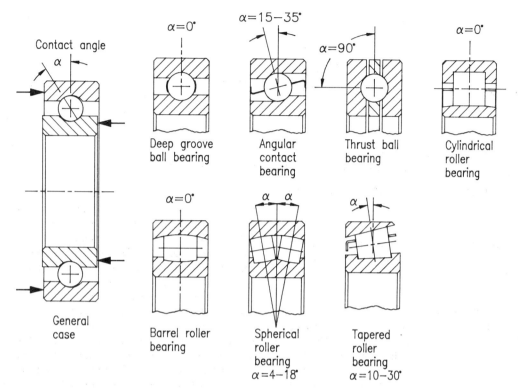

Fig. 7.2 Examples of normal contact angle

7.2). If axial loading is not constantly applied, two angular contact bearings should be installed so that they can be adjusted against each other.

(iii) Self-aligning ball-bearing
This bearing type consists of a double ball race, the outer raceway being a common spherical track (Fig. 7.1). The inner ring and balls form a common unit which can freely align about the bearing centre, up to 5 per cent allowed misalignment. The bearing is thus incapable of supporting a moment load. The bearing is not affected by shaft misalignment or shaft deflection. Due to the sphered outer raceway and the low conformity, a given load will produce relatively high contact stresses, hence load-carrying capacity and axial thrust capacity are limited.

The external type of self-aligning bearing does not suffer from the limited load capacity handicap because alignment is achieved by grinding a spherical surface on the external diameter of the ring and matching it to a similarly ground surface of a housing. However, there is then a mounting space penalty, but a single row of balls is possible.

(iv) Cylindrical roller-bearing
In the various types, rollers are guided on one raceway between retaining lips and are linked by the cage to form a unit with the ring (Fig. 7.1). The other raceway may be separable, depending on the type. Although not supplied in self-lubricated form, this type of bearing will still function very well when grease lubricated. Light axial loads up to 10 per cent of radial load can be supported but when limiting thrust conditions are approached, lubrication may become critical. Capable of heavier duty than the radial ball-bearing, the cylindrical roller bearing has a much higher dynamic load rating. Maximum misalignment is 0.1 per cent.

To prevent (a) high contact stresses at the ends of the rollers, and (b) provide some protection from the effects of slight misalignment, it is usual to crown the rollers slightly (Fig. 7.3), otherwise referred to as a logarithmic profile. Crowned raceways may be used in lieu of crowned rollers. As a further refinement, slight tapering of the raceway flanges and slight sphering of the roller ends help to improve significantly axial load-carrying capacity.[3] The end contact geometry so produced helps to promote the lubricant film.

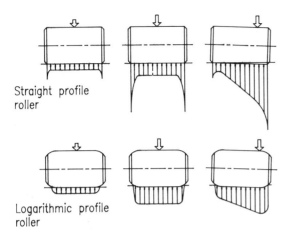

Straight profile roller

Logarithmic profile roller

Fig. 7.3 Effect of straight and logarithmic profiled rollers on the counterface stress distribution. Courtesy of SKF (UK) Ltd

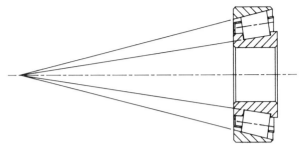

Fig. 7.4 Tapered roller-bearing

(v) Needle roller-bearing

This is a particular type of cylindrical roller-bearing used as a floating bearing, i.e. no axial load. The rolling elements have a large length:diameter ratio ($>3.5:1$). Because of the small roller diameter, these bearings are particularly useful where radial height is limited (Fig. 7.1). Three basic types of needle roller bearing are available: (1) needle rollers and cage assembly only; (2) needle roller-bearing with drawn-cup outer race; (3) bearings with machined and hardened rings. Needle rollers and cage assemblies (1) run as independent bearing units directly on the shaft and in the housing provided that the shaft and housing can be suitably hardened and ground. In this context it is an inexpensive bearing. Drawn cup bearings (2) are bearings of small radial section height with outer ring and cage made by metal-forming methods, the thin heat-treated outer ring formed of accurately controlled sheet steel. However, these must be rigidly supported to exploit the full load-carrying capacity.

If thrust loads need to be contained, this must be met by an additional ball-bearing. Higher speeds are possible where the needles are caged since they are then guided, tend not to skew so that running tolerances can be more closely controlled. Shaft misalignment or any tilting force can produce clutching.

(vi) Tapered roller-bearing

This consists of an inner ring with two lips, a separable outer ring and a complement of truncated cone-shaped rollers (Fig. 7.1). Due to the inclined position of the tapered rollers, the contact lines between races and rollers intersect at a common point on the bearing axis (Fig. 7.4), therefore ensuring a true rolling action. A cage retains the set of rollers and the inner ring as a unit.

This type of bearing will handle a wide range of radial/thrust load combinations but, as with angular contact ball-bearings, cannot accept pure radial load. In that case it is necessary to fit two bearings on a shaft, axially adjusted. Compared to the angular contact, the tapered roller has a higher load capacity but less flexibility to cope with misalignment, allowable up to 0.1 per cent. Because of inner–outer race contact angle differences, there is a force component present that drives the rollers against the guide flange. Due to this resultant friction, this bearing type is not suitable for high-speed operation without special lubrication and/or cooling.

As with cylindrical rollers, tapered rollers or raceways are frequently crowned to relieve peak contact stresses at the ends of the rolling contact zone.

(vii) Thrust ball-bearing

This consists of two grooved washers, shaft and housing, plus a caged set of balls (Fig. 7.1). Under axial load and favourable conditions, the balls are guided in the apex of the groove. The bearing becomes unsuitable for use at high speed and low thrust when centrifugal forces load the balls outward.

Thrust ball-bearings are used where axial forces are too high to be transmitted by radial bearings or where a rigid axial guidance is required. With combined loads, however, it is more usual to favour angular contact bearings which are simpler to mount and more suitable for high speed.

7.2.2 Materials

The stress levels in roller-bearings limit the choice of materials to those with a high yield and high creep strength. Important considerations in the selection of suitable materials include the following: high impact strength; good wear resistance; corrosion resistance; dimensional stability; high endurance under fatigue loading and uniformity of structure.

Steels have gained the widest acceptance as rolling contact materials as they represent the best economic and performance compromise. The through-hardening steels are listed in Table 7.2. The 535 grade is most popular (formerly EN31), but is not satisfactory at elevated temperatures due to loss of hardness and fatigue resistance. High-speed tool steels containing principally tungsten and molybdenum are superior to other materials at elevated temperatures, assuming adequate lubrication. Heat treatment to give a satisfactory carbide structure of optimum hardness is essential to achieve the maximum rolling contact fatigue life.

Steel-making practice will influence bearing life, i.e. electric arc steel-making and vacuum degassing applied to commercial bearing steels produce a high level of steel cleanliness and good fatigue life for general applications. For greater material reliability secondary processes such as electroflux refining or vacuum-arc refining produce ultra-clean steels.

Carburizing is an effective method of case hardening and is used for some types of rolling element and raceways (Table 7.2) and also for certain bearing sizes and sections that pose through-hardening and heat treatment difficulties. A deep case with the correct structure supported by a satisfactory core can be as effective, if not tougher than a through-hardened structure.

For use in a corrosive environment, martensitic stainless steels are used (Table 7.2). These generally have a poorer fatigue resistance than the through-hardening steels at ambient temperatures, but are preferred for elevated temperature work.

7.2.3 Cages

An important factor in the design of most rolling element bearings is the retainer or cage which provides a positive separation of the rolling elements. Without this, constantly variable rolling element velocities both in and out of the loaded zone will cause direct impact and rubbing of the rolling elements at higher speeds. A retainer will limit and absorb these dynamic forces while guiding and spacing the rolling elements. However, the running properties of a cage are strongly influenced by the design of its pockets and the method of rolling element retention. In certain bearing types the cage has the additional function of holding the rolling elements and one of the bearing rings together to form an inseparable unit.

Three basic types of cage are used (Fig. 7.5): (1) ball riding or roller riding; (2) inner ring land riding; (3) outer ring land riding.

The element riding types of cage (1) are usually of inexpensive manufacture and are not used in critical applications. Race (2) is driven by a force between the cage rail and inner ring land, as well as by the rolling elements, whereas (3) is retarded by the cage rail/outer ring land drag force. Traditionally, cages have been stamped out of sheet or machined from solid, but to an increasing degree, cages for rolling bearings are now injection moulded, typically glass-filled nylon. Not only does this allow for an increased element complement as less space is taken up within the raceway but moulding precision allows for reduced friction, improved toughness and smoother running. Various cage types are illustrated in Figs 7.6 and 7.7.

7.2.4 Seals

The function of a bearing seal is the containment of lubricant and the exclusion of contaminants. They are essential features of most pieces of rotating and reciprocating machinery and a large range of types and categories are available. The design of a bearing seal is governed by what the seal is expected to do, i.e. nature of material to be excluded, type of lubricant, surface speed. Designs range from a simple baffle or felt strip to a complex system of fingers, lips, labyrinths, springs, shields, etc.[4]

Seals can be divided into two broad categories:

1. *External shields and seals* — devices used external to the bearing itself and where renewal is often possible without replacing the bearing;
2. *Built-in shields and seals* — sealing incorporated into the bearing, typically throw-away types where space is limited.

Within each category there are two basic types of seal:

1. *Non-rubbing or contact-free seals.* A narrow gap between shaft and housing provides a simple, and

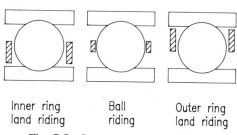

Fig. 7.5 Cage types

Table 7.2 Summary of rolling element bearing materials

Type	Specification	Maximum operating temp. (°C)	%C	%Si	%Mn	%Cr	%Ni	%Mo	Other	Notes
Through hardened carbon steel	SAE 52100[a]	160	0.98/1.10	0.15/0.35	0.25/0.45	1.35/1.60				Most popular types, used for majority of ball- and roller-bearings
	535A99[b]	160	0.95/1.10	0.10/0.35	0.40/0.70	1.20/1.60				
	M50[c]	310	0.80	0.25	0.30	4.0		4.25	1% V	Intermediate high temperature steels, widely used
	M10[d]	430	0.85	0.30	0.25	4.0		8.0	2% V	
	BM1[d]	450	0.80	0.30	0.30	4.0		8.0	1% V; 1.5% W	
Carburized steel	805M17	180	0.14/0.20	0.10/0.35	0.60/0.95	0.35/0.65	0.35/0.75	0.15/0.75		Most used grade
	665M20	180	0.17/0.23	0.10/0.35	0.35/0.75		1.5/2.0	0.20/0.30		Increased Cr and/or Ni improve hardenability on thicker sections
	SAE4320	180	0.17/0.22	0.20/0.35	0.45/0.65	0.40/0.60	1.65/2.0	0.20/0.38		
	SAE9310	180	0.08/0.13	0.20/0.35	0.45/0.65	1.0/1.4	3.0/3.5	0.08/1.15		Higher shock resistance and core hardness
Martensitic stainless steel	440C[e]	180	1.03	0.41	0.48	17.3		0.50	0.14% V	Conventional stainless grade but low hot hardness
	14Cr4Mo[f]	430	0.95/1.20	<1.0	<1.0	13.0/16.0		3.75/4.25	<0.15% V	High-temperature corrosion resistance
	15Cr4Mo5Co	450	1.10/1.15	<0.15	<0.15	14.0/16.0		3.75/4.25	2% W; 5% Co	

[a]SAE steel classification.
[b]BS 970.
[c]AISI steel classification.
[d]BS 4659.
[e]AISI steel classification.
[f]Proprietory steel.

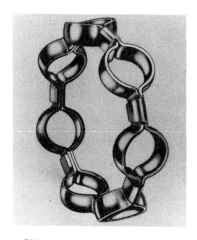

1. Ribbon cage
Ball piloted, two-piece pressed strip, tightly crimped and spot welded/riveted together. General-purpose application

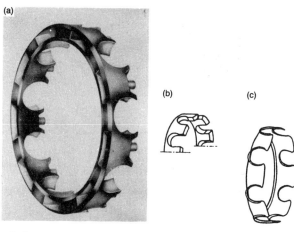

2. Crown cage
One-piece snap-in cage. Types illustrated are (a) plastic injection moulded, (b) machined, (c) pressed from strip

3. One-piece cage
Tend to be expensive. Either (a) machined from bronze, steel or reinforced phenolic, (b) sintered metal, or (c) plastic injection moulded as two halves and staked or ultrasonically welded together

4. Stamped cage
A land riding cage for roller bearings piloted on the outer ring ribs. Roller retention features formed into cage
See Fig. 7.7; cylindrical and taper roller-bearings

5. Machined and pinned cage
Bronze or brass roller guided cages with cylindrically bored pockets. Pinned or staked outer ring

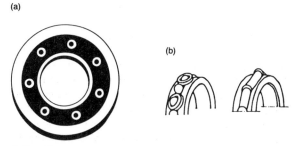

6. Special cage
Either (a) solid lubricant filled polymer, or (b) cut lengths of individual PTFE tube or ring separators

Fig. 7.6 Examples of ball- and roller-bearing cages. 1, 2(a), 3 and 5 courtesy of the Torrington Company, Connecticut. 6(a) Courtesy of Bemol Corporation

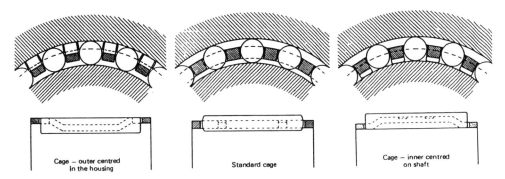

Fig. 7.7 Needle roller-bearing cage types. Courtesy of SNR Division Nadella, Rueil Malmaison, France

in many cases, sufficient seal. The gap can be made quite narrow because run-out and bearing wear effects are minimal with rolling element bearings. Typically the gap varies from 0.1 to 0.3 mm. Since there is no friction except for lubricant drag in the sealing gap, there are no speed restrictions.

2. *Rubbing or contact seals.* These vary from simple felt strips to precision lip seals. In all cases an intimate contact between surfaces provides a barrier against lubricant loss, but shaft surface finish has a critical effect on seal performance and being a rubbing contact, shaft speed is limited to avoid thermal effects. Through-feed shaft grinding can introduce surface spiral effects acting as a micro-scale auger through the contact zone.

Examples of the different seal types are illustrated in Fig. 7.8. All these are intended as suitable for grease lubrication. This is by far the more usual form of lubrication found in small machine applications today.

7.2.5 Bearing identification coding

Each rolling bearing is described by a code sequence which defines the bearing construction, bearing dimensions, clearances, type of lubrication, etc. Although the code structure follows essentially the same format between manufacturers, the actual alpha-numeric descriptors will almost certainly vary. Consequently, although standard preferred sizes exist throughout the bearing industry, interchangeability tables are normally published to define equivalent codes.

The basic bearing code structure is outlined in Fig. 7.9. Deviations from standard or normal construction are indicated by the prefix and suffix codes of individual manufacturers.

(i) Boundary dimensions

A rolling bearing is a ready to mount and interchangeable machine element. The standardization of boundary dimensions or bearing size is outlined in BS 292.[5] Formerly the terms extra light, light, medium and heavy were applied to bearing diameters and narrow and wide to the widths. These terms are clearly insufficient and the standardization codes now refer to a two-digit group numerical series, width plus diameter, indicating the dimensional series.

In the dimensional plans, several external diameter and width dimensions are assigned to each bearing bore diameter (Fig. 7.10). Various bearings of the same design may then be constructed having the same bore but different load-carrying capacities. The dimensional series reference both a width series and a diameter series, essentially defining the bearing section (Fig. 7.10). Thus by maintaining the section ratio, a range of bearing sizes are possible (Fig. 7.11). The width series is referenced by a number 0 to 6 and the diameter series by 8, 9, 0, 1, 2, 3 or 4 (8 is the lightest radial section, 4 the heaviest).

Generally the last two digits of a bearing designation indicate the bore size. These are defined in Table 7.3.

Table 7.3 *Bore code for radial bearings*

Nominal bore d (mm)	Code
<10	Nominal bore
10	00
12	01
15	02
17	03
>17	Nominal bore ÷ 5

ROLLING ELEMENT BEARINGS

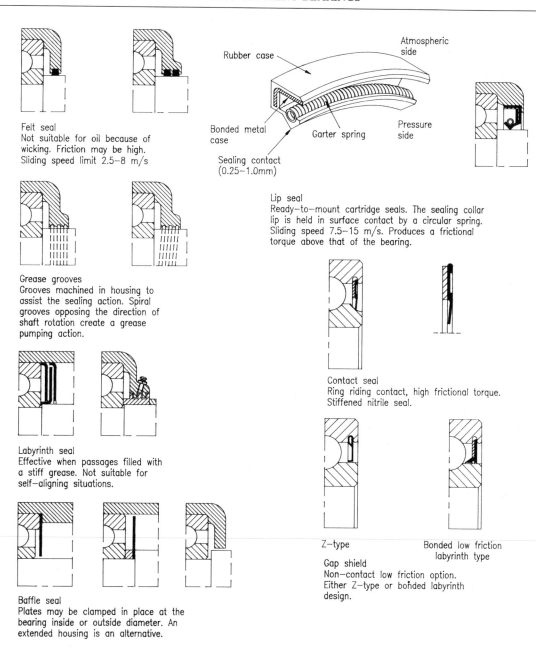

Felt seal
Not suitable for oil because of wicking. Friction may be high. Sliding speed limit 2.5–8 m/s.

Grease grooves
Grooves machined in housing to assist the sealing action. Spiral grooves opposing the direction of shaft rotation create a grease pumping action.

Labyrinth seal
Effective when passages filled with a stiff grease. Not suitable for self-aligning situations.

Baffle seal
Plates may be clamped in place at the bearing inside or outside diameter. An extended housing is an alternative.

Spring seal
Enclosed version of the baffle plate.

Lip seal
Ready-to-mount cartridge seals. The sealing collar lip is held in surface contact by a circular spring. Sliding speed 7.5–15 m/s. Produces a frictional torque above that of the bearing.

Contact seal
Ring riding contact, high frictional torque. Stiffened nitrile seal.

Z-type / Bonded low friction labyrinth type
Gap shield
Non-contact low friction option. Either Z-type or bonded labyrinth design.

Fig. 7.8 Examples of seal and shield types. Contact and gapshields are built into the bearing; other types are external

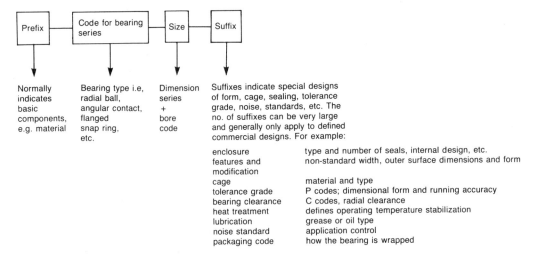

Fig. 7.9 Basic rolling element code structure

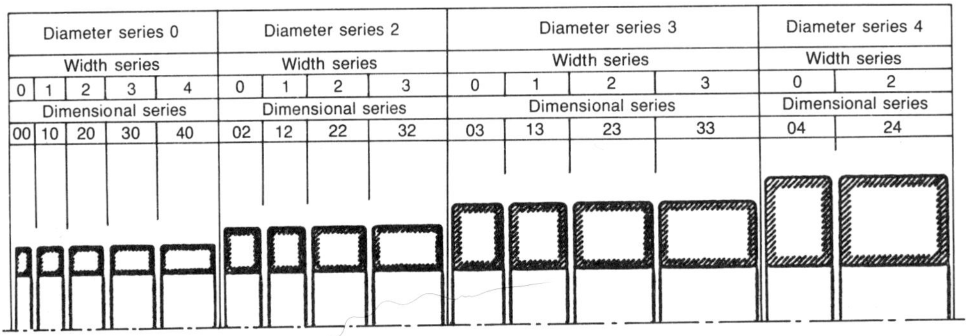

Fig. 7.10 Excerpt from the dimensional plan for radial bearings, constant bore diameter, variable section

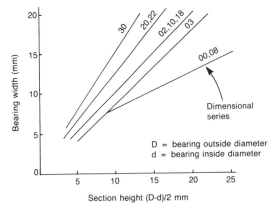

Fig. 7.11 Examples of bearing section dimensions possible for a given dimensional series

Table 7.4 *Tolerance class equivalents*

ISO class	ISO alternative code	ABEC	AFBMA	BSI
0	Normal (P0)	1	20	Normal
6	P6	3	12	—
5	P5	5P	—	EP5
4	P4	7P	—	EP7
2	P2	9P	—	EP9

Notes: ISO = International Standards Organization (ISO 492)
BSI = British Standards Institute (BS 292)
AFBMA = Anti-Friction Bearing Manufacturers Association
ABEC = Annular Bearing Engineers Committee of AFBMA

(ii) Precision class tolerances

Table 7.4 defines the equivalence of standards used by the main Bearing Standards committees in the USA and Europe, all of which may be quoted by a supplier depending on the manufacturing base. Using the International Standard Organization (ISO) standard, tolerance class 0 refers to bearings of standard accuracy which satisfy the majority of quality demands of industry. However, where high accuracy and quiet running are essential, tighter tolerance classes are available 2, 4, 5 and 6 (Table 7.5). Class 2 is the tightest tolerance (Table 7.5. In tolerance classes 4 and 5, size and axial run-out

Table 7.5 *Tolerances for metric series radial ball bearings*

(a) Inner ring

ISO tolerance class	Nominal bore diameter d mm		Bore diameter				Radial runout	Side runout with bore	Axial runout	Width		
			Mean variation		Actual variation					Deviation		Variation of individual ring
	Over	Incl.	High	Low	High	Low	Max.	Max.	Max.	High	Low	Max.
0	0.6	2.5	0	−8	+1	−9	10	—	—	0	−40	12
	2.5	10	0	−8	+2	−10	10	—	—	0	−120	15
	10	18	0	−8	+3	−11	10	—	—	0	−120	20
6	0.6	2.5	0	−7	+1	−8	5	—	—	0	−40	12
	2.5	10	0	−7	+1	−8	6	—	—	0	−120	15
	10	18	0	−7	+1	−8	7	—	—	0	−120	20
5	0.6	2.5	0	−5	0	−5	3.5	7	7	0	−40	5
	2.5	10	0	−5	0	−5	3.5	7	7	0	−40	5
	10	18	0	−5	0	−5	3.5	7	7	0	−80	5
4	0.6	2.5	0	−4	0	−4	2.5	3	3	0	−40	2.5
	2.5	10	0	−4	0	−4	2.5	3	3	0	−40	2.5
	10	18	0	−4	0	−4	2.5	3	3	0	−80	2.5

(b) Outer ring

ISO tolerance class	Nominal outside diameter D mm		Outside diameter						Outside surface runout with side	Radial runout	Groove runout with side	Width		
			Mean deviation		Actual variation							Deviation		Variation of individual ring
					Open		Sealed shielded							
	Over	Incl.	High	Low	High	Low	High	Low	Max.	Max.	Max.	High	Low	Max.
0	2.5	6	0	−8	+1	−9	+4	−12	—	15	—			
	6	18	0	−8	+2	−10	+5	−13	—	15	—			
	18	30	0	−9	+2	−11	+6	−15	—	15	—	Same as for inner ring width		Same as for inner ring width
6	2.5	6	0	−7	+1	−8	+3	−10	—	8	—			
	6	18	0	−7	+1	−8	+3	−10	—	8	—			
	18	30	0	−8	+1	−9	+4	−12	—	9	—			
5	2.5	6	0	−5	0	−5			8	5	8			5
	6	18	0	−5	0	−5			8	5	8			5
	18	30	0	−6	0	−6			8	6	8			5
4	2.5	6	0	−4	0	−4			4	3	5			2.5
	6	18	0	−4	0	−4			4	3	5			2.5
	18	30	0	−5	0	−5			4	4	5			2.5

Note: Tolerance limits in μm.

of inner and outer rings have been included. All dimensions are referenced to 20 °C.

(iii) Bearing clearance

Clearance or play of a rolling element bearing is defined as the amount of possible displacement of one ring relative to another in the radial or axial directions (Fig. 7.12). For precision running, the bearing should have very limited radial clearance when at the operating temperature. Radial clearance groups are defined by a suffix consisting of C + number, i.e. C1 to C5 (Table 7.6). The normal radial clearance group bears no suffix.

7.3 Design factors

7.3.1 Contact angle

The contact angle, or the nominal contact angle, is the angle between a plane perpendicular to a bearing axis (a radial plane) and the line of action of the resultant of the forces transmitted by a raceway to a rolling element (e.g. Fig. 7.2). The contact angle influences both the axial and radial characteristics of a bearing. The contact angle may also change with bearing load, and hence with the deformation of the bearing components. The nominal contact angle α applies to the load-free bearing in which the rolling elements are in stress-free contact with the raceways. The operating contact angle applies to the loaded bearing. It can be seen that the axial load-carrying capacity increases with α.

The following comments are relevant:

1. With the tapered roller-bearing, the slope of the inner ring raceway differs from that of the outer ring raceway, the value of α being measured at the separable ring.

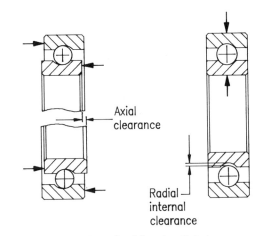

Fig. 7.12 Definition of axial and radial clearance

2. As a rule lower contact angles are used for light axial loads, high speed and high contact angles are selected when high axial loads and/or axial rigidity are major requirements.
3. For an axially loaded deep groove ball-bearing, α varies with the ratio of axial load (F_a) to the basic static load rating (C_0) (Fig. 7.13), depending on the radial clearance.

7.3.2 Preload

Preload is the application of a permanent axial load to a bearing in order to eliminate free radial and axial movement or bearing clearance. The two types of clearance are illustrated in Fig. 7.12. Preload is usually employed in thrust bearings, taper roller-bearings and angular contact bearings. A light preload is sometimes

Table 7.6 Clearance groups and radial internal clearance values for single row radial bearings with cylindrical bore

ISO code	Group 2		Normal group		Group 3		Group 4		Group 5	
Alternative codes	C2 1 dot		CN 2 dot		C3 3 dot		C4 4 dot		C5 5 dot	
Bore diameter d (mm)										
Over / Incl.	Min.	Max.	Min.	Max.	Min.	Max.	Min.	Max.	Min.	Max.
6 10	0	7	2	13	8	23	14	29	20	37
10 18	0	9	3	18	11	25	18	33	25	45
18 24	0	10	5	20	13	28	20	36	28	48
24 30	1	11	5	20	13	28	23	41	30	53

Note: Clearance values in micrometres.

ROLLING ELEMENT BEARINGS

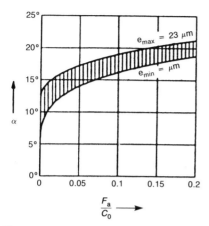

Fig. 7.13 Variation of contact angle α in an axially loaded deep groove ball-bearing. From Eschmann et al, *Ball and Roller Bearings*. Copyright © 1985 John Wiley & Sons Ltd. Reprinted by permission of J Wiley & Sons Ltd

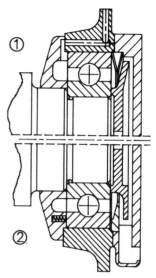

Fig. 7.14 Spring preloading of deep groove ball-bearings: (1) with spring washers, (2) with helical springs. From Hjertzen et al, Rolling element bearings. *Tribology* Series 8, Elsevier 1983. Courtesy of Elsevier Science Publishers BV

used on deep groove ball-bearings where accurate running is important. The ideal setting for a bearing would be with neither endplay nor preload in most general applications. However, temperature effects and deflections under load rarely make this possible. As a guiding rule, preload should not be employed unless there is a definite reason for so doing.

Specific reasons for the preloading of bearings are as follows:

1. To increase the overally rigidity of the whole design and reduce deflections from externally applied loads;
2. To provide assurance that angular contact bearings and tapered roller-bearings do not run free of load that could give rise to ball skidding;
3. To reduce running errors arising from small differences in the rolling elements;
4. To reduce or eliminate fretting between the housing and the raceway;
5. To reduce bearing noise;
6. To alter the natural frequency of the assembly and eliminate vibration or chatter;
7. To prevent damage by shock loading. Preload reduces dynamic peak load effects;
8. To achieve a better element load distribution when under heavy loading
9. To increase running accuracy by positive tracking;
10. To prevent hammering of the tracks, particularly with taper roller-bearings.

Examples of spring preloading are shown in Fig. 7.14. A useful rule of thumb here should be 4.5 N loading per millimetre of bearing bore. In the case of angular contact bearings, application of preload tends to maintain the original contact angle and also raises the speed at which the effects of ball skidding become significant. Hence preload increases as speed increases. An approximate preload value for any particular speed can be obtained from

$$\text{Preload} = \left(\frac{n}{1000}\right)^2 \left(\frac{C_0}{1000}\right)^2 k$$

where n = r.p.m., C_0 = basic static load rating (N) (see also section 7.3.3.(ii) and k = constant (Table 7.7)

7.3.3 Load ratings

(i) Basic dynamic load rating

This is one of the parameters characterizing the load-carrying capacity of a rolling element bearing, and is an indicator of the fatigue life expected under given running conditions. It is defined for radial bearings as that load of constant magnitude and direction which a sufficiently large group of identical bearings can endure for a basic rating life of 10^6 rev, calculated as follows:

C = basic dynamic load rating

$\quad = f_c (i \cos \alpha)^{0.7} Z^{0.67} D_w^{1.8}$ (for ball-bearings, radial loading)

Table 7.7 Values of preload constant k

Contact angle (°)	k
15	0.0065
20	0.0095
25	0.0135
30	0.0170
40	0.0255

$$= f_c(il_{eff} \cos \alpha)^{0.78} Z^{0.75} D_w^{1.07} \text{ (for roller-bearings, radial loading)}$$

and

$$= f_c Z^{0.67} D_w^{1.8} \text{ (for thrust ball-bearings, } \alpha = 90°)$$

where f_c = a tabulated factor which depends on the geometry of the bearing components, the accuracy of piece-part manufacture and the material (complex factor) (see Table 7.8), i = number of rows of balls/rollers in the bearing, l_{eff} = effective length of roller, i.e. without the corner radii, Z = no. of balls or rollers in a single row bearing or number per row of a multi-row bearing, D_w = rolling element mean diameter and α = nominal bearing contact angle.

Over the past 30 years there have been significant advances in material quality, dimensional control, surface finishing and geometric form to the extent that dynamic load ratings of modern rolling bearings will exceed the calculated ratings above. When standards were originally introduced, load ratings related to good-quality acid open-hearth steel. Today the norm is a much cleaner steel (section 7.2.2). Hence the result is a higher basic dynamic load rating quoted in manufacturers' catalogues, i.e.

$$C_r = b_m C$$

where b_m = a material and manufacturing quality factor (MMQ) and C_r = effective dynamic load rating.

The factor b_m was introduced by the ISO and is derived from manufacturing test data. Values are typically in the range 1.1–1.35 (Table 7.9). There may be some variation in b_m factors claimed by different bearing manufacturers. Some bearings, e.g. cylindrical roller thrust, do not benefit from cleaner steel.

(ii) Basic static load rating
This is defined as the maximum load which can be applied to a bearing without creating internal damage in terms of permanent (plastic) deformation of rolling elements and raceways, manifest as unacceptable running noise. Experience has shown that a total permanent deformation of 1/10 000 of the rolling element diameter, at the centre of the most heavily loaded rolling element/raceway contact, can be tolerated in most bearing applications without the bearing operation being impaired. The basic static load rating, C_0, is that radially acting load for

Table 7.8 Summary of f_c and f_o factors for some rolling element bearings types

$\dfrac{D_w \cos \alpha}{D_{pw}}$	Deep groove and angular contact bearings		Self-aligning ball-bearings		Thrust ball-bearings		Radial roller-bearing
	f_c	f_o	f_c	f_o	f_c^a	f_o^a	f_c
0.06	49.1	15.9	18.6	2.2	62.9	56.7	76.9
0.08	52.8	16.3	21.1	2.3	68.5	55.1	81.2
0.10	55.5	16.4	23.4	2.4	73.3	53.5	84.2
0.12	57.5	15.9	25.6	2.4	77.4	51.9	86.4
0.14	58.8	15.4	27.7	2.5	81.1	50.4	87.7
0.16	59.6	14.9	29.7	2.6	84.4	48.8	88.5
0.18	59.9	14.4	31.7	2.7	87.4	47.3	88.8
0.20	59.9	14.0	33.5	2.8	90.2	45.7	88.7
0.22	59.6	13.5	35.2	2.9	92.8	44.2	88.2
0.24	59.0	13.0	36.8	3.0	95.3	42.7	87.5
0.26	58.2	12.5	38.2	3.1	97.6	41.2	86.4
0.28	57.1	12.1	39.4	3.2	99.8	39.7	85.2
0.30	56.0	11.6	40.3	3.3	101.9	38.2	83.8
0.32	54.6	11.2	40.9	3.4	103.9	36.8	—
0.34	53.2	10.7	41.2	3.5	105.8	35.3	—
0.36	51.7	10.3	41.3	3.6	—	—	—
0.38	50.0	9.8	41.0	3.7	—	—	—
0.40	48.4	9.4	40.4	3.8	—	—	—

[a] $\alpha = 90$, hence diameter ratio becomes D_w/D_{pw}.

Table 7.9 *Summary of material and manufacturing load rating quality factor (b_m) for different bearings*

Rolling element	Bearing type	Radial	Thrust
Ball	General purpose	1.3	1.3
	Angular contact	1.1	—
Roller	Cylindrical roller	1.1	1.0
	Needle roller with hardened raceways	1.1	1.0
	Drawn cup needle roller	1.0	—
	Tapered roller	1.1	1.1
	Spherical roller	1.2	1.2

Note: Conventional refining = 1.

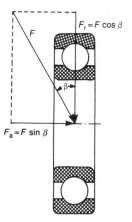

Fig. 7.15 Combined loads acting on a radial ball-bearing

which the calculated Hertz contact pressure reaches a value of 4600 N mm^{-2} for self-aligning ball ball-bearings, 4200 N mm^{-2} for all other ball-bearings and 4000 N mm^{-2} for roller-bearings.

Here C_0 is evaluated as follows:

$C_0 = f_0 i Z D_w^2 \cos \alpha$ (for radial ball-bearings)

$C_0 = f_0 Z D_w^2$ (for thrust ball-bearings ($\alpha = 90°$))

$C_0 = 22 \times i Z l_{eff} D_w \cos \alpha$ (for radial roller-bearings)

where f_0 = factor which depends on the geometry of the bearing components and the stress level (see Table 7.8).

7.3.4 Equivalent loads

(i) Dynamic equivalent radial load

In loading analysis, all bearing loads are converted to an equivalent load P permitting a direct comparison of the converted loading with the published bearing rating. The equivalent load P is defined as that constant stationary radial load under which a rolling bearing would have the same life as it would attain under actual conditions of loading.[6] To reduce a given set of conditions giving rise to an oblique load F acting at an angle β to the radial plane (Fig. 7.15), to that for a bearing under pure radial load, the axial load must be corrected by a multiplier Y which is dependent on the contact angle α and is generally >1. A factor X is also introduced into the radial load to satisfy rotational conditions, so that in terms of axial and radial components

$P = XF_r + YF_a$ (for ball- and roller-bearings)

where P = equivalent constant radial (or axial) force, F_r = radial load, F_a = axial load, X = a radial load factor and Y = a thrust factor.

Figure 7.16 is an example of a single-row angular contact, ball- or roller-bearing under combined loading showing the dependence of the quotient P/F_r on the load ratio F_a/F_r. The straight lines are in fact a simplified approximation to the curve derived from fatigue theory, defined by $P = XF_r + YF_a$ and intersecting at e where

$e \approx 1.5 \tan \alpha$

Both X and Y are determined by the location and inclination of the straight line, the e constant merely defining the transition point from one straight line approximation to another. Table 7.10 defines X and Y values for various bearing types according to the F_a/F_r ratio. A number of points arise from Table 7.10:

1. In single-row radial bearings, axial forces in the range $F_a/F_r \leq e$ have no influence on the

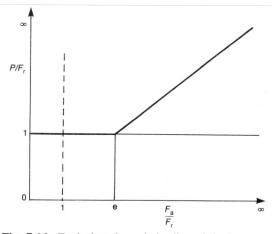

Fig. 7.16 Equivalent dynamic loading of single row rolling element bearings

Table 7.10 X and Y factors for radial and thrust rolling element bearings

Bearing type	Relative axial load F_a/C_0	e	Single-row bearings $\frac{F_a}{F_r} \leq e$		$\frac{F_a}{F_r} > e$	
			X_1	Y_1	X_2	Y_2
1(a) Deep groove ball-bearings with normal radial clearance (CN) (b) Angular contact bearings $\alpha = 5°$	0.014 0.028 0.056 0.084 0.11 0.17 0.28 0.42 0.56	0.19 0.22 0.26 0.28 0.30 0.34 0.38 0.42 0.44	1	0	0.56	2.30 1.99 1.71 1.55 1.45 1.31 1.15 1.04 1.00
2(a) Deep groove ball-bearings with C3 radial clearance (b) Angular contact bearings $\alpha = 10°$	0.014 0.029 0.057 0.086 0.11 0.17 0.29 0.43 0.57	0.29 0.32 0.36 0.38 0.40 0.44 0.49 0.54 0.51	1	0	0.46	1.88 1.71 1.52 1.41 1.34 1.23 1.10 1.01 1.00
3(a) Deep groove ball-bearings with C4 radial clearance (b) Angular contact bearings $\alpha = 15°$	0.015 0.029 0.058 0.087 0.12 0.17 0.29 0.44 0.58	0.38 0.40 0.43 0.46 0.47 0.50 0.55 0.56 0.56	1	0	0.44	1.47 1.40 1.30 1.23 1.19 1.12 1.02 1.00 1.00
4. Angular contact bearings	$\alpha = 20°$ $\alpha = 25°$ $\alpha = 30°$ $\alpha = 35°$ $\alpha = 40°$ $\alpha = 45°$	0.57 0.68 0.80 0.95 1.14 1.33	1	0	0.43 0.41 0.39 0.37 0.35 0.33	1.0 0.87 0.76 0.66 0.57 0.50
5. Thrust ball-bearings[a]	$\alpha \neq 90°$	1.25 tan α	b	b	c	1
6(a) Self-aligning ball-bearings (b) Taper roller-bearings (c) Spherical roller-bearing (d) Radial roller-bearing	$\alpha \neq 0°$	1.5 tan α	1	0	0.4	0.4 cot α

[a] Thrust (axial) load acting in one direction only
[b] Unsuitable for use at $F_a/F_r \leq e$
[c] 1.25 tan α (1 − 2/3 sin α)

equivalent dynamic load ($X = 1$, $Y = 0$) hence axial forces need only be taken into account in the range $F_a/F_r > e$.
2. Radial bearings are exceptional in that the contact angle α varies with the applied load hence varying the multiplier Y. Thus there would be a family of lines (Fig. 7.16), with F_a/C_0 as the characteristic value.
3. In angular contact bearings with $\alpha > 20°$, α changes little with axial load so in Table 7.10 dependence on axial load is not taken into account.
4. For self-aligning ball, taper roller, radial roller and

spherical roller-bearings, X and Y factors are directly derived from the contact angle.
5. For radial roller-bearings where $\alpha = 0°$, the X factor is 1 and Y factor is 0 and theoretically thrust load will have no effect on fatigue life. However, the ability of radial roller-bearings to support axial loads when $\alpha = 0°$ varies considerably with the actual bearing design and its use.
6. Thrust loading on cylindrical roller-bearings above a nominal value locating thrust requires excellent lubrication, a stabilized radial load plus limited misalignment.

A summary of the use of equivalent dynamic load formulae is given in Table 7.11.

(ii) Other calculations for equivalent load
In many bearing arrangements, load and speed may well change periodically, either randomly or in a controlled cycle. In these cases the equivalent dynamic load must be calculated for the given load and speed values, i.e. that constant load and average speed that results in the same fatigue life as the variable loads and speeds. If the individual forces act obliquely on the bearing, then prior to calculating the equivalent load of the work cycle, the equivalent loads P_i, must be calculated for the individual forces F_i (as section (i)).

(a) Constant speed, radial load varying with time. For a load varying from P_{max} to P_{min} gradually over a period of time, the following can be used for a fairly accurate calculation of the equivalent load:

$$P = \frac{P_{min}}{3} + \frac{2P_{max}}{3}$$

(b) Constant speed, radial load varying in steps. For a load P_i varying stepwise over a total time period T

$$P = \sqrt[p]{\left(\frac{q_1 P_1^p + q_2 P_2^p + \ldots q_N P_N^p}{100}\right)}$$

where $q_i = \Delta t_i/T \times 100$ = percentage time duration of each particular speed step, N = no. of time periods, Δt_i = individual time period and p = life exponent: 10/3 for rollers (line contact) and 3 for balls (point contact).

(c) Constant speed, varying bearing load. If the load is dependent on the time T in a definable function P_i over a time period T, the general value of the equivalent bearing load can be defined by

$$P = \sqrt[p]{\left(\frac{1}{T}\int_0^T P_i^p(t)dt\right)}$$

(e.g. Fig. 7.17). A graphical solution to the above is as follows:

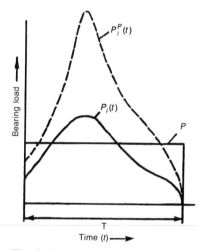

Fig. 7.17 Equivalent bearing load

Table 7.11 *Equivalent dynamic load formulae*

		Kind of load	
Type of bearing	Radial	Thrust	Combined radial–thrust
1. Deep groove ball-bearing	$P = F_r$	$P = Y_2 F_a$	$P = 0.56 F_r + Y_2 F_a$
2. Angular contact ball-bearing	$P = F_r$	$P = Y_2 F_a$	$P = 0.56 F_r + Y_2 F_a; \alpha = 5°$
			$P = 0.46 F_r + Y_2 F_a; \alpha = 10°$
			$P = 0.44 F_r + Y_2 F_a; \alpha = 15°$
			$P = 0.41 F_r + 0.87 F_a; \alpha = 25°$
			etc.
3. Thrust ball	—	$P = F_a$	$P = 1.25 \tan \alpha (1 - \frac{2}{3}\sin \alpha) + F_a$
			$\alpha \neq 90°$
4. Taper roller	$P = F_r$	$P = 0.4 \cot \alpha F_a$	$P = 0.4 F_r + 0.4 \cot \alpha F_a$

1. Plot $P_i(t)$;
2. Calculate and plot $P_i^p(t)$;
3. Measure area under $P_i^p(t)$ and divide by T to give an average P_i^p;
4. $^p\sqrt{}$ to get P.

(d) Various radial loads and speeds acting cyclically. In certain cases a bearing may be subjected to a series of cyclically varying loads and speeds. If the factors load, speed and relevant duration q per cent are known, the equivalent dynamic load is obtained from

$$P = \sqrt[p]{\left(P_1^p \frac{n_1 q_1}{n_m 100} + P_2^p \frac{n_2 q_2}{n_m 100} + \cdots\right)}$$

and the mean rotary speed n_m from

$$n_m = \frac{n_1 q_1}{100} + \frac{n_2 q_2}{100} + \cdots \text{ min}^{-1}$$

and n_i = speed at constant load P_i and duration q_i.

(iii) Equivalent static load

Some applications subject non-rotating bearings to high radial, thrust or combined radial/thrust loads, i.e. statically stressed. This refers to stationary conditions or when subjected to small oscillations. Static thus relates to bearing operation and not to the type of load. Permissible loads are governed by the static load rating C_o or the total permanent deformation of element and raceway (section 7.3.3(ii)). The load may be constant or variable.

A static equivalent load is defined as the static radial load which, if applied to the bearing, would produce the same maximum raceway contact stress as that which occurs under the actual conditions of loading. The static equivalent load P_0 for radial-type bearings under combined radial and thrust load is

$$P_0 = X_0 F_r + Y_0 F_a$$

where X_0 = radial load factor and Y_0 = thrust load factor.

A summary of thrust and radial load factors is given in Table 7.12. The equivalent static load for thrust bearings $\alpha \neq 90°$, is calculated from

$$P_0 = F_a + 2.3 F_r \tan \alpha$$

when statically stressed. The results of this approximation are sufficiently accurate when $F_r < 0.44 F_a \cot \alpha$.

To determine whether a bearing is sufficiently adequate for a given load, a safety factor S_0 is used: where

$$S_0 = \frac{C_0}{P_0} \quad \text{and} \quad S_0 \not< 1$$

Table 7.12 *Radial factor X_0 and thrust factor Y_0 for statically stressed radial bearings*

Bearing type		Single-row[a]	
		X_0	Y_0
Radial deep groove ball-bearings[a]		0.6	0.5
Radial angular contact ball-bearings	$\alpha = 20°$	0.5	0.42
	$\alpha = 25°$	0.5	0.38
	$\alpha = 30°$	0.5	0.33
	$\alpha = 35°$	0.5	0.29
	$\alpha = 40°$	0.5	0.26
Radial self-aligning ball-bearings		0.5	$0.22 \cot \alpha$
Radial spherical roller-bearings		0.5	$0.22 \cot \alpha$
Radial tapered roller-bearings		0.5	$0.22 \cot \alpha$

[a]P_0 must always be equal to or larger than F_r.

For rolling element bearings in general:

1. Vibrationless operation; normal operation with low requirements for smooth running; bearings with only slight rotational movement ($S_0 \geq 1$);
2. Normal operation with higher requirements for smooth running ($S_0 \geq 2$);
3. Operation with distinct shock loads ($S_0 \geq 3$)
4. Bearing arrangements with high requirements for high accuracy and smooth running ($S_0 \geq 4$).

7.3.5 Rating life

(i) Basic rating life (L_{10})

'Life' is defined as the number of millions of revolutions, at constant speed, that 90 per cent of the bearings would complete or exceed before the first evidence of fatigue failure developed. In theory, if a bearing is correctly selected for its duty, correctly mounted and suitably lubricated then most failure causes are eliminated except for material fatigue. Typically this is shown by Fig. 3.10.

Basic rating life is given by

$$L_{10} = \left(\frac{C_r}{P}\right)^p 10^6 \text{ rev}$$

or in operating hours

$$L_{10} = \frac{16\,666}{n} \left(\frac{C_r}{P}\right)^p \text{ hours}$$

where n = operating speed (min^{-1}), p = life exponent: 3 for radial and angular contact ball-bearings and 10/3 for radial roller-bearings.

Bearing fatigue life for normal engineering applications is taken at a reliability level of 90 per cent. Reliability is interpreted by means of the Weibull distribution (section

3.2.3(iv)) to give a pictorial representation of the bearing fatigue life distribution pattern.

(ii) Adjusted rating life (L_a)

Use of the basic rating life (L_{10}) above applies to non-critical applications where the demands on the bearing are relatively low, and the application does not warrant an involved analysis apart from a functional check and fatigue life calculation. However, the development of conventional rolling element bearings is an ongoing process as demonstrated by the current generation of bearings having improved performance over previous bearings of equivalent size.[7] Advances are usually demonstrated by 'downsizing' the bearing necessary to carry a specific load; for example, a modern bearing is only about one-third of the weight of a traditional 1940s design. Hence to take better advantage of the improvements in modern bearings, the adjusted rating life (L_a) was introduced. Briefly, apart from load and speed, this takes into account (1) reliability other than 90 per cent, (2) the bearing material and (3) the quality of lubrication.

The extended life equation,[8] is as follows:

$$L_a = a_1 a_2 a_3 L \; 10^6 \text{ rev}$$

where L_a = adjusted rating life for non-conventional material, operating conditions and required reliability, L = basic rating life in 10^6 rev (L_{10}), a_1 = life adjustment factor for a reliability other than 90 per cent, a_2 = life adjustment factor for non-conventional materials and a_3 = life adjustment factor for operational/lubrication application conditions.

However, there are limitations to this extended life calculation. It is quite difficult to make allowances for all operating conditions by the adjustment factors, hence the estimate of life under special conditions may not be that reliable. The calculated life expectation should not therefore be overrated.

(a) Reliability factor (a_1). To achieve reliabilities other than 90 per cent, the usual approach is to obtain a factor by which the L_{10} life can be modified, i.e.

$$\text{Ratio } \frac{L}{L_{10}} = a_1$$

where L = life at the reliability required.

From a Weibull graph, a plot of L/L_{10} vs reliability (per cent probability of survival) can be obtained from

$$\frac{L}{L_{10}} = a_1 = \left[\frac{\ln (1/Sx)}{\ln (1/0.9)} \right]^{1/\epsilon}$$

where ϵ = slope of Weibull plot and Sx = probability of survival. A typical plot is shown in Fig. 7.18. In practice, there is a minimum life L_0 (section 3.2.3). By a suitable choice of L_0 it is possible to derive a curve to fit the percentage of failures below the L_{10} point and hence establish more practical reliability factors, a_1 for levels >90 per cent, i.e.

$$a_1 = (1 - x) \left[\frac{\ln Sx}{\ln 0.9} \right]^{1/\epsilon} + x$$

where x = the factor of L_{10} equal to minimum life.

Based on fatigue life tests, the life adjustment factors for reliability levels >90 per cent are given in Table 7.13. These were established by the Anti-Friction Bearing Manufacturers Association (AFBMA) and incorporated into ref 8.

(b) Material factor (a_2). At one point material quality refining improvements to basic 1 per cent C, 1.4 per cent Cr steel were incorporated into a_2, but later abandoned in favour of b_m. Hence a_2 for normal bearing steels is

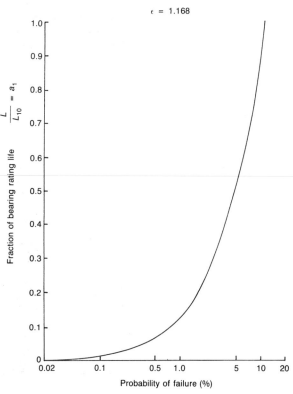

Fig. 7.18 Graph of reliability factor a_1 at 90 per cent level and below

now taken as equal to 1. Either way the result is a fatigue life improvement in modern bearings.

However, it should be noted that the use of an improved material does not overcome any deficiency in bearing lubrication or abnormal operating conditions. Values of a_2 would be obtainable from manufacturers in cases of non-conventional materials, but typically

SAE 52100/BS 535A99 $a_2 = 1$
AISI 440C $a_2 = 0.5$

(c) Operating conditions factor (a_3). Apart from correct installation aspects, the reliability of rolling bearings greatly depends on the adequacy of lubrication, adequate sealing arrangements if the bearing is self-contained and non-contamination of the lubricant. Life is prematurely terminated only by some disorder, e.g. contamination, lubricant starvation, too high a running temperature, all of which accelerate the fatigue mechanism.

Lubrication, in terms of the a_3 factor, has a profound effect on bearing performance. A good clean lubricant with adequate viscosity will dramatically improve service life.[7] For normal purposes, lubrication is defined as the degree of surface separation requiring an adequate oil film in order to avoid rolling element—raceway contact. The principal factors influencing oil film thickness are as follows:

1. Surface velocities;
2. Radius of curvature of surfaces;
3. Lubricant viscosity;
4. Contact pressure;
5. Surface roughness/surface finish.

The effect of surface texture on lubrication is defined by the film thickness/roughness ratio as

$$\Lambda = \frac{h}{\sqrt{(\sigma_1^2 + \sigma_2^2)}}$$

$$= cd_m^{0.43} n^{0.78} v_1^{0.86} \quad \text{(for radial ball-bearings)}$$

$$= cd_m^{0.6} n^{0.7} v_1^{0.8} \quad \text{(for roller-bearings)}$$

where h = film thickness, Λ = effectiveness factor (ehl), σ_1 = surface finish of body 1, σ_2 = surface finish of body 2, d_m = mean bearing diameter (mm) = $(d + D)/2$, n = speed (min^{-1}), v_1 = kinematic viscosity of lubricant at operating temperature (mm^2 s^{-1}) and c = constant = 60×10^{-6} for rollers.

Bearing reliability is related to Λ, life as a function of film percentage being shown in Fig. 7.19, from which the following equivalences are found to apply.[9]

1. $\Lambda = < 0.8-1.0$: boundary lubrication conditions apply;
2. $\Lambda = 1-2$: applicable to most bearings in service representing the achievement of partial elasto-hydrodynamic lubrication (ehl); to ensure reliable bearing operation select v_1 to give $\Lambda > 1.2$;
3. $\Lambda > 3$; represents complete surface separation, i.e. very favourable lubrication conditions providing there are no application factors suggesting otherwise, e.g. low material hardness.

Although Λ was originally used by AFBMA to define

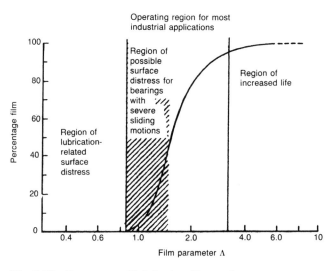

Fig. 7.19 Percentage of lubricating film vs Λ

a life factor in the American Society of Mechanical Engineers (ASME) reliability assessment procedure,[1] the ISO procedure[8] simplifies the number of parameters involved, the equivalent life adjustment factor a_3 being given in Fig. 7.20 where $a_3 = f(\Lambda)$.

An equivalent approach adopted by some bearing manufacturers is to regard the factor a_2 as a constant (typically unity as in (b) above) and combine the material factor with a_3 to form the product a_{23} and to determine lubrication effectiveness in terms of the viscosity ratio K. This is expressed as follows:

$$K = \frac{\nu}{\nu_1} = \text{actual lubricant kinematic viscosity at 40 °C/minimum kinematic viscosity for adequate lubrication at the bearing operating temperature, i.e. to produce minimum fluid film separation such that } \Lambda = 1$$

$$= \frac{\Lambda}{\Lambda_1} \text{ when } \Lambda_1 = 1$$

A typical oil thickness in the contact zone is 0.1–1.0 μm. The application of the viscosity ratio is outlined in refs 7 and 10. In assessing a_3 (or a_{23}), the value of ν_1, the speed-dependent rated viscosity necessary to maintain an ehl film, is determined from the mean bore diameter, d_m, and the speed n (Fig. 7.21). For a given mean bearing diameter and speed, the value of the minimum viscosity requirement, ν_1, is greater for roller element bearings than for ball element bearings. The actual viscosity of the oil, or grease base oil, in a bearing at temperature is calculated using Fig. 7.22 on the basis that viscosities are normally specified by oil or grease manufacturers referenced to 40 °C. Below Λ (or K) = 1, the a_3 (or a_{23}) factor may be improved by the addition of extreme pressure (EP) additives to the oil, e.g. curve b in Fig. 7.20). Alternatively increase ν.

To summarize, the theory and practical experience on which a_3 (a_{23}) is based is as follows:

1. A rolling bearing will only run without damage as long as the rolling elements and raceways have no metal/metal contact.
2. Preconditions for a separating lubricant film are high initial viscosity, distinct pressure–viscosity properties and a sufficiently high rolling speed.
3. The minimum oil thickness should be at least as large as the mean value of the conforming surface roughnesses.
4. The condition of the lubricant, and in particular the type and amount of any contaminant, will have a substantial influence on the rating life.
5. Aim for minimum viscosities at operating temperature:[8] 13 cS for ball-bearings, 20 cS for roller-bearings, plus a low speed limit of $10^4/D_{pw}$, where D_{pw} = pitch diameter of ball or roller set. Factor $a_3 < 1$ if below these limits.

7.3.6 Speed limitations

Most bearing manufacturers specify limits for their products and these serve as a useful guide for the majority of normal bearing applications. The published speed limits are of value only when considered alongside other operational factors, but not every application will function satisfactorily at such a speed. Load, lubrication and temperature will influence performance. Bearing operation at the catalogue speed limit demands good lubrication, moderate load and a contained thermal environment. The term 'speed limit' is, however, a flexible value inasmuch that with careful design application, it is possible under certain circumstances to exceed it.

If the speed of a rolling element bearing is gradually increased, a stage will be reached where smooth uniform operation no longer occurs. Inertial forces of the rolling elements and cage as well as small out-of-balance and out-of-round variations will lead to unsteady running, lubrication conditions will deteriorate, friction and wear increase causing the bearing to run hot.

The speed limit greatly depends on the bearing and the entire bearing arrangement, but manufacturers' catalogues

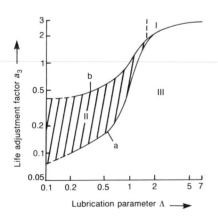

Zone
I Transition range to unlimited life for complete surface separation by lubricant film. Preconditions are total cleanliness and moderate loading
II High degree of cleanliness. Suitable additives in lubricant
III Unfavourable operating conditions. Contaminated and/or unsuitable lubricant

Fig. 7.20 Life adjustment factor a_3

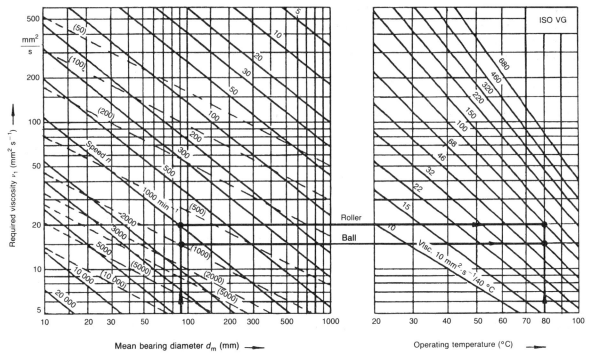

Fig. 7.21 Selection of minimum viscosity for a given bearing speed and bore

Fig. 7.22 Viscosity at operating temperature referenced to the 40° standard (based on ISO viscosity grade classification)

will generally distinguish limiting speeds both for oil and for grease lubrication. The expression

Mean bearing diameter × shaft rpm

is often quoted as the parameter on which speed performance can be based (Table 7.14), but it takes no account of oil viscosity, the rolling element diameter or load. The $n \times d_m$ factor is also influenced by the type of cage (Table 7.14). Decreasing cage weight leads to smaller imbalance forces and therefore to higher speed. Cages made of acetal or nylon have an advantage in that respect while polymeric cages generally will dampen vibrations and noise.

The speed factors quoted in Table 7.14 are not definite limits but are indicative of speeds that normal bearings correctly mounted may reach, assuming:

1. Pure radial loading
2. Moderate loads;
3. Rigid shaft and housing;
4. The correct operational clearance;
5. Good lubrication.

For higher speeds it is possible to improve lubrication with improved heat transfer or use of bearings with special cages. However, when operating at high speed, it is essential to use a sufficiently adequate radial load, plus preload if applicable, to maintain rolling contact without slip, spin or skidding of the rolling elements and also to minimize running noise.

Table 7.13 *Life adjustment factors* (a_1)

% reliability level	a_1
90	1
95	0.62
96	0.53
97	0.44
98	0.33
99	0.21

Note: 100% reliability is statistically impossible.

7.3.7 Friction

Low friction and moderate lubrication requirements are two important advantages of rolling element bearings. The

Table 7.14 *Limiting speed factors for various rolling element bearings*

	nd_m factor $\times 10^6$	
	Grease lubrication	Oil (mist) lubrication
Radial ball-bearing, snap fit phenolic (or acetal) cage; crown type, inner ring location	0.7	0.85
Radial ball-bearing, solid phenolic (or acetal) cage, inner ring location	1.0	1.6
Radial ball-bearing, two-piece or crown-type metal cage, ball riding	0.38	0.38
Cylindrical roller, machined brass cage, roller locating	0.53	0.72
Cylindrical roller, machined brass cage, inner ring located	0.35	0.54
Angular contact ball-bearing, phenolic cage, outer ring located		
15° contact angle	1.0	1.6
20° contact angle	0.95	1.5
25° contact angle	0.87	1.35
30° contact angle	0.5	0.76
Tapered roller-bearing	0.2	0.4
Needle roller-bearing	0.28	0.47[a]

Notes: $C/P \geq 10$; n = shaft speed, rotating shaft (min^{-1}) or $n = \frac{1}{2}$ outer race speed if fixed shaft; d_m = mean bearing diameter = $D + d/2$ (mm). Expected life 100 000 hours. Single-row bearings only.
[a]Max. value, small bearings 0.32.

resistance to rotation is composed of rolling, sliding and lubricant friction as follows:

Rolling losses. Elastic deformation of the raceways and elastic hysteresis as a result of the rolling contact.

Sliding losses. Cage/element rubbing contact plus some slippage in the actual contact zone.

Lubricant losses. As a result of oil viscosity, hydrodynamic action, churning and general working action.

Total running resistance will normally be small in comparison with the transmitted force. When starting an anti-friction bearing, some dry friction is normally encountered such that starting torque will be higher than the rotating torque. In the case of grease lubrication, high friction must be expected if the bearing cavities are completely filled with grease and there is not enough room for the sides to take up the excess. How large the friction may rise can be deduced from the fact that the danger of overheating will arise in excessively lubricated bearings. If, however, excess can escape from the bearing, low friction can be expected, even with grease.

For rolling elements bearings:

$$\mu = \frac{M}{F \times d/2}$$

where M = total frictional torque of bearing (N mm), F = resulting bearing load = $\sqrt{(F_r^2 + F_a^2)}$ (in N) and d = bearing bore.

Frictional torque

$$M = M_0 + M_1$$

where M_0 = torque of load-free bearing, i.e. lubricant-dependent losses and cage/element friction and M_1 = load-dependent component of total frictional torque, where

$$M_0 = f \times 10^{-7}(\nu n)^{0.67} d_m^3$$

and f = coefficient depending on bearing type and lubrication method (Table 7.15), ν = operational oil, or grease-based oil, viscosity (mm^2 s^{-1}) and n = speed (min^{-1}).

$$M_1 = \frac{\mu_1 F \times d_m}{2}$$

where μ_1 = friction factor depending on bearing design and relative bearing load (Table 7.15) = $K_f(P_0/C_0)^y$ and K_f, y = constants.

Due to contact area curvature, the friction factor μ_1 in ball-bearings is proportional to $(P_0/C_0)^y$. The magnitude of the y exponent depends on the spinning friction component. For ball-bearings with small spinning friction $y = 0.5$, and for bearings with a larger spinning friction component, e.g. angular contact, $y = 0.33$. For roller-bearings, μ_1 is constant.

Although the above relationships will permit a fairly accurate estimation of bearing frictional torque, the practical exceptions are as follows (1) in a number of cases mixed friction may occur; (2) at start-up, frictional torque may exceed the calculated values and (3) additional torque occurs when rubbing seals are installed.

Resulting from the frictional torque and operating speed, the rate of heat energy developed, W_F watts, is given by

$$W_F = 1047 \times 10^{-4} nM$$

7.4 Lubrication

The variety of lubricants available for rolling element

Table 7.15 *Summary of friction factors*

Bearing type	Coefficient of friction μ [a]	f Oil circulation	f Grease lubrication	μ_1 [b]	K_f	y
Deep groove ball-bearings	0.0015–0.003	1.5–2	0.7–1	—	0.0009	0.5
Angular contact, 20°	0.0015–0.002	1.5–2	0.7–1	—	0.001	0.5
Angular contact, 30°	0.0015–0.002	1.5–2	0.7–1	—	0.001	0.33
Angular contact, 40°	0.0015–0.002	1.5–2	0.7–1	—	0.0013	0.33
Roller-bearings (cylindrical)	0.001–0.003	2–3	1.5–2	0.0005	—	—
Needle roller-bearings	0.002	6–12	3–6	0.0005	—	—
Spherical roller-bearings	0.002–0.003	4–6	2–3	0.001	—	—
Tapered roller-bearings	0.002–0.005	3–3.5	1.5–2	0.001	—	—
Ball thrust bearing	0.003	3–4	1.5–2	—	0.0012	0.33

[a] Values apply to a load angle β (Fig. 7.15). Lower values refer to mainly radial loads, larger μ's refer to radial bearings and axial loading.
[b] Values assume lubricant separation.

bearings is very large and there are probably hundreds of commercial brands in the market-place. Hence the question of lubricant selection may at first sight seem a daunting task, but in actual fact rolling bearings are not, as a rule, very demanding as regards lubrication. Even in the mixed friction region (Figs 6.1, 6.10) they will work satisfactorily. The primary objectives of a lubricant are as follows:

1. To reduce friction and/or wear.
2. To act as a coolant. In some systems this may be a vital lubricant function to remove frictional heat without the assistance of any other form of cooling.
3. To remove wear debris and contaminants or to prevent other contaminants from entering the system.
4. To protect metals against corrosion. Mineral oils for example are a very effective corrosion preventative.

Grease may be used satisfactorily in all but the most critical applications, oil lubrication being more effective where speeds/loads/temperatures are high. In general, oil utilization requires a more complex design, whereas grease is more easily retained in any assembly, sealing is easier and housing designs simpler and less expensive. Accordingly there is an increasing tendency to use grease. Present-day quality greases have high reliability and wide performance ranges. In that context, this section will concentrate on grease as a lubricating medium, being more appropriate to the design of small machines and consumer goods.

The majority of grease-lubricated bearings now operate with so little trouble that they tend to be taken for granted. Problems that do arise are more likely to do with a mechanical aspect of the design rather than the limitations of the grease.[11] In a sense, grease is a stored oil supply and the simplest form of lubrication is the use of only

Table 7.16 *Advantages and disadvantages of grease lubrication*

Advantages
1. Simple and economic
2. Convenient to apply and retain. Serves as a seal against intrusion of moisture and foreign matter
3. Grease protects the working surfaces of a bearing by adhesion to them, thus prevents rusting during idle periods when oil tends to drain away
4. Fit for life in many applications
5. Easier to retain in a housing than oil without risk of leakage
6. Convenient to handle

Disadvantages
1. Unsuitable for high speeds. Churning at high temperature will cause grease breakdown
2. Prone to trap debris and contaminant
3. Not suitable for bearings where heat must be dissipated. There is no coolant flow to remove heat where bearing speed is high or bearing heavily loaded

a very small quantity of oil in the bearing.[12] The advantages and disadvantages of grease as a lubricant are outlined in Table 7.16.

7.4.1 Grease constitution and properties

Greases are semi-solid materials composed of a thickening agent and a liquid lubricant plus one or more additives such as an antioxidant, a corrosion inhibitor plus a possible EP additive to give better overload performance. Many additives used in lubricating oils are equally effective in grease. Solid lubricants such as graphite and MoS_2 can confer important advantages in greases, especially in high-temperature lubrication situations when the solids take over should the grease dry out or oxidize.

The liquid lubricant is usually a mineral oil, but in certain cases synthetic fluids such as silicones or esters are used to enhance high- or low-temperature properties. Most common thickening agents are metallic soaps, e.g. calcium, sodium or lithium, but in order to make a more temperature-stable product, silica, clay and soap/salt complexes, etc. are increasingly used. Some typical greases are described in Table 7.17. When selecting a grease for a particular application, some parameters used to describe performance are given under the headings below.

(a) Consistency (stiffness). Grease consistency or stiffness is primarily determined by the percentage thickener (crystalline fibres) and the base oil viscosity. Hence a grease of a given consistency may be compounded in many ways by varying these criteria. In practice greases may vary from semi-fluids, hardly thicker than a viscous oil, to solid greases almost as hard as softwood. Equal stiffness greases are not necessarily equal in performance. Medium-consistency greases are normally favoured in rolling bearing applications.

(b) Worked penetration. This is a performance value obtained after the grease has been manually or mechanically forced through a perforated plunger a number of times to simulate service conditions of churning or working. Worked penetration is grouped into a numerical hardness scale of NLGI numbers (National Lubricating Grease Institute) where the higher the NLGI number the stiffer the grease. For roller-bearings, grease of NLGI number 2 or 3 is mostly used (Table 7.17).

(c) Stability. Mechanical stability is the ability of a grease to resist structural breakdown resulting from the shearing forces present when a grease is worked in a bearing. Chemical stability is the ability to resist chemical change.

(d) Channelling (or clearability). A channelling grease forms a stable wall or channel as the rolling elements churn through the grease, the oil from the grease being thought to feed the raceways from the grease wall. Channelling-type greases normally fall into the NLGI 3 or 4 classification. Some smooth greases channel very little, i.e. they slump or flow back readily into the voids left by the rotating elements. If the grease flows back too rapidly, this impedes the elements, retaining torque and increasing bearing temperatures, hence decreasing lubricant life. Rough-textured greases channel very

Table 7.17 *Summary of typical principal rolling-bearing greases*

No.	Grease thickener	Base oil	NLGI no.	Drop point °C	Temperature range min./max. (°C)	Remarks
1	Calcium soap	Mineral	2	100	−12/60	Good water resistance. Fibres up to 100 μm length
2	Sodium soap	Mineral	3	210	−10/150	Absorbs water. May soften and flow out of bearing. Fibres up to 100 μm length
3	Lithium soap	Mineral	2	190	−30/105	Universal grease. Emulsifies with water only to a very limited extent. Fibres up to 100 μm length
4	Lithium soap	Synthetic	2	—	−60/130	Low-temperature, high-speed grease
5	Calcium complex	Mineral or synthetic	2	—	−25/130	Multipurpose, water repellant grease for high temperatures and loads
6	Polyurea	Synthetic or mineral	—	238	−30/170	Water stable. Suitable for high temperatures, loads and speeds
7	Bentonite	Mineral	3	>250	−20/200	Water resistant. Suitable for temperatures at low speed. Corrosion and oxidation inhibited. No smell. Primary particles plate like 0.02/0.05 μm in size

rapidly and the bearing temperature hardly rises, but this is undesirable in practice since such greases also tend to wind out of the bearing seals. An intermediate grease is clearly optimum.

7.4.2 Performance

The bearing design and operating conditions will determine the lubricating performance of the grease, i.e. the degree to which the grease is stressed.

The most difficult type of cage to grease lubricate are those riding on the inner ring. Cages centred on the rolling elements may be run at a higher speed as may outer race riding cages. With outer race rotation, the grease is subjected to centrifugal loading which could exceed the yield value of the grease, causing it to flow continuously into the bearing causing overheating. Figure 7.23 illustrates approximate but realistic maximum speeds for various bearing and cage configurations using a lithium stearate grease.

The ability of a grease to lubricate depends on its retention properties, i.e. it must have a 'drop point' well above the operating temperature (Table 7.17). If the grease softens too much, it will slump and ultimately liquefy. At high temperature a grease gelled by a treated clay, such as bentonite, would be used rather than a metal soap.

As temperature increases, grease life shortens (Fig. 7.24). For a lithium soap grease, the useful life of a single charge of grease is reduced from several years at ambient to a matter of weeks at 100–130 °C. Prolonged exposure at high temperature causes darkening and hardening of the grease, together with the formation of gummy or lacquer-like deposits on the bearing surface.[11]

Some applications may require lubrication at subzero temperatures. Normally there is no problem down to −15 to 25 °C (Table 7.17), although the grease will stiffen to some degree. However, where the torque requirement becomes excessive, a synthetic oil-based grease should be used.

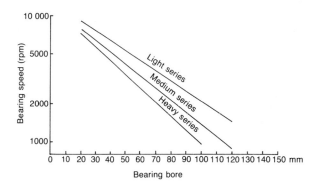

Multiplication factors applied to the diagram above for different types of bearings and cages

Bearing type	Cage centred on inner race	Pressed cages centred on rolling elements	Machined cages centred on rolling elements	Machined cages centred on outer race
Ball	As diagram	1.5–1.75	1.75–2.0	1.25–2.0
Cylindrical roller	As diagram	1.5–1.75	1.75–2.0	1.25–2.0
Taper and spherical roller	0.5	—	—	—

Bearings mounted on adjacent pairs } 25% reduction
Bearings on vertical shafts } maximum speed

Bearings with rotating outer } 50% reduction
and fixed inner races } maximum speed

Fig. 7.23 Approximate maximum speeds for grease lubrication; lithium stearate grease. From Scarlett N A *Proceedings of the Institute of Mechanical Engineers*, **182**(3A): 585 (1967–68). Reprinted by permission of the Council of the Institution of Mechanical Engineers

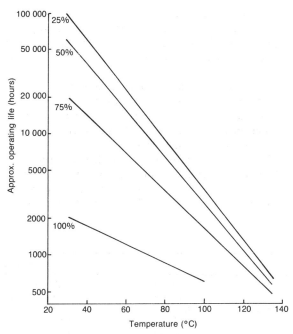

Fig. 7.24 Variation in operating life of a single charge of lithium stearate grease as a percentage of maximum speed rating at given operating temperature. From Scarlett N A *Proceedings of the Institute of Mechanical Engineers*, **182**(3A): 585 (1967–68). Reprinted by permission of the Council of the Institution of Mechanical Engineers

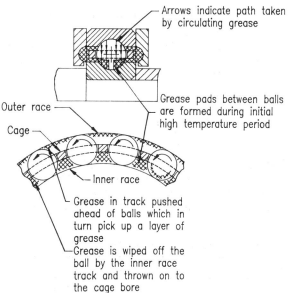

Fig. 7.25 Mechanism of grease redistribution in ball-bearing assembly. From Scarlett N A *Proceedings of the Institute of Mechanical Engineers*, **182**(3A): 585 (1967–68). Reprinted by permission of the Council of the Institution of Mechanical Engineers

7.4.3 Grease lubrication mechanism

Experiments have shown that grease (and base oil) in the housing or shield recesses of a well-run bearing does not migrate into the bearing. It does, however, provide a very close-fitting seal and prevents the escape of grease/oil from the bearing. Only grease which is immediately adjacent to and in contact with the raceways contributes to lubrication. A thin layer of grease on the working surfaces is structurally broken down under shear to a low-viscosity base oil material which acts as the lubrication medium.[11]

When commencing to run a grease-packed bearing, some is rapidly displaced from the bearing. Other grease is pumped between the inner and outer races via the balls rotating in the cage bore, creating considerable turbulence and generating heat. During the process, grease is splashed on to the cage bore between each ball where it adheres and forms pads of grease (Fig. 7.25). The formation of excess grease into static pads takes some hours to complete, as indicated by the period of running before a normal settled temperature is reached.[11]

Unsatisfactory greases do not form complete pads, but circulate continually with the bearing temperature staying high. After a long period of running, the pads of grease will harden as the oil is centrifuged out.

7.4.4 Grease packing

Effective bearing running-in and subsequent operation depend on both the quality and the quantity of the grease. Too much grease gives overheating and lubrication breakdown, too little leads to lubricant starvation and bearing damage.

In radial ball-bearings fitted and two shields or seals, only approximately one-third of the free space should be filled with grease. As a general guide with open bearings and inner ring rotation, the pack should be a half to three-quarters full, but housings should be dimensioned to accommodate the grease expelled. If anything, grease packs should always err on the low side.

7.5 Application factors

7.5.1 Bearing noise

Rolling element bearings can act as a noise source, but if this is a problem it can be contained to a large degree by clearance specification and preload (section 7.3.2). Quiet running is especially important for small motors used in domestic appliances; the bearing noise can be dampened by use of specially filtered stiff grease. See also section 16.4.1(i).

7.5.2 Bearing fits

Rolling bearings need to be fixed on the shaft and in the housing in three directions, axially, radially and circumferentially. Radial and circumferential fixing is achieved by interference fitting the bearing raceways, axial location by preload (section 7.3.2). In selecting a fit, the following should be considered: (1) press fits may affect radial clearance; (2) ease of assembly and disassembly; (3) bearings should be well supported over the circumference to utilize the load-carrying capacity fully.

If it is not possible to provide both rings with an interference fit, a sliding fit is possible on either shaft or housing depending on the load direction and the rotating member (Table 7.18). An alternative for bearing fits is the use of anaerobic adhesive (section 4.5.2).

For bearings in the normal radial clearance group, the clearances occurring during normal mounting with standard fits are taken into consideration in the recommended table of fits (Table 7.19). Large amounts

Table 7.18 *Selection of seating fit*

Rotating member	Radial load	Shaft seating	Housing seating
Shaft (stationary outer)	Constant direction	Interference fit	Sliding fit
Shaft	Rotating	Sliding fit	Interference fit
Shaft or housing	Combined constant direction and rotating	Interference fit	Interference fit
Housing (outer)	Constant direction	Sliding fit	Interference fit
Housing (outer)	Rotating	Interference fit	Sliding fit

Table 7.19 *Shaft and housing seating limits for metric series bearings, bore up to 30 mm*

				Radial ball-bearings				Cylindrical roller-bearings			
Shaft	Interference fit	Light radial load <0.07C	Grade Limits	h_5 −5 / 0	−6 / 0	−8 / 0	−9 / 0	j_6 +6 / −2	+7 / −2	+8 / −3	+9[a] / −4
		Normal radial load >0.07C <0.15C	Grade Limits	j_5 +3 / −2	+4 / −2	+5 / −3	+5 / −4	k_5 +6 / +1	+7 / +1	+9 / +1	+11 / +2
		Heavy radial load >0.15C	Grade Limits	k_5 +6 / +1	+7 / +1	+9 / +1	+11 / +2	m_5 +10 / +4	+12 / +6	+15 / +7	+17 / +8
	Sliding fit		Grade Limits	g_6 −4 / −12	−5 / −14	−6 / −17	−7 / −20	g_6 −4 / −12	−5 / −14	−6 / −17	−7 / −20
Housing	Interference fit	Normal radial load	Grade Limits	M_6 −1 / −9	−3 / −12	−4 / −15	−4 / −17	M_6 −1 / −9	−3 / −12	−4 / −15	−4 / −17
		Heavy radial load	Grade Limits	N_6 −5 / −13	−7 / −16	−9 / −20	−11 / −24	N_6 −5 / −13	−7 / −16	−9 / −20	−11 / −24
	Sliding fit		Grade Limits	J_6 +5 / −3	+5 / −4	+6 / −5	+8 / −5	J_6 +5 / −3	+5 / −4	+6 / −5	+8 / −5
		Range of bearing bore (mm)		up to 6	6 to 10	10 to 18	18 to 30	up to 6	6 to 10	10 to 18	18 to 30

[a] Tolerances in μm.

of interference in fitting practice can cause bearing clearances to decrease significantly or even disappear. The thermal conditions of bearing operation may also affect the diametral clearance, the relative expansion/contraction of the shaft, housing and bearing components depending on the shaft and housing materials together with the magnitude of the thermal gradient. The radial clearance groups C3, 4 and 5 (Table 7.6) are necessary for some assemblies, not because the bearing needs a large radial clearance but because clearance is reduced by fit and temperature considerations.

7.5.3 Housings

The bearing housing should be reasonably robust and rigid in its support for the bearing. A variable housing stiffness created by webs, ribs and reliefs may provide a distorted bearing support under load, thereby affecting the load distribution. Housings may be separate components fastened to a machine frame or they may be an integral part of the machine, i.e. the machine frame itself.

A housing is only as stable as the supporting structure. For effective support the housing itself should have a solid path to a rigid foundation. If bolted to a flexible machine frame or plate undergoing deflection, the bearing may be subjected to distortion loading for which it was not designed. Even though deflections may not harm the housing, transmitted to the bearing, damage is possible. Design of the housing should be such that shaft deflection under load should not oppose the deflected angle of the housing.

The raceways within a housing are frequently located axially by abutting against a shoulder. A radius at this point is essential for the avoidance of stress concentrations and ball races are provided with a radius or chamfer to allow space. Because the corners of the race are formed before the functional surfaces are finish-ground, the cross-section of the race edge does not reveal a truly circular arc but rather one of intermediate profile (Fig. 7.26).[1]

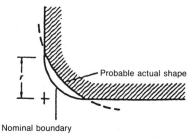

Fig. 7.26 Specification of bearing corner radius

A radius r is therefore specified which represents an imaginary boundary beyond which the race material does not penetrate.

7.6 The design selection procedure

In designing a rolling element bearing application, the basic job of the design engineer is to calculate the operating loads on the bearing, specify the bearing geometry necessary for an adequate life and to ensure proper functioning and survival of that bearing. The calculation of rolling element bearing loads and the selection of rolling element bearings may be adequately accomplished in the majority of cases from the application of simple mechanics and access to a reasonable bearings catalogue. By and large it is a factual, well-understood process.

The calculation procedure outlined in this section is based upon the implications and correct interpretation of the data presented in previous sections of this chapter. The procedure normally applies to applications where operating conditions are generally well defined and where some degree of calculation refinement is possible to take better advantage of the modern quality bearing, particularly taking into account the lubrication conditions within the bearing. This latter aspect is generally ignored where the demands on the bearing are not critical. The selection procedure algorithm is outlined in Fig. 7.27, described under the headings below.

(i) Specify the requirements

Certain basic information should be available at the start of the selection procedure to define the boundaries on which the design can proceed. The typical input parameters would be as follows:

1. The required life — which in some cases may exceed the design life of the machine.
2. Operating speed — and if known at this stage the speed spectrum if the operation is non-continuous.
3. Available space — the bore diameter is often set by the shaft size allowable.
4. Rotation — i.e. whether a shaft or a wheel bearing design is required.

(ii) Definition of the application

This requires an analysis of the bearing loading plus a definition of bearing type and size:

1. *Loads and directions*. Determine the axial and radial load components (F_a and F_r) on the bearing relative

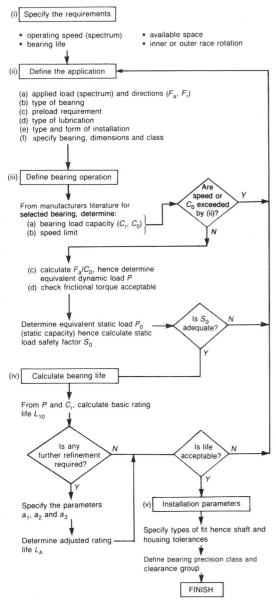

Fig. 7.27 Standard rolling element bearing design procedure

load is a radial one resulting from one, or a combination of, the following:
(a) weight of machine parts supported by the bearing;
(b) tension due to belt or chain pull;
(c) starting or stopping or reversing torque;
(d) centrifugal forces from out-of-balance eccentric loads or cam actions.

However, thrust loads may also arise from vertical shafts, gear-wheel thrust loads and thermal expansion of shaft or housing. Shock or impact loads are difficult to assess, but should be estimated if the risk is likely to exist. Three-bearing shafts should be avoided since slight misalignment in practice could impose destructive bearing loads if on short centres.

2. *Bearing type*. Once loads are known, choose appropriate type (section 7.2.1). Operating speed may also influence choice.
3. *Preload*. Choice of bearing type based upon application detail will then define any preload requirement (section 7.3.2). Adjust axial loading.
4. *Lubrication*. The basic decision is for oil or grease (section 7.4, also Table 7.15). Simplest solution is a sealed radial ball-bearing, i.e. fit for life (section 7.2.4(ii)). Other options require sealing to be designed into the application. Consider also manufacturability practicalities.
5. *Installation*. General design outline should be understood at this point and the manner of bearing retention.
6. With the above parameters defined, the bearing can be identified (section 7.2.5), although clearly this is an iterative process in practice throughout the application definition activity.

(iii) Bearing operation

When deciding on the bearing size for a certain application, loads must be related to the carrying capacity. Operating speed and frictional torque also need to be checked. The assessment of operating suitability is as follows:

1. *Load-carrying capacity*. Bearing catalogues give two basic load ratings, static (C_0) and dynamic (C_r). These values may also be calculated (section 7.3.3). These basic ratings are based on two different failure criteria:
   (a) *Static load rating*. This refers to the application of shock loads when stationary or at slow running speed and is the maximum equivalent bearing load (section 7.3.3(ii)).
   (b) *Dynamic load rating*. This is based on surface

to the load angle (Table 7.10). Loads exerted on shafts by machine components should be translated into bearing reactions by use of static beam solutions. Normally the shaft is considered as a simple beam with two supports (bearings). Such analysis is widely covered in strength of materials texts. In most applications, shafts are horizontal and the principal

fatigue and defines the operating life limit of the bearing (section 7.3.3(i)).

Catalogue estimates are based on the number and size of the rolling elements fitted, hence the bearing potential may be manipulated by adjusting these parameters. Increasing the number of balls or rollers increases the number of stress repetitions per revolution, but decreases the load per element such that the net result is an increased load capacity, hence improved life. However, increasing the number of elements also infers reducing size. For example, standard load capacity for ball-bearings $\propto$ (ball diameter)$^{1.8}$.

2. *Limiting speed*. Check application against catalogue data for type of lubrication and bearing construction (section 7.3.6). Limiting speed factors given in Table 7.14.
3. *Equivalent dynamic load* (P). For complex duty, other than purely radially loaded, it is necessary to reduce loading conditions to those of an equivalent imaginary load P. With F_a and C_0 known (section 7.6(ii)), determine relative axial load F_a/C_0, hence e (Table 7.10). Radial and thrust factors X and Y are also listed in Table 7.10, hence calculate P using Table 7.11. Adjust for fluctuating load conditions (section 7.3.4(ii)).
4. Knowing speed, equivalent load and type of lubrication, check frictional torque acceptable (section 7.3.7).
5. *Equivalent static load* (P_0). Used where life is not the determining factor in the design (section 7.3.7(iii)). With F_r and F_a known, P_0 can be calculated from the X_0 and Y_0 values in Table 7.12. With C_0 also known, the static load safety factor S_0 can be determined and assessed (section 7.3.4(iii)).

(iv) Life calculation

With P and C_r known, the basic rating life (L_{10}) follows (section 7.3.5(i)). For a typical application with a shaft rotating 500–3000 rpm, C_r/P must be at least 10 to achieve a respectable life. In more critical applications, the extended life calculation requires the following input (section 7.3.5(ii)):

a_1 — life adjustment factors >90 per cent given in Table 7.13.
a_2 — material factor, for hardened steel = 1.
a_3 — the more critical operating conditions factor — either approach via the film thickness/surface roughness ratio Λ (Fig. 7.20), or by ν estimate requirement of bearing surface speed; a_3 (or a_{23}) then based on viscosity ratio K once *actual* viscosity of lubricant at the operating temperature is known (Figs 7.21, 7.22).

An aspect so far ignored is the question of stacked races. The introduction of a second row of rolling elements does not double the bearing life as might be thought, nor is load sharing likely to be equal. If a single radial ball-bearing is replaced by a stacked pair, life is increased by a factor of ~4.3 (~5 for roller bearings). Alternatively design for a double-row bearing.

(v) Installation

If the calculated life is acceptable, bearing precision class, clearance group and fits should be established (section 7.2.5) based on the following guidelines:

1. General rule is to provide an interference fit to the component experiencing the rotating load (Table 7.18).
2. Fatigue failure can be accelerated by misalignment and other shortcomings of the assembly and fitting.
3. Bearing manufacturers apply correct tolerances to bearings, and the user must ensure that correct tolerances are applied to shaft diameters and housing bores.
4. Bearing elements operate with small running clearances. In production assemblies, internal clearances will vary slightly, with every build since every component in the assembly is made to a tolerance (Table 7.19).

7.7 Other rolling element products

Although the conventional rolling element bearing finds use in the vast majority of bearing applications, there are none the less a few important additional products which have been developed out of rolling element theory which either address a drawback of a conventional bearing or attempt to combine some additional feature. Some of the more successful of these are discussed in this section. Because of the specific features offered by some of these products, inclusion within other sections of this book would have been equally justified, but are included here simply because of the rolling element common denominator.

7.7.1 Plastic raceways

These bearing types are an analogue of single-row radial bearings except that the raceways and cage are moulded

Table 7.20 *Advantages and disadvantages of plastic radial ball-bearings*

Advantages
1. Low cost in quantity plus opportunity to mould in additional design features
2. Low weight, inertial consideration
3. High corrosion resistance. Stainless steel balls allow for bearing to run immersed in process fluid if required
4. Quiet in operation
5. No oil present
6. Very good low temperature ability; will operate at −50 °C
7. A cheap alternative to stainless steel bearings where hygiene conditions apply

Disadvantages
1. Inferior load–speed ratings
2. High loads will dent tracks
3. Radial bearings cannot take axial loading; axial loads will cause indentations on raceway shoulders
4. Limited off-the-shelf size ranges
5. Not a precision bearing
6. A real lack of meaningful quantitative design data available to the designer

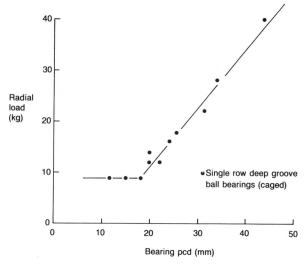

Fig. 7.28 Load-carrying capacity of acetal plastic bearings at low speed

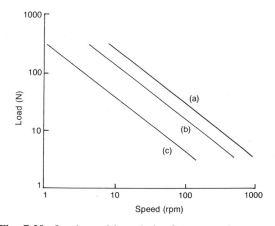

Fig. 7.29 Load-speed boundaries for an acetal thermoplastic ball-bearing assuming steady-state temperatures of (a) 120 °C, (b) 80 °C and (c) 40 °C. From Hilyard N C et al *Proceedings of the Institute of Mechanical Engineers* **199**(C4): 339 (1985). Reprinted by permission of the Council of the Institution of Mechanical Engineers

thermoplastic and the balls invariably stainless steel. They offer advantages over conventional bearings in the terms of cost, weight and corrosion resistance, but have inferior load/speed ratings (Table 7.20). As with a number of polymers used in tribological applications, design application is for the most part based on trial and error testing over an extended time interval.

By virtue of the materials used and the counterformal contact conditions, performance is inherently limited relative to other types of bearing. However, there are many light-duty applications where this type of bearing finds extensive use. In ambient conditions of low load, the maximum safe speed is approximately 1000 rpm. Safe working load is a function of the bearing diameter (Fig. 7.28). To improve these rather low values, further bearings may be necessary to spread the load. Acetal is the polymer most widely used for rolling element raceways (see also Ch. 5).

In continuous operation, load and speed are limited by frictional temperatures developed in the contact zone, hence the thermal stability of the polymer. Operation limits of acetal in terms of speed and loading are defined for various temperatures in Fig. 7.29.

Normal failure processes present with plastic bearings are surface fatigue and progressive wear over a period of time. With glass-reinforced raceways, material failure will occur by surface flaking.[13] At high contact stresses, true brinelling will occur if the yield strength of the subsurface material is exceeded. With thrust or impact loads, similar effects can occur on the race shoulder.

Under very heavy loads and speeds, a rapid plastic flow type of wear called 'smearing' can occur, which can result in redistribution of race material within the bearing.

7.7.2 The roller clutch

The roller, or one-way, clutch is similar in appearance to drawn-cup needle bearings, but is designed to transmit

LOCK FUNCTION
Shaft drives gear
clockwise (white arrows)

or gear can drive shaft
counter-clockwise
(black arrows)

OVERRUN FUNCTION
Shaft overruns in gear
counter-clockwise (white arrows)

or gear overruns on shaft
clockwise (black arrow)

Fig. 7.30 Principle of the roller clutch. Courtesy of the Torrington Company, Connecticut

torque between a shaft and housing in one direction of rotation but allowing free overrun in the opposite direction. The principle is outlined in Fig. 7.30 for the two modes of operation, the overrun mode and the lock mode. Either the shaft or the housing can be the drive input. In the overrun mode, because rollers need to rotate unloaded, the clutch will only allow a very light radial load. With heavy radial loads, either external support is required or a combined clutch/bearing unit.

Modes of operation are depicted in Fig. 7.31. Operation is controlled by the direction of clutch or shaft rotation with respect to the locking ramps. In the overrun mode, the relative rotation between the housed clutch and the shaft causes the rollers to move away from their locking positions. In the lock mode the relative rotation between the housed clutch and the shaft is opposite to that in the overrun mode. The rollers, assisted by leaf-type longitudinal springs, become wedged between the locking ramps and the shaft, hence transmitting torque. The interior ramps which control the clutch operation are formed during the cup draw operation.

The following application points apply to the roller clutch:

1. Should operate directly on to a hardened shaft, approximately 750 kg mm^{-2}. Unhardened shafts are seldom satisfactory.
2. Clutches are designed to transmit pure torque loads. Manufacturers' recommendations should not be exceeded and should include intertial forces on load take-up.
3. Oil is the preferred lubricant, alternatively a mineral oil loaded soft grease. Since clutch lock-up depends on static friction, thick grease will retard roller engagement and encourage roller slip.
4. Clutch engagement is at rates up to 200 per minute. Higher rates are achievable with a light oil rather than grease lubrication. At high engagement frequencies the drive should be via the shaft.
5. Minimal backlash. Will increase if grease lubricated.
6. Performance relies on strict conformance to recommended shaft and housing tolerances. For proper functioning it is often necessary to determine optimum fitting tolerances empirically, especially if low-modulus housings (e.g. aluminium) are used.

7.7.3 The ball nut

The ball nut is a low-cost, low-friction device for converting rotary motion from a rotating power source into linear movement along plain shafting (Fig. 7.32). It consists of a precision-ground outer race with a low

Clearance between the rollers and cup ramps is exaggerated in these drawings.

Fig. 7.31 Mode of operation of roller clutch (a) in lock mode, (b) in overrun mode. Courtesy of the Torrington Company, Connecticut

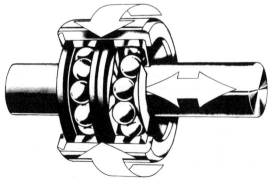

Fig. 7.32 Section through tandem ball nut. Courtesy of RHP Bearings, Newark

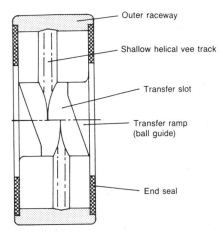

Fig. 7.33 Section of ball nut. Courtesy of RHP Bearings, Newark

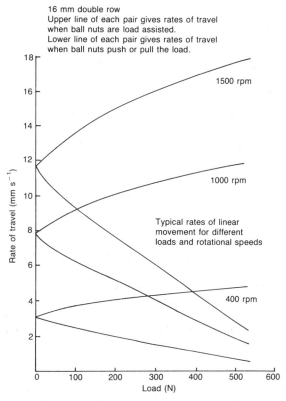

Fig. 7.34 Rate of linear movement load and direction dependency. Courtesy of RHP Bearings, Newark

contact angle, equal angle shallow vee helical pathway, interposed by a transfer slot that connects the two ends of the helix (Fig. 7.33). The transfer slot ($\sim 20°$ of arc) is so designed that only one ball is ever out of load and the ball in the slot is then free to transfer from the end to the beginning of the helix via the internal face cams.

Because the unit is full complement with the cage, the end covers act as seals to retain the balls once the shaft is removed. Due to the operational requirements of this device, units are supplied matched with a hardened shaft ($\sim 700 \text{ kg mm}^{-2}$). For greater load capacity and increased stability, nuts are standard tandem or triple units (Fig. 7.32). Some specific features are as given under the headings below.

(a) Rate of linear movement. The movement of the nut itself is controlled by the epicyclic behaviour of the rotating balls and depending upon the nature of lubricated rolling friction, the extent of ball spin and Hertzian contact theory. Consequently the ratio of rotary/linear movement will depend on load/speed conditions (Fig. 7.34). Although the load/speed relationship is unique to a given size of nut, load/speed behaviour will depend upon the magnitude of the axial load, the direction of the axial load, the speed of shaft (or nut) rotation and the bore diameter (shaft size).

Either the shaft or nut can be driven, the displacement per revolution is the same in either case. The single helix raceway has a standard lead of 1.0 mm, but due to the epicyclic ball traction, the relative no-load axial displacement is reduced to a pitch of approximately 0.5 mm. Because of this fine pitch characteristic, the device can be driven at speeds up to 300 rpm.

(b) Axial security. The bearing itself is a three-point contact device. The helical track is a very shallow, wide-angle feature, hence if an attempt is made to push the nut along the shaft without relative rotation, the balls will lock. There is therefore no free backlash.

If the nut is driven on to a hard stop, it will slip and not cause any override damage. On reversing the direction of rotation, the ball nut will resume normal functioning.

The main advantages of the ball nut unit can be summarized briefly as follows:

1. No limit to the fineness of pitch possible;
2. Can direct drive from a motor, no gearbox is necessary;
3. Statically secure with no backlash;
4. Slips on overload;
5. Clean, safe shaft.

7.7.4 Linear motion rolling bearings

(i) Recirculating ball bushings

These were introduced to overcome the problems of

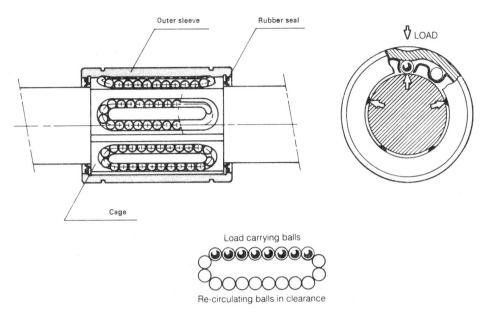

Fig. 7.35 Construction and operation of linear ball-bearing. Courtesy of Thomson Industries, Port Washington, NY

lubricated sliding action. Operating on a hardened steel shaft, ball bushings provide for a linear situation many of the low-friction and minimal wear characteristics of radial roller-bearings. Ball bushings contain three or more oblong circuits of recirculating balls (Fig. 7.35). Each circuit has the balls in one of its straight sides in bearing contact between inner ball bushing bearing sleeve surface and shaft. The load is actually rolled along linearly on balls in this part of the circuit. Balls in the remainder of circuit are free to roll in the sleeve clearance provided in the return track. A retainer or cage within the sleeve guides the balls along their proper path. Retainers are usually made from pressed steel, reinforced nylon or machined from solid. Operation is quieter with a nylon cage. The ball rolling surface will be hardened and ground. In some bushings, indents on the back indicate the loaded ball track positions and by correct circumferential positioning during installation, load-carrying capacity may be increased (Fig. 7.36).

The advantages of ball bushings compared to sliding plain bushes can be summerized as follows:

1. A constant and lower value of friction coefficient;
2. Inherently less wear, hence longer life and enhanced reliability;
3. Precision alignment, accurate support and no run-out;
4. Very little lubrication is required;
5. Clean, compact design;
6. Higher speeds of operations;
7. Greater load-carrying capacity.

A number of variants of the original ball-bushing design now exist in relation to different design solutions as follows:

1. Self-aligning with a central rocking fulcrum;
2. Slotted sleeve hence adjustable diameter possible;

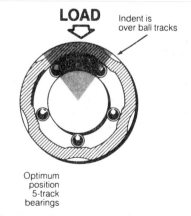

Fig. 7.36 Circumferential position for maximum rolling load capacity. Courtesy of Thomson Industries, Port Washington, NY

3. Open, rigid type for use on supported shafts;
4. Ball spline, also enabling a high torque transmission in addition to linear movement;
5. Grooved shaft.

A slight amount of radial preload is generally permissible, but excess preload will cause rough operation and may well damage the bearing or the shaft surface. Hence with press fitting, too tight a press fit must be avoided. Split bushings are split longitudinally and hence are adjustable to allow preloading when mounted in adjustable diameter housings — the simplest way of achieving this is by tightening until a resistance is felt on the shaft.

(ii) Recirculating roller-bearings

This type of linear bushing is used for high accuracy guidance. The design of the bushing is such as to provide a high degree of rigidity necessary for the precise control of movement. The unit is a full complement roller-bearing with a precision ground and hardened H-section steel body (Fig. 7.37). In the loaded zone the rollers are guided by their faces and rotate without lateral reaction. Roller recirculation is ensured by two lateral containment guides.

(iii) Linear bearing life

For ball bushings

$$L_{10} = \left(\frac{C}{F_r}\right)^3 5 \times 10^4 \, \text{m}$$

and for recirculating roller bearings

$$L_{10} = \left(\frac{C}{F_r}\right)^{10/3} \times 10^5 \, \text{m}$$

where C = basic dynamic load rating, i.e. the constant stationary load which a linear bearing can theoretically endure for a basic rating life of 10^5 m (rollers) or 5×10^4 m (balls), F_r = radial load and L_{10} = nominal life. (m) = maximum travel summation without fatigue for 90 per cent of bearings in a group of the same type and tested under the same conditions.

The calculated life figures apply to a shaft finish of 0.035 μm Ra and hardness 580/680 kg mm^{-2}. When the stroke length and frequency are constant quantities, then

$$L_h = \frac{833 \times L_{10}}{Hn}$$

when L_h = life (hours), H = stroke length (m) and n = frequency of oscillation (min^{-1}).

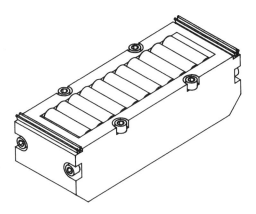

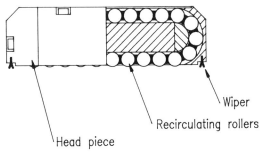

Fig. 7.37 Recirculating roller linear bearing

References

1. Barwell F T *Bearing systems, principles and practice*. OUP 1979
2. Harris T A *Rolling bearing analysis*. Wiley 1984
3. Bearing technology. *Eureka* **9**(10): 101 (1989)
4. Eschmann P et al *Ball and roller bearings*. Wiley 1985
5. BS 292 (ISO 15) *Specification for dimensions of ball and roller bearings* 1982
6. BS 5512: 1977
7. Herraty A G New methods for the selection of roller bearings. *Proceedings of the 7th Cheltenham Bearing Conference*. MEP 1986
8. ISO 281. International Standards Organisation 1978
9. Wood R et al Roller bearings, load ratings and life. *Proceedings of the 7th Cheltenham Bearing Conference*. MEP 1986
10. Albert M Load carrying capacity of rolling bearings. *Certif. Eng* **57**(3): 38 (1984)
11. Scarlett N A Use of grease in roller bearings. *Proceedings of the Institution of Mechanical Engineers* **182**(3A): 585 (1967/68)
12. Lansdown A R Selection of lubricants. In *Tribology*, Series 8. Elsevier 1983: 223
13. Marshek K M Failures in plastic ball bearings. *Wear* **52**: 141 (1979)

CHAPTER 8
DRIVE MOTORS

8.1 Introduction

Ideally, the motor selected for a particular drive application should comply with the following:

1. Start under all specified conditions;
2. Require a minimum start current;
3. Be able to be accelerated to the rated full load speed within the limits set by the requirements of the application;
4. Have maximum efficiency at the rated load;
5. Have adequate overload capability;
6. Be electrically and thermally safe;
7. Be cost effective.

Practical motor applications are generally complicated by variable operating conditions, i.e. variations in the magnitude and duration of the rated load, but successful motor application depends primarily on selecting a motor that satisfies the kinetic requirements of the driven machine without exceeding the temperature and torque limitations of the motor. Power requirements may generally be met by various types of a.c. or d.c. motor, but performance application and system requirements will narrow down the selection process.

A comprehensive guide to the many types of small motors and their variants available would run to several volumes; consequently this chapter will only consider in detail the more common types of fractional horsepower single-phase small rotational motor. Basic knowledge of electrical machines and power electronics is assumed.

8.1.1 A.c. motors

The a.c. motor family is shown schematically in Fig. 8.1. The most common type of a.c. motor is the cage induction motor, variants distinguished by different arrangements for starting. A single-phase design without the aid of a starting device will have no 'inherent' torque. Basically the induction motor is an a.c. transformer with a rotating secondary, the starter sets up a magnetic field which reacts with the current-carrying conductors of the rotor to produce rotational torque by induction action. Induction motors are fairly easy to procure when compared to other motor types.

Where installations require continuous running drives at a controlled speed, the synchronous motor is used, the only economic small motor forms being the reluctance or permanent magnet types (Fig. 8.1), both built into standard induction motor frames.[1] A summary of the applications, advantages and disadvantages of a.c. motor types is given in Table 8.1.

8.1.2 D.c. motors

D.c. motors are used in a wide variety of industrial applications because the torque—speed relationship can be varied to almost any useful form. The types available are outlined schematically in Fig. 8.2. Continuous operation is commonly available over a speed range of 8:1 and, depending on the power supply, can temporarily deliver three to five times their rated torque without stalling. The rating of a d.c. motor is determined by its heating effect.

D.c. motors often incorporate integral feedback devices within the frame itself, and some brushed types now have options for built-in electronic brush wear indicators together with forced-air cooling. Brushed d.c. motors have the ability to operate at variable speeds and high start torque, hence where there is a rapid stop—start operation or changing velocity profile, the d.c. motor is usually a natural choice. However, if frequent stop—starts are required, brush life may be as short as 1000 hours.

Table 8.1 *Summary of AC motor types*

	Split phase induction	Capacitor start induction	Permanent split capacitor
Areas of application	Widely used for domestic appliances, office machinery and industrial fractional hp applications	To overcome high inertial loads, e.g. pumps, compressors machine tools, conveyor belts etc.	Usually used for direct drive fan and blower applications (capacitor start and run)
Performance summary			
Duty	Continuous	Continuous	Continuous
Supply	a.c.	a.c.	a.c.
Reversibility	At rest only	At rest or during rotation	At rest or during rotation
Speed	Fairly constant	Fairly constant	—
Start Torque	130–200% of rated	Up to 300% of rated torque	50–100% of full load
Start Current	High	Normal	Low
Advantages	Good general purpose low cost drive	Greater locked rotor torque than PSC type, i.e. greater acceleration	Speed control possible by voltage control
	Good efficiency and starting torque, rapid acceleration times	After start winding removed from circuit, performance identical to PSC	Used for long and frequent start applications
	Fairly constant speed		Quiet
	Moderate efficiency	Improved start torque characteristics made possible by capacitor action	High efficiency
			Smooth performance
		Capacitor enables reduction in start current	No centrifugal switch
		Fairly efficient	
		Suitable for frequent starts	
		Generally quiet running	
Disadvantages	Special switching required for high speed reversal	Whilst an optimum capacitor value can enhance motor performance, an incorrect value can seriously affect performance	Tend to overheat if lightly loaded
	Not suitable for frequent stops and starts, windings will heat up resulting in loss of torque		Should be run at near the rated load point
	High starting current, 5–10 times run current	No speed flexibility	Locked rotor torque is low compared to other cap. mtrs.
	Unsuitable for accelerating high inertial loads		Not recommended for belt driven loads
	Aux. winding designed for very short duty, if stays in circuit >1–2 secs., winding can be damaged		Most expensive of the capacitor motors
	No speed flexibility		Usually the capacitor value is a compromise between start and run modes
	Low power factor		

DRIVE MOTORS

2 Capacitor (start & run)	Shaded pole induction	Reluctance synchronous	Hysteresis synchronous
Used where high locked rotor torque is required	Used in countless consumer and industrial applications, i.e. fans, blowers, pumps, hair dryers etc.	Timing and constant speed output applications. Drives for various mechanisms	Electric clocks, self-recording devices, synchronous servo systems, tape reels, record turntables, gyros etc.
Continuous a.c. At rest or during rotation — >200% of rated Medium to high	Continuous a.c. Unidirectional Fairly constant 50–100% of rated torque Low	Continuous a.c. or PWM a.c. As non-synchronous Constant 50–100% of pull-in torque —	Continuous a.c. Preferably at rest Constant 100% rated Low
Exceptionally quiet High torques during start and acceleration Use of two capacitors helps to improve the power factor at run speed and preserve efficiency Normally designed for continuous operation	Simple and economical drive Amenable to high volume low cost production Reliable Quiet and vibration free Rugged and inexpensive Reasonably constant speed drive, load variation within rated range doesn't significantly affect speed Can be speed controlled by simple series resistance	Constant speed feature Speed dependent on the frequency of the supply and no. of poles Speed control possible with variable frequency drives Ordinary squirrel cage can be used with this type of drive Moderately efficient	Special motor performance characteristics associated with its rotor design Robust and reliable drive Quiet and smooth in operation No preferred direction for synchronizing Efficiency can be as good as or better than conventional induction motors Low surge currents; 1/10 the start current of other synchronous motors Can be designed for very high shaft power in small sizes Constant speed In the ideal hysteresis motor, torque, current and power remain uniform from stall to synchronism
Capacitor value chosen is normally 'tuned' for one particular load value	Low efficiency, hence use where heat can be tolerated or use air cooling Low start and running torques Use only for low duty applications Enclosed versions only possible in smaller ratings Poor power factor Motor leakage fluxes high, rotor bars must be skewed	If load too high, motor won't obtain synchronism As the ratio of load inertia to rotor inertia increases, the load torque that can be synchronized decreases rapidly Increased acceleration of the rotor is required at the critical point as it approaches the rotational speed of the field Usually larger and costlier than a non-synchronous motor for a given power rating Until synchronism reached, separate starting must be employed Expensive Poor power factor Less efficient than induction motors	Sensitive to non-ideal performance conditions Loss of efficiency results from voltage variations Low supply voltage sharply reduces torque output

Permanent magnet synchronous	Universal motor	Induction torque motor
Combination of best features from induction and permanent magnet motors	Popular in domestic appliances and power tools (a.c. commutator motor)	Designed for use in slow speed or controlled tensioning applications
Continuous a.c. or PWM a.c. Reversible 6000 : 1 range Up to 300% of rated	Continuous or short time a.c. or d.c., SCR control common Usually unidirectional Varies with load >175% of rated torque High	Intermittent Varying power input — Low; 6 or more poles High —
Able to handle industrial loads Very high efficiencies possible Reduced size and cost per unit power Good power:weight ratio High pull-out torque compared to pull-in torque Good start torque and smooth accelerating torque Reduced cooling requirement Near linear torque:speed characteristics Small electrical and mechanical time constants	Highest hp per unit weight per unit cost of any single phase a.c. motor Only motors capable of >3600 rpm on single phase a.c. Higher start torque than any other motor of equivalent physical size Speed adjustable over broad range, usually 4000–10 000 rpm Negligible performance difference between 50 & 60 Hz Efficiency fairly high due to the operating speeds Modern designs tend to favour a.c. operation rather than a true universal characteristic High power factor	Will deliver max. torque under stalled or locked rotor conditions, maintainable for long periods Rugged, requiring minimum of service Designed as simple reversible
Expensive Sophisticated drives control and feedback devices required	Poor speed regulation and load variation in open loop mode Speed changes pronounced since armature and field are connected in series High speeds may cause bearing life problems Motors built for one direction of rotation should never be reversed With no load, tends to run-away, speed only limited by windage and commutation Brush life can be low, hence maintainance requirement Very noisy, noise emission greater than other a.c. machines RFI suppresion filters required in supply line Moderate to low efficiency	Possible temperature rise Fail safe brake may have to be used to limit temperature if load has to be held safe over a long period

DRIVE MOTORS

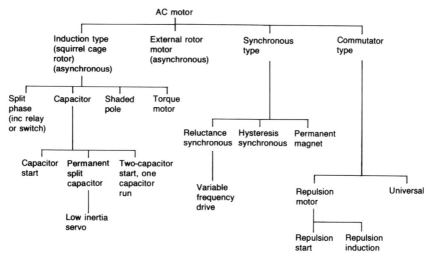

Fig. 8.1 Family tree of a.c. motors

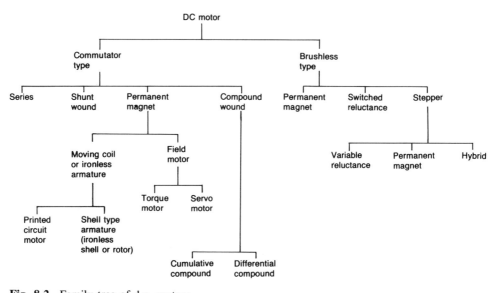

Fig. 8.2 Family tree of d.c. motors

Obtainable speeds are up to 20 000–30 000 rpm, with some brushless motors exceeding this. Steppers are limited to approximately 6000 rpm, but lower speeds are more typical.

A summary of the applications, advantages and disadvantages of d.c. motor types is given in Table 8.2.

The class of fractional horsepower (FHP) d.c. motors which utilize electromagnets to generate the starter magnetic field are called wound-field motors. When the electromagnets are energized in conjunction with the armature, the motors are known as self-excited motors. The class of motors which utilize permanent magnets differs from wound-field types in that no external power is required in the stator structure.

8.1.3 Production of torque

The force on a conductor carrying a current (i) in and lying in a magnetic field of flux density B (between two poles of a magnet) will be Bi per unit length provided that the directions of B and i are at right angles to one

Table 8.2 *Summary of DC motor types*

	Series wound	Shunt wound	Permanent magnet
Areas of application	Battery powered vehicles, general traction duty	Suitable for application where constant speed is needed at a given control setting	Most expanding segment in the 0–50W market, used in numerous products, i.e. cars, consumer electronics, office m/c's etc.
Performance summary Duty Power Supply Reversibility Speed Start Torque Start Current	Continuous or short time a.c. or d.c. Usually unidirectional Variable with load >175% of rated torque High	Continuous d.c. At rest or rotating Fairly constant 125–200% of rated Normal	Continuous d.c. At rest or rotating Fairly constant and adjustable >175% of rated High
Advantages	Good starting and acceleration torque Efficient Marginal speed control easily obtained with field diverter resistance	Earliest and most reliable of d.c. motor types Will operate from rectified mains Speed stability in open loop mode is quite good Can control speed by varying either field current or armature voltage Brush life is generally good Const speed characteristics enhanced by SCR controls Good starting torque Efficient Fairly load independent	High starting torque enables inertial loads Efficiency moderate to high Simple to reverse and dynamically brake High torques at low speed Linear torque/speed curve and ease of electronic control Use of rare earth magnets has reduced overall size and increased scope of application Low cost, fairly cheap With rapid acceleration/deceleration, ideal for stop/start applications No electrical power required to generate stator field
Disadvantages	With no load on motor, speed can run-away because torque/speed curve rises steeply Connect to load Expensive	Brush life adversely affected by reversing whilst rotating +ve brush wears faster than the other Dynamic braking is severe on brushes Expensive Not widely used	Cannot operate continuously at high torques since serious overheating may result Transmission of excess torque to rest of system Overload may cause partial demagnetization Compromise between a compound wound and a series wound motor Limited to 100 V d.c. because of field demagnetization problems Speed regulation good as long as voltage fairly constant and loads moderate High voltage produces excess arcing of brushes and commutator Brush life decreases with higher commutator surface speeds Limited to small ratings

DRIVE MOTORS

DC servo	Printed circuit motor	Shell armature motor
Used as prime movers where starts and stops must be made quickly and accurately	For incremental motion application in competition with steppers. Thin compared to diameter	Used for servo applications and incremental motion situations
Continuous d.c. At rest or rotating Variable with load High High	Continuous d.c. At rest or rotating Adjustable up to 5× rated High	Continuous d.c. At rest or rotating Adjustable >175% of rated High
Operate at speeds from stall to 2K–4K rpm and from zero torque to several times cont. torque Substantial transient over-load capability Low rotor inertia Rugged brushes and commutator to handle high powers Low electrical time constant High torque linearity with current input High torque/inertia ratio Fast response at all speeds due to rapid current rise in armature Excellent low speed characteristics	Suitable for intermittent duty (positioning servo) where torque smoothness required Speed control possible within a single rev. Low inertia high acceleration drive (high torque/inertia ratio) Armature has no 'wound' windings Moderately efficient Even running possible at speeds as low as 1rpm Fast response capacity to deliver high instantaneous torques Excellent power/volume ratio Silent operation Simplicity High linearity, torque increases linearly with input current With some types, constant torque from rest to high speed Reliable, high brush life with multiple brushes	Very low inertia and high acceleration Low inductance and thus low electrical time constant High torque/inertia ratio Max speed, 10K–15K rpm
Expensive, requires directly mounted tachogenerator Requires closed loop electronic control system Generally special designs with particular characteristics — performances can vary widely	Fragile construction of thin armature usually limits application to controlled situation Windings arranged across a very large radius Radius factor contributes substantially to armature M of I	Low thermal time constant for armature and long thermal time constant for housing Armature can easily overheat High speed 'whip' Fragile construction, hence should only operate in a controlled environment Expensive, employ only where unique performance characteristics required

Compound wound	VR stepper	PM stepper
Largely disappeared from the marketplace	Digital control of machine functions. Positional control of printers, plotters etc.	Digital control of machine functions. Positional control of printers, plotters etc.
Continuous d.c. At rest or rotating Fairly constant and adjustable >125% of rated Normal	Continuous switched d.c. At rest or rotating Adjustable N/A Normal	Continuous switched d.c. At rest or rotating Adjustable N/A Normal
High start torque No overspeed at low load Efficient Combination of series and shunt characteristics Fairly flat speed characteristics at the rated load	Unique performance characteristics Steppers are excellent positioning drives, performance and cost wise Can replace mechanical devices e.g. clutches with improved reliability Accuracy Suitable for high speed/high performance operation Robust rotor Lower cost stepper Pulse rates variable over wide range	Unique performance characteristics Steppers are excellent positioning drives, performance and cost wise Can replace mechanical devices e.g. clutches with improved reliability Accuracy Compact low weight low cost unit Detent torque up to 15% of holding torque available Can be used in open loop mode, ideally suited for μP control
Brushgear maintainance Expensive	Use only for light load applications Fairly low torque and inertial capability Prone to some resonance problems No detent torque when coils not energized Not efficient Fixed angle devices Tend to 'ring' on stopping, hence some means of damping usually required When inertial loads are being driven, speed ramping is usually required	Fairly low torque and speed capability Not efficient Fixed angle devices Limited ability to handle large inertial loads Tend to 'ring' on stopping, hence some means of damping usually required When inertial loads are being driven, speed ramping is usually required Rough low speed operation without microstepping

Hybrid stepper	Brushless d.c.	Switched reluctance
Digital control of machine functions. Takes over where PMDCs leave off.	Wide and varied, first choice when life requirement >10K hours. Used in robotics, office machinery, m/c tools etc.	Low cost brushless drive with wide speed range. Application domestic appliances and industrial drives.
Continuous switched d.c. At rest or rotating Adjustable N/A Normal	Continuous 3ph. alt. polarity d.c. At rest or rotating Adjustable >175% of rated High (rotor position feedback required)	Continuous switched d.c. — Continuous control possible Up to 200% of rated Low (rotor position feedback required)
Unique performance characteristics Steppers are excellent positioning drives, performance and cost wise Can replace mechanical devices e.g. clutches with improved reliability Accuracy Combines best features of VR and PM types High performance at reasonable cost Fine resolutions available, 1.8 deg. steps being common	No brushes or commutator to limit life Reliable and relatively maintainance free drive Long life motors, operational life at rated power not limited by wear considerations Very low electrical noise Speeds to 60K rpm not unusual Simple low cost versions now popular in small fan drives Can sit idle for years without performance loss Peak torque maintainable up to high speed	Simple low inertia rotor Stator is simple to wind Torque-speed characteristics can be tailored to application more easily than many other motors Most losses appear in stator which is easily cooled Rotor losses very small hence no rotor cooling problems as with other motors Torque is independent of the polarity of the phase current High starting torques without the problem of surge currents High speeds possible Serious alternative to d.c. + inverter drives
Not efficient Fixed angle devices Limited ability to handle large inertial loads Tend to 'ring' on stopping, hence some means of damping usually required When inertial loads are being driven, speed ramping is usually required	Electronically complex compared to other d.c. systems Can't be reversed by simple power source polarity change Excess coasting, special circuits can be added at cost Copper losses are high, not very efficient in low voltage systems Electronic commutation required can only be used with purpose built electronic drive Very expensive compared to conventional PMDC drives Limited economically to small motor sizes	Cannot equal the power density or efficiency of PMDC in small frame sizes Torque ripple (pulsed or non-uniform nature) may produce acoustic noise, especially the larger size motor Ripple current in d.c. supply tends to be large, which may cause a.c. line harmonics if rectified a.c. fed No smaller than an induction motor designed to the same specification Shaft position sensing required. Correct choice of switching angles, current chopping etc. in control strategy is of crucial importance Cannot start from sinusoidal a.c.

another. The direction of movement is by the left-hand rule, at right angles to both B and i. The two common rotor geometries satisfying these requirements are shown in Fig. 8.3, the resultant electromechanical torque being M_e.

Consider any two pole machine, e.g. Fig. 8.4. Magnetic forces both attractive and repulsive, will tend to bring the magnets into alignment, the total force and corresponding torque will vary with the torque angle. The effect is the same whether coil excited or permanent magnets, the rotational torque considered due to the mutual reactions of the poles. On rotor rotation, if the magnetic force distribution is assumed sinusoidal, then

$$M_e = K F_1 F_2 \sin \delta$$

where δ = torque angle between stator and rotor axis, K = fn of the air gap dimension, F_1 = peak stator magnetic force and $F_2 \sin \delta$ = quadrature component giving rise to peak tangential torque.

In a d.c. machine, the torque angle δ is a maximum at 90°, but in an a.c. motor the angle is a function of load so the term 'load angle' tends to be used. Now the magnetic force F is usually referred to as the magnetomotive force (m.m.f.) where F_1 is due to the stator coil(s), F_2 due to the rotor coil(s), effectively the current flow enclosing the magnetic circuit.

8.2 Terminology and formulae

Before consideration of the more common types of electric drive motors, it is worth while defining the fundamental equations of motion used in drive analysis, together with a definition of the motor performance characteristics used. Firstly, the following equations are valid for constant acceleration and velocity:

Linear motion

1. $v = u + at$
2. $s = ut + \frac{1}{2}at^2$
3. $v^2 = u^2 + 2as$
4. $P = Fv$
5. $F = ma$
6. $W = Fs = \frac{1}{2}mv^2$
7. —
8. —

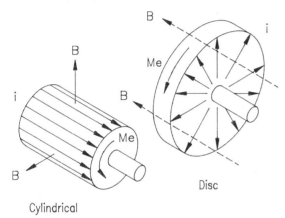

Fig. 8.3 Basic unit length rotor geometries of electromagnetic motors

Rotary motion

1. $w_2 = w_1 + \alpha t$
2. $\theta = w_1 + \frac{1}{2}\alpha t$
3. $w_1^2 = w_2^2 + 2\alpha\theta$
4. $P = Tw$
5. $T = Fr = J\alpha$
6. $W = T\theta = \frac{1}{2}Jw^2$
7. $w = 2\pi n/60 = 0.105n$
8. $J = mk^2$

where u = initial velocity (m s^{-1}), v = final velocity (m s^{-1}), a = acceleration (m s^{-2}), s = distance travelled (m), w_1 = initial velocity (rad s^{-1}), w_2 = final velocity (rad s^{-1}), α = acceleration (rad s^{-2}), θ = angular travel (rad), n = rpm, P = output power (W), J = polar moment of inertia (kg m^{-2}), F = force (N), m = mass (kg), W = work (J), T = torque (N m), r = radius arm (m) and k = radius of gyration (m).

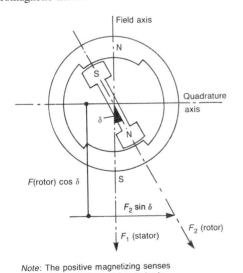

Fig. 8.4 Generalized production of torque

(i) Load

This is all the electrical and mechanical quantities that signify the demand to be made on a rotating machine at any instance. Types of load occurring in practice are shown in Fig. 8.5. Load types of curve (c) are typically inertial. Some load torques depend on the position of the motor shaft, e.g. a crank, and some loads exhibit a torque depending on the position of the load itself.

(ii) Referred load (and inertia)

Various parts of a geared drive system rotate at different speeds. To calculate the required torque for such systems, all rotating component inertias must be reduced back to the motor output shaft.

Now equating the power either side of the gear

$$T_1 w_1 = T_2 w_2 \quad \text{and} \quad T_2 = T_1 w_1 / w_2$$

where T_1 = load torque, T_2 = referred torque, w_1 = load speed, w_2 = motor speed and $w_1/w_2 = n_1/n_2$ = gear ratio.

The referred inertia is obtained by equating the stored kinetic energy, i.e.

$$J_1 w_1^2 = J_2 w_2^2$$

hence

$$m_2 k_2^2 = m_1 k_1^2 \left(\frac{n_1}{n_2}\right)^2$$

where J_1 = load inertia, J_2 = referred inertia, k_n = radius of gyration and m_n = mass.

Assuming 100 per cent gearbox efficiency and ignoring gearbox inertia, acceleration of the load is

$$\frac{T_{MP}}{\frac{w_2}{w_1}\left[J_M + \frac{J_L}{(w_2/w_1)^2}\right]}$$

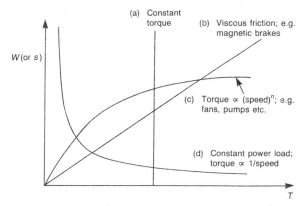

Fig. 8.5 Typical load curves that occur in practice

where T_{MP} = peak motor torque (max. requirement), J_L = inertia of the load and J_M = motor inertia.

Between extremes of the w_2/w_1 ratio there is a value of the gear ratio that gives maximum acceleration for fixed values of T_{MP}. The optimum value can be obtained by equating the differential of the above expression to 0 giving

$$\frac{w_2}{w_1} = \sqrt{\frac{J_L}{J_M}}$$

which is a well-known result that makes referred inertia equivalent to motor inertia. The maximum acceleration of the load is therefore

$$\alpha_{max} = \frac{1}{2} \frac{T_{MP} w_1}{J_M w_2}$$

(iii) Duty cycle

Many machines, e.g. fans and blowers, have continuous steady loads over long periods. In other cases the load may be applied intermittently or cyclically, in which case the duty cycle should be specified in order to obtain the motor rating. The duty cycle is a fixed repetitive pattern of loading over a given period of time, expressed as ratio of on time to cycle period:

rms load = peak load $\sqrt{\text{cycle period time}}$

The duty rating of a drive motor is normally classified as continuous rating or short-time rating. However, where this is not possible, various duty types can be specified (Table 8.3) based on the cyclic duration factor.[1]

(iv) Torque definitions

Determining the power to drive the machine is only the first step in sizing the motor. It is then necessary to examine the basic motor performance information contained in the torque/speed curves to determine whether the motor has enough starting torque to overcome static friction, to accelerate the load to full speed and to handle the likely overload. Torque definitions are as follows:

1. *Locked rotor or static torque.* The minimum torque that a motor will develop at rest for all angular positions of the motor under rated conditions.
2. *Breakdown torque.* The maximum torque a motor will develop under rated conditions without an abrupt drop in speed.
3. *Full-load torque.* The torque necessary to produce rated power output at full load speed.
4. *Acceleration torque.* At a particular speed, the accelerating torque available may be described by

Table 8.3 *Types of motor duty*

	Cyclic duration factor % (CDF)	Motor duty specification
S_1 Continuous duty	100	Cont. S_1
S_2 Short time operation at load (short time rating)	N/A	$S_2 \times$ min[a]
S_3 Sequence of identical on/off duty cycles, a constant load period (c) and a de-energized period (r)	$\dfrac{c}{c+r} \times 100$	S_3 (CDF)%
S_4 As S_3 but including a significant period of starting (d) at higher torque	$\dfrac{d+c}{d+c+r} \times 100$	S_4 (CDF)% $\times J_m \times J_{EXT}$[b]
S_5 As S_4 but including a period of rapid electrical braking (b)	$\dfrac{d+c+b}{d+c+b+r} \times 100$	S_5(CDF)% $\times J_m \times J_{EXT}$
S_6 Sequence of duty cycles, a period of operation at load (c) and a period of operation at no-load (a)	$\dfrac{c}{c+a} \times 100$	S_6 (CDF)%
S_7 As S_6 but including a braking period (b)	100	$S_7 \times J_m \times J_{EXT}$
S_8 As S_7 but with one or more load periods (c_1, c_2, etc.) corresponding to different speeds of rotation	$\dfrac{d+c_i}{d+\Sigma b+\Sigma c} \times 100$	$S_8 \times J_m \times J_{EXT} \times P \times n \times$ (CDF)%[c,d]
S_9 Non-periodic duty; variable loading at variable speeds including overload	N/A	Take full load on basis of overload concept

[a] duration of duty.
[b] J_m = moment of inertia of the motor; J_{EXT} = referred moment of inertia of the load.
[c] P = motor power, n = motor speed.
[d] list load, speed and CDF for each speed condition.
[e] Symbols: c = time period of operation at constant load, r = duration of rest period, d = start-up time period, b = period of electrical braking.

$$T_{acc} = T_M - T_L - T_F$$

where T_M = delivered motor or electromechanical torque, T_L = load torque and T_F = frictional load.

The torque developed by the motor must exceed the load torque during starting and acceration if normal full-load speed operation is to be obtained. When the mechanical load characteristic intersects the motor torque–speed characteristic, there is a balanced operating speed. The load torque is often taken as constant, but this is not necessarily the case (Fig. 8.5). Torque definitions are illustrated for an induction motor by Fig. 8.6 for a constant torque load. The case for a torque load increasing with speed is illustrated by d.c. motor behaviour in Fig. 8.7 with the difference $T_M - T_L$ being absorbed by the inertial torque $J\alpha$ and the frictional load T_F in accelerating the motor and its coupled mechanical load. Here the speed continues to rise until $T_M - T_L \to 0$.

There is no universal practice in the assignment of torque and speed axes although it is the more usual practice for d.c. motors to have the torque axis drawn horizontally. This convention is retained for performance curves in this chapter.

(v) Acceleration

The time required for a motor to run up to speed from standstill (neglecting friction) is determined by the equation

$$T_M - T_L = J \frac{dw}{dt}$$

from which

$$t = J \int_{w_1}^{w_2} \frac{dw}{T_M - T_L} \quad \text{(sec)}$$

is the time required to change speed from w_1 to w_2. Hence

$$t = \frac{2\pi m k^2}{60\, T} (n_1 - n_2) \quad \text{(sec)}$$

where T = net accelerating (or decelerating torque), ($n_1 - n_2$) = speed change and mk^2 = total rotating system inertia referred back to motor shaft.

If T_M and T_L can be defined algebraically then the above may be solved analytically. However, as is often the case, T_M may not be represented by an algebraic

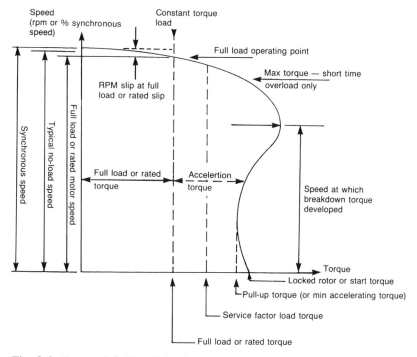

Fig. 8.6 Torque definition (induction motor example)

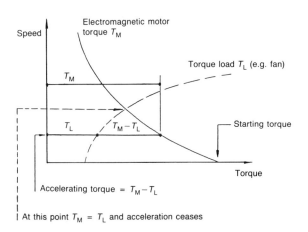

Fig. 8.7 DC motor and load torque–speed characteristics

expression, the run-up curve can be obtained graphically (Fig. 8.8).

(vi) Motor efficiency

This is the ratio of mechanical output to electrical input and represents the effectiveness with which the motor converts electrical energy into mechanical energy output. In so doing, the motor occurs losses which are converted into heat:

$$\text{Efficiency } (\eta) = 1 - \left[\frac{\text{losses}}{\text{input}}\right] = 1 - \left[\frac{P_{\text{loss}}}{P_{\text{in}}}\right]$$
$$= 1 - \left[\frac{P_{\text{loss}}}{VI}\right]$$

where the bracketed term may be called the deficiency. Thus for a motor of full load efficiency 0.72, there is a deficiency of 0.28. Losses can comprise the following:

1. *Electrical losses* — due to winding resistances and (if fitted), brush contact;
2. *Iron losses* — stator and rotor losses in the magnetic field due to hysteresis and eddy current effects;
3. *Mechanical losses* — due to friction in bearings and brushes (where used) and windage effects from rotor rotation and fitted cooling fans.

In general:

1. A change in efficiency as a function of load is an inherent characteristic of motors. Maximum efficiency is obtained at the rated load.
2. Efficiency of motors at the rated load increases as the motor rating increases, i.e. the larger the motor, the inherently more efficient.

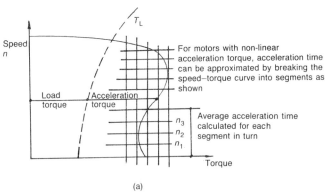
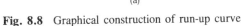
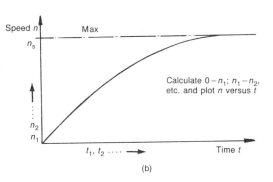

(a)   (b)

Fig. 8.8 Graphical construction of run-up curve

3. For the same power rating, higher-speed motors generally have higher efficiencies than motors with lower-rated speeds (increasing no. of poles for induction motors).

(vii) Power factor

The power factor of an a.c. electrical motor is simply a way of stating what fraction of the apparent power supply is real or active power in producing torque. In the general case of a.c. supplying to a complex impedance, voltage and current differ in phase by an angle $\theta°$, then

$$P = VI \cos \theta$$

where P = average active power (W) delivered or absorbed by the circuit, VI = apparent power supplied, θ = phase angle associated with the equivalent impedance and $\cos \theta$ = the power factor.

In a circuit containing resistance only, current and voltage are in phase, $\theta = 0°$ hence $P = VI$. An induction motor operates with a lagging power factor.

8.3 a.c. induction motors

With induction motors the rotor is constructed from a series of steel laminations, each punched with slots or 'cages' along its periphery. When the laminations are stacked together and riveted, the slots are filled with uninsulated copper or aluminium conductors or 'bars', short-circuited on themselves with end rings of the same material. Standard practice is to pour in molten aluminium, the one-piece casting then termed a 'squirrel-cage' (Fig. 8.9). The primary winding (stator) is connected to the power source and the shorted secondary (rotor) carries the induced secondary current. Torque is

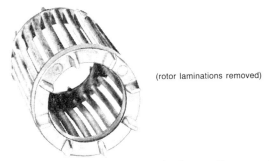

(rotor laminations removed)

Fig. 8.9 AC induction rotor squirrel cage. Courtesy of Bodine Electric Company, Chicago

produced by the action of the rotor currents on the air gap flux, and the rotor accelerates until stable conditions are reached.

The induction motor operates in much the same way as a commutator motor, except that rotor currents are induced by the changing magnetic field of the stator which rotates with the a.c. supply. With the stator poles supplied at f hertz, there will be a rotation of the pole and resultant m.m.f. axis at a speed of $n_s = 2f/p$ rev s^{-1}, where n_s is the synchronous speed and p the number of pole pairs. It is not essential for torque production that stator and rotor magnetic axes are stationary in space.

In a synchronous motor, the rotor is provided with a field system of N and S poles when the rotor is in synchronism with the rotating stator field, hence δ is fixed. Here δ increases from 0 at no load to ~ 90 at approximate maximum operating torque.

Similarly in an asynchronous motor, the angular displacement of the m.m.f. varies with load, but in this case there are two rotating fields, the stator-field speed n_s and the rotor-field speed $(n_s - n)$, where n is the rotor speed relative to the stator. The motion of the rotor-field with respect to the rotor must make up the difference between n_s and n so that the two magnetic axes are kept in step.

In a non-synchronous a.c. induction motor, the rotor will always operate at some speed less than synchronous; this lag of the rotor behind the rotating magnetic field is called slip, s, (e.g. Fig. 8.6) expressed as

$$\text{Percentage slip} = \frac{(\text{synchronous rpm} - \text{actual rpm})}{\text{synchronous rpm}} \cdot 100$$

$$= \frac{(n_s - n)}{n_s} \cdot 100$$

The maximum power factor of an induction motor usually occurs at a slip slightly greater than the full load slip, decreasing somewhat at loads significantly below the rated torque. For a given number of poles, the power factor increases with the rating.

To produce torque initially, some means must be employed to create a rotating field and start the rotor moving. The particular method employed determines the type of induction motor, i.e. split-phase induction motor, capacitor types of induction motor and shaded-pole induction motor.

8.3.1 Split phase motor

An induction device with a main and auxiliary winding connected in parallel on the stator, but the auxiliary start winding arranged electrically at 90° to the main winding. Torque is approximately equal to

$$I_m I_s \theta_R$$

where I_m = main winding current, I_s = start winding current and θ_R = relative phase angle.

The start winding is designed with high resistance and low reactance, or has a resistor in series to create a phase shift and induce torque to cause initial rotation and acceleration. The effect of resistance on the starting characteristic is shown in Fig. 8.10. After the motor has obtained ~70 per cent of rated speed, the auxiliary winding is automatically disconnected from the circuit by means of a centrifugal switch or relay, the motor then continues to run on the single oscillating field established by the main winding. The motor connection diagram and torque/speed curve are also shown in Fig. 8.10.

8.3.2 Capacitor types

These are split-phase motors with capacitors used in series with the auxiliary windings; three types as follows: (1)

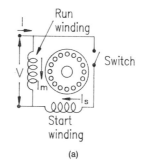
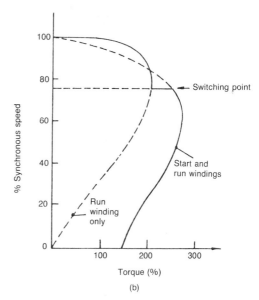

Fig. 8.10 Split-phase induction motor: (a) electrical configuration; (b) speed–torque curve

capacitor start (CS); (2) permanent split capacitor (PSC); (3) two-capacitor motor (CSCR). A summary of the behaviour of these is outlined in Table 8.1.

(i) Capacitor start motor

A greater phase difference can be obtained with a split-phase motor if a capacitor is substituted for the resistor in the auxiliary winding. For economic reasons, the capacitor will be as small as necessary to ensure adequate start torque. Most usually, the value offered is 300 per cent of full load torque, typically 20–30 μF for a 100 W motor to 60–100 μF for a 750 W motor. This type of motor produces a greater locked rotor and accelerating torque per ampere than does the split-phase motor, i.e. for the proper selection of capacitor, torque can be

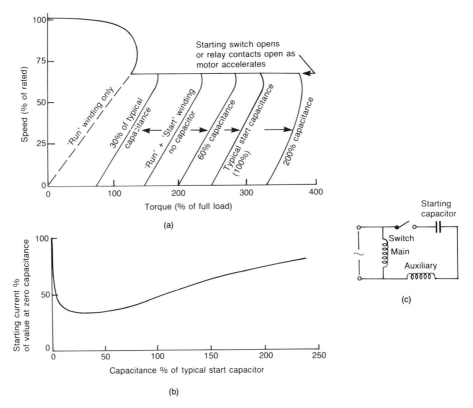

Fig. 8.11 Capacitor start induction motor; typical start torque and start current as a function of capacitance for a FHP motor: (a) speed–torque curve; (b) variation of start current with capacitance; (c) electrical configuration

increased and start current decreased (Fig. 8.11). The connection diagram is also illustrated in Fig. 8.11. After the start winding is removed from the circuit, performance is identical to the split-phase motor.

(ii) Permanent split capacitor

Similar to a CS motor but now the auxiliary winding is permanently connected in series with a capacitor (Fig. 8.12). Constantly energized main and auxiliary windings produce a high power factor and the motor runs with less noise. Choice of capacitor is a compromise to optimize both run and start characteristics, such that any improvements in starting are at the expense of run performance (Fig. 8.12). The more inefficient the run performance, the greater the risk of overheating.

The circuit (Fig. 8.12) is a phase-splitting circuit, creating a two-phase winding from a single-phase supply. This produces the necessary time-phase displacement between currents in the two windings, the auxiliary winding leading the main winding by 90°.

(iii) Two-capacitor motor (two capacitor start, one capacitor run)

The use of two capacitors helps to preserve the efficiency and quietness of the PSC motor as well as improving the start characteristics. Basically it overcomes the compromises made in PSC designs and allows for a 5–30 per cent increase in breakdown torque and 5–20 per cent increase in locked rotor torque. Ideally the capacitance for running should be about one-third of that for best starting. The effect of the running capacitor on the motor is to give a marked improvement in performance, efficiency and power factor plus a reduction in losses. Too high a start capacitor may, however, affect acceleration (e.g. Fig. 8.12). The torque–speed characteristic for the two-value capacitor motor is shown in Fig. 8.13 along with the connection diagram.

8.3.3 Shaded-pole induction motor

This is a simple, robust single-phase motor suitable for

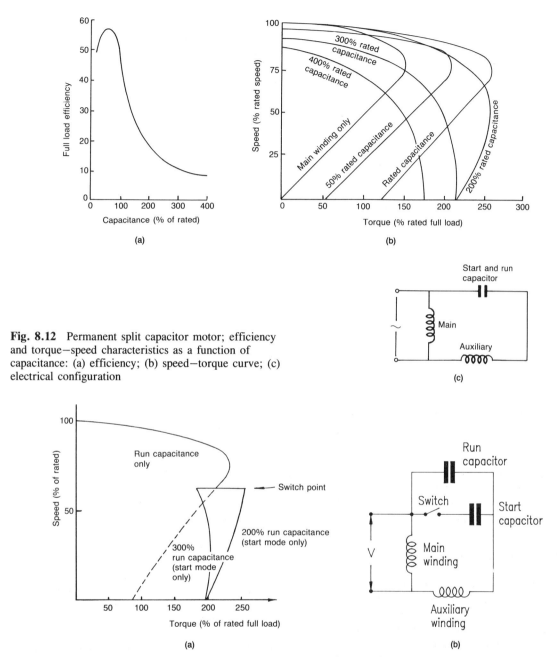

Fig. 8.12 Permanent split capacitor motor; efficiency and torque–speed characteristics as a function of capacitance: (a) efficiency; (b) speed–torque curve; (c) electrical configuration

Fig. 8.13 Two-capacitor motor; torque–speed curve and connection diagram: (a) speed–torque curve; (b) electrical configuration

the small power motor range. It has a squirrel cage rotor and an elementary form of field is produced by the interaction of the stator primary current and the current induced in a stator 'secondary' or auxiliary circuit, i.e. a solid ring of copper or brass called a shading ring (Fig. 8.14), a short-circuited strap wound on a portion of the pole. The shaded-pole motor is adaptable to salient-pole, skeleton or distribution winding. Figure 8.14 shows a salient-pole version having two windings on the stator. The shading coil causes the reaction necessary to give the frequency. As with squirrel-cage induction motors, speed is determined by the number of pole pairs. The rotor design enables the motor to 'lock into step' with the field, hence slip → 0 as the motor approaches synchronous speed.

A synchronous motor cannot start itself, and since acceleration from zero speed requires slip until synchronism is reached, start methods common to induction motors, i.e. split phase, capacitor start and shaded pole, can be employed. The electrical characteristics of these motors cause them to switch automatically to synchronous operation. There are three types of non-excited synchronous motor: (1) reluctance motors, (2) hysteresis motors and (3) the permanent magnet synchronous motor.

Typical torque–speed curves are given in Fig. 8.15 for types (1) and (2). Before considering synchronous motors in detail, some definitions are necessary:

Pull-in torque. The torque developed when pulling the connected inertial load into synchronism, during the transition from slip speed to synchronous speed as the motor changes from induction to synchronous operation. Usually the most critical period in starting a synchronous motor.

Pull-out torque. The maximum sustained torque the motor

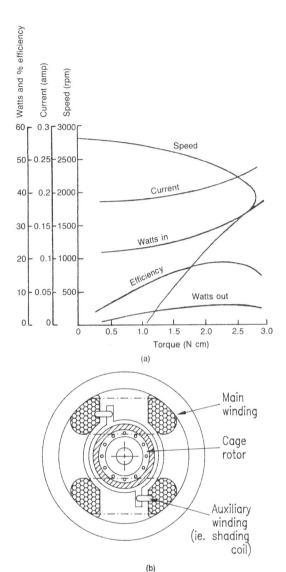

Fig. 8.14 Shaded-pole induction motor; configuration and performance of a typical two-pole, 40 W fan motor: (a) performance curves; (b) cross-section of motor

motor its starting torque. However, the motor produces only moderate torque during acceleration and run, has a low efficiency and consequently is only used for light duty applications.

8.3.4 Synchronous motors

Synchronous motors are inherently constant-speed devices. They operate in absolute synchronism with line

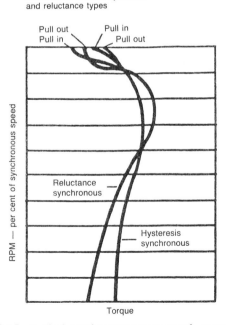

Fig. 8.15 Typical synchronous motor speed–torque curves. Courtesy Bodine Electric Company, Chicago

develops at synchronous speed for one minute at rated frequency.

To ensure that motor characteristics can be matched to the load requirements, accurate values of the load inertia and torque requirement should be known. Mechanical power output is given by the product of rotational speed and torque (section 8.2). If w_s is the synchronous speed of the motor in radians per second then the actual speed is

$$w = w_s(1 - s)$$

and

$$Tw_s = \frac{n_s}{n} \times P$$

where P = power produced by rotor and Tw_s = power output in synchronous watts.

(i) Reluctance synchronous motor

These resemble conventional squirrel-cage induction motors except that the rotor has salient poles, which on acceleration then approach the rotational speed of the field; a critical point is reached where there is an increased acceleration and the rotor pulls into synchronism with the stator field (Fig. 8.15). Instead of settling down to operate with some degree of slip, as in the ordinary induction motor (Figs 8.11, 8.12), reluctance torque accelerates the rotor above its induction motor speed into synchronism, provided the motor is not heavily loaded. The purpose of the cage is to provide starting and the motor can be designed for SP, CS and PSC configurations.

The connection diagram for a capacitor reluctance motor is shown in Fig. 8.16, consisting of two single-phase windings with mutual displacement of half pole pitch on the stator, main and auxiliary. The rotor is a salient-pole version to provide both synchronizing and starting torque. Too high a torque load will prevent synchronization, though a somewhat higher load may be applied once synchronization has been achieved.

(ii) Hysteresis synchronous motor

While the stator is of a conventional-wound induction motor type, the rotor consists of a heat-treated permanent magnet alloy cylinder securely mounted to the shaft by end-caps or over a non-magnetic (e.g. aluminium) insert (Fig. 8.17). Unlike other synchronous motors, the hysteresis motor has a perfectly round and symmetrical rotor with no salient poles, hence no preferred synchronization direction. Below synchronous speed, torque is primarily produced by a hysteresis effect. Instead of the permanently fixed poles of the reluctance synchronous design, hysteresis poles are induced by the rotating magnetic field, the poles shifting around the periphery of the rotor to reflect the relative speed differential between the stator field and the rotor during acceleration. Once the rotor reaches synchronous speed, the rotor poles take up a fixed position leaving the rotor magnetized and essentially running as a permanent magnet motor.

Performance curves are given in Fig. 8.17. Because torque is developed up to synchronism, the motor has the ability to bring high inertial loads up to synchronous speed. The motor is started as an induction motor.

(iii) Permanent magnet synchronous motor

It is a sub-variety of reluctance and hysteresis synchronous types with a permanent magnet embedded in the rotor (Fig. 8.18). Alternatively a cage assembly and permanent magnet may be placed on a common shaft. The squirrel-cage winding brings the rotor up to near synchronism when the inductive effect falls off, then accelerates from slip speed into synchronism under the influence of the permanent magnet, rotating exactly in step with the a.c. being fed to the stator. The main problem is in the magnet design to enable effective transfer from inductive to permanent magnet rotor operation.[2]

When used with the correct type of controller, permanent magnet (PM) synchronous motors are called a.c. servomotors or a.c. brushless motors. They resemble d.c. brushless motors in that torque is directly proportional to current rather than to torque angle. However, the servomotor controller adjusts the torque angle as the speed changes to produce maximum torque per ampere.

The PM synchronous motor acts as a synchronous reluctance motor if the magnets are left out or demagnetized.

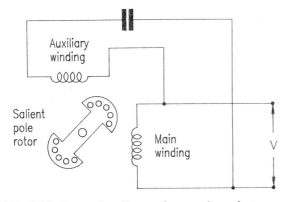

Fig. 8.16 Connection diagram for capacitor reluctance motor

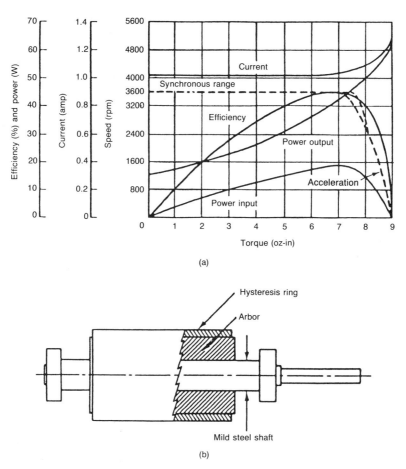

Fig. 8.17 Hysteresis synchronous motor; rotor construction and performance curves: (a) performance curves (1 oz in = 0.706 N cm); (b) rotor section. Courtesy of E H Werninck

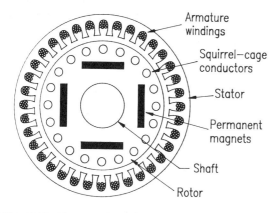

Fig. 8.18 Cross-section of permanent magnet synchronous motor

8.3.5 Special-purpose a.c. motors

(i) The external rotor motor

Basic features are a low noise level, high efficiency, constant speed, hence generally better performance than conventional motors of equivalent size. The mechanical design is shown in Fig. 8.19. The outer housing (rotor) rotates while the internal stator contains the primary winding. Asynchronous motors can be either permanent capacitor, shaded-pole or eddy current types. Some torque–speed curves are given in Fig. 8.19. The inertial effect of the rotor is useful for stabilizing speed, both magnetic flux variations and load transients.

The motor is used where the flywheel effect of the rotor may be of value in providing a smoother drive and gentler

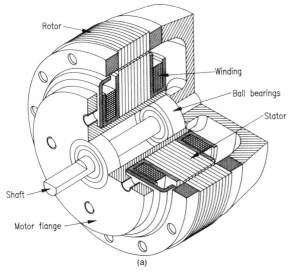

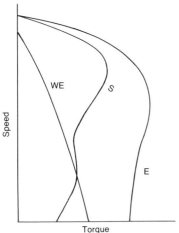

Key: S = Shaded-pole motor
E = Capacitor-start capacitor-run motor
WE = Eddy current rotor motor (soft iron rotor without slots)

(b)

Fig. 8.19 External rotor motor: (a) general construction; (b) performance curves of different types of external rotor motor

starting conditions. Major usage is for fan applications where the fan blades are directly attached to the external rotor, but also for more precise applications, i.e. typewriters, turntables, recorders.

(ii) Torque motor
Normally of the PSC type, the a.c. torque motor is used typically in one of the following applications:

1. Where the motor is stalled and no rotation is required, i.e. a constant torque loading.
2. Where the motor shaft rotates only a few degrees or revolutions, i.e. used as an actuator to clamp, close a valve, operate a switch.
3. Where the shaft must rotate at very low speed for all or part of a cycle, e.g. tape drives.

When specifying a torque motor, output is expressed as torque (rather than watts or power) against a given duty cycle. The duty cycle defines a necessary cooling period to avoid motor overheating. Some typical torque–speed curves for an a.c. induction torque motor are shown in Fig. 8.20. Performance curves that are close to a straight line approach the 'ideal' for torque motor service.

8.4 Induction motor speed control

Since the majority of speed-control applications these days are based on the use of electronics, it is probably appropriate to introduce this section on speed control by outlining types and behaviour of compact power semiconductors (SCRs). Power SCRs can be used for the control of d.c. and a.c. power, for the rectification of a.c. to d.c. and the inversion of d.c. to a.c. Such devices are not ideal, only a limited amount of power can be dissipated by thyristors, triacs, power transistors, etc. hence they must be operated as a switch, either open (maximum circuit voltage) and virtually no current, or closed, rated at maximum circuit current and small forward voltage drop. Triggering methods are not covered in this section. Control circuitry waveforms are not pure d.c. nor sinusoidal a.c., but the average voltage with the possibility of also controlling the fundamental frequency. Because such fast switching devices can operate at any point in the cycle of mains frequency, harmonics may be generated that are fed back into the mains supply. The main types of SCR device are as follows:

1. *Thyristor*. Most commonly used; a controlled rectifier, blocking current in one direction but allowing current to pass in the other provided a gate signal is applied. The process of interrupting current flow and restoring the thyristor to its non-conducting or blocking state is known as commutation. The techniques used to achieve commutation are the major difference between thyristor circuits.
2. *Triac*. Behaves as two thyristors connected back to back, i.e. in reverse parallel. In fairly wide use but limited to a.c. power control, specifically using phase control techniques. Characteristics are very similar

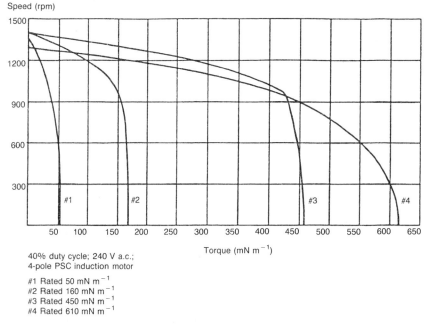

Fig. 8.20 AC torque motor; typical torque–speed characteristics

40% duty cycle; 240 V a.c.;
4-pole PSC induction motor
#1 Rated 50 mN m^{-1}
#2 Rated 160 mN m^{-1}
#3 Rated 450 mN m^{-1}
#4 Rated 610 mN m^{-1}

to the thyristor, either positive or negative pulses to the gate will trigger the triac in either direction.

3. *Gate turn-off (GTO).* As a replacement for the conventional thyristor, the GTO thyristor can be switched off as well as on by the appropriate high current pulses applied to the gate. This removes the need for complex commutation circuits.

4. *Power transistor.* The bipolar or field-effect transistor (FET) power transistor is controlled like an ordinary transistor with a continuous signal applied to the gate (base). The transistor only passes current when the base current is applied and ceases to do so when the current is removed. Current gain is typically ~10, but with power control work, it is used as a switch. Like the GTO they require no auxiliary commutating circuits and can switch at fast speeds. They are less rugged than the thyristor when switching inductive loads.

8.4.1 Speed control

Most induction drives run at constant speed, but in those applications where a variable speed is necessary, the appropriate equation (section 8.3) is as follows:

$$n = (1-s)n_s = (1-s)\frac{f}{p}$$

where n, n_s are in revolutions per second. Hence speed control is possible by either varying the number of pole pairs, p, changing the supply frequency, f, or controlling motor slip, s.

(i) Changing number of poles

Changing the number of stator poles will only produce a discrete speed change, e.g. two to four poles by a reconnection of the stator windings. Usually limited to two widely spaced speeds, requiring that a portion of the winding be idle during the operation of one or more speeds, therefore resulting in motor inefficiency and performance reduction for any given frame size.

(ii) Slip control

Suitable for shaded-pole and PSC motors, slip control allows the slip of the motor to increase by decreasing the motor volts. Since the torque of an induction motor $\propto V^2$,[3] small changes in voltage give significant torque changes. To obtain effective speed control:

1. Match the motor closely with the load;
2. Match the rotor resistance to the torque–speed characteristic of the load;
3. Use a rotor designed for high slip which will aid in obtaining maximum speed change.

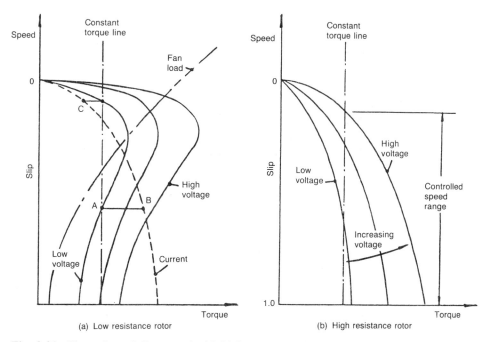

Fig. 8.21 Illustration of slip control with high and low resistance induction motor rotors

Changing the speed—torque characteristic such that a different value of *s* is required for any given load torque is a simple technique but often wasteful in terms of power and motor capacity. At increasing values of slip, increasing values of rotor current result, thus motors running at high percentage slip are inefficient and tend to overheat. With a fan load and low resistance rotor (Fig. 8.21(a)), the torque requirement decreases with speed, voltage is reduced at low speeds so the system efficiency is reasonable. With a constant torque load, however, such a system is unstable, i.e. the motor can be controlled to run at point A, but the current is much higher at B than at point C where the motor would normally run at that torque. With a high-resistance rotor, on the other hand (Fig. 8.21(b)), the constant torque load can be speed-controlled by voltage variation to give a stable system although current is still high and the motor would need to be derated at low speed.

There are several methods of changing rotor slip, as follows:

1. *Tapped winding.* Mostly used with shaded-pole or PSC motors, a change of motor impedance through use of portions of total winding. The number of speeds is determined by the number of winding taps.
2. *Series resistance* Used with shaded-pole or PSC motors.
3. *Variable voltage transformer.* Used in place of a series resistor to reduce voltage across the winding. With the PSC motor, by reducing voltage across the main winding only (Fig. 8.22), more stable operation is provided at low speed by maintaining full voltage across the capacitor winding.
4. *Winding change.* Applicable only to a PSC motor, and used for two speeds only, the function of the main and capacitor windings being switched for high and low speeds. An efficient technique but does require the motor windings to be tailored to the loads.
5. *SCR phase control.* This is variable a.c. from an a.c. supply. Method of control is only suitable for use

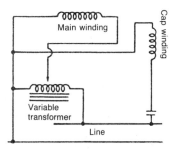

Fig. 8.22 Variable voltage transformer method in a PSC motor

ELECTROMECHANICAL PRODUCT DESIGN

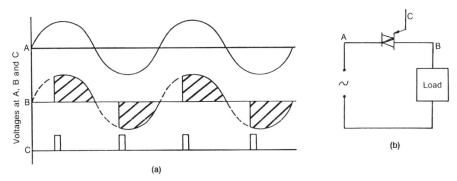

Fig. 8.23 Basic triac circuit for controlling a.c. + voltage waveforms: (a) voltage waveforms; (b) electrical diagram

on a standard low-resistance rotor motor with a fan-type inertial load. By controlling the phase relationship of the trigger to the zero axis crossing of a.c. current, the amount of power transmitted through the SCR can be varied (e.g. Fig. 8.23). This technique is phase control. By using the motor back e.m.f. as feedback control, if the motor slows down, the SCR will automatically fire sooner in the cycle (Fig. 8.24).

(iii) Frequency change

In theory, this is a simple form of speed control (variable f) but in practice with a single-phase induction motor it is quite complex and costly to implement. It has the advantage of providing stepless speed changes over a wide range, usable with both synchronous and non-synchronous motors, but the motor must be designed for the frequency range. With a robust squirrel cage and maximum flux maintained, a maximum torque is available at all frequencies (e.g. Fig. 8.25). There are two basic systems for obtaining variable frequency, based on either

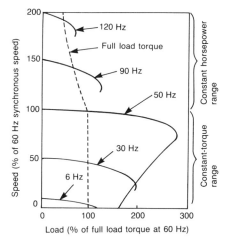

Fig. 8.25 Adjustable frequency drive; torque−speed characteristics with voltage boost at lower frequencies

a fixed or variable d.c. link voltage (Fig. 8.26). The a.c. supply in this context is the 'link' between the constant and variable frequency parts of the systems.

1. *d.c. link inverter.* Operates from a variable d.c. bus supply, obtained by rectifying a phase-controlled a.c. mains supply and then inverted. This is the quasi-square wave inversion (Fig. 8.26), with typically a frequency range 5−200 Hz. At low frequencies (Fig. 8.25), 'jerky' torque pulsations and mechanical resonances are likely.
2. *Pulse width modulation (PWM).* Using a fixed level d.c. from rectified mains, the voltage is broken by the PWM inverter to produce a fixed number of pulses per cycle, in which case the voltage is determined by the length of each pulse (Fig. 8.26). Because of the limitations and cost of current

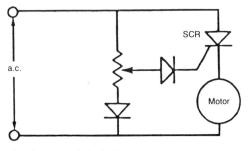

Feedback control circuit using counter EMF of the motor as feedback control voltage.

Fig. 8.24 SCR feedback control

DRIVE MOTORS

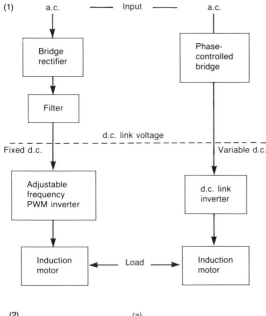

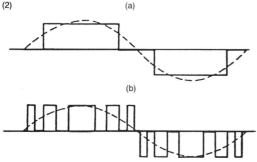

Static frequency-changer output-voltage waveforms:
(a) quasi-square wave inverter (d.c. input);
(b) pulse-width-modulated inverter (d.c. input).

Fig. 8.26 Variable frequency control techniques: (1) techniques; (2) output voltage waveforms

regulation schemes, it is more usual in a.c. drives to find PWM control applied to the voltage rather than the current. The requirement for high-speed chopping may impose upper limits of 100 Hz or so. Although the waveforms are far from sinusoidal, the motor inductance smoothes the current wave shape, dampening the harmonics considerably.

8.4.2 Dynamic braking

Apart from using mechanical friction devices, induction motors may be electrically or dynamically braked. This section will assume a non-energized rotor, and in general the braking techniques would be applicable to either synchronous or non-synchronous motors. There are three commonly used methods as given under the headings below.

(i) Reverse current braking ('plugging')
This is achieved by reversing the direction of the rotating field by switching any two stator connections. The slip for this condition is

$$s = \frac{-n_s - n}{-n_s}$$

where $-n_s$ indicates a reversal of the rotating field.

For a pre-plug slip of 0.05, the slip immediately following the reversal is 1.95. However, although this is the fastest method of stopping an induction motor, high supply currents result and any motor continually started and reversed must be specially designed. At zero speed, before the motor can rotate in the opposite direction, power is removed.

(ii) Rheostatic braking
Often called d.c. braking, obtained by removing the stator a.c. power and substituting d.c. The stator is then similar to the field winding of a shunt motor and the squirrel-cage rotor similar to a shorted armature in braking mode; in essence the motor behaves as a d.c. generator on short circuit, the load inertia being the prime mover. Rotational energy is dissipated in the form of rotor heating and the motor quickly brought to a stop. The d.c. wattage level should be limited to what the motor can dissipate without overheating if left on continuously.[4] The technique is applicable to shaded-pole and split-phase permanent capacitor induction motors. Either the main winding is used or, in the case of a permanent capacitor motor, d.c. applied to the second winding also with a three-pole switch to give series or parallel switching arrangements (Fig. 8.27). Without such wiring modifications, the capacitor will prevent use of the auxiliary winding since it blocks d.c.

(iii) Capacitor shorting
This is limited to PSC motors of the high-slip, non-synchronous or hysteresis synchronous types. The procedure is to short the capacitor, placing both the main and capacitor windings across the a.c. supply (Fig. 8.28). This process then eliminates the rotating field, but the two windings need to be equivalent for this technique to work.

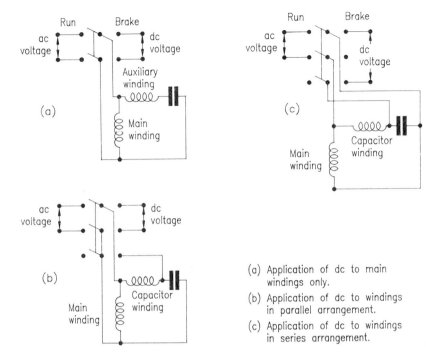

(a) Application of dc to main windings only.
(b) Application of dc to windings in parallel arrangement.
(c) Application of dc to windings in series arrangement.

Fig. 8.27 PSC motor; rheostatic braking alternatives

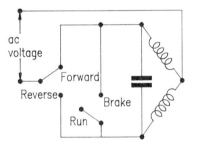

Fig. 8.28 PSC motor: capacitor shorting method of motor braking

8.5 a.c. commutator motors

8.5.1 The universal motor

The basic circuit, construction and principle of operation of a universal motor are essentially the same as a d.c. series motor (e.g. Fig. 8.29). However, in contrast to solid poles of small d.c. series motors, the entire magnetic circuit of a universal motor is laminated to minimize eddy current and hysteresis losses. A shunt winding is not used because its high inductance would introduce a large phase angle between field and armature currents, reducing torque and efficiency.

For a given voltage, a.c. operation draws less current, develops less torque and hence runs at a lower speed than d.c. operation (Fig. 8.30). On a.c. operation, the power factor tends to be poor, 0.4–0.7, because of the large inductance in the windings, but this is of little significance at low ratings. Commutation on a.c. current is substantially poorer than on d.c. and thus the brush life is correspondingly less, although extended length brushes can be designed for a brush life of 3000–5000 hours. Although these motors are usually supplied in unidirectional rotation, bidirectional motors are possible with a split or double field winding, but reducing the available power in a given frame. There is also some initial speed differences in bidirectional usage until the brushes seat adequately on the commutator in both directions.[5]

There are two types of universal motor, compensated and uncompensated, the latter being less expensive and simpler in construction. Where a compensation winding is incorporated (Fig. 8.30), it not only neutralizes armature reactance but also aids in commutation. The compensation type has generally better torque–speed characteristics and a.c.–d.c. differences are not substantially different at higher speeds. Speed can be adjusted over a broad range by means of rheostatic control, adjustable autotransformer or electronic control.

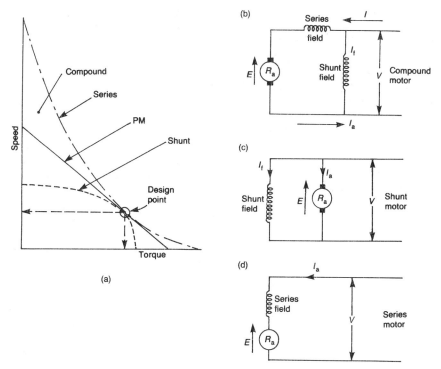

Fig. 8.29 Wound-field d.c. motors; connection diagram and torque–speed characteristics: (a) torque–speed curves; (b) electrical configuration of compound motor; (c) electrical configuration of shunt motor; (d) electrical configuration of series motor

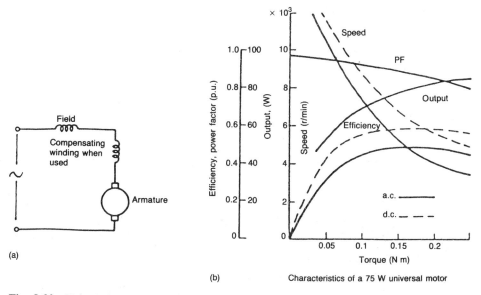

Fig. 8.30 Universal motor; torque–speed curve and connection diagram: (a) eletrical configuration; (b) performance curves. From Avallone E., Baumeister J., *Mark's Standard Handbook for Mechanical Engineers*, McGraw-Hill 1987. Reproduced with permission of McGraw-Hill Inc.

8.5.2 Repulsion motors

The repulsion motor is very similar to the d.c. series motor but with the rotor energized inductively instead of conductively. The commutator rotor winding is designed for a low working voltage. The brushes are permanently short-circuited and shifted circumferentially, the brush axis displaced from the field axis, to give the effect of two stator windings. The induced e.m.f. and current in the rotor winding give rise to a magnetic force having a component 90° to the field axis such that a torque is produced. Repulsion motors are used where a high starting torque is required, again similar to the d.c. series motor.

8.6 d.c. motors

In its simplest form, the brush axis is usually set for maximum torque at right angles to the field axis (Fig. 8.31). The two-pole armature has a current distribution such that half the armature carries current in an opposite direction to the other half, the dividing axis being the peak m.m.f, $F_2 \sin 90°$, hence

$$F_2 \sin 90° = \text{current per conductor} \times \frac{\text{no. of conductors}}{2}$$

where the m.m.f. per pole is half this.

Increasing the number of commutator segments will reduce the torque variation as a function of angular position.

As δ is constant, the torque is only proportional to the armature current, I_a, and the field produced flux

$$T \propto \phi I_a$$

$$T = \left[\frac{Z\phi p}{2\pi}\right] I_a$$

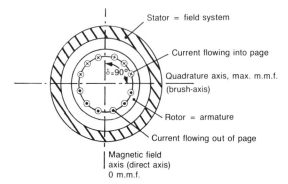

Fig. 8.31 Basic d.c. brushed motor schematic

where ϕ = magnetic flux per pole, p = no. of pole pairs, Z = no. of conductors per parallel path of winding and where

$$\left[\frac{Z\phi p}{2\pi}\right]$$

is constant for a given motor design and is called the torque constant K_T (torque sensitivity or electromechanical conversion coefficient) i.e. $T = K_T I_a$.

Note that the above is gross torque; the net shaft output torque is less than this by a small percentage which is absorbed by the windage brush contact and friction losses, etc. Now the d.c. motor is a variable-speed motor consisting of an armature and conductors rotating in a fixed magnetic field, the conductors move alternatively under a N and S pole, the generated and commutated voltage opposing the d.c. supply voltage. The power balance equation becomes

Electrical power = mechanical power ± armature copper loss (converted power)

$$VI_a = EI_a \pm (R_a I_a \times I_a)$$

where V = terminals e.m.f., E = induced e.m.f. and R_a = armature resistance.

Equating mechanical power (EI_a) to the torque–speed product gives

$$T = \left[\frac{E}{w}\right] I_a$$

where w = angular rotational velocity = $2\pi n$ rad s^{-1}, or

$$w = \frac{V - R_a I_a}{K_T} \quad \text{and} \quad E = K_T w$$

Note: in SI units the torque constant is numerically equal to the back e.m.f. or motor voltage constant $K_E = E/w$, where E is the back e.m.f., the torque constant K_T expressed in newton metres per ampere and the voltage constant expressed in volts per radian per second. This arises from the fact that the torque produced by a motor with a given winding and physical configuration is directly related to the voltage it produces when externally driven or used as a generator. Torque is a function of current and speed is a function of voltage.

The value of speed is determined by the point at which the motor torque is balanced by the load torque (Fig. 8.7).

8.6.1 Wound field motors

Of the d.c. motors that have electromagnetic excitation, they are classified according to the source of the excitation current, i.e. series, shunt or compound (Fig. 8.2). The torque−speed characteristics of the different field systems are shown in Fig. 8.29 along with connection diagrams. See section 8.5 for interpretation of motor equations. In a shunt motor, the sum of the armature and field currents $I = I_a + I_f$. In a series machine the same current flows in both armature and field windings (Fig. 8.29).

(i) Shunt-wound d.c.

The strength of the field is not appreciably affected by load change, hence speed stability in open loop mode is fairly good. In this respect it is the nearest d.c. motor to the induction motor. However, armature reaction can act to weaken the magnetic field at loads >200 per cent of rated.

(ii) Series-wound d.c.

When the motor starts, the current is at a maximum; consequently the magnetic field density is a maximum, hence high starting torque (Fig. 8.29). As the motor accelerates, the current and flux both reduce. The motor will operate from a.c. or d.c. Unlike the shunt-wound motor, the torque developed I_a^2 and the total resistance of the armature circuit now include the field coil resistance.

(iii) Compound-wound d.c.

These motors have both a series and a shunt winding and characteristics are largely intermediate between a shunt and a series motor. When the series winding aids the shunt winding the motor is termed a 'cumulative compound motor', and when the series winding opposes the shunt winding, the motor is termed a 'differential compound motor'. The compound motor exhibits high start torque and relatively flat torque−speed characteristics at the rated load. Because of elaborate circuitry needed to control compound motors, they have largely disappeared from the FHP market-place.

8.6.2 Permanent magnet motors

Since the stator magnetic field of a d.c. permanent magnet motor (PMDC) is generated by permanent magnets, no power is used in the field structure, and the stator field remains essentially constant at all levels of armature current, hence the speed−torque curve is linear (Fig. 8.29). With modern rare-earth magnets, the stall torque will tend to be higher and the speed−torque curve more linear than for a comparable wound field motor. Because of the high coercive strength of PMs, PMDC motors are typically one-quarter the diameter of a wound field motor.

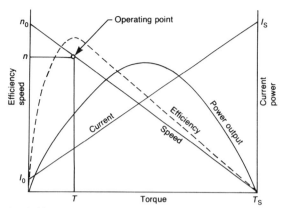

Fig. 8.32 PMDC motor operating characteristics

There are two basic types, the iron rotor motor and the ironless rotor motor that we shall consider.

Traditionally, the PM motor performance is defined by curves that show the relationships between torque, speed, input current, efficiency and power output (see also section 8.2). A typical diagram of the motor operating characteristics is shown in Fig. 8.32. Motor efficiency rises rapidly until it reaches a maximum at torque T and speed n. Beyond this point, efficiency decreases and output power increases until the point of maximum power is reached. Increased load decreases both power and efficiency as the stall conditions are approached. Maximum power output is at one-half the stall torque.

As a general rule

$$P = \frac{\pi T n}{30} \quad \text{and} \quad \eta = \frac{\pi T n}{30 V I}$$

where P = power output, T = torque, n = speed (min^{-1}), V = applied voltage, I = current, I_0 = no-load current, I_{st} = stall current and η = efficiency.

Maximum efficiency occurs at relatively high speeds and is a function of the frictional torque/stall torque ratio, i.e.

$$\text{Max. } \eta = \left[1 - \sqrt{\frac{I_0}{I_{st}}}\right]^2$$

The no-load speed is a function of the supply voltage and is reached when E becomes nearly equal to V. No-load current I_0 is a function of frictional torque. Additional motor characteristics are defined under the headings below.

(a) Motor constant (K_m). Often overlooked, this is used during motor sizing as a figure of merit of the motor

power:torque ratio. It is a constant for a given motor regardless of different voltage windings, i.e.

$$K_m = \frac{T_s}{\sqrt{P_s}} = \frac{T}{\sqrt{P_L}} = \frac{K_T}{\sqrt{R}}$$

where T_s = stall torque, P_s = power input to motor when stalled at peak torque, P_L = power loss at given load torque T and R = motor terminal resistance.

(b) Motor regulation constant. Basically indicates the slope of the torque–speed curve. For a given torque–speed condition, the motor with the lowest regulation value will dissipate the least power. Motor regulation represents the ability of the motor to transform electrical power into mechanical power, i.e.

$$\text{Regulation constant} = \frac{1}{\text{damping constant}} = \frac{w_0}{T_s}$$

$$= \frac{R}{K_T^2} = \frac{1}{K_m^2}$$

where w_0 = no-load speed.

For a given motor, a family of speed–torque curves can be plotted that correspond to various terminal voltages (Fig. 8.33).

(c) Mechanical time constant (t_m). This is a parameter defining motor response, the time needed by the motor to reach 63 per cent of its no-load speed under the set operating conditions. In $4t_m$, the rotor will have reached >99 per cent of no-load speed. It is a mechanical load characteristic taking into account electrical (R), magnetic (K_T) and mechanical (J) parameters, a product of motor regulation and rotor inertia, i.e.

$$t_m = \frac{RJ_m}{K_T^2} = \frac{Rw_0 J_m}{T_s}$$

where J_m = rotor inertia. When connected to a load, t_m will change, since

$$J_t = J_m + J_L$$

where J_L = load inertia and J_t replaces J_m in the above equation.

(d) Electrical time constant (t_e). This describes the rate at which current can change in the motor windings, i.e. time to rise to 63 per cent of maximum:

$$t_e = \frac{L_a}{R_m}$$

where L_a = armature inductance and R_m = motor impedance at stall.

(e) Thermal characteristics. As the motor is operated, power losses are dissipated in the armature, resulting in temperature rise. The limiting factor in most d.c. systems is the thermal rating of the motor since most motors can be operated at high power levels for short periods if needed. In incremental motion systems, the velocity

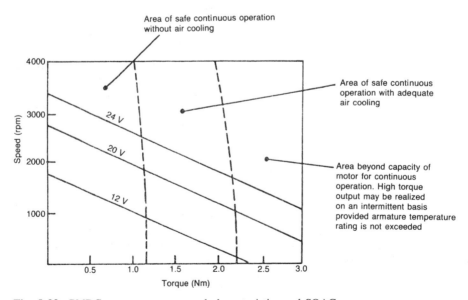

Fig. 8.33 PMDC motor; torque–speed characteristics and SOAC

profile chosen to achieve the required positional change will affect the armature heat dissipation, e.g. a parabolic velocity profile produces minimum dissipation compared to triangular and trapezoidal profiles.

In continuous operation, values of speed and torque can be limited so that the resulting temperature is acceptable, e.g. the safe operating area boundary for continuous operation (SOAC) (e.g. Fig. 8.33). Power losses are assumed to be dissipated in the armature (also valid where losses are dissipated in the housing) such that taking the electrical analogy

Thermal difference $(\Delta \theta °)$ = rate of heat dissipation $(I^2 R)$ × thermal resistance (of rotor)

The thermal time constant is the time (in seconds) taken by the armature (or stator) to reach 63 per cent of the temperature rise corresponding to a given constant dissipated power. Such values indicate comparisons.

The maximum continuous torque depends on the dissipated power, i.e.

$$T_{max} = K_T I_{max}$$
$$= K_T \sqrt{\frac{P_{diss}}{R_{max}}}$$
$$= K_T \sqrt{\frac{\Delta \theta °}{R_{max} R_T}}$$

where $\theta °$ = temperature, P_{diss} = dissipated power, R_{max} = rotor resistance at $\theta °$ max, $\Delta \theta °$ = rotor temperature rise and R_T = thermal resistance of rotor at $\theta °$ ambient (see Ch. 15).

The temperature rise of copper windings can be determined from resistance values of the wound rotor (or stator) as follows:

$$\theta ° \text{ max} = \left[\frac{R_{max}}{R_{amb}} \right] (\theta ° \text{ ambient} + 235) - 235$$

where R_{amb} = rotor resistance at $\theta °$ ambient.

(i) Permanent magnet type

The most expanding sector of the PMDC motor market is in the range 0–50 W; these motors are used for example in cars, consumer electronics and office machines. The scope of application has increased significantly with rare-earth magnets now being used in place of traditional ferrous or ceramic magnets, and at the same time, because of higher coercivity, reducing weight by ~30 per cent and reducing motor diameter without reducing torque. The newer magnets will also permit different designs, i.e. the inside-out motor with the magnet on the rotor and windings on the stator, the result being a smaller, lighter motor with lower rotor inertia and greater heat dissipation.

(ii) Shell armature type

As a variation of the permanent magnet motor, the shell motor, or coreless motor, is a low inertia design which utilizes a stationary iron core within an annular armature (Fig. 8.34) to increase torque rather than speed. The rotor consists of a coil former fixed to the spindle on which the coils are wound. The stator is a steel alloy cylindrical magnet located inside the rotor. The motor housing completes the magnetic circuit. The commutators of some types are equipped with zener diodes or capacitors for heavy-duty applications or extended life. The ironless rotor removes most of the disadvantages of small PMDC motors; ironless rotor motors are characterized by

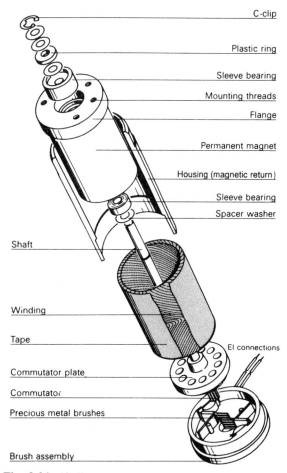

Fig. 8.34 Shell armature type of permanent magnet d.c. motor. Courtesy of Interelectric AG, Sachseln

excellent acceleration, mechanical time constant, start torque and low speed stability. The shell-type armature has a high torque/moment of inertia ratio, providing acceleration capabilities up to 10^6 rad s^{-2}.

The field coils are usually thin overlaid copper wire loops, bonded to provide a rigid frame. This technique totally removes magnetic hysteresis, hence the motor cannot cog and thus provides the excellent low speed–torque characteristics. Efficiencies of > 80 per cent are readily achieved, thus suitable for a battery power source.

(iii) Disc armature type

Variously referred to as a printed circuit motor, pancake motor or flat motor, the disc armature motor is a low inertia device in which the conventional copper armature windings are formed into a flat disc form and then moulded in epoxy resin. Either there is no separate commutator, in which case the brushes bear directly on to the armature conductors, or in general-purpose motors the armature windings lead out to a conventional barrel commutator.

The absence of iron in the armature not only makes it light but also results in very small values of inductance, hence linear torque–speed characteristics. In conventional brushed motors, torque modulation is produced by reluctance changes as slots pass under each pole tip, but this is absent with disc armature motors since the armature does not contain any magnetic material. This, plus the high number of commutation periods in a narrow commutating zone with face commutation, eliminates cogging and gives smooth torque down to zero speed. Compared to a cylindrical servo motor, and hence limited number of commutation segments, it is the ability of the disc motor to run at speeds as low as 1 rpm and by virtue of evenness of torque for any armature position, the disc motor can be considered for incremental applications in competition with steppers.

The basic geometry involves a ring of permanent magnets of alternate polarity to provide the axial field, either:

1. Two rings of magnets flank the armature in high-performance servo drives to provide the constant field (e.g. Fig. 8.35), the magnetic path closed by the steel end frames. The magnetic field is created by excitation coils, permanent magnets or a combination of the two to avoid the tendency of high start currents to demagnetize the magnets.[6]
2. In general-purpose motors, magnets are only mounted on one side of the rotor, the front plate of the motor then forming the return path for the magnets.

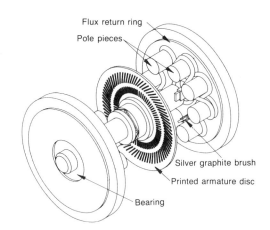

Fig. 8.35 Exploded view of a typical printed motor. Courtesy of E H Werninck

3. The yoke is carried by the armature. This increases rotor inertia but decreases the effective air-gap length.

The axial field produced by high coercivity magnets allows a shallow form of motor construction, because of their short length. Magnets may be sector rather than circular in shape to increase flux density in the outer radius where it is most effective. With face commutation, multiple brushes are normally used to minimize current into each brush, increasing both brush life and motor performance.

(iv) d.c. servos

These are typically high-performance motors with lightweight low-inertia armatures that respond quickly to excitation changes, usually used as prime movers in computers and control mechanisms where stops and starts must be made quickly and accurately. Whereas servo types used to be wound field motors, today the servomotor types would include permanent magnet motors, disc armature motors, shell armature motors and possibly steppers in some instances.

Each servomotor type would have its own characteristics such as inertia, shape, cost, weight, resonance, speed, etc. and although of similar torque ratings, other physical and electrical constants may vary considerably. The greater torque-to-interia ratio and low armature inductance of these motors give mechanical and electrical time constants of a few milliseconds. A great advantage of d.c. servomotors used in closed-loop control is that full use can be made of their high initial peak torque since the drive is self-compensating.

Most servo systems are complex and the requirements

vary, i.e. to control acceleration, to control velocity, to control position, hence a servo system must be specified very carefully, establishing the performance specification, determining the criticalities and setting up the tolerances. In general, however, a servomotor should have the following desirable characteristics:

1. A high continuous torque rating for a given frame size and electrical power dissipation;
2. The ability to operate both at low speed and high torques for accurate tracking with minimal cogging;
3. A high peak torque rating giving high acceleration rates with load-compatible armature inertia values plus a high thermal capacity if totally enclosed;
4. A rugged mechanical construction, to eliminate unwanted structural resonance, incorporating a directly mounted tachogenerator to provide a stable velocity feedback signal;
5. Brushes and commutator adequately designed to handle high torques and high current.

The following parameters are relevant to servomotor performance when evaluating compatibility:

1. *Torque-to-inertia ratio.* The higher this ratio, the faster the acceleration. Inertia referred back to the motor shaft is typically 30 per cent of rotor inertia.
2. *Motor bandwidth.* This is a measure of the motor's ability to respond to small rapid changes in input voltage, i.e.

$$\text{Bandwidth (rad s}^{-1}) = \sqrt{\frac{K_E K_T}{J_t L_a}}$$

where J_t = total inertia and L_a = armature circuit inductance (H). All servos have a bandwidth of >100 and in general it should be three times the bandwidth of the digital controller.

3. *Commutation limit.* Repeated high torque demands speed-up brush wear. Excessive high torque (commutator current) can cause flashover between adjacent commutator bars resulting in burnt areas and, in PM motors, demagnetization. Thus the machine acceleration requirements should be translated into the SOAC limits (Fig. 8.33).
4. *Heat dissipation.* The considerable start torque of servomotors can be exploited, but it is advisable to monitor the armature temperature which depends on the thermal capacity of the motor. The high coercivity of modern magnets permits servomotors not only to be started and reversed on full voltage but also, when required, to be given short current pulses in excess of the normal start current.

A servo system in its most elementary form consists of an amplifier, a control actuator and a feedback device, i.e. a closed loop system (Fig. 8.36). The feedback device is most commonly a tachometer or encoder. The output variable is measured, fed back and compared to the desired input function; any difference between the two is a deviation from ideal, the deviation is amplified and used to correct the error. However, the system response in that case depends on the closed loop; the response may be overdamped, underdamped or critical (Fig. 8.37), hence special care should be taken in the design of the closed loop in order to achieve the correct result. Control

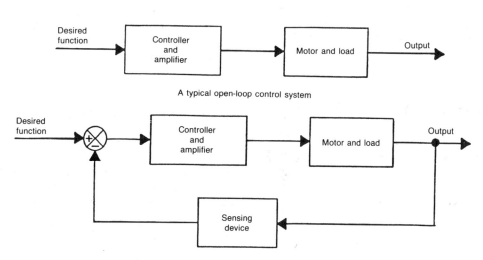

Fig. 8.36 Open and closed loop control

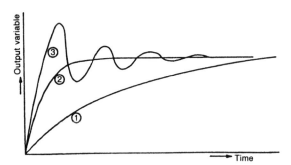

Fig. 8.37 Response of a system to a step input command; curve 1 = overdamped response; curve 2 = critically damped response; curve 3 = underdamped response

theory and servo theory are subjects outside the scope of this book, and input/output relationships and transfer functions are methods used in determining servo system stability. The various types of closed-loop control systems are as follows:

1. *Velocity control system.* In this system the motor velocity is controlled to follow a signal describing the desired velocity profile.
2. *Position control system.* The objective in positional control is to control the angular position of the shaft, either locking the shaft at the desired position or rotating it at a given rate.
3. *Torque control.* In some cases, it is required to keep the motor torque constant, i.e. by providing a constant rotor current. This can be achieved as in Fig. 8.38, where the input voltage $V_i = -I_d R$ is compared with the feedback voltage $V_f = I_p R$ when input to a high gain amplifier. If the output current I_p is different from the desired current I_d, the differential is amplified and used to correct the situation.
4. *Hybrid control system.* In some applications the advantages of various control modes can be combined to establish a desired servo performance by switching the system from one control mode to another. For example, alternating between a velocity control mode, where movement is to a desired velocity profile, and positional control mode where stopping is done by positional control for greater accuracy.
5. *Phase locked system* (PLS). Exhibits excellent speed regulation and inherent insensitivity to parameter changes. The basic elements are shown in Fig. 8.39. When the system is phase locked, the frequencies of the command and feedback signals become identical. As long as the system remains phase locked, the motor velocity follows the frequency of the command signal independently of any error condition such as amplifier gain drift, so guaranteeing a motor speed regulation which is as good as the stability of the command frequency. Operation is as follows: encoder pulsed output is input to the frequency divider which provides an output pulse train. The phase frequency detector compares feedback and reference frequencies and provides an output voltage proportional to the phase error. The loop will settle with motor and reference frequencies equal, but offset in phase by a fixed amount large enough to maintain the motor speed once the PFD output has been amplified. During acquisition of the phase lock condition, feedback and reference frequencies are unequal, the PFD output, positive/negative, will then cause the motor to accelerate or decelerate towards the lock speed.

8.7 Brushed d.c. motor speed control

8.7.1 Speed control

One reason for the continued use of d.c. motors is the simple and economic speed control possible over a very wide range. Speed and torque can be described by

$$\text{rpm} = \frac{K_1(V_a - I_a R_a)}{\phi} \quad \text{and} \quad T = K_2 \phi I_a$$

where V_a = armature voltage, I_a = armature current, R_a = armature resistance, ϕ = field flux, T = torque and K_1, K_2 = constants.

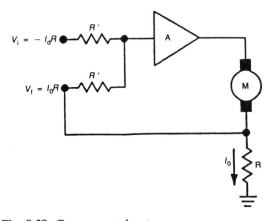

Fig. 8.38 Torque control system

DRIVE MOTORS

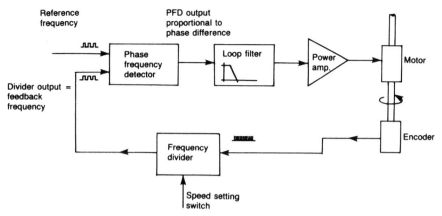

Fig. 8.39 Phase locked control system

Hence speed may be controlled by varying the applied voltage, the flux or the armature resistance. Armature current is proportional to, and varies with, the load, but it is not usually considered as a method of speed control. The various methods of speed control are summarized in Table 8.4. Field weakening, reducing the field current from its rated value, is a simple method of marginal speed control. By varying armature voltage, on the other hand, using SCR or chopper circuits, produces a more overall technique of speed control.

(i) Field weakening
The field current (and thus flux) is varied by adding resistance to the field circuit, in series in the case of the shunt motor and in parallel with the field winding in the case of the series motor (Fig. 8.40). The base speed of a d.c.-wound field motor is the speed at which the motor runs when full armature voltage is applied at full field strength. A d.c. motor can be operated below base speed by reducing armature voltage, but to operate above base speed, the motor field must be reduced by reducing field current while armature voltage is kept at the rated supply voltage. However, the field can only be weakened within limits if unstable speeds and armature overheating are to be avoided. Since motor output torque is proportional to field strength, weakening the field to increase speed reduces output torque.

(ii) Armature resistance control
Basically this is a resistance in series with the armature, increasing R_a, which can be used with series, shunt and PM motors, producing motor speeds below the base speeds. In the series motor, the field winding current is also affected by the armature series resistance, producing a greater effect on the speed–torque relationship than the other two (Fig. 8.41). Technique is essentially opposite to field weakening, the field winding in this case remaining excited at the line voltage. The higher the resistance the greater the drop in speed as load is increased, hence a series resistor will have its maximum effect on the motor start torque.

(iii) Armature shunt resistance
This can be added to a shunt or series motor along with the armature series resistance (Fig. 8.42) to give light load speeds down to 10 per cent. It slows the motor to creep speed by diverting a large portion of the line current around the armature. This system provides a positive slow-down or brake action when moving from high-speed operation to an armature shunt resistance point. It can also be used to brake dynamically.

Table 8.4 Brushed d.c. motor control alternatives

Shunt	Wound field motors Series
Field control (field weakening)	Field control (field weakening)
Armature resistance control (shunt or series)	Series resistance control (a.c. or d.c.)
Thyristor control (armature resistance control)	Shunt resistance control (a.c. or d.c.
	SCR control (a.c. supply or pulsed d.c.)

Permanent magnet motors

Armature resistance control (as shunt)
Armature voltage control (SCR control)

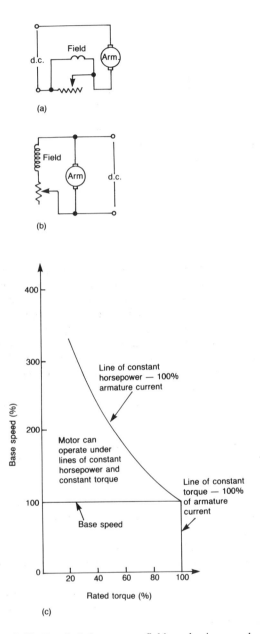

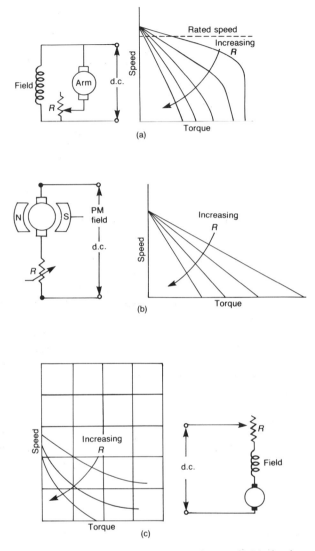

Fig. 8.41 Armature resistance speed control: (a) simple series armature resistance—shunt motor; (b) armature resistance control circuit for a typical PM motor; (c) armature resistance control for a series motor

Fig. 8.40 Brushed d.c. motor; field weakening speed control: (a) series motor, (b) shunt motor, (c) increase of speed by reducing field current and effect on torque

(iv) Armature voltage (SCR) control

As with induction motors (section 8.4), thyristors can be used to control d.c. currents from a d.c. supply, but since there is now no voltage reversal, the thyristor has to be turned off by other means, e.g. a capacitor in parallel

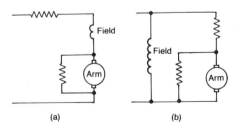

Fig. 8.42 Armature shunt resistance speed control: (a) series motor; (b) shunt motor

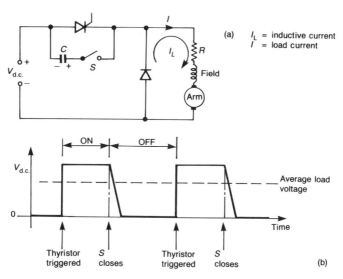

Fig. 8.43 SCR control of a d.c. motor: (a) basic 'chopper' circuit showing use of flywheel diode; (b) voltage waveform

(Fig. 8.43). This represents a rudimentary chopper circuit. In Fig. 8.43 a switch S is used to bypass the current, but in practice the gating circuit would be another thyristor. With an inductive load there is an advantage in using a freewheeling diode in parallel (Fig. 8.43): (1) to smooth the current in the load inductance; (2) to prevent large voltages appearing in the circuit due to interruption of the inductive current — the motor inductance discharges during the OFF period.

The d.c. power would be switched through an SCR operating at ~100 Hz. The mark–space ratio (ON:OFF time) can be changed by the firing circuit, thereby controlling the mean voltage V_a applied to the motor. There is no detectable speed change due to current and torque fluctuation since the electrical transients are much faster than the mechanical ones. For a fixed field current, the speed is nearly proportional to the duty cycle ratio. Starting d.c. motors supplied from chopper circuits is achieved by suitable adjustment of the firing angle so as to provide a gradually increasing V_a as the motor speed increases.

A d.c. motor can be fed from an a.c. supply through thyristors, T_1 and T_2 (Fig. 8.44), rectifying the current and (by delaying the commutation angle with controlled gate pulses) furnishing a variable voltage supply to the armature. The field current is obtained from the same a.c. supply. With armature voltage feedback, good speed holding to very low speeds (<1 r.p.s.) is possible. The phase angles of the trigger pulses are varied with respect to the a.c. line voltage to provide phase control of the

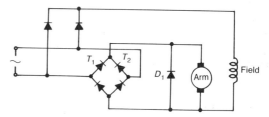

Fig. 8.44 Single-phase d.c. motor speed control from a.c. supply

rectified supply. Since only half the bridge is controlled by thyristors in Fig. 8.44, it is known as a half-controlled bridge.

8.7.2 Dynamic braking

In critical applications a brake or electromechanical clutch would be used, but where braking can be less precise, dynamic braking is possible. Electrical braking possesses definite advantages over its mechanical counterpart; there is no additional component, hence no wear element and control of the braking torque is more reliable.

(i) Reverse current braking ('plugging')
Either the field or armature connections are reversed in order to reverse the torque. In practice, however, the field winding has a much greater time constant than the armature so it is usual to reverse the armature.

The dynamic equation is

$T_m = T_L + J\alpha$ (neglecting frictional load)

but in this case, both the load torque T_L and the braking torque T_m oppose the motion and deceleration occurs, hence

$$T_m + T_L = -J\alpha = -J\frac{dw}{dt}$$

and

$$t = -J \int \frac{dw}{T_m + T_L}$$

Plugging can be used to the point where the armature passes through zero speed in its attempt to reverse itself. It is a severe method of braking because, at the instant of reversal, voltage across the armature is approximately twice normal, resulting in excess heating and brush arcing. In the case of PM motors, the coercive force of the magnets may be exceeded. Consequently, extra external armature resistance is normally included to reduce the braking current.

Electric braking can be described by second quadrant operation (Fig. 8.45); for example, a shunt motor operating at point A will be plugged across to point B, the speed reducing to point C. On the same basis, a series motor operating at E will go to F then G.

(ii) Rheostatic braking
This implies operating the motor as a generator, thereby developing its own resistance load. Dynamic braking is accomplished by disconnecting the armature from the power source and placing either a short or a current-

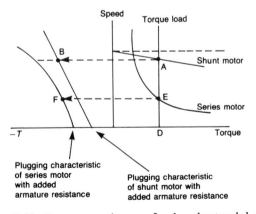

Fig. 8.45 Torque–speed curves for d.c. shunt and d.c. series motors for normal running and reverse current braking

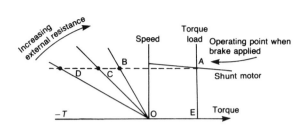

Fig. 8.46 Torque–speed curves for d.c. motor; rheostatic braking

limiting resistor across the armature while the field coils remain energized. The armature current reverses, torque reverses, and the motor will try to reverse. Speed in the forward direction rapidly decreases and along with it the back e.m.f. At the point of reversal, zero speed, generated voltage is zero, no current can flow in the armature and the motor stops.[5] Braking rate is controlled by the value of the shunt resistor. Braking torque is proportional to armature current, hence the smaller the external resistor the shorter the stopping time (Fig. 8.46). Some resistance is usually recommended to limit the severity of the braking action especially with gear motors. This method is appropriate for a series-wound motor if run on a.c. because of repulsion motor behaviour.

8.8 Brushless d.c. motors

8.8.1 Permanent magnet brushless

This is the same motor action as a brushed d.c. motor except that the supply current switching is via solid-state circuitry. A logic circuit senses the position of the rotor and controls the distribution of current to the field windings, energized in sequence to produce a revolving magnetic field. A brushless motor comprises four basic elements: an electronic commutator, a rotor position sensor, the stator and the rotor. In a brushless d.c. motor, the roles of the stator and rotor are reversed when compared to a conventional d.c. motor; the stator now becomes the armature, resembling a typical induction motor stator and the rotor provides the magnetic field (permanent magnet) (Fig. 8.47). In using an 'inside-out' construction brushless motors place the windings close to the stator surface so improving heat dissipation and maximum torque capability.

Typically the rotor houses a three-phase winding, the switching of the motor phases controlled by positional sensors, usually digital Hall sensors, located inside the

DRIVE MOTORS

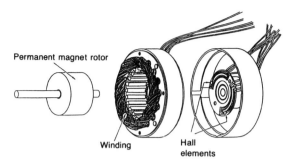

Fig. 8.47 The d.c. brushless motor in exploded and sectional views. From Kenjo T. et al., *Permanent Magnet and Brushless d.c. Motors*, Oxford Clarendon Press 1985. By permission of Oxford University Press

motor. These sensors ensure that excitation of the electromagnetic armature field always leads the permanent magnet field to create torque. Power transistors drive the armature windings at a specific motor current and voltage level and must withstand large voltage transients and current surges. Some motors use optics and photocells to commutate the current but this is less common. The essential parts of brushless d.c. motor control are shown in Fig. 8.48.

Torque is proportional to the back e.m.f., which in a single winding is the same thing as driving the rotor externally. The result for three phases is shown in Fig. 8.49, a simple sinusoidal distribution which when commutated will produce torque ripple. This may be acceptable in less precise applications, but ripple can be reduced by any of the following:

1. Increasing the number of commutator cycles per revolution. This means using smaller and smaller portions of the peak of the back e.m.f. waveform thereby limiting torque-ripple excursion and is analogous to the number of commutation bars on a brush motor.
2. Modify the shape of the back e.m.f waveform so that it is more trapezoidal, i.e. flatter during the commutation period.
3. Use a motor having a sinusoidal back e.m.f. waveform with a sine wave drive. More expensive in terms of the drive system and positional sensing, but results in very smooth torque output. Based on the classical a.c. synchronous motor, but complex circuitry is required to synthesize and control all three sinusoidal currents.

It is usual to shape the back e.m.f. waveform for each

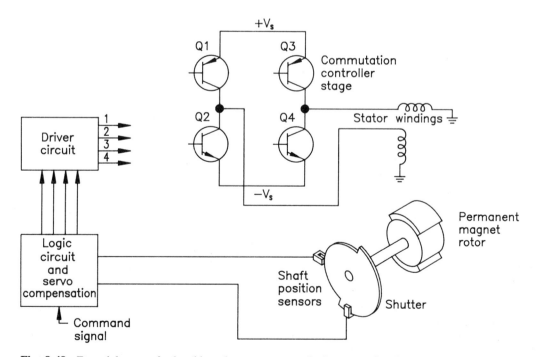

Fig. 8.48 Essential parts of a brushless d.c. motor control. Courtesy of Reliance Motion Control, Eden Prarie

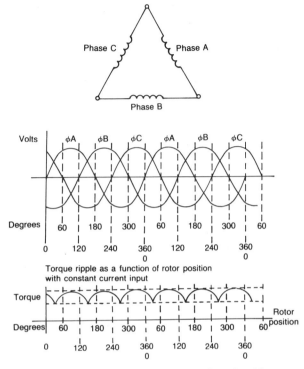

Fig. 8.49 Back-emf and torque ripple from brushless d.c. motor with sinusoidal windings (for a two-pole motor, electrical degrees = mechanical degrees). Courtesy of Inland Motor, Radford VA

phase so that constant torque is produced for constant current over this range. Ideally each phase should generate 120° of flat top back e.m.f. per half-cycle, an alternative terminology for the brushless d.c. motor being the trapezoidal-type drive. The simplest practical brushless d.c. motor circuit is shown in Fig. 8.50, a half-wave control with a conduction angle of 120°. As can be seen, each winding is used one-third of the time and hence the logic control is simple. To reverse the motor, all logic functions are shifted 180 electrical degrees. In a conventional d.c. motor, the supply polarity would be reversed, this being a basic difference between the two motor types.

A more common d.c. system is three-phase full-wave control (Fig. 8.51). Current flows into two of the three phases at all times at each 60 electrical degrees, one of the two phases is switched off and the third one switched on. By commutating twice as often during one revolution by using the negative as well as the positive half of the back e.m.f. waveform, produces a sequence of six phase combinations (Fig. 8.51). Each phase is active for two consecutive working strokes of 120 electrical degrees (twice per cycle).

8.8.2 The stepper motor

Steppers are quite unlike other motors. They are widely used where incremental or repeatable movement is required, converting digital information into discrete changes of position. However, while the stepper motor has simple positional control and reliability, it has introduced the need for electronics, control signals appearing as pulse trains rather than analogue voltages. A stepping motor can be defined as a reversible brushless d.c. motor, the rotor of which rotates in discrete angular increments when its stator windings are energized in a programmed manner. Rotation occurs because of the magnetic interaction between the rotor poles and the poles of the sequentially energized stator phases.[7] The rotor has no electrical winding but rather salient, projecting or magnetized poles. The basic characteristics and features of a stepper motor may be summarized as follows:

1. When energized, the motor will move and come to rest in a given number of steps, or detent positions, i.e. the motor is an electromagnetic actuator. Equivalent electromechanical devices would be rotary solenoids or ratchet and pawl drives.
2. For rotation to occur, the motor phase currents have to be changed in a controlled manner. The majority of drive circuits are microprocessor controlled, the widespread use of digital control having widened the application of steppers.
3. When energized the motor has a restoring torque (holding torque) which returns the rotor to a fixed position. May be used for positioning and maintaining a load.
4. The stepper is a passive, pulsed torque static device not relying on commutator or eddy currents for its operation. Consequently performance is more predictable and reliable.
5. Power consumption is not dependent on load conditions. It will consume the same power whether it is moving a load, holding a position or stalled.
6. The torque—speed curve, hence velocity—displacement characteristics for the motor depend on the electronic drive control.
7. The motor is not totally silent, an audible noise is present which is a function of the driving pulse rate.
8. A stepper accelerates and decelerates with each control pulse, even when rotating at 'full' speed, i.e.

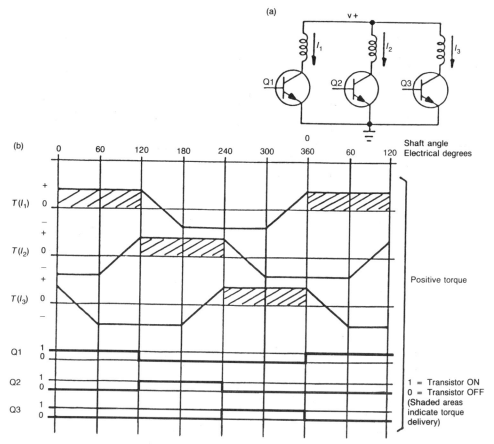

Fig. 8.50 A three-phase, half-wave brushless motor controller, $T(I)$ shows the torque functions for each winding: (a) electrical configuration; (b) torque function. Courtesy of Reliance Motion Control, Eden Prairie

there is always an inherent velocity ripple on the output shaft.
9. Load inertia should be limited to $\not> 4 \times$ rotor inertia for a high-performance (fast) system and $\not> 9 \times$ rotor inertia for a low-performance system. For optimum power transfer, hence maximum acceleration, the motor inertia should match the load inertia.
10. Most step motors can operate as a bidirectional synchronous motor.

As discussed in section 8.5, when a simple bipolar stator is excited by a direct current, the maximum holding or static torque is developed at 90°, i.e. when the magnetic axes of rotor and stator are in quadrature. Basic stepper theory can be illustrated by Fig. 8.52, a two-phase stepper with a two-pole permanent magnet rotor. Rotating the stator magnetic fields sequentially and reversing polarities (Fig. 8.52), produces a rotor angular step movement of 90°, i.e. step position. The change of rotor position with time is shown in Fig. 8.53, the faster the switching rate, the shorter the time period. Figure 8.53 basically demonstrates an underdamped response. Once stopped, the motor will resist dynamic movement up to the value of the holding torque. In practice, 2-, 3-, 4- and 5-phase steppers are common with multiple phase excitation and/or differing numbers of rotor teeth to give step angles 0.75–90°. The most common is 1.8° per step, 200 steps per revolution. In all cases, by sequencing the voltage applied to the coils, stator and rotor teeth are forced into alignment and the motor made to step. Each pulse advances the rotor shaft and latches it magnetically to the point to which it is stepped.

$$\text{Step angle} = \frac{360}{Ny}$$

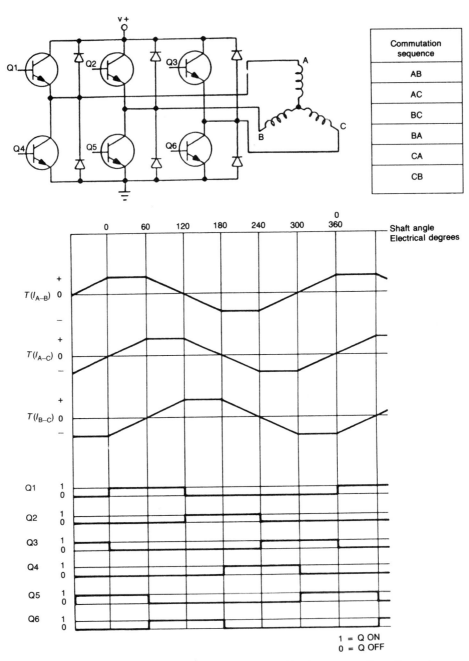

Fig. 8.51 Controller configuration, torque function and logic sequence for a three-phase, full-wave brushless d.c. motor. Courtesy of Reliance Motion Control, Eden Prarie

and no. of steps per revolution = pNy

where N = no. of phases, y = no. of teeth on rotor, p = no. of stator coil pairs energized at a given time

Shaft angular displacement = step angle × no. of pulses supplied

and

DRIVE MOTORS

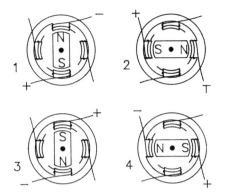

Fig. 8.52 Stepping action of a single stack PM stepper motor

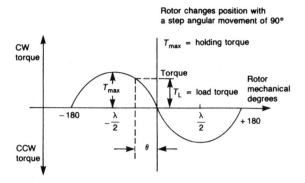

Fig. 8.54 Torque–displacement curve for a two-pole, two-phase stepper

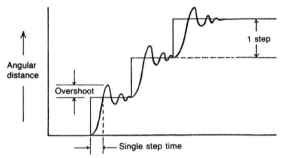

Fig. 8.53 Variation of step position with time

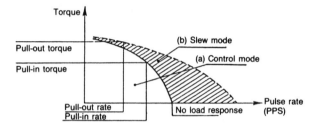

Fig. 8.55 Stepper motor; typical speed–torque curve. Courtesy of Astrosyn International Technology plc, Chatham

Motor speed (steps s^{-1}) = $\dfrac{NV}{3LI}$

= maximum speed with winding current reaching 95 per cent of rated motor current current

where V = supply voltage, L = inductance (H) and I = rated motor current.

The positional error at each step is typically ±5 per cent and the error is not cumulative. The static torque produced by the motor is proportional to the coil current and the misalignment between the stator and rotor teeth:

$T = J\alpha T_{max} \sin \theta_s$

where T_{max} = maximum torque or holding torque and θ_s = step displacement angle (Fig. 8.54).

From Fig. 8.54 the torque in the alignment position is zero (stator field and rotor field parallel) and a maximum at ±90°. Thus there are 'x' stable positions where an armature will come to rest when a given coil is energized. Output torque availability decreases with increasing shaft speed (Fig. 8.55), there being two modes of operation:

1. *Control mode.* For operation in the control mode, or response range, the stepper can be started, stopped or reversed on a single step impulse. In this region the motor can be successfully used as an open loop positioning device.
2. *Slew mode.* This mode of operation is achieved by accelerating from the control mode, thus extending the speed range. The motor runs in synchronism but cannot be stopped, started or reversed. The motor must be decelerated back to the control mode to achieve this. The technique of accelerating into and decelerating out of the slew range is called ramping.

The power availability from a stepper is the key to its ability to stop or start on load without error, but in spite of normally low efficiency (Table 8.2), one of the main features of a stepping motor is its relatively high torque at low speed. Some additional features are necessary to define stepper performance:

1. *Holding torque.* If an attempt is made to rotate the motor shaft by hand, the magnetic field will oppose the rotation. The required force is termed the 'static torque' and is a function of displacement angle (Fig.

8.54). The holding torque quoted by manufacturers is used as a machine—machine figure of merit, but in practice the motor cannot drive this amount of load torque since this is a maximum level achieved at only one relative rotor displacement position.

2. *Dynamic torque.* The useful torque that a stepping motor can develop is always less than the holding torque. As speed increases, the torque reduces (Fig. 8.55).
3. *Pull-in torque (control mode).* This is the maximum torque at which an energized stepper will start driving a specified load and run in synchronism without losing steps, on application of a fixed pulse rate (Fig. 8.55). The maximum torque—pulse rate pull-in curve is also referred to as the error-free-start-stop (EFSS) curve.
4. *Pull-out torque (slew mode).* The maximum torque that can be applied to a stepper driven at a given pulse rate under specified drive conditions without missing steps (Fig. 8.55). Further speed demand will result in lost steps. The critical step rate is termed the pull-out rate (of synchronism).
5. *Instability.* Instability of a stepping motor's performance can occur under certain circumstances. The mass moment of the rotor and its load, together with the magnetic stiffness, forms a spring system which causes:
   (a) a resonance at low stepping rates;
   (b) hunting around the required speed at high step rates (most pronounced with multi-phase motors operating in their slew range).

Steppers have a low natural frequency (150—250 Hz) and when operating in this range, can cause the rotor to overshoot and oscillate (ring) (e.g. Fig. 8.53). This may cause the motor to miss steps, and such unstable areas are generally indicated by dotted lines in manufacturers' performance curves. Applied friction will have the effect of reducing overshoot and primary resonance. Inertial loads can usually be accelerated through the resonance region. The primary resonance effect can also result in torque resonance at critical harmonics, typically 800—1500 Hz.

6. *Microstepping.* This is an electronic proportioning technique to divide a motor's fundamental step angle into smaller sub-steps, i.e. steps/step, from 2 to 125. These motor types drive both phases continuously applying currents in sine/cosine ratio. Advantages of microstepping are:
   (a) improved positional resolution;
   (b) reduces torque ripple;
   (c) eliminates resonance in most cases.
7. *Load inertia.* The effect of load inertia is to shift the torque—speed curve. Manufacturers' data will typically reference the pull-in torque curve (Fig. 8.56). The greater the torque/inertia ratio, the less the overshoot at each step.

The stepping motor is one of a four-part system (Fig. 8.57). The d.c. power supply provides the desired voltage and the control logic generates the start, accelerate, run, reverse, stop, etc. signals according to a program algorithm. The driver distributes the current (drive power) to the coils in the right order and at the correct level. The motor's dynamic behaviour will depend, for a given frame size, on the particular driver used. Types of driver are as follows:

1. *R/L drives.* These have a resistor placed in series with the phase winding to limit the motor current to its rated value when the motor is stopped or running slowly (Fig. 8.58(a)). The rise time of the winding current is improved at the expense of power dissipation in the series resistor. The number of

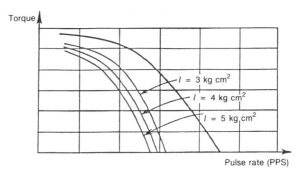

Fig. 8.56 Stepper motor; effect of load inertia. Courtesy of Astrosyn International Technology plc, Chatham

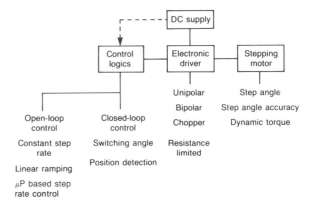

Fig. 8.57 Elements of the stepping motor system

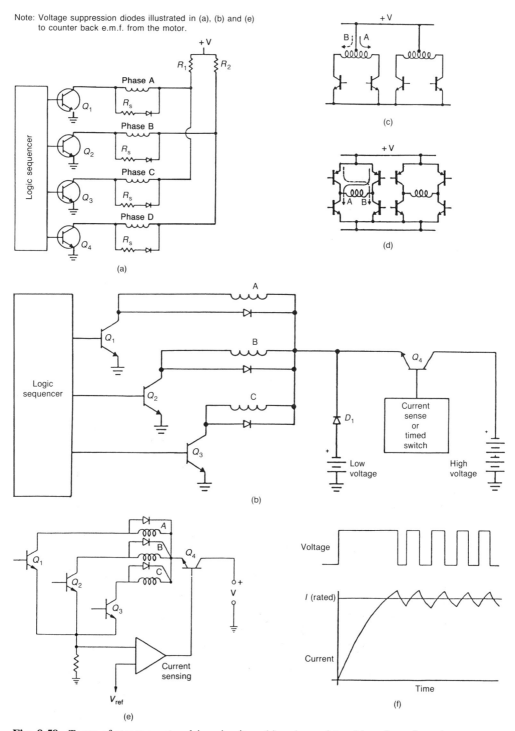

Fig. 8.58 Types of stepper motor drive circuitry: (a) series resistor driver for a four-phase step motor; (b) simplified dual voltage driver; (c) unipolar drive; (d) bipolar drive; (e) simplified chopper driver; (f) chopper voltage and current waveforms

resistors required can be minimized by placing them in a common return lead or in a power supply lead (e.g. Fig. 8.58(a)). Often a compromise between high-speed performance and complexity. In this case

$$t_e = \frac{L_a}{R_{total}} = \frac{L_a}{R_{series} + R_m}$$

i.e. decreasing the time constant from L_a/R_m to L_a/R_{total}.

2. *Dual voltage.* Referred to as a dual voltage or bi-level drive, this is a more efficient driving method using high voltage only to build current in the windings and then to switch to a lower voltage. A simplified circuit for a three-phase dual voltage, single-phase-on driver is shown in Fig. 8.58(b). When a motor is stepped, a high-voltage switch Q4 plus a phase switch (Q1, Q2 or Q3) are turned on. After the current has reached the rated value, sensed or timed, the high-voltage switch is turned off and the current is maintained by the low-voltage supply.

3. *Unipolar drive.* Each stator coil of a motor is provided with a centre tap which is connected to one side of the supply, say positive. The direction of current flow through a coil is then determined by the end to which the negative supply line is connected via a switching device. Switching coil halves results in the magnetic poles of the stator being reversed (Fig. 8.58(c)). Allows for a simple drive circuit and power supply.

4. *Bipolar drive.* The stator coils have no centre tap, so to reverse current flow through the coil, both supply lines are switched (Fig. 8.58(d)) generally by a bridge-configured circuit. Bipolar drives produce more torque at low speed than unipolar drives (Fig. 8.59), but electronics are more complex. Alternatively they allow a reduction of system power, up to 50 per cent.

5. *Chopper drive.* A current-limited driver using pulse width or frequency modulation to control the average current to the step motor windings (Fig. 8.58(e)). The chopper driver applies a high voltage at the beginning of each step but prevents the current exceeding its rated value by sensing and chopping it, i.e. the power transistor Q4 is ON/OFF controlled by the current sensing network (Fig. 8.58(f)).

The other type of chopper control system controls the chopping frequency at approximately 4 kHz, while the voltage 'on' time is varied to provide current control. These drives produce increased high-speed performance but there is a tendency for the circuit to generate electrical noise.

Moving on to speed control, this can either be open loop or closed loop to match system performance and cost

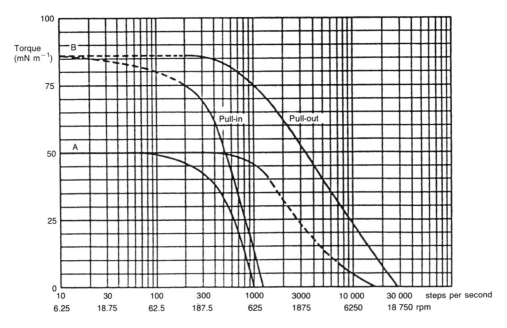

A: unipolar motor, B: bipolar motor.

Fig. 8.59 Stepper motor; comparison of unipolar and bipolar drive performance

requirements. Traditional open loop control is adopted in 95 per cent of stepper motor applications, mainly because it is cheaper and simpler than with encoders included. Numerous sequence logic or controller configurations have been published, and the following three are typical examples in order of increasing elegance:

1. *Constant step rate control.* In its simplest form, a constant step rate applied to the motor until the load reaches the target position. A schematic is given in Fig. 8.60(a). The clock is turned on by a start signal, causing the motor to run at the clock frequency until turned off by a stop signal. Operation is in the control mode and, allowing for resonance difficulties, the system will produce good performances at minimum cost.

2. *Linear ramped control.* This allows for higher motor speeds by increasing the step rate with time, into and out of the slew range. A schematic is shown in Fig. 8.60(b). The oscillator is voltage controlled via an analogue circuit, i.e. the step rate is controlled via R. Resistor R_A adjusts the integration time constant, hence the initial acceleration rate. The deceleration rate is adjusted via R_D, the start of deceleration controlled by a preset counter (timer). The linear ramp, however, only provides an optimum velocity profile if the load torque is constant, but in practice, motor torque decreases with speed increase thereby reducing the system's acceleration capacity significantly. Thus either
   (a) restrict the acceleration rate to the level required for high-speed operation, or
   (b) restrict operation speeds to the near-constant region of the torque–speed curve.

   An experimental acceleration (and deceleration) ramp is possible with an RC combination charging up from a constant voltage source.

3. *Microprocessor control.* Can provide a velocity profile close to the ideal (Fig. 8.60(c)). The block diagram schematic is fairly simple, since all control functions are performed in the software algorithm. This offers the maximum performance of all open-loop modes. The control algorithm is usually based on the empricial simulation of a particular motor-drive characteristic.

One of the major advantages of a stepper motor is its inherent ability to operate in an open-loop control system, but it does mean the high-speed performance is usually limited and acceleration loads set conservatively. In a closed-loop system, higher speed and smoother operation are possible with accurate position control. In Fig. 8.61 a closed loop controller matches the motor excitation

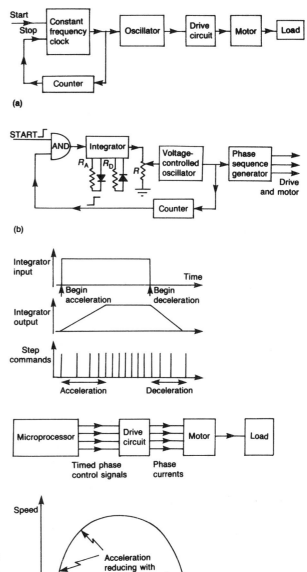

Fig. 8.60 Types of stepper motor control: (a) constant stepping rate open-loop control; (b) linear ramped open-loop control; (c) microprocessor based open-loop control and optimum velocity profile

timing to the load conditions and is able to give a near-optimum velocity profile with consequent rapid load positioning. Since the load position is monitored direct, there is no possibility of losing synchronism between step commands and the rotor position.

ELECTROMECHANICAL PRODUCT DESIGN

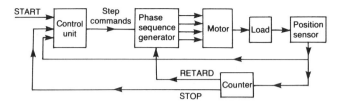

Fig. 8.61 Closed-loop control of a stepping motor

The three common types of stepping motor are: (1) variable reluctance, (2) permanent magnet, and (3) hybrid.

(i) Variable reluctance (VR) stepper

This motor uses a multi-toothed soft iron rotor, having a different number of teeth from the stator, which is caused to rotate to the position of minimum reluctance (position of minimum resistance to the flow of magnetic flux) in the stator's magnetic field. Energizing coil A in Fig. 8.62 causes the rotor to align with the stator field as shown, i.e. a 15° step angle with four coils and four phases. Increasing the number of stator and rotor teeth will reduce the step angle. The rotor of a VR stepper has practically no residual magnetism hence there is no magnetic force to hold the rotor in position once the stator supply is switched off. Also, because the rotor has little inertia, it will tend to resonate. Two basic types are available, single stack and multiple stack.

(ii) Permanent magnet stepper

This consists of a permanent magnet rotor and wound stators (e.g. Fig. 8.63). Since the rotor is a permanent magnet, the poles are fixed, and their number is inherently limited. Enlarging the rotor diameter to provide for a large number of poles results in a drastic increase of rotor inertia. The stator assembly comprises two or more stators. By reversing the direction of current flow, N and S poles can be transposed. Thus reversing the current flow through successive stator coils creates a rotating magnetic field which the rotor follows. Figure 8.63 illustrates a unipolar-driven four-phase motor in which when phases P and R are energized, the rotor assumes the upper position. If S1 is then operated, R and Q are energized and the rotor moves through 90°. Thus operating S1 and S2 alternately, the rotor will rotate in 90° steps. With a third switch position to turn off phases, half-step operation is possible (Fig. 8.63).

Recent developments in magnet technology has resulted in new designs with low-inertia thin disk rotors and improved resolution performance.[8]

(iii) Hybrid stepper

This motor combines the features of VR and PM motors. It is a widely used motor, the most commonplace style being the 42 mm square, 200-step motor found in almost every floppy disk drive. The active rotor is an axially magnetized cylindrical permanent magnet. Soft iron cups or gear-like hubs are mounted on each end of the rotor and surround the magnet (Fig. 8.64). The stator teeth or rotor teeth on each end of the magnet must be offset half a tooth pitch from each other in order to prevent the rotor teeth on both ends of the magnet from lining up with the same stator teeth. The hybrid design will feature either (a) a dissimilar number of teeth on rotor and stator to provide maximum damping, e.g. 48–50 to reduce the number of harmonics within the operating cycle, hence reduced resonance, or (b) the same number of teeth on rotor and stator to provide maximum torque, i.e. 50–50.

Both the above result in a step angle of 1.8° with a four-phase design. The rotor is made to move one step at a time by properly sequencing the drive currents. Because the rotor is a permanent magnet, there is always some detent or residual torque.

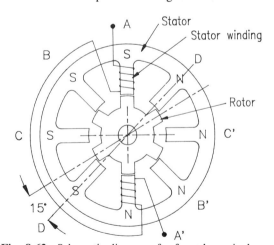

Fig. 8.62 Schematic diagram of a four-phase single-stack variable-reluctance step motor with one phase at a time energized; only the 'A' windings are shown for clarity

DRIVE MOTORS

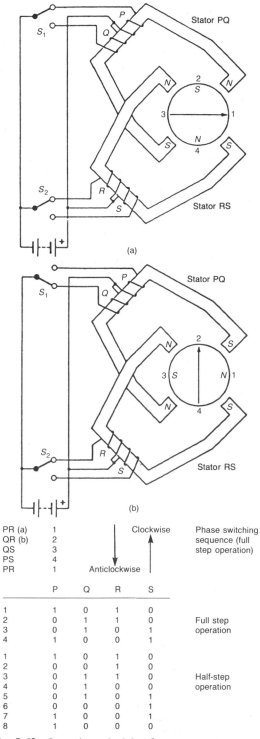

	P	Q	R	S	
1	1	0	1	0	
2	0	1	1	0	Full step
3	0	1	0	1	operation
4	1	0	0	1	
1	1	0	1	0	
2	0	0	1	0	
3	0	1	1	0	
4	0	1	0	0	Half-step
5	0	1	0	1	operation
6	0	0	0	1	
7	1	0	0	1	
8	1	0	0	0	

Fig. 8.63 Operating principle of permanent magnet four-phase stepper

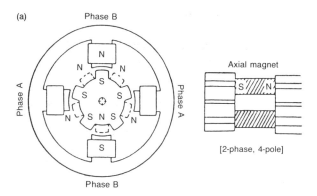

The rotor position is shown with phase B energized. When phase A is energized the direction of rotation is controlled by the polarity of the two A phase poles. If the right-hand pole is north the rotor will rotate clockwise. If the right-hand pole is south the rotor will rotate counterclockwise.

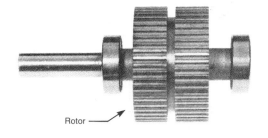

Fig. 8.64 Hybrid stepper motor: (a) simplified explanation of 18° stepper action (20 steps per rev.); (b) example of 1.8° step angle motor components

8.8.3 Switched reluctance

The switched reluctance (SR) motor is a direct derivative of the VR single-stack stepper motor but unlike the stepper motor, it can produce continuous torque at any rotor position and speed. It is a double salient, singly excited motor, i.e. it has salient poles on both the rotor and stator but only one member (usually the stator) carries the windings. The rotor has no magnets, conducting circuit or cage, but is built up from a stack of salient pole laminations (Fig. 8.65). The stator poles are energized by a switched d.c. current.

Alternative terms, VR motor, brushless reluctance motor or commutated reluctance motor are sometimes used.[9] The SR drive is fairly new compared to its competitors so it has not yet reached full potential. It is capable of providing high levels of performance over a wide range of conditions with the correct motor design, power electronics and control strategy. The motor is capable of high overall efficiencies and good load control yet it has a simplicity and ruggedness of construction only previously associated with the induction motor. It is used where a brushless drive is required, a wide speed range and a cost saving over PM brushless.

In producing torque, the magnetic flux set up by the current in the stator coils tends to pull the rotor poles into alignment with those of the stator (Fig. 8.65). The direction of torque is independent of the direction of current flow and the current needs only to be switched, not reversed as with some other motors. To achieve torque in one direction, the current must be switched ON and OFF at appropriate angular positions of the rotor, e.g. $\theta_1 \to \theta_2$ (Fig. 8.65). Coil inductance increases with increasing overlap of stator and rotor poles and when current flows into regions where inductance is increasing, positive torque is produced. Negative torque is produced if the current is flowing in the region of decreasing inductance, $\theta_3 \to \theta_4$.

Variation in torque and power is achieved at higher speeds through appropriate selection of switch-on and switch-off angles. If switching angles are held constant and speed is allowed to increase, the motor will exhibit the runaway characteristic of a series (universal) motor as a consequence of flux magnitude being proportional to 1/speed, torque decreasing as $1/(\text{speed})^2$.

The power supply can be from a battery, but more usually from rectified mains through a d.c.-link inverter. A basic power circuit configuration, two switches per phase, is shown in Fig. 8.65, the type of electronic switch could be thyristor, triac or power transistor. To self-start in either direction, at least three phases are required, e.g. a three-phase 6:4 motor is illustrated in Fig. 8.66.

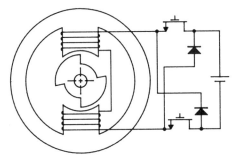

Basic power circuit, 1 pole pair, single phase

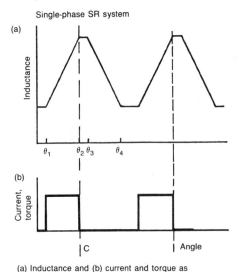

(a) Inductance and (b) current and torque as functions of angle turned through by rotor

Fig. 8.65 Switched reluctance motor; basic motor configuration, communication and circuit

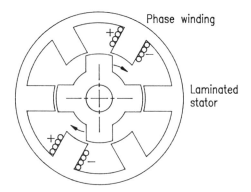

Fig. 8.66 Cross-section of three-phase 6:4 motor. From Muller T.J.E., *Brushless Permanent Magnet and Synchronous Reluctance Motor Drives*, Oxford Clarendon Press, 1989. By permission of Oxford University Press

The ideal current waveform is a rectangular pulse coinciding with the rising inductance (Fig. 8.65). However, depending on the control strategy, in order to maintain a current within prescribed limits at low speed when the self-e.m.f. of the motor is much smaller than the supply voltage, it has to be forced to an approximate rectangular shape by chopping or by pulse width modulation (pwm) in the controller (e.g. Fig. 8.67). At high speed, the current waveform takes up a natural shape determined by the speed, turn-off and turn-on angles, the applied voltage and the rate of change of inductance. Unlike other motors, the current waveform in an SR motor may vary widely over the speed range (Fig. 8.67). It is convenient for the logic design to use one transistor primarily for commutation (Fig. 8.65), the other for regulation (chopping). At the end of the conduction period, both switches are forced off.

The generic form of the torque–speed curve for an SR motor is shown in Fig. 8.68. For speeds up to the base speed (base point), a 'constant' torque–speed characteristic is achievable by varying current up to a maximum at the rated voltage ($\equiv w_b$, Fig. 8.68). If the firing angles are kept fixed, torque decreases $\propto 1/(\text{speed})^2$, but if the conduction angle θ (dwell angle) is increased (e.g. Fig. 8.67), a constant power characteristic (increasing current and torque) can be maintained up to a point P (Fig. 8.68). Advancing the turn on or conduction angle can only be up to the point where the total conduction angle equals the rotor pole pitch. With other d.c. motors, the torque/ampere ratio

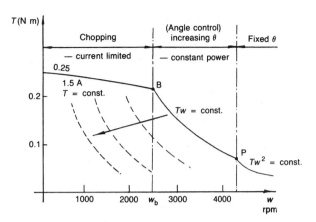

Fig. 8.68 SR motor; form of the torque–speed curve (characteristic of small SR motors)

is approximately constant, but with the SR motor there is no simple relationship.

8.9 Motor selection criteria

The range of modern motion control applications is virtually unlimited (Tables 8.1, 8.2). Loads have widely differing requirements, the commonest being for speed control with varying degrees of precision and accuracy, particularly in, for example, automated processes and office machinery. In some cases it is the steady-state operation that is important, e.g. fan or pump drives, whereas in other cases such as robotics, tape drives and actuators, dynamic performance is important because of the need to minimize time taken to perform actions or effect control operations. In these cases, torque/inertia ratio is an important parameter.

It is not possible to set down a formal procedure to select the 'optimum' motor for all control applications since the duty cycle, type of control, space limitations, etc. all impose restrictions which are individual to a particular situation. However, it is possible to approach all applications in a prescribed manner, determine the relevant application factors and narrow down the drive choice in most instances to a general motor type. Because of the number of drive types now available, this in itself is a significant task. There are a large number of possible factors, and these are given under the headings below.

(i) Motor characteristics

1. Operating speed requirement;

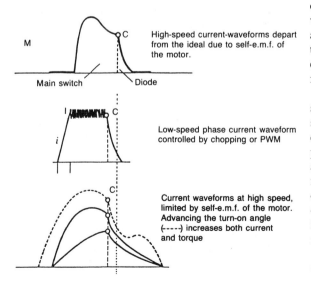

Fig. 8.67 SR motor; current waveforms

2. Life requirement and maintenance interval possible;
3. Efficiency;
4. Torque requirement, starting, synchronizing, etc.;
5. Mechanical aspects of available space envelope, environment, noise level, weight, etc.;
6. Compliance with national and industry standards including radio frequency interference (RFI), (IP) rating etc.;
7. Overload rating and duration.

(ii) Load requirements

1. Preliminary analysis of the operating cycle to determine the most critical position as regards motor selection, i.e. periods of high acceleration/deceleration;
2. Dynamic acceleration torque as a function of speed, normal full-load torque and any likely overload condition;
3. Analysis of the starting conditions, inertia to be accelerated and/or synchronized;
4. Behaviour of load to any transient irregularities;
5. Gearbox or direct drive; optimum gear ratio;
6. System friction.

(iii) Control requirements

1. Power input availability (voltage, current, regulation, etc.);
2. Motor operating voltage, current and variation required;
3. Open or closed loop;
4. Forward/reverse operation;
5. Motoring and/or braking requirement;
6. Torque or servo performance required;
7. Degree of position or speed control required;
8. Complexity and availability of controller type.

Having decided, from these factors, the general motor type, i.e. a.c. or d.c., it is then necessary to undertake a more detailed analysis to determine which particular variant (Tables 8.1, 8.2) will meet the detailed requirements. For example:

1. Identify the optimization criteria consistent with acceptable performance, e.g. cost, availability, etc.
2. Detailed analysis of the identified critical portions of the operating cycle to produce estimates of motor size.
3. Use of these estimates together with the requirements of other elements in the system to determine the most suitable method of coupling the load.
4. If a feedback sensing is required, and not integral with motor, define type and accuracy specification.
5. Optimization of the selected drive system in terms of the identified factors to produce a final specification for the motor. The essence of motor selection is to consider the whole drive system as part of the process, i.e. the control circuit, couplings, gearbox, encoder, relative to the reliability and cost targets required of the design configuration.
6. Confirmation of a suitable motor type. Obtain manufacturer's data on torque, speed, duty cycle, cost, etc.
7. Confirmation of voltage/current drive requirements based on the motor performance requirement together with control electronics if working within a closed or a feedback sensing system.

8.9.1 Motor selection process

The general criteria for all system elements have already been summarized. We will now take this one step further and deal with a generalized d.c. motor application. This particular drive has been chosen as it provides most of the design and selection factors relevant in motor selection. For performance equations relevant to a specific drive type, however, reference should be made to the sections on particular motors. The process starts by defining a motion specification, e.g.:

1. Movement profile, continuous, intermittent ...;
2. Type of control, velocity or positional;
3. Distance travelled;
4. Time allowable plus possible constraints;
5. Duty cycle;
6. Acceleration rate and slew speed;
7. Accuracy of speed and allowable variations;
8. Positional accuracy requirement;
9. The nature of the drive system, e.g. gearbox, pulley, leadscrew, coupling stiffnesses.

(i) Sizing the motor

Described stepwise as follows:

1. Calculate the maximum speed w (rad s^{-1}).
2. Select a motor, or a range of possibles, that can meet the required speed.
3. Calculate the reflected inertia J of the load and drive components (kg m^{-2}) (see section 8.2(ii)).
4. Calculate the peak torque, T_{MP}, during acceleration where $T_{MP} = J\alpha + T_F$ (N m), T_F = frictional torque (section 8.2(iv)) and $J = J_{motor} + J_{load} + J_{coupling}$, etc.
5. Check that the motor can provide the required peak

torque. If not, continue an iterative process (2) to (5) by increasing motor size.

6. Define the following motor parameters (section 8.6.2): K_T, the motor torque constant (N m A^{-1}), R_a, the armature resistance (ohms), J_m, the rotor inertia (kg m^{-2}), K_E, the back e.m.f. or motor voltage constant (V rad^{-1} s^{-1}) and R_T, thermal resistance (°C W^{-1}).

7. Check that motor can operate within its thermal specification as follows:

$$I_{\text{acceleration}} (I_{\text{max}}) = \frac{T_{\text{MP}}}{K_T}$$

$$I_{\text{deceleration}} = \frac{\alpha J - T_F}{K_T}$$

$$I_{\text{run}} \text{ (constant speed)} = \frac{T_F}{K_T}$$

from which

$$I_{\text{rms}} = \sqrt{\frac{\Sigma I_i^2 t_i}{\Sigma t_i}}$$

and

Temperature rise = $I_{\text{rms}}^2 R_a R_T$ (°C)

Check that the motor can handle this temperature. Depending on the movement profile, an individual portion of the profile may become dominant in heat dissipation (section 8.6.2(e)).

(ii) Defining the power requirement

After motor selection, the next step is to calculate the peak voltage and current necessary to drive the motor. Here I_{max} can be obtained from (7) above.

$$V_{\text{max}} = R_a I_{\text{max}} + K_E w_{\text{max}}$$

(iii) Defining the transmission

Although the design of the drive system is known, there are nevertheless a number of factors that will need to be checked relative to the movement specification once the motor type and size have been defined. These are stiffness, backlash, dimensional constraints, efficiency, noise.

These are mostly items which are covered elsewhere in this book, but the following general points can be made:

1. In a system involving a flexible or timing belt drive, system stiffness clearly decreases with increased belt length.
2. With timing pulleys, the power transmission capability is a function of number of teeth in contact with the belt.
3. The natural frequency of the drive

$$f = \frac{1}{2\pi} \sqrt{\frac{\text{torsional stiffness}}{J}} \quad (\text{Hz})$$

must be well above the operating frequency.
4. A motor-mounted encoder may operate at higher speed, not be subject to backlash effects and allow a lower resolution requirement, but it is then not such a fully closed loop as positioning at the load. A load-mounted encoder gives best feedback for load control, but since the drive is usually geared down in some form, it also requires higher resolution. However, it is pointless specifying a high-resolution encoder if backlash in the drive system gearing is noticeable.

References

1. BS 4991: 1987 *General requirements for rotating electrical machines*
2. New AC motor reaches 95% efficiency. *Eureka* 7(10): 64 (1987)
3. Werninck E H (Ed) *Electric motor handbook*. McGraw-Hill 1978
4. Hale R Dynamic braking of electric motors. *Eng. Designer* May 1980: 27
5. Solowski W Series wound motors. *Power Trans. Design* 30(7): 53 (1988)
6. The Lynch traction motor. *Eureka* 9(10): 66 (1989)
7. BS 5000: Pt 60: 1982 *Rotating electrical machines of particular types or for particular applications*
8. Ormond T Modern stepper motor control system. *EDN* Jan 1987: 85
9. Muller T J E *Brushless permanent magnet and reluctance motor drives*. Clarendon Press 1989

CHAPTER 9

DRIVES, BELTS AND COUPLINGS

9.1 Introduction

This chapter deals with a number of the more common types of driveline component used for the transmission or transfer of movement (power) between shafts. Each type of mechanical drive connection will have specific features that often dictate selection in a given design application, although the final choice is generally based on a number of considerations. The operating characteristics of several drives will be discussed. The subject of gears is left until Chapter 10.

9.2 Belts

The origin of the present-day flexible belt lies in the rope drives and flat leather drives of the 1800s, the fundamentals of operation still remaining the same. Belts provide an efficient means of torque transfer between shafts in a wide variety of applications and have major advantages over other power transmission methods. Such advantages are the wide selection of speed ratios, cost, efficiency and reliability. We shall consider three main types of belt, i.e. flat belts, light duty vee belts and synchronous toothed belts.

Both vee belts and flat belts creep, hence they are not suitable for drives requiring synchonized input and output. Precise speed ratios are possible with synchronous belts and these find significant usage in office machinery, computer peripherals, etc.

9.2.1 Flat belts

Generally used where high speed rather than high torque transmittance is important. Because of lowish mass, minimum belt height and a centre of gravity close to the pulley surface, they do not develop significant centrifugal tension at high speed, i.e. they are less affected by centrifugal force than, say, the vee belt. The belts depend on high tension to maintain traction, hence the support shafts and bearing must be stable. Shaft alignment may need to be controlled to prevent belts 'walking off'. The advantages and disadvantages of flat belts are summarized in Table 9.1.

Table 9.1 *Advantages and disadvantages of non-synchronous belt drives*

Advantages
Design flexibility, basic geometry easily defined in pitch terms
Belt sag rarely encountered hence great variety in centres possible
Parallel shafts not necessary, will also tolerate some misalignment
No lubrication required
Can slip under gross overload
Shock and vibration absorbent, quiet
Belt failure decouples drive (safety aspect)
Wide range of speed variation possible
High mechanical efficiency, >98%
Little heat generated through hysteresis
High power transmission possible
High speeds
Good strength to weight ratio

Disadvantages
Periodic tension checks and adjustment required
Uneconomic at low speeds
Max. reduction ratio 6 : 1
Creep plus speed reduction of drive shaft
Cannot maintain precise angular relationship between shafts
Not tolerant to temperature extremes
Deterioration on exposure to contaminants
Damaged belts must be replaced rather than repaired
Rely on frictional grip

DRIVES, BELTS AND COUPLINGS

Numerous types of belt are available today for different types of application, the generic class of belt basically dependent on the belt construction i.e.:

1. *Elastic belt.* These are unsupported elastomeric stretch bands used for lightly loaded low-duty application. Typically made of silicone, polyurethane or a synthetic elastomer blend. Truely elastic, only initial tensioning required in service.
2. *Semi-elastic belt.* These have the features of unsupported stretch belts while maintaining a fabric, woven cord or helically wound cord substrate for support (e.g. Fig. 9.1). Used in a variety of small multi-pulley drives, this belt is fairly inexpensive and can be 'tuned' to any spring rate. Pulley take-up adjustment may or may not be required. The traction surface is either an elastomer impregnated directly onto the substrate, (e.g. polyurethane) or it can be neoprene or a rubberized fabric bonded to the substrate. The tensile member is usually polyamide, cotton or polyester cord.
3. *Non-stretch belt.* These are endless belts that can be used with small pulleys and at high speed, made very thin (0.4–1.5 mm thick) with a continuous, or near-continuous, nylon or polyamide tensile member (e.g. Fig. 9.2). The traction surface is a bonded elastomer. In some instances thin steel belts are a possibility. These run on pulleys without camber, but shafts must be perfectly aligned with the pulleys running true. They have the advantage of no permanent stretch.

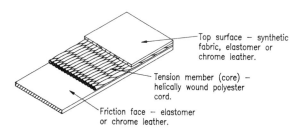

Fig. 9.1 Structure of semi-elastic flat belt

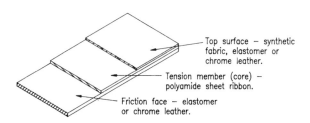

Fig. 9.2 Structure of non-stretch flat belt

(i) Fundamentals of flat-belt power transmission

The force that produces torque with a belt drive acts on the rim of the pulley causing it to rotate. When a drive is transmitting power the belt pulls or belt tensions are not equal (Fig. 9.3). There is a tight side tension T_T and a slack side tension T_S, the difference, i.e. the effective or net pull, is the maximum effective tangential pull exerted by the belt on the pulley rim

$$\text{Effective pull} = T_e = T_T - T_S = \frac{P}{v} = \frac{P}{D_p n}$$

where P = power transmitted, v = belt speed, D_p = pitch diameter (Fig. 9.4) and n = rpm.

Depending on the values of T_T and T_S, the same effective belt tension can be created at different shaft loadings (e.g. Fig. 9.3).

If the transmitted power and belt speed are known, the above equation can be used to determine the effective pull. However, when the belt is initially fitted (or at rest), belt tensions will be equal, i.e. T_0 each side. When power is applied to one pulley, the tension of the two free lengths changes to T_T and T_S respectively, such that

$$T_T - T_0 = T_0 - T_S$$

if the belt is elastic.

To determine actual values of dynamic belt tension, a second relationship, the tension ratio T_T/T_S, is required. At the point of *impending* belt slip, the tension ratio is described:

$$\frac{T_T}{T_S} = e^{\mu\theta}$$

where μ = belt/pulley coefficient of friction and θ = belt wrap or arc of belt/pulley contact (rad).

The higher the ratio the closer the belt is to slipping, i.e. the belt is loose. The prevailing coefficient of friction will depend on the physical condition of the belt surface and pulley surface but modified by any surface contamination present. At increased values of θ, a drive can

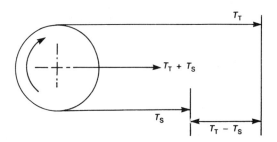

Fig. 9.3 Flat belt tension inequality

241

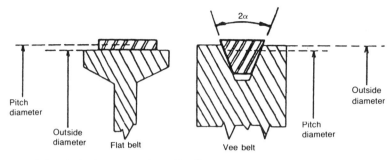

Fig. 9.4 Pitch diameter and outside diameter

operate at a higher tension ratio and, relatively speaking, transmit more power. A low-tension ratio on the other hand means more slack-side tension compared to tight-side tension so the belt is less likely to slip but the effective pull is lower. If there is a recommended tension ratio for a given belt, then

$$\text{Effective tension} = T_T \left(1 - \frac{1}{K}\right)$$

and

$$T_T = \left[\frac{2K}{K+1}\right] T_0$$

where $K = T_T/T_S$ (smaller pulley).

Working tensions are important in the way they affect belt life, i.e. strain fatigue, but in addition there are other tensions that develop as a belt passes around a drive loop, i.e.

Bending tension. Occurs as the belt flexes in tension and compression around a pulley, the extent depending on the pulley radius.

Centrifugal tension. At high speed the belt is subject to the centrifugal effect of belt inertia passing over the pulleys such that the peak belt tension at the pulley is given by

$$T_p = T_T + T_B + T_C$$

where T_B = bending tension, T_C = centrifugal tension = Wv^2/g, W = weight of belt/unit length and g = acceleration due to gravity.

Figure 9.5 gives an idea of the repetitive stressing of a belt when transmitting power. Neither the bending nor the centrifugal tension effects are imposed on the pulleys, only on the belts. Peak tension is empirically related to belt life (Fig. 9.5).

One of the major features of the non-elastic, high-modulus flat belt is high efficiency, i.e. low energy loss in service as it flexes around the drive loop. The hysteresis can be represented as in Fig. 9.6, the width of the belt being proportional to the damping factor of the belt material.[1]

(ii) Practical aspects of flat-belt power transmission
(a) Drive geometry. The practical equations describing the basic drive geometry (Fig. 9.7), are as follows:

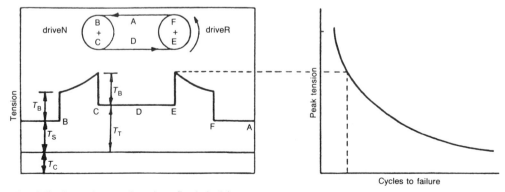

Fig. 9.5 Operating tensions in a flat belt drive

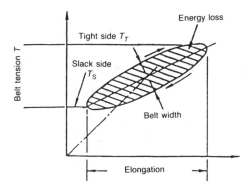

Fig. 9.6 Representation of flat belt hysteresis loss

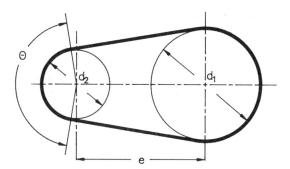

Fig. 9.7 Flat belt drive geometry nomenclature

$\theta°$ (error $\not> 5$ per cent) $= 180 - \dfrac{60(d_2 + d_1)}{e}$

where d_1 = diameter of the driving pulley, d_2 = diameter of the driven pulley and e = centre distance.

Belt length $l = 2e + \left[\dfrac{\pi}{2}(d_2 + d_1)\right]$
$+ \dfrac{(d_2 - d_1)^2}{4e}$

approximately, or for greater accuracy

$l = \left[2e \sin\dfrac{\theta}{2}\right] + \left[\dfrac{\pi}{2}(d_1 + d_2)\right]$
$+ \left[\dfrac{\pi}{360}(180 - \theta)(d_2 - d_1)\right]$

and for a crossed drive

$l = 2e + \left[\dfrac{\pi}{2}(d_1 + d_2)\right] + \dfrac{(d_1 + d_2)^2}{4}$

Note that the effective pulley outside diameter is used to determine belt length, whereas pitch diameter (Fig. 9.4)

is used to determine belt tensions and drive speed. For a flat belt to run true, some crowning of the pulleys, especially the smaller drive pulley, is recommended. Unless unusual tracking problems exist, crowning should be kept to a minimum. The general recommendation is that the pulley diameter at the crown should exceed the diameter of the pulley at the edge by 1.5–2.5 per cent of pulley width. Pulley width should be 5–10 per cent wider than belt width.

Where pulley pitch diameters and centre distances are given, the length of a belt can be easily calculated. For most applications, however, a standard length would be used, hence requiring a centre distance calculation for locating the pulleys:

$e = 0.25b + 0.25\sqrt{[b^2 - 2(d_2 - d_1)^2]}$

where $b = l - 1.57(d_1 + d_2)$ and l = length of belt.

(b) Belt creep. Flat belts exhibit 'creep', the result of the tension difference between the tight side and the slack side of the pulley. Hence because of elongation under tension, a greater length of belt leaves the drive pulley than returns to it, so that there is always a difference between the peripheral speed of the driving pulley and that of the driven pulley. Even if the tension ratio is much less than the limiting value, the difference is small, but nevertheless it exists, typically ~ 0.5 per cent for most drives. The peripheral speed is identical to that of the oncoming belt so the driven shaft will run slower.

The effect of creep is to reduce the ratio of the pulley speeds ω_2/ω_1 from d_1/d_2 to

$\dfrac{d_1}{d_2}\left[1 - \dfrac{(T_T - T_S)}{AE}\right]$

where A = belt cross-sectional area, E = modulus of elasticity of belt and ω = angular velocity.

Creep does not take part over the whole of the belt/pulley arc of contact, but is limited to an arc defined by

$\dfrac{T_T}{T_S} = e^{\mu\beta}$

where $\theta > \beta$ (see Fig. 9.8).

Over the remainder of the arc $(\theta - \beta)$ there is no change in tension and the inside surface of belt and pulley rim will have identical speeds. At the tension ratio limit, then $\theta = \beta$.

(c) Belt stretch. Whether classed as elastic or inelastic, all elastomer/polymer belts stretch to some extent (e.g. Fig. 9.9). Belt stretch results not only from static loads,

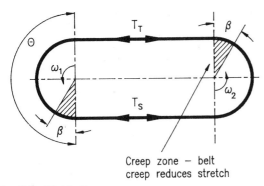

Fig. 9.8 Flat belt creep

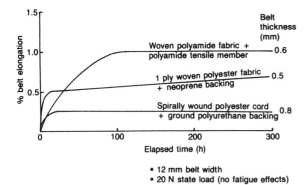

- 12 mm belt width
- 20 N state load (no fatigue effects)

Fig. 9.9 Stretch behaviour of some semi-elastic flat belts

i.e. the initial drive tension/belt elongation required for initial installation, but also from dynamic service load variations (see section 9.2.1(i)), i.e. a result of fatigue (Fig. 9.5). Few manufacturers supply installation belt-stretch information based on the post run-in condition.

In time, therefore, it is likely that the tension ratio will decrease, so to restore the effective pull, either the shaft centres need to be adjusted or some form of spring-loaded or gravity idler employed. The main effect of the idler is to prevent tension on the slack side from diminishing so rapidly (Fig. 9.10), and, depending on the design of the belt path, also increase the arc of contact on the driver pulley. In Fig. 9.10, with a dead weight and jockey pulley pivoted at the drive pulley centre

$$T_0 = \frac{Wl}{2a \cos(\psi/2)}$$

where ψ = belt angle subtended by idler pulley, W = weight, l = lever arm, Wl = gravity torque arm and a = perpendicular distance of line of action from the pivot.

The increase in mean tension produced by the idler will

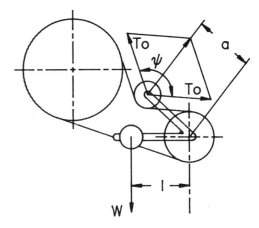

Fig. 9.10 The gravity idler

also cause some belt stretch. Because of the reverse bending of the belt as it passes over the idler, and the increase in cycling frequency, it means that belt conditions are more severe in a short-centre than in a long-centre drive.

(d) Belt operation. It goes without saying that the life and performance of a flat belt depend on how it is used. The following is a summary of the more important aspects:

1. Flat belt applications are nominally designed at a 2:1 tension ratio for equal-diameter pulleys.
2. Other service aspects being equal, if pulley diameters are reduced, T_T and T_C remain constant but T_B increases and belt life reduces.
3. With an increasing load transmit requirement, T_T and T_S increase as the load is increased, thus T_p increases and belt life decreases.
4. When reducing the belt centre distance, belt length decreases and, assuming all other parameters stay constant, belt life decreases since the number of stress reversals/unit time increases. If the arc of contact on the smaller pulley also decreases, belt tension will increase, further lowering life.
5. Too small a pulley accelerates flexural fatigue failure.
6. When pulley diameters are reduced 10 per cent, belt life is reduced by ~50 per cent.
7. When the power requirement is increased 10 per cent, belt life is reduced by ~50 per cent.
8. Speed ratios should not exceed 6:1.
9. Most cases of short belt life can be traced to extreme mechanical flexing action unsuitable for that belt type.

10. Mechanical failure occurs when initial belt tension (extension when fitted) is too high.
11. With polyamide core belts, humidity changes may cause unfavourable expansion in run-in belts.
12. If belt surfaces wear or contaminate with use, the effective coefficient of friction will decrease; hence to maintain an effective pull, shaft loading is increased, thereby accelerating deterioration and decreasing fatigue life (Fig. 9.5).

(iii) **Flat-belt design procedure**

The objective here is to size a belt correctly for the required application in terms of length, width and initial stretch. However, firstly a certain amount of data need to be available from the application requirement and/or belt manufacturer's literature, before starting the design procedure. Typically:

1. The power to be transmitted, i.e. the tangential force on the driver × speed.
2. Belt type and rating from the manufacturer's guidelines.
3. Pulley diameters.
4. Belt/pulley friction.

First find the arc of contact: see section 9.2.1(ii)(a) to determine. Then determine the belt tension: given μ and θ, check that the belt tension ratio at the point of slip is $\geqslant 2$. Knowing the power requirement (demanded power) a design tension ratio of 2, then values of T_T, T_S ($T_T - T_S$) and T_0 can be established. (*Note.* For polyamide core belts, it is usual to start with a tension ratio of 2.5.)

The belt width can be determined from

$$\text{Belt width} = \frac{(T_T - T_S)}{T_R} C_2$$

where C_2 = a service factor[2] and T_R = rated belt power/unit width.

For the majority of small machine applications, the value of C_2 can be taken as unity (Table 9.2). Rather than calculate individual values of T_T and T_S as above, what is done in belt manufacturers' design catalogues is to modify the original value of T_e in the power transmission equation, $P = T_e v$, by a factor C_1, i.e.

$$\text{Belt width} = \frac{T_e C_2}{C_1 T_R}$$

where C_1 = an arc contact factor (small pulley).

The arc contact factor above basically reduces the frictional torque capability of the belt in line with the reduction in the belt contact angle below 180° and increases it above 180°. At wrap angles below 180°, therefore, the 'effective pull requirement' is 'increased'. The C_1 factor is estimated from the power relationship

$$C_1 = A_r^m$$

where A = contact area ratio based on contact area at 180°, wrap = unity and m = exponent ≤ 1, the more elastic the belt, the greater is m and the more contact area sensitive the material (Table 9.3).

The next step is to determine the belt length (see section 9.2.1(ii)(a)). Then determine the belt elongation. Knowing the required belt tension and belt type, consult the manufacturer's information for elongation characteristics, or alternatively measure the load extension properties (e.g. Fig. 9.9). Then establish installation elongation of belt.

Adjust the belt length if fixed centres: the reduction in belt length to achieve the required belt tension with fixed pulley centres can be calculated as follows:

$$l_s = l - \frac{l\epsilon}{100}$$

where l_s = shortened length, ϵ = percentage elongation required and l = unmodified length.

Check shaft bearings. Given that shaft load is known, refer to Chapters 5, 6 or 7 depending on the type of bearing.

Table 9.2 *Service factors for various belt types*

Type of belt	Duty class	Soft start hours of duty/day			High torque start hours of duty/day		
		≤ 10	10–16	>16	≤ 10	10–16	>16
Flat belt	1	—	1.0	—	—	1.2	—
	2	—	1.1	—	—	1.4	—
	3	—	1.2	—	—	1.6	—
Vee belt	1						
	2	1.0	1.1	1.2	1.1	1.2	1.3
	3	1.2	1.2	1.3	1.2	1.3	1.4
High torque	1	1.0	1.4	1.5	1.2	1.7	1.8
	2	1.2	1.3	1.4	1.5	1.6	1.7
	3	1.3	1.4	1.6	1.6	1.7	1.9
XL/L	1	1.0	1.4	1.5	1.2	1.7	1.8
	2	1.5	1.6	1.7	1.8	1.9	2.0
	3	1.6	1.7	1.9	1.9	2.0	2.2
Modified curvilinear	1	1.0	1.2	1.4	1.2	1.4	1.6
	2	1.4	1.6	1.8	1.6	1.8	2.0
	3	1.5	1.7	1.9	1.7	1.9	2.1

Notes: Duty class definition: Class 1, very light duty, e.g., business machines, domestic appliances, typewriters, printers. Class 2, light duty, e.g. mixers and agitators, blowers, pumps and compressors. Class 3, medium duty, e.g. printing machinery, presses and punch tools, conveyors, mixers and granulators.

Table 9.3 *Arc of contact correction factors*

$\dfrac{d_2 - d_1}{e}$	Arc of small pulley of small pulley (degrees)	C1 factor for v-belts	C1 factor for flat belts	
			Woven polyester cord tension member	Polyamide film tension member
0	180	1.0	1.0	1.0
0.1	174	0.99	0.98	0.98
0.2	169	0.97	0.96	0.96
0.3	163	0.96	0.93	0.94
0.4	157	0.94	0.91	0.92
0.5	151	0.93	0.88	0.91
0.6	145	0.91	0.86	0.88
0.7	139	0.89	0.83	0.86
0.8	133	0.87	0.80	0.84
0.9	127	0.85	0.77	0.82
1.0	120	0.82	0.74	0.80
1.1	113	0.80	0.71	0.77
1.2	106	0.77	0.67	0.74
1.3	99	0.73	0.63	0.72
1.4	91	0.70	0.59	0.68
1.5	83	0.65	0.55	0.65

Notes: d_2 = diameter of large pulley; d_1 = diameter of small pulley; e = centre distance of the drive.

9.2.2 Light duty vee belts

Rubber vee belts first appeared in the 1920s, developing high driving forces with belt tensions considerably less than required by flat belts, also reducing load on shaft bearings. Vee belts make use of the wedging action between the belt and the pulley sheave.

There are two lines of vee belt produced:

1. *Standard wrapped*. Vee belts are formed of cord and fabric impregnated with rubber (Fig. 9.11), where the cord is either cotton, a synthetic or steel. In the standard belt there is a protective envelope encasing the section (Fig. 9.11). These are not available cogged since the wrap is difficult to manufacture in cogged form.
2. *Premium construction*. Alternatively known as raw-edged or non-wrapped construction. None of the cross-section is allocated to the wrap, thus the total section consists of working tensile material and hence these belts are high power rated. However, the lack of a wrap may make the belt more vulnerable to damage. Belts of non-wrapped construction can be made with teeth, cogs or moulded notches on the underside (Fig. 9.12). Cogged belts permit more severe bends over smaller diameter pulleys (typically 25 per cent diameter reduction).

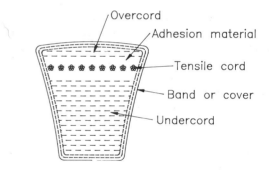

Fig. 9.11 Basic vee belt construction

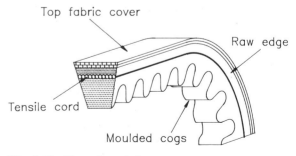

Fig. 9.12 Cogged vee belt

DRIVES, BELTS AND COUPLINGS

Table 9.4 *Summary of Vee-belt parameters*

	Pitch width W_p (mm)	Top width W_t (mm)	Belt height h (mm)	Min. small pulley dia. (mm)	Average deflection force (N) at centre of belt span for speed range	
					0–10 m s^{-1}	10–20 m s^{-1}
Wedge belt SPZ	8.5	10	8	67–95	12–18	10–16
SPA	11	13	10	100–140	22–32	18–26
Vee belt 2L	—	4	3	20	—	—
Y(3L)	5.3	6	4	38	—	—
Z	8.5	10	6	56	6.5–9	5.5–7.5
A	11	13	8	75	12–18	10–15

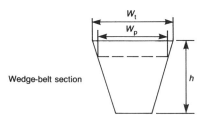
Wedge-belt section

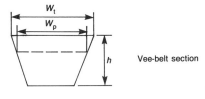
Vee-belt section

Apart from the form of construction, there are two forms of section:

1. *Vee belt*. Also known as the classical belt, a belt in which the trapezoidal section is characterized by a relative height of 0.73 × pitch width (Table 9.4).
2. *Wedge belt*. Also known as narrow vee belt, a belt in which the trapezoidal section is characterized by a relative height of 0.95 × pitch width (Table 9.4).

The size and operating parameters for light-duty vee belts are given in Table 9.4, including the FHP range. While the advantages and disadvantages of non-synchronous drives in Table 9.1 still apply, there are some specific qualifiers when comparing vee belts to flat belts, i.e.

1. Vee belts provide the highest power in the smallest package.
2. Vee belts inherently track better than flat belts.
3. The thickness of vee belts limits the bend radius, thus fairly large diameter pulleys are required and hence greater space requirement.
4. Because of heavier construction, the vee belt generates high centrifugal forces that place relatively low limits on top speed.

(i) Fundamentals of vee belt power transmission

Essentially these are the same as for the flat belt (section 9.2.1(i)), except for the following aspects. The fundamental tension ratio for vee belts is

$$\frac{T_T}{T_S} = \exp(\mu\theta \operatorname{cosec} \gamma)$$

where cosec γ = a wedging factor and γ = one-half the included pulley angle, Table 9.4.

The wedging factor takes into account the force between the belt and the groove surface that occurs because of the wedging action of the belt in the groove. This increases the exponential exponent, allowing operation at higher average tension ratios than flat belts, i.e. vee-belt drives are normally designed at 5:1 tension ratio and up to 9:1 is possible if the drive is equipped with automatic tensioning. The higher tension ratio translates directly into lower operating tension and bearing loads for vee belts. The flat-belt rim may be regarded as the limiting case of a grooved pulley section, where $2\gamma = 180°$. Note that as with flat belts, the tension ratio is not a function of belt size, only belt/pulley contact area.

Because of the heavier section, bending and centrifugal forces become significant with vee belts, the bending force in particular being sensitive to pulley diameter. The forces present in a moving vee belt are shown as in Fig. 9.13, where

$$T_p = T_T + T_{B1} + T_C \quad \text{and} \quad T_B \propto \frac{1}{D_p}$$

where D_p = pitch diameter of pulley and T_{B1} = bending stress around small pulley, from which

$$\frac{1}{N} = \frac{1}{F_1} + \frac{1}{F_2}$$

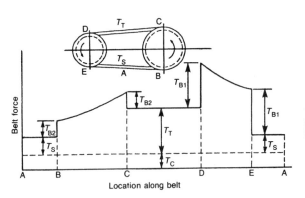

Fig. 9.13 Forces in a moving vee belt

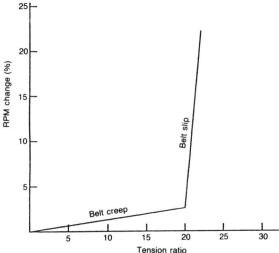

Fig. 9.14 Percentage change in revolutions per minute between driven and driver pulleys as a function of tension ratio (creep and slip)

where F_1 = no. of force peaks a belt can sustain passing over the smaller pulley, F_2 = no. of force peaks a belt can sustain passing over the larger pulley and $1/N$ = proportion of belt life 'consumed' by one complete passage around both pulleys.

To allow for fatigue effects in practice, a belt length correction factor C_3 is introduced which, in conjunction with the arc of contact factor, adjusts the belt rating required to take into account the number of stress reversals/unit time, i.e.

$$\text{Required belt rating} = \frac{\text{demanded power}}{C_1} \times C_2 C_3$$

Values of C_3 are normally quoted by belt manufacturers.

Considerable levels of static charge may be built up on the moving belt, even if shaft and pulley are technically earthed, hence in many environments, belts are specified as antistatic to avoid the risk of periodic discharge. To meet this criterion,

$$R \not> \frac{5 \times 10^6 \times l_x}{8I}$$

where $I = 2h \sec(\gamma/2)$ = sum of the lengths of the two sloping sides of the belt, l_x = distance between belt contacts, R = electrical resistance (Ω) and h = nominal belt height, Table 9.4.

(ii) Practical aspects of vee-belt power transmission

Like flat belts, vee belts creep. The extent of the creep, hence velocity variations between driver and driven shaft, depends on the belt tension ratio (Fig. 9.14). If the tension ratio becomes too high (tension too low) the belt will slip on the pulleys (Fig. 9.14). Because vee-belt materials and cross-sections are universally standard, belt tensions are also standard for a given rated belt, and average conditions of use modified only by belt centrifugal effects at high speed. Consequently, when referred back to the static installation belt tension (Table 9.4) the maximum belt deflection at centre span should be 0.015 mm per millimetre of span. However, the belt tensions in Table 9.4 only reflect the average situation. With new belts, the drive should be tensioned to the higher value, since tension falls rapidly in the early stages of run-in.

If tensioning is required with a vee-belt drive, tensioning pulleys must be fitted with a grooved pulley on the inside of the drive on the slack side positioned as close to the larger pulley as possible. Flat tensioning pulleys are only possible with classical belts, positioned close to the small pulley and of an equivalent diameter.

Some further practical aspects of vee-belt operation are as follows:

1. The included pulley groove angle is typically 34–38°, the wedging action increasing the tractive force.
2. Sufficient clearance must be provided at the base of the groove to prevent the belt bottoming out as a result of wear.
3. A vee belt is normally designed to ride with the top surface approximately flush with the top of the pulley groove.
4. For velocity ratio calculations, pulleys are specified by their pitch diameters, hence belt lengths need to be calculated in terms of pitch length.
5. Belt efficiency is typically 70–96 per cent.

(iii) Vee-belt design procedures

This procedure is similar in many respects to the flat-belt design procedure except that in this case, the belt is very much a standardized product, hence more straightforward. The initial data requirement, from either the application or manufacturers' literature, is the power to be transmitted by the belt, the speed ratio and pulley diameters. Subsequently:

1. Determine belt type (and rating) from manufacturer's literature.
2. Find the arc of contact; see section 9.2.1(ii)(a) to determine θ.
3. Calculate belt length; see section 9.2.1(ii)(a) and then choose nearest standard pitch length.
4. Calculate centre distance (see section 9.2.1(ii)(a)).
5. Determine belt length correction factor, C_3, from manufacturer's literature or ref. 2.
6. Determine arc of contact correction factor, C_1 (small pulley) from Table 9.3.
7. Calculate the required belt rating (watts) from section 9.2.2(i) using the appropriate correction factors, C_1, C_2 and C_3. Check rating still acceptable for the belt type chosen.
8. Knowing belt speed, refer to Table 9.4 for the required deflection force. Adjust pulley centres to provide the required belt deflection, 0.015 mm per millimetre of span. The latter is actually a set-up operation with a force gauge on the application.

9.2.3 Synchronous belts

Synchronous belts, or timing belts, are basically flat belts with a series of evenly spaced teeth on the inside, or on both circumferences, thereby combining the advantage of the flat belt with the positive grip feature of chains. The first synchronous belt was produced in 1946 for sewing machines, synchronizing needle and bobbin movement. Belt construction consists of a tensile member, a backing, teeth and facing (Fig. 9.15), as follows:

1. *Tensile member.* Originally steel cord but soon changed to fibreglass for greatly improved fatigue (flex) life. Fibreglass is most common today although polyester cord is sometimes used and, for maximum capacity, nylon fibre. The tensile member controls length stability.
2. *Backing.* Neoprene or polyurethane.
3. *Tooth.* Neoprene or polyurethane. The higher the hardness the less the tooth deformation, but the belt is then more rigid and less easy to bend.
4. *Facing.* Normally present as a nylon fabric (duck)

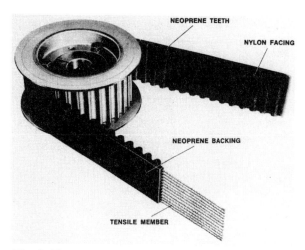

Fig. 9.15 Synchronous belt construction. Courtesy of Gates GmbH, Aachen

covering the belt wearing faces. Provides a wear-resistant surface that becomes highly polished with prolonged service.

Only two components have a significant effect on synchronization, the tensile member and the teeth. Some advantages of synchronous belts are as follows:

1. Belts are thin and flexible, so operate well on miniature drives and in high-speed/small pulley applications.
2. Positive engagement. No slippage or creepage frictional effects as with flat belts or vee belts.
3. Do not stretch.
4. The required belt tension is low, hence low bearing loads.
5. Speed is transmitted uniformly, there is no chordal rise and fall of the pitch line as with the roller chain (section 9.3(i)).
6. Can transmit fairly high loads over a wide speed range at low noise level without lubrication.

The drawbacks of synchronous drives are the higher cost relative to other drive types (belts and pulleys), the demand for fairly accurate alignment of pulleys and the possibility of resonances. Synchronous belts are available with a number of tooth shapes (Fig. 9.16), and within those profiles, a variety of pitches (Table 9.5). Larger pitches are available,[3] but they relate to heavier-duty applications outside the scope of this book. The traditional, or classical, tooth profile is trapezoidal (Fig. 9.16). However, this profile has a limited load capability due to line stress concentrations produced during belt/sprocket meshing. To overcome this the high torque

Table 9.5 *Summary of neoprene synchronous belt characteristics*

Belt type	Pitch (mm)	Distance of pitch line to belt tooth bottom (mm)[a]	Clearance distance (mm)	Allowable working tension (T_a) for 25.4 mm belt width (N)[b]	Tooth profile
Minipitch MXL[c]	2.03	0.25	—	142(95)[d]	Trapezoidal
40DP[e]	2.07	0.18	—	100[f]	Trapezoidal
Minipitch	2.07	0.25	—	142(95)	Trapezoidal
XL[c]	5.08	0.25	—	182(122)	Trapezoidal
L[c]	9.53	0.38	0.14	250	Trapezoidal
T5	5	0.5	—	570[g]	Trapezoidal
T10	10	1.0	—	2281[g]	Trapezoidal
Mini HTD[c]	3	0.38	0.17–0.24	267	Curvilinear
Mini HTD[c]	5	0.57	0.20–0.36	445	Curvilinear
HTD[c]	8	0.68	0.38–0.51	470	Curvilinear
Mini GT	3	—	0.025–0.13	—	Modified curvilinear
Mini GT	5	—	0.056–017	—	Modified curvilinear
GT	8	0.8	0.22	—	Modified curvilinear
RPP	3	0.38	—	—	Parabolic
RPP	5	0.57	—	—	Parabolic
RPP	8	0.67	—	—	Parabolic

[a] Pulley DD = pitch dia. − twice this distance.
[b] No correction for T_c; apply belt width factor for different widths.
[c] Most popular belt types.
[d] Urethane belt ( ).
[e] Designed to run on a 40DP spur gear.
[f] Polyester tension member.
[g] Steel cord tensile member.

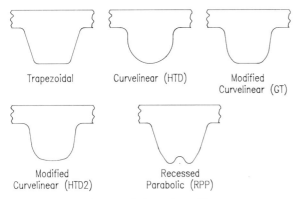

Fig. 9.16 Synchronous belt tooth profiles

drive profile (HTD) was introduced, reducing tooth distortion and increasing torque transmissibility by up to 30 per cent with the full rounded profile (Fig. 9.16). Advantages of the high-torque profile are as follows:

1. Proportionally deeper tooth, hence tooth jump is less likely.
2. A light construction and correspondingly reduced centrifugal effect.
3. A larger tooth contact area, hence smaller tooth unit pressure.
4. Greater shear strength due to larger tooth contact.
5. Compared to a trapezoidal profile, a nearly constant strain distribution across the belt (Fig. 9.17).

However, because of the nature of this profile form, substantial belt/sprocket clearance is required for correct meshing. In the modified curvilinear design (GT) the tooth form is midway between the trapezoidal and curvilinear shape and to improve the load/life characteristics, nylon fibre is chosen as the tensile member. The GT belt is generally stated to outperform chain drives, vee belts and other timing belts in numerous applications.[4] To improve the meshing of high torque belts on small pulleys, a further development is the HTDII belt, a modified GT profile with slightly radiused crest (Fig. 9.16). A change of backing material to nitrile in this belt type also improves oil and heat resistance.

Timing belts are also available in double-sided designs. This offers a number of design possibilities on lighter-duty applications; some machine configurations are best served by belts able to deliver power from both top and bottom surfaces, making it possible to change the direction of more than one pulley with only one belt. Typically XL

DRIVES, BELTS AND COUPLINGS

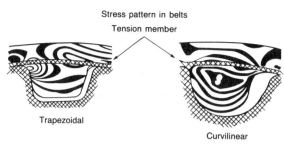

Fig. 9.17 Stress pattern in belts

and L sections are available, known as DXL and DL. Standard double-sided belts moulded with nylon facing on both sides will have the same power rating as a comparable, single-sided belt, i.e. capable of transmitting 100 per cent of the maximum-rated load from either side of the belt or in combination where the sum of the loads exerted on both sides does not exceed the maximum rating of the belt.

(i) Fundamental aspects of synchronous belt power transmission

In terms of drive geometry, belt length and centre distance equations are the same as for the flat belt (section 9.2.1(ii)(a)). However, the tooth profile of timing belts is an involute curve similar to a spur gear tooth but unlike the spur gear, the OD of a timing pulley is manufactured smaller than its pitch diameter, thus creating an imaginary pitch diameter larger than the pulley itself (Fig. 9.18). The theoretical pitch diameter lies within the tensile member. In terms of number of pulley pitch lengths:

$$e = \tfrac{1}{2}p\,[(N_b - N_2) + k(N_2 - N_1)]$$

where

$$k = \frac{1}{\pi}\left[\tan\left(\frac{\pi}{4} - \frac{1}{2}\phi\right) + \phi\right]$$

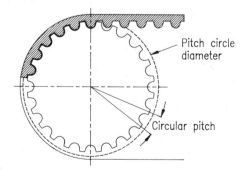

Fig. 9.18 Drive parameter definition

N_b = no. of teeth on belt, p = circumferential pitch length of belt, N_2 = no. of teeth on larger pulley, N_1 = no. of teeth on smaller pulley and ϕ = one-half the wrap angle (small pulley) (rad).

It is generally recommended that the minimum number of teeth in mesh be six, a belt loading which is generally taken to satisfy the maximum shear strength of the belt per unit width. If this is less than six, belt width needs to be increased to compensate. The teeth in mesh is given by

$$\text{TIM} = \frac{\phi N_1}{\pi}$$

or approximately

$$\text{TIM} = N_1\left[0.5 - \frac{(D_2 - D_1)}{6e}\right]$$

where D_2 = pitch diameter of large pulley and D_1 = pitch diameter of small pulley. (*NB*. Fractional teeth do not count.)

As with vee belts, centrifugal effects increase belt tension during rotation, but additionally in the case of the synchronous drive geometry, radial forces also increase belt tension, i.e.

Total belt tension = effective tension + bending stress
                     (working tension)
+ radial load + centrifugal load

Total tension in a synchronous belt increases as the torque is applied, and the interaction between belt and pulley teeth develops a radial force (Fig. 9.19), the radial vector force tending to lift the belt tooth out of the space between the pulley teeth, hence there is an increase in total belt tension. As a result, dynamic bearing loads for synchronous belts and vee belts tend to be equivalent. As the belt tightens, the tension ratio goes down. The torque-carrying capacity of synchronous belts is not directly related to belt width, i.e. it decreases disproportionately with the narrower belts due to the inefficiency of the tensile members at the edges of the belt, i.e. the narrower the belt, the smaller the ratio of full load-carrying tension cords relative to the total number of cords in the belt.

Although synchronous belts are widely used as positive drives, there are a number of sources of error affecting precise registration, in particular:

1. Belt elongation under load;
2. Tooth deformation under load;
3. Incorrect belt tension;
4. Pitch errors in belt or pulley;
5. Tooth/groove backlash.

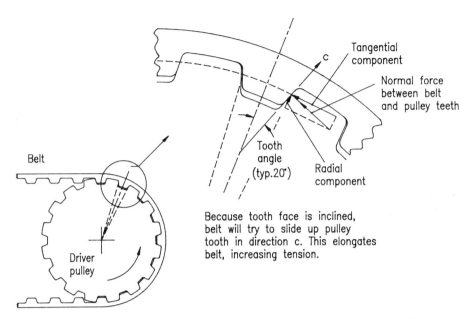

Fig. 9.19 Synchronous belt and pulley forces. Belt exerts circumferential force against pulley tooth to cause rotation. Due to inclined tooth face, belt attempts to move in direction *c*, elongating belt and increasing belt tension

Consequently the driven pulley may slightly lag behind the driver. Tooth clearance is the main cause of system backlash and each type of profile has its characteristic clearance (Table 9.5). Because of their deep profile and meshing behaviour, high torque belts have the largest clearance.

(ii) Practical aspects of synchronous belts

(a) Belt tracking. Because timing belts have a natural tendency to run to one side or another during operation, one pulley in the system must be flanged otherwise the belt will run off. Clearly crowning is not suitable for a positive drive belt. Belt tracking will always occur to one side, but the side thrust is negligible and does not cause edge wear or flange climb if the drive is correctly aligned. Drive factors contributing to side travel are:

Misalignment. Any belt will always climb to the high or the tight side.
Tensioning. Sideways travel can generally be altered or modified by changing tension.
Pulley taper. Moulded pulleys have draft, hence the belt will climb to the high side of the taper.
Plane of centres. Vertical drives have a greater tendency for sideways belt movement.

Belt factors contributing to side travel are as follows:

Cord twist. Depending on the direction of cord twist of the tensile member, the belt may track left or right.
Cord lay. The lay of the tensile member in the belt can work with or against cord twist. Most belts are wound left to right from the top of the mould, the shorter the belt the greater the helix angle and the greater the tendency of cord lay to make the belt track. Thinner backings pronounce the effect.
Facing. Its weave and manner of application may also affect the direction and intensity of tracking.

A no-tracking situation is unlikely in production belts although changes in belt construction may help. Some belts may track neutrally by chance, but there is no guarantee if a belt will track or not and in which direction.

(b) Noise. Synchronous belts inherently produce some noise because of the belt passage over an uneven surface. Increasing belt tension tends to increase belt noise. The belt teeth entering the pulleys at high speed act as a compressor, creating noise. Some noise is also the result of rubbing against the flange. A softer tooth stock (special quiet construction) will reduce noise, but at the expense of belt wear rate. Various modifications to pulleys and belts have been tried over the years to reduce the noise level. One improvement that appears to work is to provide

holes in the belt itself, but this will be at the expense of belt rating. As a refinement of the curvilinear belt, a modified profile with a crest indentation has been introduced to ease air discharge and reduce belt noise, allowing a softer meshing of belt and groove. This type is known as a recessed parabolic profile belt (RPP) (Fig. 9.16), having the increased tooth depth of the GT/HDT type belts but the side tooth shaping of the trapezoidal belt. (*Note*: HTD and GT are trade marks of the Gates Rubber Co., RPP is a trade mark of Pirelli Transmissions.)

(c) *Power rating*. A comparison of the power ratings for classical, high torque and parabolic profile belt styles is given in Fig. 9.20. In practice, for a given belt, the manufacturer's literature will either quote the power, torque or tension rating relative to a specific sprocket size and rotational speed. Such data are plotted for two pulleys in Fig. 9.21. Although an allowable belt working tension is quoted in Table 9.5, ratings are normally published relative to a flex life of approximately 3000 hours, so taking into account the fatigue effects of belt speed and belt load, as given below.

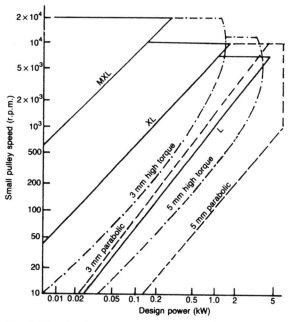

Fig. 9.20 Synchronous belt power transmission comparison chart

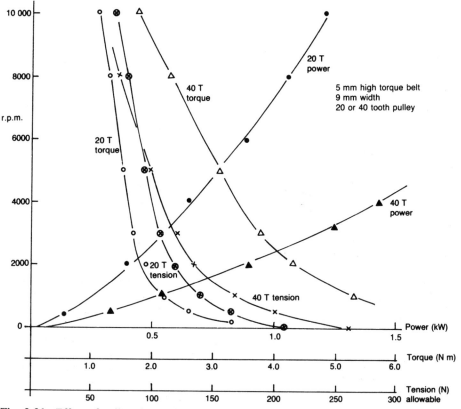

Fig. 9.21 Effect of pulley size on belt performance ratings

1. Allowable belt tension decreases with speed increase for a given pulley. The corresponding working tension decreases as centrifugal force loss increases.
2. Allowable belt tension is greater the greater size of sprocket, i.e. the less the degree of articulation necessary hence lower fatigue loads. Likewise load per tooth is improved.

For design purposes, belt selection is based on the rated power of the driven unit corrected by a number of operational and configurational factors such that

Design power = rated power × S

or

Peak design torque = rated torque × S

or

Allowable belt tension = total belt tension × S

where

S = total correction factor = total service factor
$\times C_4 C_3 = (C_2 + C_5 + C_6 - C_7) C_4 C_3$

where C_2 = basic service factor (Table 9.2), C_3 = belt length correction factor (high torque belts only); refer to manufacturers' literature, C_4 = TIM factor, C_5 = speed-up factor, C_6 = idler factor; for an idler add 0.2 to the basic service factor and C_7 = intermittent operation factor; if operation is intermittent or seasonal deduct 0.2 from the basic service factor.

Values of C_4 and C_5 factors are given in Table 9.6. As can be seen, belt selection is very much an empirical process, based upon manufacturers' tables of rated power or torque and then suitably factored to determine the actual belt size that will provide the standard flex life.

(d) Design guidelines. Some general pointers, applicable to all timing belts, are as follows:

1. Always design with some reserve capacity and use overload safety factors if unsure. Belts should be rated at 1/15–1/20 of their ultimate tensile strength (UTS).
2. For mini-pitch belts, the smallest recommended pulley will have 10 teeth. The minimum recommendation increases with increasing pitch size.
3. The pulley diameter should not be smaller than the belt width.
4. Belts with fibreglass tensile members should not be subject to sharp bends or rough handling.
5. Belts are in general rated to yield a minimum of 3000 hours' useful life if designed in the application correctly.
6. Efficiency is ≥95 per cent.
7. Timing belt installation should be a snug fit, neither too tight nor too loose. The positive grip of the belt eliminates the need for high initial tension; preloading is not necessary.
8. Shafts and pulleys must be in good alignment.
9. It is important that a rigid framework support the pulleys. A non-rigid frame causes a variation in centre distance, belt slackness and possibly tooth jumping.
10. Although belt tension requires little attention after initial installation, provision should be made for some centre distance adjustment for ease in installing and removing belts.
11. Recommended belt tension, as a general rule of thumb, is such that the belt deflects 0.015 mm at centre span for every 1 mm of span.

(iii) Timing belt design procedure

Similar in many respects to flat belts and vee belts, the selection procedure can be outlined as follows:

1. State the design requirements, in particular: (a) the type and description of the driver; (b) rated power of the driver; (c) the torque requirement of the driven load; (d) the speed reduction or step-up ratio; (e) conditions of loading and usage; (f) the allowable range of centre distance.
2. Determine the total service factor (see section 9.2.3(ii)(c)), i.e. decide values of C_2, C_5, C_6 and C_7 (Tables 9.2 and 9.6).
3. Select design power rating, belt type and belt pitch from Fig. 9.20 (see also section 9.2.3(ii)(c)).
4. From the known transmission ratio, determine the pulley combination from catalogue sizes.
5. Determine belt length from the pulley selection and approximate centre distance (see sections 9.2.1(ii)(a) and 9.2.3(i)). Adjust centre distance to obtain a standard catalogue belt length item.

Table 9.6 *Summary of correction factors*

(1) Speed-up factor C5		(2) TIM factor C4	
Speed-up range	C5	Teeth in mesh	C4
1–1.24	0	≥6	1
1.24–1.74	0.1	5	1.25
1.75–2.49	0.2	4	1.6
2.5–3.49	0.3	3	2.5
>3.5	0.4	2	5

DRIVES, BELTS AND COUPLINGS

Table 9.7 *Belt width correction factor*

Belt width		Correction factor
mm	in	
6.4	0.25	0.15
7.9	0.31	0.21
9.5	0.375	0.28
11.1	0.44	0.35
12.7	0.5	0.42
15.9	0.625	0.57
19.1	0.75	0.71
22.2	0.875	0.86
25.4	1	1
38.1	1.5	1.56
50.8	2	2.14
63.5	2.5	2.72
76.2	3	3.36

Note: trapezoidal section belt; all pitches.

6. Calculate the TIM factor C_4 and, if appropriate, determine C_3 from manufacturers' data.
7. Using the total correction factor S (section 9.2.3(ii)(c)), correct the power, torque or tension requirement.
8. Determine the belt width for the required power, torque or tension depending on the form of the catalogue ratings available. If values are quoted only for a standard width of belt, a width factor then has to be used (Table 9.7).
9. Performance ratings are normally catalogued as a function of pulley diameter and rotational speed. However, if this is not the case and a single allowable tension figure is quoted, due allowance within this must be made for centrifugal tension loss if appropriate (sections 9.2.1(i) and 9.2.3(i)).
10. As a final check, ensure compatibility of pulley sizes, belt length and belt width with the general design requirements of (1).

9.3 Chain drives

A chain drive may be considered intermediate between belt and gear drives in that it has features in common with both. Chain drives are used where a positive drive is required, but the action is such that it cannot be used where precise timing is a requirement. The basic power chain used in the majority of small machine applications is the precision roller chain. Other types of metallic chain are available, i.e. inverted silent tooth, double pitch, detachable and pintle chains (Fig. 9.22). These tend to

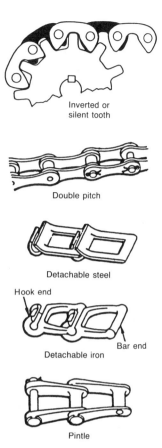

Fig. 9.22 Alternative types of chain

have specific application such as high power take-off usage as in the case of the inverted tooth or in low-cost, less precise drives for the remainder.[5] The newer style of chain, sometimes referred to as cable drive chain, combines the features of link chain and synchronous belt. Based on tough elastomers, these find application where there is a low loading and a reduced noise requirement. The advantages and disadvantages of chains for speed and load transmission, compared to alternatives such as belts or gears, are summarized in Table 9.8.

(i) Precision roller chains

Roller chains are designed for smooth, free-running operation at high speed and high power. A basic simple strand chain may be rated at approximately 370 kW but a service factor must be applied to the nominal load when impact conditions exist. The type of construction ensures a low loss, quiet operation. The rollers themselves turn on bushings that are press fitted to the inner link plates, the pins are prevented from rotating in the outer link plates

Table 9.8 *Chain drives — advantages and disadvantages*

Advantages	Capacity and service life maintainable at service extremes
	Positive drive ratios
	Available in a wide range of accuracies from precision to non-precision
	Can be used with arbitrary shaft centre distances
	Can be assembled around obstacles
	Favourable load and life ratings, reliability superior to that of belts
	Readily connected to form almost any length
	Can be removed and replaced more easily than belts
	More compact than a similarly rated belt drive
	Do not stretch or slip, hence maintain constant speed ratios under widely varying load conditions and needing only occasional adjustment
	Efficiency of a chain drive is 85–90%
	The range of chain availability is very wide, chains can be used for large and small amounts of power
	A large speed reduction is possible
	A greater tolerance to the installation itself than gears
Disadvantages	Non-precision types of chain do not provide an exact close fit between sprocket and link so do not articulate smoothly
	Compared to belts, chains do not provide any overload or jam protection since there is no slippage
	The weight of a metallic chain compared to a light elastomer belt produces penalties in acceleration time i.e. poor efficiency in stop–start applications
	Chains are noisier than belts
	An idler sprocket is generally necessary to remove excess slack from the chain as it wears
	Lubrication is an important feature in the successful design application of a metal roller chain
	Susceptible to contamination

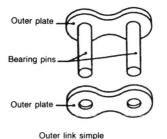
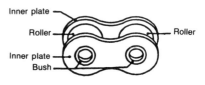
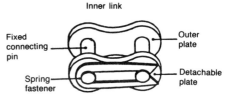

Fig. 9.23 Typical roller chain components

by the press-fit assembly (Fig. 9.23). Rated values depend on adequate lubrication. Self-lubricating types using sintered metal oil-impregnated bushes are available but useable only at low speed.

At low speed, the power rating is determined by the fatigue life of the link plates, expressed as

$$0.18 \times 10^{-6} N_1^{1.08} n_1^{0.9} p^3 \text{ kilowatts}$$

where N_1 = no. of teeth in smallest sprocket, n_1 = rpm of smallest sprocket and p = pitch (mm). At higher speed, the power rating is determined by the roller bush fatigue life, expressed as

$$\frac{56 \, k_1 N_1^{1.5} p^{0.8}}{n_1^{1.5}}$$

where k_1 = 29 for 6.4 and 9.5 mm pitch chain.

Normal chain wear is caused by the articulation of the pins in the bushings, i.e. the link swings through an angle β on engagement with the sprocket (Fig. 9.24). Taking the articulation effects into account, an empirical relationship between the life of the chain and dimensional factors is expressed as follows:

$$L^* = \frac{\pi \, n_1 \, T}{k_2 l_p} \left[1 + \frac{N_1}{N_2} \right]$$

where T = total chain tension at smallest sprocket (N) = 2 × torque (Nmm)/pitch diameter (mm), l_p = chain length in pitches, N_2 = no. of teeth on larger sprocket and k_2 = constant 0.49 for 6.3 mm pitch chain and 0.76 for 9.5 mm pitch chain. From the calculated value of L^*, chain life may be estimated from Fig. 9.25.

When the number of teeth on a sprocket is small, the driven shaft of a roller chain drive may be given a pulsating or jerky motion. As a chain unwraps from a sprocket, it unwraps from two different diameters; one is the true pitch circle (radius R) and the other is the

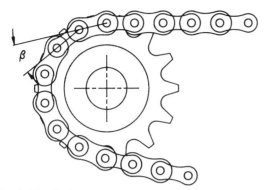

Fig. 9.24 Chain articulation

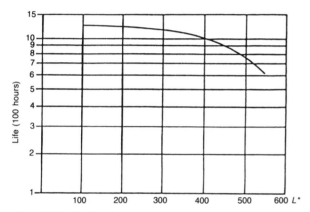

Fig. 9.25 Chain life conversion graph

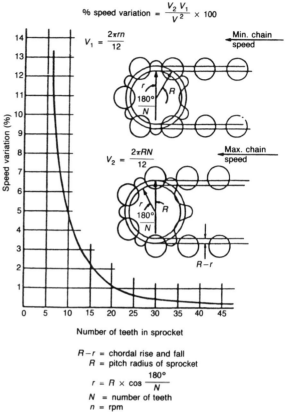

Fig. 9.26 Chain chordal action and speed variation

inscribed circle within the pitch polygon (radius r) (Fig. 9.26). The chain alternates between these two diameters as each tooth and each tooth gap passes through the tangent position. The result is a rise and fall of the chain strand as well as a speed fluctuation. The chordal action is strictly a function of the number of teeth on the sprocket (Fig. 9.26). By using an odd number of teeth on smaller sprockets, this will give smoother operation because the vertical movement of the upper and lower parts of the chain are in phase.

The average speed of the sprocket cannot be obtained using the pitch diameter, but rather the length of chain passing round the sprocket in unit time:

$v = pNn$

and v = average chain velocity.

(a) Design practice. A number of points can be made regarding good design application of roller chains, summarized as follows:

1. The small sprocket wrap angle should be at least 120°, if too low, for example 90°, the sprocket will tend to ratchet within the chain.
2. For a drive ratio of 3:1 or less, there will always be a 120°, or greater, wrap angle on the small sprocket.
3. The centre distance should not be as long as to produce excessive catenary pull, i.e. a centre distance 30–80 times chain pitch is generally ideal.
4. Very short centre distances subject the chain to frequent articulation and high wear.
5. Engagement of chain rollers with sprocket teeth results in impact, the magnitude of the impact reduced with an increase in the number of sprocket teeth, a reduction in drive tension or speed.
6. Service factors may need to be applied. Use same ratings as vee belts in Table 9.2.
7. To minimize torque pulsations, at low speed the smallest sprocket should be 12 tooth, while at high speed, >25 tooth.

8. Plastic sprockets do not have the load capacity of metal ones, but are both quieter and require only minimal lubrication.
9. Tension idlers should be no smaller than the driver and preferably engaging in slack span with a minimum chain engagement of three teeth.
10. The greater the number of teeth on a sprocket, the lower the angle of articulation, the greater the possible chain load, the less the wear and the less the speed variation.

(b) Centre distance. Calculated as follows:

$$e = \frac{p}{4}\left[l_p - \frac{(N_1 + N_2)}{2}\right]$$
$$+ \sqrt{\left[\left(l_p - \frac{N_1 + N_2}{2}\right)^2 - 8C\right]}$$

where e = centre distance (mm), $C = (N_2 - N_1)^2/2\pi$ and l_p = no. of chain pitches, i.e.

$$= \frac{2e}{p} + \frac{N_1 + N_2}{2} + \frac{Cp}{e}$$

(c) Chain tension. The proper fit of a chain may be obtained either by adjusting sprocket centres or by use of an adjusting idler. When the chain is correctly tensioned, the total mid-span movement (double amplitude) in the slack span should be 4–6 per cent of the span lengths for normal drives. For high speed, impulsive or reversing loads, the total movement should reduce to 2–3 per cent.

(d) Drive arrangements. A number of drive geometries are illustrated in Fig. 9.27, correct and incorrect. The drive direction should be such as to keep the upper span of the chain in tension. Slack in the upper strand of a long centre-distance drive could cause it to hit the lower strand as the chain elongates. Extremely long spans should be avoided at high speed in case of 'whip' and in the case of low ratio drives, sprocket size should be adequate.

With short centres, slack should be in the lower strand. Slack in the upper strand could force the chain out of engagement. Horizontal centres are preferred for a chain drive, the centres horizontal or inclined >45°. The slack strand may be either side, but the lower side is preferred. Vertical centre drives should be avoided if at all possible since they must run fairly taut, so requiring frequent tensioning as the chain wears. When the line between centres is >45° to horizontal, the slack should be in the strand with the lesser angle to the horizontal.

In multi-sprocket or serpentine drives, each additional sprocket adds additional chain articulation, making it difficult to maintain good design criteria hence resulting in reduced chain life and increased noise.

(ii) Plastic chains

Varieties available include traditional designs with discrete links that connect to form almost any length, plus the newer design types that include one, two or three parallel strength members. The newer style chains, sometimes called cable drive chains, combine the features of a link chain and a synchronous belt, and are generally plastic links moulded on to steel wire ropes. On single strength member chains, the plastic links or drive pins are either dowel or cross-shaped. The various types are illustrated in Fig. 9.28, together with a mechanical description and performance summary.

The advantage of plastic chains is low noise with no lubrication requirement. The pitch-to-pitch moulding tolerance can be tightly controlled to eliminate play between teeth and cogs. Clearly the load capability of cable chain is much less than roller chain, but nevertheless is still adequate for many applications (Fig. 9.28). The working load of the chain, based on shared load per pin, is limited by the shear strength of the wire/plastic pin interface. The maximum tensile strength of the cable is limited to the cable join quality, nominally reduced by 50 per cent if spliced in the field. These two factors therefore dictate the design application.

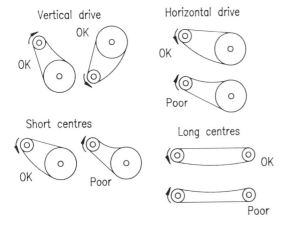

Fig. 9.27 Preferred drive arrangements

9.4 Flexible couplings

Flexible couplings are designed to compensate for angular misalignment, parallel offset, axial float or any

DRIVES, BELTS AND COUPLINGS

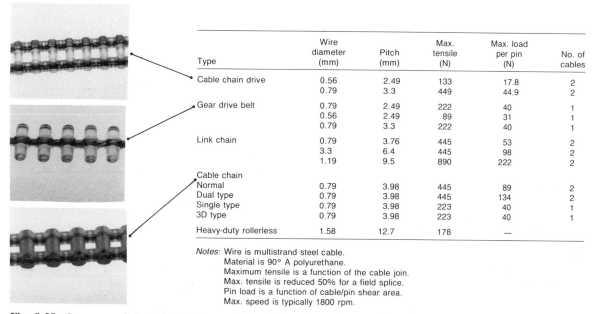

Type	Wire diameter (mm)	Pitch (mm)	Max. tensile (N)	Max. load per pin (N)	No. of cables
Cable chain drive	0.56	2.49	133	17.8	2
	0.79	3.3	449	44.9	2
Gear drive belt	0.79	2.49	222	40	1
	0.56	2.49	89	31	1
	0.79	3.3	222	40	1
Link chain	0.79	3.76	445	53	2
	3.3	6.4	445	98	2
	1.19	9.5	890	222	2
Cable chain					
Normal	0.79	3.98	445	89	2
Dual type	0.79	3.98	445	134	2
Single type	0.79	3.98	223	40	1
3D type	0.79	3.98	223	40	1
Heavy-duty rollerless	1.58	12.7	178	—	

Notes: Wire is multistrand steel cable.
Material is 90° A polyurethane.
Maximum tensile is a function of the cable join.
Max. tensile is reduced 50% for a field splice.
Pin load is a function of cable/pin shear area.
Max. speed is typically 1800 rpm.

Fig. 9.28 Summary of plastic chain characteristics. Illustrations courtesy of W.M. Berg Inc., East Rockaway, NY

combination of these (e.g. Fig. 9.29). Misalignment couplers compensate for errors elsewhere in the drive system. When shafts are connected, it is seldom practical to align them perfectly and accurately or to calculate simply the loads and stresses when there are more than two bearings on a shaft. Alignment is also affected by temperature, bearing wear, frame deflections, etc. hence the practicality of a flexible connection.

There is a large availability of flexible coupling types, and this in itself can make selection difficult. This section will reference the more usual types available for small machine usage. Some designs offer specific advantages in terms of system maintainability, and with proper sizing and selection the coupling can also be designed to provide overload protection.

Coupling life is clearly infinite if there is no misalignment. Some all-metal couplings can withstand only slight misalignment while many elastomeric types can withstand 4° or more of angular misalignment. If the misalignment capability of a coupling is exceeded, and depending on the specific coupling design, the reaction force may be enough not only to damage the coupling but shaft support bearings as well.

All misalignment couplers are capable of routeing the transmission through an angle. When this is the limit of capability, the coupler is referred to as a single engagement coupler. Parallel misalignment requires a double engagement coupler (Fig. 9.30). For a given offset, the

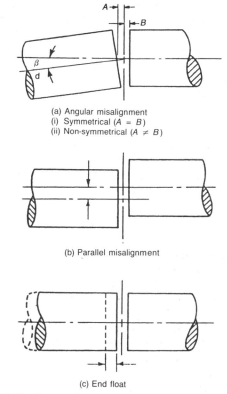

(a) Angular misalignment
(i) Symmetrical ($A = B$)
(ii) Non-symmetrical ($A \neq B$)

(b) Parallel misalignment

(c) End float

Fig. 9.29 Types of misalignment

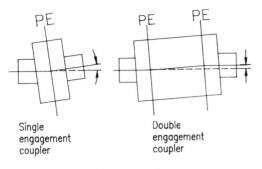

Fig. 9.30 Types of coupler engagement

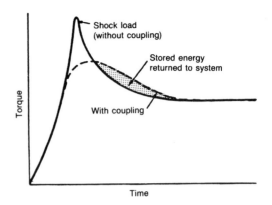

Fig. 9.31 Shock load reduction

angles can be reduced by increasing the distance between the planes of engagement (Fig. 9.30), i.e. providing a larger coupler. A severe offset can be alleviated with two single engagement couplers, providing there is no orbiting of the link shaft: couplers with a defined plane of engagement produce a Cardan when paired; couplers which compensate for parallel misalignment by radial displacement of hubs operate on the Oldham system.

Although some couplings offer a torsionally stiff unit, it should not be overlooked that the coupling is connecting two rotating masses, hence producing an assembly capable of vibration. In common applications the resonant frequency is given by

$$f \text{ resonant (natural frequency)} = \frac{1}{2\pi} \sqrt{\left(C_t \left[\frac{J_{in}+J_{out}}{J_{in} J_{out}}\right]\right)} \text{ hertz}$$

where C_t = torsional stiffness of coupling (Nm rad^{-1}) and J = moment of inertia of input and output (kg m^{-2}). In practice the resonant frequency should be at least twice as great as the excitation frequency of the drive.

There are three categories of flexible coupling that compensate for the out-of-line torque conditions, as follows:

1. *Transfer using elastomer.* These can also be divided into torque transmission in compression (e.g. the spider coupling) and torque transmission in tension (e.g. the rubber disk, elastomer sleeve or the double loop coupler). These types are torsionally 'soft', i.e. the construction offers torsional resilience, which can be varied with the grade of material, the application 'tuned' to absorb vibration, cushion shock, etc. (e.g. Fig. 9.31). They are not used for a precise translation of motion requirements and generally have limited torque capacity per unit size. Ambient temperature is more critical with elastomeric couplers.
2. *Transfer using metallic flexures.* These are couplers with relatively unrestrained misalignment compensation based on the flexible shaft system (e.g. the bellows, helical beam and leaf couplers). Fairly precise in the translation of motion.
3. *Transfer by sliding contact.* These couplings will involve either sliding, rotational or pivotal action within the coupler design in order to compensate for misalignment (e.g. the universal joint, Oldham, gear or mechanical jaw couplings). Lubrication is required in all metal couplings with sliding contact.

A summary of types of coupling and their performance limitation is given in Table 9.9.

9.4.1 Types of flexible coupling

(a) Spider coupling. Torque transmission is via a one-piece elastomeric spider design to absorb shock loads and vibration also (Fig. 9.32). It consists of a pair of hubs with inward-facing lugs which engage via the elastomeric spider held in compression and providing the misalignment compensation. It is a well-known coupling, but with wide variations in design from one manufacturer to another. Different insert materials, e.g. nitrile, polyurethane, allow the stiffness characteristics to match the application. The harder the elastomer, the greater the torque rating.

(b) Rubber sleeve coupling. In its basic form this is a rubber tube forced over and clamped to plain shaft ends. It can be manufactured in many variants, e.g. a rubber tyre or a splined tube. The splined tube type (Fig. 9.33), is a rubber tube with internal splines to engage a matching form on each hub. Sleeves can be cut to various lengths and fitted internally with a metal sleeve support to prevent collapse of the rubber under load.

Table 9.9 Summary of the performance characteristics of misalignment couplers

| Type | Category | Shaft dia. (mm)[a] | Misalignment mode | | | | Torsional deflection (min. of arc/Ncm)[c] | Max. r.p.m. | Constant angular velocity I/O | Material of construction |
			lateral Allowable parallel offset (mm)	angular Max. angular misalignment (deg)	longitudinal Max. longitudinal displacement (mm)	Max. unidirectional torque (Nm)[d]				
Spider (6 web)	Elastomer transfer	6	0.2	0.5	—	1.0	—	7,000	no	metal lugs, nitrile spider
		12	0.2	0.5	—	2.3	—	7,000		polyurethane
Double loop	Elastomer transfer	6	2.4	10	4.8[b]	0.035	—	3,600		"
		12	4.8	15	7.5[b]	0.33	—	3,600		"
Helical beam	Metallic flexure	6	0.18	5	±0.1	2.1	220			aluminium
		12	0.38	7	±0.1	14.4	35			"
Bellows	Metallic flexure	6	0.13	7	—	0.68	0.35			brass
		12	0.5	10	—	2.0	0.037			"
Flexible leaf	Metallic flexure	6	0.45	8	0.58	116	3.2			Be/Cu leaves
		12	0.75	8	1.14	310	0.44			"
Flexible Schmidt	Polymer flexure	6	0.4	1.5	allowable	0.7	—	30,000	yes	acetal torque member
		12	0.5	1.5	allowable	1.0	—	22,000	yes	"
Gear	Sliding contact	12	0.3	±1	±1	20	—	14,000		
Multijaw (6 jaw)	Sliding contact	12	0.8	2	2	3.0	—	10,000		
Multijaw (12 jaw)	Sliding contact	12	0.8	2	1.5	1.0	—	9,000		
Oldham	Sliding contact	6	1.5	—	—	0.85	1.02	250	yes[e]	acetal or nylon torque disks
		12	4.0	—	—	15.7	0.06	250	yes[e]	"
Uni-Lat	Sliding contact	6	1.3	10	—	0.27	1.63		yes[e]	acetal annular bearing rings
		12	1.3	10	—	3.14	0.24		yes[e]	"
Schmidt coupling	Sliding contact	12	2.0	0.5	1.0	112	—	3,200	yes	steel + needle bearings

Notes: [a] performance details refer to the most compact coupler size for the given shaft dia.
[b] without angular or parallel offset
[c] torsional stiffness increases with coupling size
[d] running torque
[e] parallel misalignment only

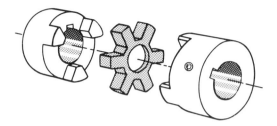

Fig. 9.32 Spider coupling. Illustration courtesy of Huco Engineering Industries Ltd

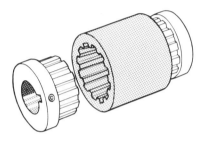

Fig. 9.33 Sleeve coupler. Illustration courtesy of Huco Engineering Industries Ltd

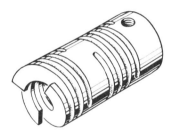

Fig. 9.34 Helical beam coupler. Illustration courtesy of Huco Engineering Industries Ltd

(c) Helical beam coupler. This is manufactured from a one-piece material achieving compliance in all three modes by means of a helix machined through the wall of the material (Fig. 9.34). The flexing elements are designed to allow the same torque capacity in either direction, providing uniform bearing loads and consistent operation in either handed installation. They are manufactured with a one or three start helix, the three start has a higher level of torsional stiffness making it easy to align and balance, thus suitable for high speeds. There are five characteristics defining the helical coupler, i.e. torque, angularity (bending moment), parallel misalignment (radial load), torsional stiffness and spring rate. These are affected by bore, coil thickness and no. of coils.

(d) Bellows coupling. This consists of a convoluted sleeve with a hub attached at each end (Fig. 9.35). The convolutions are compliant in the angular, radial and axial planes but rigid in torsion. The greater the number of convolutions, the greater the ability of the coupling to compensate for shaft misalignment, but the number of convolutions is limited by the accuracy requirement of the transmission. Generally used in small light-duty equipment for large amounts of shaft misalignment combined with low radial loads.

(e) Gear coupling. Basically this is a pair of hubs, each with an end gear form engaging a steel or plastic sleeve with a matching internal form. A curvature of the hub teeth permits slight angular displacement relative to the sleeve and this movement gives the coupler its angular and parallel misalignment compensation plus a small degree of end float. The coupling is torsionally stiff with a high torque transmission capability (Fig. 9.36).

(f) Multi-jaw coupling. This is a rapid type of shaft coupling consisting of two halves with mating teeth or abutments which lock together forming a jaw-like

Fig. 9.35 Bellows coupling. Courtesy of Hydroflex Caradon Mira Ltd, Cheltenham

Fig. 9.36 Gear coupler. Illustration courtesy of Huco Engineering Industries ltd

Fig. 9.37 Multijaw coupler. Illustration courtesy of Huco Engineering Industries Ltd

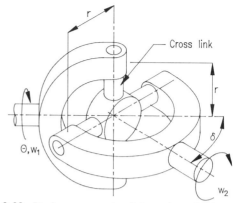

Fig. 9.38 Basic components of the universal joint

connection. Simplistically it is a spider coupling without the elastomer, the number of teeth variable depending on the manufacturer (Fig. 9.37). In misalignment situations, the interlocking abutments are in sliding contact with one another during use, and the resulting friction is basically the only bending load transmitted to the shaft. Allows compliance in all three planes and usually moulded in thermoplastic to reduce wear.

(g) The universal joint. Also known as Hooke's or Cardan's joint, the universal joint is a kinematic linkage used to connect two shafts having permanent angular misalignment, i.e. intersecting centre lines or axes. This joint cannot be used to compensate for offset shaft alignment or axial play. The single universal joint produces a variable velocity ratio, i.e. the output shaft will have a non-constant angular velocity for a constant input velocity, i.e.

$$\omega_2 = \frac{\omega_1 \cos \delta}{1 - \sin^2 \delta \sin^2 \theta}$$

where ω_1 = input shaft angular velocity, ω_2 = output shaft angular velocity, δ = angular misalignment and θ = angular displacement of the input shaft.

The universal joint is illustrated schematically in Fig. 9.38. In terms of relative shaft lead or lag, the speed of the output shaft as a function of angular rotation is given in Fig. 9.39. Differentiating ω_2 with respect to t gives the angular acceleration of shaft 2 as

$$\alpha_2 = \frac{d\omega_2}{dt} = \frac{\omega_1^2 \sin^2 \delta \, \cos \sin 2\theta}{(1 - \sin^2 \delta \sin^2 \theta)^2}$$

As a result of output velocity variations, vibration may be induced into the driven equipment; to counter this

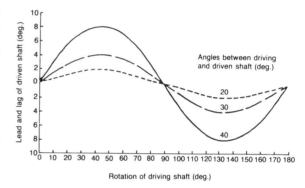

Fig. 9.39 Variation of output shaft speed with respect to driver shaft speed for universal joint shaft misalignments of 20, 30 and 40°

effect, shaft misalignment should be limited to 15° although this type of joint can tolerate up to 40°. A constant velocity ratio between input and output shafts may be achieved using two joints providing that both input and output shafts are in the same plane and that the angles between the driver shaft, the driven shaft and the connecting shaft are equal (Fig. 9.40).

(h) Oldham coupler. Also known as the sliding disk coupler. Comprised of two hubs, each with a tenon machined on each end plus a central disk with matching slots in each face (Fig. 9.41). By orienting these slots at 90° to each other the disk is able to slide radially in respect of one hub without disengaging from the other, hence giving the ability to compensate for lateral offsets. It is recognized for high torque transmission and misalignment

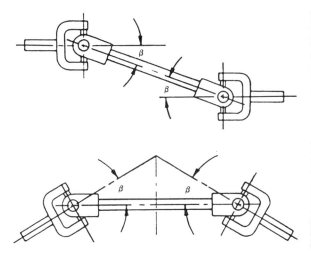

Fig. 9.40 Two arrangements of a double universal joint system for achieving a constant velocity ratio

Fig. 9.42 Universal lateral coupling. Illustration courtesy of Huco Engineering Industries Ltd

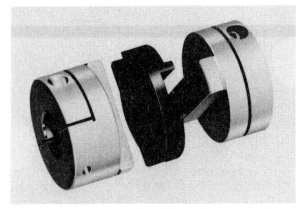

Fig. 9.41 Oldhams coupling. Illustration courtesy of Huco Engineering Industries Ltd

capability at low speed. The actual performance characteristics are governed by the design, dimensions and material of the torque disk. Being a single-stage coupler, the Oldham coupler can be manufactured to a very short overall length, hence is suitable for applications where there is a space constraint.

(i) Universal lateral coupling. This combines the principle of the Oldham coupling with that of the universal joint within a small envelope, giving the coupler both parallel and angular misalignment capability of a high order. The functioning principle is of a sliding bearing, not a flexure, the journals displace laterally to accommodate parallel misalignment and pivotally to accommodate angular offsets. The journals can both pivot and slide laterally within the bearings (Fig. 9.42). The coupling gives a constant velocity ratio for parallel offset conditions (no angular displacement), the amount of misalignment governing the wear rate, hence life, of the coupler. Use is aimed at bidirectional and backlash-free positional control applications. Since there is no radial restraint in the coupling, always support shafts as close to the coupling as practical. Always use singly, never as a Cardan.

(j) Schmidt coupling. Essentially consists of three rotating disks, each connected in series by a pair of parallel links, the first link pair arranged at 90° to the second link pair. Each link locates and pivots on drive pins on the disks. The kinematics of parallel orientated links gives a constant angular velocity between radially offset shafts. The allowable parallel misalignment is determined by the length of the coupling links. The centre disk will radially deflect by a similar amount and should be allowed for in a design application. As a variation, the centre disk and links can be replaced by an acetal element, also providing some degree of angular misalignment. This is a compact mechanical coupling with the minimum of offset reaction forces.

9.5 The drive shaft

A common machine element is a shaft used to transmit torque. Usually a gear, pulley or sprocket is attached to the shaft resulting in a bending moment, as well as a torque, acting on the shaft. The relationship of the combined stress to the material fatigue characteristics should therefore be considered. If the torque and bending moment are effectively constant, then the shaft may be designed by replacing the torque T by the equivalent torque T_E, i.e.

$$T_E = \sqrt{(M^2 + T^2)}$$

where M = bending moment and T_E is the equivalent torque that would produce the same shear stress in the shaft as the combined actual torque and bending moment, but the power transmitted will be a function of T and not T_E. Some basic formulae are given in the following, but reference should be made to specialist texts for fatigue life, the effect of stress concentrations and situations where loads vary.

Bending only,

$$M = \frac{\pi D^3 F}{32}$$

and for a hollow shaft

$$M = \frac{\pi (D^4 - d^4) F}{32 d}$$

where D = OD of shaft, d = ID of hollow shaft and F = maximum tensile or compressive stress.

Shear stress

$$\tau = \frac{16T}{d^3}$$

where T = torque relative to two faces at a known distance and for a hollow shaft

$$\tau = \frac{10 T d}{(D^4 - d^4)}$$

Torsional stiffness C_t(rad) $= \dfrac{Tl}{JG}$

where G = shear modulus = τ/γ, γ = shear strain, J = polar moment of inertia and l = distance between the two torque faces. It is sometimes more convenient to write

$$K_t = \frac{T}{C_t} = \frac{GJ}{l}$$

where K_t = torsional spring constant = torque/unit rotation for a finite shaft length.

It is standard practice to limit the torsional deflection for machinery shafting to 0.26° per metre length. Shafts that permit an excessive angular displacement may contribute to vibration.

9.5.1 The flexible shaft

The function of the flexible shaft is to transmit rotary power or motion from its driving element through a curve path to the driven element. It is used in applications where connection between two points by convenient shafting is not simple or feasible, i.e. for handling difficult drive problems. There are two types of flexible shaft: (1) where the shaft is designed to transmit power in one direction of rotation; (2) where the shaft is primarily designed to transmit motion for remote control applications that can be worked rotationally in both directions.

Shafts are constructed of 1–12 layers of wire which are helically wound on a mandrel (Fig. 9.43). Each layer is wound in an opposite direction, the top layer determines whether the cable 'lay' is left or right. Maximum torque occurs when the flexible shaft is rotated in the direction that tightens the outer layer of wires. The casing supports the rotating core and acts as a bearing surface at the same time. Appropriate end fittings terminate the drive (Fig. 9.44).

Flexible drive shafts are used primarily in the 1000–3600 r.p.m. speed range, but up to 10 000 r.p.m. is possible with the smaller diameters. Power transmission capability is illustrated in Fig. 9.45. The torsional effect on deflection of a flexible core varies proportionately with the torque applied, and to reduce this it is advantageous to operate a shaft at high speed; hence in the case of step-down gearing for example, the flexible shaft should be on the driver side. Usually, charted speed figures (e.g.

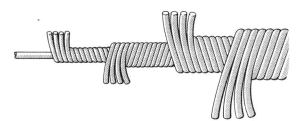

Fig. 9.43 Flexible drive shaft construction

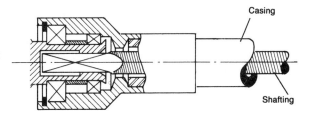

Fig. 9.44 End connection for the flexible drive shaft

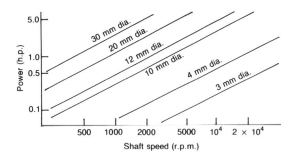

Fig. 9.45 Flexible shaft power transmission capability (straight)

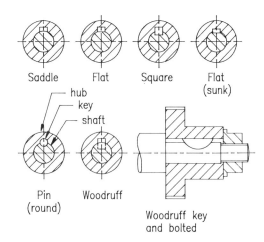

Fig. 9.46 Types of key

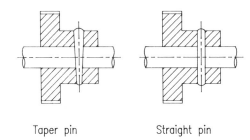

Fig. 9.47 Examples of pin fixings

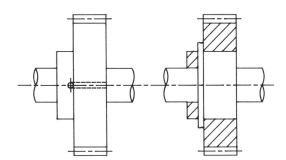

Fig. 9.48 Spring pin modification

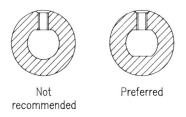

Fig. 9.49 Use of set screws for shaft attachment

Fig. 9.45) are for a relatively straight-drive application. When operating in a curve at load, a speed reduction must be made to avoid heating effects. At the minimum radius of curvature, the maximum speed should be reduced to approximately one-third of the regular speed. With a 3 mm diameter shaft for example, the minimum operating radius is 64 mm.

9.5.2 Shaft attachment techniques

There are many ways of attaching gears and pulleys, etc. to drive shafts. The technique chosen will much depend on the cost, torque, alignment and concentricity requirements of the design commensurate with manufacturability. Some of the more common methods are summarized as follows:

1. *Keying*. This is an effective method of shaft fastening, has good torque capacity, can be easily disassembled but has higher cost implication and is usually limited to assemblies requiring a fixed timing position. Some types of shaft key are illustrated in Fig. 9.46. With plastics, rounded keys should be used in preference to flat keys to avoid stress concentration.
2. *Pins*. Spring pins, tapered pins and groove pins are often used as shaft fasteners, working best when placed as close as possible to the gear or pulley centre (e.g. Fig. 9.47). Pins are considered in more detail in Chapter 14. With moulded gears, etc. a U-slot can be moulded in (Fig. 9.48).
3. *Set screws*. Often used for light application, but are not particularly ideal. Not only do they tend to loosen with reverse loadings, threads in plastic may well strip or creep (loosen) in use. Where set screws are used with D-drives, use a set screw relief flat opposite the D (Fig. 9.49).
4. *Tapered collar*. Apart from the more obvious

DRIVES, BELTS AND COUPLINGS

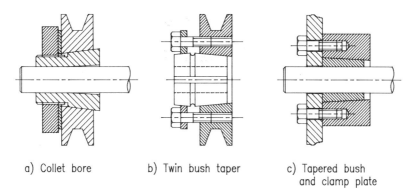

a) Collet bore   b) Twin bush taper   c) Tapered bush and clamp plate

Fig. 9.50 Some examples of tapered shaft fastenings

fastening methods of keys, pins or screws, maintaining gear/pulley/shaft concentricity is often a major additional requirement of many drives and, at the same time, allowing ease of assembly. In order to minimize run-out, it is usual to use a tapered shaft collar, tightened up against the component by means of a wedging action to secure a strong frictional grip on both diameters. For very high torque requirements, it may be possible in some cases to use D-bore versions. Tapered fastening systems include the tapered collet, the twin-tapered bush and clamp plate and conical locking units (Fig. 9.50). Numerous types exist in the market-place, but the general principle is the same.

5. *Concentric fasteners*. Typical of this type of shaft fastening are tolerance rings (Ch. 14), and the split hub and clamp collar (Fig. 9.51). However, the design and application of hub clamping devices should be such as to ensure run-out is not adversely affected.

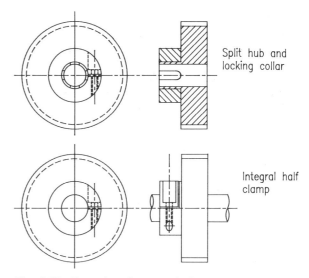

Fig. 9.51 Examples of concentric fastenings

References

1. Morf, R.E. Flat belts. *Mach. Des*, 9 March 1989.
2. BS 3790 : 1981.
3. BS 4548 : 1987.
4. Wallin A.W. Selecting synchronous belts for precise positioning. *Power. Trans. Des*. **2**: 65 (1989).
5. Mechanical fixed speed drives. *Eureka* **9**(9): 192 (1989).

CHAPTER 10

GEARS

10.1 Introduction

Toothed gears give the most compact means of transmitting torque, speed and direction. They may be considered as one of the basic machine mechanisms. However, where the shaft axes are widely spaced, belts or chains may sometimes be more suitable. There are a number of different gear types, each suited to a particular situation. In practice gears are not only used in pairs but in more complex interactions such as gear trains, epicyclic systems, etc. In transmitting motion and power between shafts, the general advantages of gears can be said to be as follows:

1. High efficiency, usually >95 per cent;
2. Ability to cope with almost any angle of input and output shaft;
3. A synchronous drive, an exact ratio between input and output speeds, increasing or decreasing;
4. Torsionally stiff and smooth in transmission.

But there are some drawbacks, mainly:

1. Cost; machined gears can be expensive compared to chains or belts;
2. Unable to tolerate shaft misalignment at high loading;
3. The need for lubrication and possibly some cooling;
4. Noise and vibration with some designs.

Historically, numerous rules and national standards have existed for gear design and gear rating, the approaches to standards making having varied with the type and responsibility of the authority concerned. As a result, different rating formulae have been offered that give different results in practice. It will be apparent from a reading of current gear engineering texts that there are two main approaches to gear-tooth stressing, i.e. that of AGMA (American Gear Manufacturers' Association) and that of the ISO (International Standards Organization).

Although nomenclature is different[1] and the full formulae do not produce identical results in the absolute sense, they are nevertheless quite similar at the simplest level. The position of the British Standards Institution (BSI) would seem to be the adoption of the ISO approach,[2,3] although some standards still exist based on a 'speed-factor' approach.[4,5] In the full ISO procedure, the complexity of the full formulae is somewhat intimidating. However, this does provide a framework of gear design that we will use to determine gear stresses. Since this selection is aimed at the design engineer employing gears for a particular mechanism rather than the engineer designing a gear from scratch, some of the more complex modifying factors and independent parameters are left out as not necessary for providing adequate comparative assessment.

In discussing gear basics, the simpler geometrical case of the spur, or straight-toothed parallel axis, gears will be generally used. Other gear forms are considered separately and although details and geometry vary, the main design concerns are similar.

10.2 Gear types

Although there are many different types of gear, there are only four basic classes, spur, helical, bevel and worm. A summary of the main features of these is given in Table 10.1. For the purposes of small machine design, we will consider the basic gear form within each group, as given under the headings below.

(a) Spur gears. Spur gears connect parallel shafts and are characterized by straight teeth parallel to the centre line of the bore. If both meshing gears have external teeth (Fig. 10.1) the shafts rotate in opposite directions,

GEARS

Table 10.1 *Summary of the main features of common gear types*

Types	For	Against	Other
Spur	Most widely used gear type Simple, relatively low cost drive	Has the least load capacity of all gear types, i.e. limited tooth/tooth overlap Noisier than other gear types External spur gears have small contact ratio	Teeth straight and parallel to shaft axis. Use rack instead of wheel for linear motion.
Helical	Capable of carrying higher loads than equivalent spur gears Transmit smooth and quiet motion Suitable for very high peripheral speeds	Costlier than spur gears Shaft thrust bearings required to cater for helix angle end thrust (axial load) No end-thrust with a double helical gear, but an expensive solution Crossed axis helical gears only suitable for light duty service Axial displacement (thrust) results in increased angular backlash Not as satisfactory as spur gears for critical timing applications	Increased tooth overlap plus increased tooth section thickness in the plane of rotation. Teeth wind helically around gear axis.
Straight bevel	Crowning avoids load concentrations at ends of teeth	Shaft support needs to be rigid to maintain proper tooth contact Introduces thrust on gears acting away from apex Need careful adjustment to get correct tooth contact distribution	Usually right-angled drives but any inter-connecting shaft angle can be accommodated. Teeth formed in conical blanks.
Worm	Simplest way of achieving large-velocity ratios Connects non-intersecting shafts Worm and wheel provides the most cost-effective means of obtaining the maximum reduction ratio for a given centre distance A large contact angle provides a high load capacity in spite of worm/wheel sliding action	Efficiency lower than other gear types, high heat generation Self locking, i.e. gear not able to drive worm Worm is nearly always case-hardened steel, wheel is bronze	Used only for speed reduction, cannot be back driven. Shafts nearly always at right angles. Gear has one or more teeth in the form of a screw thread. Worm is nearly always case-hardened steel, wheel is bronze.

Fig. 10.1 Spur gears. Courtesy of Reliance Gear Co. Ltd., Huddersfield

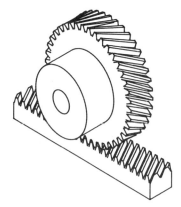

Fig. 10.3 Helical gear and rack

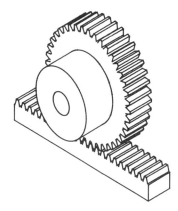

Fig. 10.2 Spur gear and rack

Fig. 10.4 Straight bevel gears. Courtesy of Reliance Gear Co. Ltd., Huddersfield

whereas if the larger gear has internal teeth, the shafts rotate in the same direction. The smaller gear is traditionally called the 'pinion', the larger gear the 'wheel'. If, instead of a wheel, a rack is used (Fig. 10.2), linear motion is produced.

(b) Helical gears. Helical gears connect parallel shafts, their teeth cut on a cylinder but at an angle (helix). In a gear set, the helix for the pinion and the gear must be of opposite hand (Fig. 10.3). A variation of the helical gear is the crossed helical gear or spiral gear in which shafts are not parallel but skewed.

(c) Bevel gears. The simplest type is the straight tooth bevel gear (Fig. 10.4). The teeth are formed in conical blanks with connecting shafts whose axes intersect. If the gear ratio is 1:1, the gears are termed 'mitre gears'. As a variation of the basic form, spiral bevel gear teeth may have any form of spiral or developed curve in the longitudinal direction, e.g. hypoid, spiroid.

(d) Worm gears. Worm gears connect non-intersecting shafts, normally at right angles (Fig. 10.5). The worm itself resembles a screw thread in appearance and can be either handed. With large reduction ratios it is impossible for the gear to drive the worm, thereby providing a self-locking feature. Compared with the other gear types there is a large tooth slide : roll ratio.

Fig. 10.5 Worm gear unit. Courtesy of WHH & Raine Transmissions Ltd., Tamworth

10.3 Gear materials

Various materials are used for manufacturing gears, depending both on the method of manufacture and the application to which the gear will be put. The choice ranges from plastics to hardened steels. The application, advantages and disadvantages of various gear materials are summarized in Table 10.2. With gears of useful load capacity, high pressures have to be carried on fairly small contact areas, hence relatively hard materials are generally required to carry such pressures without surface deformation or fatigue. However, having said that, the use of thermoplastic gears in small machine drives is a rapidly growing area of application, not only satisfying load requirements but also providing production advantages that cannot be met by metal, Table 10.2. In consequence of the significant usage of thermoplastic gears, and the different design criteria that apply, a section of this chapter is specifically devoted to plastic gearing.

In general terms, the required properties of gear materials may be summarized as follows:

1. Material must have adequate fatigue strength in bending to carry the required loading on the teeth roots for the design lifetime. Uniquely, with polymers, failure may also occur at the pitch point.
2. Material should have adequate resistance to surface fatigue in order to carry the tooth surface loadings (Hertzian). Surface fatigue (pitting) is a common mode or failure in gears that have a limited amount of sliding (spur, helical, bevel). A material with a high surface fatigue resistance normally has a high elastic limit, i.e. high hardness at the surface and of adequate depth.
3. Material of adequate toughness and ductility to withstand any impact and shock loads imposed in use.
4. Material of sufficient wear resistance under the required loading conditions in conjunction with the material selected for the opposing gear material with the defined degree of lubrication present.

10.4 Gear terminology

The terminology applied to gears covers a wide range of items and concepts specific to gear technology that need to be explained at this point. The mechanics of tooth meshing define the basic gear geometry that, translated into the basic tooth form, defines the gear system. The various factors influencing gear terminology together with the mechanical relationships between different parameters are considered in the following sections.

(i) Basics of tooth engagement

Gears are toothed wheels of multilobed cams used for transmitting uniform motion from one shaft to another by means of positive contact of successively engaging teeth. To achieve this, tooth profiles must be of conjugate form and long enough to ensure continuous action such that the angular velocity is constant at all times. Two basic curves can be used to define gear profiles, cycloidal and involute, although the cycloidal form was more difficult to manufacture and is now mainly of historical interest.

The involute curve can be constructed graphically by wrapping a piece of string around a cylinder and then tracing the path a point on the string makes as it is unwrapped from the cylinder (Fig. 10.6). When applied to gearing, the cylinder around which the string is unwrapped is referred to as the base circle. Here A is any point on the involute curve and the line AB is equal to the arc subtended by the angle β (Fig. 10.7), and the involute of ϕ:

$$\text{inv } \phi = \tan \phi - \phi \text{ radians}$$

where ϕ = pressure angle and A = the pitch point, OA = the pitch circle radius and BA = line of action.

A basic characteristic of the involute is that any line drawn tangent to the base circle is normal to the involute

Table 10.2 *Summary of gear materials*

Material type	Typical usage	Advantages	Disadvantages	Commercial specification	Surface hardness kg mm^{-2}	U.T.S. MN m^{-2}	Max. Hertz contact stress σ_{Hlim} MN m^{-2}	Surface load capacity factor C^* MN m^{-2}	Max. bending stress σ_{FO} MN m^{-2}
Zinc base die casting	Lightly loaded situations	Possible to make large numbers of components consistently to drawing Smooth finish and capable of fine detail reproduction Wide application, most types of gear possible About same strength as phosphor bronze machine cut gears Low melting points, good castability	Creep will occur at room temperature under load Strength is stress rate dependent at room temperature The higher the aluminium content, the greater the strength but at the expense of cost Some lubricants will corrode castings	BS1004A BS1004B	90 —	280 320	260 300	3 —	35 —
Brass (machined)	Light duty	Ease of manufacture	Expensive Poor wear resistance	61–64% Cu	190–240	—	690	20	—
Phosphor bronze (cast)	Wormwheels typically	Low friction Work hardens in use Compatible with hardened steel Suitable for high working loads and slow speed only with good lubrication	Low loading only	BS1400PB2C sand cast chill cast	70–140 125–175	220–310 270–340	— 205	9.3 11.6	49 63
Sintered metals (powder metallurgy route)	Low and medium duty spur and straight bevel gearing	Cheap to make a large quantity of simple gearing High accuracies possible, no machining required Flexibility of alloy choice, enables alloys unobtainable by other routes Blind corners possible Allows for lubricant impregnation	Expensive for small quantities Only straight tooth forms possible by this method	Iron Fe, 0.6%C Fe, 5%Cu Fe, 3%Cu, 0.6%C 316 Stainless	65[a] 120[a] 95[a] 140[a] 85[a]	180 min. 250 min. 280 min. 330 min. 380 min.	— — — — —	— — — — —	— — — — —
Heat treated steels	High duty industrial gears	Low cost and good mechanical properties Wear resistant to a degree Good compatibility with other mating gear materials Good machinability	Not suitable for high duty or critical applications	Normalized 080M40 080M40 Q&T Normalized 070M55 070M55 Q&T	150–180 180–230 200–250 200–250	540 min. 615 695 770–850	720 760 825 895–965	21 16 19 23–26	145 166 235 275
Heat treated alloy steels	High duty worms and pinions	Wide variety of through hardened steels to give desired properties Increased load capacity Amount of alloying depends on section size	Machining difficult in high strength steels More alloying addition and more expensive Some tendency to distort when heat treated	Through hardened and tempered 150M36 150M19 605M36 150M36 708M40 817M40 826M31 826M40	[b] R S T	200–250 225–295 255–320	700–850 775–925 850–1000	825 895 965	19 23 26	235 255 275

Case hardened steels	High duty applications	When ground, used in more critical applications Carburizing gives a deep, tough, high strength, high hardness layer Compressive surface residual stresses aid contact loading and bending fatigue	Some heat treatment distortion possible Added cost Special conditions may be required for running in	665M17 665M23 655M13 659M15 835M15	750 min. 710 min.	695–850 1310	1930–2000 2130	105–114 130	480 515
Nitrided steels	High duty industrial applications	Reduced level of distortion compared to carburizing Higher surface hardness than carburizing	Quite expensive process Tends to deteriorate at high load if not well supported by the case	772M24 897M39	850 min. 850 min.	695–925 1230	1450–1580 1720	59–72 83	250 320
Plastics (injection moulded)	Light to medium duty applications, non-critical	Relatively low cost High resilience and damping capacity Low noise, quiet operation, often reductions of 10 dB compared to metal gears. Ease and speed of manufacture, no finishing operations Low friction ability to operate with no or little lubrication Lightweight, low inertia Filled plastics give a higher performance potential. Wear resistant to a degree. Capability for designing built-in complex features, e.g. hubs, inserts, snap fits etc. Improved load sharing between teeth. Colour coding possible to eliminate assembly errors. Low elastic modulus may be used to tune out torsional vibrations.	Maximum operating speed linked directly to heat generation Low load carrying capacity compared to metal gears Mould tool costs are high Dimensional stability may be affected by shrinkage, thermal expansion and moisture absorbtion. May be affected by certain lubricants. Sections may provide stress raisers, cross-pins, keys and D-holes etc., need careful consideration. Low rigidity. Typically only around 10% the capacity of equivalent hardened steel gears.	Nylon 6/6 Acetal copolymer Acetal homopolymer Nylon +30%GF +15%PTFE Nylon +30%GF Polycarbonate +30%GF +15%PTFE	— — — — — —	60 73 69 160 230 120	— — — — — —	— — — — — —	28 31 26 50 55 38

Notes: a = Apparent surface hardness
b = Tensile range to BS 970: 1983

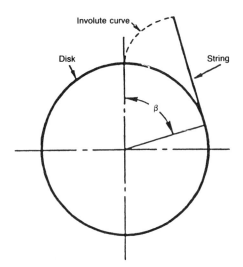

Fig. 10.6 Construction of the involute curve

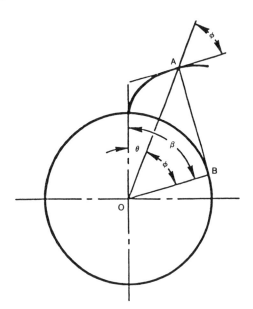

Fig. 10.7 Definition of pressure angle

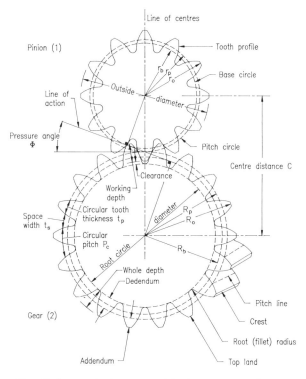

Fig. 10.8 Tooth elements and basic geometry of meshed standard involute spur gears

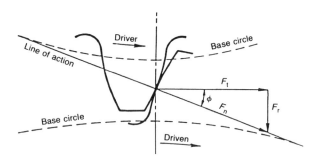

Fig. 10.9 Resolution of forces at the pitch point

at its point of intersection, hence the common normal to any surface in contact with the involute must be tangent to the base circle: this is called the line of action (Fig. 10.8).

From Fig. 10.8, the pressure angle ϕ is the angle between the line of action and the perpendicular to the gear centres and also the angle between the transmitted tooth force (useful component) and the normal force (Fig. 10.9), i.e.

$$F_t = F_n \cos \phi$$

where F_t = transmitted force, i.e. the tangential component representing the useful work done in transmitting the load from the driving to the driven shaft, F_n = normal force, driving force acting along line of action and F_r = work done attempting to separate components, dissipated as heat.

The greater the centre distance (Fig. 10.8), the greater

is ϕ, and the greater are the radii of the pitch circles, thus the pressure angle and the pitch circle cannot be determined precisely until the gears are in contact. With the involute gear profile, slight changes in centre distance (tolerances, bearing wear, etc.) are possible without affecting the angular-velocity ratio. The base radius is fundamental and cannot be changed for a given involute.

(ii) Standard involute gear system

The pressure angle, tooth thickness and length of tooth can all be varied to obtain optimum gear drive performance. However, the cost is often prohibitive in designing and manufacturing gears, hence standard interchangeable gears to an agreed profile are usually employed off the shelf. For interchangeable involute gears they must have the following:

1. The same pitch;
2. The same pressure angle;
3. The same addendum (Fig. 10.8);
4. A circular thickness equal to half the circular pitch — after the addendum is specified, the working depth is known, then clearance is the remaining dimension specified as a function of the pitch.

The basic gear nomenclature and tooth element geometry are illustrated for a meshed spur gear set by Fig. 10.8. Definitions of terminology are given in Table 10.3.

The standard pressure angle is 20° for modern involute gears,[2] although historically 14.5° and 25° angles were used and these will still be manufactured to some extent. The 20° pressure angle form is recognized as meeting the great majority of applications and is defined in British and International Standards in terms of the basic rack (Fig. 10.10). The basic rack is a section of the tooth surface of a gear of infinitely large diameter on a plane normal to the tooth surfaces. The profile of the basic rack is used as the basis of a system of cylindrical gears, having either straight or helical teeth. The inclination of the side of a rack tooth to its radial centre line is, in general, its pressure angle. Specific size dimensions are obtained by multiplying by m, the module. The basic rack will vary for each tooth form system.[6]

In the metric gear, the module is used to specify the size of teeth on a gear; the metric module is the standard pitch diameter divided by the number of teeth, i.e. the smaller the module the smaller the tooth size. In the metric module, the addendum of the gear is specified in millimetres and all other dimensions are set in proportion to the addendum. For non-metric gears, the term traditionally used is diametral pitch (DP), i.e. the number of teeth per inch of standard pitch diameter. Standard modules together with the equivalent DPs are given in Table 10.4. The module is thus a dimension (expressed in millimetres) equal to the reciprocal of the diametral pitch (Table 10.3).

All dimensions in a standard system of interchangeable gears can be defined in terms of pressure angle and pitch. This can be seen from Table 10.3 which is a summary of all basic gear tooth formulae in module format. For interest, the DP format is also included for spur gears.

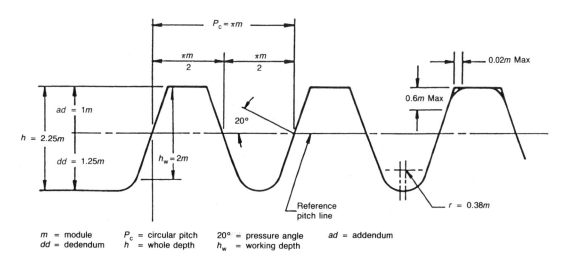

Fig. 10.10 The basic rack — cut gear tooth form

Table 10.3 Common gear formulae

Term	Symbol	Definition	(DP)	Spur gear (Module)	Helical gear(b) (Module)	Worm (Module)	Wormwheel (module)	Bevel pinion (module)	Bevel gear (module)	Notes
Diametral pitch	P_d	The ratio of number of teeth to the pitch diameter in inches; i.e., the number of teeth per inch. of p.c.d.	$\frac{N}{d}$	—	—	—	—	$\frac{n}{d}$	$\frac{n}{d}$	N = no. of teeth d, D = pitch circle diameter α = helix angle
Circular pitch	P_c	Distance between corresponding points on adjacent teeth measured at the p.c.d.	$\frac{\pi}{P_d}$	πm	πm_n		$\frac{\pi D \cos\lambda}{N_g}$	πm	πm	
Module	m	Pitch diameter (in mm) divided by the no. of teeth; i.e. no. of mm of p.c.d. per tooth	$\frac{25.4}{P_d}$	$\frac{d}{N}$	$\frac{d \cos\phi}{N}$	$m \cos\lambda$ (a)	$m \cos\lambda$ (normal value)	—	—	m = axial pitch/π
Addendum	a	The radial distance between the p.c.d. and the gear OD, i.e. the height of the gear tooth outside the pitch circle	$\frac{1}{P_d}$	m	m_n	m	$m(2\cos\lambda - 1)$	—	—	
Dedendum	b	The radial distance between the p.c.d. and the tooth root, i.e. tooth depth below the pitch circle	$\frac{2.25}{P}$ if $\lambda \geq 20 P_d$ $\frac{2.4}{P_d}$ if $\lambda < 20 P_d$	1.25–1.40m	1.25–1.40m	min. = $m(2.2\cos\alpha - 1)$ max. = $m(2.25\cos\alpha - 1)$	$m(1+0.2\cos\lambda)$	$h - a_p$	$h - a_g$	h = whole depth of tooth
Whole depth	h	The whole tooth space depth, i.e. $a + b$	$h - a$	$a + b$	$a + b$	$2.2 m_n$	$2.2 m_n$	2.19m+0.05	2.19m+0.05	
Working depth	h_w	Depth of engagement of the two mating gears; the sum of the addenda of mating gears	$\frac{2}{P_d}$	$2a$	$2m_n$	$2m_n$	$2m_n$	$2m$	$2m$	
Clearance	c	Positive difference between whole depth and working depth	$h - h_w$	0.25m	—	$0.2m\cos\lambda$ $-0.25m\cos\lambda$	$0.2m\cos\gamma$ $-0.25m\cos\gamma$	—	—	
Base circle (root circle diameter)	D_b	The circle from which the involute tooth form is generated	$P_d \cos\phi$	$\frac{\cos\phi}{m}$	—	$\frac{L}{\pi \cot \lambda_b}$		—	—	ϕ = pressure angle λb = base lead angle; i.e. lead angle on the base circle (complement of the base helix angle) θ = semi-included apex angle
Outside diameter	d_o/D_o	Extreme tip diameter of the gear teeth	$d + \frac{2}{P_d}$ $= d + 2a$	$d + 2m$ $= m(N+2)$	$d + \frac{2}{P_d}$ $= d + 2a$	$P_d + 2a$ $= m(q+2)$	$P_d + 2.4m$ (min) $P_d + 3m$ (max)	$(d+2)a_p \cos\theta_p$	$(D+2)a_g \cos\theta_g$	

276

GEARS

Property	Symbol	Description						
Pressure angle (angle of obliquity)	ϕ	Angle between common normal to two teeth in contact and the common tangent to the pitch circles	20°	20°	$\tan\phi = \sec\alpha\,\tan\phi_n$	$\cos\phi_n = \dfrac{\cos\lambda b}{\cos\lambda}$	—	20°	20°
Circular tooth thickness	t_{sd}	Thickness of a single tooth measured along the pitch circle	$\dfrac{\pi}{2P_c}$	$\dfrac{m}{2}$	$\dfrac{\pi m}{2}$	$\dfrac{P_c}{2}$	—	$\dfrac{\pi m}{2}$	$\dfrac{\pi m}{2}$
Lead	L	Distance measured parallel to the axis by which each tooth advances per revolution	—	—	$\dfrac{\pi d}{\tan\alpha}$	nP_c	—	—	—
Centre distance	C	Distance between gear axes	$\dfrac{N_1+N_2}{2P_d} = \dfrac{d+D}{2}$	$\dfrac{m(N_1+N_2)}{2}$	$\dfrac{\sec\alpha(N_1+N_2)}{2P_c}$	$0.5m\left(N_w+\dfrac{P_d}{m}\right)$	—	$\dfrac{\sec\alpha(N_1+N_2)}{2P_c}$	—
Helix angle	α	Angle between gear tooth and gear axis	—	—	$\tan\alpha = \dfrac{d\pi}{L}$	$\tan^{-1}\dfrac{nm}{P_d}$	—	—	—
Lead angle	λ	In worm gearing, the complement of the helix angle	—	—	—	$\tan\lambda = \dfrac{nm}{d} = \dfrac{L}{d\pi}$	—	—	—
Cone distance	A_0	The distance between the pitch cone apex and the pitch diameter (measured along the cone)	—	—	—	—	—	$\dfrac{D}{2\sin\delta_g}$	$\dfrac{D}{2\sin\delta_g}$
Dedendum angle	δ	Angle between gear or pinion root and the pitch angle	—	—	—	—	—	$\tan^{-1}\left(\dfrac{b_p}{A_0}\right)$	$\tan^{-1}\left(\dfrac{b_g}{A_0}\right)$
Face angle	θ_f	Angle between the gear or pinion face and the axis of rotation	—	—	—	—	—	$\theta_p + \delta_g$	$\theta_g + \delta_p$
Face width	F_w	The length of tooth measured along the pitch cone	—	—	—	—	—	≯ $10m$ or $\dfrac{A_0}{3}$	≯ $10m$ or $\dfrac{A_0}{3}$

ϕ_n = normal or true pressure angle

n = no. of thread starts on the worm
P_c = pitch of the worm

Notes: a Gives value of the normal module, m in this case is the axial module
b For helical gears, normal values apply to P_d, m and P_c, i.e. measured normal to tooth helix

General notes: Certain symbols applied generically, e.g. p.c.d. P_d refers to the particular gear in which the formula appears. Formula apply to external gears only.

Convention used: smaller gear is the pinion (subscript p), the larger gear is the wheel (subscript w or g)

Table 10.4 *Standard module equivalents*

Standard module (m)	Nearest approx. diametral pitch (P_d)	Actual equivalent diametral pitch (P_d)
0.5	48	50.8
0.6	40	42.3
0.8	32	31.8
1 [a]	24	25.4
1.25[a]	20	20.3
1.5[a]	16	16.9
2 [a]	12	12.7
2.5[a]	10	10.2

[a] Preferred modules.

(iii) Planes and angles

The geometry of spur, helical and worm gears, which have cylindrical pitch surfaces, is considered relative to (a) a transverse plane perpendicular to the axis, (b) the axial plane containing the axis and (c) in the case of helical gears, the normal plane perpendicular to the tooth helix (Fig. 10.11). When a three-dimensional basic rack tooth is intersected by these planes, the inclination of the sides of the rack teeth on those plane sections are the transverse pressure angle, the axial pressure angle and the normal pressure angle. The inclination of a tooth helix to the axial plane (Fig. 10.11) is the helix angle. For a worm gear, the complement of this angle, the lead angle, is more commonly used.

10.5 Gear application factors

10.5.1 Centre distance

From Fig. 10.7, it can be seen that the centre distance of two mating gears is half the sum of the base circle diameters divided by the cosine of the pressure angle, i.e.

$$C = \frac{D_{b1}+D_{b2}}{2 \cos \phi} = \frac{d+D}{2} = \frac{m(N_1+N_2)}{2}$$

and

$$\phi = \cos^{-1}\left[\frac{D_{b1}+D_{b2}}{2C}\right]$$

where d = pitch circle diameter of pinion, D = pitch circle diameter of gear, D_{b1} = base circle diameter of pinion, D_{b2} = base circle diameter of gear and C = theoretical, standard or close mesh centre distance. Alternatively, if the standard pitch diameters and the pressure angles are known, the operating pressure angle can be obtained

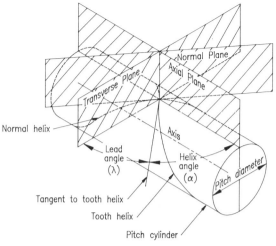

Fig. 10.11 Transverse, axial and normal planes

$$\phi_1 = \cos^{-1}\left[\frac{(d+D)\cos \phi}{2C_1}\right]$$

and

$$C_1 \cos \phi_1 = C \cos \phi$$

where C_1 = operating centre distance and ϕ_1 = operating pressure angle.

For involute gears the standard pressure angle is 20° and this would apply to the gear pair at the standard centre distance, but making the centre distance other than standard ϕ may be varied. Centre distance variations will have no effect on theoretically correct gear tooth action. In applications calling for high gear efficiency, the operating pressure angle should be kept as low as other factors permit, i.e. the centre distance can be manipulated to give the best possible 'drive'. The above formulae do not apply to non-standard gears where tooth thicknesses differ.

The close mesh centre distance assumes perfection in the gears and bearings used and in the stability of the housings. In practice, tolerances, run-out and environmental changes will cause the close mesh centre distance to vary. However, the minimum operating gear centre distance specified must be greater than or equal to the close mesh centre distance otherwise the gears may bind. Tolerances are defined for a given grade of gear as the expected error level by

Errors in teeth + gear run-out = total composite error
(tooth–tooth   (TCT)
composite error)

where for a pair of mating gears, the total composite tolerance factor is

$$\frac{TCT_1 + TCT_2}{2}$$

A summary of gear tolerances and quality grades is given in Table 10.5. The average standard is generally grade 8. Higher-quality gears may require special finishing operations and can be more expensive. For general design work, aim to use the lowest quality that the intended service will permit. The minimum operating centre distance must exceed the calculated close mesh centre distance by the total composite tolerance factor above.

10.5.2 Contact ratio (CR)

This is a numerical index of the existence and degree of continuity of gear tooth action and is equal to the average number of teeth in continuous contact. A minimum contact ratio of 1.2 will ensure smooth operation of the gear set, and since at least one pair of teeth is in contact at all times, load or motion is transferred at constant velocity. Contact ratios below 1.0 means that there is not enough involute surface available to make a timely exchange with correct angular rotation. Aside from inertial carry-over, damaging edge contact can occur which is usually associated with gear noise. As well as an indicator of smoothness of action, contact ratio also determines the degree of load sharing between teeth.

The profiles of a pair of mating teeth and the rotating contact point is shown in Fig. 10.12. The tooth of the driving gear first makes contact with the driven gear at C'. Contact continues along the line of action, through the pitch point A until the teeth break contact at C''. Contact ratio is defined as

$$CR = \frac{\text{active length of the line of action}}{\text{base pitch}} = \epsilon$$

where base pitch = distance between two involutes along the line of action and

Table 10.5 Summary of tolerance formula for metric gears

Quality grade		Tolerance on radial tooth run-out (μm)	Tooth–tooth composite tolerance (μm)	Total radial composite tolerance (μm)
	1	$0.22\phi_p + 2.8$	$0.16\phi_p + 2$	$0.32\phi_p + 4$
	2	$0.36\phi_p + 4.5$	$0.22\phi_p + 2.8$	$0.5\phi_p + 6.3$
	3	$0.56\phi_p + 7$	$0.32\phi_p + 4$	$0.8\phi_p + 10$
Higher accuracy of meshing	4	$0.90\phi_p + 11$	$0.45\phi_p + 5.6$	$1.25\phi_p + 16$
	5	$1.4\phi_p + 18$	$0.63\phi_p + 8$	$2.0\phi_p + 25$
	6	$2.24\phi_p + 28$	$0.9\phi_p + 11.2$	$3.15\phi_p + 40$
Lower accuracy of meshing	7	$3.15\phi_p + 40$	$1.25\phi_p + 16$	$4.5\phi_p + 56$
	8	$4.0\phi_p + 50$	$1.8\phi_p + 22.4$	$5.6\phi_p + 71$
	9	$5.0\phi_p + 63$	$2.24\phi_p + 28$	$7.1\phi_p + 90$
	10	$6.3\phi_p + 80$	$2.8\phi_p + 35.5$	$9.0\phi_p + 112$
	11	$8.0\phi_p + 100$	$3.55\phi_p + 45$	$11.2\phi_p + 140$
	12	$10.0\phi_p + 125$	$4.5\phi_p + 56$	$14\phi_p + 180$

Notes: $\phi_p = m_n + 0.25\sqrt{d}$ with ϕ_p in mm; d = reference circle diameter, (mm); m_n = normal module.

Definitions: Tooth–tooth composite tolerance (*TTCT*) = sum of single pitch tolerance + tooth–tooth profile tolerance (total) error. Total composite tolerance (*TCT*) = tooth–tooth error + radial run-out. From above; *TCT* = *TTCT* + radial run-out.

Gear grade: The lower the grade no. the greater the degree of meshing, i.e. the greater the accuracy of manufacture hence the greater the contact area over the tooth face and flank.

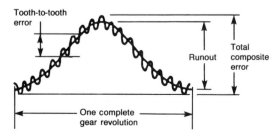

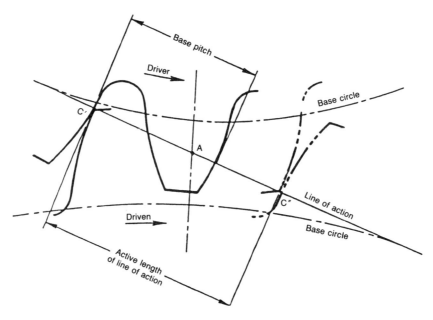

Fig. 10.12 Definition of the contact ratio

$$\epsilon = \frac{\sqrt{(r_o^2 - r_b^2)} + \sqrt{(R_o^2 - R_b^2)} - C_1 \sin \phi_1}{P_{b1}}$$

where r_o = outside radius of pinion, R_o = outside radius of gear, r_b = base radius of pinion, R_b = base radius of gear and P_{b1} = base pitch of pinion.

In Fig. 10.12 from C1 to A, the driving tooth is sliding into the driven tooth. At the pitch point A there is momentary rolling and from A to C2, the driving tooth slides out of engagement with the driven tooth. In gearing terminology: C1 to A is termed the approach action (AA); A to C2 is termed the recess action (RA), and

$$RA = \frac{d_o}{2} \sqrt{\left[1 - \left(\frac{N_1}{d_o Y}\right)^2\right]} - \frac{C_1 N_1}{N_1 + N_2}$$

$$\sqrt{\left[1 - \left(\frac{N_1 + N_2}{2C_1 Y}\right)^2\right]}$$

$$AA = \frac{D_o}{2} \sqrt{\left[1 - \left(\frac{N_2}{D_o Y}\right)^2\right]} - \frac{C_1 N_2}{N_1 + N_2}$$

$$\sqrt{\left[1 - \left(\frac{N_1 + N_2}{2C_1 Y}\right)^2\right]}$$

where d_o = outside diameter of pinion, D_o = outside diameter of gear, C_1 = operating centre distance and $Y = 1.06/m$, N_1 = number of teeth on pinion (driving gear) and N_2 = number of teeth or gear (driven gear). Hence

$$\epsilon = \frac{RA + AA}{P_b}$$

provided that the circular tooth thicknesses are greater than that given by $t = (2.33 - 0.0426N)m$.

A pair of gears should be designed to have as much recess increment as other considerations allow as this part of the gear action is mechanically less severe than the approach action. Providing a gear is speed reducing, recess action of a pair of involutes can be increased >50 per cent by increasing the addendum of the driving gear over the driven gear.

10.5.3 Interference and undercutting

For correct tooth action, the points of contact on the two mating gear teeth must lie on the involute profiles. If the addendum of one tooth is too large, however, contact may occur between the tip of that tooth and the non-involute portion of the mating tooth between the base circle and the dedendum circle. Interference therefore occurs when

Action addendum circle radius $> r_a$

where

$$r_a = \sqrt{(r_b^2 + C^2 \sin^2 \phi)}$$

and r_b = base radius, C = centre distance and r_a = allowable radius of addendum circle.

The points of tangency of the base circle with the line of action are known as pressure points, e.g. point A in Fig. 10.8, and if the addendum circle intersects the line of action outside these points, interference will occur. A number of methods are available to prevent interference as given under the following headings.

(a) Increase centre distance. Correct tooth action is maintained but the pressure angle is increased. This leads to higher tooth pressures for a given torque transmitted and backlash is correspondingly increased. Increasing the pressure angle on its own to decrease the base circle diameter results in a non-standard gear.

(b) Stub the tooth. In other words, remove the tip of the gear, but the resulting reduced contact ratio produces a rougher and noisier gear.

(c) Undercutting. If a standard gear has a small number of teeth, the base circle is greater than the root diameter (Fig. 10.8). Since no action can take place below the base circle, that part of the tooth inside the base circle is non-functional, i.e. the portion of the mating tooth which projects inside the interference point is not needed. To accommodate the tip of the mating tooth as it sweeps round, the lower portion is undercut (Fig. 10.13). Undercutting is generally undesirable since it weakens the tooth, reduces the contact ratio and may well produce a noisier and rougher gear. No undercutting can occur if the addendum does not extend inside the reference point.

For a standard tooth thickness, $\pi m/2$, minimum number of teeth for no undercutting (Nc)

$$= \frac{2}{\sin^2\phi}$$

which for the standard gear module, $\phi = 20°$, is 18 teeth. The lower the number of teeth being cut, the more the undercutting until at nine teeth, insufficient involute is left to permit proper functioning of the teeth, i.e. the contact ratio <1.

(d) Tooth correction. Modification of both pinion and wheel addenda by increasing the pinion addendum and decreasing the gear addendum, the two made as a pair to maintain the same working depth. The pressure angle, centre distance and base circles remain unaltered, but the thickness of the pinion tooth increases slightly and that of the gear decreases slightly. Since pinion teeth are weaker than gear teeth, tooth correction helps to obtain a more balanced strength. The major disadvantage with long and short addendum teeth is non-standard gears, clearly not interchangeable. The minimum circular tooth thickness to avoid undercutting is given by

$$t = (2.333 - 0.0426N)m$$

10.5.4 Backlash

Backlash is the play between mating tooth surfaces at the point of tightest mesh in a direction normal to the tooth surface when the gears are mounted in their specified position (Fig. 10.14). Essentially it is the amount by which a tooth space exceeds the thickness of the engaging tooth, providing some clearance so that the gears will not bind. Backlash is governed by the operating centre distance and the tooth thickness. If both mating gears have standard tooth thicknesses and are mounted at the standard centre distance, they will theoretically close mesh with zero backlash. In practice, either standard gear teeth are cut slightly thinner to take account of manufacturing composite tolerances and always provide some clearance at standard centres or the centre distance is varied accordingly to adjust backlash.

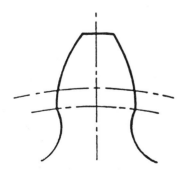

Fig. 10.13 Example of tooth undercutting

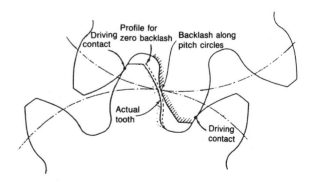

Fig. 10.14 Backlash between two gear profiles

A number of ways exist of defining backlash:

$B = 2 \Delta c \tan \phi$

where B = backlash along pitch circle and Δc = difference between close mesh centre distance and actual running centre distance.

$B_L = t_s - t_p$
$ = B \cos \phi$

B_L = linear backlash along line of action, t_s = tooth space on pitch circle (Fig. 10.8) and t_p = mating tooth width on pitch circle.

$B_\theta = \dfrac{2B}{d \text{ (or } D)}$

where B_θ = angular backlash. Alternatively,

$B_{min} = 2 \tan \phi (\Delta r_p + \Delta R_p - \text{tol } \Delta C)$

$B_{max} = 2 \tan \phi (\Delta r_p + \Delta R_p + \text{tol } \Delta C + \text{tol } \Delta r_p + \text{tol } \Delta R_p)$

where Δr_p = difference in pitch radius between close mesh centre distance and actual running centre distance for the smaller gear (pinion) and ΔR_p = difference in pitch radius between close mesh centre distance and actual running centre distance for the larger gear.

Enough backlash should always be provided for satisfactory gear operation, but excess backlash can result in instability in some dynamic situations plus positional errors in gear trains. Typical backlash for a 0.8–2.0 m gear pair after assembly would be 0.05–0.10 mm.

A number of schemes have been suggested to control backlash, and these are described below.

(a) Adjustable centre gears. This requires an adjustable shaft which is not always possible because of design and cost reasons. Always difficult to set up on the production line.

(b) Spring-loaded scissor gear. This consists of two spring halves connected by springs (Fig. 10.15). The advantage is simplicity and reliability, the disadvantage is that the springs must be given a torque larger than the load to be driven otherwise the springs will deflect, resulting in lost motion. As a result, spring-loaded gears are only used in light-duty torque applications.

(c) Locked gears. For high torque applications, split gears are assembled and the two halves riveted or screwed together, thus all backlash except that due to rotation is eliminated. The limiting torque is therefore controlled by

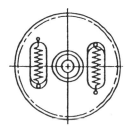

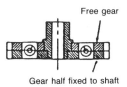

Free gear

Gear half fixed to shaft

Fig. 10.15 Spring-loaded scissor gear

the strength of the fastener. However, interchangeability is lost.

10.5.5 Surface distress and lubrication

There are two main types of surface deterioration evidenced on gear teeth:

1. *Pitting (surface fatigue).* Tends to occur at the lower end of the speed range and is common in soft or through-hardened gears. Typical of the symptom are small-sized craters in the contact faces.
2. *Wear (scuffing).* Tends to occur at the higher end of the speed range, but may well also follow pitting if the oil film breaks down locally. The wear form may be described variously, but it occurs when lubrication fails.

The zones of gear tooth distress are shown in Fig. 10.16. Full oil film thickness or rather the film thickness/surface roughness ratio[7] is important. The ideal condition of full ehl lubrication occurs at low load and high speed (Fig. 10.16). The pitting resistance of a gear is proportional to the surface load capacity (Hertzian stress level) (the permissible stress level is discussed in section 10.6). Adhesive wear appears as a thermal phenomenon, essentially due to tooth–tooth sliding friction which also results in an increase in noise level and power requirement.

Gears operate under a diversity of conditions and the methods of lubrication vary accordingly. Frequent application of a small amount of lubrication is preferable to large volumes at longer intervals. When gears run in an enclosed case, the larger gears may dip into an oil bath.

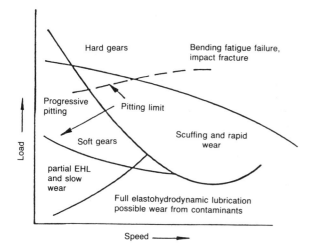

Fig. 10.16 Zones of gear tooth distress. From Fowle T.I., *Practical Gear Tribology*, in Industrial Tribology Series 8, Elsevier Science Publishers BV, 1983

At very high contact pressures, EP lubricants may be necessary. In practice, select an oil of lowest possible viscosity to minimize frictional losses consistent with speed and tooth loadings. Since large teeth tend to be more heavily loaded than small teeth, this tends to have an influence on the incidence of scuffing. Table 10.6 generalizes experience. Recommended oil viscosity grades are generally given in terms of pitch line velocity.

Lubrication of gears can be a more complex field than the lubrication of journal bearings, but because of conservative design factors, the great majority of gearing now in use is not seriously affected by lubrication deficiency. However, in very compact designs requiring a high degree of reliability at high operating speeds, stresses or temperatures, the lubricant becomes a true engineering material requiring careful consideration.

In the case of open gears, it is usual to use high-viscosity oil or grease which should maintain a protective film without being centrifuged off the gear. Such gears then need to be slow running.

10.5.6 Gear noise

Noise is often a problem with gears, the most frequent cause being the variation in tooth loading due to errors of geometry. Tooth errors can produce resonance over a wide frequency range; the predominant note, however, will reflect the frequency of tooth engagement. If noise is critical then attention should be paid to achieving greater gear accuracy and uniformity of motion, including shafts and bearings.

The following points are useful in making gears as quiet as possible:

1. Specify the finest pitch allowable for the load conditions;
2. Use coarse pitch gears to produce a lower pitch sound;
3. Use a low-pressure angle. Relatively speaking a 14.5° pressure angle is half as noisy as a 25° pressure angle;
4. Use a modified profile with root and tip relief;
5. Increase the working depth of the teeth, also resulting in a larger contact ratio;
6. Increase backlash;
7. Use high-quality gears with a fine surface finish;
8. Balance the gear set;
9. Use a non-integral gear ratio if both pinion and gear are made from a hard steel;
10. Make sure that critical speeds are at least 20 per cent apart from the operating speed or speed multiples of the mesh frequency;
11. Ground gears are usually quieter than machine-cut gears;
12. Parallel drives are normally quieter than angled-drive gears;
13. Type of gear influences noise; gears are noisiest in the order: spur (noisiest) > helical > worm (least noisy). With helical gears, best overall noise reduction ~35° helix angle.

Table 10.6 *Lubrication factors relating to spur, helical, and bevel gears*

Pitch line speed (m s^{-1})	Oil viscosity range cSt @ 40°C
0.5	460–1000
1.3	320–680
2.5	220–460
5.0	150–320
12.5	100–220

Module (mm)	Risk of gear scuffing
1.25	Very low
2.5	Only at high speeds with thin oil
5	At moderate speed even with medium viscosity oil
10	At low speeds even with heavy oil

10.6 Gear rating

There are a number of established standards for rating metal gears such as BS 436,[2] AGMA 218[8] and ISO 6336 covering basically two aspects of loading: (1) resistance to tooth surface fatigue and pitting, and (2) resistance to tooth breakage.

One further aspect of gear application, the resistance to lubricant breakdown, wear and scuffing is difficult to cover in great detail, although lubrication implications are introduced into the coverage of (1) above by classification societies. Thus gear rating is determined primarily in terms of tooth contact stress and tooth bending stress, calculations normally performed separately for wheel and pinion. The fundamental formulae for calculating tooth bending and surface contact stresses are derived from Lewis bending and Hertz contact equations, modified by the application of correction factors to allow for stress concentrations, dynamic and load distribution effects. For satisfactory service, the beam strength and the wear strength must equal or exceed the design load. The permissible contact stress is calculated from the material endurance limit (modified by configurational and system factors), which has to exceed the actual stress calculated from the nominal tangential force (modified by gear geometry factors). The gear tooth torque capacity is calculated from the permissible bending stress, load modifying factors and gear geometry. This needs to exceed the torque requirement of the gear pair.

It should be noted that beam strength and wear strength have little in common so it is possible to short cut the design verification process if one or the other is decided as the most critical in a particular case. For example, if gear is non-metallic, beam strength is critical; if both gears are steel, wear is generally critical unless the gears are subject to some abuse.

The rating of the various types of common gear are discussed in the following sections.

10.6.1 Spur gears

(i) Contact stress

This is evaluated using the Hertz equation for contacting cylinders, the radii of the cylinders made equal to the radius of curvature of the involute tooth form at the point of contact and the resultant contact stress is given by

$$\text{Contact stress} = \sqrt{\left[\frac{F_t}{bd} \frac{1}{(1-\nu_1^2)/E_1 + (1-\nu_2^2)/E_2} \frac{4}{\sin 2\phi} \frac{u+1}{u}\right]}$$

where b = face width (mm), u = gear ratio, d = reference (pitch) diameter (mm), ν = Poisson's ratio, E = Young's modulus (MN m^{-2}) and F_t = nominal tangential force at the pitch circle (N). For the actual Hertzian contact stress, σ_H, the above equation is modified in form to give

$$\sigma_H = Z_H Z_E Z_\epsilon \sqrt{\left(\frac{F_t}{bd} \frac{u+1}{u} K_A K_M\right)}$$

where Z_H = zone factor which accounts for the influence of tooth flank curvature at the pitch point on Hertzian stress and converts the tangential force to a normal force at the pitch cylinder, i.e.

$$= \sqrt{\left(\frac{4}{\sin 2\phi}\right)}$$

if gears at standard centres; Z_E = elasticity factor which accounts for the influence of specific material properties, E and ν, on the Hertzian stress, i.e.

$$\sqrt{\left(\frac{1}{\pi[(1-\nu_1^2)/E_1 + (1-\nu_2^2)/E_2]}\right)}$$

and for steel/steel gears Z_E = 189; Z_ϵ = contact ratio factor which accounts for the load-sharing influence of the contact ratio and the overlap ratio on the specific loading = $\sqrt{[(4-\epsilon)/3]}$ and ϵ = contact ratio (section 10.5.2) = length of contact along line of action base circle pitch. Finally, K_A = application factor which is an empirical adjustment modifier to account for load fluctuation in use, i.e. for uniform loads, K_A = 1.0, for light shock loads K_A = 1.25, for medium shock loads K_A = 1.50, for heavy shock loads K_A = 1.75, and K_M = load distribution factor (see section 10.6.1(ii)).

The permissible contact stress, σ_{Hp}, is given by the contact stress endurance limit modified by a series of configurational and system factors:

$$\sigma_{Hp} = \sigma_{Hlim} Z_{LV} Z_R Z_M Z_W Z_N / S_{Hmin}$$

where S_{Hmin} = safety factor in surface contact stress. This value should reflect the confidence in actual operating conditions and material factors achieved. The recommended values are as follows: for normal industrial applications, S_{Hmin} = 1.0–1.2; for high reliability and critical applications, S_{Hmin} = 1.3–1.6. σ_{Hlim} = contact stress endurance limit, i.e. the maximum contact stress (Hertzian) that can be sustained for an infinite no. of cycles without contact fatigue damage (pitting). Values of σ_{Hlim} are given in Table 10.2. Where a range of values appears, the higher the module, the larger is σ_{Hlim}, the larger the pitch circle diameter, the lower is σ_{Hlim}. Z_{LV} = speed and lubricant factor (Fig. 10.17). This affects the lubricant film thickness which in turn affects the allowable contact stress. Z_R = surface roughness factor (Fig. 10.17). Consider alongside Z_{LV}. Z_W = the work hardening factor is the increase of surface durability (work hardening) produced from meshing a wheel of <400 kg mm^{-2} hardness with a hardened steel pinion (Fig. 10.17). Z_N = life adjustment factor. This takes into account the increase in permissible contact stress allowable if the required number of stress cycles of the gear is less than the material fatigue limit (Fig. 10.17). Z_M = material quality factor (Fig. 10.17). This results from improved material quality and mechanical properties as per the following:

Quality C. Normal requirement for materials and heat treatment used for lightly loaded gears in non-critical applications. No special inspection.

Quality B. Requirements and materials for the majority of industrial gears, i.e. uniform microstructure with no surface or subsurface cracks.

Quality A. Maximum quality requirements for materials and heat treatments for critical applications, i.e. fine uniform surface structure, no inclusions or material discontinuities.

The allowable or permissible contact stress σ_{Hp} will be required to exceed the calculated contact stress σ_H, making due allowance for contact ratio. If the contact ratio is not used in respect of the actual stress, the design may well be over-conservative.

The power transmission in the surface durability approach is given by

$$P = \frac{F_t \pi d n}{60 \times 10^6} \quad \text{and} \quad M = \frac{60 \times 10^3 P}{2n}$$

where n = rotational speed (rpm), P = power capacity (kW) and M = actual torque (Nm).

Note: Because of the numerous approaches and standards that have been applied to gearing, some explanation of the calculated contact stress equation derivation is worth while since the design engineer may encounter alternative forms of equation. In general terms, the intensity of tooth loading per millimetre of face width may be written

$$F_i = \frac{F_t}{b}$$

from which a loading factor, K, has been derived such that

$$K = \frac{F_i}{d}\left(\frac{u+1}{u}\right)$$

The factor K was widely known as the Lloyds K-factor,[9] its use as a loading criterion based on the fact that the Hertzian contact stress is roughly equivalent to $\sqrt{K}$ corrected to allow for the finite curvature of the mating gear teeth. Although the units of K are in MN m^{-2}, K-values are not visualized as direct stress but rather an arbitrary criterion of tooth surface loading. The value F_t/bd is sometimes referred to as the surface load capacity factor C^* (Table 10.2). The first 'step' in converting K into a meaningful tooth surface loading is by means of the contact stress conversion factor C_K, i.e.

$$S_C = K C_K$$

where $C_K \equiv (Z_H)^2$ and S_C = contact stress, whence for gears at standard centres and $\phi = 20°$

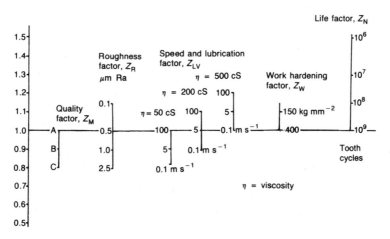

Fig. 10.17 Summary of contact stress modification factors

$$S_C = 6.22C^* \left(\frac{u+1}{u}\right) \text{ MN m}^{-2}$$

However, to obtain a correct value of surface compressive stress (Hertzian) the above is modified by a number of factors, i.e. moduli of elasticity (Z_E), some allowance for contact ratio (Z_c) and the application environment in which the gear is put (K_A).

(ii) Bending stress

A gear tooth is a short cantilevered beam (Fig. 10.18), the normal force F_n resolved into components F_t and F_r at the centre of the tooth Q, locating the critical section. The tangential force F_t produces a bending stress distribution of maximum value:

$$\sigma_b = \frac{K_b 6 F_t h^*}{b t_r^2}$$

where K_b = bending stress concentration factor, b = face width, h^* = height of critical section, t_r = tooth root thickness and σ_b = maximum bending stress at root.

There is clearly also a small radial compressive stress, but since fatigue failures start on the tension side it is common practice to ignore the compressive stress. Taking the tooth geometry into account, Lewis (1897) developed the well-known equation for tooth root bending stress:

$$\sigma_b = \frac{K_b F_t}{bmY}$$

where Y is the Lewis form factor, a function of the gear tooth geometry.

For simple spur gears, it has become more convenient to define a geometry factor J where

$$J = \frac{Y}{K_b} \quad \text{or} \quad = \frac{1}{Y_F Y_S}$$

as in ref. 2 where Y_F = tooth form factor and Y_S = stress concentration factor.

The geometry factor J evaluates tooth shape, fillet configuration, stress concentration in the fillet area and load distribution and is usually presented as a function of the number of teeth in the gear pair.[10] Differences in gear types are reflected in the values of J. Procedures for evaluating Y_F and Y_S separately are overly complex[2] and not of straightforward applicability. The basic bending stress equation applicable to the highest point of single pair contact (HPSPC) is as follows:

$$\sigma_F = \frac{F_t}{bmJ} K_M K_A$$

where J = a geometry factor (Table 10.7), σ_F = actual root bending stress, K_A = application factor (section 10.6.1(i)), K_M = load distribution factor which accounts for inaccuracies in bearing bore location leading to misalignment of axes, errors due to gear tooth inaccuracies and load/thermal distortions or deflections. It may be separated[2] into non-uniform load distribution across the face width and in the transverse plane (meshing direction), i.e.

$$K_M = K_{F\beta} K_{F\alpha}$$

where $K_{F\alpha}$ = transverse load factor and $K_{F\beta}$ = face load factor.

However, calculations of $K_{F\alpha}$ and $K_{F\beta}$ can become overly involved[2] hence it is convenient to adopt the following generalized values: accurate mounting, low bearing clearances precision gears, $K_M = 1.3$; less accurate gears, less rigid mountings, $K_M = 1.6$; poor alignment, less than full face contact exists, $K_M = 2.0$.

The *permissible root bending stress*, σ_{FP}, is calculated as follows, the calculation based on modifying the fatigue

Table 10.7 *Geometry factor J for standard spur gears*

Number of teeth for which J is desired	Teeth in mating gear						
	12	17	25	35	50	85	Rack
15	0.25	0.25	0.25	0.25	0.25	0.25	0.25
16	0.25	0.25	0.25	0.25	0.25	0.25	0.25
17	0.29	0.29	0.29	0.29	0.29	0.29	0.29
18	0.30	0.31	0.32	0.32	0.32	0.32	0.32
19	0.31	0.32	0.32	0.33	0.33	0.34	0.36
20	0.31	0.32	0.32	0.34	0.34	0.35	0.37
22	0.32	0.33	0.34	0.35	0.36	0.36	0.38
24	0.33	0.34	0.35	0.36	0.37	0.37	0.39
30	0.35	0.37	0.38	0.39	0.39	0.40	0.42
40	0.38	0.39	0.40	0.41	0.42	0.43	0.45
60	0.40	0.42	0.43	0.44	0.45	0.46	0.48
80	0.42	0.43	0.45	0.46	0.46	0.48	0.50
125	0.43	0.45	0.46	0.47	0.48	0.49	0.52
275	0.44	0.46	0.48	0.49	0.50	0.51	0.54

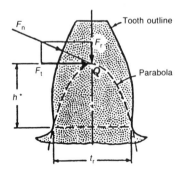

Fig. 10.18 Gear tooth bending stress

endurance stress to account for the effects of surface finish, life required and material quality:

$$\sigma_{FP} = \frac{\sigma_{FO} Y_N Y_R Z_M}{S_{Fmin}}$$

where σ_{FO} = basic endurance limit for bending stress based on the reversed fatigue limit. Values of allowable stress are given in Table 10.2. If gears are subject to reversed loading, use 70 per cent of the values in Table 10.2. Y_N = life factor for bending stress that accounts for an increase in permissible stress if the number of stress cycles is less than the endurance life. If the S/N curve for the material is known, this can be the basis for the life factor. Alternatively use the values in Table 10.8. Y_R = surface condition factor for bending stress accounts for the reduction of endurance limit due to flaws in the material and the surface roughness of the roots (Fig. 10.19). Z_M = material quality factor (see section 10.6.1(i)). S_{Fmin} = minimum demanded safety factor or alternatively known as the reliability factor. The allowable stresses for standard gear materials will vary with material quality, heat treatment, etc., σ_{FO} and Z_M. However, S_{Fmin} is sometimes introduced to ensure high reliability or in some cases to allow for a calculated design risk. Values are as follows:

Normal design = 1.25
High reliability = >1.6
Fewer than 1 failure in 100 = 1.0
Fewer than 1 failure in 3 = 0.7

10.6.2 Helical gears

These are used in place of spur gears to give a smoother drive between parallel shafts since the engagement between helical teeth is gradual, i.e. an initial point contact becoming a line contact of increasing length as the gear rotates. The instantaneous line contact of a spur gear tends to produce a shock effect that can result in

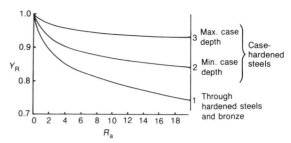

Fig. 10.19 Values of surface conditioning factor Y_R

rough operation. The transverse section of a helical gear tooth is identical with that of an involute spur, but the teeth are inclined to the gear axis, the helix angle α (Fig. 10.20). However, the helix angle results in a thrust load, F_{thrust}, in addition to the usual tangential and separating loads (Fig. 10.21) such that

$$F_t = F_n \cos \phi_n \cos \alpha$$

and

$$F_{thrust} = F_t \tan \alpha = F_n \cos \phi_n \cos \alpha$$

where ϕ_n = normal pressure angle.

The thrust load needs to be accommodated by the appropriate use of thrust bearings or washers. The direction of axial thrust is illustrated by Fig. 10.22. In order for two helical gears on parallel shafts to mesh, they must have the same spiral angle and be of opposite hand (Fig. 10.22). Large helix angles may produce excessive end thrust while small angles may require excessive face width or fine pitch to achieve full overlap. In general, helix angles up to $\approx 20°$ are satisfactory and where end thrust can be tolerated, the helix angle can increase to $45°$.

(i) Contact stress

As with spur gears, the actual contact stress, σ_H, is given by

Table 10.8 Life factor for bending stress, spur and helical gears

Number of cycles	Gear hardness			Case carburized
	160 kg mm^{-2}	250 kg mm^{-2}	480 kg mm^{-2}	
≤ 1000	1.6	2.4	3.4	2.7
10 000	1.4	1.9	2.4	2.1
100 000	1.25	1.5	1.7	1.6
10^6	1.1	1.1	1.2	1.2
10^7	1.0	1.0	1.0	1.0

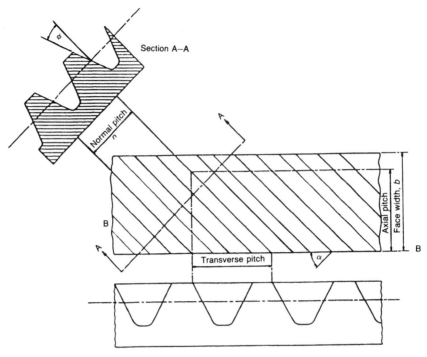

Fig. 10.20 Definition of helical gear tooth geometry

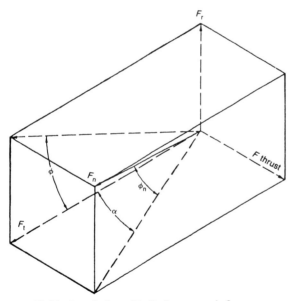

Fig. 10.21 Resolution of helical gear tooth forces

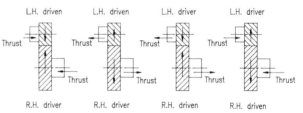

Fig. 10.22 Direction of axial thrust in helical gears

account the effects of the helix angle and so overall produce a lower contact stress, as follows:

Z_ϵ = contact ratio factor for helical gears

$$= \sqrt{\left[\left(\frac{4-\epsilon}{3}\right)(1-\epsilon) + \frac{\epsilon_\alpha}{\epsilon}\right]} \text{ for } \epsilon_\alpha < 1$$

or

$$= \sqrt{\frac{1}{\epsilon}} \text{ for } \epsilon_\alpha \geq 1$$

where ϵ_α = the overlap ratio which is the quotient of overlap angle and angular pitch or alternatively the quotient of face width and axial pitch = $b \tan \alpha/(\pi m)$ (see Fig. 10.20) and

$$Z_H = 2\sqrt{\frac{\cos \alpha}{\sin 2 \phi_{(t)}}}$$

$$\sigma_H = Z_H Z_E Z_\epsilon \sqrt{\left(\frac{F_t}{bd_1} \frac{u+1}{u} K_A K_M\right)}$$

but in this case, Z_H and Z_ϵ are modified to take into

for gears at standard centres where $\phi_{(t)}$, etc. is the transverse pressure angle, plane BB (Fig. 10.20).

The permissible contact stress, σ_{Hp}, is given by the same formula as for spur gears.

(ii) Bending stress

The formula for the root bending stress, σ_F, is the same as that for spur gears except that the geometry factor J is modified by a helix factor:

$$J = \frac{YC_\alpha}{K_b} \text{ or } = \frac{1}{Y_F Y_S Y_\alpha} \text{ (as in ref. 2)}$$

where C_α and Y_α are helix factors.

The helix angle factor for bending stress accounts for the fact that the conditions of tooth root stress in helical gears are more favourable as a result of the inclined line of contact compared to spur gears. Values of geometry factor for helical gears are given in Table 10.9, but these apply only to an overlap ratio ≥ 2. Overlap ratio is a measure of the degree of tooth engagement on a single helical gear. Good helical gear design assumes a value ≥ 2, but in those cases where face width is narrow or the helix angle is low, the gear can be best thought of as intermediate between spur and helical and a midway value of J chosen (Tables 10.7, 10.9).

Values of K_M, the load distribution factor, are also reduced to take into account the influence of tooth form, as follows: accurate mounting, low bearing clearances, precision gears, $K_M = 1.2$; less accurate gears, less rigid mounting, $K_M = 1.5$; poor alignment, less than full face contact existing, $K_M = \geq 2.0$.

The permissible bending stress σ_{Fp} for helical gears is given by the same formula as for spur gears (section 10.6.1(ii)).

10.6.3 Bevel gears

These are described as gears cut from conical blanks, i.e. based on pitch surfaces that are frustums of cones having positive driving and sliding in the same direction at the same time. There are four types, straight tooth, Zerol, spiral and hypoid, but for applicability to small machine design, we shall concentrate on the simpler straight bevel. The name is derived from the fact that the teeth are cut straight (Fig. 10.4). Bevel gear pitch cones and axes must intersect at the same point, although shaft intersect angles and gear pitch angles can vary. The terminology and physical characteristics of bevel gears are illustrated in Fig. 10.23.

The action of bevel gears is the same as spur gears, but a given pair of bevel gears will have a larger contact ratio and will run more smoothly than a pair of spur gears with the same number of teeth.

Bevel gears are inherently non-interchangeable and usually operate therefore as matched pairs, hence there is an advantage in addendum modification. The standard pressure angle is 20°,[5] the basic rack as Fig. 10.10. From Fig. 10.24

Shaft angle = sum of pitch angle of pinion and gear

$$\gamma = \theta_p + \theta_g$$

where θ = semi-included angle at apex (pitch angle) and

$$\tan \theta_g = \frac{\sin \gamma}{N_1/N_2 + \cos \gamma} = \frac{N_2}{N_1} \text{ if } \gamma = 90°$$

$$\tan \theta_p = \frac{\sin \gamma}{N_2/N_1 + \cos \gamma} = \frac{N_1}{N_2} \text{ if } \gamma = 90°$$

In applying the basic rack to bevel gears, the pinion is cut with a long addendum, the gear with a short addendum as follows:

Addendum of pinion $= m(1+k)$

Addendum of gear $= m(1-k)$

where

$$k = 0.4 \left[1 - \left(\frac{N_1 \sec \theta_p}{N_2 \sec \theta_g} \right) \right]$$

Table 10.9 Geometry factor J for standard helical gears, 10°, 20° and 30° helix angles

No. of teeth in gear for which J required	No. of teeth in the mating gear				
	20	30	50	75	150
20	0.46/0.47/0.43	0.47/0.48/0.43	0.48/0.49/0.45	0.49/0.5/0.46	0.5/0.51/0.47
30	0.51/0.51/0.47	0.52/0.53/0.48	0.54/0.54/0.49	0.55/0.55/0.50	0.56/0.56/0.51
60	0.57/0.56/0.5	0.58/0.58/0.51	0.6/0.5/0.52	0.61/0.6/0.53	0.62/0.61/0.54
150	0.6/0.6/0.53	0.62/0.61/0.54	0.64/0.63/0.55	0.65/0.64/0.56	0.66/0.65/0.57

[a] Values of J for the three helix angles given in the format 10°/20°/30°.
[b] Above values only valid if overlap ratio >2.

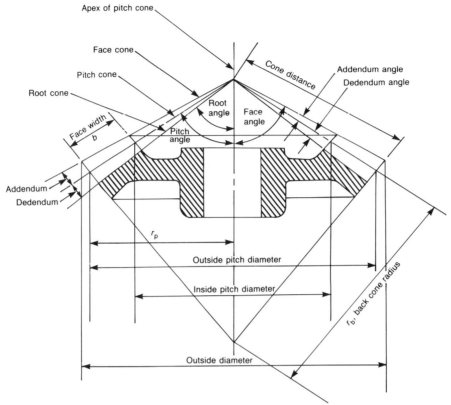

Fig. 10.23 Bevel gear terminology

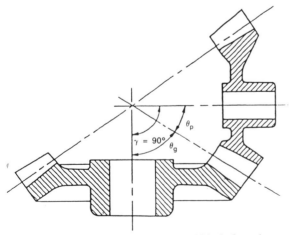

Fig. 10.24 Regular bevel gears with 90° shaft angle

(i) Contact stress

In a similar format to spur and helical gears,

$$\sigma_F = Z_H Z_E \sqrt{\frac{F_i}{d_v}\left(\frac{d_v + D_v}{D_v}\right) K_A K_M}$$

where

$$F_i = \frac{2M_p}{db}\frac{C}{C-b}$$

and C = cone distance (Fig. 10.23), M_p = torque on pinion, $d_v = d \sec \theta_p$, $D_v = d \sec \theta_w$ and

$$Z_E = \sqrt{\left(\frac{1.5}{\pi[(1-\nu_1^2)/E_1 + (1-\nu_2^2)/E_2]}\right)}$$

with K_M, K_A values as section 10.6.1(i) and $Z_H = \sqrt{(4/\sin 2\phi)}$ where $\gamma = 90°$, but in the general case Z_H values for helical gears may be used as in section 10.6.2(i).

Values of allowable stress may be calculated using the procedure in section 10.6.1(i).

(ii) Bending stress

The actual bending stress at the root is given by

$$\sigma_F = \frac{F_i K_M K_A K_S}{mJ}$$

where K_S = size correction factor for bevel gears, i.e.

Table 10.10 Geometry factor J for straight bevel gears having a 20° pressure angle

No. of teeth in gear of interest	No. of teeth in the mating gear					
	20	30	40	50	70	100
20	0.2	0.23	0.24	0.24	0.25	0.28
30	0.2	0.23	0.25	0.26	0.28	0.3
40	0.2	0.23	0.25	0.27	0.3	0.33
50	0.2	0.23	0.25	0.27	0.31	0.34
70	0.21	0.24	0.26	0.28	0.33	0.36
100	0.23	0.26	0.28	0.3	—	0.38

Module = 1 2 3 4
K_S = 0.5 0.53 0.58 0.63

K_M and K_A as section 10.6.1(i) and F_i as section 10.6.3(i); values for geometry factor J are shown in Table 10.10.

The geometry of bevel gears is more complicated than parallel axis gearing and is more dependent on the manufacturing tools used, consequently the geometry factor chart is not as uniform and smooth. Values of allowable bending stress may be calculated using the procedure outlined in section 10.6.1(ii).

10.6.4 Worm gearing

The worm is basically a single or multi-start screw, a true involute gear with one or multi-teeth, which meshes with a gear wheel whose teeth are inclined to the axis at the load angle of the screw (Figs 10.5, 10.25). The combination of a relatively large helix angle of the worm and a small diameter results in the appearance similarity to a screw thread, the same terminology applying to both. If the worm meshes with a helical gear, or spur gear on crossed axes, there is only kinematic point contact hence only light loads are possible. To increase gear capacity and give line contact and a greater degree of conformity with the gear wheel, the teeth are concave, as in Fig. 10.25. Referring to Fig. 10.25:

Pitch of wheel   = pitch of worm
Lead of worm, L = distance a point on the gear or the helix would advance parallel to the axis of the gear during one revolution of the worm
  = nP
  = $\pi n m$

where n = no. of thread starts, P = pitch of worm and m = axial module = axial pitch/π and

$\lambda + \alpha = 90°$ for a worm

where λ = lead angle, = the angle at the pitch diameter between a thread and the plane of rotation and, α = helix angle.

Because the worm has no operative pitch cylinder, what is referred to is an arbitrarily selected cylinder to which other dimensions are referenced such that

$$q = \frac{\text{worm pitch diameter}}{\text{module}} \equiv \text{an integer if possible}$$

i.e. the pitch diameter of a worm ≡ spur gear with q teeth from which

$$C = \frac{d + D}{2} = \frac{m(q + N_2)}{2}$$

The basic rack is to BS 436,[2] (Fig. 10.10), the standard worm thread surface being involute helicoids with ϕ_n = 20° at the pitch circle diameter.

Because there is sliding contact present, the capacity of a worm drive reduction unit is usually limited by the rate of heat dissipation. The efficiency of the drive, η, is given by

$$\frac{\tan \lambda \cos \phi_n - \mu \tan \lambda}{\cos \phi_n \tan \lambda + \mu}$$

where μ = coefficient of friction.
If M is the torque on the worm (Fig. 10.25) then

$$F_a = F_t \cot \lambda = \frac{C \cot \lambda}{d/2}$$

(i) Contact stress

The general load capacity of worm gears is given by the following formula:[4]

$$M_c = S_c^* Z_H D^{1.8} m$$

where $S_c^* = S_c X_c$ and M_c = permissible torque (Nm), Z_H = zone factor, X_c = speed factor, D = pitch circle diameter of the worm wheel (mm), S_c = surface stress factor, the allowable contact stress in the worm wheel teeth and m = axial module (mm).

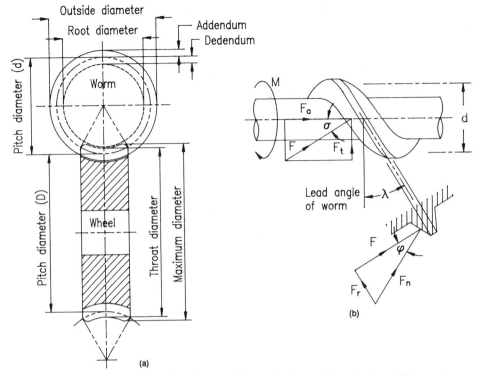

Fig. 10.25 Basic worm gearing elements: (a) terminology; (b) resolution of forces. From Dudley D, *Handbook of Practical Gear Design*, McGraw-Hill Inc. 1984. Reproduced with permission of McGraw-Hill Inc.

The above relationship has been developed more by empirical experience than by rational derivation. Because of the value of the exponent of D, the expression

$$F_t = 2XS_c ZD^{0.8} m$$

where F_t = transmitted load after taking worm set efficiency into account, cannot be converted into the corresponding value of Hertzian contact stress, thus it is more reasonable in the case of worm gears to calculate the tangential load capacity than to calculate stress and compare calculated to allowable stress. Efficiency may well not be known at the design stage, hence initial estimates will have to be made. Values of X can be obtained from Table 10.11 where

Sliding velocity (rubbing speed) = $\dfrac{\pi d n}{\cos \lambda}$

Values of S_c and Z are obtained from Tables 10.12 and 10.13 respectively. The available output power from the gear set is given by

$$P = \dfrac{M n Z_N}{9550}$$

Table 10.11 Contact stress speed factor X for worm gears

R.P.M.	Rubbing speed (m s^{-1})					
	0.1	0.5	1.0	3.0	5.0	10.0
1	0.7	0.58	0.53	0.45	0.41	0.35
4	0.6	0.52	0.47	0.39	0.37	0.31
10	0.55	0.47	0.42	0.36	0.33	0.28
40	0.48	0.4	0.37	0.32	0.28	0.24
100	0.4	0.33	0.3	0.26	0.24	0.20
400	0.3	0.22	0.24	0.18	0.17	0.14

where Z_N = life adjustment factor

$$= 3\sqrt{\left(\dfrac{27\,000}{1000 + L_c}\right)}$$

where L_c = total equivalent running time (hours).

(ii) Bending stress

The bending strength of worm threads is generally much higher than strength of worm wheel teeth, hence the

Table 10.12 Stress factors for worm gears

No.	Material	BS spec.	Bending stress factor S_b	Surface stress factor, S_c running with			
				1	2	3	4
1	Phosphor bronze chill cast	BS 1400	63	—	6.2	6.9	12.4
	Phosphor bronze sand cast	BS 1400	49	—	4.6	5.3	10.3
2	0.4%C steel normalized	080M40	138	10.7	—	—	—
3	0.55%C steel normalized	070M55	173	15.2	—	—	—
4	Carburized steel	—	276	48.3	—	—	15.2

Table 10.13 Values of worm gear zone factor Z for contact stress

No. of worm starts	Value of d/m ratio										
	6	6.5	7	7.5	8	8.5	9	9.5	10	11	12
1	1.045	1.048	1.052	1.065	1.084	1.107	1.128	1.137	1.143	1.160	1.202
2	0.991	1.028	1.055	1.099	1.144	1.183	1.214	1.223	1.231	1.250	1.280
3	0.822	0.890	0.989	1.109	1.209	1.260	1.305	1.333	1.350	1.365	1.393
4	0.826	0.883	0.981	0.098	1.204	1.301	1.380	1.428	1.460	1.490	1.515

former need not be considered. Usually the rating of all kinds of worm gear is based on surface durability; tooth strength is not normally much of a risk unless abnormally small threads or teeth are used. However, the strength of a worm wheel tooth is not amenable to exact calculation so is handled on an approximate basis. The number of worm wheel teeth in contact nominally alternates between two and three with a variable total length of contact line. The calculation is treated in a similar way to contact stress,[4] such that

Permissible torque $M_b = S_b X_b m l_r D \cos \lambda$

where

l_r = length of curved root of tooth in axial section (Fig. 10.26).

$= 2r_r \dfrac{(\theta°)}{57.3}$

and

$\sin \theta = \dfrac{b_w}{2r_r}$

where b_w = face width of worm wheel, r_r = root radius of worm wheel, θ = semi-subtended angle, S_b = bending stress factor (Table 10.12), X_b = speed factor (Fig. 10.27), from which the transmitted load on the wheel

$F_t = 2 S_b X_b l_r m \cos \lambda$

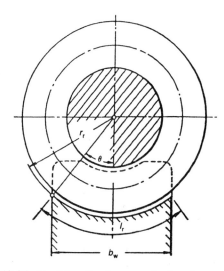

Fig. 10.26 Root length of worm wheel teeth

where F_t is after taking worm set efficiency into account.
The life adjustment factor in this case is

$$\sqrt[7]{\left(\dfrac{26\,200}{200+L_c}\right)}$$

10.7 Plastic gears

Plastic gears are being used by industry in increasing

Fig. 10.27 Table of bending stress speed factor, X_b, for worm gearing

RPM	Speed factor, X_b
1	0.65
5	0.60
10	0.56
20	0.52
50	0.47
100	0.42
200	0.37
500	0.32
1000	0.27
2000	0.23
5000	0.19
10 000	0.13

amounts as the economic and technical advantages of non-metallic gears become more and more appreciated. The advantages and disadvantages of plastic gearing are outlined in Table 10.2. Plastic gears are well suited to motion control where low power is transmitted, but in order to design for plastic gears, or to use them to replace metal gears effectively, the implications of material differences need to be understood. In particular there are two main areas of difference between metal and plastic gear behaviour, as given in the headings below.

(a) Elastic modulus The low elastic modulus of plastics allows for a greater tooth deflection under load and the probability of a larger contact ratio effectively increasing the load-bearing capacity of the gear. The susceptibility of plastic gears to stress concentrations is also reduced together with a reduction in dynamic loading during operation. Misalignment and tooth error loadings are less as the teeth are able to deform to compensate. Tooth overlap can be improved by the use of helical teeth, but at the expense of introducing axial thrust.

(b) Temperature effects These are very important with plastic gears. With increase of temperature, the decrease in modulus reduces contact stresses and increases load sharing, but the load-bearing capacity decreases. Low thermal conductivity, together with internal hysteresis effects (tooth bending) and frictional losses (tooth sliding) cause gear life to decrease with increase in pitch line velocity. In order to run plastic gears at high speed, some approaches have been as follows:

1. Mate a plastic gear with a metal gear, the metal gear acting as a heatsink to offset temperature effects;
2. Design plastic gears with a metal inset;
3. Provide an airflow directly on to the gear mesh;
4. Provide a lubricant as a coolant;
5. Select materials that are less thermally sensitive;
6. Limit the speed or duty cycle;
7. Design for as fine a gear pitch as possible since sliding velocities are lower for smaller teeth;
8. Use combinations of the above.

Two conventional methods are used to form plastic gears, injection moulding and gear cutting, the economics of the application usually dictating the former.

(a) Injection moulding Generally confined to less than 100 mm pitch diameter because of shrinkage, but advantageous as a process inasmuch as additional features can be incorporated into the tool. The advantages outlined in Table 10.2 tend to relate to injection-moulded gears. Not all plastics can provide all the advantages quoted nor are they all handicapped by the disadvantages. Moulded gears have generally harder teeth than cut plastic gears and the features of fine pitch teeth can be controlled to provide high contact ratios. A number of materials can be moulded that cannot ideally be cut, i.e. glass-reinforced materials.

(b) Conventional cutting Such gears produced commercially will rarely depart from the standard metal tooth forms. Cut plastic gears should be limited to nylons, acetals and laminated phenolics. Poor performance generally results from gears cut in reinforced plastics.

10.7.1 The tooth form

The basic rack for cut gears is as Fig. 10.10. However, with a moulded plastic gear (or powder metal gear for that matter), the designer is freed from standard tooth form restrictions and beneficial modifications can be introduced at no extra cost, in particular a full fillet radius and tooth tip relief. Such a modified rack is shown in Fig. 10.28. A gear with the basic form of the moulded system will function perfectly well with a mating gear having the standard cut form.

(a) Full fillet radius It is normal for the teeth of heavily loaded metal gears to be given a full fillet radius between two teeth at the root as this can increase fatigue strength by up to 20 per cent. This is specified for moulded gears to increase stiffness and reduce stress concentrations since plastics are notch sensitive. It preserves material and strength in the root area where most often gear failure occurs.

(b) Tip modification If a tooth deflects during rotation and the trailing edge is out of register with the oncoming

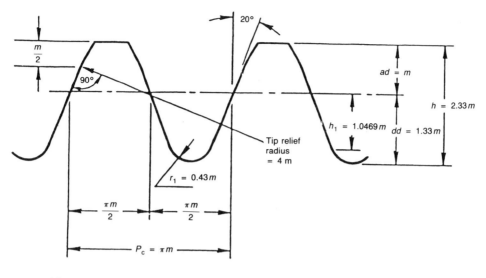

Fig. 10.28 Basic rack; plastic moulded gear standard

m = module P_c = circular pitch 20° = pressure angle
ad = addendum dd = dedendum h = whole depth
h_1 = depth of straight portion of dedendum to point of tangency with root radius

tooth of the mating gear, this can result in interference which can cause noise, tooth wear and abnormal motion. This can be remedied by means of 'tip relief' or trimming the tips of the teeth from half-way up the addendum (Fig. 10.28). Since plastics flex to a greater extent than metals, all plastic teeth should have this modification. This should not affect the strength of the gear.

10.7.2 Temperature

The relevance of temperature to gear performance has already been mentioned, and in order to design for plastic gearing, the operating gear temperature needs to be known, both surface temperature relevant to contact stresses and bulk temperatures relevant to bending stress and operating centre distance. However, calculation of temperature rise is not an exact science and depends on a number of variables. For open gears,[3] temperature is given by the following:

$$T_1 = T_0 + \frac{136 P_T (1+u)}{5+N_2} \left[\frac{1.71 \times 10^4 K_a}{b N_1 (v m_n)^{K_m}} \right] + 5$$

where T_1 = pinion temperature (°C), T_0 = ambient temperature (°C), u = gear ratio, b = face width (mm), m_n = normal module (mm), v = pitch line velocity (m s^{-1}), P_T = transmitted power (kW), K_a = experi-

mental factor to account for material temperature (Table 10.14), which determines bulk or surface temperature determination, K_m = gear material constant: nylon = 0.75; acetal = 0.40 and μ = coefficient of friction (Table 10.14).

Table 10.14 *Summary of plastic gear temperature factors*

Gear combination	Bulk temperature constant	Surface temperature constant
Nylon/steel	1	10
Nylon/nylon	2.4	15
Acetal/acetal	2.5	70

Factor K_a for dry running or initial grease lubrication

Gear combination	No lubrication	Initial greasing
Nylon/steel	0.25	0.11
Nylon/nylon	0.18	0.11
PBT/steel	0.25	—
PBT/PBT	0.22	—
Acetal/steel	0.21	—
Acetal/acetal	0.18	—
Nylon/acetal	0.10	—

Coefficient of sliding friction, μ

Note: For other polymer frictional values, refer to Chapter 5.

10.7.3 Design allowances

Failure analyses of plastic gearing indicate that the main problems are backlash, contact conditions, thermal expansion or moisture absorption, or a combination of these. Hence due consideration should be given to these in the initial design.

Certain plastics are hygroscopic, e.g. nylon, which expand slightly with moisture uptake. In the case of applications where the gears will not be exposed immediately to conditions of high humidity, the expansion of most plastics is small and gradual and can be discounted, because it is offset by the equally small and gradual shrinkage that occurs as mould stresses are relieved. Temperature has been considered in section 10.7.2 and gear tolerance in section 10.5.1. Thus, after moulding and stabilization, the total amount by which the minimum operating centre distance must exceed the calculated close mesh centre distance is given by

$$\Delta C = \frac{TCT_1 + TCT_2}{2} + C\left[T_1 - T_0\left(\frac{\alpha_1 N_1}{N_1 + N_2}\right.\right.$$
$$\left.\left. + \frac{\alpha_2 N_2}{N_1 + N_2} - \alpha_H\right) + \left(\frac{M_1 N_1}{N_1 + N_2}\right)\right.$$
$$\left. + \left(\frac{M_2 N_2}{N_1 + N_2}\right)\right] + \frac{TIR_1 + TIR_2}{2}$$

where ΔC = required increase in centre distance, T_1 = gear operating temperature (°C), α_1 = coefficient of linear expansion of pinion (m m^{-1} °C^{-1}), α_2 = coefficient of linear expansion of gear, α_H = coefficient of linear expansion of housing, M_1 = expansion due to moisture pick-up of pinion (m m^{-1}), M_2 = expansion due to moisture pick-up of gear, TIR = maximum allowance run-out of bearings, and values for M are as follows (m m^{-1}): acetal 0.005; nylon 0.0025; 30 per cent glass-filled nylon 0.0015; polycarbonate 0.005.

It is worth noting that with non-hygroscopic materials, the effect of differential thermal expansion in the above formula can be countered by also making the housing from the same material. The value of C calculated above will also need to be increased to provide backlash (section 10.5.4); the chief problem is knowing what backlash value to use in a specific design. Values for nylon are given in Fig. 10.29. Alternatively start with the tight mesh centre distance and adjust empirically.

10.7.4 Lubrication

All gearing works best when lubricated and the optimum plastic gear performance does include an occasional oiling

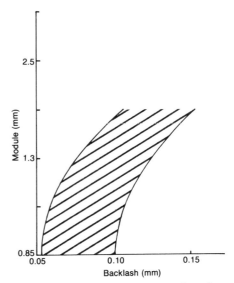

Fig. 10.29 Suggested backlash range values for nylon gears

and, at the very least, greasing before initial operation. Regular lubrication can improve gear life tenfold over no lubrication. It has been found that mineral-based greases provide good lubrication for nylon, acetal and most types of plastic.[3] Because of its low surface energy compared to plastics (section 4.2.1), silicone oil will act as a lubricant, and along with PTFE, can be dispersed within the moulded component, being then self-lubricating (sections 5.2.5(ii) and 5.3.4). Aggressive lubricants and those containing extreme pressure additives should be avoided. Amorphous thermoplastics may prove more of a problem to lubricate (section 4.2.3).

10.7.5 Plastic gear rating

The procedures used for metal gears have already been discussed in section 10.6. While the fundamental formulae are still based on Lewis bending and Hertzian contact equations, the modification factors used for metal gears do not necessarily have the same influence on stresses in plastic gears and additional factors such as temperature and moisture need to be considered. The stressing formulae used for plastic gears[3] are derived by modifying and simplifying those for metal gears[2] but, as described in section 10.7.2, the determination of the gear temperature is essential for quantifying the mechanical parameter used in bending and contact stress equations. Principally the following analysis refers to the basis rack. For modified profiles refer to ref. 3.

(i) Contact stress

In some design practices, contact stresses are ignored on the basis that bending stresses dictate the basis of the design. However, from ref. 3, contact stresses can be calculated in the same format as with metal gears except that variable Young's modulus, Poisson's ratio and permitted contact stress are introduced to allow for temperature effects, the tooth flank temperature relevant to surface durability. For contact stress

$$\sigma_H = Z_H Z_E Z_\epsilon \sqrt{\left(\frac{F_t}{bd} \cdot \frac{u+1}{u} K_A\right)}$$

where Z_H, Z_E, Z_ϵ and K_A are as in section 10.6.1.

The effect of temperature on both Young's modulus and Poisson's ratio is given in Fig. 10.30. The permissible stress is given by

$$\sigma_{Hp} = \frac{\sigma_{Hlim}}{S_{Hmin}}$$

where values for S_{Hmin} are as in section 10.6.1. The contact stress endurance limit σ_{Hlim} for nylon and acetal is given in Fig. 10.31. As with metals, the allowable permissible contact stress σ_{Hp} is required to exceed calculated contact stress σ_H.

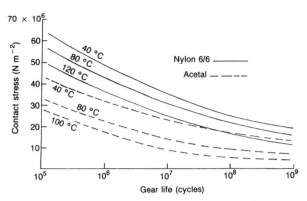

Fig. 10.31 Stress endurance limit versus gear life for nylon and acetal polymers as a function of temperature

(ii) Bending stress

A modified Lewis equation is used for plastic gears to calculate the root bending stress:

$$\sigma_F = \frac{F_t}{bm} Y_F Y_\epsilon K_A$$

where Y_F = tooth form factor, a geometry factor that takes into account the nominal bending stress for application of a load at the highest point of a single tooth pair contact with values as follows:

No. of gear teeth = 18 20 25 40 60 80 100
Y_F = 2.90 2.80 2.63 2.40 2.28 2.23 2.19

The above values relate to standard module unmodified gear teeth.

Y_ϵ = contact ratio for bending stress = $0.2 + 0.8/\epsilon$ and K_A values are in section 10.6.1. The permissible root bending stress σ_{FP} is given by

$$\sigma_{FP} = \frac{\sigma_{Flim}}{S_{Fmin}}$$

where values of S_{Fmin} are as in section 10.6.1. The basic fatigue endurance limit for bending stress σ_{Flim}, for nylon and acetal as a function of bulk gear temperature is given in Fig. 10.32, based on the life requirement of the particular application.

10.8 Gear transmissions

Gears provide a very durable form of mechanical drive and any combination of two or more meshing gears that are used to change the output characteristics (e.g. speed,

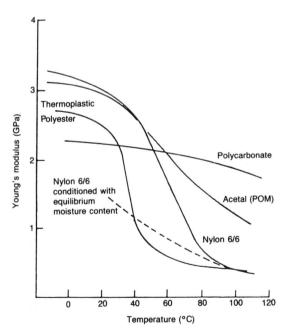

Fig. 10.30 Effect of temperature on Young's modulus and Poisson's ratio of plastics

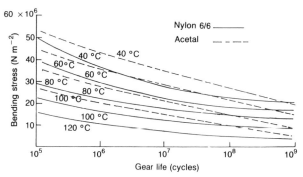

Fig. 10.32 Bending stress endurance limit versus gear life for nylon and acetal polymers as a function of temperature

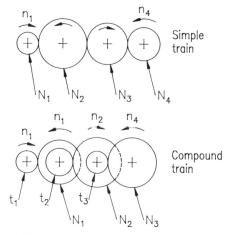

Fig. 10.33 Gear trains

direction, torque) relative to the characteristics of the prime mover, is referred to as a transmission. This section will consider the important aspects of the more common types of gear combination.

10.8.1 Gear trains

For two gears to maintain constant angular velocity ratio, the gears must mesh to produce conjugate action (section 10.4). Velocity ratio is defined as the ratio of the angular speed of the driven gear to the angular speed of the driver gear — alternatively the ratio of output speed to input speed, i.e. <1 when the pinion is the driver and >1 when the gear is driving. Thus

$$\frac{w_2}{w_1} = \frac{n_2}{n_1} = \frac{N_1}{N_2} = \frac{d_1}{d_2}$$

where w = angular velocity (rad sec^{-1}), n = rotational speed (min^{-1}), N = no. of teeth, d = pitch circle diameter, w_2/w_1 = velocity ratio and N_1/N_2 = gear ratio.

Direct gear trains, in which all the gears rotate on stationary axes, can be arranged in one of two basic ways. In the simple train (Fig. 10.33), the velocity ratio depends on the number of teeth in the first and last gears of the train, the remaining gears are idler gears and the train is an idler gear train. With an odd number of idlers, the first and last shafts rotate in the same direction and vice versa with an even number of idlers (including zero). From Fig. 10.33

$$\text{Gear ratio of a simple train} = \frac{n_1}{n_4} = \frac{N_4}{N_1}$$

The disadvantage of a simple train is that for large-velocity ratios, the overall size of the gear train is large, so in order to use large-velocity ratios economically, a compound gear train is used (Fig. 10.33). In this case, the train is effectively two simple trains in series and the intermediate gears must be considered when the velocity ratio is calculated, i.e.

$$\text{Velocity ratio} = \frac{\text{product of number of teeth on driven gears}}{\text{product of number of teeth on driving gears}}$$

i.e.

$$\frac{n_1}{n_4} = \frac{N_1 N_2 N_3}{t_1 t_2 t_3}$$

10.8.2 Epicyclic gears

An epicyclic gear unit consists of three coaxial torque transmitters, the sun pinion, the annulus gear and the planet carrier (Fig. 10.34). The sun meshes with a number of planet pinions mounted on a planet carrier and enveloped by an internal annulus gear. The gear ratio is a function of the ratio of teeth on the annulus to teeth on the sun pinion, depending on the choice of reaction member. The term 'epicyclic' serves to describe a motion of a gear that rolls around another (its centre describing a circle), an epicycloid curve being generated by a point on the surface of a planet gear as it rotates around the sun.

The terms 'epicyclic' and 'planetary' gears are often confused. The term 'planetary' describes two coaxial gears connected by a number of fixed position intermediate gears, i.e. an epicyclic gear with fixed carrier is a planetary construction. The epicyclic chain is commonly arranged to form a reduction gear, output speeds given in Table 10.15 for the various choices of

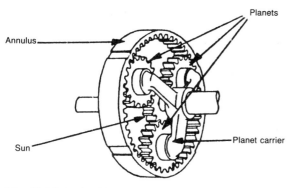

Fig. 10.34 Epicyclic gear unit. Courtesy David Brown Corporation plc

$$\frac{n_{a/s}}{n_{c/s}} = 1 - \frac{n_{a/c}}{n_{s/c}} = -\frac{N_s}{N_a} \quad \text{(Table 10.15)}$$

Epicyclic and plantary gear trains are usually designed so that two or more pairs of gears are in contact at all times, the more planet gears the greater the torque capacity of the system. If equally spaced planet gears are used, the following relationship must be true:

$$\frac{\text{No. of sun gear teeth} + \text{no. of annulus gear teeth}}{\text{No. of planet gears}}$$

$$= \text{an integer}$$

If the above cannot be satisfied, it is possible to use unequally spaced planet gears. With the annulus stationary, the maximum allowable velocity ratio can be calculated from the following:

static reaction member. For gears having comparable pinion diameters, the epicyclic gear is smaller in size than the parallel shaft gear, and the epicyclic gearing principle therefore offers a volumetric space saving together with the lower pitch line velocities and gear inertias that result from the reduced sizes of the gear elements.

$$\frac{n_s}{n_c} \max = \frac{2N_s - 4}{N_s[1 - (\sin 180/x_p)]}$$

where x_p = no. of planet gears.

10.8.3 Harmonic drive

By using equations for relative rotational speeds

$$n_{a/s} = n_{a/c} + n_{c/s}$$

and

The harmonic drive is essentially a unique type of constant ratio gear drive offering a high reduction ratio, 50:1 or

Table 10.15 *Summary of epicyclic gear output options*

Driver (input)	Follower (output)	'Earthed' member	Speed of follower	Direction of rotation	Speed ratio	Type
Sun	Carrier	Annulus	$n_s \left[\dfrac{N_s}{N_a + N_s}\right]$	Unchanged	Reduced	Planetary gear
Annulus	Carrier	Sun	$n_a \left[\dfrac{N_a}{N_s + N_a}\right]$	Unchanged	Reduced	Solar gear
Carrier	Sun	Annulus	$n_c \left[\dfrac{N_s + N_a}{N_s}\right]$	Unchanged	Increased	Planetary gear
Carrier	Annulus	Sun	$n_c \left[\dfrac{N_s + N_a}{N_a}\right]$	Unchanged	Increases	Solar gear
Sun	Annulus	Carrier	$n_s \dfrac{N_s}{N_a}$	Reversed	Reduced	Star gear[a]
Annulus	Sun	Carrier	$n_a \dfrac{N_a}{N_s}$	Reversed	Increased	Star gear[a]
Sun annulus	Carrier	—	$\dfrac{n_a N_a + n_a N_s}{N_a + N_s}$ [b]	—	—	—

Notes: n_s = rotational speed of sun; n_c = rotational speed of carrier; n_a = rotational speed of annulus; N_s = no. of teeth on sun gear; N_p = no. of teeth on planet gear; N_a = no. of teeth on annulus gear.

[a] not strictly epicyclic, simple planetary gear train.
[b] n_s is negative if sun rotates in opposite direction and vice versa since N is proportional to gear diameter (constant pitch), then $N_p = (N_a + N_s)/2$.

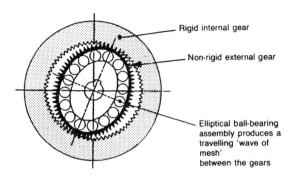

Fig. 10.35 Harmonic drive

more. It is more efficient than worm gearing and is much more compact than gear trains used to achieve the same ratio. The three basic components (Fig. 10.35) are a rigid outer circular gear, a flexible gear and an elliptical wave generator. The flexible gear normally has two teeth less than the rigid gear. In operation the wave generator rotates and elastically deforms the flexible gear so that teeth are fully engaged on the major axis and fully disengaged on the minor axis of the ellipse.

Three modes of working are possible:

1. If the outer gear is fixed and input rotation is via the wave generator and output is taken from the flexible gear, then the gear ratio is

$$\frac{N_F}{N_R - N_F}$$

where N_F = no. of teeth on flexible gear and N_R = no. of teeth on rigid gear. The output shaft rotates in the opposite direction to the input shaft.

2. If the flexible gear is fixed and input rotation is via the wave generator and output taken from the outer gear then the gear ratio is

$$\frac{N_R}{N_R - N_F}$$

The output shaft rotates in the opposite direction to the input shaft.

3. In the third mode of operation, the wave generator is fixed and the input applied to the rigid gear. The output is taken from the flexible gear or vice versa. In either case, the drive provides a near unity ratio.

With the flexible gear fixed and the rigid gear the output, input and output shaft rotation will be in the same direction.

10.9 Gear selection process

The problem facing the design engineer is to determine a gear application design that will be able to carry the power required for the lifetime required. The gears therefore must be of sufficient size, of sufficient hardness and of adequate quality to do the job required. The design procedure is to keep bending stresses and surface compressive stresses within allowable limits and frequently some refinement to the detail is necessary to meet all the operating limits. The size and material of a gear are controlled by the allowed stresses in both the driver and driven elements.

All formulae involved in rating the application are ultimately based on a comparison between a calculated and an allowable stress.[11] When all other quantities which enter into both calculated and allowable stresses such as speed, power and torque are included, the terms can be transposed in a variety of ways. The process of stress comparison may be pursued at various levels depending on the nature and amount of information available and the modifying factors included. The stresses calculated in gear design are not necessarily true stresses, e.g. aspects like stress concentration and residual stresses make it difficult to obtain true values. However, gear stress formulae are a necessary and valuable design tool, a yardstick on which to base the initial design; field experience is normally required to give the final answer on performance.

There are probably many ways to approach the design of a gear pair application to give satisfactory results. The process does become one of trial and error, though since size, tooth form and gear dimensions must be stated before actual loads and stresses can be determined, but choice of assumed values will improve with experience. A summary of the criteria affecting gear design and gear application is given in Table 10.16. The following procedure usually works well.

(i) Input the operating conditions

Specify the application by defining the following parameters:

1. The transmitted power requirement/the required torque;
2. Input speed;
3. Gear ratio/velocity ratio;
4. The operating environment/lubrication;
5. The life requirement;
6. Allowable speed variation;
7. Torque and speed duty cycle/variations.

Any kind of rating formula requires a statement of the

Table 10.16 *Gear selection parameters*

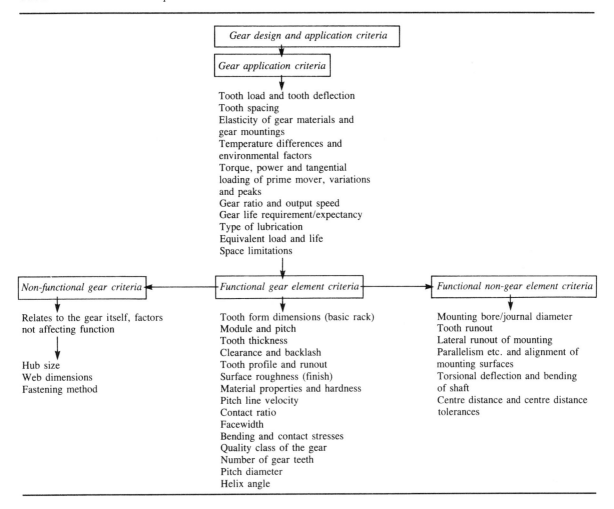

loading criteria. Inasmuch as a case of a constant continuous load at constant speed is rare, the general situation is one of torque/speed/duration that needs to be reduced to an equivalent condition of torque and speed acting for an equivalent life.[10] This may be defined by a duty cycle or by equivalent load and life,[10] the analysis of complex conditions outside the scope of this chapter.

(ii) Define the gear dimensions

Once the input/output requirements are known, initial sizing is then possible. It is assumed that the standard ISO basic rack and 20° pressure angle[2] apply in all cases. The following mechanical aspects of the gears can be stated:

1. Type of gear;
2. Module;
3. Face width;
4. Number of teeth/gear diameters;
5. Pitch circle diameter;
6. Class of gear;
7. Helix angle, if applicable.

In general, the more teeth a pinion has the quieter it will run and the better its wear resistance. It is also sometimes desirable to obtain a hunting ratio between gear and pinion teeth (i.e. no common factor between number of gear teeth). This tends to equalize wear and improve angular positional accuracy, but the pair may take longer time to run in.

Table 10.17 *Summary of unit load factors in strength and surface durability*

Application	Gear material	K factor N mm^{-2}	Unit load (N mm^{-2})
Small commercial; pitch line speed <5 m s^{-2}	Plastics	0.25–0.35	—
Small mechanism; pitch line speed <2.5 m s^{-2}	Zinc alloy or brass	0.10	—
General-purpose industrial drive	220 kg mm^{-2}	1.4	38
	320 kg mm^{-2}	2.1	48
	680 kg mm^{-2}	5.5	68

Notes: Valid for 10^7 cycles.
Values apply to uniform torque only; reduce values by 40% when moderate shock loads present.

(iii) Define the gear set
This particular set of parameters is defined as follows:

1. The tangential loading;
2. Gear materials and hardness/unit loading and K factor;
3. Centre distance/centre distance tolerance;
4. Operating pressure angle;
5. Pitch line velocities;
6. Tooth profile and run-out/TCT;
7. Backlash.

In order to make a preliminary choice of gear material, the index criteria to tooth loading may be used. These indicate the intensity of loading the teeth is trying to carry and what the design is achieving from the mesh. These indexes are based on unit load for strength and the K factor (section 10.6.1(i)) for surface durability and relate to the geometry and quality of the application.

Unit load is defined by F_t/bm_n with a $C/C-b$ factor introduced with bevel gears (section 10.6.3(i)). A summary of load indices is given in Table 10.17. Having chosen the basic material type, the centre distance/centre distance tolerance/TCT/backlash scenario may be defined. In the case of plastics, the temperature rise will first be required (section 10.7.2).

(iv) Test the application
From the information derived from (i)–(iii) above:

1. Determine actual bending and contact stresses;
2. Determine allowable contact and bending stresses;
3. Check both gears.

By use of the appropriate modification factors, i.e. application factor, safety factor, load distribution factor, calculate the tooth stresses, determine whether strength or surface durability is predominant and compare to the allowable values for the materials chosen and the life required. Where the results indicate an under- or over-design situation, the process will be iterative, adjusting gear dimensions, and/or possibly materials, to achieve the required gear set performance output.

References

1. Dudley, D.W. *Handbook of practical gear design*. McGraw-Hill 1984.
2. BS 436 : 1970 *Spur and helical gears*.
3. BS 6168 : 1987 *Non-metallic spur gears*.
4. BS 721 : 1983 *Worm gearing*.
5. BS 545 : 1982 *Bevel gears*.
6. Janninck, W.L. *Basic spur gear design. Gear Technology* Nov. 1988: 36.
7. Fowle, T.I. 'Practical gear tribology' *Tribology*, Series 8, *Industrial Tribology*, Elsevier 1983.
8. AGMA 218 : 1982 *Rating the pitting resistance and bending strength of spur and helical involute gear teeth*.
9. Young, I.T. 'Gear rules and standards'. *Tribology Int.*, 10(5): 257 (1977)).
10. Drago, R.J. *Fundamentals of gear design*. Butterworths 1988.
11. Merritt, H.E. *Gear engineering*. Pitman 1971.

CHAPTER 11

CLUTCHES, SOLENOIDS AND BRAKES

11.1 Introduction

This chapter deals with electromagnetic actuators, principally clutches, brakes and solenoids. There are numerous types of mechanical clutch and brakes described in the literature,[1] but these do not usually find application in small machines. Exceptions are the ball-nut and the one-way clutch which are discussed in section 7.7, and the spring-wrap clutch/brake included in this chapter.

All electromagnetic actuators function via the energization of a wound coil, and since the electromagnetic principles are the same, some discussion is devoted to the origin of magnetic pull at the start of this chapter. Traditionally the performance and specification of brakes and clutches have been presented differently from solenoids, mainly because of the transfer of kinetic energy in the former but only of force in the latter. As with all electromechanical systems, performance limits are largely established by thermal aspects; in the case of electromagnetic clutches and brakes, armature movement is small but frictional heating significant. Solenoids, on the other hand, require high power for long stroke travel and can overheat electrically if incorrectly applied.

11.1.1 Magnetic pull

In a magnetic circuit, a source of m.m.f. in the form of the coil causes a magnetic flux ϕ to flow through a circuit with reluctance R_m (i.e. magnetic resistance) as follows:

m.m.f. (in ampere turns) $= IN = R_m \phi$

where ϕ = flux in magnetic circuit (Wb), R_m = reluctance (ampere turns per weber), I = current flowing through windings (i.e. amperes through each coil turn) and N = no. of coil turns.

See also section 8.6 for application to electric motors. If the m.m.f. generated by the coil (referred to electrically as a solenoid) is uniformly distributed over the full length l_m of the magnetic flux path of uniform cross-sectional area (i.e. uniform magnetic resistance) then the m.m.f. per metre is termed the magnetic field strength or magnetizing force H, i.e.

$$H = \frac{\text{m.m.f.}}{l_m} = \frac{IN}{l_m}$$

or m.m.f. = magnetizing force $\times$ length of magnetic circuit.

Now the magnetizing force H (in ampere turns per metre) and the flux density B (in Wb m^{-2}) are linked:

$$B = \mu_0 \mu_r H$$

where μ_0 = constant, = permittivity of free space = $4\pi \times 10^7$, μ_r = relative permeability of the material within the magnetic flux path ($\mu_r = 1$ in air by definition, 10^2–10^5 for ferromagnetic materials), μ_a = absolute permeability = $\mu_0 \mu_r$ and

$$R_m = \frac{1}{\mu_0 \mu_r a}$$

where a = cross-sectional area of the magnetic circuit (m^2).

The B–H or magnetizing curve for certain ferromagnetic materials is given in Fig. 11.1. Initially an increase in H produces an increase in B, but when the curve passes over the knee into the saturation region, flux density increases only marginally. Operation above the saturation point reduces the efficiency of the design so equipment should be designed to operate below the knee.

Magnetic energy can also be expended in a non-magnetic material, such as air. In the case of an air gap

$$B = \mu_0 H$$

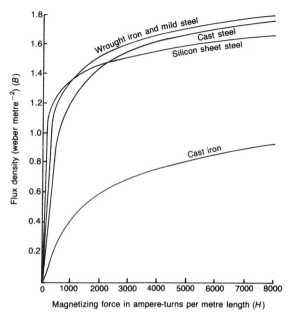

Fig. 11.1 Typical $B-H$ or magnetizing curves for certain ferromagnetic materials

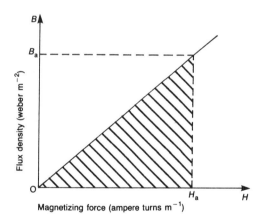

Fig. 11.2 The $B-H$ curve of an air gap

i.e. a linear relationship (Fig. 11.2); the energy expended per cycle is represented by the area below the $B-H$ curve for the air gap. Thus

Energy per unit volume of air gap = $\tfrac{1}{2} B_a H_a$

where H_a = field strength across air gap and B_a = flux density across the mating surfaces.

The two opposing surfaces forming the air gap will be attracted to each other, this force being the basis of electromagnetic actuators, clutches and solenoids. An iron path is much more efficient than air, but air is needed to permit movement. In each case a force $F(N)$ will act either to attract a moving armature towards a fixed coil yoke or alternatively hold an armature firmly against it (e.g. Fig. 11.3). Now

$$F = \frac{B_a^2 \, a}{2\mu_0}$$

$$= \frac{\mu_0 \, a I^2 N^2}{2 l_a^2}$$

where F = magnitude of force, a = surface area traversed by the magnetic flux (m^2) and l_a = air gap. In other words, the force of the solenoid is inversely proportional to the distance between the poles squared. A major objective in the design of a solenoid is to provide an iron path capable of transmitting maximum magnetic flux density with a minimum energy input. Another is to obtain the best relationship between the variable ampere turns and the working flux density in the air gap. In an electromagnetic clutch or brake, torque is a function of both the magnetic pull and the friction face material.

11.2 Clutches and brakes

Clutches and brakes are often similar in design, performing similar but opposite functions within a mechanical system. The function of a brake is to resist motion (converting kinetic energy to frictional heat), while that of the clutch is to transmit motion in a controlled manner. A clutch therefore becomes a brake if it can decelerate and/or stop the driven component. Because of the similarity of application, we shall consider clutches and brakes together with any differences indicated where appropriate. Applications may be classified into one of the following:

1. *Occasional stop–start.* Typically a cycle rate <5 per minute. The most common type of application and one that provides least complications where disconnection of the prime mover is at regular intervals.
2. *Cyclic stop–start.* This situation always requires careful evaluation since torque, referred inertia, peak thermal input, response times and facing life are all relevant. Dynamic torque is important where controlled stop–start times are required. Heat dissipation is clearly important in periodic braking.
3. *Non-slip.* Where there is a positive physical connection between the drive and driven member, engagement normally made when both members are stationary, or rotating at zero differential speed. In

CLUTCHES, SOLENOIDS AND BRAKES

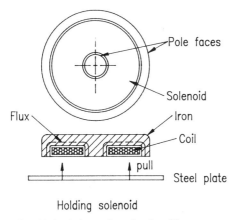
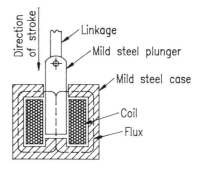

Fig. 11.3 Schematics of solenoid types

the case of brakes, a positive lock would prevent further movement once stopped.

4. *Continuous slip*. Typically continuous tension device type applications where continuous thermal dissipation occurs. Ability to perform satisfactorily and reliably within the required temperature envelope is important.

There are three generic types of clutch that find application in machine design as follows:

1. *Magnetic drives*. These may be used as light-duty clutches, couplings or brakes. They work on the principle of coupling via magnetic forces generated in an air gap. Most are friction free and some may be operated through a partition.
2. *Disc clutches and brakes*. Typically these are two rotating discs with friction material applied to one face. The method of actuation can be varied, mainly electromechanical actuation where torque is transmitted across an air gap via the creation of an electromagnetic field. Units are also available with multiple friction discs.
3. *Wrap spring drives*. Here one end of a coil spring is anchored to the driving member, the free end riding closely on the driven member and with differential rotation present, the coil spring tightens up to lock the driving and driven members together. Normally electromagnetically actuated, primarily used as one-way clutches but may be used in clutch/brake combination.

Other types of clutch and brake exist that find application in small machine design, e.g. the roller clutch, detent clutch, viscous clutch, disc brakes, band brakes. The roller clutch is discussed in section 7.7.2 and the use of drive motors for dynamic braking is discussed in sections 8.4.2 and 8.7.2.

11.3 Clutch and brake application factors

11.3.1 Clutch torque

A simple type of disk clutch/brake is shown in Fig. 11.4, and in its simplest form consists of a steel disk with friction material attached in an annular, or segmented angular format. Torque is transmitted through axial loading, the torque capacity being given by

$$T = 2\pi \int_{r_i}^{r_0} \mu p' r \, dr$$

where μ = coefficient of friction, p' = clamping load on an elemental area of friction material, and assuming μ to be constant

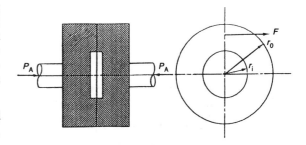

Fig. 11.4 Simple disc clutch

$$T = \tfrac{2}{3}\mu P_A \frac{(r_0^3 - r_i^3)}{(r_0^2 - r_i^2)}$$

where P_A = axial load and $F = \mu P_A$ where F = frictional force.

Assuming a uniform wear rate and uniform pressure over the contacting surfaces and applying to the point of impending slip, then

$$T = \frac{\mu P_A}{2}(r_i + r_0)$$

and for multiple disk clutches

$$T = \frac{\mu P_A n}{2}(r_i + r_0)$$

where n = number of pairs of contacting surfaces. Good design practice is to make $r_0/r_i \approx 1.5$.

The torque performance of a typical electromagnetic clutch is shown in Fig. 11.5, showing the torque transmission capacity as a function of slip at full and 50 per cent of applied current.

11.3.2 Torque requirements

Dynamic torque is important in cyclic applications. The average amount of torque required to accelerate a part from rest to a rotational speed of n r.p.m. in t seconds, or decelerate a part from n r.p.m. to rest in t seconds is given by

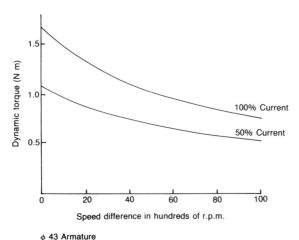

ϕ 43 Armature
110Ω; 5 W coil
Torque at 0 speed differential = static torque

Fig. 11.5 Example of electromagnetic friction clutch torque–speed characteristics

$$T(\text{av}) = \frac{2\pi mk^2 n}{60 t}$$

where mk^2 is the referred inertia of the driven parts back to the clutch/brake speed (see section 8.2); or in general terms

$$T = \frac{mk^2 \Delta w}{t}$$

where w = speed difference in radians per second.

If the start/stop times are short, the above torque requirement may exceed the available transmitted torque, e.g. from the main drive motor. Most electric motors have a pull-out torque approximately twice their running torque (see Ch. 8), which generally allows for quick start requirements. Available drive torque can be defined as

$$T = \frac{PS}{n}$$

where S = service factor (overload) = motor pull-out torque/theoretical rated torque and P = rated motor power.

In practice the available drive torque should reflect frictional losses in the system plus any gear train ratios prior to driving the clutch. If frictional losses are disregarded, then the power delivered by the clutch equals the power of the prime mover. The dynamic torque capability of the clutch should reflect this.

To accelerate any driven part of a machine from rest to a given velocity, or to decelerate from a given velocity to rest, energy must be added or removed. Neglecting windage and friction losses, then this energy is the kinetic energy possessed by the driven components moving at a fixed velocity, i.e. (from section 8.2):

$$W = \tfrac{1}{2} mk^2 w^2$$

where $w = 2\pi n/60$, then for a clutch/brake situation, energy dissipated during the take-up on-time is given by

$$W = \frac{T_R J w^2}{T_R \pm T_L 2}$$

where T_R = dynamic torque (Fig. 11.6) and T_L = constant load torque, static torque.

A load torque T_L restricting acceleration or braking is designated negative while a load torque assisting braking is designated positive, and when repeated at a cyclic rate of f Hz, energy supplied,

$$W = \tfrac{1}{2} J w^2 \frac{f T_R}{T_R \pm T_L}$$

= heat dissipated at the cyclic rate

CLUTCHES, SOLENOIDS AND BRAKES

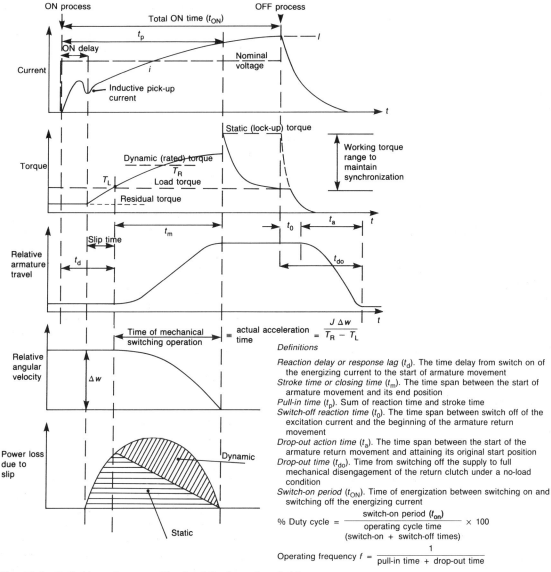

Fig. 11.6 Definition of monostable clutch/brake coil switching

This then becomes important in determining whether a given size of clutch or brake is capable of dissipating the heat (energy) required to cycle at a given load. For a given load, the heat dissipation capability of the clutch/brake is given as follows:

$$W \text{ per second (av)} = \frac{t_1}{t_1+t_2} W \text{ per second at r.p.m 1}$$
$$+ \frac{t_2}{t_1+t_2} W \text{ per second at r.p.m. 2}$$

where t_1 = time (s) at r.p.m. 1, t_2 = time (s) at r.p.m. 2, r.p.m. 1 = starting speed and r.p.m. 2 = maximum speed.

Values of W per second at speeds 1 and 2 are obtained from manufacturers' literature. Figure 11.7 illustrates the heat input, speed and temperature relationship for a typical electromagnetic faceplate clutch.

If dissipation capacity is greater than heat to be dissipated, then the particular cyclic torque application is acceptable. For a constantly slipping brake/clutch

ELECTROMECHANICAL PRODUCT DESIGN

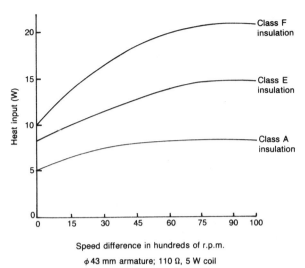

Fig. 11.7 Example of permissible thermal input to electromagnetic friction clutch for different coil insulation classes

$$W = 2\pi Tn$$

where n in this case is a speed differential.

Having calculated the energy dissipation by engagement (W), an alternative approach adopted by some manufacturers is to provide curves of W_{max} vs cyclic rate for a given clutch or brake (e.g. Fig. 11.8). Knowing the required cycling rate, $W_{max} > W$ for acceptable operation. Permitted values of W_{max} are again based on the allowable temperature of coil and friction material in relation to heat dissipation characteristics of the unit.

11.3.3 Response time

In many control applications, clutch or brake response speed can become a key design consideration. Response times are conditioned by the unit construction, principally governed by the electrical and mechanical time constants of the coil/armature on energization. The major factors affecting response time are as follows:

1. The current build-up time lag in an inductive winding corresponding to the electrical time constant (L/R — see also section 8.6.2) defined by

$$i = I\,(1\,-\,\exp\,[(-R/L)t]$$

where i = instantaneous current (A), t = instantaneous time, L = coil inductance, R = coil resistance and I = steady-state current. During the transition period the increase in current is exponential, hence when $i = I$ $i = V/R$. Similarly, when the clutch coil is de-energized, the decay in current is given by

$$i = I\exp\,[(-R/L)t]$$

The current behaviour is depicted in Fig. 11.9. Increase of the volume of copper in the coil increases clutch power, but also increases the electrical time constant. Pull-in time is reduced by either reducing the time constant L/R using a series resistor R_v such that the time constant becomes $L/R + R_v$ or by retaining the same time constant and increasing I. A common method increasing instantaneous current is by using a parallel resistor–capacitor network in series with the coil (Fig. 11.10); a high initial voltage is passed through the coil when the circuit is energized and capacitor C discharged. This situation

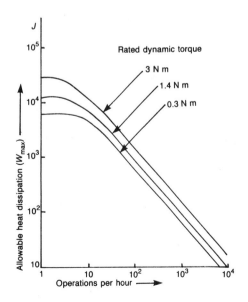

Fig. 11.8 Electromagnetic clutch–brake heat dissipation

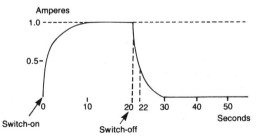

Fig. 11.9 Solenoid current–time curve

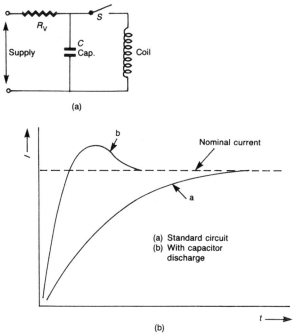

Fig. 11.10 Reduction of pull-in time with resistance–capacitor circuit: (a) electrical diagram; (b) current–time profile

occurs for a brief period only, once the capacitor is discharged the voltage falls to the solenoid voltage as determined by the value R_V. A typical current–time curve for capacitive discharge is shown in Fig. 11.10.

2. The inertial delay is defined by the mechanical time constant which largely depends on the mass of the armature and its travel, all electrical aspects remaining fixed. This is not always straightforward to calculate and is more easily established experimentally. Correct and careful application design can minimize inertia and total clutch travel.
3. The restoring spring force opposing the applied excitation force. Clearly a trade-off between rapid engagement or disengagement.

Response time definitions are illustrated in Fig. 11.6. However, since magnetically actuated devices depend on ampere-turns, response times diminish at high temperatures where increase coil resistance reduces current and power. Clutches and brakes are normally rated at a 35°C operating temperature above which the clutch should be derated in accordance with manufacturers' instructions. Below 35°C, the clutch can be slightly uprated on published performance figures. Operating frequency can normally be increased by reducing the electrical time constant, e.g. high-speed excitation at switch-on and counter excitation at switch-off.

11.3.4 Disengagement

When the clutch or brake is de-energized, a voltage spike may be induced as a result of the coil inductance. This reverse voltage can be very high and may cause damage to the coil and the circuit. Thus disengagement (or dropout) times have to be considered together with surge protection circuitry. Suitable protection devices are resistors, capacitors, varistors, diodes, etc. The following methods may be used:

1. *Resistor/diode.* In parallel with the coil. The values of resistor and the voltage peak are inversely proportional to each other, but as the resistance increases, disengagement time decreases. A parallel resistance of six to seven times the coil resistance is typical and rated ~25 per cent of coil wattage. Using a diode in series with the resistor eliminates the added current draw, but has no effect on drop-out time.
2. *Zener diode.* Used in series with a diode and in parallel with the coil. Actual switch-off delay influenced by the voltage of the zener, typically eight times the coil supply voltage for best performance. Provides faster release than the resistor/diode option.
3. *Varistor (non-linear resistor).* Used for fast brake/clutch disengagement in parallel with the coil. Unlike (1) and (2) may be used on a.c. as well as d.c. circuits for operating voltages typically up to 30 V. Disengagement current/time curves are typically referenced to the varistor circuit as the manufacturers' norm.

It should, however, be pointed out that the surge protection circuitry described does not necessarily lead to a complete elimination of interference as required by international/EEC legislation (Ch. 17).

11.4 Types of clutch and brake

11.4.1 Magnetic drives

There are three types, magnetic particle, hysteresis and eddy current. Since torque is developed electromechanically rather than derived from mechanical friction, there are no wearing parts. The only exception to this are

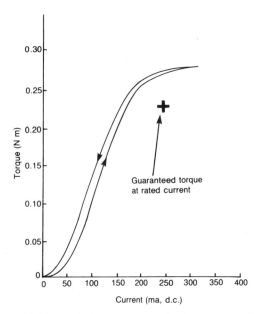

Fig. 11.11 Typical torque–control current curves for a hysteresis clutch

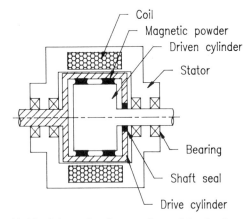

Fig. 11.12 Schematic of magnetic particle clutch

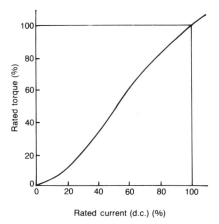

Fig. 11.13 Magnetic particle brake — variation of torque with coil excitation current

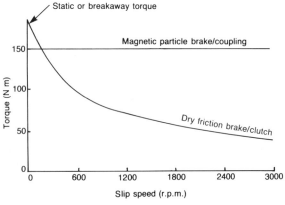

Fig. 11.14 Comparison of torque–slip curves for large dry friction and magnetic particle couplings

the containment seals necessary in magnetic particle clutches and brakes. Relationships between electric field intensity, magnetic induction, magnetic field intensity, etc. determine the torque — applied current characteristic (e.g. Fig. 11.11). However, with magnetic drives, there are no performance standards. Particular constructional variations between manufacturers can have a strong effect on performance characteristics in terms of magnetic field differences. General behaviour is summarized in Table 11.1.

(i) Magnetic particle brakes and clutches

These consist simply of two surfaces separated by finely divided dry magnetic powder (stainless steel) or a powder–oil slurry (Fig. 11.12). The cantilevered drum, or output shaft, is the inner member which is enveloped by a cylindrical chamber, the input member, attached to the input or driver shaft. The magnetic powder is contained between the two. When d.c. energized, a field coil produces a magnetic field that causes the powder to coalesce, becoming a rigid mass, orientated along the lines of flux so providing a physical coupling capable of transmitting torque.[2] Both shear and tensile stresses in the magnetic links resist relative motion. In practice torque is proportional to the magnetic field and therefore

Torque varies with the coil excitation current (Fig. 11.13)

Torque remains constant, independent of slip (Fig. 11.14)

Virtual instantaneous disconnection of the drive can be achieved by breaking the excitation current. A slipping clutch will generate heat in the clutching space due to particle–particle friction, but since the friction characteristics of the iron dust are relatively temperature independent, a very stable torque–current relationship prevails and precise modulation of torque level is possible. However, where clutches are to be slipped, they may need to be derated.

(ii) Hysteresis clutches and brakes

These comprise an electromagnetic coil encircling a high retentivity iron ring with a reticulated pole structure (Fig. 11.15), together with a permanent magnet in the form of a drag cup rotating between the inner and outer pole structure. The pole structure is keyed to the input, the permanent magnet to the output. Magnetic field variation is accomplished by reticulating the inner and outer walls of the outer member to produce an alternating set of N and S poles (Fig. 11.15), when the outer member becomes magnetized. The hysteresis brake differs from the clutch only in that the outer member becomes fixed, i.e. the pole structure is a stator in the gap of which the drive-actuated permanent magnet cup rotates.

To engage the clutch, the coil is d.c. energized generating a steady flux field to induce opposite poles in the rotor. When the rotor cuts the field of the input assembly, the resultant discrete poles generated in the rotor ring by the input field lead the poles of the input assembly field. The lead angle depends on the excitation of the stator and hysteresis losses in the rotor. Attractive forces between the induced poles in the rotor ring and those in the input assembly cause the stator and rotor to 'lock' magnetically, producing torque that opposes the rotation. Rotation is in synchronism up to a load where the drive starts to slip. Instead of pulling out, the rotor will then be subject to asymmetric magnetization and continue to maintain the drive with increasing slip (Fig. 11.16). The shape of the decreasing torque portion of the curve to the right of the critical point reflects both the change in the hysteresis loop with temperature increase plus heat transfer characteristics of any cooling system. Power loss here $\propto$ (slip speed).[2]

Torque is a function of the d.c. current applied to the device (Fig. 11.11), but is different when the control current is increasing from when it is decreasing due to hysteresis in the magnet materials. Torque control is within 1 per cent when the control cycle is repeated.

Because of the nature of hysteresis clutches and brakes, unavoidable residual magnetism in the output rotor results in de-energized drag. This effect can be minimized by applying a reversed polarity input after each normal application of the device.

(iii) Eddy current clutches and brakes

These have a similar physical construction to hysteresis clutches and brakes except that the rotor is now made of a magnetically soft material and toothed or generated in the form of magnetic segments to promote the formation of eddy currents and restrict hysteresis. Such a unit will only develop torque when differential motion is present, i.e. there is slip. The strength of the generated eddy currents, hence generated torque, increases in proportion to (rotor field strength)^2 and almost linearly with slip (e.g. Fig. 11.17). Eddy current couplings are well suited

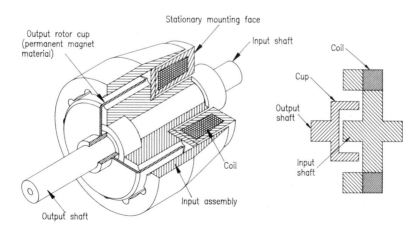

Fig. 11.15 Schematics of hysteresis clutch and components

Table 11.1 Comparison of brake and clutch type characteristics

	Eddy current drives	Hysteresis drives	Wrap spring drives	Magnetic particle clutches	Electromagnetic friction drives	Tooth clutch
Positive features	Maximum torque at maximum slip	Self-starting against full load	Suitable for both light and heavy drives	May be used as a clutch or brake, either direction and also in continuous mode	Favourable torque:size ratio	Can be run and disengaged at high speed
	Can carry temporary overload	Can operate at 100% torque up to the slip point	Power capacity depends on the amount of force the springs exert	For a given clutch, torque is directly proportional to excitation current	Attractive cost	Small size and high torque capability
	Well suited to speed control applications	Synchronous up to the slip point then asynchronous	Direction of rotation depends on direction in which coils are wound	Can be used as a torque limiting device	Vary clutch torque by varying clutch voltage	Large number of teeth provide good load sharing plus a short/fast engagement stroke
	Less efficient high slip units can develop relatively high starting torques	Can accommodate overload with slip	Unidirectional; positive load engagement	Torque is speed independent for all practical purposes	Ideally suited for start–stop applications	Overload protection can be provided with torque limiting teeth option
	Different characteristics obtained by modifying pole designs	Long life, no friction surface	Self-energizing	Small size delivering high torque	AC or DC operation	Backlash free clutching is possible
		Low residual drag	Transmits torque with positive displacement	Since they slip continuously, may be used as bidirectional actuators	Single or multiple-disc faces	Some differential engagement speeds are acceptable when flexible torsion elements are fitted to input and output
		Low power consumption	Simple, economical rotation control devices	Inherently fast response times	Can operate wet or dry	Can be run in oil
		Initial performance maintainable indefinitely, high repeatability	Various types of actuator allowable, e.g. mechanical, pneumatic, electromagnetic, etc.	Fluid ones operate more smoothly than dry ones	Popular; ease of control	
		Wide speed range	High torque:size ratio	Torque accurate and repeatable	Relatively low power requirements	
		Smooth operation due to rotor symmetry	Time to speed not inertial or frictional load dependent within torque capacity	Silent, maintenance free operation		
		High heat dissipation capability	Consistent time to speed approx. 0.003–0.006 s	High torque:inertia ratio		
		High torque:signal linearity below saturation except near zero				

CLUTCHES, SOLENOIDS AND BRAKES

Negative features	Asynchronous No holding torque at standstill — must have slip to develop torque Limited to long time cycles Torque–temperature sensitive	May be subject to low speed cogging due to presence of residual coupling torque Not suitable for rapid reversal duty cycles Limited to long time cycles Response times up to 0.5 s depending on size Larger per unit torque than other devices hence slower response times	Limited movements, spring susceptible to fret Tangs under shear; weak point in their design — ensure design application is correct Medium speeds and cycles Cannot modulate torque take up; design the load connection to allow for this	Inherent zero excitation drag due to seal friction + magnetic particle friction Hysteresis can result in de-energized drag Limited by heat dissipation capability and low inertia loads Powder may 'wear' in a slipping application	Ensure correct thermal dissipation for the application Not used for precise torque control unless feedback circuit employed Cannot use in some constant slip applications Do not function well where there is a substantial amount of slip during the useful life	Can only engage at speeds up to 60 r.p.m.; ideally at standstill or zero differential speed Wear occurs when engaged at speed Tooth form is critical to performance Any shocks created during engagement may overload clutch Only rated at a static torque value, no dynamic torque rating If used in overload mode, sensing system required to de-energize clutch	In line drives coupling where driven components required to follow driver precisely with engagement at zero speed differential
Applications	Speed and torque control Oscillation dampers Magnetic tape drives	Torque limiting and shock load cushioning Uniform torque permits use in tension control devices Use limited to small instrument devices; low power gain Feed control, take up control	Incremental rotation control with single rev. units See text for details of individual types and application Indexing, positioning, stop/start and intermittent feed type applications	Spooling, tensioning and positioning devices Continuous variable speed control Rapid cycling situations Protective torque limiters, energized to transmit a predetermined value and slipping on sudden excess demand Can step large loads faster than servos or steppers	In line drives coupling		

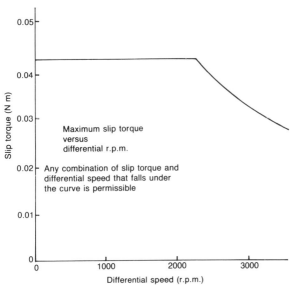

Fig. 11.16 Torque–speed curve for hysteresis drive

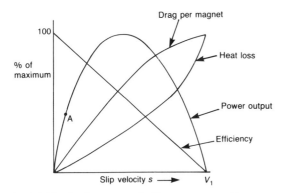

Fig. 11.17 Performance of eddy current couplings

Slip speed = $w_{in} - w_{out}$

and

Fractional slip $s = \dfrac{w_{in} - w_{out}}{w_{in}}$

where w_{in} = input shaft speed and w_{out} = output shaft speed. Since input torque ≡ output torque, input power = Tw_{in} and output power = $T(1-s)w_{in}$ and neglecting losses

Slip power = $T(w_{in} - w_{out}) = sTw_{in}$

and

Efficiency = $\dfrac{\text{output power}}{\text{input power}} = 1 - s$

11.4.2 Electromagnetic clutches and brakes

Electromagnetically operated clutches and brakes provide a simple and adequate means of mechanical torque control in low- and medium-power systems. Various forms of construction are available. For general operation, when windings are energized, the magnetic field causes the central member to move axially and make strong physical contact with a mating surface and thus develop the required locking action up to the rate torque limit. General behaviour is summarized in Table 11.1.

(i) Friction clutches and brakes

Construction varies from a simple shoe to single or multiple discs. The more common single plate type is shown in Fig. 11.18. The clutch unit comprises three basic parts, a stator, a rotor and an armature. The stator, incorporating the coil, is always stationary (stationary field) and may be attached to the machine or any grounded component. The rotor and armature both rotate and either of these may be connected to the driven shaft. The rotor (clutch) or stator (brake) incorporates the friction disc. When the stator coil is energized, the rotor's magnetic poles directly attract the armature plate into contact and the drive is taken up. A flux is generated by the bearing-mounted stationary field coil which unidirectionally permeates the armature to force-lock the system and effect drive. In principle, when energized, the magnetic circuit operates without an air gap, thereby effecting a strong axial adhesion force.

On removal of the excitation current, a leaf spring connecting the armature to the shaft fastening disengages the two members. The stiffness of this spring usually determines the operating times, e.g. a stiff spring increases the engagement time but reduces release time.

to speed control applications, but for maximum clutch efficiency the amount of slip must be small (Fig. 11.17). It is desirable to design the drive to work on a rising part of the power curve, e.g. at A, since slip velocity is then low and efficiency high. Because there would be no relative slip, the eddy current brake has no holding torque at standstill.

Performance can be assessed by

Torque = Dxr

where x = no. of magnets, D = unit drag per magnet pair, r = mean radius of magnets, and $D \propto$ slip velocity,

and in an analogy with the induction motor

Fig. 11.18 Electromagnetic friction clutch. Courtesy of Simplatrol Ltd, Bedford

Fig. 11.19 Tooth clutch: stationary field. Courtesy of ZF Friedrichshafen AG

This type of disc/armature operating arrangement affords a backlash-free operation and zero drag torque and is used where high relative speeds are combined with long idling periods or where the fixing arrangement is for vertical operation. An alternative design, a disc or discs operating on a splined hub, is preferred for high cyclic operations with short response times.

Electromagnetic friction clutches and brakes have two things in common: (1) the use of friction to transmit motion; (2) the use of an electromagnet to control the normal force applied to the friction faces.

The holding power of the clutch depends on the number of laminations, the size of the laminations and the coefficient of friction of the faceplate material. The friction coefficient decreases with slip velocity due to temperature rise (e.g. Figs 11.5 and 11.14). Above ~200°C, some resin-bonded friction materials exhibit a marked decrease in friction coefficient, i.e. they tend to fade. Attention to engagement time, inertial take-up and clutch pressure should minimize heat dissipation. Because of the negative $d\mu/dv$ (Fig. 11.14, see also section 5.2.6), this will assist torque take-up.

The theory of friction clutches and brakes is essentially dealt with in the discussion of application factors (section 11.2.1).

(ii) Tooth clutches

These are similar in concept to the frictional clutch in that the armature is similarly driven but torque in this case is transmitted by a series of fine teeth on the outer circumference of the poles (Fig. 11.19). Because of the mechanical interlocking, stationary field toothed clutches can transmit a higher torque than single friction faced clutches of the same physical size.

The actual tooth form is the key to performance. However, because of the interlocking feature, it is only possible to engage the clutch at zero or low differential speeds although the clutch may be run and disengaged at high speed. For ease of engagement, choose a design where one tooth ring is non-ferrous, e.g. cupro-nickel, so that the flux path does not pass through the teeth and crown—crown engagement is prevented. An accurate machined tooth fit is essential for long-life performance.[3]

Tooth forms available are as follows (Fig. 11.20):

1. Symmetric 30°, bidirectional drive with engagement at slow relative speeds;
2. Sawtooth profile, permits engagement at higher speed

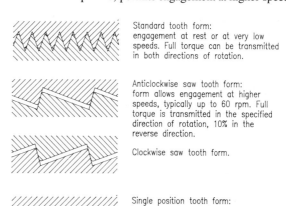

Fig. 11.20 Common types of tooth style

but the drive is then unidirectional and needs to be specified;
3. Single position, only engages with input and output synchronized at a fixed angular position;
4. Multiple position, will engage in regular fractions of a revolution, e.g. up to eight positions equispaced at 45°;
5. Trapezoidal, often the standard form with a small amount of circumferential play;
6. Torque limiting, teeth can be cut at a shallow angle to ensure that the clutch disengages at a predetermined overload torque.

Tooth clutches may also be used, say in an emergency situation, as overload detent clutches, because their torque is limited by the axial force holding the toothed jaws together. For example, from Fig. 11.21, the maximum torque that can be transmitted is estimated by

$$T = P_n r F_n \tan(\gamma + \beta)$$

where r = radius of tooth ring, $\beta = \tan^{-1}\mu$, μ = friction coefficient, $F_t = F_n \tan(\gamma + \beta)$, γ = tooth angle, P_n = normal load on single tooth and F_n = axial load on single tooth.

11.4.3 Wrap spring clutches and brakes

The basic spring clutch consists of three elements, the input hub, the output hub and the spring (Fig. 11.22). The internal diameter of the spring is smaller than the outside diameter of the two hubs and when the spring is forced over the two hubs, rotation in the direction of the arrow (opposite to the wind of the spring) wraps it down tightly on the hubs, connecting them in a positive engagement (Fig. 11.22(a)), i.e. the increased friction between spring and hubs tends to resist relative rotation. Relative rotation of one hub in the other direction, however, tends to loosen the helix and so disengages the clutch (Fig. 11.22(b)), i.e. the clutch will overrun or freewheel. If one hub is held from rotating, the other hub can rotate in the opposite direction as the wind of the

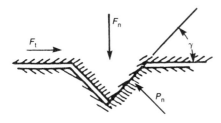

Fig. 11.21 Forces acting on single wedge tooth

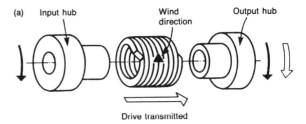
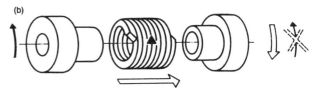
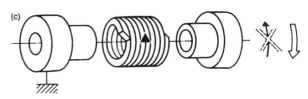

Fig. 11.22 Basic components and operation of wrap spring clutch/brake. Courtesy of Warner Electric, Bishop Auckland

spring but is prevented from rotation in the same direction (Fig. 11.22(c)). Estimates of torque can be made from the following:

$$T_t = EI_A r_h \left(\frac{1}{r_1} - \frac{1}{r_2}\right)\left(e^{2\pi N\mu} - 1\right)$$

where T_t = transmitted torque, E = elastic modulus of spring material, I_A = spring wire moment of area, r_1 = neutral radius of spring when free, r_2 = neutral radius when spring in tight contact with hubs, N = smallest number of spring turns on either hub, μ = spring/hub friction coefficient and r_h = hub radius.

The above equation holds for spring tensile stresses below the yield point to ensure adequate life. The limiting torque is found to satisfy the inequality, assuming rectangular section wire:

$$T_{max} \leq bd^2 \frac{1.05}{2r_h}\left(r_2 - r_1 - \frac{t}{2}\right)$$

where d = thickness of wire in radial direction and b = thickness of wire in the axial direction.

A number of forms of wrap spring clutch exist, described as follows

CLUTCHES, SOLENOIDS AND BRAKES

(i) Overrunning clutch
In its basic form, the wrap spring clutch performs as an overrunning clutch (Fig. 11.22). When the input hub is driven in the direction of the arrow, the spring tightens on the output hub so transmitting torque. Rotation in the opposite direction opens the spring, releasing the output hub and the load is free to overrun (Fig. 11.22(b)).

(ii) On–off clutch
Clutch control is via a modified clutch tang which when released causes the spring to wrap down (Fig. 11.23). When the tang is externally held, preventing it from gripping the input hub, the spring opens providing instant disengagement and the load coasts to a stop (Fig. 11.23(a)). The system should be fairly stiff to prevent the output rotating backwards. The output can continue rotating in the normal drive direction (overrun).

(iii) Single revolution clutch
If a second tang, at the output end of the spring, is permanently fastened to the output hub, it provides both an angular reference to the input tang and a braking function, so allowing the clutch to perform as a single revolution or multiple stop clutch. The wrap down is achieved as before, the double tang preventing the load from overrunning when the input tang is re-engaged. Provided the load inertia does not exceed the sheer strength of the spring tang, stopping accuracy is between 2 and 5°. Braking torque is limited to ~20 per cent of maximum clutch torque. In Fig. 11.23(b), the actuator is illustrated with an integral solenoid. The stop collar connects to the input tang.

(iv) Brakes and clutch/brakes
This is a combination of two clutch springs as in (iii) above, operated by the same control collar, but with one spring clutch (of opposite hand) used to brake the rotation of the load. Here the control tang holds the brake spring open to permit rotation, but when the control collar is stopped it wraps down on the brake hub to brake the load. The clutch brake will stop and start an equal load to a positional accuracy of ±0.5°.

A minimum kinetic energy is needed to declutch a clutch/brake unit, this energy provided by the output inertia prior to stopping. The following inequality is used to check whether enough inertia is connected to the output shaft:

$$J_a = 10t \frac{(T_c + T_0)}{n} - (J_c + J_{out})$$

where J_a = relative inertia, J_c = clutch inertia (manufacturer's literature), J_{out} = inertia reflected at the

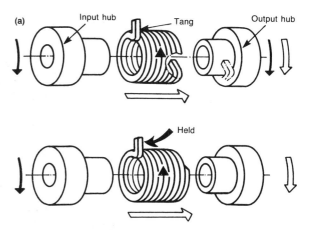

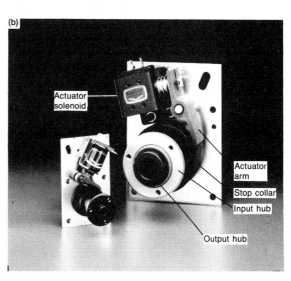

Fig. 11.23 Wrap spring clutch applications: (a) basic components and operation of an ON–OFF clutch; (b) single revolution spring wrap clutch. Courtesy of Warner Electric, Bishop Auckland

output shaft, T_0 = maximum load torque of system applied to the output shaft, T_c = torque required to fully activate the clutch (manufacturer's literature) and t = deceleration time (manufacturer's literature). If $J_a \leq 0$, application is satisfactory; if J_a is positive, the clutch/brake lifetime and stopping accuracy may well be affected; there are three possible solutions:

1. Increase speed;
2. Reduce the system torque;
3. Install additional inertia to the output shaft but also recheck the dynamic clutch torque capability.

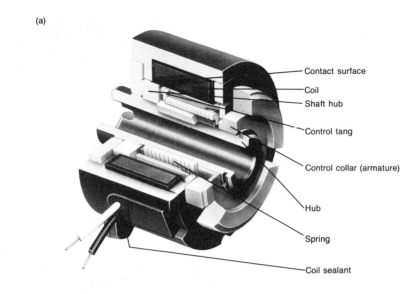

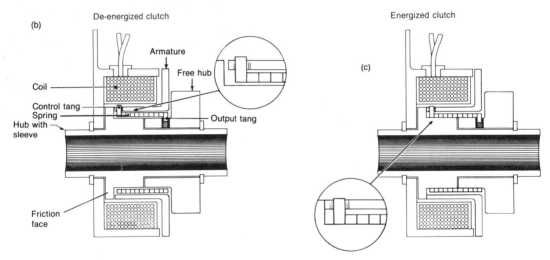

Fig. 11.24 Components and operation of an electromagnetic spring wrap clutch: (a) cut-away view of clutch; (b) de-energized clutch; (c) energized clutch. Courtesy of Warner Electric, Bishop Auckland

(v) Electromagnetic spring wrap clutch

The magnetic spring clutch uses the same spring type as in (iii) above, but with the internal diameter of the spring larger than the diameter of the two hubs (Fig. 11.24). In this design, the control tang of the spring is connected to the armature (Fig. 11.24). The output tang is connected to the free hub. Energizing the coil causes the spring to wrap down. Magnetic force is only necessary to maintain a tight spring grip for total torque transfer. Time to speed depends upon the wrap-down angle and input speed. Operation is very similar to a friction clutch but is capable of higher torque output for the same package size.

11.5 Clutch and brake design procedure

11.5.1 Clutches

The following process is divided into three areas. However, because such a wide range of application conditions exist, the process described is very general with as broad an applicability as possible.

(i) Preliminary decisions

Some early factors that should be considered prior to the

analytical process are as follows; in order to specify the application:

1. Is bidirectional output required? If so this would rule out some clutch types such as spring wrap units.
2. Is the clutch to be d.c. or a.c. driven? Reaction time is faster in a.c. units although a.c. coils may hum.
3. Is the output system stiff, i.e. it should not rotate backwards once the clutch is disengaged.
4. Type of load take-up required, i.e. rapid start or smooth take-up with slippage.
5. Likely environmental conditions. For example, friction clutches are less suitable for high humidity conditions if exposed.
6. Define the clutch control action, e.g. continuous slip, occasional stop–start.
7. Where is the clutch to be positioned? In a speed reduction system, a reduced torque requirement on the high-speed side allows a small size clutch, but a larger clutch on the low-speed side dissipates heats more effectively.
8. Is allowance for shock in elements downstream of the clutch necessary?
9. What is the differential speed when clutch engagement required?
10. Any size constraints likely?

(ii) Define system parameters

Because applications will vary significantly, it is firstly necessary to determine which parameters are likely to be important (e.g. Table 11.2). The breakdown here is based upon the type of clutch control required. Taking the more involved case:

1. Calculate the maximum and minimum speed at the clutch shaft taking into account all operating conditions;
2. Calculate the reflected inertia, mk^2, at the clutch output (see section 8.2);
3. From (i) and (ii), estimate the dynamic torque range requirement, hence duty cycle, having first defined the acceleration time required;
4. Estimate or measure system friction in any elements downstream of the clutch. System load torque = dynamic torque + frictional torque;
5. If a cyclic stop–start situation, calculate the heat dissipation (section 11.3.2). Similarly if application is continuous slip;
6. Calculate number of operations required during average life of machine.

(iii) Define the clutch

1. Initially select a suitable clutch unit based upon: (a) size; (b) torque capability, i.e. the total clutch torque requirement of maximum dynamic torque plus system load torque (usually manufacturers only declare a static torque value so assume dynamic torque ≈ 80 per cent static torque); (c) speed capability; (d) required power input; (e) acceptable mounting arrangement; (f) environmental specification; (g) service life.
2. Having short-listed or selected a particular clutch, a number of performance checks will be required to ensure satisfactory operation, in particular: (a) check that the required acceleration time is compatible with the clutch response and engagement times. If disengagement time is also important, it may be necessary to trade off clutch engagement/release characteristics; (b) check that the thermal dissipation characteristics of the clutch are adequate for application (section 11.3.2); (c) check life of unit is not adversely affected if clutch is working >50 per cent of its thermal capacity. Check that manufacturer's thermal ratings are applicable to actual operating speed; (d) if clutch working raises average coil temperature, check that any power decrease is not detrimental to response time.

Table 11.2 *Relative importance of performance parameters when based on application type*

	Occasional start–stop	Cyclic start–stop	Constant slip
Static torque	Important	Could be important	Not applicable
Dynamic torque	Sometimes important	Important	Not applicable
Heat sink capacity	Sometimes important	Not applicable	Not applicable
Heat dissipation	Not applicable	Important	Important
Inertia	Sometimes important	Important	Not applicable
Response time	Not applicable	Could be important	Not applicable
Disk wear rate	Sometimes important	Could be important	Important
Speed limit	Check	Check	Check

11.5.2 Brakes

As far as brakes are concerned, the design procedure is analogous to that described for clutches, but in this case speeds are reducing and loads decelerating. However, in this case, any frictional torque in the system will assist braking such that

Total braking torque = maximum dynamic torque −
 minimum system load torque
 (including clutch drag)

11.6 Solenoids

The energy produced within a solenoid can either be used to provide a straight linear pulling action, or it may be converted to provide rotary action. In either case, the magnetic pull force increases as the air gap is reduced between armature and core. The available types of solenoid may be described as follows:

1. *Single-action linear solenoid*. A linear stroke motion from the start position to the end position. The return action is effected by some other external force/mechanism. Depending on the solenoid design, the force can be applied in a 'push' or 'pull' sense. An example of a push solenoid is given in Fig. 11.25.
2. *Double-acting linear solenoid*. This in effect is two solenoids back to back, the stroke obtained by energizing in one of the two opposing directions from neutral. Return to neutral is provided by some other force/mechanism.
3. *Mechanical latching solenoid (bistable)*. These are fitted with an internal latching mechanism for holding the plunger against the pull of the load. The unit effectively operates by pulsing, and since the duty cycle is reduced, coil power can be applied to produce higher pull forces than conventional solenoids of the same frame size. A reverse polarity pulse unlatches the plunger.
4. *Keep solenoid*. These are fitted with a permanent magnet so that no power is necessary for holding loads in the pulled-in position. Because of the consequently reduced duty cycle, for a given size of coil, wattage can be increased. The plunger is released by supplying a pulse in reverse polarity to that for the pull action.
5. *Rotary solenoid*. Utilizes rotary motion from a neutral position when the solenoid is energized. Standard travel is 25−95°. Return action is via some other external means. An example of a rotary solenoid is given in Fig. 11.26.
6. *Reversing rotary solenoid*. Rotary motion is from one end position to the other when energization occurs. There is no neutral position, the end position one way is the start position for the other direction.

The magnetic principles outlined in section 11.1.1 are applicable to solenoid coils, together with the discussion of response time in section 11.3.3 and disengagement in section 11.3.4. With solenoids in particular, one of the more commonly used methods to decrease response time is by means of a switchable series resistor controlled by the mechanical movement of the armature (Fig. 11.27). When the circuit is energized, full voltage V_a is delivered across the coil, bypassing resistor R_v. Movement of the armature towards its final position opens the switch in parallel to R_v, reducing the solenoid voltage as determined by R_v. The hold-in resistor value can be determined from Fig. 11.27. The resistance value

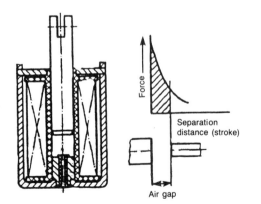

Fig. 11.25 Flat face armature and core face

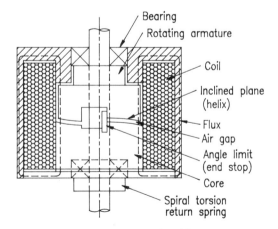

Fig. 11.26 Section of rotary solenoid

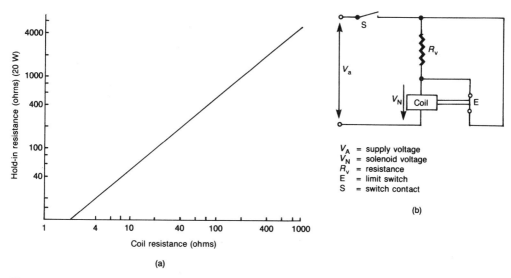

Fig. 11.27 Value and circuit details of switchable series resistor: (a) value of hold-in resistor; (b) electrical diagram

is selected such that when the winding reaches the maximum temperature, the current is limited to a permissible value. Values in Fig. 11.27 reflect an insulation class E winding. For other classes of insulation, use the following to estimate R_v:

$$R_v = R_1 \frac{(V_a - V_n)}{V_n}$$

where R_1 = coil resistance at limiting temperature, V_n = solenoid voltage and V_a = supply voltage.

For continuous duty, hold-in resistor circuits are commonly used to provide higher start forces than normally obtainable at continuous duty. Pulsing the power may also provide higher solenoid loading without the risk of solenoid overheating.

11.6.1 Solenoid rating and duty cycle

The flux in the working gap can be increased or decreased by varying the electrical input to the coil (section 11.2.4). However, as the electrical input increases, the coil heats up in proportion to (current)^2. In practice the electrical input depends on the number of ampere-turns of the coil and the solenoid duty cycle such that the coil temperature, as determined by the insulation class of the windings, is not exceeded. In other words, the flux generated by the excited turns is limited by the heat that a given solenoid can dissipate. The most common class of insulation available for coils is class E, i.e. a maximum working temperature of 120°C. Given a maximum working ambient temperature of 35°C, manufacturers therefore provide standard solenoid rating data based on a coil temperature rise of 80°C.

The higher the solenoid input power, the higher the solenoid force and speed of operation, but the greater the coil heating, the higher the input power and its time of application. In many situations, the solenoid is briefly energized then turned off for a period so allowing it to cool down; in other words, the solenoid takes more current for a given wire size so increasing its power. The standard way of expressing the allowable solenoid rating is by means of the duty cycle, e.g. as in Fig. 11.6. However, even at the intermediate ratings there is still a time limit that the coil can safely carry the current and this will be stated in data sheets. Excess current can lead to burn-out.

Each solenoid type/size will have a maximum power rating for continuous operation, i.e. its basic rating normally referred to a 20°C coil resistance. Different application voltages are possible by varying the wire gauge of the coil. However, because it is usual to maintain the same number of ampere-turns, the power rating is also maintained constant. For a given wire gauge, the allowable increase in power for a non-continuous duty cycle may be expressed as

$$\text{Allowable input power (W)} = \frac{\text{power rating at 20°C}}{\text{duty cycle}}$$

Note that solenoid power is defined by watts for d.c. operation, but in terms of VA rating for a.c. operation. In any application though, a portion of the input power is dissipated in heating the coil, hence as the coil resistance increases the actual solenoid power decreases such that

$$\text{Solenoid coil power (W)} = \frac{\text{power rating at 20°C}}{\text{duty cycle} \times \text{resistance factor}}$$

where resistance factor = $R_T/R_{20} = 1 + 0.003\,93\,(T - 20)$, R_T = coil resistance (Ω) at working temperature and R_{20} = coil resistance (Ω) at 20°C.

Knowing the available solenoid power together with the duty cycle, the solenoid pull can be determined. Figure 11.28 illustrates the force at zero stroke (close-up) for tubular-type pull solenoids. Individual solenoid ratings are also given in Fig. 11.28.

A similar situation exists with rotary solenoids, but in this case the magnetic pull is given in terms of torque (Fig. 11.29(a)). However, in this case, the end of stroke is not a fixed position and is determined by the working angle range required (Fig. 11.29(b)). The same rules on rating, ampere-turns and duty cycle apply as with linear solenoids.

11.6.2 Linear solenoids

A linear solenoid in its simplest form consists of a frame, a wire coil and a plunger, the iron path around the coil concentrating the magnetic field. It is normally possible to energize a linear solenoid with either a.c. or d.c.; the pros and cons of both are outlined in Table 11.3, and the differences in force/stroke characteristics are represented diagrammatically by Fig. 11.30. It is also possible to alter the force/stroke characteristics by modifying the shape of the plunger face and core location. The available configurations are as described in the headings below.

(i) Flat face armature and flat core face

Generally the most common arrangement (Fig. 11.25), the magnetic air gap corresponds to the stroke of the solenoid armature (section 11.1.1). A sharply rising force results at the end of the stroke (Fig. 11.31), depending on duty cycle. These are generally used where a small stroke and a high end force are required at the expense of medium stroke performance. As discussed in section 11.6.1, the smaller the duty cycle the higher the allowable solenoid power and the shorter the stroke times (Fig. 11.31).

(ii) Conical armature and core face

Typically as in Fig. 11.32. In most cases stroke = length of the cone portion. The force/stroke curve is determined by the change in surface area of the air gap with stroke for a given cone angle α. In that case the force/stroke curve can vary from near horizontal (small α) to steeply increasing as $\alpha \to 90°$, the flat armature case, but again depending on duty cycle. Figure 11.31 is a comparison of force/stroke behaviour for the same solenoid with flat end conical plunger faces.

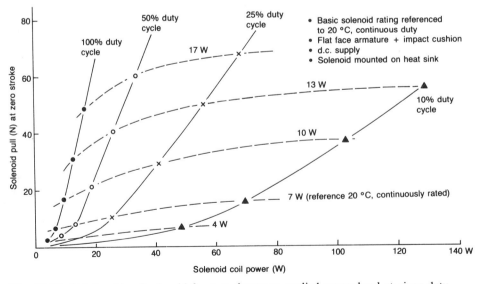

Fig. 11.28 Relationship of solenoid force at closure to applied power level at given duty cycle for tubular solenoids

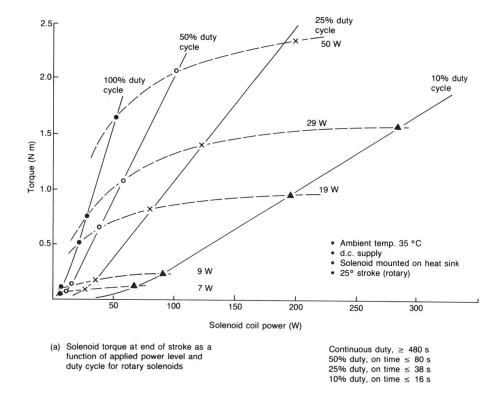

(a) Solenoid torque at end of stroke as a function of applied power level and duty cycle for rotary solenoids

Continuous duty, ≥ 480 s
50% duty, on time ≤ 80 s
25% duty, on time ≤ 38 s
10% duty, on time ≤ 16 s

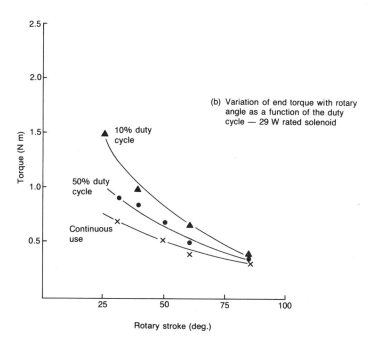

(b) Variation of end torque with rotary angle as a function of the duty cycle — 29 W rated solenoid

Fig. 11.29 Rotary solenoid performance characteristics

Table 11.3 *Comparison of the pros and cons of energizing voltage type applied to linear solenoids*

a.c. supply	d.c. supply
Greater long stroke force, gaining this advantage from its inrush current, the longer the stroke the greater the current surge	Good short stroke performance
	Slower acting than a.c.
	Hold at zero stroke is strong and quiet
Fast closing	Can be stopped anywhere along stroke without overheating or hum
On repeated cycling, repeated current surges increase coil heating	
Compared to d.c., not so sensitive to power loss when hot	No d.c. inrush current
Resistance change with temperature the same as d.c. but the inductance of the coil does not change hence effect of temperature on pull not so marked	Hold characteristics can be varied significantly by modifying the shape of the plunger and core stop
	It is possible to fit an 'O' ring shock absorber internally to prevent high closing impacts
Tend to hum when closed	Sensitive to coil temperature; a 70 °C rise in coil temperature causes a 30% reduction in coil current and hence a reduction in pull
Hum can worsen with wear or with dirt on mating faces	
If prevented from closing, solenoid will be noisy and rapidly overheat	Solenoid power reduction — temperature effect greater on d.c. than on a.c.
Characterized by high VA input when plunger prevented from seating	
Need to introduce a spring into the plunger linkage so that the plunger can seat if the mechanism jams	
To obtain quiet operation, fit a shading ring and ensure that faces of plunger and core have a smooth finish and are flat and square	
Ensure linkage design enables square contact of plunger and core	
Larger sizes are laminated to prevent excessive losses due to eddy currents	
If cannot cushion plunger impact, try to prevent impact noise ringing throughout machine structure	
No noise problems with mechanical latching solenoids when used on a.c. or d.c.	

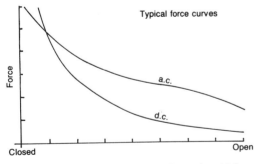

Fig. 11.30 Representation of a.c.–d.c. solenoid force–stroke differences

(iii) Flat armature into core with external conical shape

This is essentially a combination of (a) and (b) in Fig. 11.33, but in this case the conical design influences the force/stroke characteristic from near horizontal (small α) to steeply decreasing as $\alpha \to 90°$.

11.6.3 Rotary solenoids

A rotary solenoid is a very simple means of performing basic point-to-point rotary movement quickly and efficiently. It is a d.c. device rotating through a fixed angle up to 95° and holds its position as long as it is energized. Return to the start position may be inherent in the process used or controlled by a spring or some other mechanism. The action of a solenoid can be compared to that of the rotor (armature) and stator of an electric motor,[4] the shaft-mounted armature rotating between two internal end stops. The principle of the rotary solenoid is based simply on the inclined plane, such that when the solenoid is energized, the armature rotates, closing the air gap until the end stop is reached. Either the armature and core are

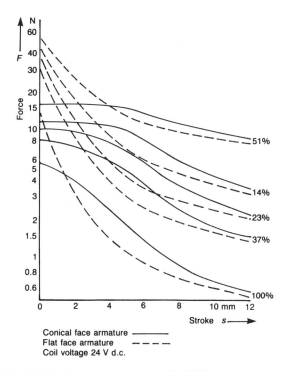

Fig. 11.31 Force–stroke curves for 7 W rated solenoids at different duty cycles

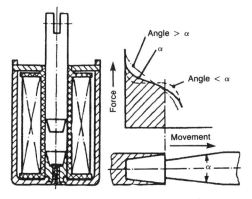

Fig. 11.32 Conical face armature and conical core face

inclined relative to the shaft centre line with an air gap in between (Fig. 11.26), or the armature faces are normal to each other and the armature is supported on three balls that travel down inclined raceways. In the latter design, there is shaft axial movement associated with the rotation.

Unlike linear solenoids, rotary solenoids are not affected by acceleration or deceleration effects due to the symmetry of forces within the solenoid. The total force

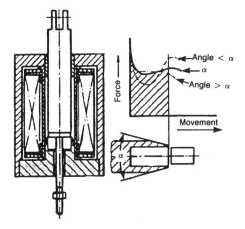

Fig. 11.33 Flat armature into external conical core

over the stroke rotation may be distributed fairly evenly simply by adjusting the angle of the ball raceways.

11.6.4 Solenoid design application

A solenoid is inherently a reliable device if applied correctly in a given elctromechanical system of which it forms part. As separate devices, solenoid reliability is quite high; the problems that occur are normally associated with the application. It is therefore worth while considering the practical problems which can be encountered.

(a) Solenoid loading The energy produced by a solenoid is constant regardless of load. If the solenoid is significantly underloaded, its life will be decreased. The best solenoid for a given application is generally the smallest one that will give sufficient pull without overheating.

(b) Friction Frictional effects in one form or another are probably responsible for the majority of failure causes (e.g. Fig. 11.34). Typically the solenoid problem is a failure to pull in or drop out, partial movement as a result of contamination and/or wear or stiffness/misalignment of an associated linkage.

(c) Contamination Generally a result of accumulation of airborne dirt and dust, particulate process matter, etc. on the solenoid/system working surfaces. The problem is aggravated by mounting the solenoid vertically, plunger up. The problem may be overcome by using a flexible

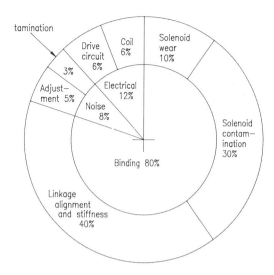

Fig. 11.34 Typical solenoid failure modes

rubber bellows or boot to provide a working seal over the solenoid and plunger. Plunger-down operation is preferred.

(d) Mounting This is important for a number of reasons. Vertical mounting is clearly preferred to avoid horizontal plunger loads and side wear. Also when mounted on the machine framework or metallic surface, the amount of heat that can be dissipated is increased. The most effective way to mount a solenoid is in such a way that the impact closing force is transferred directly to the mounting structure, an important consideration for the ultimate life of the solenoid.

(e) Alignment Often the alignment of the solenoid to its linkage is critical to prevent excess side loads on the plunger and solenoid bore. Any side loads increase friction and the probability of the assembly binding in use. Most solenoid problems are the result of linkage malfunction (Fig. 11.34), i.e. the force—stroke characteristics have not been well optimized at the design stage, the velocity ratios not ideal or a poor choice made of bearings or pivot pins. Linkages also suffer from contamination.

(f) Noise a.c. solenoids suffer noise problems when the solenoid is not seated properly, possibly as a result of contamination or poor adjustment. Not only may this cause user annoyance but also overheating and electrical failure.

(g) Residual magnetism This is the remaining force after the solenoid has been de-energized. Good solenoid design results in a negligible residual magnetism characteristic. With d.c. solenoids, the plunger need not go completely home and a thin washer used to prevent a sticking force.

11.6.5 Design procedure

The design procedure for solenoids is quite straightforward once the relationship between duty cycle, coil heating, applied power and magnetic pull is appreciated and understood (section 11.4.1). A general procedure is described under the headings below.

(i) Define the operating requirement

Specify the application as far as possible by defining the following:

1. The available voltage and power maximum (if limited).
2. Type of movement, i.e. linear/rotary and single or reverse action.
3. The length of stroke or angular rotation required based on the proposed linkage design.
4. Using the proposed linkage, calculate or measure the force—stroke requirement, bearing in mind the fairly fast solenoid actuation times. Normal expectation for both linear and rotary solenoids would be to complete the stroke in 100 ms maximum. Thus depending on whether movement is linear or rotational, force calculations should be based on the mass/inertia of the moving parts together with the estimated acceleration requirement. With a linear solenoid, if mounted vertically, the force will be ± the armature weight.
5. Determine the duty cycle, i.e. specify the maximum ON time and the minimum OFF time.
6. Determine the mounting and sealing requirement as outlined in (c) and (d) above.

(ii) Define the solenoid

1. From the force—stroke characteristics of the linkage system, decide the most appropriate end shape for plunger and core if application is for a linear solenoid.
2. With rotary solenoids, the smaller the duty cycle then it is safe to assume that the start torque ≥ end torque. However, as the duty cycle → continuous, especially at the larger angles, then start torque will be less than the final end torque.
3. Assume a coil temperature rise of $\Delta 80\,°C$ and from the expected ambient temperature determine the

resistance factor (section 11.6.1). Define a maximum 20°C power rating for solenoid.
4. Use equivalent of Fig. 11.28 (11.29(a)) to establish whether the maximum hot coil pull is adequate for the connected system, available solenoid power and proposed duty cycle.
5. If the required closing force is less than the maximum, then use Fig. 11.28 (Fig. 11.29(a)) to indicate the required basic rating for the solenoid and select solenoid.
6. It should be pointed out that while Fig. 11.29(a)) is fairly typical of rotary solenoids, Fig. 11.28 for the linear case is necessarily very dependent on the shape of the plunger end form and whether or not any impact cushion is present (d.c. type). With a linear solenoid and flat plunger end form, the rise in pull force below ~ 1 mm stroke is very significant to the extent that the zero stroke pull force may not be precise. On this basis some manufacturers do not publish force data below a 3 mm stroke length. However, while Fig. 11.28 is not necessarily typical of all manufactured linear solenoids, it does illustrate the general procedure.
7. Once the solenoid is chosen, refer to manufacturer's data sheets (e.g. Fig. 11.31), and extrapolate if necessary to ensure that the actual force−stroke characteristics do match the linkage system requirements.
8. As a summary, the more important of the parameters that influence solenoid force−stroke curves are as follows: (a) plunger geometry; (b) plunger diameter, (c) coil temperature, (d) no. of ampere-turns, (e) applied power, (f) duty cycle and (g) solenoid orientation.

11.7 Limited angle torquers

These are a special type of brushless d.c. motor constrained to produce torque only through a rotation angle of +90°, in that respect an alternative to a rotary solenoid. Since the limited angle torquer (LAT) is wound single phase, there is no need for commutation circuitry, thereby not incurring the electronics cost penalty normally associated with brushless d.c. drives (section 8.6.2). As with brushless motors, the rotor carries the field magnets and the stator supports the armature windings. The rotation angle is determined typically by the number of poles, the greater the number of poles the lower the rotation angle. Figure 11.35 illustrates a four- and a two-pole type. Change of supply polarity changes the direction of torque.

Armature windings are either as a conventional brushless d.c., in slots around the periphery of a laminated stator, or toroidally wound on a solid or laminated stator.[5] For a conventional winding, the shape of the torque−stroke curve is defined by

$$T = T_p \frac{\cos \theta N_x}{2}$$

where T_p = peak torque, T = torque, N_x = no. of poles and θ = angle of rotation.

Toroidally wound types, however, produce a trapezoidal waveform (Fig. 11.36), maintaining a constant torque over a fairly wide operating angle. The cosine formula above still applies to the torque values in the roll-off portions of the curve (Fig. 11.36). Compared to rotary solenoids, torque values are not particularly high, i.e. 0.015−1.0 Nm peak torque. However, LATs are quiet in operation.

Performance calculations are made with the same set of equations applicable to d.c. motors (section 8.6.2).

Fig. 11.35 2- and 4-pole types of LAT. Courtesy of Muirhead Vatric Components Ltd, London

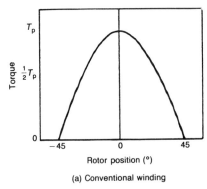
(a) Conventional winding

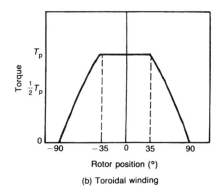
(b) Toroidal winding

Fig. 11.36 Torque–position characteristics of limited angle torquers depending on type of winding

References

1. Orthwein, W.C. *Clutches and brakes*. Marcel Dekker 1986.
2. Pedu, J.C. 'Magnetic particle clutches and brakes', *Power Trans. Design*. July 1987.
3. Spear, G.S. Tooth clutches, *OEM Design* Oct. 1985: 63.
4. Lawson, M. Rotary solenoids offer point to point angular motion, *Drives and Controls* **3**(4): 17 (1987).
5. Fleisher, W.A. Brushless motors for limited angle rotation, *Drives and Controls* **6**(2): 35 (1990).

CHAPTER 12

SPRINGS

12.1 Introduction

Many machine components are utilized on the basis of strength where deflection is very often not an issue. In the design of springs, however, a large elastic deflection is as important as strength. The main uses of springs are as follows:

1. *To apply force.* The major industrial case, e.g. to provide the operating force in brakes and clutches, to provide a clamping force, to provide a return load, to keep rotational mechanisms in contact, make electrical contacts, counterbalance loading.
2. *To control motion.* Typically storing energy, e.g. wind-up springs for motors, constant torque applications, torsion control, position control.
3. *To control vibration.* Used in essence for noise and vibration control, e.g. flexible couplings, isolation mounts, springs and dampers (see Ch. 16).
4. *To reduce impact.* Used to reduce the magnitude of the transmitted force due to impact or shock loading, e.g. buffers, end stops, bump stops.

In practical situations, springs are often used to provide more than one of the above functions at the same time. Because of superior strength and endurance characteristics under load, most springs are metallic. However, other resilient materials, e.g. polymers, are used where special properties such as a low modulus and high internal damping capacity are required. Apart from the more common types of spring, this section will also consider less well-known types such as gas springs, elastomer springs, etc. The discussion of wave springs, snap rings, disk springs and load washers is dealt with in Chapter 14.

12.2 Spring materials

The most common choice of metallic spring material, and with generally the best stock selection, is carbon steel, whether in wire or flat form. In the majority of spring applications, significant deflection is necessary, but however good the spring material, there are limits over which it can be expected to work consistently and exhibit a reasonable fatigue life. For correct design application there are occasions where for example temperature, corrosion or conductivity requirements dictate the use of a different material, e.g. stainless steel or a non-ferrous material.

Copper and copper alloys do not allow high working stresses but do have certain advantages over steel, e.g. electrical/thermal conductivity, non-magnetic properties, that on occasion may be important. Beryllium copper is the most well known of the group and obtains its spring characteristics from a precipitation hardening heat treatment. Failure to undertake the correct heat treatment procedure and so provide the optimum material structure will produce inferior properties. For those spring applications involving property requirements outside the above material capabilities, e.g. high-temperature stability, subzero ductility, high nickel and titanium alloys could be specified.

A material description of the more common types of springs is given in Table 12.1. Materials intended for high static or dynamic loadings should be quality screened for structural inclusions, surface defects and in the case of carbon steels, surface decarburization.

12.3 Mechanical aspects

Basically the life of a spring depends on six main factors:

1. The maximum working stress;
2. The working range;
3. Material structure and surface condition;
4. The stress application (type of loading, i.e. tension/compression/torsion);

Table 12.1 *Summary description of conventional spring materials*

Wire material	Condition	Type	Composition	Max. working temp. °C (5% creep at 400 Nm after 1000 hours)	Application	Corrosion resistance
Cold drawn spring wire (BS 5216)	Stress relieved	Static duty Dynamic duty music wire	0.45/0.85% C 0.55/0.85% C 0.70/1.0% C	160 140 120	Most widely used in industry. Not satisfactory for high or low temperatures. Music wire is popular	Not corrosion resistant. Performance fair at normal ambient conditions Improve by using: (a) galvanized wire (b) epoxy paint or (c) blackodize and oil
Pre-hardened carbon and low alloy steel (BS 2803)	Stress relieved	Carbon steel, low alloy steel	0.55/0.75% C 0.46/0.72% C 0.35/1.1% Cr 0.30% V max.	170 170	General-purpose materials for all types of spring. Low alloy types are equivalent to music wire at lower cost	
Annealed carbon and low alloy steel (BS 1429)	Hardened and tempered	Carbon steel, low alloy steel	0.70/1.0% C 0.46/0.6% C 0.5/1.1% Cr	175 220	Principally used for conditions involving high stress/shock loads/high/low temperatures	
Cold drawn austenitic stainless (BS 2056)	Stress relieved	Austenitic 316, precipitate hardened 301	18% Cr/8% Ni 17% Cr/7% Ni	300 250	Suitable for high temperature. Hardenable only by cold work.	Good to excellent
Martensitic stainless steel (BS 2056)	Hardened and tempered	Martensitic 420	12/14% Cr	250	Useful at high temperature, but embrittle below zero.	Not as resistant as austenitic stainless
Beryllium copper (BS 2873)	Precipitate hardened	—	1.7/1.9% Be	150	Can withstand high stresses, but is the most expensive of metallic springs	Good to excellent

5. The spring index (section 12.5.1), where applicable;
6. The working environment.

In order to specify a spring for a particular application, the designer will normally be aware of approximate loads, the forces involved and the operating environment. Having determined a material and its permissible stressing, calculations can then establish spring dimensions. Safe working stress guidelines exist for particular materials (Table 12.2). These figures assume working loads, deflecting 25–75 per cent, without buckling and with the spring properly stress relieved. Material properties are outlined in Table 12.3.

Most spring failures result from too high forces creating too high material stresses for too many deflections. While spring steels have a fatigue limit, non-ferrous materials do not, hence non-ferrous materials would not be recommended for unlimited fatigue life situations.

Working stresses may be increased by improved material quality, improved surface condition and by either pre-stressing or shot peening.

Pre-stressing (setting) is carried out by manufacturing a spring larger than the required final size then compressing to solid a number of times until the desired length is achieved, thereby introducing a degree of work hardening. It is an important operation with compression springs; it avoids permanent set when solid and also improves fatigue performance or, alternatively, allows a 5–15 per cent increase in working stress.

Shot peening, or vapour blasting, are effective methods of increasing working stresses under fatigue conditions, increasing the allowable levels outlined in Table 12.2 by 20–30 per cent. The process introduces a surface residual compressive stress thereby inhibiting surface crack nucleation. Tension springs are not suitable for this treatment.

Table 12.2 *Design stress guidelines*

Recommended design stress (N mm^{-2})							
		Compression springs			Torsion springs		
Spring material	Type of service	0.5 mm dia.	1.0 mm dia.	2.0 mm dia.	0.5 mm dia.	1.0 mm dia.	2.0 mm dia.
Music wire	L	1200	1070	960	1690	1520	1360
	A	1080	980	870	1460	1310	1160
	S	900	800	760	1280	1150	1040
Cold drawn spring wire	L	1100	1000	900	1270	1200	1000
	A	1000	900	810	1120	1070	910
	S	830	760	670	970	920	800
H and T carbon steel	L	1100	1030	920	1800	1590	1440
	A	1000	920	825	1510	1380	1230
	S	820	770	690	1330	1220	1080
H and T low alloy steel	L	1360	1300	1140	1820	1680	1550
	A	1150	1100	970	1540	1470	1320
	S	930	900	790	1380	1290	1170
Austenitic stainless	L	1040	960	840	1450	1340	1190
	A	935	860	760	1240	1130	1030
	S	780	720	620	1100	1030	900
PH stainless	L	1130	1090	1000	1580	1510	1410
	A	1020	980	900	1360	1300	1200
	S	860	820	760	1200	1150	1060
Martensitic stainless	L	640	630	625	980	970	965
	A	580	570	570	890	885	875
	S	530	525	520	820	815	810
Be.Cu 3/4H Pretempered	L	685	650	615	965	935	900
	A	570	550	525	995	880	855
	S	490	475	450	835	820	800

Notes: L = light service, typically 10^3–10^4 deflections; A = average service, typically 10^5–10^6 deflections; S = severe service, $>10^6$ deflections. For extension springs, reduce recommended stresses by 10–15%.

Table 12.3 Summary of spring material properties

Material	Duty	Average tensile strength for given wire diameter, (N mm^{-2})			Elastic limit[a] as % of tensile strength		Young's modulus, E (N mm^{-2} × 10^4)	Rigidity modulus, G (kN mm^{-2})	ρ Density (kg mm^{-3} × 10^{-6})	Spring constant, Kn (circular section) (mm min^{-1} × 10^7)	Average hardness (kg mm^{-2})
		0.5 mm	1.0 mm	2.0 mm	Tension	Torsion					
Cold drawn spring wire (BS 5216)	NS	—	—	1470	70	55	20.69	79.3	7.83	2.13	380/475
	HS	2190	1900	1670	65	50	20.69	79.3	7.83	2.13	
	NS, HD	2450	2120	1870	60	45	20.69	79.3	7.83	2.13	
Cold drawn music wire (BS 5216)	HS	2560	2320	2050	70	50	20.69	82.7	7.83	2.13	435/475
	HD	2790	2470	2200	65	45	20.69	82.7	7.83	2.13	
Hardened and tempered carbon steel (BS 2803)	All	1990	1890	1720	80/90	40/50	19.65	77.2	7.83	2.13	410/530
Hardened and tempered low alloy steel (BS 2803)	All	—	2030	1870	88/90	65/75	20.34	77.2	7.83	2.13	465/530
Cold drawn austenitic stainless (BS 2056)	All	1960	1860	1670	65/75	45/55	19.31	66.9	7.97	2.01	430/490
Precipitation hardened stainless steel (BS 2056)	All	3220	2270	2070	72/78	53/57	20.34	75.8	7.67	2.01	490/530
Martensitic stainless steel (BS 2056)	All	—	—	—	65/75	45/55	20.6	79.3	7.75	2.13	490/540
Heat-treated beryllium copper (BS 2873)	All	1050/1240	1050/1240	1050/1240	73/77	50/55	11.98	44.8	8.75	1.5	360/420

Notes: [a] = Elastic limit/proportional limit/proof stress as appropriate. NS = normal static duty; HS = heavy static duty; ND = normal dynamic duty; HD = heavy dynamic duty.

12.4 Simple flat springs

As a basic and widely used spring form, the spring characteristics of a rectangular section beam experiencing bending will depend on the mode of support. Two types are possible.

(i) Singly supported

For a simple cantilever spring (Fig. 12.1), constant rectangular cross-section, at the free end:

Deflection $\delta = \dfrac{4Pl^3}{Ebt^3}$

and

Stress at the fixed end $\sigma_0 = \dfrac{6Pl}{bt^2}$

where P = load, l = length of cantilever, b = width of section, t = thickness of section and E = Young's modulus.

Since stress in the spring material is inversely proportional to width and (thickness)^2, increasing thickness reduces the material stress more effectively than width increase. For the more general case, distance x from the fixed end:

$\delta_x = \dfrac{2Pl^3 x^2}{Ebt^3 l^2}\left(3 - \dfrac{x}{l}\right)$

$\sigma_x = \dfrac{6Pl}{bt^2}\left(1 - \dfrac{x}{l}\right)$

The design procedure is best handled iteratively, but assumptions have to be made on material type, spring load and spring length and thence to evaluate the material section if the required deflection is known or vice versa. All calculations should finally be checked to ensure the maximum advisable working stress is not exceeded (Table 12.2). Apart from a flat rectangular section, square or round wire could be used; the alternative equation forms are given in Table 12.4.

Since stress varies linearly along the cantilever and inversely with width so it should be theoretically possible to achieve a state of constant stress along the spring. This in fact results in an isosceles triangle, width b at the clamped end and zero width at the load point. Not a particularly practical proposition, but what is more common is the truncated triangle or trapezoidal spring (Fig. 12.2).

Under these conditions, for constant material thickness:

$\delta = \dfrac{6Pl^3}{BEbt^3}$

where $B = 3b_0/(2b_0 + b_1)$, b_0 = material width at fixed end and b_1 = material width at free end.

If it is required to produce an increasing rate cantilever spring, it is possible to provide the cantilever with a curved stop which allows each portion of the spring to reach, but not exceed, the design stress as it rolls around (Fig. 12.3). The radius of the stop, r, is given by

$r = \dfrac{Et}{2\sigma_D}$

where σ_D = design stress and t = material thickness.

As progressive parts of the spring contact the stop and

Table 12.4 *Summary of flat spring performance*

Spring wire section		Deflection unknown	Deflection known		Stress
Simple support	Flat	$\delta = \dfrac{4Pl^3}{Ebt^3}$	$bt^3 = \dfrac{4Pl^3}{E\delta}$		$= \dfrac{6Pl}{bt^2}$
	Square	$\delta = \dfrac{4Pl^3}{Ea^4}$	$a^4 = \dfrac{4Pl}{E\delta}$		$= \dfrac{6Pl}{a^3}$
	Round	$\delta = \dfrac{4Pl^3}{Ed^4}$	$d^4 = \dfrac{4Pl^3}{E\delta}$		$= \dfrac{6Pl}{d^3}$
Support both ends	Flat	$\delta = \dfrac{Pl^3}{4Ebt^3}$	$bt^3 = \dfrac{Pl^3}{4E\delta}$		$= \dfrac{3Pl}{2bt^2}$
	Square	$\delta = \dfrac{Pl^3}{4Ea^4}$	$a^4 = \dfrac{Pl^3}{4E\delta}$		$= \dfrac{3Pl}{2a^3}$
	Round	$\delta = \dfrac{Pl^3}{4Ed^4}$	$d^4 = \dfrac{Pl^3}{4E\delta}$		$= \dfrac{3Pl}{2d^3}$

Notes: a = side of square; d = wire diameter; l = length of cantilever; P = load; δ = deflection; t = thickness of section; b = width of section; E = Young's modulus.

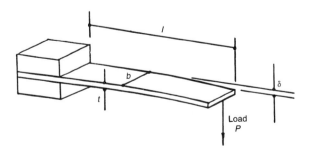

Fig. 12.1 Simple cantilever spring

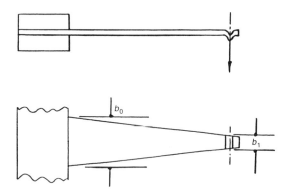

Fig. 12.2 Constant stress cantilever

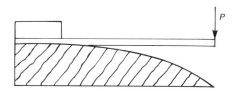

Fig. 12.3 Increasing rate cantilever assembly

reach the design stress, deflection continues as the active length decreases, thereby increasing the rate.

In general application of cantilevered springs, the frequency of the load should ideally be less than one-fifteenth the natural frequency of the spring or at least not a harmonic between 1 and 15. The natural frequency of a rectangular leaf spring is given by

$$f_n = \frac{1.015 t(Eg/\rho)^{0.5}}{l^2}$$

where g = acceleration due to gravity and ρ = material density.

(ii) Supported both ends

For this case, Fig. 12.4, the stress and deflection formulae are modified as per Table 12.4, load P applied centrally.

12.5 Helical springs

These types of spring are so designated since the spring coil as a whole is compressed or extended in length along its central axis. The wire is loaded principally in torsion, the shear stress due to the axially applied load is small.

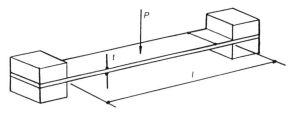

Fig. 12.4 Spring supported both ends

Hence, following formulae reflect the helical spring material in pure torsion.

12.5.1 Helical compression springs

A compression spring is a helically wound open coil, formed as a constant-diameter cylinder that offers resistance to a compressive force applied axially. The primary spring characteristics that are useful in designing a circular section wire compression spring are as follows:

Spring index $c = \dfrac{D_m}{d}$

Spring rate $R = \dfrac{P}{\delta_w}$

Free length $L_f = L + \delta_w$

Solid height $H_s = L_f - \delta_s$

Deflection $\delta = \dfrac{8nD_m^3 P}{Gd^4}$

Torsional stress $S = \dfrac{8DK_1 P}{d^3}$

where D_m = mean coil diameter = OD $- d$ = ID = d, OD = spring outside diameter, ID = spring inside diameter, P = working load, d = wire diameter, δ_w = working deflection, L = working length under working load, δ_s = deflection distance to solid, n = number of active coils, G = modulus of rigidity of spring material and K_1 = Wahl stress correction factor.

The torsional stress, S, above, is the working stress at the working load P. This should not exceed the guidelines in Table 12.2. The Wahl stress correction factor K_1 is introduced specifically at low values of the stress index to allow for the higher stresses on the inside of the coil caused by wire curvature. It is usual to identify the factor K_1 as follows:

$$K_1 = \frac{4c - 1}{4c - 4} + \frac{0.615}{c}$$

For large values of c, this correction factor can reasonably be ignored. A further stress correction factor K_2, known as Wood's factor, should also be applied to the deflection formula. This is expressed as

$$K_2 = \frac{2c^2 + c - 1}{2c^2}$$

The deflection formula then becomes

$$\delta = \frac{8nD_m^3 P K_2}{Gd^4}$$

The main factors influencing the design of helical springs are the elastic limit in torsion and the rigidity modulus of the material, the stiffness being proportional to d^4 and to $1/D_m^3$. Hence both d and D_m significantly affect spring performance.

(i) Spring index

With values of $c < 4$ there are often manufacturing problems with the high forces required to form the wire, while the higher the spring index, the more the dimensional variations occurring during manufacture. For spring selection, it is usual to choose a spring index between 5 and 10 since the spring is less prone to distortion both during manufacture and assembly. If the required spring rate is known together with the maximum working load, the optimum material condition can be chosen with Fig. 12.5.

(ii) Working deflection and solid height

We have defined the working deflection δ_w as the deflection corresponding to the working load P, giving a spring working length of L (Fig. 12.6). The solid deflection, δ_s, is defined as the difference between the

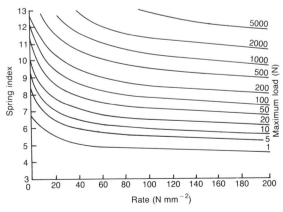

Fig. 12.5 Optimization of spring material condition for compression springs

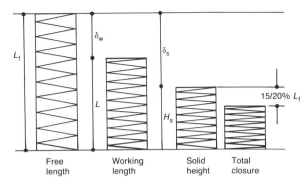

Fig. 12.6 Schematic of loaded spring conditions

no-load and solid heights, hence the clash allowance is given by

$$r_c = \frac{\delta_s - \delta_w}{\delta_w}$$

The clash allowance typically is a margin of extra deflection between working deflection and complete bottoming. An allowance of 15–20 per cent is satisfactory for most applications. Clearance between maximum deflection and solid length is necessary with compression springs because it is impractical to manufacture a spring in which all the coils are solid at the same load and therefore the relation between load and deflection is not always linear beyond 80–85 per cent deflection.

(iii) Spring ends

The number of active coils in a spring are those actually 'working'. There are four basic types of compression spring end (Fig. 12.7), the form of these affecting solid height, number of active coils, total coils, free length and seating characteristics of the spring. Table 12.5 outlines the dimensional formulae as applied to the four end types.

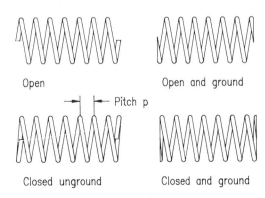

Fig. 12.7 Helical spring end forms

Table 12.5 *Dimensional configuration formulae for compression springs*

Spring characteristic	Open	End Type		
		Open and ground	Closed	Closed and ground
Pitch (p)	$\dfrac{L_f - d}{n}$	$\dfrac{L}{N}$	$\dfrac{L_f - 3d}{n}$	$\dfrac{L_f - 2d}{n}$
Solid height (H_s)	$d(N + 1)$	dN	$d(N + 1)$	dN
Active coils (n)	N	$N - 1$	$N - 2$	$N - 2$
Total coils (N)	$\dfrac{L_f - d}{p}$	$\dfrac{L}{p}$	$\dfrac{L_f - 3d + 2}{p}$	$\dfrac{L_f - 2d + 2}{p}$
Free length (L_f)	$pN + d$	pN	$pn + 3d$	$pn + 2d$

Note: d = wire diameter.

As a result of end coil effects, the spring rate tends to lag over the first 20 per cent or so of the deflection range, this being taken up with seating down the spring ends firmly. Together with a spring rate increasing for the final 20 per cent or so of the deflection range as the coils progressively close out, this leaves a central deflection zone where the rate is linear.

(iv) Natural frequency

Springs which are subject to rapid variations in displacement may show harmonics or surging frequencies corresponding to the natural frequency of the spring. Suddenly applied forces often cause a surge wave to occur and if constantly repeated, can excite the natural period of vibration to such an extent that coil clashing may unseat the spring. Harmonics above the 15th do not usually have to be considered. In other words, the loading frequency of the spring wherever possible should be less than one-fifteenth the natural frequency of the spring. In any case, the higher the natural frequency the better. Natural frequency of an unloaded spring can be calculated from the following:

$$f_n = \frac{d}{2nD_m^2}\sqrt{\frac{10^3 G}{2}} \text{ cycles per second}$$

$$= \frac{D_m}{d}\sqrt{\frac{P}{\delta_w M}} \text{ cycles per second}$$

where ρ = density and M = mass of active spring = $0.25\pi^2 n D_m \rho$.

(v) Diameter changes

The outside diameter of a compression spring increases when the coils are deflected as a result of the coil slope changing. Allowance for this should be made if the compression spring is constrained within a guide tube. When deflected to solid, the new spring diameter becomes

$$D^* = \sqrt{\frac{D_m^2 + p^2 - d^2 + d}{\pi^2}}$$

where D^* = new diameter and p = pitch.

(vi) Other wire sections

Use can be made of square and rectangular section material where spring location space is limited. As a guide, it is possible to obtain approximately one-third more load from a square-section spring than from an equivalent made in round wire. For the general case,

$$\text{Deflection } \delta = \frac{PnD_m^3}{Gbt^3\beta}$$

where b = width of wire (longest side), t = wire thickness (shortest side) and β = De Saint Venant coefficient (Table 12.6). Or, for the specific case of square wire,

$$\delta = \frac{PnD_m^3}{(0.18)Gt^4}$$

For calculation of stress, the general case is given by

$$\text{Torsional stress } S = \frac{PD_m K_1}{\mu bt^2}$$

where μ = De Saint Venant coefficient (Table 12.6) and for the square wire case,

$$S = \frac{PD_m K_1}{(0.416)t^3}$$

An important consideration in the design of square and rectangular section springs is the calculation of solid length H_s. As square or rectangular wire is coiled the wire cross-section wire deforms slightly to a trapezoidal

Table 12.6 *Helical compression spring coefficient factors for square and rectangular sections*

b/t	1.0	1.2	1.5	2.0	2.5	3.0	5.0	10	∞
μ	0.416	0.438	0.462	0.492	0.516	0.534	0.582	0.624	0.666
β	0.180	0.212	0.250	0.292	0.317	0.335	0.371	0.398	0.424

Notes: b = width of wire; t = thickness of wire; μ, β = De Saint Venant coefficients.

(keystone) shape which modifies the solid height. Material thickness after coiling, t^*, can be determined approximately by

$$t^* = 0.48t \left(\frac{\text{OD}}{D_\text{m}} + 1 \right)$$

To overcome this material section distortion during coiling it is possible to start with trapezoidal section wire and form it inside out. However, such a concept has limited applicability since the initial wire shape will only apply to a particular spring design.

(vii) The design method

The two basic formulae contain seven variables (δ, P, n, D_m, G, d and S) which clearly prevents a one-step process. In practice it is mainly an iterative procedure of manipulating the basic formulae with close reference to manufacturers' catalogue standards in order to identify possibilities. However, as the following outline demonstrates, some basic estimations from the designer are required initially, e.g. spring length at working load, space available, end type, spring material, in order to start off the process. Such design technique is ideally suited to a computerized solution. Whatever method is used the approaches cover the same ground, descending from theoretical strength of material relationships. The following represents a generalized procedure for round wire as a typical example:

1. Specify initial design requirements: (a) material type; (b) spring rate (R); (c) working load (P); (d) working length (L).
2. Calculate spring operating window. Determine: (a) safe working stress (S) (Table 12.2); (b) working deflection (δ_w) (e.g. Fig. 12.6); (c) deflection to solid (δ_s) e.g. δ_w + 15 per cent; (d) solid height (H_s); (e) free length (L_f).
3. Specify additional data: (a) mean coil diameter (D_m) — make D_m = OD to obtain trial value; (b) type of spring end (Fig. 12.7); (c) wire diameter (d).
4. Calculate spring design geometry. Determine: (a) spring index (c); (b) total number of coils (N); (c) number of active coils (Table 12.5); (d) pitch (Table 12.5).
5. Check results acceptable to application: (a) calculate actual diameters from parameters determined in (4); (b) check torsional stress still acceptable (Table 12.2).

12.5.2 Tapered helical compression springs

With the tapered or conical spring, each coil is of a different diameter (Fig. 12.8), so giving the spring a variable rate. The variable rate could also be realized in a straight-sided spring using a variable pitch. The stress imposed by any load causing a deflection is also variable from coil to coil. However, for conical spring design purposes it is the maximum stress occurring in the largest active coil which is the most relevant, i.e.

$$\text{Maximum stress} = \frac{8PD_1 K_1}{d^3}$$

where D_1 = mean diameter of largest active coil and K_1 = Wahl correction factor.

The above holds true since stress is proportional to coil diameter, but the stress in any coil can be deduced by using the appropriate diameter (Fig. 12.8). Deflection under a constant load will vary. In the case of a compression spring, first the largest coil will bottom, then the next largest and so on (Fig. 12.9). Thus

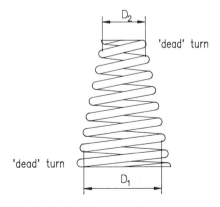

Fig. 12.8 Conical spring geometry

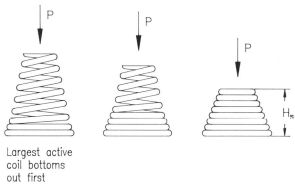

Largest active coil bottoms out first

Fig. 12.9 Conical spring compression behaviour

Deflection $\delta = \dfrac{8nPD_2^3}{Gd^4}$

where D_2 = mean diameter of smallest active coil and n = number of active turns (Table 12.5).

The solid height of a tapered spring is geometrically less than that of a straight-sided spring as individual coils 'stack' to some degree (Fig. 12.10). Thus the active solid height of the spring is given by

$H_s = n\sqrt{(d^2 - x^2)}$

For the general case of rectangular wire conical springs, the basic formulae are as follows:

$S = \dfrac{4.5D_m P K_1}{2t^2 b}$

$= \dfrac{0.225nD_2^3(t^2 + b^2)}{t^3 b^3 G}$

where t = small dimension of section and b = large dimension of section.

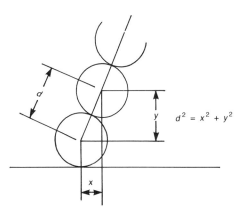

Fig. 12.10 Stacking geometry of conical spring

Although conical springs require special production facilities and hence cost more than regular compression springs, they do have useful features as follows:

1. Variable increasing rate if pitch constant;
2. A small solid height is possible if each active coil is designed to fit within the next;
3. Generally good lateral stability and less tendency to buckle;
4. As each coil bottoms, natural frequency of the spring increases, thereby reducing resonance.

12.5.3 Tension springs

Tension or extension springs absorb and store energy by offering a resistance to an axial tensile load. The same basic design formulae for compression springs also apply to tension springs except that when extension springs are close coiled in hard-drawn or oil-tempered wire (Table 12.7), there is invariably some initial tension present. This initial tension gives a variable load to the spring before the coils begin to open. When this initial tension is overcome, the spring rate follows a linear law. The actual amount of tension that can be wound into a tension spring depends both on the spring index (typically <10), and the coiling method used. the initial tension load, P_B, can be calculated from the following:

$P_B = \dfrac{\pi q d^3}{8D_m}$

where q = initial tension stress.

Table 12.7 *Initial tension present in coiled tension springs*

Spring index D_m/d	Initial tension stress (Nn mm^{-2}) (q)
3	172.3
4	155
5	137.8
6	124
7	111.6
8	99.9
9	89.6
10	79.9
11	73
12	66.1
13	60.6
14	54.4
15	48.2

Notes: D_m = mean coil diameter; d = wire diameter.

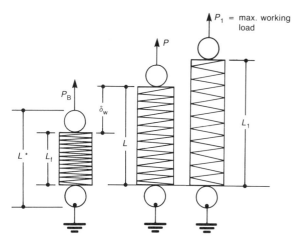

Fig. 12.11 Schematic of spring loading

Corresponding maximum values of q are given in Table 12.7. Alternatively, with reference to Fig. 12.11,

$$P_B = \frac{L_1 P - L P_1}{(L_1 - L)}$$

where L_1 = maximum working length and P_1 = maximum working load. Hence

$$\delta = \frac{8 n D_m^3 (P - P_B)}{G d^4}$$

$$S = \frac{8 D_m K_1 (P + P_B)}{d^3}$$

Spring rate $R = \dfrac{P_1 - P}{L_1 - L} \equiv \dfrac{P_1 - P_B}{\delta_w}$

Free length $L_f = L - \delta_w$

Maximum deflection $= \dfrac{P_1 - P_B}{R}$

The similarity of these equations with the helical compresssion case will readily allow solutions to alternative wire sections.

(i) Spring ends

A wide variety of extension spring ends are possible. These can include reduced and expanded eyes, loops, hooks, square ends, etc. Characteristically the end is a loop when the opening is less than one wire size, but a hook when the opening is larger. The double loop is preferred to wire diameters <1.5 mm and a full loop for diameters above this. Under fatigue conditions, cone ends are advisable.

Tension springs have no end coils that are dead, all coils

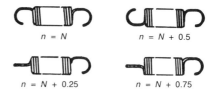

Fig. 12.12 Effect of relative end angular position on number of coils (n = number of working coils; N = number of complete coils)

whether full or partial, are active. Consequently, the relative angular positions of loops or hooks will affect the number of active or working coils (Fig. 12.12).

Most extension spring failures occur in the end zone where sharp bends and stress concentrations may occur. To maximize spring life, the path of the wire should be smooth and gradual as it is formed to the end shape, a minimum bend radius of 1.5 × the wire diameter is recommended. For a spring with central loops of the same diameter as the spring:

$$L^* = (n - 1)d + 2(OD - d)$$

where L^* = initial free length of spring.

Calculations to determine exact stresses in loops are somewhat complicated by the difficulties in determining bend radii on some designs. However, in the case of the regular hook (full loop) (Fig. 12.13) the hook experiences a principal bending stress at A and a torsional stress at B. These stresses are higher than those experienced by the spring coils in the body, the principal reasons why hooks break first and why recommended torsional stresses in tension springs are lower than in compression springs (Table 12.2). From Fig. 12.13:

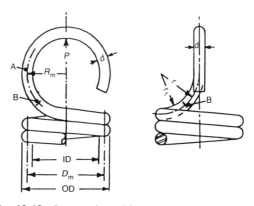

Fig. 12.13 Stresses in end loops

Stress in bending at A $= \dfrac{5.1 P D_m^2}{d^3 \, \text{ID}}$

Stress in torsion at B $= \dfrac{5.1 P R_m r}{d^3 r_1}$

(ii) Design method

The only significant difference between the design cases for compression or tension helical springs is that in the case of helical extension springs, initial tension needs to be assessed and included in the method. The following represents a generalized procedure for round wire:

1. Specify the initial design requirements: (a) material type; (b) spring rate (R); (c) working load (P); (d) working length (L).
2. Calculate the spring operating window. Determine: (a) safe working stress (S) (Table 12.2); (b) working deflection (δ_w); (c) free length (L_f), assuming $L = 85$ per cent of maximum working length; (d) maximum length (L_1); (e) maximum load (P_1); (f) initial tension (P_B), assumed to be approximately 15 per cent of working load P.
3. Specify additional data: (a) mean coil diameter (D_m); (b) end type (Table 12.5).
4. Calculate spring design geometry. Determine: (a) wire diameter, based on L_1; (b) spring index (c); (c) number of coils (n); (d) initial overall length (L^*).
5. Check results are acceptable to the application: (a) ensure initial tension (P_B) acceptable to load requirement; (b) calculate actual coil diameters from parameters determined in (4) and ensure result is practical; (c) check torsional stress is still acceptable (Table 12.2).

12.6 Torsion springs

These are subject to torque about the central axis of the spring coil. This induces a bending stress in the spring wire, tensile and compressive, hence Young's modulus of elasticity governs deflection. The springs are assumed to be anchored at one end and torqued at the other.

12.6.1 Helical torsion springs

With coiled springs of this type (e.g. Fig. 12.14), the following effects are observed when torqued:

1. If the spring is 'wound up': (a) the diameter of the helix decreases; (b) the number of coils increments; (c) the body length increases.

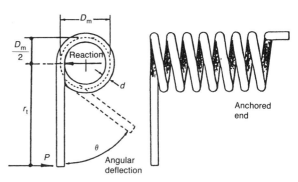

Fig. 12.14 Helical torsion spring

2. If the spring is 'wound down': (a) the diameter of the helix increases; (b) the number of coils decrements; (c) the body length decreases.

Normally the spring is fitted over a shaft or spindle, hence due care must be taken to ensure that there is adequate allowance for the size changes that may occur. It is unusual to use a torsion spring in 'unwind' mode because of two factors:

1. The vast majority of torsion springs are manufactured from hard-drawn or heat-treated wire and the residual stresses that result from the original forming operation are tensile on the inner surface of the coil. Hence, when wound up, these stresses reduce but increase if wound down. Thus if it is necessary to design the spring such that it is unwound by the applied torque, a low-temperature stress relief should be given before use.
2. The nature of the moment arm r_t in Fig. 12.14 will increment to $[r_t + D_m/2]$.

The fundamental formulae for designing helical torsion springs are based upon simple beam theory. Load is applied as a torque, Pr_t, which either turns the spring while winding or retains its position. In the case of springs of circular wire section

Stress in body of spring $S = \dfrac{32 P r_t K_4}{\pi d^3}$
(ignoring residual stress)

where K_4 = stress correction for heavily close coiled springs $= (4c - 1)/(4c - 4)$ and

Spring index $= \dfrac{D_m}{d}$ (4–14 preferred)

Angular deflection $\theta = \dfrac{64 \, P r_t L_t}{E d^4}$ radians

where length of active material $L_t = \pi D_m n + y$ and y

= the effective wire length between points of load application and the body of the spring. Hence

$$\theta = \frac{1.17 \times 10^3 \, Pr_tL_t}{Ed^4} \text{ degrees}$$

or

Angular deflection per coil $\alpha = \dfrac{3.67 \times 10^3 \, Pr_tD_m}{Ed^4}$ degrees

No. of turns when loaded N_f $= n + \dfrac{\theta}{360}$

and

Final mean diameter $D_{mf} = \dfrac{D_m n}{N_f}$

After deflection, the inside diameter of the spring may be given by

$$\text{ID*} = \frac{n(\text{ID})}{n + \theta/360}$$

The torsional spring rate R_t may be defined as

$$R_t = \frac{\text{torque}}{\theta} = \frac{Ed^4}{3.76 \times 10^3 \, D_m n}$$

Note that although the constant here is the theoretically correct 3.76×10^3, empirical testing shows 4×10^3 to produce better correlation with practice, allowing for an increase in n and a decrease in ID. As a design guide, springs made with less than 3 coils tend to buckle and are difficult to assess, whereas with ≥ 30 coils, loads may not achieve uniform coil deflection.

(i) Spring end types
It has previously been mentioned that the length L of the wire in the spring effectively includes the lengths of wire between the points of load application and the body of the spring. This is particularly so when different types of fixing are considered.

Torsion springs are also made handed, in other words the applied torque should normally be in the direction that winds the coil: loaded clockwise = wound left hand and vice versa. The spring in Fig. 12.14 is right handed.

(ii) Other wire sections
As with tension/compression springs, use may be made of square and rectangular wire sections. For a material of section bt, the modified basic formulae are

$$S = \frac{6Pr_tK_5}{b^2 t}$$

where b = radial depth of section, t = axial thickness, and

$$K_5 = \frac{3c - 1}{3c - 3}$$

for closely coiled springs.

$$\theta = \frac{2.16 \times 10^3 Pr_tD_m n}{Eb^3 t} \text{ degrees}$$

A practical example of a helical tension spring of square section wire is the spring wrap clutch (section 11.4.3).

(iii) Design method
As with other spring design calculations performed manually, this is an iterative process of manipulating the eight basic variables (P, r_t, D_m, d, S, θ, L_t, n) in the formulae outlined. The following is an example of a typical procedural method for round wire:

1. Specify the initial design requirement: (a) material type; (b) specified torque (load, moment arm); (c) type of anchorage.
2. Calculate the wire size. Determine: (a) maximum stress (S) (Table 12.2); (b) Young's modulus (Table 12.3); (c) a suitable wire diameter (d) initially excluding K_4; adjust to a standard wire size if necessary.
3. Specify additional design requirements: (a) wire diameter (d); (b) OD or ID (gives D_m); (c) total working deflection (θ).
4. Calculate spring performance characteristics: (a) spring length (L_t); (b) number of coils required (n); (c) spring index (c); (d) torsional spring rate (R_t).
5. Check results are acceptable to the application: (a) recalculate true spring stress, including K_4, based on results from (4); (b) adjust nearest no. turns to give required relative leg locations; (c) no. of turns when fully wound up (N_f); (d) maximum solid length; (e) check mean diameter when wound up (D_{mf}); (f) check shaft/hole clearance to avoid binding.

12.6.2 Spiral torsion springs

Springs coiled into a spiral form from flat strip are known as spiral springs. If during operation the springs remain separate from each other, the term 'hair spring' is sometimes used. If the spring is tightly wound and restrained such that coils touch, this is commonly referred to as a clock spring. We will only consider the former as relevant to small machines.

Springs of this type (Fig. 12.15) made of rectangular section material, deliver a linear torque for the first 360° of rotation. Above this, the coils tend to wrap down on

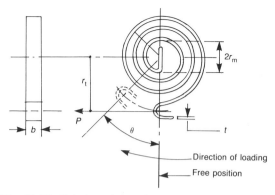

Fig. 12.15 Spiral torsion spring

to the central shaft or pin, torque increasing significantly from thereon.

For a load P and moment arm r_t,

Angular deflection $\theta = \dfrac{12 L_t r_t P}{E b t^3}$ radians

where b = width of coil strip, t = thickness of coil strip, L_t = total length of active material, and

Torsional spring rate $= \dfrac{\text{torque}}{\theta} = \dfrac{b t^3 E}{12 L_t}$

The stresses imposed on a spiral torsion spring are in bending, hence from standard beam theory

Stress $S = \dfrac{P r_t 6 K_5}{b t^2}$

where $K_5 = (3c - 1)/(3c - 3)$, $c = 2 r_m / t$ and r_m = minimum radius of curvature at centre of spiral.

12.7 Tensator springs

Tensator springs,[1] or as they are sometimes known, zero-rated springs, are coiled lengths of pre-stressed spring steel strip. The manufacturing technique provides a uniform curl tendency in the strip so that when relaxed, it forms a tightly wound coil. Performance of these spring types is made possible by preforming each portion of the flat band to the same curvature in such a way that it can be straightened without permanent deformation. The pre-stressing process controls the magnitude and distribution of residual stresses and so accounts for the high elastic deflections that can be obtained. While the design formulae given in this section are accurate, the design and manufacture of such springs are of a specialized nature, hence there is limited design scope outside the standard product range. These springs are basically of two types:

1. *Constant force*. Obtained when the outer end of the spring is extended tangentially to the coiled body of the spring (Fig. 12.16);
2. *Constant torque*. Obtained when the outer end of a spring is attached to another spool of larger diameter and caused to unwind in a direction the same as or reverse of the original wind (Fig. 12.17).

(i) Constant force springs

Used in a counterbalance type of situation, the spring produces basically constant force/deflection characteristics regardless of extension and number of turns. Because the material is pre-stressed heat-treated steel strip, it has limited fatigue performance, the smaller the coil diameter and the larger the load, the lower the life. Most design life applications would be 3000–20 000 cycles. Using beam theory

$$P = \dfrac{E b t^3}{12 C r_n^2}$$

or

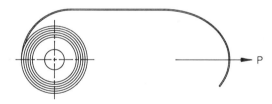

Fig. 12.16 Constant force spring

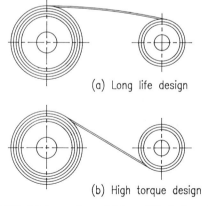

Fig. 12.17 Motor springs: (a) long life design; (b) high torque design

$$P = \frac{Ebt S_f^2}{26.4}$$

where $S_f = t/r_n$ and

Material stress $S = \dfrac{Et}{2r_n}$

where b = width of coil, t = thickness of coil strip, C = correction factor for cross curvature in extended form = 2.2, r_n = natural radius of coil in its free state and S_f = stress factor related to expected life (see Table 12.8).

When a spring is assembled into a mechanism, it is fitted over a free-running bush which is of diameter r_2 slightly greater than r_n such that (Fig. 12.18).

Storage drum radius = $r_2 = 1.15 r_n$

and

Length of strip in spring (L_t) = deflection required (δ) + $10 r_2$

i.e.

$L_t = \delta + 10 r_2$

and

Table 12.8 *Stress factors applied to tensator spring fatigue life*

Stress cycles	S_f (load)	S_f (torque)
5K	0.027	0.026
10K	0.023	0.023
20K	0.018	0.019
40K	—	0.014
50K	—	0.012
75K	0.013	—
150K	0.009	—

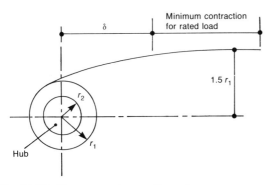

Fig. 12.18 Tensator spring configuration

$r_1 = 0.56 + 9.5 r_2 + 4 r_2^2$
$L_t = \pi n (r_2 + nt)$

From the above relationships, together with Table 12.8, it is straightforward to determine the rated load for a given fatigue life and spring dimensions to achieve the required deflection. The following characteristics define the constant force spring:

1. Diameter build-up on long springs needs to be allowed for;
2. Long deflections possible;
3. Unstable if not supported;
4. Smooth operation, very low inter-coil friction;
5. Rated load reached after $\delta = 2.5 r_1$ (Fig. 12.18);
6. Constant retracting force.

(ii) Constant torque motor springs

If a constant torque spring is fitted to a freely rotating hub, then wound on to a larger, freely rotating mandrel, it will be found that on release, the spring recoils itself on to the smaller mandrel. There are two types of configuration: the low-torque, long-life A-motor, and the higher torque lower life B-motor (Fig. 12.17) (reverse wound). During winding, the middle 85 per cent of the spring will give an almost constant torque.

Although the A-motor design gives a long life, to obtain a usable torque, the physical design size is often prohibitive, hence this type is generally not preferred. The B-motor on the other hand is a more effective use of material, but limited typically to 35 000 cycles. The design formulae are as follows (Fig. 12.19):

Torque $T = \dfrac{Ebt^3 r_3}{24} \left(\dfrac{1}{r_n} + \dfrac{1}{r_3} \right)^2$

and

$S = \dfrac{Et}{2} \left(\dfrac{1}{r_n} + \dfrac{1}{r_3} \right)$

and an optimum value of $r_3 = 3t/S_f$ with

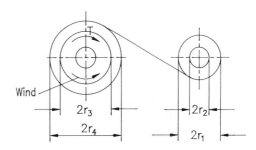

Fig. 12.19 Spring motor configuration

$$S_f = t\left(\frac{1}{r_n} + \frac{1}{r_3}\right)$$

and optimum centre distance = $r_1 + r_2 + 30t$.

Also, the following formulae are useful design approximations for establishing spring dimensions:

$r_1 = 0.56\sqrt{(L_t t + 4r_2^2)}$
$r_4 = 0.56\sqrt{(L_t t + 4r_3^2)}$
$L_t = 11(2r_3 N + tN^2)$

where N = number of working revolutions of large mandrel.

The following characteristics define the constant torque B-motor:

1. Near constant output torque, low hysteresis loss;
2. Low life;
3. High torque output;
4. Smooth in operation, no inter-coil friction;
5. Should not be used above 95 °C.

In cases where additional torque is required within a given space, multiple band motors may be employed (e.g. Fig. 12.20). The torque will be increased proportionately by the number of spring bands used in the system and has some torque/space advantage over a single spring.

12.8 Elastomer springs

Elastomeric materials, and in particular natural rubbers, have been used in engineering for many years. Their natural behaviour presents them as obvious spring material candidates, useful where long life, arduous duty and inherent damping requirements are involved. While various types of rubber mount can technically be classified as springs, this section will only consider elastomers under compression as a more conventional spring situation.

(i) Material aspects

Under continuously loaded static situations, elastomers are very often modified by stress relaxation, creep and permanent set. Creep is an increase in deformation (deflection) with time under a constant load, and stress relaxation is a decrease in force with time at constant deformation. A similar relaxation occurs when the imposed load or deformation is removed. The amount of deformation not recovered is known as set or permanent set. Both creep and relaxation effects result from internal molecular movements so there should exist a physical correlation between them (Fig. 12.21). Stress relaxation and creep rates vary considerably with the composition and type of elastomer, but in any case will be most rapid

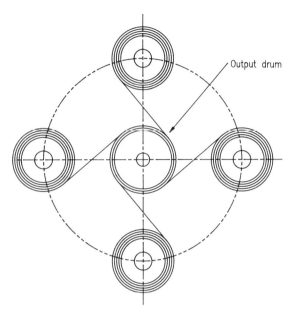

Fig. 12.20 Example of a multiple band motor

during the first few weeks under load. In Fig. 12.21, deflection follows a simple physical law, i.e.

Creep ∝ log time

For many design applications, the requirement is for acceptable load deformation characteristics throughout the required lifetime of the component. However, at sufficiently low temperatures all elastomers become stiff and rigid, and at very high temperatures viscous flow will occur. The useful temperature range, where the required 'rubbery' characteristics predominate (Fig. 12.22), will depend on the elastomer.

With moderate extension, natural rubber containing minimal filler produces little hysteresis. It is a preferred material for vibratory applications where the hysteresis can provide a small but desirable measure of damping without the danger of serious self-heating. Elastomer springs are normally designed to give a frequency ratio (i.e. the ratio of the disturbing frequency to the natural frequency) of approximately 3, corresponding to an attenuation factor of about 10 (section 16.5.2). The attenuation factor is defined as the amplitude of the disturbing vibration at one end of the spring to the amplitude of vibration at the other end of the spring.

(ii) Application

Elastomer springs can be used individually, in the form of a cushion (in parallel) or stacked serially. An example

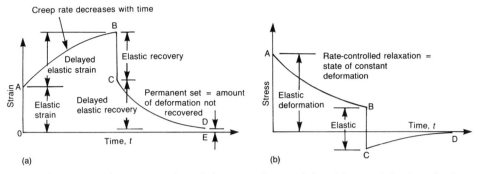

Fig. 12.21 Schematic representation of elastomer characteristics: (a) creep behaviour (load applied at $t = 0$, load removed at B); (b) stress—relaxation behaviour (load applied at $t = 0$, load removed at B)

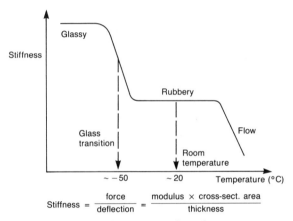

Fig. 12.22 Schematic representation of elastomer/temperature behaviour

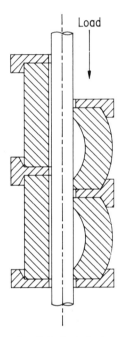

Fig. 12.23 Example of stacked elastomer spring

of stacked springs is shown in Fig. 12.23. However, since the material is essentially incompressible, ($\nu = 0.5$), bulging will occur under pressure (Fig. 12.24). If the material is compressed 20 per cent, the diameter increases approximately 20 per cent. Annular springs produce more favourable stress characteristics than solid material, but when loaded the bore ends reduce in diameter, tending to choke any guide bolt if present, and the centre bore widens (Fig. 12.24). A central guide is required in some circumstances to avoid the risk of buckling with high aspect ratios and large deflections. Larger spring travel can be achieved by stacking elastomer springs; the individual springs should, however, be separate from each other by means of spring collars such that each spring can bulge out individually.

Compared to metal springs, the choice and use of elastomer springs have a number of advantages/disadvantages. These can be summarized as follows.

Advantages:

1. Elastomer springs have a high energy storage capacity;
2. They require no maintenance;
3. Easy and inexpensive to manufacture;
4. High resilience with inherent hysteresis serving to dampen resonant vibration;
5. Can be designed to give different stiffness characteristics in different directions or even non-linear behaviour.

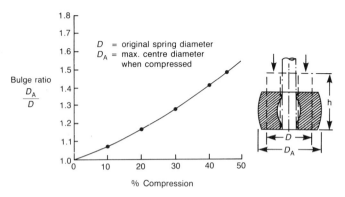

Fig. 12.24 Elastomer spring bulging

Disadvantages:

1. Viscoelastic material with mechanical properties both time and temperature dependent;
2. Narrower environmental operating window than metals;
3. High static loads can impart dimensional changes.

(iii) Mechanical design

The stiffness of rubber in compression, when the loaded surfaces are prevented from slipping (i.e. bonded or mechanically located) depends upon the shape factor s and compression modulus E_c such that the compression stiffness K for a circular block is given by

$$K = \frac{P}{\delta} = \frac{E_c A}{t}$$

where

$$E_c = E(1 + 2ks^2)$$

$$s = \frac{D}{4t} = \text{shape factor}$$

$$= \frac{\text{loaded area}}{\text{force-free area}}$$

$$\text{Compressive strain} = \frac{\delta}{t} = e$$

and P = compressive load, δ = linear deflection, t = thickness, D = diameter, A = cross-sectional area, E = Young's modulus and k = compression factor (0.5, high hardness to 1.0, low hardness).

It is generally not advisable to rely on frictional location of the springs since slippage is possible when $s > \mu/2$ where μ is the frictional coefficient.

While the above relationships generally hold true for strains up to approximately 10 per cent, above 10 per cent and no slip, strain is approximated to by $e(1 + e)$. For static load applications, deflection should be limited to 10 per cent if stress relaxation is not to be a problem. Figure 12.25 illustrates the effect of shape factor and hardness on compressive load for polyurethanes at 10 per cent deflection. As a guide, to avoid overloading, polyurethane springs should not be compressed more than 40 per cent and rubber springs not more than 30 per cent.

As previously discussed, bored elastomer springs perform better than solid pads, and in this case compression stiffness is given by

$$K = \frac{P}{\delta} = \frac{4E}{3} \frac{(D^2 - d^2)}{4t} \left(1 + k\left[\frac{D - d}{4t}\right]^2\right)$$

$$= \frac{4ED_m b}{3t}\left(1 + \frac{kb^2}{4t^2}\right)$$

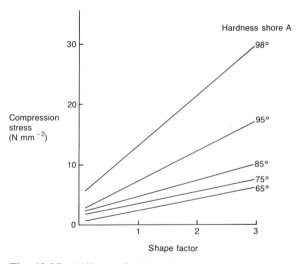

Fig. 12.25 Stiffness of polyurethane in compression

SPRINGS

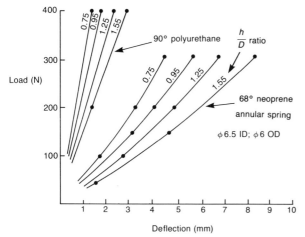

Fig. 12.26 Force–deflection curves for two spring materials as a function of height:outside diameter ratio

where D = external diameter, d = internal diameter, D_m = mean diameter = $\frac{1}{2}(D + d)$ and b = radial width of section = $\frac{1}{2}(D - d)$.

Therefore, from the above formulae it should be possible to estimate the force–deflection characteristics of a particular elastomer. Curves for 16 mm diameter polyurethane and neoprene annular springs at various aspect ratios are shown in Fig. 12.26.

12.9 Plastic springs

Because of lower mechanical properties, it is not possible to replace a metal spring with a plastic one without specific design adjustment, carefully taking the properties of the plastic into account. Even so, plastics cannot compete with steel springs directly, but there are a number of plastic spring applications now evolving which tend to use a particular material feature to achieve acceptable performance. Some types of plastic springs are considered in the following.

(i) Snap-fit hooks

Snap joints are frequently designed into plastic mouldings for ease and rationalization of the assembly or for component retention, etc. The requirement is for minimal actuations, typically one, the types of assembly being the flexural hook, snap-ring joint or torsional snap joint.[2] The basis of the flexural hook design is the simple flat spring (section 12.4). In Fig. 12.27,

Maximum percentage deformation in edge fibres ϵ = $\dfrac{3ft \times 100}{2l^2}$

where f = lug height = max deflection, t = thickness, l = length and assembly force P

$$= \frac{Ebt^3 f}{4l^3}\left(\frac{\tan\alpha + \mu}{1 - \mu\tan\alpha}\right) \text{ newtons}$$

where α = angle (use α_1 to calculate assembly force, use α_2 to calculate pullout force) and μ = coefficient of friction.

Permissible percentage distortion for plastics (ϵ) will vary from 1.5 per cent (filled resins) to 5 per cent (unfilled resins). If repeated assembly is anticipated, the strain level should be reduced by 1 per cent. This can be done by increasing beam length, reducing thickness or reducing lug height. In designing such a snap fit, the following points apply:

1. Minimum radii should be 0.5 mm.

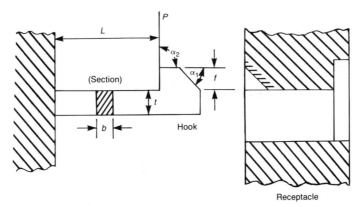

Fig. 12.27 Plastic snap-fit joint

2. The lug should be dimensioned so that shrinkage, part distortion, heating and loading in operation do not disturb the functioning of the snap fit.
3. Stressing in use and assembly forces should not act in the same direction.
4. Arm brackets should be considered to prevent the escape of the hook after joining.
5. Permanent deformation of the joint may occur if the height of the undercut is too large, hence strains too high.
6. If the snap hook fractures as a result of overload during assembly, cross-section area should not be increased, the joint should be made more flexible.
7. Lead angle should not exceed 60 per cent.
8. A conically tapering section is best (Fig. 12.28), reducing the outer strain level in the hook. Also it increases stability of the flexing member once assembled.[2]

(ii) Helical compression springs

Coil springs are torsionally loaded (section 12.5.1), requiring a high shear modulus to function acceptably. Since the shear modulus of plastics is typically one-eightieth that of steel, plastics on their own lack torsional stiffness and would be very limited in performance in a helical spring application. To be of potential use, plastics must be fibre filled with the fibre optimally orientated to enable a high shear modulus and a reasonably allowable strain level. It is possible to achieve a shear modulus of $10-20 \text{ kN mm}^{-2}$ with glass fibre and $10-50 \text{ kN mm}^{-2}$ with carbon fibre.[3]

(iii) Sulcated springs

Using reinforced plastic in a spring application in bending rather than in torsion has evolved the sulcated spring[3] (Fig. 12.29). Materials such as glass-filled plastics have excellent flexural stiffness. The spring stiffness characteristic can be tailored in the X, Y and Z planes to suit by varying the aspect geometries. Non-linear load–deflection characteristics can be achieved by changes in section form. Such springs may be produced by moulding, hot compression or resin transfer techniques.

(iv) Disc springs

These are made from glass fibre reinforced epoxy or polypropylene and use a dished profile that maximized load capacity and provides an even stress distribution on the periphery, avoiding high strains in the edge fibres. The design and performance characteristics are shown in Fig. 12.30. Features such as limit stops can be included in the design.

Springs may either be stacked or nested to give stiffer spring rates. Variable spring rates can be obtained by nesting springs at different thicknesses in such a way that they bottom against each other progressively as the load is applied.[4]

12.10 Gas springs

A gas spring is a hydropneumatic ram having the characteristics of a compression spring. It is a compressed

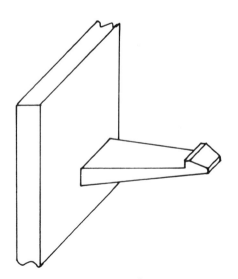

Fig. 12.28 Snap-fit tapered section

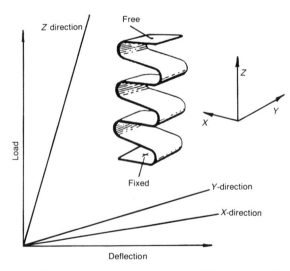

Fig. 12.29 Sulcated spring; relative stiffness in X-, Y- and Z-axes

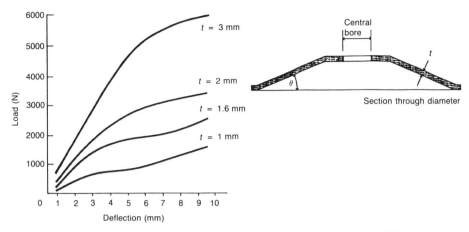

Fig. 12.30 Load–deflection characteristics for 150 mm diameter springs, different thickness. Courtesy of TBA Composites, Rochdale

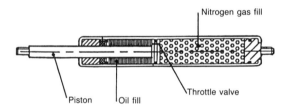

Fig. 12.31 Gas spring. Courtesy of Alrose Products, Stamford

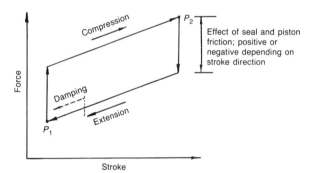

Fig. 12.32 Force–stroke characteristics of a gas spring

nitrogen-filled cylinder with an axial piston rod (Fig. 12.31). The piston has a gas throttle which controls the gas flow and allows equalization of pressure on both sides of the piston. The throttle controls piston speed, allowing free flow on piston compression but resisting flow on extension. Gas springs have the following performance characteristics:

1. The spring force increases as the spring shortens, but the force increases only a small amount for a given compression compared to other spring types.
2. The force of compression increases linearly with stroke (Fig. 12.32).
3. Rate of movement is controlled in extension (Fig. 12.31).
4. There is a hydraulically cushioned zone at the end of the extension stroke provided by free oil in the piston. To be effective the gas spring must be mounted rod downwards within ± 60° of the vertical.
5. Standard struts are undamped in compression.
6. Force in compression depends on the relative volumes of cylinder, piston rod and oil volume present.
7. As the spring extends, the force it exerts will diminish by some 20–50 per cent.

Gas spring applications are basically concerned with a lid or cover rotating about a fixed point from closed to open positions. Design application is concerned with the resolution of moments taken about the pivot point (Fig. 12.33). Since perpendicular distances from line of action of both weight (cover) and spring will alter as the cover rotates, the forces will need to be checked at all relevant positions to ensure that the application requirements are met. In Fig. 12.33,

$$P_1 L_3 = W L_1 + H_1 J$$

and

$$P_2 L_4 = W L_2 + H_2 J$$

ELECTROMECHANICAL PRODUCT DESIGN

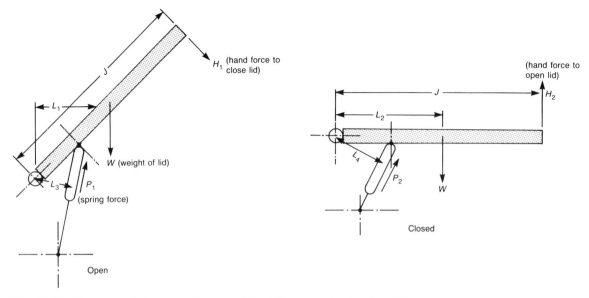

Fig. 12.33 Gas spring design example; cover/lid rotation, open to closed positions

where H_1, H_2 = applied force to close or open cover, W = weight of cover, P_1 = extended force (nominal force), J = length of cover, P_2 = force fully compressed, L_1, L_2 = length of cover mass moment arm, L_3, L_4 = length of moment arm, gas spring. For the cover to be held open $P_1L_3 > WL_1$ and $P_2L_4 < WL_2$ for the cover to stay closed, where P_2/P_1 = compression ratio.

Gas springs are usually quite powerful so can be fitted fairly close to the pivot point. If a high force spring is used, the mounting and pivot points should have adequate strength and rigidity.

12.11 Shape memory alloy springs

Shape memory is a physical metallurgy phenomenon that has been known for some 50 years.[5] The shape memory effect (SME) is demonstrated by the ability of a component to accept mechanical distortion at low temperatures and then return to its original shape when the temperature is raised above a critical point. The effect is reversible unless the shape memory alloy (SMA) is restrained in some way. Unlike bimetallics, SMAs change shape at a specific temperature and also produce a usable force and stroke. For best performance, nickel/titanium alloys are normally used.[6] Typical transformations are in the range -10 to $+100$ °C with an 8 per cent shape memory. There are four types of shape change event possible:

1. *Free recovery.* Simple recovery, i.e. heat is used to recover the shape of a bent SMA. One-way effect.
2. *Constrained recovery.* The shape change is limited by a second component, e.g. pipe clamps, compression joints.
3. *Work production.* Shape change is used to produce a mechanical movement and/or exert a force, e.g. thermally sensitive actuators, springs, valves and levers. Can be considered as an alternative to solenoids, linear short stroke devices, etc.
4. *Energy storage.* This is an event where the material displays pseudo-elastic properties below the critical temperature where small shape changes are caused by small temperature changes, i.e. the metal exhibits the behavioural characteristics of elastomers.

When used in wire form as a spring ((3) above), deformation and production of force can be exploited to the maximum. Figure 12.34 illustrates a simple compression actuator required to push or pull. Thermal energy to drive the mechanism can be supplied by convection, radiation or process fluid conduction.

Normal spring design formulae may be used for SMAs (section 12.5.1), but the following parameters and restrictions will apply:

1. Preferred actuation range is $+20$ to $+80$ °C with a repeat accuracy of ± 2.5 °C;

SPRINGS

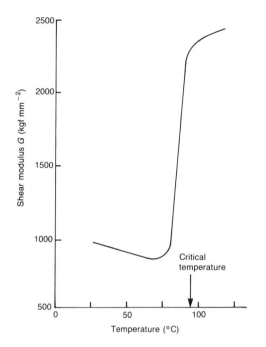

Fig. 12.34 Effect of temperature of the shear modulus of SMA

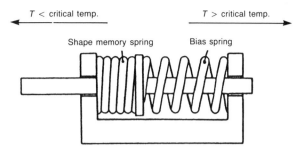

Fig. 12.35 Two-way actuator using shape memory alloy (illustrated in low-temperature condition)

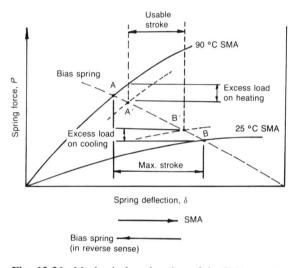

Fig. 12.36 Mechanical explanation of the SMA actuator design principle

2. Maximum recovery stress (torsional shear) can be up to 300 N mm^{-2}, but typically a maximum force of 50 N;
3. Above the critical temperature, $G \approx 22 \times 10^3$ N mm^{-2} (Fig. 12.35);
4. Below the critical temperature, $G \approx 7 \times 10^3$ N mm^{-2};
5. Use a spring index, c, of 6 (5 absolute minimum);
6. Depending on the fatigue life required, shear strain should not exceed the following: 6 per cent for a few cycles, 1–1.5 per cent for 10^4 to 10^5 cycles, 0.5–0.8 per cent for $\geq 10^7$ cycles.

The working explanation of the actuator illustrated in Fig. 12.34 is given in Fig. 12.36, illustrating the balance of forces.[7] Empirical load–deflection curves are plotted (approximate straight lines) for the SMA at typical 90 and 25 °C temperatures with the bias spring plotted in the reverse sense. The full stoke AB is reduced in practice to A'B' to allow for excess forces at the ends of the stroke: AA' of SMA over bias spring at A'; BB' of bias spring over SMA at B'.

In a practical bias spring application, the design process can be summarized as follows:

1. Obtain manufacturer's performance data on SMA spring sizes at temperatures above and below the critical. Alternatively, knowing wire sizes available, calculate spring geometry.
2. Plot as in Fig. 12.36, together with excess load requirements at either end of the stroke and the maximum stroke length requirements.
3. Check that at the points identified, shear strains in the SMA do not exceed those recommended for the given life requirement.
4. Given the bias spring force–deflection requirements at two points, can then iteratively determine bias spring geometry, making initial assumptions of spring and wire diameter (section 12.5.1).
5. Knowing the number of coils in the bias spring, hence length, the actuator assembly can be specified.

References

1. Tensator is a registered trade mark of Tensator Ltd.
2. Shonewald H Snap joints in use. *Kunststoffe* **79**(8): 732, (1989)
3. Hendry J C et al Carbon fibre coil springs. *Materials and Design* **7**(6): 330 1986
4. *Eureka* May 1988: 35
5. Miles P Shape memory materials. *Engineering* **229**(7): 29 (1989)
6. Hornbogen E Shape memory, 3 usable effects in one material. *Design Eng.* May 1990: 67
7. Bowyer M J Design and application of Ni–Ti SMA springs. *Eng. Designer* **14**(6): 15 (1988)

CHAPTER 13

SURFACE ENGINEERING

13.1 Introduction

It was recognized a while ago that the majority of engineering failure processes, i.e. fatigue, corrosion and wear, originate at the surface of a component[1] and that in most cases different material properties from those of the bulk material are required to address these problems in the context of cost-effective manufacturing. Surface coating technology was predicted as a major growth area[1] and indeed has become so over the last 10 years,[2] seeing the evolution of a number of unique process developments into the manufacturing engineering environment. In the early 1980s, the interdisciplinary subject of surface engineering arose:[3]

> Surface engineering involves the application of traditional and innovative surface technologies to engineering components and materials in order to produce a composite structure with properties not obtainable in either the base or surface material. Frequently the various surface technologies are applied to existing designs of engineering components, but ideally surface engineering will involve the design of the component with a knowledge of the surface treatment to be employed.

Over the years, engineers have devoted much time and effort into developing ways to protect and enhance metal and plastic surfaces and to provide special characteristics. Recently, many new techniques have been developed which enable the surfaces of relatively low-grade materials to be 'engineered' to meet the increasing cost-effective demands for products with greater reliability, longer life and improved performance. However, there is a need to consider the requirements and characteristics of the finishing process at an early stage in the design process, to regard it as an essential element of the production cycle and not as an afterthought.[4,5] Most benefit is therefore derived from a design that considers the essential features of the finishing process. Because the subject of surface engineering, surface finishing or surface coating may have been neglected historically, it does present opportunities to reduce manufacturing costs and improve product quality more so than many other facets of the manufacturing cycle.[6] Since few companies have anyone specifically assigned to the technology, the responsibility and understanding lie with the design engineer.

Having accepted that the effective use of coatings, treatments, etc. can provide worthwhile product improvements, the subject is very diverse, but with no recognized overall designs selection procedure. This chapter outlines the more readily available coating and finishing procedures, with the design-related aspects of each process emphasized in the individual sections. The use of published quantitative performance data is generally not a particularly reliable approach; the adequacy for the job has to be experimentally tested since reproducible standard tests do not exist to cover all the circumstances likely to be encountered in practice. Corrosion testing is reasonably well defined and meaningful, but there is no standard wear test that reproduces practical tribological phenomena. The range of surface engineering processes readily available to manufacturing industry is outlined in Fig. 13.1.

Whatever the process chosen, there are a number of common considerations:[5]

1. Good-quality base metal and sound construction of the article should apply; the coating should not be relied upon to cover up flaws.
2. All surfaces to be coated need to be adequately cleaned first and the design should ensure that all surfaces are accessible for surface preparation.
3. Sharp edges and corners, deep recesses and blind holes should be avoided.

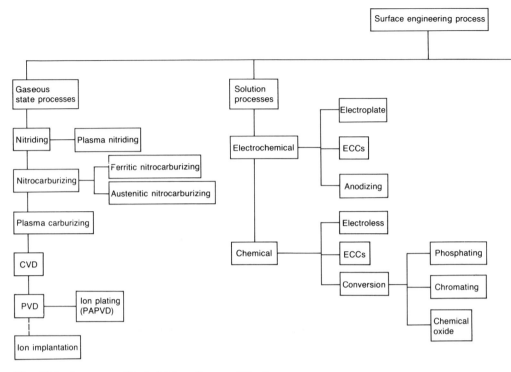

Fig. 13.1 Summary of industrial surface modification processes

4. Aim to design clean smooth lines in order to ensure uniform finishes. This also avoids accumulation of corrosive agents, e.g. in crevices, pockets, folds.

13.2 Electrochemical and electroless treatments

Probably the group of surface-coating processes offering the greatest diversity of properties and applications are those based on electrochemical and electroless treatments, all providing a means of enhancing the properties of a material in a precise and controllable manner. These treatments include the electro and electroless deposition of metals, alloys and composites as well as the production of conversion coatings by anodic oxidation (anodizing). These technologies are very important to industry. To be able to cover inexpensive and standard base materials with a thin coating of specific properties extends their use to applications which may otherwise prove very expensive.

Electroplating generally has held sway as one of the most important ways of creating metallic coatings to modify surface properties, i.e. to improve corrosion and wear resistance, to impart desirable electrical or magnetic properties or to achieve a purely decorative finish. Composite systems are co-deposits of particulate matter ranging from organic polymers to metal alloys and refractory materials, generally aimed at improving wear resistance, but coating properties will vary significantly depending upon the supporting plated matrix, the type of co-deposit and its particle size and distribution. Electroless deposition is a rapidly developing area of finishing, producing uniform thickness deposits, evenly coating the most complex metal or plastic surface.

13.2.1 Electroplating

Electroplating is a solution treatment application process of metallic coatings on to electrically conductive surfaces. Electrodeposits tend to vary in thickness over the surface of an article due to variation of current density with article shape and its positioning relative to the anodes. In order to obtain a specified minimum thickness, a much higher average thickness needs to be applied. Apart from improving the plating process, a number of straightforward design rules should be observed.[5] In general, sharp changes in contour should be avoided as should deep recesses, i.e. countersink any holes and round off leading edges.

SURFACE ENGINEERING

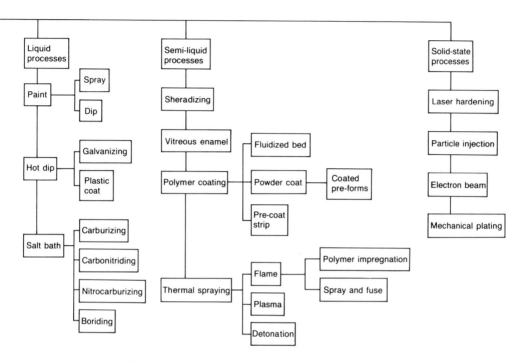

Articles are preferably electroplated before assembly, thus due allowance should be made on threads and close tolerance fits. With external metric threads, a diameter factor of ×4 applies, i.e. a 25 μm deposit increases the effective thread diameter by 0.010 mm. The factor will vary with the thread form and clearly depends on the flank angle. However, the deposit thickness on external threads is not uniform and the effect of deposit thickness variation is to increase the compound effective diameter rather more than might be expected from a geometrical factor. Suggested thread diameter allowances for plating are given in Table 13.1. With internal threads, the deposit is usually restricted to the first few threads, hence there is no usual requirement to tap oversize.

The base metal may influence the type of deposit suitable for a given application; for example, zinc provides corrosion resistance on steel, but it is not suitable for that purpose on copper and copper alloys. Electrodeposited metals tend to be finer grained, more brittle, harder and usually more wear resistant than wrought alloys. Many deposits have internal stresses which can be minimized by the choice of process conditions. A summary of types of electrodeposit, properties and uses is given in Table 13.2. Some wear guide properties of electroplates are shown in Fig. 13.2.

Table 13.1 *Allowances by which diameters of external threads should be reduced from basic size to allow for plating thickness*

	Major diameter allowance (mm)	
	≤8 mm	>8 mm
Minimum risk of interference		
Deposits with thickness to BS 3382 (11)	Standard thread	Standard thread
Deposits with thickness greater than specified in BS 3382 up to 1.5 μm	0.051	0.061
Deposits greater than 1.5 μm and up to 2.5 μm	0.081	0.061
Deposits greater than 2.5 μm and up to 3.3 μm	0.122	0.102
With some selective assembly		
Deposits up to 1.0 μm	Standard thread	Standard thread
Deposits greater than 1.0 μm up to 2.5 μm	0.051	0.061
Deposits greater than 2.5 μm and up to 3.0 μm	0.081	0.061
Deposits greater than 3.0 μm and up to 3.8 μm	0.122	0.102

Table 13.2 Summary of electrochemical coatings

Plating	Base material	Thickness μm	BS specification	Properties	Uses
Aluminium	steel, titanium, zinc	25	—	No risk of hydrogen embrittlement. Deposits may be anodized and dyed. Gives bright finish on low grade aluminium and zinc base alloys. Coating very pure. Deposit harder than commercial aluminium.	Corrosion protection of steels and fasteners. Wave guides, parabolic mirrors and aerospace components.
Black chrome	nickel undercoat over steel, copper or aluminium	up to 12.5	—	Deposit is 75% chromium, remainder is oxygen.	Optical instruments. Useful selective absorption properties.
Black nickel	nickel plated steel	0.15	—	Thin deposits contain nickel and zinc sulphides. Typically 50% Ni, 7% Zn & 15% S. Protective non-reflective coating. Early coatings had low humidity resistance. Can be stabilized with a chromate conversion film.	Optical instruments and solar collectors. Provides non-reflective decorative effect.
Brass	steel, zinc alloys	0.5–25	—	Requires lacquer to prevent tarnishing. Alternative to chromium plating for decorative finish. Vary colour of deposit by altering plating conditions.	Earliest commercially plated alloy. Steel fittings, decorative purposes etc. Pre-vulcanizing coating on steels; copper content must be in the range 72–78%
Chromium	nickel plate, steels, copper alloys, aluminium, titanium	0.25–1.0 (decorative) 10–500 (engineering)	1224	Load bearing ability (assuming substrate support) Hard wear resistant deposit, exhibits low-friction non-stick quality. Chromium is virtually unwettable and lubrication may well need to be considered. Can be ground and polished to a mirror like finish. Working temperature up to 650°C. A low temperature process; can be deposited onto a variety of base metals. Process modification can produce a porous deposit where oil retention is required. To take full advantage of the high hardness of chromium electroplate, deposit directly onto steel without any intermediate layer. Fatigue strength is reduced by hard chromium plating, minimized by heat treatment. Deposit is in a stressed condition so exhibits a cracked appearance, 1–2 μm. Satin and black finishes available. Below 0.5 μm, coating is porous. Coating brightness depends on the condition of the nickel layer.	Heavy deposits of chromium are known as 'hard' although hardness doesn't differ substantially from decorative coatings. Most widespread wear resistant electroplate. Low friction applications against most commonly used engineering materials.

				Resistance to corrosion is proportional to the thickness of the underlying nickel layer. More recent micro-cracked chromium has a widespread network of very fine cracks instead of the more localized coarse cracks in the conventional deposit; improves pitting/corrosion resistance.	
Cobalt	steel	up to 500		Expensive, hence a restricted use. Excellent high temperature wear resistant properties. Stronger than soft nickel deposits. Inherent tensile stress in many cobalt deposits may approach very high values. Heat treatment at 400°C increases squareness of hysteresis loop and decreases coercive force of some coatings. Can be alloyed with nickel or tungsten.	Coating metal forming dies. Wear resistant alloys cobalt-tungsten and cobalt-molybdenum can be electrodeposited. Soft magnetic coatings with low coercivity.
Copper	steel zinc alloys	up to 500	—	Can be bright, satin coloured or bronzed as deposited. Lacquer to provide corrosion and wear protection. Rarely has deposit high internal stress.	pcb's, electro-forming and electrotyping. Stopping off for carburizing treatment. Undercoat for other plating processes, e.g. nickel and chromium. Drawing lubricant.
Gold	brass copper alloys zinc alloys silver	0.5–20	4292	Resistant to tarnish and oxidation. Can be coloured by addition of alloying elements e.g., nickel, cobalt, silver and copper. Porosity possible with thin deposits.	Electrical contacts, components and connectors. Decorative finish. Gold alloys developed for wear inhibition.
Lead	steel	2–25	—	Less porous than hot-dipped coats of similar thickness. Resistant to sulphuric acid. Coating easily worn or damaged. Corrosion protection indefinite against many substances. Minimal internal stresses; plated articles will withstand considerable deformation before coating fracture.	Not for decorative purposes. Nuts and bolts. Equipment handling brine. Overlays on plain bearings. Steel strip.
Nickel	steel copper alloys zinc alloys	7–30	1224	In conjunction with hard chrome for the production of decorative nickel-chrome finishes. Excellent ductility. Even heavy deposits are porosity free. Will tarnish under outdoor conditions, tend to overlay with a thin chromium flash deposit. Fairly hard deposit, 150–400 Kg mm^{-2}; Machinable. Service life of coating influenced by its physical properties. For decorative purposes, bright, semi-bright or satin finishes available.	As undercoatings for chromium deposits. Decorative and protective coating. Hydraulic components, electroforming etc.

Metal	Substrate	Thickness		Properties	Applications
Palladium	copper alloys	—		Bright nickel deposits contain sulphur and are generally unsuitable for engineering use. General usefulness as a coating.	Suitable for sliding contacts under high mechanical pressure. Pcb edge connectors. General electronics connectors.
Platinum	silver, nickel copper	up to 10		Low cost, low stress deposits. Fairly hard, low contact resistance.	Decorative applications. Useage declining in favour of rhodium.
Rhodium	nickel silver	up to 125		Deposit tends to have internal stress.	Slip rings. Telecommunications switching equipment. Sliding contacts operating infrequently.
Silver	nickel–silver anodized aluminium steel brass copper alloys	5–25	4920	Highly wear resistant. Hard, non-tarnished aesthetic coating. Exeptionally high corrosion resistance. High reflectivity. Readily forms a tarnish film in a sulphur containing atmosphere.	Earliest decorative electroplate application. Electrical contacts, bearings and slip-rings. Reflectors, cutlery etc. Stop off for nitriding treatment.
Tin	steel	2.5–30 (8–13 on brass)	1872	Poor adhesion to brazing or solder. High reflectance. Excellent resistance to alkalis and detergents. Included as a continuous production line process. Deposit is dull, overcome by subsequent flow brightening (momentary surface melting after deposition). One of the easiest metals to deposit. Plating solutions have high throwing power.	Food industry generally, (metal is non-toxic) Electrical equipment (for solderability).
Zinc	steel	5–25	1706	Uniform thickness, effective and cheap method of corrosion protection. Good adhesion and post-formability. Bright zinc plating and various passivation treatments enable attractive finishes. Basically non-porous. Specify subsequent heat treatment to minimize hydrogen embrittlement. For extra corrosion protection, specify a chromate passivation.	Protective coating for steel. Wide range of components, domestic appliances, fixtures and fittings etc. Frequently used as a base for paints. Alloys provide overlays on journal bearings.
Zinc-Nickel	steel	7.5	—	Twice the corrosion resistance of ordinary zinc as plated. Average 12% nickel in deposit. Can be conventionally chromated. Attractive semi-bright appearance. Good thermal stability, uniform distribution of nickel.	Automotive etc. High salt spray corrosion resistance @ 5–15% Ni.

Zinc-Cobalt	steel	10	—	0.8% cobalt in deposit. Up to three times the corrosion resistance of zinc plate when chromated. Hardness 170–190 Kg mm^{-2}.	
Anodizing	aluminium (titanium) (magnesium)	5–10 (decorative trim) 1–25 (conventional anodizing) 25–150 (hard anodizing)		Ordinary anodizing hardness 200–300 Kg mm^{-2}, hard anodizing 350–650 Kg mm^{-2}. Films are uniform in thickness, have good heat transfer characteristics and are thermal shock resistant. Long term exposure to 500°C acceptable. High electrical resistance; withstand voltages >1000 Vd.c. Film acts as an electrical insulator. High corrosion resistance, can be improved further by chromate sealing. Freshly formed anodic film will absorb a wide range of water soluble organic dyestuffs. Cannot be applied to assembled or riveted parts since any trapped electrolyte is corrosive. Different type of aluminium alloy should not be anodized at the same time. Hard anodizing may lower the fatigue strength of some alloys by up to 50% Heavily worked surfaces tend to be grey rather than silver after anodizing. The more highly alloyed the component the darker and rougher the coating appears whatever the pretreatment or anodizing process. Cast surfaces anodize poorly compared to machined surfaces. A hard anodized finish is often rough so may be lapped or ground. Coatings craze on bending or as a result of excessive thermal fluctuations, the thicker and smoother the original coat, the more crazing is apparent.	Reflective finish, used generally for decorative, protective or wear resistant purposes. Plastic injection moulds. Gears, valves, slides etc. Architectural fittings. Electrical shielding.

ELECTROMECHANICAL PRODUCT DESIGN

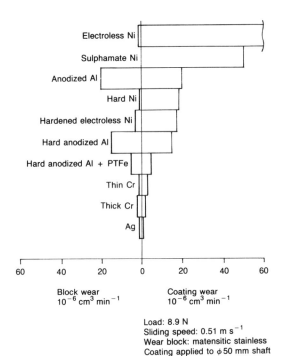

Fig. 13.2 Wear rate of coatings sliding against a hardened stainless steel counterface. Copyright ASTM. Reprinted with permission

Certain materials, especially cold-worked or high-strength steels, absorb atomic hydrogen during processing in aqueous solutions and become subject to delayed brittle failure well below the yield strength, i.e. hydrogen embrittlement. The risk can be reduced with a low-temperature stress relief before and after plating.

(i) Chromium electroplate
Of all the various electroplates available, chromium plate probably has the longest and most effective history as a low-temperature coating process. For engineering purposes the plating thickness is greater than that used purely for decorative purposes and is typically called hard chrome.

Chromium plate is brittle and cracks spontaneously at $\sim 0.5\,\mu$m thickness, cracking occurs repeatedly in layers, but successive crack patterns do not usually coincide. However, substrate distortion can cause macrocracks which can produce unsightly corrosion sites. Uncracked chromium acts as cathodic areas while the underlying nickel exposed at the crack sites acts anodically to generate electrolytic action (see Table 13.2). The so-called microcracked chrome-plated system was introduced to overcome the corrosion problem. In this system a ductile $20\,\mu$m layer of semi-bright nickel is first applied to the steel followed by a $10\,\mu$m layer of bright nickel and a $0.8\,\mu$m layer of chromium on top, deliberately microcracked. Galvanic corrosion is then limited because the anode:cathode ratio is much more favourable than when macrocracks appear.

13.2.2 Electroless plating

Electroless plating is the deposition of metals by controlled autocatalytic reduction, i.e. the process continues automatically. Metal ions are reduced to the metal by a chemical agent, the surface of the workpiece catalysing the reaction so that a coating is produced on it. The general features of electroless plating may be summarized as follows:

1. The process requires no power supply;
2. The ability to coat evenly the most complex of surfaces;
3. Can be applied to non-conductors with a suitable pretreatment, i.e. pre-etching followed by sensitizing with salt solutions;
4. More expensive than electroplating, the high cost and relative inefficiency of reducing elements plus unavoidable solution losses;
5. Possible to incorporate fine particles, e.g. SiC, Al_2O_3, polytetrafluoroethene (PTFE) in electroless deposits, but then stabilizers need to be added to prevent particles acting as nuclei for metal deposition in their own right.

The most important industrial coatings are based on nickel (Table 13.3), but electroless plating need not be confined to nickel, i.e. copper, cobalt, tin, gold and palladium are also capable of electroless deposition. Nickel–phosphorus is one of the important compounds, phosphorus imparts a passive surface film on to the nickel so that compared to electroplated nickel, corrosion resistance is better. When the coating contains 10–12 per cent phosphorus, it offers excellent corrosion resistance to steel. Post-plating heat treatment may be carried out to reduce hydrogen embrittlement, improve adhesion or increase hardness, typically at 190–230 °C. Maximum hardness is obtainable from 1 hour at 400 °C, but heat treating for maximum hardness will produce a contraction of the coating that results in microcracking. Such microcracking may well affect strength and lower corrosion resistance.

Table 13.3 *Summary of electroless plating types*

Electroless coating	Hardness (kg mm^{-2})	Thickness (μm)	Properties	Uses
Nickel–phosphorus (7–13% phosphorus)	450–650	10–25	Deposition of electroless nickel gives a chemically adherent bond ensuring no separation from substrate Working temperature up to 550°C Usually pore free, excellent corrosion resistance No levelling of rough substates. Heat treating >200°C precipitates nickel phosphide and increasing room temperature hardness Wear resistance may equal chromium at best and less wear of the counterface Ability to cover complex shapes with uniform coatings, e.g. threads No problem with throwing power.	Most suitable for general engineering Corrosion-resistant in many environments Numerous industrial components Tools and surgical instruments Providing sanitary coating for food processing and packaging machinery Textile machinery parts Electromagnetic shielding Parts are made solderable by electroless nickel plating Low coercivity useful for fast switching memory devices
Nickel–boron (0.5–0.7% boron)	650–750	—	Grey in colour, lacklustre Partly amorphous, partly crystalline Heat treatment for 1 hour at 400°C raises hardness to 1200 kg mm^{-2} due to precipitation of nickel borides Higher temperature capability than nickel–phosphorous Wear resistance lower than nickel–phosphorous Salt spray corrosion resistance inferior to nickel–phosphorus	Deposits <1% B have good electrical conductivity and solderability so used on electronic components Uses generally as for nickel–phosphorus coatings
Cobalt–boron (1.7% boron, 1% carbon and 0.05% nickel)	700 as deposited 1000 after heat treatment	Up to 0.5	Effective protection against the corrosion of steel Low coercivity possible with a low phosphorus content Magnetic properties are critically dependent on preparation conditions and deposition variables	Magnetic properties, wear resistance, electrical and thermal conductivity applications

13.2.3 Electrodeposited composite coatings

The properties of electrodeposits can be modified by the incorporation of solid particles. These particles are added to the plating bath and kept in constant suspension, becoming trapped by the metal ion during deposition to produce the composite. The idea of composite coatings originated as an inherently better controlled alternative to flame-sprayed coatings. They can be developments of electrolytic, electroless and anodizing processes. There are four general application categories of electrodeposited composite coatings (ECCs) as follows:

1. *Wear resistance.* Wear properties are strongly influenced by the co-deposited particulate, typically 30 per cent by volume and 1–10 μm size. Examples are Al_2O_3, TiC, WC, SiC and CrC_3.
2. *Dry lubricant.* Basically soft particles with a low

shear strength dispersed in the metallic matrix. Examples are MoS_2, graphite and PTFE in electroless nickel.
3. *Heat-treatable metals/alloys*. Particles are co-deposited and then the composite is heat treated. Typical examples are Cr+Ni, B+Ni and Al+Co.
4. *Mechanical properties*. Improves room temperature and elevated temperature strength. Typical examples are Al_2O_3 in Au and TiO_2 in Ni.

A summary of commercially available composite coatings is given in Table 13.4. Although a very wide matrix metal/composite phase choice is possible, nickel tends to be the predominant matrix material. Ultimate properties depend not only on the particle and matrix materials but also on particle size and the consistency or the manner and the amount of particles entering the coating.

(i) Electrolytic composite coatings
These will have excellent adhesion to the substrate but some base materials may require a subsequent low-temperature de-embrittlement treatment. To date chromium carbide, silicon carbide and PTFEs have been successful, but it is not possible to achieve uniform particle incorporation on shapes with blind holes or re-entrant angles. Solid lubricants can be incorporated, e.g. or re-entrant angles. Solid lubricants can be incorporated, e.g. PTFE (Table 13.4), but molybdenum disulphide has not been successful as the desired lubrication properties are not achieved. The matrix material is limited to those that are efficient in plating terms, i.e. nickel or cobalt. Chromium is unlikely as a matrix as there is too much gas evolution at the anode and cathode. The stimulus for the development of ECCs includes the need for high-strength and wear-resistant materials with high elastic moduli, especially at high temperature. Lightweight particles are of special interest because of the prospect of high moduli-to-density ratios.

(ii) Electroless composite coatings
These overcome the disadvantage of electrolytic plating in that a coating of uniform thickness can be deposited on all surfaces; however, coating thickness is limited to a maximum of 60 μm. The most successful commercial process is electroless nickel + PTFE since this inherently avoids the problem of the particles themselves becoming coated in the plating solution.

(iii) Multistage composite coatings
This is another way to achieve a composite, typically creating a porous coating matrix and then impregnating with or infusing a low friction polymer, typically a fluorocarbon (Table 13.4). The composite matrix may either be anodized aluminium (section 13.2.4) or porous nickel. The porous films can also be impregnated with a range of substances, e.g. oils, waxes, silicones.

13.2.4 Anodizing

The process of anodizing is normally applied to aluminium and its alloys and involves passing an electric current through the component (anode) in a dilute acid to convert the aluminium surface into a strongly adherent layer of aluminium oxide. The process equipment resembles that for electroplating. The porous nature of some anodic films allows for production of coloured coatings by the use of organic dyes or deposition of metallic pigments, and a wide range of decorative and protective finishes are possible. The anodized surface is finally sealed in boiling water or steam to hydrate the film, seal the surface and increase corrosion resistance.

Applications of anodizing are summarized as follows:

1. *Decorative*. To protect polished and other finished surfaces and to provide colour. For decorative work the majority of aluminium alloys are satisfactory, but for maximum brightness and clarity use previously polished super-purity aluminium. Colour dyes can give brilliant effects, matt or satin finishes may be applied to minimize imperfections in surface structure. Referred to as dyed anodizing.
2. *Wear resistance*. A hard, brittle oxide is produced on a soft substrate, hence only light loadings with no impact conditions. Since sealing reduces wear resistance by ~20 per cent, one may wish to avoid water sealing. Referred to as hard anodizing.
3. *Corrosion resistance*. Depends on the thickness of the anodic film and the effectiveness of the subsequent sealing process. Widely used as a protection against general atmospheric and marine environments. Anodizing does not resist strong acids or alkalis.
4. *As a base for paints and other organic finishes*. In their unsealed condition, anodic oxide coatings are an excellent base for paints, lacquers, resins, etc. The good protection obtained in this way is due to the porous and highly adsorptive nature of anodic coatings, their chemical inertness, electrical inertness and increased wear resistance. Not normally used purely as a basis for paint as conversion coatings are cheaper.
5. *Specialist applications*. Those associated with

Table 13.4 *Electrodeposited composite coatings*

Matrix	Composite phase	Thickness (μm)	Hardness (kg mm^{-2})	Properties	Uses
1. *Electrolytic*					
Nickel (elnisil)	Silicon carbide[a] (0.5–3.0 μm particle size)	100	Apparent 550 particle 2400	Superior wear profile to hard chrome; Excellent oil-retention properties; Carbide uniformly distributed in nickel deposit; Carbide is 5% by weight or 30% by volume	Cylinder bores (air cooled); Extrusion screws, injection moulds, tools generally
Nickel (elniflon)	PTFE (7–8 μm particle size)	Min 12	300	PTFE 15% by volume; Suitable for low-temperature applications down to $-55°C$; PTFE will act as a running-in lubricant	Pneumatic valves and cylinders; Water pumps
2. *Electroless* (autocatalytic)					
Nickel	Silicon carbide[b] (0.5–4.0 μm particle size)	To 20	1600	20–24% by volume of carbide; Dull appearance but can be polished; Heat treatment improves wear resistance; Lower friction than electroless nickel alone	Metal-forming dies; Abrasive wear applications, e.g. paper handling; Resin working equipment
Nickel (niflor)	PTFE	5 (60 max)	250	Up to 25% PTFE by volume; Nickel acts as a binder for the PTFE; Uniformity of coating; Can increase hardness to 400 kg mm^{-2} after heat treatment at 300°C for 1 hour; Low friction coefficient, self-lubricating and non-slip; Corrosion resistance not as good as normal electroless nickel	Mould tools; Valves, pumps and machine tools; Instrument mechanisms; Stainless steel threading; Pneumatic cylinders
3. *Multistage*					
Nickel (nedox)	Infusion of low-friction polymers	—	—	As deposited porous nickel is infused with PTFE, etc.; High flexibility down to 80°C; Upper limit 260°C	Food industry self-cleaning operations; Resists salt-water corrosion
Anodized aluminium (tufram)	Low-friction polymer	—	—	Self-lubricating down to $-220°C$; Maximum operating temperature 300°C; Process is an extension to normal anodizing; Imparts previously unobtainable levels of wear resistance, corrosion resistance, etc.; Advantages of hard anodizing with low friction polymers	

[a] Dispersed particles may also be alumina, boron, carbon, titania or tungsten; these are dispersed in the electrodeposition bath and become positively charged, are attracted to the cathode and encapsulated with the electrodeposited metal.
[b] May also be alumina, boron carbide, diamond, chromium carbide, tungsten carbide, etc.; general procedure is to suspend particles in stabilized electroless nickel solution by mechanical or air agitation so that they are randomly included during the formation of the deposit.

specific properties of the anodized coating, i.e. electrical or thermal insulation. The oxide is an electrical insulator, the breakdown voltage approximately proportional to the film thickness.

The general properties of the anodized finish are given in Table 13.2. Most aluminium alloys can be provided with a moderately good anodic oxide film provided that protection is the objective and appearance is of no great concern. A specific breakdown of alloys suitable for different types of anodizing is given in Table 13.5, the chemical composition of the alloy having a decisive influence on the appearance of the film.

As an extension to normal anodizing, a low-friction surface can be produced by impregnating and sealing the porous oxide with either PTFE or a solid lubricant. The depth and density of surface porosity vary with the rate and conditions of film growth, anodizing conditions and type of electrolyte. However, the pores are not interconnecting, hence attaching the polymer demands the production of a deeply etched surface with undercut cavities to provide the mechanical keying effect. The non-

Table 13.5 *Guide to selection of finishes on cast and wrought aluminium alloys*

Material designation[a]	Nominal composition								Suitability for		
	%Al	%Cu	%Mn	%Si	%Mg	%Ni	%Cr	%Fe	Protective anodizing[b]	Decorative anodizing	Bright anodizing
LM0	99.5								VG	VG	VG
LM2	Rem	2.0		10.0					M	U	U
LM4	Rem	3.0		5.0					G	M	U
LM5	Rem				5.0				VG	VG	G
LM6	Rem			12.0					M	U	U
LM9	Rem			12.0	0.4				M	U	U
LM10	Rem				10.0				VG	M	U
LM12	Rem	10.0		2.5					M	M	U
LM13	Rem	1.0		11.0	1.0	1.5			M	U	U
LM16	Rem	1.0			0.5				G	M	U
LM18	Rem			5.0					G	M	U
LM20	Rem		0.5	12.0				1.0	G	M	U
LM22	Rem	3.0	0.5						G	M	U
LM24	Rem	3.0		8.0				1.3	M	M	U
LM25	Rem			7.0	0.5				G	M	U
LM26	Rem	3.0		10.0	1.0				M	U	U
LM27	Rem	2.0	0.5	7.0					G	M	U
LM28	Rem	1.5		19.0	1.0	1.0			U	U	U
LM29	Rem	1.0		23.0	1.0	1.0			U	U	U
LM30	Rem	4.5		17.0	0.5				U	U	U
1080A	99.8								E	E	E
1050A	99.5								E	VG	VG
1200	99.0								VG	VG	VG
2014A	Rem	4.25	0.75	0.75	0.5				M	M	U
2031	Rem	2.0		0.8	0.9	1.0			M	M	U
3013	Rem		1.25						G	G	M
3015	Rem		0.6		0.5				G	G	M
4043A	Rem			5.0					G	M	U
5005	Rem				1.0				E	VG	G
5056A	Rem				5.0				G	G	M
5083	Rem		0.7		4.5		0.15		G	G	M
5154A	Rem		0.3		3.5				VG	VG	G
5251	Rem				2.25				VG	VG	G
5454	Rem		0.75		2.7				VG	VG	G
6061	Rem			0.6	1.0		0.2		VG	G	M
6063	Rem			0.4	0.75				E	VG	G
6082	Rem		0.7	1.0	1.0				G	G	M
6463	99.8			0.4	0.75				VG	VG	VG

Notes: E = excellent; VG = very good; G = good; M = moderate; U = unsuitable.
[a] BS 1490 or BS 1470.
[b] Appearance of no concern; selection also applies to hard anodizing.

stick anodized aluminium surface is self-lubricating down to −220 °C with an upper operating temperature of 300 °C.[7]

13.2.5 Plating of plastics

A number of plastics can now be plated, enhancing both the physical characteristics and the visual appearance of a plastic item leading to a more expensive metallic appealing material from the consumer's point of view. With additional ease of production, lightness and corrosion resistance make plated plastic able to replace metals in a number of applications.

The choice of the correct plastic is important; unsuitable materials choice can result in expensive failures. Some fire-retardant grades are easy to plate, others difficult — the fire-retardant additives often make it difficult to obtain an initial mechanical 'key' for the coating. A few plastics can be made conductive for direct electroplating, e.g. carbon filled, but added conductive filler is generally found to degrade ductility and strength. The highest volume plastics used for plating are ABS and amorphous resins. It is more difficult to plate crystalline resins such as acetal or nylon. Commericially plateable plastics are as follows: ABS, including ABS/polycarbonate alloys, polypropylene, polystyrene, nylon, polyester, acetal and polycarbonate.

As a general rule, the flexural strength and modulus of the component are increased by plating, even possibly the heat distortion temperature, but at the expense of some loss in ductility. In this respect there is a negative aspect of plating in that snap-fit lugs may become too stiff to flex properly and break off. The answer is to plate selectively. While adding the visual dimension of plating is worth while on the right product, there are a number of design guidelines that should be considered. These reflect good moulding practice as well as a successful plating process outcome:

1. Weld lines, parting lines and gate marks will emphasize their presence through the plating.
2. Large flat surfaces should be slightly domed or textured in order to distract from the effect of slight surface irregularities.
3. The mould tool must be of vacuum-melted steel and polished to Ra 5.0 or finer.
4. The minimum wall thickness = 1.5−4 mm, ideally 2.3−3 mm.
5. Gates should be 50 per cent larger than normal.
6. Thickness variations, if unavoidable, should be gradual and uniform.
7. Large weight: surface area variations should be avoided due to shrinkage considerations.
8. Ribs should be no more than 50−60 per cent of the wall thickness they support at the point of intersection to avoid sink marks. Imperfections become very obvious after plating, emphasized by the reflectivity.
9. The radii of internal and external microsections should be as generous as possible to allow for a good flow of moulding material.
10. External radii should not be less than 1 mm and internal radii not less than 0.5 mm.
11. The width of recesses should be at least twice their depth.
12. If a part needs to bend, selectively plate but design in a groove or undercut if there is no natural edge.
13. Consider the temperature range of use. Thermal expansion coefficients can differ by four to eight times so causing cracking of the plating at elevated temperatures and blistering at very low temperatures.
14. Ejector pins should be positioned where sections are thickest in order not to create stress at the ejection point. Stressed areas in general will alter the performance and quality of the plating.
15. Screw-fed injection moulding machines give the best result. Raw material should be pure and dry.
16. It may be necessary in some cases to temper the part in water immediately after ejection, especially if large section differences exist that would create stresses if too rapidly cooled.
17. Avoid silicone-based mould release agents.

Whether the final coating is electroplated or autocatalytic, the initial stages in the plating process are common to both. Firstly the surface of the plastic is etched in chromium−sulphuric acid and then dipped into a solution of palladium−tin chloride which deposits a monomolecular surface layer to act as a catalyst for the next coating of ∼1 μm of electroless copper or nickel. The surface is now conductive and can be treated as a metal. If the adhesion of the initial electroless coating is not perfect, subsequent platings may delaminate.

(i) Electroplating

Following the thin layer of electroless nickel, it is usual to electroplate with copper/nickel layers and then finally chromium or gold. The thickness of each layer and type of metal used depends on the final application. Four types of service condition are given in Table 13.6 relative to layer thicknesses of the copper−nickel−chromium system. A summary of plastic electroplate performance is given in Table 13.7, a typical coating cross-section in Fig. 13.3.

Table 13.6 *Application specification for plated plastics*

Service	Application	Plating thickness (μm)		
		Copper	Nickel	Chromium
Very severe Denting, scratching and abrasive wear likely plus corrosive environment and temperature cycling.	Faucets, marine hardware, external auto	15	23–30	0.25
Severe Occasional exposure to environment or corrosive solution plus some temperature cycling.	Lawn furniture, cookware hospital equipment	15	20–25	0.25
Moderate Mainly indoor use, occasional heat and water exposure.	Kitchen ware, reflectors, controls	15	10–15	0.25
Mild Indoor use with minimal wear, moisture or temperature cycling.	Decorative and novelty items	15	7	0.13–0.25

Table 13.7 *Metallic coatings on plastics*

Coating	Thickness (μm)	Plastic substrates	Properties	Uses
1. *Electroplating* Chromium, copper, nickel, brass, gold	12–20	ABS mostly, but also nylon, acetal, PPO, polypropylene, mineral filled nylon, DMC, etc.	Impact resistance of plastic may be lowered and stiffness increased Unequal bending may crack metal coat Range of textures achievable, polished to satin Can be selectively plated Advantage in weight saving and cost compared with metallic alternative Watertight Excellent resistance to ageing, UV, weathering, etc.	Decorative purposes Copper and nickel for RFI shielding All sectors of manufacturing industry, trims, handles, appliances, knobs, plumbing items, etc.
2. *Electroless plating* Copper, nickel	0.3–10		Serves as an initial conductive film on the plate in plastics industry Generally not used as coatings on their own as more expensive per weight of deposited metal Functional EMI shield coats are used alone	EMI shielding

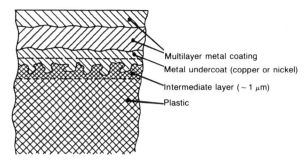

Fig. 13.3 Cross-section of plated plastic

(ii) Electroless plating

A summary of properties, etc. is given in Table 13.6. The same set of advantages/disadvantages as in section 13.2.2 apply. If plastics are plated exclusively by the electroless method, there tends to be a bright nickel deposit, $5-10\ \mu m$ thick.

13.3 Chemical conversion coatings

In many cases where surface protection is required, simple immersion treatments can be employed, chemically converting the surface into metal phosphates, oxides or chromates. Such chemical conversion coatings are generally known as metal pre-treatments, although specific conversion coatings do have value in reducing friction and interfacial contact during running-in of lubricated components. It is virtually standard practice to prepare steel, zinc-coated steel, zinc and zinc alloys or aluminium and aluminium alloys for painting with a prior conversion treatment.

13.3.1 Phosphating

This is a conversion process in which a metal surface reacts with an aqueous solution of heavy metal acid

Table 13.8 *Phosphating applications*

	Application	Nature of coating	Comment
(a)	Preparation for painting	Amorphous iron phosphates Zinc and iron phosphates	For general rustproofing preparation for painting, i.e. external surfaces for outdoor exposure Lightweight iron phosphate gives a lower corrosion resistance Overall corrosion resistance of phosphate–paint system is much superior to paint alone
(b)	Corrosion resistance	Iron and zinc phosphates Manganese and iron phosphates	Original application for phosphate coatings Invariably immersion-type operations apply Finish with grease or wax Heavier iron and manganese coatings provide higher corrosion resistance than zinc phosphate Zinc processes tend to dominate as cheaper and easier to control
(c)	Cold working	Zinc and iron phosphates	Tube and wire drawing, deep drawing and extrusion Used to facilitate cold heading and sheet stamping operations Partially react coating with soap solution to impregnate surface with insoluble heavy metal soap Zinc soaps have better lubricating properties Alternative is to use mineral oil with phosphate esters or stearates in powder form
(d)	Lubrication improvement	Manganese and iron phosphates $4-8\ \mu m$ thick coat rising to $12\ \mu m$	Principal value is in the running-in of new machinery but benefits secured may ensure survival for long periods subsequently Because of porosity, will absorb lubricant and maintain the bearing surface under load. After drying the coating, lubricate with either mineral oil, soluble oil, colloidal graphite or molybdenum disulphide grease Better coats are based on the harder manganese phosphate Alloying elements in steel should not exceed 5% Case-hardened surfaces can be treated Coating is highly resistant to adhesive wear Coating may be buffed if close tolerances required As with all aqueous processes, there is some danger of hydrogen embrittlement Used for gears, shafts, pistons, etc.

phosphates to produce an adherent layer of insoluble phosphate complex. Treatments may also be known as Parkerizing, Walterizing or Bonderizing. Used mainly on mild or low alloy steels but also on zinc and aluminium alloys to some extent, the principal phosphates involved are those of iron, zinc and manganese. Application is by immersion, spray or brushing, depending on the situation. The types of application for phosphating are summarized in Table 13.8. The major application is, however, the pre-treatment of mild steel prior to painting where the coatings improve the adherence and corrosion resistance of the paint, minimizing corrosion should the paint film be damaged.

Zinc or zinc-coated steel is readily coated, but care should be taken to ensure the coating is not over-thick or poor paint adhesion occurs. Hot-dipped galvanized steel is more difficult to treat because of aluminium alloying of the zinc layer. Steel strip coating lines typically use an amorphous lightweight iron phosphate coating, while on zinc-coated strip, either an amorphous coating of zinc phosphate or a conversion coat based on an alkali oxide.

13.3.2 Chromating

The formation of chromate coats is normally associated with the treatment of zinc or aluminium, but it may also be applied to magnesium, copper, brass, tin or silver. The coatings form tarnish- and corrosion-resistant films in their own right or act as pre-treatments for organic coatings. The process is basically a simple dip procedure, the protective value increasing up to thicknesses of 18 μm. However, the films are not very strong, so if unpainted are usually protected by lacquering. The corrosion resistance of phosphate and oxide films is also enhanced by the use of a chromate coating.

Chromating is an alternative to anodizing in providing corrosion resistance, but unlike anodizing, fatigue properties are not affected. The types of commercial processes available are given under the headings below.

(i) Chromate process

1. Produces a pale gold to orange-brown coloured coating depending on coating thickness;
2. Provides a higher corrosion resistance than the chromate–phosphate process (below);
3. Because chromium is present in the non-toxic trivalent state, may be used as a pretreatment for lacquered aluminium used in the food industry.

(ii) Chromate–phosphate process (Alocrom)

1. Coating solution is a mixture of acids containing chromates, phosphates and fluorides;
2. Chromium is present in the toxic hexavalent state;
3. Coatings blue-green to olive-green in colour. The deeper the colour the higher the coating weight;
4. Preferred as a base for organic finishes due to thickness achievable.

(iii) Dacrotizing

1. Applies coatings of zinc and aluminium flakes in a passivating chromate mix. Colour of coat is silver-grey metallic;
2. Corrosion resistance enhanced from both sacrificial protection provided by the zinc plus the effective passivation of both zinc and substrate (mild steel);
3. Coating 8–10 μm thick but requires a 300 °C cure;
4. Coating has limited ductility. Adequate elasticity is present for springs and clip-type applications;
5. Suitable for parts difficult to coat by electrodeposition.

13.3.3 Oxide coatings

Natural oxide films are generally not adequate in themselves for providing adequate corrosion and wear protection, hence a number of methods are used for thickening the oxide layer and retaining porosity to act as a key for organic coatings or as a base for oil or wax impregnation. The electrolytic anodizing process has already been considered (section 13.2.4). This section deals with some chemical oxide-thickening processes.

(i) Chemical oxidation

1. Based on an alkaline chromate solution and applied to aluminium and aluminium alloys;
2. Thickens and etches the surface oxide layer, hence ideal paint preparation;
3. Black in colour but depends on alloy composition.

(ii) Chemical anodizing

1. A process of forming the hydroxide chemically on aluminium and aluminium alloys then dehydrating;
2. Film is thick, weather resistant and white in colour, typically 9–13 μm thick;
3. Colours possible;
4. Forms oxide film in a shorter time span than electrolytically.

(iii) Blackodizing

1. A process of chemically blackening steel to give a deep black lustrous finish of oxide Fe_3O_4. Has aesthetic appeal;
2. Used either as a corrosion-resistant surface on its own, a basis for organic coatings or as an anti-wear coating on gears, shafts, etc.;
3. Thickness of coat is 1 μm, surface finish very good;
4. A more controlled coating than phosphating and with superior properties;
5. Adherence in bending very good.

13.4 Vapour coating processes

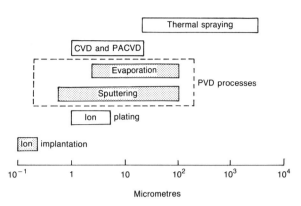

Fig. 13.4 Effective depth of treatment for various surface engineering processes

Of the more non-standard treatments to apply adherent layers of hard materials, are the vacuum-coating processes, e.g. chemical vapour deposition (CVD) and physical vapour deposition (PVD), together with lower-temperature plasma-assisted versions. Ceramics and cermet coatings would normally only find application for wear resistance. In corrosion-only applications, aluminium can be applied by ion vapour deposition to give surface protection, a serious alternative to electroplates.

However, since some vapour-coating processes require high substrate temperature, the bulk material clearly must be able to withstand the processing steps with the properties of the substrate not impaired. It is of no use using a coating for corrosion protection if the fatigue properties of the component are affected by the coating process. Some of these coatings are 'line of sight' processes so the design of the component should enable adequate coating. The advantages of ion and plasma-assisted coating processes are that the bulk material properties are much less affected. Advantages and disadvantages of the various coating types and techniques are given in Table 13.9.

As a conclusion to this section, the most recent technique, ion implantation, is considered as a means of introducing an interstitial element, nitrogen, into the surface layers. Coating thicknesses obtained by different methods are illustrated by Fig. 13.4.

13.4.1 Chemical vapour deposition

This process involves the formation of a solid layer from the gaseous phase via a chemical reaction. By analogy with electroplating, it is sometimes known as gas plating. By definition, CVD is a chemical process and hence for controlled deposition, endothermic reactions are preferred. The pressure in the coating chamber is slightly above or below atmospheric. The simplest source material is one which is a gas at room temperature and pressure, e.g. for deposition of tungsten carbide, the chemical mix is benzene vapour, tungsten hexafluoride gas and hydrogen.[8] A summary of the process is given in Table 13.9. Methods of energy input are one of the following:

1. *Thermal CVD* — traditional substrate heating methods, e.g. resistance.
2. *Plasma CVD* — advantage over (1) in that reduced substrate heating required with a plasma-assisted process, known as plasma-enhanced CVD (PECVD) or plasma-assisted CVD (PACVD) — coatings are generally amorphous and of good quality.
3. *Laser CVD* — in the development phase, general surface heating can be overcome by optimizing a laser beam to heat only a thin layer or specific focused area of surface.

Plasma CVD processes are available commercially for the coating of tools and dies producing a triphase layer of titanium carbide—titanium carbonitride—titanium nitride in the thicknesses 3, 1 and 4 μm respectively. Plasma processes have also been developed for nitriding and carburizing. These have advantages over conventional salt bath or gas processes. In the plasma nitriding CVD process, the chemical mixture is of titanium chloride (g), hydrogen, methane and nitrogen. The resultant coating has high hardness, wear resistance and fatigue resistance. In the plasma carburizing process the gas is a methane—hydrogen mixture, and the resulting carburized layer is highly uniform for most geometries with minimal distortion after subsequent hardening.

Table 13.9 Summary of vapour deposition processes

Process		CVD	PVD	Ion plating	Ion implantation
Advantages		Large range of elements and compounds can be deposited, metallic ceramic and polymeric materials Some layer properties obtained by this method may be unique Ability to vary the composition from a high degree of purity in the case of elements to a carefully controlled stoichiometry for polycrystalline and amorphous layers Compressive internal stresses in layer Variable within wide thickness limits e.g., 0.1–100 µm Coatings very dense Plasma-assisted CVD lowers the deposition temperature Ability to produce triphase composite coatings in simultaneous or successive steps	Coating below 500°C avoids subsequent rehardening in tool steels No dimensional change in precision tooling By controlling deposition conditions, coating properties can be tailored High throwing power Sputtered coatings produce the more adherent coat Shows promise for the surface treatment of aluminium and its alloys	Uniform coating distribution ideal for threaded components Coatings have excellent adhesion, even when substrate and coating do not alloy Can give good coverage on surfaces out of line of the vapour source 350–450°C coating temperature preferred; in some cases as low as 200°C Most dominant commercial process for titanium nitride Expanding range of coatings Dense coating microstructure No edge run-off or build-up in recesses High deposition rates Can be applied to certain plastics Smooth surface finish	Essentially a cold process 100–150°C; virtually eliminates substrate metallurgical change Requires less rigorous pre-clean than PAPVD Significant wear resistance despite the shallow depth of implantation No distortion No dimensional change No change in surface appearance
Disadvantages		Some coatings do not adhere well to substrate especially if deposition temperature is low Most depositions involve high substrate temperatures around 1000°C, hence suitable substrate materials severely limited, distortion inevitable Substrate cleanliness is very important Steel substrates require a thermal stress relief after coating All CVD hard coatings are brittle, liable to fracture on sharp edges	Costly in terms of plant and energy Metallic coatings deposited at low temperatures are slightly porous, columnar and poorly bonded to substrate Coating adhesion can be increased by post-coating heat treatment	Initial multistage pre-clean essential Costly Clean vacuum required and process control needs to be good	Line of sight process
Coating materials		Titanium nitride (1–10 µm for wear resistance) titanium carbide/carbonitride/nitride triphase, silicon, carbon, nickel, tungsten, boron nitride, tungsten carbide, alumina, chromium carbide, silicon carbide	Titanium nitride (2–4 µm) silver, lead, cobalt, nickel, chromium	Titanium nitride (1–5 µm for wear resistance) titanium carbide, aluminium, copper, silver, gold, lead, cobalt–chromium alloys, aluminium bronze	
Applications		Mostly for metal cutting tools and replaceable carbide tips Unlubricated rolling elements in inert environments	Optical, electronic and decorative applications Coloured interference coatings on lenses Metal coating of steel strip	Protection of high strength steels against corrosion without risk of embrittlement Wear-resistant coatings Soft metals as solid lubricating films Injection mould tools Tools, punches, dies and cutters	

13.4.2 Physical vapour deposition

Both PVD and CVD deposit coatings from the gas phase by what is essentially an atom-by-atom process, but in this case the process is conducted in a vacuum with the source metal either evaporated or sputtered on to the target component. The common feature of PVD processes is the ionization of the deposition species, increasing energy and allowing lower deposition temperatures to be used. Again the commercial process is mainly directed towards titanium nitride. The main processes are as follows:

1. *Evaporation.* The source material can be evaporated either by resistance heating, radiation, laser beam, electron beam or arc discharge. Direct, line of sight evaporation is usually considered effective and reliable for metals, but is difficult for ceramic materials because of their high melting-point, low vapour pressure and tendency to dissociate. Reactive methods are often used to obtain stoichiometric carbide, nitride or oxide coatings, the metal evaporated in the presence of the appropriate gas.
2. *Sputtering.* This process is superior in terms of throwing power compared to evaporation, but the deposition rates are much lower. Cathodic sputtering of the source takes place in an inert atmosphere at low pressure under applied potential. The material removed by ion bombardment of the source condenses on the component target to form the coating. The composition of the sputtered film is the same as the source.

PVD treatments have also been applied to aluminium substrates. Chromium and copper films on LM13 have proved successful in simple wear tests as well as titanium nitride coatings.[9] Materials for coating should be selected which do not temper below 400 °C and that do not contain zinc.[10] For some PVD systems there are also size limitations.

13.4.3 Ion plating

Ion plating is an ionization or plasma-aided PVD technique (PAPVD) in which deposition is accompanied by simultaneous ion bombardment. The usual way of providing ion bombardment is to make the substrate the cathode in a low-pressure plasma discharge, the ion bombardment continuing during the deposition.[11] Ion bombardment prior to deposition removes surface contamination and roughens the surface, both aspects helping to maximize the mechanical bond. The vaporized coating can be provided by evaporation, sputtering, electron beam, etc. If a reactive gas species is introduced into the vacuum chamber, a compound coating, e.g. titanium nitride, can be deposited and the process is known as reactive ion plating.

Of the number of coatings possible by vapour deposition processes, titanium nitride has emerged as the most popular, not because of superior properties *per se*, but because:

1. It is easy to deposit by ion techniques;
2. It has an attractive golden appearance;
3. It retains its high hardness and chemical stability at high temperature;
4. Increases the life of cutting tools, for example, range from $\times 4$ to $> \times 20$;[12]
5. Low frictional coefficient, ~ 0.14, virtually environment independent.

Another commercial application of ion plating is the ion vapour deposition (IVD) of aluminium on to steels, titanium and copper alloys as a corrosion-resistant non-toxic alternative to cadmium. Other applications for IVD are increasing the conductivity on insulating surfaces, electromagnetic interference (EMI) shielding, etc. As a post-coating operation, the coat is glass bead peened to densify the structure and give a smooth finish. Aluminium itself is not readily electrodeposited hence the attraction of this technique. Similarly it may be applied as a coating on stainless steel.

13.4.4 Ion implantation

This is an innovative surface treatment originally developed to introduce one or more ion species into semiconductor devices, but the wider engineering application has been to implant nitrogen into the surfaces of steel, aluminium, beryllium copper, titanium, chromium plating, etc. to provide significantly increased wear resistance. Impressive benefits are reportedly gained without any surface physical or dimensional change. It is a low-temperature process involving the migration of nitrogen ions below the surface, and in the case of chromium plate, the precipitation of chromium nitrides.[13] Because it is a non-equilibrium process, solubility limits may be exceeded with or without subsequent precipitation, i.e. ions can be incorporated without developing a set of diffusion conditions or consideration of chemical constraints. Much less total energy is involved in the process, although individually the ions have high energy for penetration and modification. It is essentially a complementary, but simpler, process to PAPVD requiring a less rigorous pre-clean. Treatment time is 2–10 hours.

The process has potential in a number of areas, e.g. it has been applied to stainless steel where conventional case hardening methods seriously lower the corrosion resistance of the steel and for hardening short-run soft aluminium tools. Some small amount of machining/refurbishment is also possible without destroying the wear properties.

13.5 Thermal and thermochemical treatments

Surface heat treatment processes are employed to improve the performance of engineering materials by enhancing surface hardness, improving wear and fatigue resistance and in some cases corrosion resistance, while maintaining a ductile core. Less costly steels are used to maintain the required surface condition. Most thermochemical treatments apply to steels and cast irons but some may be applicable to non-ferrous metals. In ferrous materials, the treatments fall into three categories:

1. Selective surface hardening treatments, e.g. flame and induction hardening;
2. Low-temperature thermochemical treatments, e.g. nitriding;
3. High-temperature thermochemical treatments, e.g. carburizing.

Case hardening may be achieved in a number of different ways in the above; the simplest means is by producing the core in the correct condition by prior heat treatment followed by selective treatment of the surface by flame or induction hardening. The laser is an alternative energy source.

Thermochemical treatments are interstitial hardening processes in which one of the elements carbon, nitrogen, oxygen or boron is introduced into the surface layer of a component, either in solid solution or as in interstitial phase via solid, liquid, gas, fluidized bed or plasma transfer media at elevated temperature. Ferritic thermochemical treatments, e.g. nitriding and mitrocarburizing, which are carried out below the austenitic transformation temperature, harden by means of a fine dispersion of alloy nitrides and carbonitrides in the matrix. Treatments carried out at higher temperatures, e.g. carburizing, require the components to be quenched to develop the case hardness.

13.5.1 Selective surface hardening

In the conventional methods of induction and flame hardening, steel components are subjected to a selective heating so that the surface is austenitized to a controlled depth prior to quenching. Information regarding normal heat treatment processes may be found in general metallurgical texts. Heat treatment parameters are known to exert a considerable influence on the wear resistance of steels through the variation in microstructure,[14] but in the first instance it is usual to accept that wear resistance follows a similar course to hardness variation. The following guidelines may be applied:

Adhesive wear. To ensure the initial wear decreases into the mild wear regime after the running-in process is complete, the following hardness levels are suggested: 350–450 Vpn for a hypoeutectoid steel (≤ 0.8 per cent C) >250 Vpn for a hypereutectoid steel (>0.8 per cent C). For mild wear to prevail from the start of sliding, a hardness level >700 Vpn is required.

Abrasive wear. Large increase in wear resistance occur when $H/H_a \geq 0.5-0.6$, where H = normal steel hardness and H_a = abrasive hardness. This simple relationship becomes more pronounced as the microstructure of the steel becomes more complex.

While conventional techniques enjoy continued use, there is a growing interest in the use of lasers and the electron beam as alternative energy sources for heat treatment purposes, so extending the scope of selective surface hardening. These can be regarded as sources of finely controllable high energies compared to conventional heating, the important characteristic being the intense energy flux which promotes very fast heating rates,[15] outlined below. When using a conventional heat source, the relatively low energy density allows the heat to be conducted into the bulk of the component while the surface is reaching the required temperature, thus heating the surface inefficiently. A high-energy laser beam heats the surface rapidly, reducing the time for heat conduction to the bulk.

(i) Laser transformation hardening

With steels, a thin surface layer is rapidly heated above the austenitizing temperature in the short time the beam is incident. Once the beam moves on, this surface area is quenched by the conduction of heat into the still cold bulk component material. The quench rate is usually fast enough to harden without the need of an external coolant. Because of the initial high reflectivity of a machined surface, it is often necessary to improve efficiency by matt painting the surface, graphiting the surface or shot-blasting. Reflectivity does, however, decrease with increase of surface temperature. The process in operation is shown in Fig. 13.5.

SURFACE ENGINEERING

Fig. 13.5 Laser hardening process. Courtesy of Midlands Electricity plc

This treatment offers the following advantages:

1. Treatment can be localized to a required area.
2. The thermal profile is controlled, the affected zone is confined to a thin surface layer hence minimum component distortion.
3. No surface disruption except for a very slight volume increase; no after-machining if at all.
4. No quenchant required except on very small parts.
5. Treatment can be carried out on the finished article.
6. Any areas that can be seen optically can be treated.
7. The power–density distribution can be controlled to allow shaping of the hardness profile.
8. Process speeds are relatively high, interaction times are a fraction of a second.
9. Rapid quench rates lead to a fine-grained surface structure.
10. Case depth 1 mm.
11. The martensitic transformation will induce surface compressive stresses, so improving fatigue life as well as wear resistance.

(ii) Laser surface melting
A similar process to laser transformation hardening except that a near focused beam is used and the surface to be melted shrouded in inert gas. Produces moderate to rapid solidification rates, a smooth surface finish and fine, near homogeneous structures. Again, as there is little thermal penetration, results in little distortion.

(iii) Laser surface alloying
Similar to laser surface melting with another material added to the melt pool by prior electroplating, vacuum-evaporated coating, pre-placed powder coating, powder blowing, wire feed, etc. Most materials can be alloyed into different substrates and the high quench rate ensures minimal segregation. Very thin fast-quenched alloy regions are therefore possible with interesting future application.

(iv) Particle injection
Generally achieved by either pre-placing a powder on the surface or propelling powder into a laser-generated melt pool, the powder in question being a hard ceramic, i.e. titanium carbide, silicon carbide, tungsten carbide or alumina. The hard particles must be wetted by the metal matrix and have a strong bonding to it, thereby improving hardness and wear resistance via the dispersed ceramic phase. The process shows promise for the hardening of aluminium and aluminium alloys.[9,16]

(v) Electron beam hardening
Unlike lasers, the electron beam process works wholly in a vacuum. However, since the kinetic energy of the electrons is transmitted directly to the surface atoms of the workpiece, there is no need to coat components prior to treatment. Can harden selectively to a depth of 1.5 mm.

13.5.2 Low-temperature thermochemical treatments

(i) Nitriding
Applied via the gas phase or in a salt bath, the mechanism is the diffusion of nitrogen into a steel surface at temperatures below the austenitic transition to form alloy nitrides and carbonitrides. An effective process for increasing hardness, wear resistance and fatigue resistance. A summary of the process features is given in Table 13.10.

(ii) Plasma nitriding
This is a thermochemical treatment performed in the plasma of a current-intensive glow discharge. Traditional nitriding processes have been in the liquid or gaseous media, but plasma nitriding is now a widely implemented

Table 13.10 *Summary and features of nitriding processes*

Process	Features of process
Nitriding	Nitrided steels have excellent abrasion resistance, high hardness and improved fatigue performance Steels must contain nitride-forming elements such as aluminium, titanium, chromium, molybdenum and vandaium Quenching not required to develop surface hardness Less distortion than carburizing or carbonitriding Axisymmetrical components preferred for this process Steels must be heat treated before nitriding in order to obtain desired core properties Harder and thinner cases compared to carburizing, i.e. 0.4–0.6 mm and 800–1000 kg mm^{-2} Case may be brittle
Gas nitriding	Ammonia atmosphere diluted with hydrogen Lengthy process, used on low alloy steels
Plasma nitriding	By adjustment of process parameters, blind holes and long bores can be nitrided Since nitriding potential is governed by electrical parameters, plasma nitriding can be carried out at lower temperatures than gas nitriding Fully viable industrial process Fast process with greater control and reproducibility of the nitrided structure Appreciable improvements in wear resistance, fatigue life and surface lubricity may be obtained Minimal distortion, cooled via forced convection

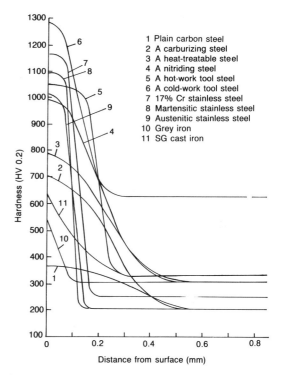

Fig. 13.6 Typical hardness profiles from a range of plasma nitrided materials. Courtesy of Klockner Ionen, Leverkusen

technology, using the glow discharge plasma to bring about the mass transfer of nitrogen from a low-pressure gas atmosphere into the component.[17] Alternatively known as ion nitriding or glow discharge nitriding. One feature of the process is the removal of material from the cathode by sputtering which continues through the whole treatment and allows the workpiece surface to be cleaned and depassivated. The process can be used to harden cast irons, carbon steels, stainless steels and titanium alloys. In the case of stainless steels the adherent surface oxide film is firstly removed by the sputtering action of active particles present in the discharge.[17] The process produces compound layers that are hard, thin, dense and adherent, the surface finish and metallurgical structure being superior to the salt bath and gaseous treatments. A summary of plasma-nitriding process features is given in Table 13.10. Typical hardness profiles from a range of plasma-nitrided materials are shown in Fig. 13.6.

(iii) Ferritic nitrocarburizing
This is a general term adopted for nitrocarburizing carried out below the iron–nitrogen eutectoid temperature, i.e. in the ferritic range below 590 °C. It is basically a non-distorting nitriding process carried out at 'stress–relieving' temperatures in the presence of a surface carburizing reaction and specially designed for treating non-alloy steels. A number of salt bath treatments have been available for some time, e.g. Tenifer, Tufftriding, Sulfinuz, but these are mostly based on cyanide salts. During the early 1980s gaseous techniques became available (e.g. Nitrotec) together with an improvement in corrosion resistance, the characteristics produced by a flash oxidation immediately after processing and before cooling. The features of the nitrocarburizing processes are given in Table 13.11.

In the Nitrotec process,[18] a single heat treatment can confer improvements in fatigue strength, corrosion resistance and tribological properties. Nitrogen diffuses inwards to a depth dependent on treatment time at temperature, the improvement in fatigue properties directly related to the depth of nitrogen diffusion. The surface compound layer possesses an inherent porosity which can act as an oil retention layer or a key for the application of an organic coating. To develop the required mechanical properties without distortion, a water/emulsion quench is used from 550°C, but for

Table 13.11 *Summary and features of nitrocarburizing processes*

Process	Features of process
Nitrocarburizing	Involves the diffusion of mostly nitrogen, but some carbon into surface; layer thickness 20 μm Process produces a thin wear-resistant compound layer at surface + a nitride diffusion zone up to 1 mm Surface hardness 500–650 kg mm^{-2} If part is quenched after treatment, most of the nitrogen in the diffusion zone is retained in solid solution, increasing hardness Some porosity in outer layer helps with oil retention Treatment becoming widely used, particularly for upgrading low carbon non-alloy steel parts Ability of process to enhance corrosion resistance is a bonus now exploited both in molten bath and gaseous treatments; key is controlled oxidation conducted straight after nitrocarburizing Fatigue strength increases with depth of nitrogen diffusion layer
Salt bath	Based on molten cyanide salts Various forms of treatment known as Tufftriding, Sulfinuz, Sulfocyaniding etc. To ensure coating uniformity, bath temperature and composition need to be closely controlled
Ferritic nitrocarburizing	Treatment aimed at improving the wear, fatigue and corrosion resistance of a range of low-carbon steels Nitrocarburizing followed by quenching into an oil/water emulsion at 70–80°C increases yield strength and produces an aesthetic blue/black oxide film Oxide film needs to be <1 μm if exfoliation is to be avoided Surface has good oil retention properties and good finish Organic sealant may be introduced into micro-porous compound layer Negligible shape distortion If steel contains nitride-forming elements, will produce a nitrided case behind the surface compound layer
Austenitic nitrocarburizing	Process provides a back-up case by initial low temperature carburizing Depth of case may be varied as required Low distortion by starting quench at 700°C Uses plain carbon steels Tends to replace conventional carburizing and carbonitriding processes

aesthetic/corrosion protection purposes prior to quench, a short oxidation of the outer part of the compound layer to black Fe_3O_4 is allowed. The application of this process may be summarized as follows:[18]

1. Use of the tribological properties of the microporous surface layer to avoid the need for sintered bearings.
2. The increased yield strength of the treated components to reduce the component section size.
3. The aesthetic black appearance plus enhanced corrosion resistance to eliminate the need for more expensive finishing processes.
4. Using the controlled quench technique to enable the heat treatment and surface modification of distortion-prone components.

(iv) Austenitic nitrocarburizing

To obtain a back-up case with the ferritic nitrocarburizing process, it is necessary to employ a nitridable alloy steel. If working with a mild steel, the austenitic process provides the back-up with an initial low-temperature carburizing. The process is thus a combined treatment[19] producing gaseous carburizing plus a nitrided compound layer and a standard case of 0.75–1.5 mm. By final quenching from 700 °C, distortion is controlled. The features of austenitic nitrocarburizing can be summarized as follows:

1. Combines benefits of the compound layer (as in ferritic nitrocarburizing) with a supporting hard case;
2. Improved mechanical core properties;
3. Low distortion compared to conventional case hardening;
4. Enhanced tribological characteristics and corrosion resistance.

13.5.3 High-temperature thermochemical treatments

(i) Carburizing

Normally carried out at temperatures in the range 800–925 °C in controlled furnace atmospheres or salt baths (Table 13.12). It is a process in which a steel surface is enriched with carbon, but the introduction of carbon does not in itself produce a hardened case — the treated component has to be subsequently oil quenched. Rapid carburizing is possible using a fluidized bed, the rapid heat transfer and temperature uniformity enabling greater productivity.

(ii) Plasma carburizing

The plasma is created in a methane/hydrogen gas mixture

Table 13.12 *Summary of features of carburizing processes*

Process	Features of process
Carburizing	A process in which a steel surface is enriched with carbon at temperatures $>723\,°C$ Layer thickness 0.25–0.4 mm, hardness 700–900 kg mm^{-2} Core will have lower hardenability so retains toughness Normally a subsequent temper at 150–200°C is required Compressive stresses developed in case during carburizing increases fatigue strength Applicable to plain and low alloy steels Some distortion may occur during treatment Only thermochemical treatment that can provide a deep case, used for heavily loaded components Can produce a wide range of case depths as required so can select load-carrying capacity in conformal situations
Pack carburizing	Cannot quench directly, need to reaustenitize hence process is very energy inefficient Generally useful for one-offs; simplest and cheapest
Salth bath	Rapid heat-transfer characteristics of molten salt bath make it ideal for a large number of small components High rates of production possible for medium to shallow case depth
Gas carburizing	Accounts for the bulk of carburizing carried out Uses a specially prepared carburizing atmosphere
Plasma carburizing	Component heated by resistance heating in a low-pressure argon and hydrocarbon atmosphere Applied potential ionizes gas and produces a glow discharge Activation of carburizing atmosphere causes rapid carbon mass transfer Materials exhibit very steep carbon gradients at short treatment times Offers advantage of vacuum carburizing Reduces carburizing times due to rapid establishment of a high carbon concentration. High carburizing rate lends itself to a boost-diffuse approach Quench in oil for minimal distortion Distortion is less compared to gas carburizing. Directly attributable to a more uniform case.

at low pressure and a d.c. glow discharge. The physical aspects of the process are similar to that for plasma nitriding (section 13.5.2(ii)) except that the process temperature is now ~900 °C. The type of discharge produced has a characteristic visible appearance or glow that totally envelopes the workpiece and arises from the excitation of ion species within the discharge. The glow follows the surface contours of the component uniformly so that the resultant case is correspondingly uniform, the uniformity in itself reducing subsequent distortion.

A steep carbon profile is obtained in about half the time of traditional gas carburizing, and since the surface layer is highly carburized, a subsequent diffusion treatment ('boost-diffuse') is carried out to adjust the carbon profile to be more acceptable before oil quenching.

(iii) Carbonitriding

Usually applied to carbon and low alloy steels to produce relatively shallow cases (Table 13.13). The introduction of nitrogen into the carburized layer improves hardenability, hence the resultant shape distortion is less than for carburizing. The process results in property improvements very similar to carburizing.

(iv) Boriding

Also known as boronizing. This can be either a salt bath treatment or applied from the gas phase, a thermochemical

Table 13.13 *Summary and features of carbonitriding processes*

Process	Features of process
Carbonitriding	Variant of the carburizing process in which up to 0.5% of nitrogen as well as carbon, is introduced into the steel surface increasing hardenability More uniform hardness case, 600–850 kg mm^{-2} on plain and low alloy steels compared to carburizing Cases are relatively shallow, 50–75 μm, and the core is unaffected by treatment Similar property improvements as with carburizing Lower process temperature than carburizing, 750–900°C, hence distortion is somewhat less Needs to be quenched, direct harden into quenching medium
Salt bath	Process is complex, control is not particularly good Regular use within industry Lower operating temperature than salt-bath carburizing
Gas carbonitriding	Process introduces ammonia into a carburizing atmosphere

Table 13.14 *Summary and features of boriding processes*

Process	Features of process
Boriding (boronizing)	Thermochemical diffusion treatment where boron is diffused into a metal substrate, typically steel, to form a metal boride layer A high-temperature process of several hours' duration Risk of distortion Surface layers have a high melting-point, hardness, wear and corrosion resistance Thick layers are friable No appreciable strengthening of the core by the process — most suited to medium carbon and low/medium alloy steels Hardness is 1500–3000 kg mm^{-2}, layer thickness 50–150 μm Lack of use in the UK
Pack boriding	Packed with boron carbide and activator compounds
Paste boriding	Several paste coatings required to provide adequate boriding Expensive process
Salt bath	Thermochemical and electrolytic process, both must be used above 850°C Can introduce chromium, aluminium, and zirconium salts for boro-chromizing, boro-calorizing and boro-zirconizing respectively
Gas boriding	Used to combine boriding and hardening in one operation using a carrier gas

process involving the simultaneous or consecutive diffusion of boron into the surface. The treatment gives high corrosion resistance and high wear resistance (Table 13.14). If a salt bath is used, chromium, titanium, aluminium and zirconium may also be introduced and co-diffused, the process then known as multi-component boriding.[20]

13.6 Vitreous enamelling

Vitreous enamelling is a coating of glass/ceramic which has been fused into the base metal in a form of chemical bonding. It is found as a finish in many industries and is a sound choice where thermal and corrosion resistance are required. Because of its additional aesthetic qualities it is mainly used in the domestic appliance industry, but there are numerous industrial applications, some of which are listed in Table 13.15.

The process is also known as porcelain enamelling, particularly so in the USA, where the enamel tends to be applied by the 'dry' process where the powdered frit is applied directly on to the red-hot component. Other techniques involve the spray, powder or electrophoretic coating of the component prior to fusing.

For vitreous enamelling today, the design requirements are understood[21] and if the guidelines are adhered to, the opaque vitreous coatings rarely fracture unless the substrate is severely deformed. All vitreous colour enamel is from inorganic pigments which are highly stable, virtually lightfast and colour reproducible in manufacture. The available range of shades will also allow almost any colour to be matched. Logos and lettering can also be incorporated into the finish enamel by screen printing, offset or slide transfer techniques.

In recent years, numerous enamel frit formulations have been produced to cater for various industrial requirements, e.g.

1. A hard non-continuous base on aluminium to provide a key for subsequent fluorocarbon impregnation and produce a low friction surface;
2. A crystalline, microporous matt coating to absorb and oxidize food soil in the self-cleaning oven liner;
3. A single-coat enamel for domestic appliances providing high-temperature and high thermal/mechanical shock resistance;
4. Vitreous enamel PCBs.

A summary of vitreous enamelling, process features, advantages and disadvantages is given in Table 13.15. The main materials usually enamelled are sheet steels, grey iron castings and aluminium components.

(i) Sheet steel

The design rules for sheet steel are as follows:

1. Avoid closed or folded seams as these present pre-enamelling cleaning problems.
2. To avoid enamel burn-off and flaking on corners, the minimum bend radii should be 5 mm (two coats) and 3 mm (one coat).
3. Product rims should be rolled and left open to permit drainage, any burn-off is then also concealed.
4. Flanges and broken edges are better than sharp edges and corners.
5. Beaded edges improve product rigidity, folded edges may distort during fusing.
6. Flanges provide rigidity and flatness to a panel and should be continuous wherever possible. Holes and cut-outs in flanges promote stress and weakness.
7. Avoid cut-outs and swaged tapped holes.
8. Stresses around deep-drawn profiles should be eased with generous radii to blend into flat surfaces.

Table 13.15 *Summary of vitreous enamelling*

Hardness	>1100 kg mm^{-2}
Heat resistance	Up to 400°C, 550°C with certain materials Thermal shock resistance 195–260°C
Applications	Domestic appliances, industrial products, panels and signs, high-temperature pcbs, tanks, ducting and heat exchangers
Process factors and properties	Enamel is chemically keyed to the substrate, coating will not fracture unless the steel base is deformed Firing temperature is 750–900°C for steel, 680–780°C for cast iron If coating cast iron, require high-quality castings with no chill, blowholes, porosity, etc. Vitreous enamelling grades of sheet steel exist; low carbon steels are essential for trouble-free single-coat enamelling Enamel is electrostatically sprayed or electrophoretically applied to form a slip. After drying the enamel is fused at approx. 850°C. The base enamel coat contains cobalt oxide which produces a dark blue/black colour and results in a very strong bond with steel. The second or colour coat is applied over the base coat Correct choice of substrate is a crucial factor in component design for enamelling Can apply enamels directly on to steel sheet in one coat process with improvement in sag and damage resistance Can also enamel some aluminium alloys, e.g. LM6. Die castings not recommended. Commercial grade copper and copper alloys also suitable substrates
Advantages	Enhances product appearance; cost is usually economic when product life and durability are considered Resistant to acid, alkali, solvent and boiling water Almost unlimited range of colours available (frits) Colour fastness greater than 20 years; unaffected by UV light or heat Up to 100% gloss achievable; also obtainable from high gloss to full matt Smooth surface, literally as smooth as glass Easily cleaned, very high levels of hygiene are readily attained. Can withstand abrasive cleaners Resistant to cigarette burn Finish coats formulated to give a high degree of opacity Performs well if irradiated
Disadvantages	Adequate rigidity of component is necessary if only coated on one side — need to resist stresses caused by thermal expansion differences if the coating is thick Impact or mechanical strain (substrate) can cause fracture of the glass layer Enamel will flow during firing and may cause build-up on edges and in corner traps. Depends how components are stacked in the furnace Cast iron 'grows' during the enamelling process and so due allowance must be made for this Distortion is the chief difficulty in enamelling sheet metal parts having continuous flat surfaces. This is caused largely by an inappropriate relationship between the overall size of the product and the thickness of the section — minimize by correct design Steel sheet with too high a carbon content can result in pin-holes and black speck enamel defects

9. Joints should not be soldered or brazed as neither withstand the fusing temperature.
10. Weld joints wherever possible from the face to be enamelled. Position at a distance from a corner since smooth dressing of the weld is essential.
11. Brackets, lugs and stiffeners should not be riveted. Use spot welds and minimum show-through. Material should be thinner than the product to avoid distortion during welding.
12. Holes should always be 1.5 mm larger than the fastener. Avoid overtightening screws in enamelled plates.
13. Thickness of enamel in assemblies must always be allowed for and sufficient clearance provided between mating parts. Tolerances should be such as to avoid distortion and enamel failure.
14. Enamel-free areas can be provided for earthing tags, e.g. by wiping the dry enamel when the cover coat is in the bisque state or weld on stainless steel tags or lugs from which enamel can be easily removed.

(ii) Cast iron

The design rules for iron castings are as follows:

1. Most important is the provision of adequate and

uniform metal thickness so that distortion is minimized during the firings.
2. Castings should have generous radii on all corners. Sharp corners are difficult to coat and may well result in enamel runoff during fusing to give dark lines showing through on pale coats.
3. Cast iron grows during the enamelling process, typically 0.7–1 per cent linearly. Due design allowance should be made for growth plus enamel thickness.
4. Always enamel fresh from the foundry. Castings with porosity and blowholes are unsatisfactory, and cracked castings should not be enamelled.
5. The main enamelled face should be face down when cast since rising slag is then confined to less important faces.
6. With tapped holes, the first thread should be counterbored/countersunk to prevent the enamel fouling.
7. Lustre, hammered or ornamental enamels can be used to obscure surface irregularities.

(iii) Aluminium

The design rules for aluminium components are as follows:

1. The general preceding comments for sheet and cast components apply.
2. Adequate component rigidity and thickness is required to withstand fusing temperatures 500–600 °C.
3. Edge finishing and hole blanking possible after fusing on to thin gauge material.
4. Sheet is usually enamelled in a single coat to provide better thermal and mechanical shock resistance. Initial sheet quality, however, must be good since a single thin coat cannot mask surface damage.
5. If a casting, pressure diecasting structure is not advised, gravity casting structures are preferred.

13.7 Sprayed surface coatings

Spray coating is basically a process of depositing finely divided particles in a semi-molten condition to produce an adherent, layered coating on a substrate. The deposit is created from solid rod or powder which is heated, atomized and then projected in the molten or semi-molten condition on to the surface of the workpiece. On impact, particles flatten and overlap to form a coat, the degree of interlocking depending on the velocity of the particle and its plasticity. Depending on the process used, particle velocity can vary from 50 to 700 m s^{-1}. Deposit structure is not homogeneous and cohesion is due predominantly to mechanical interlocking, some point-to-point metallic fusion and some oxide–oxide sintering.

Sprayed surface coatings can be used for wear resistance, corrosion resistance, oxidation resistance, electrical conductivity and also for restoration of dimensions (parts reclamation). The choice of coating type clearly will reflect the application. Some examples of spray materials and application are given in Table 13.16. However, the coating choice will also depend upon the substrate material to be sprayed, the requirements of the finished surface, the thickness of coating and the required rate of deposition.

These types of surface coating can be classified into two main groups, spray deposits and spray and fuse deposits, as given under the headings below.

13.7.1 Spraying

(i) Flame spray

The deposition of finely divided metal, intermetallic or oxide particles, the powder or rod feed melted and projected by an oxyacetylene flame on to the workpiece. Residual porosity is up to 20 per cent which allows for oil retention or polymer impregnation. Metallizing is a term used to describe a flame spray process using metal in wire form. This type of deposit can be susceptible to corrosion and fatigue loading.

(ii) Plasma spray

This process can apply a variety of materials, including refractory ceramics and even polymers. A section through the plasma gun is shown in Fig. 13.7. An inert gas (argon) flows into the arc where it is ionized to form a plasma. Depending on the geometry and power of the gun, the flame can reach supersonic speed. Powder is introduced into the plasma jet and accelerated to the target surface. This process carries less risk of degrading the coating and substrate than many other high-temperature processes because the gas in the flame is chemically inert and the target can be kept cool. Residual porosity is typically 5 per cent. Spraying in a vacuum will yield a dense void-free coating.

(iii) Detonation spray

Particularly suited to cermet coatings, the process produces the most dense and well bonded of all spray coatings. In this case the bond has metallurgical characteristics, although coating penetration into the base metal is negligible. The low porosity, typically ~1 per

Table 13.16 *Summary of sprayed surface coating materials*

Material	Application/performance
Zinc	Most commonly used material for steel protection, particularly in a marine environment. Usually followed by painting/sealing
Aluminium	Used for protection of steel, particularly under conditions of high humidity. A subsequent paint system is recommended for sealing
Copper	Used for decorative purposes and also to give high electrical conductivity to surfaces
Brass/bronze	Used for bearing applications and for decorative coatings
Stainless steel	Used to provide a corrosion resistant surface as well as for decorative coatings
Molybdenum	Some degree of metallic bonding claimed with steel, i.e. self-bonding. A reliable and wear-resistant coating. Residual oxide level probably responsible for high hardness and wear resistance.
Tungsten carbide	Because of the high hardness carbide, generally always deposited in softer matrix materials to give an optional ductility to the coating. Sprayed metal and cement is better bonded to the substrate than sprayed oxides. Believed to achieve its maximum wear resistance when plasma coated with 12% cobalt. Can achieve higher wear resistance with added alumina and titanium oxide but at the expense of ductility
Chromium carbide	Usually sprayed with a matrix of nickel/chromium alloy to obtain a dense coat and to increase ductility
Nickel/chromium spray and fuse alloy	Boron and silicon provide self-fluxing characteristics. Additions of molybdenum or tungsten increase wear resistance. Both nickel and chromium provide corrosion resistance, the higher the nickel content the greater the coating ductility
Cobalt/chromium/tungsten spray and fuse alloy	Chromium carbide provides the wear resistance, tungsten stiffens the matrix and provides high temperature strength. Chromium provides corrosion resistance and cobalt increases operating temperature. in addition to silicon and boron, nickel is added to obtain self-fluxing characteristics
Oxides	Powder form of alumina, chromia, titania and zirconia used for wearing/abrasion resistance, but typically for thermal barrier coatings

cent, does manifest itself in enhanced mechanical properties, especially wear resistance.

(iv) Coating characteristics

Standard flame spray materials include numerous steels, non-ferrous metals, hard-facing alloys, carbides, oxides and refractory metals. The sprayed particles in many deposits are very hard, which is to be expected for carbides and ceramics, but the hardness of metal coatings is sometimes surprising. The rapid quench rates produced on impact plus some degree of work hardening both contribute to high microhardnesses. The coatings are harder and more brittle than the original material, though the physical properties of the coatings can be altered within wide range limits by varying the spray technique. Sprayed metals bear some resemblance to the stronger types of sintered metal.

To achieve a high deposit density and strong mechanical bond, a close control of particle size is necessary. For example, coarse particles do not always melt completely and too many will produce a poor coating. On the other hand, undersize particles of fines tend to vaporize. The size of particles will vary from 5 to 100 μm, the smaller size ranges produced by powder systems and the large sizes by rod spraying. However, apart from the condition of the particle, the structure of the coating also depends on the velocity of the particle, the denser coatings being produced by the higher velocity systems.

In general, flame-sprayed coatings are not recommended for impact conditions or where the surface geometry is counterformal. Under conformal sliding conditions, carbides and oxides are generally recommended, e.g. chromium oxide, chromium carbide, tungsten carbide and titanium carbide are most common. While being hard, these coatings are brittle and highly susceptible to fatigue and shock loading. Thus it is common practice to include metal additives which increase shock resistance and ductility although they may lower wear resistance. This also has the added advantage of reducing porosity, so obtaining a denser coat. Typical examples are as follows: tungsten carbide — cobalt or nickel, chromium; chromium carbide — nickel, chromium; zirconium oxide — nickel.

The same comments on coating structure and deposition effects apply to detonation spraying as to flame spraying, but the nature of the process does not necessitate a binder. If extreme wear resistance is required, 100 per cent ceramic can thus be deposited. However, the coating ductility would be negligible so a binder is often added. Typical deposits for wear resistance are as follows:

Alumina

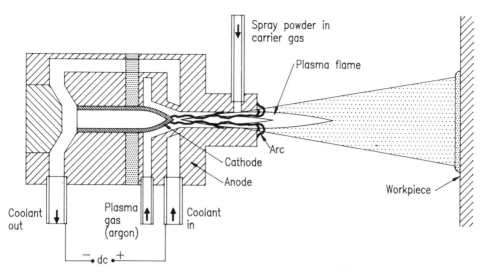

Fig. 13.7 Schematic of plasma spray gun

Alumina + titania
Chromium carbide + nickel–chromium matrix
Chromium oxide
Chromium oxide + titania
Tungsten carbide + cobalt matrix
Cobalt–nickel–chromium–tungsten

Of all the surface coatings considered, the major part of their wear resistance is due to an interstitial phase present, usually a carbide. This, however, is only the starting-point for obtaining overall wear resistance. The problem resolves itself into one of carbide shape and form, carbide size and carbide type as well as the suitability of its support within the coating structure.

Porosity may be a disadvantage in some applications if, for example, the wear process involves impact or fatigue, but in others it is an advantage. The structure of a sprayed metal coating is such that a machined face provides an ideal surface for oil retention. It is a well-known fact that surface porosity is useful in dealing with boundary lubrication conditions and in fact some surface treatments are deliberately produced with this aim in view.

13.7.2 Spray and fuse coatings

This is a unique group of deposits falling in between sprayed coatings and welded coatings. This type of deposit is applied by any of the available spray methods and is then simultaneously or subsequently fused, providing a dense uniform coat metallurgically bonded within itself and to the substrate. Application of the powder on to the surface is largely done by a spray method though other techniques are currently being introduced, for example the powder in a paste form, in a gel or bonded to tape.

For a long time it was not possible to fuse and bond sprayed coats as oxide layers formed during spraying and oxide layers on the substrate prevented the coat from coalescing properly. However, the addition of boron and silicon provided fluxing characteristics, their oxides forming low-melting-point glasses with metal oxides so cleaning up surfaces to allow metallurgical bonding. Of course it does mean that the coatings have a residual slag content, and excessively oxidized surfaces cannot be dealt with successfully.

With very few exceptions, the materials available for application by this method are hard-facing alloys and generally used in powder form. Most of the fusible alloys fall into three groups, the addition of boron and silicon characterizing these types of deposit:

1. Nickel base: Ni–Cr–Si–B and Ni–Si–B;
2. Cobalt base: Co–Cr–Si–B and Co–Cr–W–Si–B;
3. Bonded carbides; Cr–Ni–B–Si+WC and Cr–Ni–B–Si+NbC.

The addition of boron and silicon to this type of deposit not only provides self-fluxing characteristics but also provides low-melting-point eutectic phases that promote coherence between carbides, borides and matrix material during fusing. These phases are based on nickel, e.g. Ni/NiB, Ni/NiSi, Ni/CrB and Ni/FeB binary eutectics melting in the region of 1000 °C. It is thus largely a liquation effect, removing the inherent porosity of the sprayed deposit. Adequate wetting is ensured by the fluxing action. The metallurgical structure is usually a solid solution-eutectic matrix with carbides and borides present from the pre-alloyed powder. In the case of the cobalt base deposits, nickel is deliberately added as cobalt does not form low melting-point eutectics with boron or silicon below 1100 °C.

The dispersed hard phases in the deposits are borides and carbides, chiefly of chromium but also of iron, these interstitial phases being in the most part responsible for wear resistance. In this respect, composition and resulting structure are important. Large increases in wear resistance may be obtained by increasing the interstitial phase content, either by alloying or by addition of tungsten or niobium carbides to the original powder. With the incorporation of carbide granules, the finer the grade the more uniform is the wear resistance.

13.7.3 Polymer impregnation

In some flame spray applications, it may be desirable to seal off the interconnecting pore network, for example to form a more effective barrier against corrosion. A post-treatment is then given, e.g. with a phenolic or a fluorocarbon. As well as providing improved corrosion resistance, it has also been found to be a useful way of providing low-friction surfaces. By deliberately producing a porous deposit and then impregnating with PTFE, in a direct analogy with thin layer bearing materials (section 5.3.2(iv)), a low-friction surface is produced with an exceptionally low wear rate possessing none of the problems of bulk PTFE. The flame-sprayed coat is usually stainless steel, though molybdenum can be used if a harder coating is required. These types of coating were originally described as 'non-stick', but nowadays find a much wider field of application than domestic cookware. Nominal coating thicknesses are 30–60 μm.

13.7.4 Plastic substrates

Up to this point we have assumed that the substrate being spray coated is metallic or ceramic in nature. However, this need not be a limitation; the concept of spraying on to plastic is not new, but the techniques are still being evaluated and there are limitations that should be borne in mind. By definition, thermal spray coatings are applied hot, but with the correct conditions the substrate temperature can be held below 90 °C, the distortion temperature for most plastics. As a result, electric arc spraying has become the standard procedure for coating a broad range of plastics, since the oxyacetylene flame directs a lot of excess heat on to the workpiece.

The deposit can be for decorative or functional purposes, typically a thickness of 25–400 μm which will permit after-polishing and importantly gives a solid metal appearance and feel. A major application of the process is for radio-frequency interference/electromagnetic compatibility (RFI/EMC) shielding. The metals applied tend to be the lower-melting-point materials, e.g. copper, bronze, nickel, and it is usual to apply as low a melting-point bond coat first. It is essential firstly to have good surface preparation by removing any surface contamination, resins, release agents, etc. that could affect adhesion.[22] Thermal expansion adhesion is inadequate. Table 13.17 gives some idea of coating adhesion obtainable on different plastics.

13.8 Organic finishes

The term 'industrial organic finish' covers a broad spectrum of paints and coatings that are applied to a wide variety of articles, most usually in the factory as part of the production process. As true for modern finishes as it is for engineering materials, complexity renders selection and specification difficult for the product designer. In addition the terminology applied may sometimes appear confusing. 'Paint' usually applies to materials applied primarily for aesthetic effect, and any corrosion or abrasion protection also received is a secondary bonus. The term 'coating' or possibly 'protective coating' on the other hand is generally reserved for use where atmospheric protection and/or scuff protection is the primary consideration. The thickness of a coated layer also tends to be thicker than a paint

Table 13.17 *Tensile bond strengths of sprayed zinc coatings on various substrates*

Substrate	Tensile strength of coating (kg mm^{-2})[a]
Modified PPO	0.25
Polycarbonate	0.27
Polystyrene	0.21
Structural foam	0.30
ABS	0.17
Phenolic	0.32
Glass-filled nylon	0.74

[a] Coating pulled at right angles to plastic substrate. Surface lightly grit blasted prior to spraying.

layer. We shall adopt this broad generic classification of organic finishes in the presentation of this section.

Industrial organic finishes will differ from ordinary decorative types of paint finish in the following regard:

1. Drying and curing times are very much shorter in order to satisfy production process requirements. The methods used are rapidly evaporating solvents, faster drying oils, accelerated thermal curing methods, etc.
2. The resistance to chemicals and solvents is of a higher order.
3. They are often required to withstand considerable temperature fluctuations or to operate in above-average environmental ambients.
4. A high proportion of industrial finishes are applied to metals, either directly or primed. The second major industrial substrate is plastics.
5. Fabrications of different materials can be coated as one finish overall, plus coverage of welds, sealants, fillers, etc.

The object of an organic finish is to interpose a durable film between the substrate and its environment, but the finish should be considered as an integral part of any manufacturing process and the correct choice is as important as the choice of the substrate itself. Many different types of finish are available, each with differing properties and uses, but the application process used depends on a number of factors as follows:

1. *The conditions the finish has to withstand.* For good durability, the adhesion to the substrate is improved by an adhesive primer, but this does constitute a productivity limitation.
2. *The state of the surface.* Whatever coating is used, it must be applied over a thoroughly prepared surface, i.e. free of surface contaminants and corrosion products. A clean surface improves coating adhesion hence corrosion resistance.
3. *Standard of appearance.* For a smooth finish over a rough surface, filling and/or undercoating is necessary.
4. The article being finished must withstand the processing temperatures. A large number of finishing processes involve a thermal cure/dry in the range 150–400 °C. In some cases components are also pre-heated to encourage polymer adhesion and flow.

13.8.1 Paints

Paint is an extremely versatile coating, giving general surface protection plus a broad range of decorative effects, but not primarily for mechanical wear or heat resistance applications. It has very wide use in the engineering industry, is easy to apply and gives good value for money. Paints have been significantly modified over the years to produce modern types that are fast drying, have high gloss and low susceptibility to dirt retention.

Paint is a liquid suspension of finely divided polymeric solids that when applied to a surface, sets or dries to an adherent, coherent film. It is basically made up of three constituents:

1. *Vehicle.* This contains the film components (binder) — it is a natural or synthetic resin responsible for the main properties of the paint.
2. *Pigment.* This provides the paint with its colour and opacity plus any special property requirements such as rust inhibition, film reinforcement, UV stability, etc. Clear lacquers and varnishes are not pigmented.
3. *Solvent or thinner.* This is used to adjust viscosity and regulate application properties, evaporating/oxidizing as the paint dries.

As the paint dries, the above vehicles change from a liquid to a solid by one or more of the following mechanisms:

1. Oxidation (of a drying oil);
2. Polymerization through application of heat, addition of a catalyst or a combination of reactive ingredients;
3. Evaporation of solvents.

In describing the drying or curing of a paint film, the terms 'conversion' and 'non-conversion coatings' are sometimes used in industry, defined as follows:

Conversion. Refers to the conversion or cure of a fluid or resinous paint binder into a hard, tough adherent film by a polymerization reaction. The reaction can take one or more forms according to the nature of the binder and the temperature.

Non-conversion. Refers to paints that dry by solvent evaporation only, e.g. lacquers, and no cure or intermolecular reaction takes place.

Broadly speaking then, industrial paints can be divided into air-drying and stoving types. Product usage and properties are given in Table 13.18. The types of paint cure depicted by Table 13.18 are as follows:

1. *Paints that dry by oxidation.* The paint binder reacts with oxygen to form a solid film, e.g. alkyd and epoxy–ester paints. A free radical is produced that promotes cross-linking of the paint film. Polymerization speeded up by heating to ~70 °C, i.e. forced drying.
2. *Application of heat (stoving).* A stoving paint dries

Table 13.18 *Types of industrial and engineering paints*

Type	Single-coat thickness (µm)	Features	Applications
1. *Stoving Paints*			
Acrylic		Thermosetting paints based on modified acrylic resins that cross link on stoving Excellent gloss and colour retention. Hard and tough Good resistance to chemicals, grease, etc.	Domestic appliance cabinets Strip coating, coated metal General industrial use
Alkyd	25–37	Based on modified synthetic polyester resins Good gloss and colour retention Good chemical resistance Combine with urethane resin for additional hardness Not for immersion service or for use in an alkali environment	Domestic and electrical equipment General machinery Large variety of mass-produced components
Epoxy (1 part)		Resin modified with phenol formaldehyde High resistance to detergents, solvents and oils Good adhesion and flexibility, but poor colour stability Good wear resistance High gloss	Domestic and industrial equipment Domestic appliance interiors generally Exterior enamels
Silicone	20–25	High resistance to heat and corrosion When pigmented with aluminium can withstand 500°C Coloured or modified paints can withstand up to 250°C	Domestic heating equipment General application where high temperatures expected Exterior metal coatings
Vinyl	20–25	Based on vinyl resins Hard, flexible, tasteless and odourless High chemical resistance Not resistant to temperatures >150°C	Strip coated metal General containers Chemical-resistant finishes
Plastisol	Up to 1.0 mm	Vinyl resin and pigment dispersed in plasticizer On stoving, the resin dissolves in the plasticizer to form the film High flexibility, durability and corrosion resistance Smooth, high gloss	Strip coating Process fabrications, wirework, external metalwork, etc. General-purpose protective coating Noise reduction
2. *Air-drying paints*			
Acrylic lacquer	20–25	Thermoplastic paints that dry by solvent evaporation only Excellent gloss and colour retention Good chemical and heat resistance High flexibility Easily polished, low dirt retention	Cars and aircraft Clear lacquer for metals Sealers for sprayed metal Protective lacquers generally
Cellulose lacquer		Fast drying High gloss finishes with good durability Good colour and gloss retention	Cars generally Industrial products and machinery Electrical equipment General domestic items
Epoxy esters	25–37	Durable in chemical atmospheres Good adhesion to many surfaces Moderate to good chemical resistance Can be applied by brush	Known as brushing enamel General industrial work
2-pack epoxy	50–75	Chemical curing Mix immediately before use, limited pot life Require 5–7 days at room temperature to achieve maximum resistance, accelerate polymerization process by stoving High chemical and solvent resistance Good flexibility and abrasion resistance Approaches the properties of a baked-on coating more than any other type of cold-cure coating	Can produce relatively thick films General chemical-resistant surfaces

by the action of heat. Based on resin types that will only polymerize at 110–150 °C.
3. *Drying by catalytic action*. In two-pack materials, depends on the chemical reaction between a base and an activator. Mixing of the two components starts the polymerization reaction. Paint types are polyurethanes, polyesters and epoxies. However, polymerization is fairly slow and may continue for some time before optimum properties are achieved. The process can be speeded up by forced drying.
4. *Drying by solvent evaporation*. Occurs to some extent in the drying of all paints, but some dry solely by this method. In these paints the binder is already polymerized and it forms a solid film as soon as the solvent or carrier evaporates. Typically lacquers, but also in this category would be emulsion paints where particles of pigment and polymer are collodially suspended in water.
5. *Radiation cure*. Using either UV radiation or electron beam, the paint cure is effected by a 'free-radical' mechanism liberating free radicals and promoting rapid cross-linking. With the UV radiation cure, photo-initiators are included in the coating that decompose under UV to generate free radicals.

Application of industrial paint finishes can in principle be achieved by any practical brush, spray or roller technique and possibly some dipping processes if the paint is stable enough. In practice, brush application is difficult if the paint is fast drying. Lacquers can only be sprayed. The principal transfer methods are summarized as follows:

1. *Air spraying*. The conventional and widely used liquid paint application method. Compressed air is usually the primary transport mechanism moving the atomized paint particles form the spray (gun) tip to the object being coated.
2. *Airless spraying*. No air is used, but the paint is pumped under pressure to the spray nozzle. On release, atomization of the paint occurs due to the sudden drop in pressure, the droplets being carried to the component surface by their own momentum. High application rates are possible.
3. *Electrostatic spraying*. An electrostatic field is used to assist the paint deposition by giving the atomized paint droplet a negative charge. If the object to be coated is grounded, paint droplets are attracted. Compared to conventional spray it is more than a line of sight process, showing considerable paint saving since there is virtually no overspray, the electrostatic coating being able to 'wrap' itself around the article being sprayed. However, no electrostatic charge exists on the inside surfaces of articles hence the process will not adequately coat the inside; hence avoid recesses and cavities with the design. Electrostatic attraction also produces build-up of paint on edges.
4. *Electrophoretic coating (electropainting)*. This is a dipping process applicable to water-dispersible resins where the dispersed phase in the emulsion usually has an inherent negative charge. Applying a potential difference between the article to be coated (positive) and the tank will cause the changed species to migrate and coat the article. The process gives uniform thickness and good edge coverage. Autophoretic paints are a development of the process where deposition will initiate via electrochemical potentials instead of an applied potential. Able to provide paint coverage on areas that cannot be reliably painted by other means, e.g. inside box sections.

The decorative value of paint finishes is of course very high and a wide selection of colours and gloss levels are available (gloss to matt). Apart from the plain finishes, a number of decorative effects are possible, as follows:

1. *Metallic finish*. Well known from its widespread use in the car industry, aluminium particles in the paint give the finish a translucent appearance that has different shades of colour when viewed at different angles.
2. *Hammer finish*: Aluminium particles in this case form into small groups, once the paint is applied, that simulate a speckled or 'beaten' metallic effect. This pattern will also hide surface roughness and imperfections to some extent.
3. *Spattered texture*. Ideal for uneven surfaces and foamed mouldings, can be obtained by using a fleck finish or by applying conventional paints with a special spray technique.

13.8.2 Polymer coatings

Traditionally liquid paints were the prime industrial finishing material, but in the early 1970s the concept of powder coating brought a new and high-quality range of organic finishes to the metals and plastics industries. The resins used for coating are of both the thermoset and thermoplastic types, blended with pigments, flow agents, etc. The solid or powder process covers polyethylene, nylon, fluorocarbons, epoxies, etc. There are only a few products that cannot be powder coated. Application of liquid polymer dip coatings is not as common, typically restricted to PVC and PVDF (polyvinylidenedifluoride).

A summary of plastic coating types is given in Table 13.19.

The major market in polymer coating finishes is in environmental protection and corrosion control, an area where future development of polymer coatings is likely to develop. A successful coating system on mild steel under moderately severe atmospheric conditions should last at least 10 years.[23]

There are a number of factors that should be borne in mind when considering the design of components for plastic coating, as follows:

1. The assembled dimensions must allow for the coating thickness (see Table 13.19).
2. Hinges and other similar moving parts should be so designed to allow disassembly for coating purposes.
3. Avoid air traps; complex coated parts should have adequate vent holes.
4. All corners should have a minimum 3 mm radius. Coatings tend to pull back over sharp corners; threaded connections are not recommended. Sharp edges cannot be coated.
5. Most coatings require that the substrate is pre-heated in order to gelate the polymer on to the metal in the case of a liquid or to melt and flow the particles in the case of a powder.
6. Since the coating thickness may depend to some extent on the thermal capacity of the substrate, widely differing material thicknesses should be avoided.
7. Metals should be cleaned/shotblasted prior to coating application. Finishes such as plating or galvanizing do not confer any finished product performance advantage.
8. Unmachined castings may be coated provided flash is removed and the casting is not porous.
9. Because expansion coefficients of metals and polymer coating differ significantly, a primary adhesive bonding coat may be required.
10. The substrate should not be affected by the curing or pre-heat temperatures (see Table 13.19).

The application or transfer methods suitable for each type of polymer coating material are given in Table 13.19. These processes are summarized as follows:

1. *Fluidized bed dipping*. In a fluidized bed, the powder is fluidized by the air forced through a membrane placed at the bottom of the equipment. Any thermoplastic material obtainable in powder form may be applied by fluidized bed methods, the powder behaving as a liquid and so not offering any resistance to the component part entering. Depending on the powder, it may be necessary to sinter (cure) the dipped article subsequently in order to ensure total fusion.
2. *Electrostatic powder spraying*. Most widely used for epoxy and polyester powder coats. Powder is delivered in an air system to a spray gun which electrostatically charges the particles. Since the component is earthed, powder particles are attracted as they leave the spray gun, the generated powder cloud ensuring that the component receives a uniform one-coat finish. Coated articles require stoving, the resin reacts with the hardening agent as the particles melt and flow to form an even, tough film usually a harder coating than produced by wet paint systems. Avoidance of sharp edges and deep recesses equally applies to powder coats, but the danger of paint runs and puddles is minimized.
3. *Electrostatic fluidized bed*. A combination of electrostatic spray and fluidized bed. The bed consists of an array of electrostatic elements on the bottom of the shallow bed, the bed of powder is then charged. An earthed article held in the cloud collects a layer of powder more by electrostatic than by fluidized bed principles. The coated article is subsequently stoved as in (2).
4. *Flame spraying*. The general process is considered in section 13.7.1. The problem is that the high temperatures required to melt thermoplastics (100–300 °C) may also bring them close to degradation unless the process is accurately controlled.
5. *Liquid dipping*. The easiest and oldest method of applying a coating to any substrate. However, the only polymers currently processable in liquid form are PVC (plastisol) and PVDF (Table 13.19). The heated article is dipped for a given time to create the required thickness then heated to drive off the solvent.
6. *Dispersion spraying*. A spray process of a polymer resin dispersed in either an aqueous or a solvent base, subsequently followed by sintering. Typically applied to fluorocarbons.

13.8.3 Pre-formed coatings

The use of pre-painted steel or pre-coated steel for post-forming fabrication has grown steadily in the past 10 years and is likely to continue to do so. There are two routes that pre-form coatings have taken: (1) pre-coating steel sheet and coil; (2) coating of pre-cut blanks.

(i) Pre-coated strip

The positive features attributable to pre-finished sheet steel are as follows:

1. Reduced manufacturing costs, no paint shop requirement and a reduction in line rejects through consistent quality;
2. Smooth total coverage, easier to achieve on flat coil;
3. A wide range of colours and finishes available
4. Use of conventional assembly and fabrication processes without damaging the finished surface. Resistance welding, however, is not possible although spot welding may be used;
5. Most conventional forming processes are possible including deep draw and press braking.

Although there is a rapid conversion in the appliance industry to this type of manufacture, it does mean that the product may well need to be redesigned to take full advantage of the potential of coated strip. In particular, welding needs to be designed out and raw edges need to be hidden.[24] Two types of product are available, a paint coated finish and a laminate (Fig. 13.8), both on to galvanized substrates. With the laminate option, a PVC laminate (100–250 μm coating thickness) is bonded on to the galvanized base. Hot-roll embossing at the manufacturer's plant enables veneer and woodgrain simulation finishes. A summary of available coating materials and coating performance data is given in Table 13.20.

In order to establish subsequent forming limitations for each type of coating, the T-bend flexibility testing standard is used (e.g. Fig. 13.9). The T-bend fabrication performance (Table 13.20), thus forms the basis of the sheet metal component design.

(ii) Coated pre-cut blanks

Pre-coated strip or coil presents three problems as follows:[25]

1. The product needs to be redesigned in order to hide the cut edge arising from the slitting to size of the coated substrate. This may involve retooling or the use of protective trims.
2. There is a need for tight tolerance colour matching between batches of pre-coat coil.
3. The cost of the pre-coated strip.

These factors have encouraged the powder coating of pre-cut blanks and then subsequently fabricating the coated sheet. This not only removes the edge problem with blanked, notched and pierced sheet but also eliminates wastage. Powder coats have been formulated to allow bending, pressing, etc. without compromising paint adhesion, the range of materials again as in Table 13.20. However, the raw sheet still needs to be cleaned, degreased and given a zinc phosphate dip prior to coating.

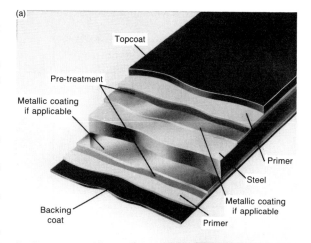

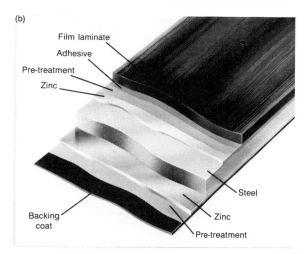

Fig. 13.8 Example of pre-coated sheet (a) and laminate (b). Illustrations courtesy of British Steel Strip Products

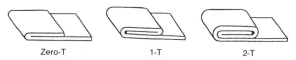

Fig. 13.9 Standard T-bend flexibility test

Table 13.19 Plastic powder coatings summary

Coating material	Application method[a]	Stoving temperature (°C)	Nominal thickness range (µm)	Working temperature range (°C)	Properties	Application
PVC powder	a,c,e	230	250–1.5 mm	−10 to +70	Good corrosion resistance Good impact strength down to −10°C Excellent edge coverage	Tough general-purpose coating External applications Tubular frames, wirework, etc.
Nylon 11	a,c,e,f	300	25–100	−50 to +100	Excellent impact strength if water quenched Reasonable chemical resistance Hard abrasion-resistant coating Components must be well radiused Coating glossy if quenched, matt if air cooled	Non-chip finish requirements Printing rollers
Polyethylene	a,c,f	200	300–2.5 mm (normally 0.7–1.0 mm)	−60 to +100	Smooth and glossy coat Soft, flexible and resistant to a wide range of acids and alkalis Retains flexibility and impact strength down to −40°C High electrical resistance Poor abrasion resistance Poor external weathering ability	Widely used as a decorative coat for wirework General cushion coating on metals
Epoxy	e,g	150	250–1.0 mm	−50 to +180	Good chemical resistance Good impact and scuff resistance	Scratch-resistant coating with ability to withstand a high temperature General coating process for manufactured components Post sintering of epoxies tends to be lengthy compared to other thermoplastics

Material	Application methods[a]		Thickness (μm)	Temperature range (°C)	Properties	Uses
PTFE	b,e	400	12–120	−80 to 250	Microporous	Non-stick, low friction coefficient type of surface coating, i.e. mould release, roller applications, etc.
PVDF	e,b,d	230	50–75 (dispersion spray) 100–350 (powder)	—	Very high outdoor weathering resistance. Excellent corrosion resistance. High UV resistance, hence excellent gloss retention. High abrasion and shock resistance. Good flexibility and impact resistance. UL 94 VO rated	Specially developed for aluminium profiles. Used where high degree of chemical or wear resistance required. Chemical plant, medical equipment, food machinery, bearing surfaces, etc.
Polyurethane	a,c,f,e	325	200–1.0 mm	−70 to +100	—	Abrasion-resistant coating with a high surface friction coefficient
PFA	b,e	400	100–300	−240 to +290	Non-stick low friction coefficient. Cryogenic stability. Tougher than PTFE. Porous-free finish. Good wear resistance at temperature	Moulds, rollers, etc. for food industry in general. Medical equipment. Reprographic rollers. Hydraulic and pneumatic pistons
Polyester	e	190	40–100	−40 to +140	Excellent detergent and chemical resistance properties. Good colour and gloss retention at elevated temperatures. Can be formulated to achieve any gloss level. Some degree of 'orange-peel' is evident on full gloss finish	Advertised as an organic replacement for porcelain in the domestic appliance industry

[a] Key to application methods: a — fluidized bed; b — dispersion spray; c — flock spray; d — dipping; e — electrostatic spray; f — flame spray; g — electrostatic fluidized bed.

Table 13.20 Summary of pre-finished steel coatings

Coating	Dry film thickness (μm)	Pencil hardness	Life expectancy (years)	Scratch resistance (gm) (BS 3900)	T-bend diameter (T=sbstr.thk.)	Max. contin. working temp. (°C)	Properties	Applications
Epoxy phenolic	8	9H	—	800	1T	120	Hard chemical-resistant coating with good fabricating properties. Not suitable for exterior use. Limited colour range	Containers and reverse side coating
Acrylic	25	H	7	2500	10T	120	Not recommended for external use. Poor fabrication performance. Good temperature and stain resistance	Heating and lighting appliances. Shelving
Deep draw polyester	15–18	H–2H	12–15	2900	0T	140	Excellent fabrication performance. Not suitable for external use	Deep drawn and severely deformed components
Polyester	25	F–H	10	2900	3T	120	High-quality domestic appliance finish. Requires care in handling	Domestic appliances in general. Heating and lighting equipment
PVDF	27	F	20	3000	4T	120	Excellent external durability. Good oil and stain resistance. Good forming properties	—
Plastisol	200	—	10–15	>3500	0T	60	Good formability, corrosion and abrasion resistance. Not for use at elevated temperature	General-purpose fabricating material
PVC laminate	100–200	—	—	>3500	2–4T	60	Good abrasion-resistant coating with good handling and fabrication characteristics. Wide range of embossed effects. Low temperature rating	General engineering applications. Domestic appliances and office furniture

13.8.4 Paint finishes on plastics

The reasons for wanting to paint plastics may be summarized as follows:

1. To improve the chemical and abrasion resistance of the base plastic;
2. To increase UV resistance;
3. To give a uniform weathering of a coloured film, i.e. unequal weathering of 'colour matched' components is always a problem;
4. More cost-effective. Uniform pigmentation distribution throughout the product is now relegated to a thin surface paint film;
5. With some plastics, especially foamed plastics, there is a need to post-finish the substrate after moulding for aesthetic/protection purposes.

While thermosetting plastics are easily overcoated with paint, some thermoplastics are very sensitive to solvent types, e.g. polycarbonates, where the inherent material strength may be completely altered if the correct paint system is not chosen. A compatibility chart for common paint types is given in Table 13.21. Although the paints will air-dry, industrial processing would be 20–40 min at 40–80 °C depending on the type of paint and plastic. Paint application will typically be as in section 13.8.1. However, with electrostatic spraying, plastics are insulators, hence the article being sprayed is made conductive, usually a dip or spray inorganic salt solution which when dried provides a sufficiently conductive surface on the substrate without interfering with paint adhesion. Paint formulations must therefore fall within a specified electrical resistance range, the object being coated maintained at ground potential.

13.9 Zinc coatings

Zinc is the most widely applied coating to iron and steel for general surface stability and corrosion resistance, and since zinc coatings are achievable by such a range of processes, a section has been devoted to the subject. There are six main methods by which zinc is used to protect steel, namely hot dip galvanizing, sheradizing, spray coating, zinc plating, zinc paint and impact plating. Of these, hot dip galvanizing and sheradizing are thermal processes involving the metallurgical reaction of zinc with iron or steel, the others are 'cold' processes not involving the formation of any interfacial alloy.

Whatever the type of zinc coating, coating life is largely independent of the method by which it was applied. In conditions of medium industrial pollution, life depends on thickness, i.e.

Average coating life (years to 5 per cent rust) = $0.0116 \times$ coating weight (g m^{-2})

and 50 μm thickness $\equiv 350$ g m^{-2}

Zinc coatings protect steel in one of three ways:

1. The coating weathers at a slow and predictable rate. Zinc corrodes more slowly than steel in most natural environments. There is an initial high corrosion rate during the early stages of exposure, but the rate slows quickly with the formation of a protective film of

Table 13.21 *Recommended paints types for common plastics*

	Recommended paint type					
Plastic	Polyurethane	Epoxy	Polyester	Acrylic laquer	Acrylic enamel	Waterborne
ABS	*	*	—	*	*	*
Acrylic	—	—	—	*	*	*
PVC	—	—	—	*	*	—
Polystyrene	*	*	—	*	—	*
PPO	*	*	*	*	*	*
Polycarbonate	*	*	*	*	*	*
Nylon	*	*	*	—	—	—
Polypropylene	*	*	*	—	—	—
Polyethylene	*	*	*	—	—	—
Polyester	*	*	*	—	—	—
Structural foam	*	—	—	—	*	*

[a] Items with asterisks are compatible

corrosion product. Corrosion then continues at a reduced rate, corrosion being proportional to exposure time.
2. The coating corrodes preferentially to provide sacrificial (cathodic) protection to any small exposed area of steel, whatever the cause. Small scratches can be sealed by corrosion products from the zinc.
3. If the area of damage is large, sacrificial protection prevents sideways creep which could undermine paint finishes.

'Galvanizing' is a generic term often applied to zinc coatings, not just to hot dip galvanizing in particular. The subject of zinc electroplate has already been dealt with in section 13.2.1 and the options of electroplate or hot dipping for sheet steel in section 13.8.3. The remaining application processes are outlined in Table 13.22.

13.9.1 Hot dip galvanizing

When a clean iron or steel component is dipped in molten zinc, a series of iron−zinc alloys are formed by the reaction. An outer layer of pure zinc sits above a relatively thick layer of hard intermetallic compound. The alloy layer is said to provide abrasion resistance and the outer zinc layer impact resistance, so producing an overall coating that is not only corrosion resistant but also has a high resistance to mechanical damage during handling and use. The formation of a fully galvanized coating will occur readily on clean component surfaces, but it will not form on areas of surface inadequately prepared. Surface defects are thus fairly obvious.

At normal bath temperatures, the reaction rate slows down after the alloy layer achieves a certain thickness, but thicker coatings can be obtained on certain alloy steels or by grit blasting prior to galvanizing. Aluminium may be added to the bath to produce a brighter and smoother coating, but this does tend to inhibit alloy layer formation.

Design considerations for hot dip galvanizing may be summarized as follows:

1. Avoid structural designs that tend to distort (stress-relieve) during the dip process.
2. To minimize stress build-up use the correct welding procedure during fabrication and avoid the use of sections with large differences in thickness.
3. Sealed compartments must not be immersed.
4. Allow for zinc thickness build-up on mating surfaces, typically 1 mm is sufficient allowance.
5. Pre-finished sheet is often post-formed, and for this reason the galvanizing conditions are controlled to give a flexible coating ∼ 25 μm.
6. Fabricated structures should be so designed that there are no narrow crevices to cause air locks or impede drainage of the molten metal.
7. Fabrications are coated on all internal/external surfaces providing both a seal and improved rigidity, reducing subsequent noise levels and also providing the opportunity for weight saving.
8. Fusion welding post-coating will destroy the protection at the weld, hence the zinc coat will need to be restored locally, e.g. by spray or paint.
9. Satisfactory welds can be made in galvanized steel, but welding speeds are slower and may require slight changes in welding procedure. Zinc oxide inhalation should be avoided.

13.9.2 Sheradizing

Articles to be coated are tumbled in a sealed barrel containing fine zinc dust at temperatures just below the melting point, i.e. 370 °C. The zinc bonds to the steel by a diffusion process which forms a hard even layer of iron−zinc compounds, the outer surface layer being zinc rich. Growth proceeds on all surfaces in contact with the zinc dust, but pre-clean is essential. The process will not bridge impurities nor will it take on parts with surface imperfections, e.g. scale or rust.

The major feature of the process is coating uniformity on all surfaces with no peak build-up on corners and threads. It is a reliable and versatile means of coating with zinc, but is only generally used for small and intricate components.

13.9.3 Impact plating

The coating is a mechanically applied zinc−iron alloy using shot-blast type of equipment, zinc−iron 'shot' is projected at high speed on to the metal substrate where impact causes a 'galvanizing reaction'. The blasting wheel speed, substrate hardness and exposure time all affect the coating speed and weight.

References

1. Halling J Tribology in manufacturing engineering. *Proceedings of the Institution of Mechanical Engineers* **192**(20): 189 (1978)
2. Mathews A The key to success for surface engineering. *Metals and Mats.* **7**(3): 150 (1991)
3. Bell T *European Journal of Engineering Education* **12**: 1 (1987)
4. Not just a pretty colour. *Engineering Digest* **47**(5): 47 (1986)

Table 13.22 Summary of zinc coatings[a]

	Hot-dip galvanizing	Sheradizing	Spray coating	Zinc paint	Impact plating
General	Bath temperature 450–465°C. 50 μm on thin steel to 150 μm+ on thick steel. Coated sheet thinner at 25 μm	Typical thickness 25 μm, other thicknesses obtained by varying the process times. Process temperature is approx. 370°C, just below the melting point of zinc	Generally 100–150 μm thickness, thicker coats clearly possible	Up to 35 μm in one coating	Film weight of 8–12 gsm
Features	Coating integral with steel. Process pre-clean treatment essential. Shot-blasting promotes a more vigorous reaction between zinc and iron, hence thicker coating results. Colour is the light grey of the zinc-rich outer layer	Dark grey colour as surface is a zinc-iron layer; matt uniform coat. Normally used on small components and fasteners. Thinner coat than hot-dip galvanizing. Diffused coating provides a chemical bond. Closely controlled coating thickness. Applied to finished articles where forming not required. Hardness of interface layers 600–700 kg mm^{-2}, surface 300–350 kg mm^{-2}	Frequently used as a base for organic finishes. Usually applied to the finished component	Use alone or as zinc base primer under conventional paints. Formulated with a high proportion of zinc dust pigment to provide electrical conductivity	After impact plating, parts have to be sealed or chromated. Plated film itself is a porous alloy film consisting of the build-up of impacted platelets
Advantages	Versatile, hence widespread usage. Only limitation in the process is the size of the galvanizing bath available. Gives protection to areas inaccessible to spray or brush finishes and avoids hidden rusting. Any discontinuities due to poor surface preparation are readily visible as 'black spots'. Coatings with little or no alloy layer are readily post-formed	An alloyed coating therefore good adhesion. Continuous and very uniform, even on threaded and irregular parts. Useful where close control of tolerances is important. Hardest of all the zinc coatings. Good key for paint. Retards the oxidation rate of iron and steel up to 400°C	Good mechanical interlocking provided part is firstly grit blasted	Good uniformity, any pores fill with reaction products. Abrasion resistance will be better than conventional paints. Suitable for anything that can be painted brush, spray or dip applied. Painted sheet can be formed. No size limitations	Innovative dry plating technology. A high performance, corrosion-resistant metal surface treatment replacing existing methods. Excellent base for painting. Cold process, no heating effects. Flexible coat will allow post-forming operations. Excellent base for bonding rubber with conventional adhesives
Disadvantages	Coatings containing the alloy layer are not easily formable. Etching treatment required before painting	Limited by barrel size	Coatings are porous but will fill up with corrosion products. Not generally suitable for small parts. Variable thickness	Performance varies. Formulation and application of paint needs careful control. Paint films are always slightly porous and sensitive to poor surface preparation	Need post treatment to seal off porosity, e.g. modified chromate, sealer or sealer + paint. Variable thickness coat with no metallurgical bond to the base steel

[a] For zinc electroplating refer to Table 13.2.

5. BS 4479:1990 *Design of articles that are to be coated*
6. Hignett J B Mechanical surface engineering. *Proceedings of the 1st International Conference on Surface Engineering*. The Welding Institute 1985
7. Hard, non-stick anodised coating. *Design Eng.* Oct 1989: 11
8. Hitchman M L Chemical vapour deposition for surface modification. *Proceedings of the Conference on recent developments in surface coating & modification processes*. Institution of Mechanical Engineers 1985
9. Bell T et al Surface engineering of light metals. *Proceedings of the ASM Conference in Heat Treatment and Surface Engineering* ASM 1988
10. Garside B et al Engineers guide to advanced surface engineering. *Metals & Mats.* **7**(3): 165 (1991)
11. Teer D G et al Principles of ion plating. *Proceedings of the Conference on Recent Developments in Surface Coatings & Modification Processes*. Institution of Mechanical Engineers 1985
12. DTI *Wear resistant surfaces in engineering*. HMSO 1986
13. Onate J I et al Tribological effects of nitrogen implantation on hard chromium coatings. *Metal Finishing*, March 1989: 25
14. Hurricks P L Some metallurgical factors controlling the adhesive and abrasive wear resistance of steels. *Wear* **26**: 285 (1973)
15. Hick A J Advances in surface heat treatments. *Proceedings of the 1st International Conference on Surface Engineering*. The Welding Institute: Brighton 1985
16. Oakley P J Laser surfacing review. *Proceedings of the 1st International Conference on Surface Engineering*. The Welding Institute, Brighton 1985
17. Bell T et al Plasma nitriding treatments for enhanced wear and corrosion resistance. In *Coatings and surface treatment for corrosion and wear resistance*. Ellis Horwood 1984
18. Dawes C et al Nitrotec surface treatment. *Heat Treatment of Metals* **4**: 85 (1982)
19. Cherry F K Austenitic nitrocarburising *Heat Treatment of Metals* **1**: 1 (1987)
20. Chatterjee-Fischer R Some new wear resistant coatings on steel. *Proceedings of the 4th European Tribology Congress*. Elsevier 1985
21. *Vitreous enamel, a performance guide*. Vitreous Enamel Development Council Ltd 1984
22. Hamilton S EMI shielding for plastic enclosures. *Design Prods. and Applications* April 1988: 43
23. Scantlebury J D Organic coating systems and their future in corrosion protection. In *Coatings and surface treatments for corrosion and wear resistance*. Ellis Horwood 1984
24. O'Toole K Pre-finished steels. *Eng. Mats. and Design*. July 1989: 29
25. McLeod K Flexible powder coatings for post-form application. *PRA Symposium on Powder Coatings* Telford, Sept. 1990

CHAPTER 14

FASTENERS

14.1 Introduction

This chapter covers the subject of fasteners and similar mechanical retention means. Background theory and tabulated information is presented to allow an understanding of the principles involved. In most cases several possibilities for fastening will be suitable for securing a joint, and the aim of this chapter, in part, is to provide an awareness of these options. Particularly in the case of fasteners, it is also worth consulting commercial application data.

Screwing components together is very often a preferred solution when designing for the assembly of several items. The reasons for this are as follows:

It is easy to design based on the application of standard fastener sizes.
The result is usually reliable since the required clamp force is much less than the screw strength.
The enabling tools and components are widely available and well known.
The joint can be disassembled easily.

Since screw fastening is relatively labour intensive, other methods of designing a joint should also be considered, particularly where cost, ease of manufacture, assembly or in cases where a large number of fasteners may be used (high volume production or large components). Better design can reduce the manufacturing effort required to produce a fastened joint provided the design resource and enabling support are available.

14.2 Threaded fasteners

While threaded fasteners are considered casually as simple joining elements, the reality of an assembled fastener represents a complex mechanical system, albeit one of the cheapest methods of joining items together. A machine fastener consists of a shank, wholly or partly threaded, plus a driving head. From this straightforward specification, a very wide range of head style and shank sizes have evolved. To ensure correctness of fit, standards exist for all commonly used items. The following terminology usefully distinguishes bolts and screws:

Bolt. An externally threaded fastener intended to mate with a nut and tightened by rotating the nut.
Screw. An externally threaded fastener intended for installation in a preformed mating internal thread and tightened (or removed) via the screw head. Applies whether or not threaded up to the head.

A threaded fastener will typically be designed to withstand axial loads (static, fluctuating or impactive) plus possible supplementary bending and shear. Torsional loads arise mainly from thread friction on torque-up. See section 14.5.1 for the principles of joint clamping. The ISO metric thread type is the one mainly considered in this chapter since its use is virtually universal in the UK and Europe.

14.2.1 Material parameters

Except for very small diameters, externally threaded fasteners are now rarely machined directly. Production nowadays is from wire or barstock of the required fastener diameter, the head being forged and the threads rolled. The rolled thread gives a better grain flow hence improved mechanical properties. Cold rolling is the most effective method for increasing the fatigue resistance of a screwed connection, due to a residual compressive stress at the thread root (notch), work hardening and improved grain flow (Fig. 14.1). Similarly, with cut threads,

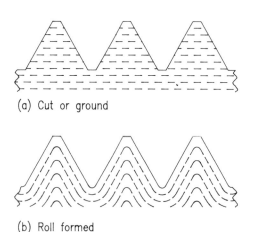

(a) Cut or ground

(b) Roll formed

Fig. 14.1 Grain flow in tooth forms

improvements can be obtained by subsequent root rolling. Any heat-treatment operation following thread rolling will obviously reduce the fatigue strength advantage. Cold-forged screws are generally manufactured either in steel, stainless steel, brass or aluminium. The minimum tensile strengths are as follows: steel = 392 N mm^{-2}, brass = 314 N mm^{-2}, aluminium alloy = 314 N mm^{-2}.

With steel bolts and screws, a strength grade system operates,[1-3] the designation consisting of two figures, one-hundredth the minimum tensile strength in newtons per square millimetre, and one-tenth the ratio between minimum yield stress and minimum tensile strength expressed as a percentage. For example, for a steel of minimum tensile strength 392 N mm^{-2} and minimum yield stress 235 N mm^{-2},

$$1/100 \times 392 \approx \text{`4'} \quad \text{and} \quad \frac{1 \times 235 \times 100}{10 \times 400} \approx \text{`6'}$$

hence the strength grade is 4.6. A summary of mechanical properties and strength grade designations is given in Table 14.1. With strength grade figures, the product of the two numbers multiplied by 10 gives the yield stress.

14.2.2 The ISO metric screw thread

The metric thread series are groups of diameter/pitch combinations. The basic profile for ISO general-purpose screw thread is shown in Fig. 14.2.[4] Two major versions of metric thread are covered by this profile standard, the general-purpose engineering M series and the MJ series.[5] The MJ series incorporates controlled radii at the root of the external thread and is intended for aerospace use and other highly stressed applications requiring fatigue resistance.

In the general-purpose series there are three pitch/diameter groupings, coarse, fine and constant (Table 14.2). The general commercial fastener is coarse pitch. The complete designation for an ISO metric screw thread comprises a designator for the thread system plus a designator for the thread tolerance class. The tolerance class designator comprises a tolerance class for the pitch diameter followed by a tolerance class for the crest diameter. If these two tolerance classes are the same, there is no need to repeat the symbols. Some examples are given below.

(i) External thread

M6 × 0.75 — 5g 6g

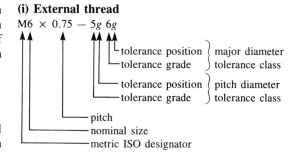

Table 14.1 *Mechanical properties of screws, bolts and nuts*

Mechanical property	Property class									
	3.6	4.6	4.8	5.6	5.8	6.8	8.8	9.8	10.9	12.9
Nominal tensile strength (N mm^{-2})	300	400	400	500	500	600	800	900	1000	1200
Hardness (kg mm^{-2}) min.	95	120	130	155	160	190	230	280	310	372
max.	220	220	220	220	220	250	300	360	382	434
Nominal yield stresss (N mm^{-2})	180	240	320	300	400	480				
Proof stress (N mm^{-2})							580	650	830	970
Elongation at fracture (%) min.	25	22	14	20	10	8	12	10	9	8

FASTENERS

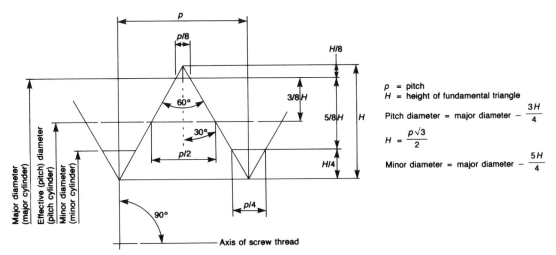

Fig. 14.2 Basic profile of ISO metric thread

Table 14.2 *General purpose metric series diameter/pitch combinations*

	Major diameter			Pitch	
Preferred	2nd choice	3rd choice	Coarse series	Fine series	Constant series
2.5			0.45	0.35	
3			0.5	0.35	
		3.5	0.6	0.35	
4			0.7	0.5	
	4.5		0.75	0.5	
5			0.8	0.5	
		5.5		0.5	
6			1.0	0.75	
	7		1.0	0.75	
8			1.25	1.0	0.75
		9	1.25		1.0, 0.75
10			1.5	1.25	1.0, 0.75
		11	1.5		1.0, 0.75
12			1.75	1.25	1.5, 1.0
	14		2.0	1.5	1.0
		15			1.5, 1.0
16			2.0	1.5	1.0
		17			1.5, 1.0
	18		2.5	1.5	2.0, 1.0
20			2.5	1.5	2.0, 1.0

(ii) Internal thread

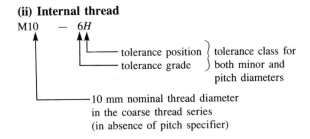

A fit between parts is indicated by the internal thread tolerance symbol followed by the external thread tolerance symbol separated by an oblique stroke, namely:

M6 + 6H/6g
M20 × 2 + 6H/5g6g

The ISO metric screw thread tolerance system provides for allowances and tolerances defined by the following:

Table 14.3 Crest diameter tolerance grades

(a) Nut thread minor diameter

Pitch (mm)	Tolerance grade				
	4 (μm)	5 (μm)	6 (μm)	7 (μm)	8 (μm)
0.35	63	80	100		
0.4	71	90	112		
0.45	80	100	125		
0.5	90	112	140	180	
0.6	100	125	160	200	
0.7	112	140	180	224	
0.75	118	150	190	236	
0.8	125	160	200	250	315
1.0	150	190	236	300	375
1.25	170	212	265	335	425
1.5	190	236	300	375	475
1.75	212	265	335	425	530
2.0	236	300	375	475	600
2.5	280	355	450	560	710

(b) Bolt thread major diameter

Pitch (mm)	Tolerance grade		
	4 (μm)	6 (μm)	8 (μm)
0.35	53	85	—
0.4	60	95	—
0.45	63	100	—
0.5	67	106	—
0.6	80	125	—
0.7	90	140	—
0.75	90	140	—
0.8	95	150	236
1.0	112	180	280
1.25	132	212	335
1.5	150	236	375
1.75	170	265	425
2.0	180	280	450
2.5	212	335	530

1. *Tolerance grade.* A series of standard tolerances applied to each of the four screw thread diameters. The tolerance grades for pitch diameters, minor internal thread diameter and major external thread diameter are given in Table 14.3.
2. *Tolerance position.* Fundamental deviation from basic size (zero line): (a) internal threads, G and H; (b) external threads, e, f, g and h. The fundamental deviations of internal and external threads can be calculated from the following empirical formulae:

$$H_{EI} = 0 \qquad h_{es} = 0$$
$$G_{EI} = +(15+11p) \qquad g_{es} = -(15+11p)$$
$$f_{es} = -(30+11p)$$
$$e_{es} = -(50+11p)$$

where p = pitch (mm), EI = lower deviation, nut (μm) and es = upper deviation, screw (μm)

Tabulated thread deviations are given in Table 14.4. Some typical examples of tolerance and positional deviation concepts for threads are shown in Fig. 14.3.

3. *Tolerance class.* A combination of tolerance grade and position, e.g. 6H, etc. A number of combinations cover the contingencies and requirements of a design as follows:
   (a) Fine quality; 5H/4h. This applies to precision threads used in high-quality production where little variation in the fit characteristic is possible.
   (b) Medium quality; 6H/6g. General-purpose threads intended for most engineering applications.
   (c) Coarse quality; 7H/8g. For use where quick and easy assembly is required, e.g. long blind holes or even with bruised or dirty threads.

Table 14.4 Tolerance position deviations from basic size

Pitch (mm)	Fundamental deviation (μm)					
	Internal thread		External thread			
	G_{EI}	H_{EI}	e_{es}	f_{es}	g_{es}	h_{es}
0.35	+19	0		−34	−19	0
0.4	+19	0		−34	−19	0
0.45	+20	0		−35	−20	0
0.5	+20	0	−50	−36	−20	0
0.6	+21	0	−53	−36	−21	0
0.7	+22	0	−56	−38	−22	0
0.75	+22	0	−56	−38	−22	0
0.8	+24	0	−60	−38	−24	0
1.0	+26	0	−60	−40	−26	0
1.25	+28	0	−63	−42	−28	0
1.5	+32	0	−67	−45	−32	0
1.75	+34	0	−71	−48	−34	0
2.0	+38	0	−71	−52	−38	0
2.5	+42	0	−80	−58	−42	0

Notes: es = upper deviation; EI = lower deviation

What will have become clear is that a very large range of tolerance deviations exist for a given thread diameter

FASTENERS

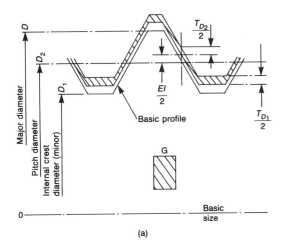

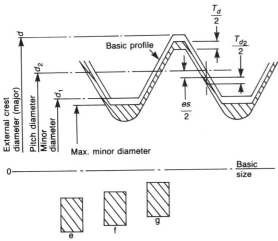

Fig. 14.3 Examples of tolerances and positional deviation concepts: (a) internal thread tolerance position G; (b) external thread tolerance positions e, f and g

Table 14.5 *Example of tolerance deviations for M8 × 1 metric thread*

Tolerance class	Pitch diameter		Minor diameter	
	Upper deviation (μm)	Lower deviation (μm)	Upper deviation (μm)	Lower deviation (μm)
Internal thread				
4H	+95	0	+150	0
5G	+144	+26	+216	+26
5H	+118	0	+190	0
6G	+176	+26	+262	+26
6H	+150	0	+236	0
7G	+216	+26	+326	+26
7H	+190	0	+300	0
8G	+262	+26	+401	+26
8H	+236	0	+375	0
External thread				
3h4h	0	−56	0	−112
4h	0	−71	0	−112
5g6g	−26	−116	−26	−206
5h4h	0	−90	0	−112
5h6h	0	−90	0	−180
6e	−60	−172	−60	−240
6f	−40	−152	−40	−220
6g	−26	−138	−26	−206
6h	0	−112	0	−180
7e6e	−60	−200	−60	−240
7g6g	−26	−166	−26	−206
7h6h	0	−140	0	−180
8g	−26	−206	−26	−306
9g8g	−26	−250	−26	−306

and pitch, e.g. the possibilities for M8 × 1 are given in Table 14.5. While it is possible to combine any of the tolerance classes for the two mating components, in practice the choice is rationalized[4] to 4h, 6g, 8g, 5H, 6H and 7H combinations, normal quality being 6H/6g. As an example the thread tolerance limits for M8 × 1 are given in Table 14.6.

As an unavoidable result of the manufacturing process, the root profiles on both threads are radiused and this root contour should not transgress the toleranced profile of the opposing part. The maximum root radius is specified as $P/8$.[4]

14.2.3 Head and drive styles

Head styles fall into two obvious categories, flush and surface proud. Within each of these there are a variety of drive styles. It is perhaps surprising that, with such a basic product, so many options do exist. It is generally the enabling requirements of sophisticated automated assembly techniques that have seen the most recent developments in screwdriving methods. Preferred head styles are summarized (Fig. 14.4), under the headings below.

(i) Flush form (countersunk)
To ensure that the screw fits flush, it is necessary to control within limits, the dimensions of the screw head and the mating countersink. For the ISO metric countersunk head,[3] the maximum or design size of head is controlled by a theoretical diameter to a sharp corner and the minimum head angle of 90° (Fig. 14.5). The minimum head size is controlled by a minimum head

Table 14.6 *Example of thread tolerance limits*

Tolerance class	Fundamental deviation	External threads						Minor diameter (min)
		Major diameter			Pitch diameter			
		Max.	Tol.	Min.	Max.	Tol.	Min.	
4h	0	8.000	0.112	7.888	7.350	0.071	7.279	6.663
6g	0.026	7.974	0.180	7.794	7.324	0.112	7.212	6.596
8g	0.026	7.974	0.280	7.694	7.324	0.180	7.144	6.528

Tolerance class	Fundamental deviation	Major diameter (min)	Internal threads					
			Pitch diameter			Minor diameter		
			Max.	Tol.	Min.	Max.	Tol.	Min.
5H	0	8.000	7.468	0.118	7.350	7.107	0.190	6.917
6H	0	8.000	7.500	0.150	7.350	7.153	0.236	6.917
7H	0	8.000	7.540	0.190	7.350	7.217	0.300	6.917

Notes: M8 × 1 metric thread; all dimensions in mm.

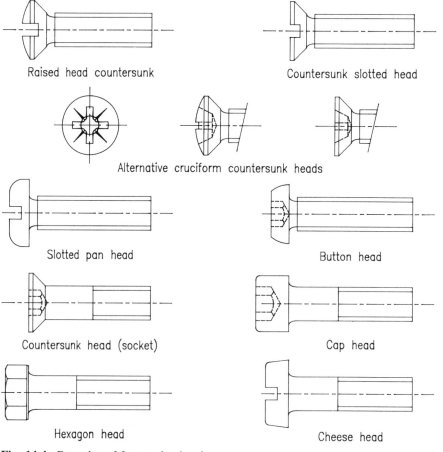

Fig. 14.4 Examples of fastener head styles

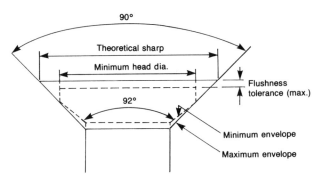

Fig. 14.5 Countersunk head design size

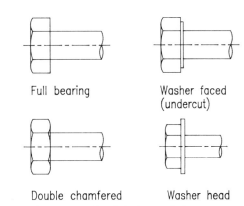

Fig. 14.6 Hexagon head design variations

diameter, the maximum head angle of 92° and a flushness tolerance, 5 per cent of the nominal thread diameter. A variation of the countersunk head is the raised countersunk head which allows greater depth of driving type plus, perhaps, improved aesthetics.

(ii) Proud form

1. *Pan head.* The pan head, when slotted (Fig. 14.4) is a flat head of twice the nominal thread size, parallel to the base of the head and radiused at the edges. In the recessed drive type, the head shape is a combination of round and pan head profile (domed) resulting in a greater head height than the slotted form.
2. *Cheese head.* Originally the most economical profile to manufacture by auto-machining methods, but not the most satisfactory head shape for cold forging. Not generally available in other than a slotted drive style. In basic form it is a cylindrical head with a flat seating. Alternatively available with a shallow domed top and deeper slot (fillister).
3. *Hexagon head.* A number of variations in the head configuration are shown in Fig. 14.6, each to some extent depending on the production process, but in essence either a washer-faced type or a full bearing head. The hexagon head style is by definition also its means of driving. For preferred sizes,[6] the following empirical size relations apply:

Distance across flats (mm) = $1.45M + 1$

where M = nominal thread diameter.

Width across corners = $1.13 \times$ width across flats

14.2.4 Drive styles

The type of drive style available can be grouped into four basic categories, summarized below.

(i) Slotted form

Still a popular way of internally wrenching a screw, but the slot has a number of disadvantages:

1. A multitude of slot widths, depths and lengths require a number of screwdrivers.
2. Screw heads easily suffer damage during driving by virtue of the high load concentration produced at screwdriver–slot contact points.
3. Screwdriver blades can slip out of the screw head and damage surrounding surfaces.
4. It is difficult to transmit high drive torques.
5. Slots are not self-aligning.

Since driver blades are usually tapered in thickness, the tendency to cam-out is a possibility. Cam-out occurs when torque is applied to drive surfaces that are not parallel to the screw axis, i.e. strictly normal to the applied torque. Some of the above problems can be overcome by using a recess screw head or enclosing or constraining the driver in position.

(ii) Recess form

There is a wide range of recessed form drive styles (Fig. 14.7), some of which have been deliberately developed for high torque transfer. Less well-known types are not always available off the shelf, being generally produced by manufacturers to order as required. Some examples are described as follows:

1. *Cruciform style.* These fasteners are self-aligning and positive drive systems. However, the curvature of the drive faces and the engagement angle of the head slots make it necessary to apply an end load to the driver when torqueing in order to avoid cam-out. In this regard, Torq-set which is used for shallow countersunk screws, Frearson with its sharp 45° recess taper and Phillips, the original production

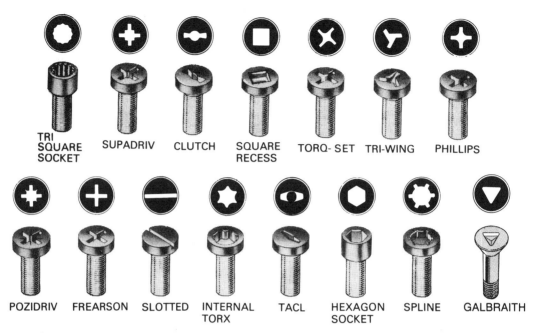

Fig. 14.7 Examples of recessed head drive styles. Courtesy of Harmsworth, Townley & Co. Ltd, Todmorden, and Courtesy of Findlay Publications

recess screw, all require significant end loading (Fig. 14.7). The Posidriv (Fig. 14.8) screw was an attempt to reduce the compound angle of the driver contact face by forcing metal forward during the forming operation. Success here was apparently limited, the Supadriv recess now superseding the Posidriv (Fig. 14.8). Apart from improved containment of metal flow, the Supadriv recess cone angle is reduced to encourage the driver to wedge in and reduce cam-out tendency.

2. *Hexagon recess drive.* This system concentrates a radial torque loading at an inefficient 60° drive angle (Fig. 14.9). This causes high stress concentrations (line contact) over a reduced wall so requires a tough grade of steel, grade 12.9 in Table 14.1. Thus fasteners with internal hexagon drives are heat treated, hence the black (thermal) oxide colour, typically oiled. Depending on the head style, the actual socket size will vary for a given key wrench size (Table 14.7).

(ii) High torque form
The division between normal recessed form and high torque is somewhat arbitrary, but the features of a high torque drive style can be summarized typically by the following:

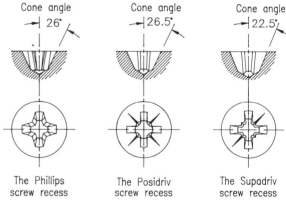

Fig. 14.8 Comparison of cruciform styles: (a) the Phillips screw recess; (b) the Posidriv screw recess; (c) the Supadriv screw recess. Courtesy of EIS Fasteners, Wednesbury

1. A number of surfaces over which to distribute the torque load.
2. A low or zero driver cam-out tendency.
3. Low or zero drive angle.
4. The necessity for a special driver.

A number of styles exist that are suitable for a high

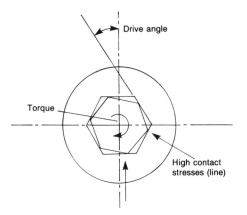

Fig. 14.9 Hexagonal socket drive system

torque driving requirement and where the drive head does not easily cam-out (Fig. 14.10). Although we are concerned here with machine screws, the ability of these head styles to easily transmit high torque loading makes them popular for self-tapping screws. The examples in Fig. 14.10 show similar features that have been developed to cope with the requirement of a high torque screw.

(iii) Security screws
The security screw fulfils an important role where it is not always possible to conceal a fastener totally, yet where access must be deterred. Essentially these are a range of screw-head styles that enable insertion and extraction of the screws with special drives and which do not allow the use of normal screwdrivers or wrenches, however ill-fitting. Such fixing devices are becoming more important with small machines where it is desired to deter entry into the product, but at the same time where expensive locking systems are not justified. Some examples are shown in Fig. 14.11.

14.2.5 Point styles

Machine screws are generally unpointed. As a result of the production method, however, the last two threads or so tend to be undersize as the screw end rolls over, leaving the characteristic shank end indentation. The 'lead' provided by the thread-rolling operation is normally regarded as producing a sufficiently good chamfer without further machining.

Cut threads are normally finished to a chamfer for sizes < M6. Other options are to end radius the screw or produce a cone point where a more positive lead is required.

14.2.6 Locking screws

The 1960s saw the advent of prevailing torque-type fasteners with nylon inserts, the most well-known example being the Nyloc nut (section 14.3.3). However, although nylon inserts are still popular, today's insert ends to be a 'patch' of fused polymeric material or micro-encapsulated adhesive. The terminology generally used to describe self-locking performance is given below:

Tightening torque. The torque required to tighten or tension a bolt or screw, typically 75 per cent proof stress.
Break-loose torque. The torque necessary to commence screw untightening.
Break-away torque. The torque necessary to put in reverse rotation a screw that has not been seated.
Prevailing-on torque. A measure of the retarding torque of a screw-locking feature during insertion.
Prevailing-off torque. Opposite of prevailing-on torque, i.e. when screw is being withdrawn.

Table 14.7 *Hexagon socket sizes for different head styles*

Key size A/F (mm)	Key size tolerance	Cap head shank dia. (mm)	C/S head shank dia. (mm)	Button head shank dia. (mm)	Set screw shank dia. (mm)
1.5	h9	1.6/2	—	—	3
2	h10	2	3	3	4
2.5	h10	3	4	4	5
3	h10	4	5	5	6
4	h10	5	6	6	8
5	h10	6	8	8	10
6	h10	8	10	10	12
8	h10	10	12	12	16
10	h10	12	14/16	14/16	20

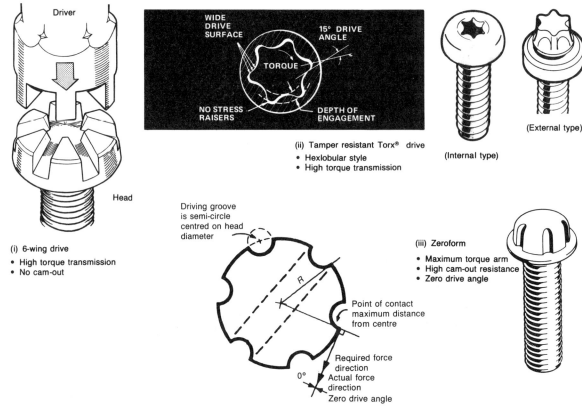

Fig. 14.10 Examples of high torque drive forms. TORX is a registered trademark of Camcar division of Textron Inc.

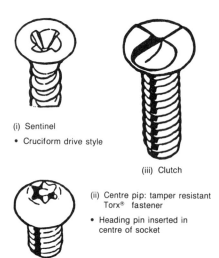

Fig. 14.11 Examples of security screws. Part (i) courtesy of European Industrial Services, Wednesbury. Part (ii) TORX is a registered trademark of Camcar division of Textron Inc.

(i) Polymer insert or patch

As the description imples, plastic inserts (typically nylon) may be incorporated into the threaded portions of bolts or screws (Fig. 14.12), their size controlled to give different degrees of frictional resistance. A nylon strip is compressed and inserted into a narrow slot milled in the screw. The self-locking is achieved by the wedge action between male and female threads which increases flank pressure on the threads diametrically opposite the insert.

Another method of retaining a polymeric thread contact is the fused patch. Either nylon or polyester material is fused on to about four to six threads of the screw around some 30–360° of circumference. With a circumferential fused band, the resultant thread infilling will also provide sealing. All types are generally quoted as 'reusable'.

(ii) Micro-encapsulated adhesive

Advances in micro-encapsulation technology have enabled a number of thread-locking adhesives to be applied in this form to fasteners over a year or so before use. As the

FASTENERS

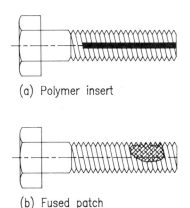

Fig. 14.12 Polymer insert and patch

screw is rotated in its mating component, the capsules rupture, releasing adhesive. Generally speaking, the break-loose torque strengths of the liquid adhesives are retained.

The most well known of the adhesive patches is the encapsulated anaerobic adhesive sealant (section 4.5.2). Cure time is inversely proportional to viscosity, typically 2–12 hours at room temperature to obtain 50 per cent of ultimate strength. However, on inactive surface such as stainless steel, cadmium plate and zinc plate, cures can be protracted. An epoxy resin two-part mixture is an alternative patch adhesive with a high break-loose torque, but a lengthy room temperature cure, typically 12–24 hours to reach 50 per cent of ultimate strength.

A recent addition is a polyacrylate type, coated initially in latex form.

14.2.7 Set screws

Also known as a grub screw, the set screw, in contrast to other fasteners, is primarily a compression device producing a strong clamp action to generally resist relative motion between assembled parts. Because of the stress concentrations of the hexagon drive form, these set screws require a toughened alloy steel (section 14.2.4). Driver key sizes are given in Table 14.7. Slotted set screws are normally supplied electroplated.

Since tensile properties are not relevant as a material specifier, hardness, as a measure of the yield stress in compression, is used instead to define property classes. Properties and characteristics of the H (hardness) range are given in Table 14.8. As with conventional screw fastenings, set screws can be divided into two groups, depending on their drive form and point style.[2,7]

(i) Point styles

Preferred point styles are illustrated in Fig. 14.13, the usual design application for each is as follows:

1. *Cup.* Probably most widely used for quick, permanent location of gears, collars, pulleys, etc. on shafts. No spotting hole is necessary, but some cutting-in action is inherent with a hexagon drive form and soft shaft. Some loss in holding power is experienced when the hardness differential between screw and shaft is < 100 kg mm^{-2}. For those situations where vibratory loosening may be a problem, knurled cup point types provide a degree of ratcheted self-locking action.
2. *Cone (or point).* Used when a permanent location of parts is desired. Because of point penetration, develops greatest axial and torsional holding power when a hardness differential as in (1) applies. Usually spotted in a hole to half its length such that sufficient cone section shear strength is available.
3. *Flat.* Used when frequent resetting is required. Flat points cause little shaft or contact damage.
4. *Dog.* Either as full dog or half dog point. Normally used where permanent collocation is required, spotted into a shaft hole. The holding power is then the shear strength of the point. Occasionally used in place of dowel if infrequent disassembly is envisaged. The dog size merely determines the length of the point itself.

Table 14.8 *Description of steel set screws*

Property class[a]	14H	22H	33H	45H
Hardness (kg mm^{-2}, min.)	140	220	330	450
%C, max.	0.5	0.5	0.5	0.5[b]
Heat treatment	N	Y	Y	Y
Drive form	Slotted only	Slotted only	—	Hex. socket only
Torque strength (N m)	—	—	—	0.9 for M3 to 310 for M20
Thread tolerance finish	6g plain	6g plain	—	5g6g black oxide

[a]The numerical part of the symbol represents one-tenth of the minimum hardness.
[b]Alloy steel.

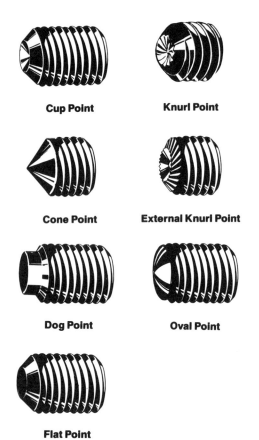

Fig. 14.13 Set screw point styles. Courtesy of Holo-Krome Ltd, Dundee

(Point styles shown: Cup Point, Knurl Point, Cone Point, External Knurl Point, Dog Point, Oval Point, Flat Point)

(ii) Performance

For evaluation of set screw performance, some definitions are required:

1. *Point penetration resistance.* The actual shaft material distortion/displacement around the screw set point, thereby inhibiting relative movement.
2. *Axial holding power.* The resistance of the assembly to relative movement along the longitudinal axis of the shaft.

An illustration of set screw fastening is shown in Fig. 14.14. Torqueing the set screw produces a frictional contact w on the shaft, an equal and opposite clamp load w plus some shaft penetration, depending on point style. Although the extent of the opposing contact area does not provide greater frictional resistance, it is clearly not good engineering practice to provide a large shaft/collar clearance tolerance. The above forces therefore provide most of the resistance to relative axial and torsional movements, excluding screw thread friction. Factors influencing screw and collar tightness are as follows:

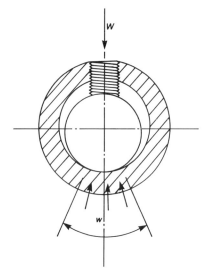

Fig. 14.14 Set screw performance

1. *Seating torque.* The greater the tightening torque, clearly the greater the torsional holding power. Needs to be viewed in conjunction with property class and drive form.
2. *Point style.* Where no holes are pre-drilled in the shaft, the point style clearly provides the greatest holding power followed by the hollow point. In relative terms, if a cup point has a holding power of 1, a cup point is 0.93, a flat is 0.86 and an oval point is 0.84. Providing a flat on a shaft provides only approximately 10 per cent more torsional holding power and no effect on the axial holding power.
3. *Length of thread engagement.* As long as there is sufficient thread engaged to prevent stripping, this has no effect on holding power.
4. *No. of screws.* Two set screws increase the torsional holding power depending on relative collocation. The holding power is doubled when axially in line, but only 30 per cent greater when diametrically opposed. On the same circumference, an optimum displacement of 60° is recommended as a compromise between maximum holding power and minimum metal between tapped holes. This displacement gives 1.75 times the holding power of one set screw. The general relationship may be expressed as follows:

$$T_2 = T_1 \frac{(200 - 0.42\phi)}{100}$$

where T_2 = holding power of assembly, (Nm), T_1 = holding power of one screw (Nm) and ϕ = displacement (deg).

14.2.8 Plastic fasteners

In the early 1960s there was practically no market for plastic fasteners in Europe, although they had been available in the USA for the preceding 10 years. However, there are many virtues of plastic fasteners and today their use is becoming widespread in industry. A number of specialized fasteners described in this chapter are plastics based. Typical uses for plastics in the fastener field are screws, nuts, spacers, washers, push clips, ratchet rivets, stud anchors, PCB insulators and expansion nuts. This is by no means an exhaustive list.

The use of a plastic fastener would favour positional location rather than a tensile clamping load. The use of nylon 6:6 is most common for screws, either machined from extruded bar stock or injection moulded directly. Typical torque figure recommendations are illustrated in Fig. 14.15. As a rule of thumb, after torqueing the joint to finger tightness, a further half-turn with a mechanical driver is enough.

Nylon, a crystalline thermoplastic, is attractive as a fastener material for the following reasons:

1. Outstanding toughness and good damping capacity;
2. Good machinability;
3. Fair creep resistance;
4. Relatively hard surface;
5. Good chemical and corrosion resistance;
6. High-temperature performance good, relative to other thermoplastics;
7. Low friction coefficient;
8. Excellent electrical insulation properties;
9. Choice of colour;
10. Low density.

However, nylons are hygroscopic, the use of the 6:6 grade being a compromise between dimensional stability and cost. Moisture absorption is a surface effect and therefore most apparent in thin sections. Typically, dimensions increase by ≈ 0.10 for each 1 per cent of absorbed moisture. The major effect, however, is in terms of strength reduction, some 75 per cent reduction in tensile strength of nylon 6:6 at a 5 per cent absorption level compared to when dry. Electrical properties are also impaired by moisture although impact resistance is improved.

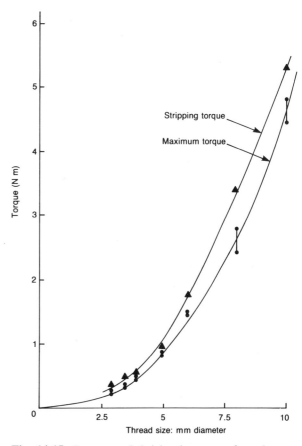

Fig. 14.15 Recommended tightening torque for nylon screws

14.3 Nuts

The function of a nut is to engineer or stress a bolt to its full strength potential, and to maintain the resulting loads throughout the life of the joint. Steel nuts may be produced by cold forging, hot forging or turning from bar, the choice of method and heat treatment being left to the manufacturer to comply with a given strength grade.[6]

14.3.1 Basic nut design

The ISO metric series coarse thread form predominates in industry.[6] For increased stress area ISO metric fine nuts may be required, provided the tolerance class is then controlled (section 14.2.2), to avoid the risk of thread stripping.

If recommended torque loads are exceeded, standard coarse pitch metric nuts are designed to give sufficient length of engagement to cause the bolt to fracture rather than the nut to strip. A bolt or screw assembled with a nut of the appropriate property class (Table 14.9) metric coarse thread, provides an assembly capable of being tightened to bolt yield without thread stripping occurring. Thread-stripping failure indicates an uneconomic use of the material of the bolt, the full core strength of which is not developed fully. The strength grade designation for steel nuts should be graded one-tenth of the specified proof load stress (Table 14.9). The proof load stress is the minimum UTS of the highest grade of bolt with which the nut is to be used, e.g. a grade 8 nut should not be used with anything less than a grade 8.8 bolt (section 14.2.1). Thus the nut material is somewhat softer than the bolt material. Nut grade will be indicated on the nut face, either as an indented code symbol based on a clock face or the grade physically stamped on the nut.

14.3.2 Nut stress distribution

The thread of a nut is subject to a force during tightening which can be expressed as two components:

1. Horizontal or radial force acting outward and tending to dilate the nut at the base.
2. Vertical or shear force acting in a line parallel to the axis of the bolt.

As the load changes from predominantly compressive in the nut to tensile in the bolt, the distribution of load by no means equally shared by all engaging threads. The result for a standard nut is shown in Fig. 14.16. This

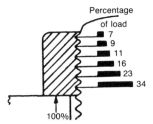

Fig. 14.16 Load distribution on threads of a standard nut

illustrates that the first thread may carry up to one-third the total load, and the stress distribution across the bolt itself results in a maximum stress at the point of thread engagement of some five times nominal. This area of concentrated load, when combined with the thread notch effect makes the area susceptible to fatigue failure. Typically the nut should have sufficient wall thickness and strength to resist the radial force trying to spread the base of the nut. However, an advantage of this radial deformation is the spreading of the load to threads further away from the abutment face.

Tapered thread forms and differential thread pitching are known to significantly improve fatigue life of a nut/bolt combination, but are difficult to employ economically. Scope does, however, exist for relieving the stress on the first few threads and a number of options are possible. Even a standard nut is greatly improved if the face is countersunk at 70° until the first thread is removed (Fig. 14.17) on either side. The thread load distribution may also be improved by tapering the effective diameter of the nut thread or bolt,[8] but this of course is a special measure, not practical on the smaller bolt size (Fig. 14.17).

Table 14.9 *Mechanical properties of steel nuts* (Nut height $\geq 0.5D < 0.8D$)

	Property class of steel nut						
	4	5	6	8	9	10	12
Proof load stress (N mm^{-2})[a]	380	500–630[b]	600–720	800–920	900–950	1040–1060	1150–1200
Hardness (kg mm^{-2})[c] min.	117	130	150	170	170	272	295
max.	302	302	302	302	302	353	353
%C max	0.50	0.50	0.50	0.58	0.58	0.58	0.58
Heat treatment	No	No	No	No	No	Yes	Yes
Property class of mating bolt	3.6	3.6	6.8	8.8	9.8	10.9	12.9
	4.6	4.6					
	4.8	4.8					
		5.6					
		5.8					

[a]Proof load = proof load stress * tensile stress area of bolt.
[b]Proof load stress increases with nut thread diameter within range indicated.
[c]Hardness increases with nut thread diameter within range indicated.

FASTENERS

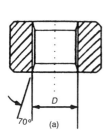 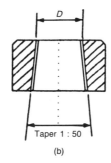

Fig. 14.17 Stress-relieving nuts

The general idea behind these methods is to increase the deflection per unit thread load of the lower threads so improving overall thread load distribution, the improvement obtained outweighing the weakening imposed on the nut and the increase in bending stress on the lower nut threads. The groove nut (Fig. 14.18)[9] is an example where metal removal from the flanks reduces stiffness on the lower nut threads. As a special measure, conical boring out of the bolt material improves the elastic thread load distribution, as does conical tapering of the nut OD, especially when this is combined with an overhung nut (or tension) design (Fig. 14.19).

14.3.3 Torque nuts

The integrity of any assembly held together by threaded fasteners depends upon the maintenance of the clamping load, hence tightening to correct torque plus a provision to ensure that the bolt/screw does not unscrew in service. There may not be that many opportunities in small

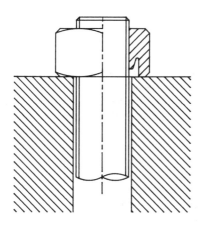

Fig. 14.18 The groove nut

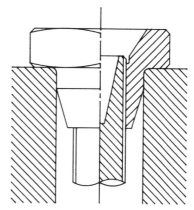

Fig. 14.19 Overhung nut and taper bored bolt

machine design where critical locking torques apply, but there may well be occasions where assemblies are factory aligned and even if dowelled, locknuts are required to ensure no deterioration during the life of the product, including transportation and abuse during use. Unscrewing, even by a few degrees, can be enough to reduce the loading noticeably. Few designers today find themselves involved in the selection and specification of nuts locked by wire, split-pins or tab washers, despite the fact that the split-pin and castellated nut is still arguably the most effective device for critical applications. However, alternative methods are typically less labour intensive and lend themselves deliberately to automated assembly.

(i) Prevailing torque nuts

A prevailing torque type nut (or stiffnut) is one which is frictionally resistant to rotation due to a self-contained torque feature, but not intended to imply permanence. The prevailing torque developed by a nut is the torque necessary to rotate the nut on its mating externally threaded component, with the torque being measured while the nut is in motion and with no axial load on the bolt. Performance specifications for class 8 and class 12 nuts are given in Table 14.10. Some of the commercially available types are as follows:

1. *Material inserts*. One way to create prevailing torque is a plastic insert to increase friction between nut and bolt. Originally a fibre material, now nylon or polyester, the most well-known example being the Nyloc nut (Fig. 14.20). It consists of a plain nut portion surmounted by a shroud, rolled over and keyed after inserting the nylon ring. The keying is to prevent rotation during use.

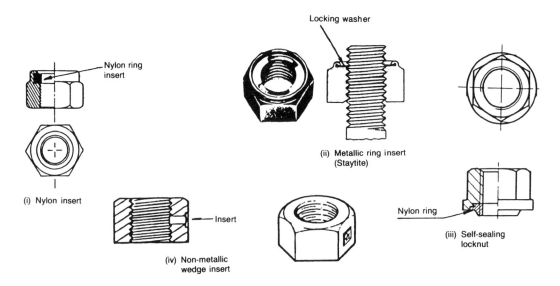

Fig. 14.20 Examples of prevailing torque nuts with material inserts. Part (ii) courtesy of Staytight Ltd, H. Wycombe. Part (iii) courtesy of Bollhoff GmbH and Co. KG, Bielefeld

Table 14.10 *Performance specification of prevailing torque nuts*

Nut size d (mm)	Clamp Load (kN)		Prevailing torque 1st Install. (N m) max.	
	Class 8	Class 12	Class 8	Class 12
M4	3.73	0.9	6.13	0.22
M5	6.08	0.29	9.9	2.1
M6	8.63	3.0	14.0	4.0
M8	16.3	6.0	25.6	8.0
M10	24.9	10.5	40.5	14.0
M12	36.0	15.5	59.0	21.0
M16	67.2	31.5	110.0	42.0
M20	192.2	54.0	171.0	72.0

The Staytite nut is of similar concept to the nylon insert nut except that it uses a stainless steel spring washer instead of the nylon ring as a locking device (Fig. 14.20). The spring pressure of the locking tongues against the thread of the bolt provide the prevailing torque.

2. *Deformed thread nuts.* These basically have a standard tapped thread, subsequently deformed over part of the nut length to provide an interference with the mating bolt around a proportion of the circumference. When the nut is assembled, the interference section is sprung to provide frictional grip as prevailing torque. There are numerous examples of this type, e.g. Fig. 14.21. Interference may be radial, axial or both.

3. *Integral spring nuts.* These incorporate spring arms or some similar thread-gripping feature at the top of the nut. Generally the arms are formed by cutting and slotting operations after tapping, the arms then depitched by deflecting downward and inward. When a screw passes through the nut the arms are forced back towards their original position, resilience of the nut material causing a locking action on the bolt threads. At least two threads should project through the arms to ensure a satisfactory prevailing torque when tightened. Some examples are given in Fig. 14.22.

(ii) Free-spinning devices

These possess no prevailing torque characteristics and rely on either the frictional locking induced between the underside of the nut (or screw) and the component surface or friction between mating threads promoted by tightening. Many of the former employ spring, toothed or serrated washers (section 14.4). Some examples of free-spinning locking nuts are shown in Fig. 14.23.

(iii) Locking systems

The ideal locking system should maintain preload without interfering with the initial setting and should be easy to dismantle for reuse. A recent system (Fig. 14.24)

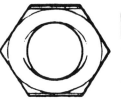

(i) Out-of round
- Controlled axial and radial deflection of upper nut cone profile
- Illustration shows dual deformation type. Trilobular deformed heads also obtainable

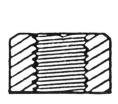 

(ii) Deformed thread
- Accurately introduced radial and axial depitching of top threads
- 3–6 indents typically useful for thin nuts

Fig. 14.21 Examples of deformed thread nuts. Part (ii) courtesy of Philidas Ltd, Pontefract

Aerotight Nut (bent beam axial deflection)

Philidas Nut (bent beam axial deflection)

Binx Nut (bent beam axial deflection)

Slotted Collar (bent beam radial deflection)

Fig. 14.22 Examples of integral spring nuts. Part (i) courtesy of The Premier Screw and Repetition Co. Ltd, Leicester. Part (ii) courtesy of Philidas Ltd, Pontefract. Part (iii) courtesy of TR Fastenings Ltd, Uckfield

overcomes the deficiencies of existing fastening systems and meets today's changing demands. The bolt has a splined extension which mates with the coaxial splines in the locking bell. This item has an internal bi-hexagonal form which, in combination with the splines, results in a vernier effect so that the bell slips over the nut in any position enabling accurate torque control. Once engaged, rotation of either component is prevented by the splines while the bell itself is retained by a separate clip, preventing axial movement.

(iv) Adhesive locking
Adhesive thread locking techniques are included in section 4.5. Adhesives are not really associated with achieving and maintaining a particular bolt loading, rather that the fabrication is vibration resistant.

14.3.4 Spring steel nuts

The first pressed metal nut was designed to back up a load-carrying nut to prevent it backing off or loosening under severe vibration conditions. Under lightly loaded conditions they can also be used as load nuts in their own right. These types of fasteners are predominantly single-thread plates. The thread engagement area is a formed helix in true relation to the pitch of the screw thread. The nuts are generally manufactured from austempered carbon steel (hardness $500-580$ kg mm^{-2}) to maximize loading from a single thread, but at the same time sufficiently resilent to behave elastically when tightened to provide a degree of vibration resistance.

These types of fasteners do have a number of quoted advantages over alternatives in some situations, summarized as follows:

1. By nature of their design, a weight saving of 65–85 per cent is possible;
2. Less space is required;
3. May be removed and reused repeatedly, as long as the coned centre has not been over-torqued;
4. Interchangeable with other locking devices and a cheaper alternative;
5. Will effectively self-clean any painted over or dirty threads;
6. Can be used for the assembly of brittle or fragile materials.

The locking principles of this type of nut are explained in Fig. 14.25. The force F generated when the nut is tightened pushes the teeth of the thread form down the bolt thread helix such that the ends of the teeth (wings) bite into the root of the thread form. This generates a

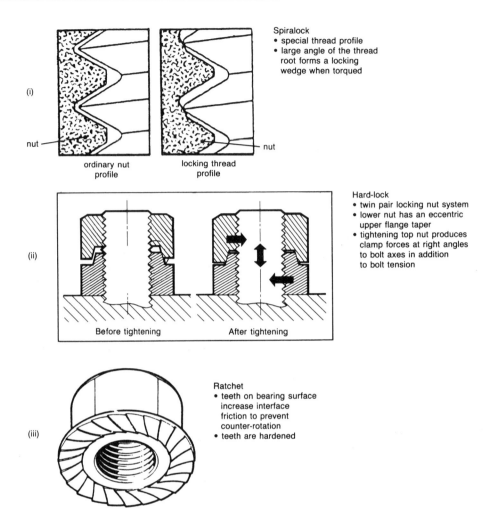

Fig. 14.23 Examples of free-spinning torque nuts. Part (i) courtesy of Microdot Inc., Fullerton, California. Part (ii) courtesy of Staytight Ltd, High Wycombe

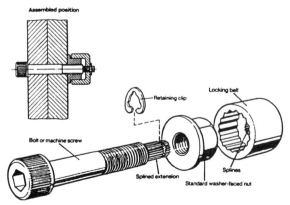

Fig. 14.24 Bell locking system. Courtesy of Findlay Publications, Horton Kirby

reproducible locking force, resolved as in Fig. 14.25, such that both bolt and nut are mutually constrained, a feature that is not true of all locking systems.

It is very difficult to compare quantitatively the performance of the different types of spring steel fastener since fundamentally they are all designed for a different role. However, a nominal comparison based on a shaft diameter of approximately 4 mm is given in Table 14.11. Some examples are given in Fig. 14.26.

14.3.5 Captive nuts

A captive or caged nut is a full threaded nut enclosed within a steel retainer and having an associated degree of float to compensate for misalignment. The nuts used

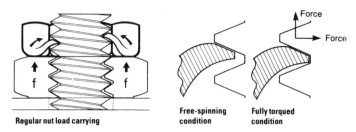

Fig. 14.25 Spring steel nut locking action. Courtesy of Jet Press Sales Ltd, Huthwaite, Notts

Table 14.11 *Comparison of spring steel nut performance*

Nut type	Stud size	Application torque (N m)	Bolt tension (N)	Push-on force (N)
Self-thread nut	M4	3.4	1112	—
Ratchet nut	d3.9	—	267	67
Regular locknut (6-notch)	8/32 UNC	1.2	667	—
Acorn (open)	M4	1.1	534	—
Wing nut (6-notch)	8/32 UNC	1.0	445	—

in these fasteners are normally square with full depth threads. The conventional cage-type nut retainer is shown in Fig. 14.27. One word of caution inasmuch that any surface to be secured clamps down on to the cage, not the panel to which it is attached. Hence, shear resistance will be entirely dependent on how well the cage clips grip the bottom panel.

An alternative form, the self-clinching cage nut, is shown in Fig. 14.27, either self-locking, where the top threads of the bush are pressed elliptical, or the normal grade. The self-clinching nuts provide positive fastening.

14.4 Washers

The significance of a washer as part of a structural fastening system is quite often overlooked in design, yet when used, washers can and do contribute greatly to the overall integrity of the fastening. A single washer is normally applied under the fastener element actually being torqued, i.e. nut or bolt/screw head. Washers act in the following ways:

1. Provide additional bearing area to spread the stress distribution, important especially in softer materials;
2. Permit bridging of oversize or large clearance holes;
3. Act as a thrust surface;
4. Provide a locking or sealing action in certain situations;
5. With certain washer designs, indicate the preload developed in a bolted assembly;
6. Provide a known friction surface during torqueing.

When using structural washers, ensure that the correct washer is specified. The washer should be compatible with the strength level of the threaded fasteners, too soft a washer may actually extrude (creep) and result in fastener relaxation. Washer, type and finish influence interfacial friction during torqueing, hence bolt-tightening torque (effort applied) should always reference washer specification in critical applications. Various types of washer are illustrated in Fig. 14.28.

14.4.1 Plain washers

Metal washers are basically divided into four categories of size, i.e. light, normal, large and extra large diameter metric series.[10] The large diameter series are useful where a greater bearing area is required. The light range of bright washers have approximately 60 per cent of the normal range thickness.

14.4.2 Spring washers

Spring washers are convenient to use in light assemblies where fastener security is required in line with compactness

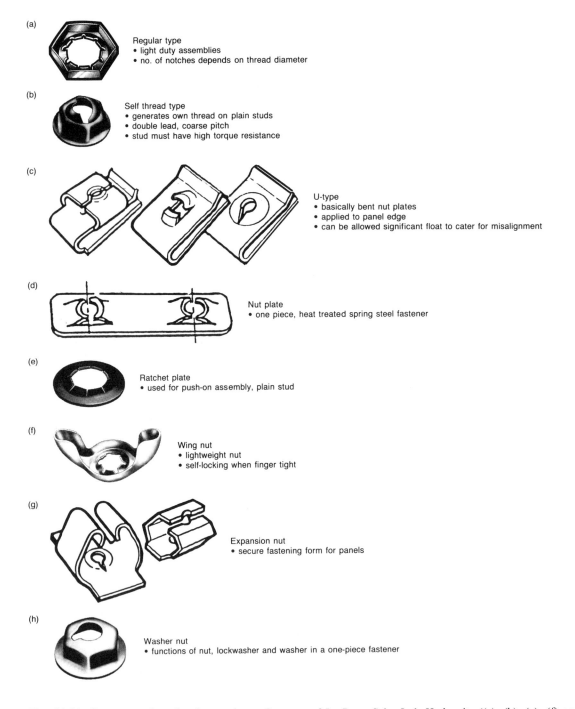

Fig. 14.26 Some examples of spring steel nut. Courtesy of Jet Press Sales Ltd, Huthwaite ((a), (b), (e), (f) and (h)) and PSM International Inc., Willenhall ((c) and (g))

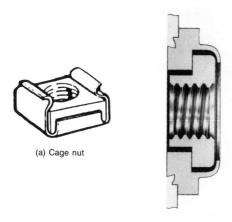

Fig. 14.27 Examples of captive nuts. Courtesy of PENN Engineering and Manufacturing Corp., Danboro, Pennsylvania (b), and PSM International Inc., Willenhall (a)

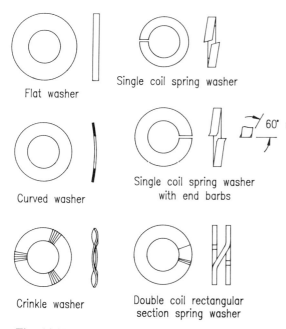

Fig. 14.28 Basic types of washer

of design. Applications include load distribution, vibration absorption, temperature compensation, electrical continuity. The basic types are curved, wavy and helical.

(i) Curved washers

The curved or single-wave washer (Fig. 14.28) is best suited to conditions requiring a maximum range of deflection when using light loads. The spring rate is nearly linear, but load tolerance is ±20 per cent. Often used to absorb rattle or radial end play. Mating surfaces should be hardened.

The approximate formula for the reactive load P is as follows:

$$P = \frac{17.8\ Est^3\ (D_0 - D_i)}{D_0^3}\ \text{newtons}$$

where E = Young's modulus (N mm^{-2}), s = deflection, (mm), t = material thickness (mm), D_0 = external diameter (mm) and D_i = internal diameter (mm).

(ii) Crinkle washers

Crinkle, wavy or multi-waved washers are manufactured in beryllium copper or stainless steel. The general form is three equispaced radial corrugations (Fig. 14.28). The characteristics of this washer type may be summarized as follows:

1. Will support only light to medium loads;
2. The spring rate is nearly linear;
3. Less expansion than a curved washer, greater reactive force for given deflection;
4. Contact points per side = no. of waves;
5. Load tolerances should not be <20 per cent.

The approximate formula for load, P, is as follows, for deflections 20–80 per cent of available height:

$$P = \frac{1.85 Esbt^3\ N^4\ D_0}{D_m^3\ D_i}\ \text{newtons}$$

where b = material radial width = $(D_0 - D_i)/2$ (mm), t = material thickness (mm), N = no. of waves and D_m = mean diameter = $(D_0 + D_i)/2$.

(iii) Helical washers

Helical spring washers fulfil a dual function of compensating for loss of preload in service while acting as a thrust face, reducing interface friction during assembly and disassembly. They are not true locking washers. There are three types available (Fig. 14.28): single-coil square section, single-coil rectangular section, double-coil rectangular section.

It is normal practice to use rectangular section lock washers, but where these washers are required for use with cap head screws, the higher collar square section is used. Helical lock washers rely on their spring compensation to achieve preload. The standard form of cutting results in a free height of washer approximately twice the thickness of the washer section (Fig. 14.28). However, a further factor is sometimes added by shaping the washer ends into 'barbs' which dig in to both the nut/bolt head and parent material, greatly resisting any tendency for the fastening to loosen. This type is generally referred to as the 'positive' type.

14.4.3 Toothed lock washers

These are normally available as either 'shakeproof washers' or 'fan-disc washers'. Toothed washers are designed to dig into the two mating surfaces, hence the hardness ratio washer/surface is important. While the load applied to shakeproof types may cause the washer to flatten out, the fan-disc washer is so designed to have overlapping teeth which cannot flatten completely, even when excessive torques are applied. The two types are illustrated in Fig. 14.29.

Toothed lock washers derive their locking function from a combination of three separate functions:

1. *Line bite* — the hardened tooth material cuts into the face of the workpiece and nut/bolt head.
2. *Spring reaction* — each tooth acts as a compensating spring.
3. *Strut action* — the teeth individually oppose the nut/bolt head tendency to loosen by rotation.

The teeth may be located on the internal diameter of the washer, internal tooth type or on the external diameter, external tooth type. For exceptional holding power, teeth can be located on both internal and external diameters. A performance comparison of the overlapping and tooth lock washer is illustrated in Fig. 14.30.

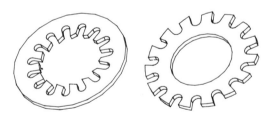
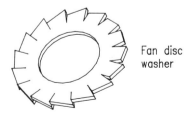

Fig. 14.29 Examples of tooth-locked washers

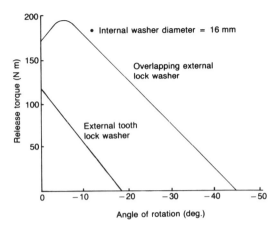

Fig. 14.30 Variation of untightening torque with angle of rotation

14.4.4 Disk springs

Disk springs are used in applications where a small controlled deflection and relatively high spring pressure is required. The disk spring, or Bellville washer is produced in conical form and derives its locking performance from the inherent spring and frictional properties of the material (Fig. 14.31).

The load-deflection characteristics of this type of disk are a function of internal diameter (D_i), external diameter (D_o), material thickness (t) and coned height (h). Numerous load-deflection curves can be obtained by varying the ratios D_0/D_i and h/t (Figs 14.31 and 14.32).

Typical curves having different h/t values are shown in Fig. 14.32 for a D_0/D_i ratio of 2.9. The character lines of a single disk obtained by changing the cone height/material thickness ratio is typically illustrated in Fig. 14.33. Although disk washers may be compressed

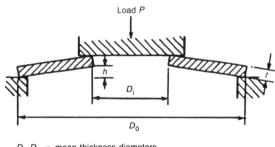

D_i, D_o = mean thickness diameters
h = coned height
t = disc material thickness

Fig. 14.31 Initially coned disc spring (Bellville washer)

FASTENERS

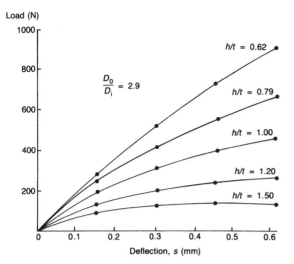

Fig. 14.32 Effect of varying h/t ratio on load/deflection characteristics of disk springs

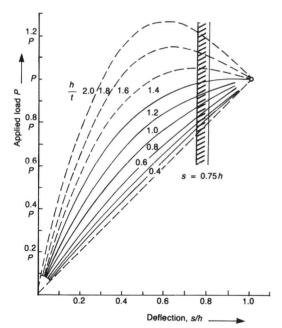

Fig. 14.33 Effect of varying h/t ratio on a given spring character

flat without damage under static load conditions, normal usage (preload plus working deflection) is for a total deflection s, of no more than 75 per cent, i.e. $s \not> 0.75h$. Loads progressively increase above the theoretical after this point. Note that disk ratios should not exceed 1.5 as this may lead to individual disks snapping inside-out.

For a single spring the performance characteristics may be summarized as follows:

1. Use for medium to heavy loadings;
2. Constant spring rate for $h/t < 0.4$;
3. Nearly constant spring rate for $h/t > 0.4 < 0.8$;
4. Decreasing spring rate for $0.8 < h/t > 1.4$;
5. Zero rate over part deflection curve for $h/t = 1.5$;
6. Least expansion under load;
7. Normal load tolerance is $\pm$ 20 per cent.

In order to produce a simplified relationship between the load and deflection characteristics, it is accepted practice to assume that: (a) the radial cross-section does not distort as it rotates about its point of support under load, and (b) that the D_0/D_i ratio $\not> 3$. Hence, in terms of deflection, the load P (Fig. 14.31) may be written as

$$P = \frac{4EsC_2}{(1-\nu^2)D_0^2}\left[\left((h-s)(h-\frac{s}{2})\right)t + t^3\right]$$

where P = load, E = Young's modulus, s = deflection, h = cone height (undeformed), t = material thickness, ν = Poisson's ratio, D_0 = outside diameter and

$$C_2 = \pi\left(\frac{\gamma+1}{\gamma-1} - \frac{2}{\ln\gamma}\right)\left(\frac{\gamma}{\gamma-1}\right)^2$$

where $\gamma = D_0/D_i$.

To simplify the design calculation, the equation may be rewritten

$$P = \frac{4Et^4C_1C_2}{D_0^2}$$

where

$$C_1 = \frac{s}{(1-\nu^2)t}\left[\left(\frac{h}{t}-\frac{s}{t}\right)\left(\frac{h}{t}-\frac{s}{2t}\right)+1\right]$$

Plots of C_2 as a function of γ (Fig. 14.34) and C_1 as function of s/t and h/t (Fig. 14.35) may be interpreted for factors C_1 and C_2. In plotting Fig. 14.35, it is assumed that when $s = h$, the washer is pressed flat and no further deflection is required. In other words the design does not require the washer 'pushed through' beyond flat. Further variations in load/deflection characteristics may be produced by stacking disks in series and/or parallel, some of which possible arrangements are shown in Fig. 14.36. For simplicity, linear spring characteristics have been assumed.

14.4.5 Preload indicating washers

These washers are designed to provide an easy but

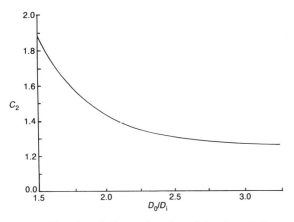

Fig. 14.34 Plot of C_2 as a function of the diametral ratio

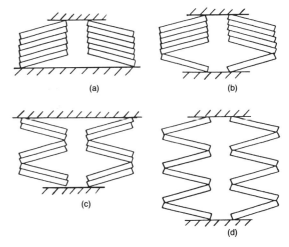

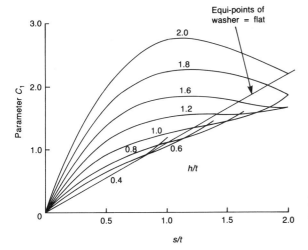

Fig. 14.35 Parameter C_1 as a function of s/t and h/t

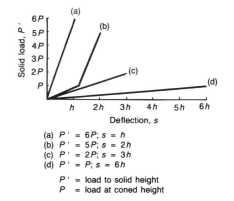

(a) $P' = 6P$; $s = h$
(b) $P' = 5P$; $s = 2h$
(c) $P' = 2P$; $s = 3h$
(d) $P' = P$; $s = 6h$

P' = load to solid height
P = load at coned height

Fig. 14.36 Load deflection characteristics of various disk spring configurations

accurate check of bolt tension when the screw or nut is being torqued. Generally, however, such washers are used in conjunction with high-tensile bolts. Examples are illustrated in Fig. 14.37 and given below:

1. *Projection type.* Basically a flat washer with a number of protrusions. It may be inspected at any time to determine tension. When the bolt is tightened the protrusions are flattened, and by measuring the gap with a feeler gauge tension can be ascertained. If used at the nut end, a special nut face washer should be used.

2. *Ring compression preload.* Consists of two hardened washers with two close tolerance steel rings sandwiched in between, the inner ring slightly thicker depending on the preload required. Bolt tension is achieved on torqueing up when the outer ring fails to rotate by hand.

3. *Sacrifical washers.* The operating principle is to cut through a dispensable washer which is pre-shaped and calibrated to provide a known shear resistance proportional to the applied load.

14.4.6 Assembled washers

To assist in mechanical assembly, it is often worth considering the use of combined screw and washer or nut

FASTENERS

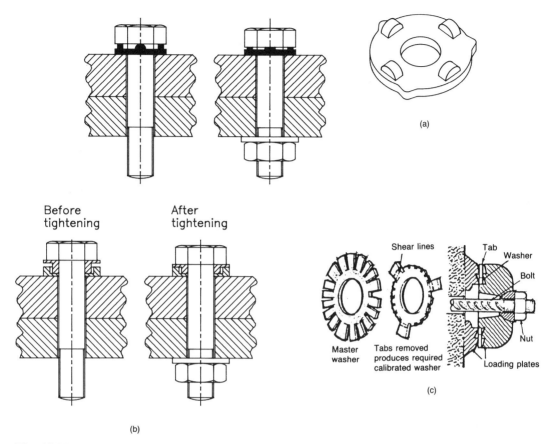

Fig. 14.37 Preload indicating washers: (a) projection type washer; (b) ring compression washer; (c) sacrificial washer. Part (c) courtesy of Findlay Publications

and washer assemblies. Typical examples are shown in Fig. 14.38.

1. *SEMS*. Available as ordinary and lockwasher screw assemblies, machine and self-tapping screws. Also SEMS ensure the fitting of the correct washer(s) every time.
2. *Rolled nylon or PTFE collar*. Strictly a sealing ring, it can be used to seal, lock and isolate vibrations. Because the outside rolled edge of the washer is stronger than the inside edge, it resists outward flow during screw tightening and the inside edge is forced inward into the screw threads and into any space between screw shank and hole.
3. *Pre-assembled nuts and washers*. Typically with conical spring washer or shakeproof washer. Washers are held in place by swaging over a projecting portion of the internal diameter and are free to rotate during preload.

14.5 Theory of nut tightening

Apart from ensuring the inherent quality of fasteners, probably no function is as important to the efficiency of a joint as is its installation torque. The action of torqueing a screw or bolt/nut combination elongates and introduces a clamp load (or preload) in the bolt itself. The clamp load does the work of maintaining joint tightness and minimizing fatigue as long as the clamp load exceeds the forces acting on the joint. For every fastener system there is an optimum torque range to develop the design clamp load.

14.5.1 Torque–tension relationship

As a threaded fastener is rotated in a joint, the fastener goes through several tightening phases. Plotting applied

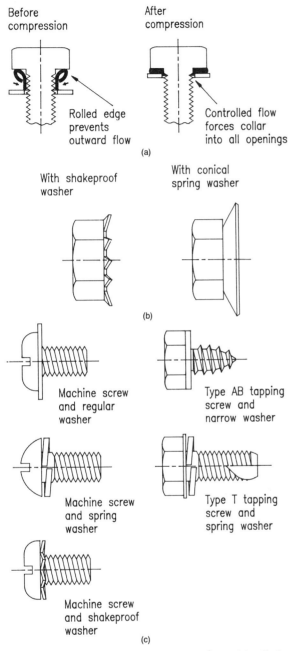

Fig. 14.38 Examples of assembled washers: (a) rolled polymer collar; (b) pre-assembled nuts; (c) SEM screws

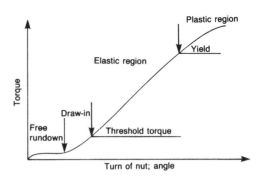

Fig. 14.39 Typical fastener torque-up with metal-on-metal joint

torque against the resulting turn usually produces a S-shaped curve (Fig. 14.39), the input work being the integral of this curve.

1. As torque is applied, the nut is run down freely and the parts are pulled together. There is turn but only prevailing torque is present, no clamp load is generated.
2. Once joint is snugged,[11] continued rotation draws in the components, flattens spring washers, smoothes the surfaces and initiates clamp loading. Build-up of torque is often linear, all parts are being loaded and the deformation is elastic.
3. Finally, some parts of the system reach its elastic limit, the curve hooks over, completing the S and torque drops off as yield and permanent deformation occur. The clamp load therefore approaches its maximum value.

Much of the externally applied torque is either converted to heat loss from nut/screw frictional resistance (40 per cent) or nut/joint interfacial resistance (50 per cent). Only 10 per cent of the work input creates the clamp load, hence frictional conditions are of major importance. Losses are even greater if the bolt twists or bends, the clamp surfaces are non-parallel, there is interference between bolt and hole, nut dilation occurs or prevailing torque nuts are used.

In the straightforward case, the following relationship can be used to give reasonably accurate results between induced preload and applied torque.

$$T = W\left[\frac{d_m}{2} \tan(\alpha + \theta) + r_n \mu\right]$$

where T = torque (N m), W = bolt load (N), d_m = mean thread diameter, α = lead angle (deg.), θ = friction angle of thread (deg.), r_n = mean radius of nut face, μ = nut face friction coefficient and p = pitch.

Tightening a nut on a threaded fastener is in effect equivalent to driving a wedge between the bolt and joint face. Unless all the frictional conditions are known, it is more usual to use a simplified form:

$T = cd_m W$

where c = torque coefficient.

The value of the torque coefficient for dry surfaces and unlubricated bolts is taken as 0.2. Using a grade 8.8 bolt and assuming a bolt tension of 75 per cent yield stress, a graph of tightening torque for the coarse metric series can be produced as in Fig. 14.40. In practice, surface conditions and lubrication affect the torque coefficient such that tightening torque correction factors need to be applied. Correction factors for different surface condition and bolt material grades are given in Table 14.12.

14.5.2 Joint tightening

At this point it is worth considering the basic principles underlying joint clamping, regardless of how the fastening is actually achieved. When a bolt is tightened it is stretched elastically and the joint is compressed. What happens when service loads are experienced depends critically on the stiffness of bolt and joint and their degree of mutual preload. The effect of preload can be shown typically by AB in Fig. 14.41. The stiffness constant K

Table 14.12 *Correction factors for surface condition and bolt grade*

(a) *Surface condition*

Nut condition	Bolt condition		
	Self	Zinc	Phosphate
Self	1.00	1.00	0.90
Zinc	1.15	1.20	1.15
Phosphate, oiled	0.70	0.65	0.75
Zinc + grease	0.60	0.55	0.55

(b) *Bolt grade*

Steel grade (BS 3692)	6.6	6.8	8.8	10.9	12.9
Correction factor	0.56	0.75	1.00	1.41	1.69

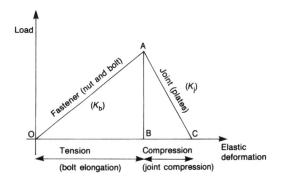

Fig. 14.41 Effect of joint preload

is defined in the same way as the spring constant for springs, i.e.

$$K = \frac{W}{s} = \frac{W}{WL/AE} = \frac{AE}{L}$$

where L = length, E = Young's modulus, A = cross-sectional area, W = load and s = deflection.

Generally, but not always, the joint is stiffer than the bolt. In Fig. 14.41 the joint is shown as approximately twice as stiff. Consider now the effect of applying an external tensile load. This will increase bolt tension and relax the compression on the joint face (Fig. 14.42). It can be seen that the external load is shared in effect between the bolt and joint in proportion to their stiffnesses.

Assuming that the initial preload is W, then on applying an external load W_e (Fig. 14.42), the result is (1) an increase in the length of the bolt W_b/K_b, (2) a decrease in deformation of the plates W_j/K_j, where K_b = stiffness of bolt and K_j = stiffness of joint. Hence

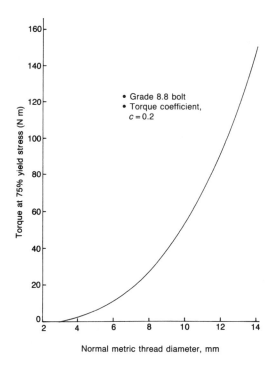

Fig. 14.40 Clamp loads for metric nuts

ELECTROMECHANICAL PRODUCT DESIGN

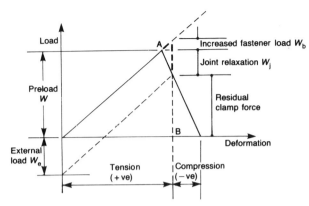

Fig. 14.42 Effect of applying an external tensile load

$$W_e = W_b + W_j$$

and

$$\frac{W_b}{K_b} = \frac{W_j}{K_j}$$

hence

$$W_b = \frac{K_b}{K_j}(W_e - W_b)$$

$$= \frac{K_b}{(K_b + K_j)} W_e$$

Now total load on bolt is

$$W + W_b = W + \frac{K_b}{(K_b + K_j)} W_e$$

And total compressive load on plates is

$$W - W_j = W - \frac{K_j}{(K_j + K_b)} W_e$$

Now if an oscillating load is applied to the joint, it will be shared by the fastener and the joint and the way it is distributed will depend on their relative stiffnesses. In Fig. 14.43 an oscillating load of half amplitude, approximately equal to half the preload, is shown applied to a comparative high stiffness joint. The result is a small increase in fastener load and a small decrease in joint compression. Merely increasing the preload does not affect the alternating stress carried by the bolt, although it will represent a reduced portion of the bolt tension and joint compression respectively. In cases where an oscillating load (hence fatigue) is likely to be a problem, and increasing the stress range of the fastener may not be desirable, it is possible to minimize the resulting stress range in the bolt by increasing its elasticity, i.e. a larger length:diameter ratio. The joint would then be relatively stiff in comparison to the fastener, i.e. $K_j \gg K_b$. An extreme ratio of 8:1 can make the bolt immune to vibration.[12]

While the stiffness of a bolt is relatively straightforward to calculate, the same is not so for the joint where structural design, metal mass and the presence of gaskets are all variables.

14.5.3 Joints in shear

In a majority of engineering situations, bolts and screws are fitted into clearance holes and carry shear load across the joint face quite successfully by virtue of the interfacial frictional forces resulting from the clamp pressure. However, for friction joints to be successful, (a) $\mu P > S$, where μ = interfacial friction coefficient, S = shear stress parallel to interface and P = bolt load and (b) thread run-out should not coincide with the shear plane between

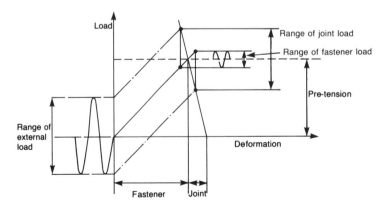

Fig. 14.43 Oscillation load applied to a stiff fastener

FASTENERS

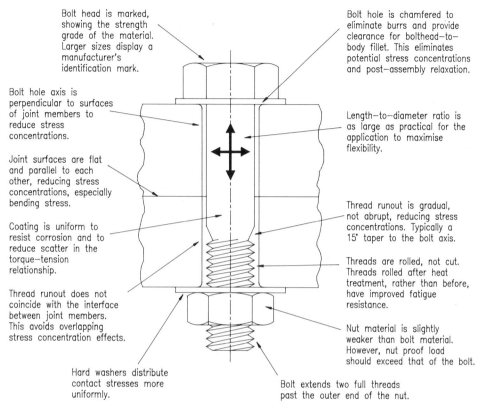

Fig. 14.44 The optimum bolted joint

joint members (Fig. 14.44). If there is periodic joint overload, fretting failure of the joint may occur. The shear strength of the bolt itself may be regarded as a final safeguard.

14.5.4 Measurement of preload

Because the mechanical efficiency of a bolted or screwed joint is low with friction taking up some 90 per cent of the applied torque, industry has learned to live with considerable variation in the resultant clamp force. Because the amount of friction generated is variable and unpredictable, torque values alone are poor predictors of bolt tension and clamp load. With the advent of quality control product demands, such variability in the tightening process may well no longer be acceptable. To control the applied tension various methods have been adopted, including the following:

1. *Torque measurement*. Most widely known technique, but as previous comments and Table 14.12 indicate, useful only as long as the locking surface conditions are known and controlled beforehand if wide scatter in the joint clamp load is to be avoided. The maximum torque is referred to as 'installed torque' or 'peak torque'.

2. *Direct measurement of bolt extension*. For example by strain gauges applied to screw or bolt, controlled measurement or by use of ultrasound. Accurate, but costly and time consuming.

3. *Purpose-designed load-indicating bolt and washer sets*. The use of a load cell under the bolt or screw head can only be justified in special circumstances,

but does provide the facility for periodic monitoring. Load-indicating washers are discussed in section 14.4.5.

4. *The turn-of-the-nut method.* After initial 'finger-tightness', the pitch of the thread gives the degree of turn necessary to stretch the bolt a calculated amount, elastically or plastically, and thus engineer a given loading. However, practical difficulties arise in knowing where the measurement of turn is to begin, typically the so-called 'snug-level'.[11] This technique does not allow for variations in joint stiffness, friction levels, etc.

5. *Taking the bolt into its yield condition.* Since this depends only on achieving yield status in the bolt, the method is not dependent on the actual torque applied or the joint stiffness. It has the advantage of high bolt tension with little variation from bolt to bolt. One way of sensing yield accurately is with an internal indicator pin.[13] At the point of yield, Poisson's ratio changes from 0.3 to 0.5, this fractional lateral contraction sensed by the induced interference of the originally loose-fit pin.

(i) Torque plus angle control

Not all of the above are straightforward or cheap or indeed applicable where productivity considerations are important. Where power tool assembly operations are justified, a torque control and monitoring process is a likely possibility. In the most useful version available, the tightening process is controlled into a defined window by measurement of both torque and angle, this shape of the torque/angle curve being monitored to ensure that both torque rate and deviations from nominal are within limits of acceptability, the so-called 'green-window'. Tightening may be in the elastic range or into the plastic yield zone, an ideal torque/angle monitored curve into the plastic range shown in Fig. 14.45. Examples of joint condition feedback from the monitoring of the torque/angle tightening relationship in the elastic range is illustrated in Fig. 14.46.

14.5.5 Thread run-out

In addition to being a stress raiser in its own right, the actual stress at the point of thread run-out on the bolt is dependent also on the relative torqued nut position. In the fully tightened position, the nut should be at least one bolt diameter from the start of run-out (Fig. 14.47). The stress at the run-out point may be reduced by either: (a) introducing a stress relief groove (Fig. 14.47), or (b) adopting a bolt design with reduced shank diameter (Fig. 14.47).

14.5.6 The optimum bolted joint

From previous discussion, it will be apparent that a large number of parameters can both affect the reliability of

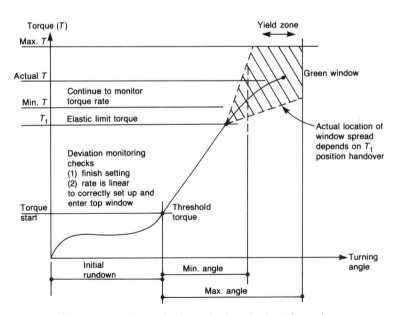

Fig. 14.45 Torque-angle monitoring: elastic+plastic deformation range

FASTENERS

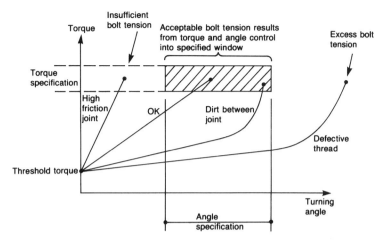

Fig. 14.46 Torque-angle monitoring of joint condition: elastic range

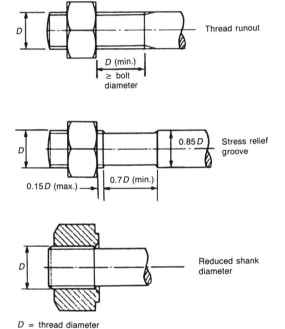

D = thread diameter

Fig. 14.47 Thread–shank stress relief options

a fastener and the clamping load obtained. Some of these are presented schematically in Fig. 14.44. Apart from ensuring the design has enough fasteners of adequate size with sufficiently sturdy joint members, fastener distribution should be uniform about the line of action of external tensile loads on the joint. Loads should not be offset. Likewise, the line of action of shear loads should pass through the centroid of the bolt pattern.

14.6 Retaining rings

A number of situations exist where item(s) are to be retained on a shaft. To satisfy specific applications many alternatives have been developed from the original basic ring types.

14.6.1 Circlips

The most frequently used basic circlip is illustrated in Fig. 14.48. The tapered section, decreasing symmetrically from beam centre to the free ends ensures that the ring maintains circularity, avoiding permanent distortion when expanded or contracted within normal working limits, i.e. approximately 10 per cent of diameter. Each end of the circlip has a hole, or in the case of the smaller external circlips a hole or a slot, to allow assembly.

Circlips are manufactured from spring quality medium carbon steel, either ausformed or quenched and tempered.[14] Their design provides for a constant pressure against the bottom of the groove, making them secure against heavy thrust loads. Variations of the basic type are shown in Fig. 14.49.

Wire form retainers or snap rings are normally auto-coiled from cold drawn spring wire of a uniform section. The gap ends are cut according to the design requirement and may be square or angled. The wire ring is available in various cross-sectional shapes (Fig. 14.49). Although thrust performance is generally inferior, use advantages over conventional circlips can be summarized as follows:

1. Ability to expand and contract over a much wider range;

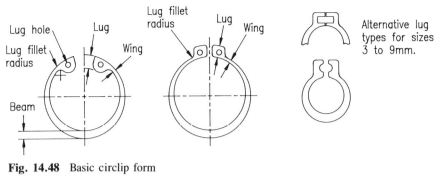

Fig. 14.48 Basic circlip form

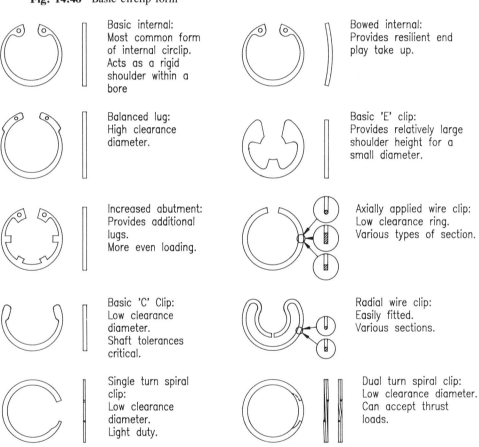

Fig. 14.49 Examples of various circlip types

2. Useful for non-standard shafts and housings;
3. Can be used where shaft diameters vary over large tolerance ranges;
4. A low clearance diameter;
5. Where quantities do not justify production tooling.

(i) Thrust load capacity

In an axially loaded situation, the ring located in a groove is subject to shear and the groove material to tensile loading. In practice it is desirable for the ring to be designed to fail first at overload, being the cheapest and most easily replaced part.

Ideally the ring should seat against a square-cornered part; if the abutting part has a corner radius or chamfer, the ring may dish under load or even be ejected. The smallest of groove wall strength, ring shear strength and loading capacity against a non-square face determines the thrust capacity of the assembly. The static thrust load capacity against a sharp edge abutment for both ring and groove may be found under the headings below.

(a) External assembly

Circlip thrust load $= \dfrac{\pi D t S \beta}{F}$

and

Groove thrust load $= \dfrac{D(D-d)Y\pi\beta}{5}$

where D = shaft diameter, d = groove diameter, t = circlip thickness (diameter in the case of wire forms), Y = yield strength of shaft, S = shear strength of circlip, β = conversion factor, i.e. fraction of circumferential wrap in contact with groove when assembled and F = safety factor, typically 4 for the majority of retaining rings but 12 for the low clearance circular wire form retainer (Fig. 14.49).

From the above, maximum thrust load capacity is obtained when the two loads are equal, but in practice, groove geometry can be adjusted to ensure that the ring fails first.

(b) Internal assembly

Circlip thrust load $= \dfrac{\pi B t S \beta}{F}$

and

Groove thrust load $= \dfrac{B(d-B)Y\pi\beta}{5}$

where B = bore diameter.

14.6.2 Grooveless rings

Either push-on fixes or grip ring types, they are used both for speed of assembly and where shaft-grooving is not preferred. Generally they are not used for precise axial positioning, rather for toys, small appliances, etc. where the shoulder provided by the retainer is not required to withstand any particular load, but merely to act as a positioning stop or locking device. Various types are illustrated in Fig. 14.50.

The inclined prongs of the push-on fixes are deflected backwards during installation. When load is applied from the opposite direction, these inclined prongs grip the shaft tightly, so providing assemblies free of end play, automatically compensating for tolerance build-up. Ideally

Fig. 14.50 Examples of grooveless shaft fixings

suited for die-cast or plastic studs, certain types catering for rivets, tubing and wire.

Grip rings are basically similar to a heavy-duty circlip, but are designed to be fitted on to, and removed from, plain shafts without a location groove. Strong circlip pliers are required to cope with the heavier gauge.

14.6.3 Tolerance rings

Tolerance rings have become accepted as fastening devices for the assembly of components requiring an interference fit, and in the majority of cases reducing part manufacturing costs. The rings are made from heat-treated strip with corrugations formed at a constant height and pitch to provide the necessary interference (Fig. 14.51). The advantages of using tolerance rings may be summarized as follows:

1. Allows for straightforward, low-cost assembly.
2. Provides an interference fit while relaxing equivalent tolerances on components to be assembled, e.g. up to $h9$ for shafts, $H9$ for bores.
3. May be used to transmit torque and, as such, can eliminate need for keys, pins, threads, locknuts, adhesives, etc. in designs.
4. Insensitive to differentials in thermal expansion coefficient, hence may be assembled and/or used at high ambient temperatures, e.g. to 100 °C.
5. Will compensate for certain production or design errors, i.e. some degree of misalignment, oversize housings, undersize shafts, etc.
6. May be used to mount light section rolling element bearings.

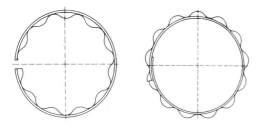

Fig. 14.51 Tolerance rings. Courtesy of Ray Engineering Co. Ltd, Bristol

The actual interference required depends on the end use of the assemblage, either for radial load capacity or for torque transfer. In most cases these are mutually exclusive. The maximum radial load-carrying capacity is with minimum deflection of the ring elements, but maximum torque capacity is with a ring element deflection of approximately 20 per cent to obtain a maximum shear contact area. When designing for a torque situation, due allowance should be made for inertial start-up effects, impulse and overload occurrences. Higher torque transmissions may also be possible by altering material thickness, wave height, pitch, etc. as well as using two rings, say, in separate grooves.

14.7 Inserted fasteners

This section represents a selection of innovative design solutions that avoid the use of nuts in component assemblies where the material being fastened is too thin or too weak to support a tapped thread. Inserted fasteners generally fall into two application groups, those inserted into a previously threaded hole, and those for insertion into drilled, cored, moulded or punched holes. It is of course equally possible for inserts to be cast or moulded *in situ*. In many cases, the strength of the insert thread and the joint of the insert of the parent material exceed the strength of the surrounding parent material. Inserted fasteners are manufactured in a variety of materials to suit the application.

14.7.1 Sheet metal inserts

These fasteners provide a means of obtaining deep, tapped holes in metal that is nominally too thin to be tapped. Likewise some of these fasteners have been adapted for use on materials that do not easily lend themselves to fastener application, e.g. printed circuit boards, certain plastics, etc. The use of such inserts also eliminates the need for locally strengthening the parent material. Additional advantages may be summarized as follows:

1. Nut remains permanently attached to parent material;
2. Can be assembled after components painted or plated, hence eliminates thread masking/cleaning;
3. Can incorporate distance piece or spacer into the design of the fastener;
4. Can incorporate a 'floating' bush or locking action.

Some available types are described under the headings below.

(i) Rivet bush

The basic type is the press or hank nut, characterized by a finely tapered and serrated tubular shank (Fig. 14.52). This shank, when riveted over, cuts into the surface of the sheet giving high torque resistance and firmly retaining the bush in position. A number of bush material/sheet material combinations are possible, but it is important to ensure adequate hardness to enable serrations to be cut in the sheet.

(ii) Self clinch

Self-clinching fasteners are a novel design of metal bush, or stud, that can be installed in a single operation. The fastener (Fig. 14.53) is inserted into a sheet metal hole and pressure then applied to embed a corrugated clinching ring. The displaced metal then flows evenly and smoothly around the fastener shank to give a positive lock with high

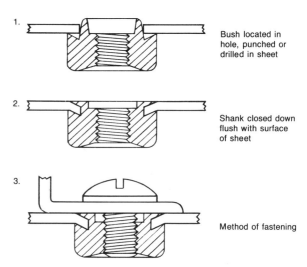

Fig. 14.52 Assembly of rivet bush. Courtesy of PSM International Ltd, Willenhall

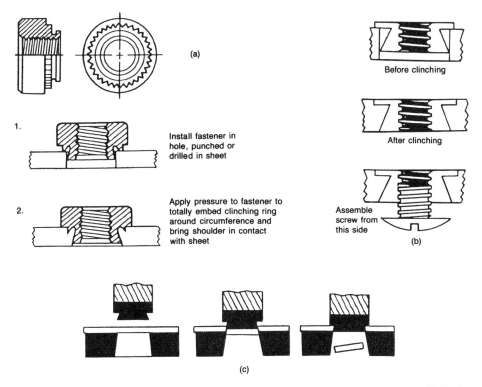

Fig. 14.53 Types of self-clinch insert: (a) self-clinch fastener and assembly; (b) self-clinch flush insert; (c) self-piercing flush insert. Courtesy of PSM International Ltd, Willenhall

push-out and torque resistances. Mounting holes must be left sharp.

Flush self-clinching fasteners are designed to provide a thread within the sheet thickness, giving a flush finish either side. Alternatively, for thin sheets, it is more usual to fit a stand-off version and so obtain a large number of active threads. Some performance values for self-clinch types are given in Table 14.13.

The self-piercing version allows installation in one operation, a hardened steel spigot acts as a punch to form the hole and the serrated clinching ring securely locks the fastener into the sheet.

(iii) Tubular locking inserts
Along with inserts generally, these represent one of the recent advances in joining processes.

(a) Cold-formed tabular rivet nut These are basically a one-piece blind fastener not requiring a nut (Fig. 14.54). It is possible to use these to clamp two sheets together as well as providing the screw thread insert to constrain a third. However, it is essential to choose a tubular rivet with a grip range appropriate to the material(s) being joined. The actual fastener installation procedure may be described as follows:

1. The tubular rivet is threaded on to the mandrel of the installation tool.
2. The fastener is inserted and the mandrel retracted so pulling the fastener shank towards the blind side of the sheet, forming a locating compression bulge in the unthreaded shank area, clinching the fastener firmly in place. This is termed the collapsing or upsetting operation.
3. The mandrel is removed and the chosen screw inserted.

A development of the rivet nut is a flat-head type with slotted shank. When upset, the shank splits to form large flaps which evenly distribute the installation and working loads over wide areas.

(b) The shear nut tubular rivet With this version the rivet contours and cross-sectional area are such that the threaded portion shears off during the upsetting operation

Table 14.13 *Performance data for various self-clinch fasteners (typical results)*

Type	Screw size	Test sheet thickness (mm)	Install force (KN)	Pushout force (N)	Retainer torque-out (Nm)
Aluminium					
Floating	M4	1.6	10	1178	17
Blind tube	M4	1.0	8.9	490	2.8
Standoff	M4	1.9	17.8	1335	—
Ring type	M4	1.4	12.3	845	3.6
Flush type	M4	1.9	8.9	1100	—
Self-locking	M4	1.4	12	1110	5.6
Steel					
Floating	M4	1.6	15	1780	22
Blind tube	M4	1.0	15.6	600	3.4
Standoff	M4	1.9	23.6	1955	—
Ring type	M4	1.4	23	1020	4.2
Flush type	M4	1.9	17.8	1100	—
Self-locking	M4	1.4	18	1330	7.9

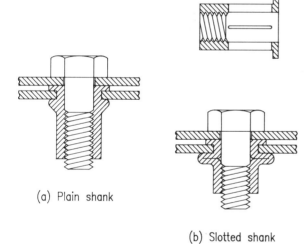

(a) Plain shank

(b) Slotted shank

Fig. 14.54 Examples of tubular rivet nuts

and is forced upward into the outer collar, expanding it to provide a strong frictional lock, tangentially and normally between all items (Fig. 14.55).

(c) The sheet edge locking insert This is not intended as a high strength fastener, as torque is limited by the shear strength/bending of the small tangs either side of the parent plate cut-out (Fig. 14.56).

14.7.2 Wire thread inserts

Wire thread inserts (Fig. 14.57) are manufactured from smooth finish stainless steel, diamond section wire with radial truncations, to a typical hardness of 450–530 kg mm^{-2}. Originally produced in the 1930s to overcome notch effects at screw thread roots, today's usage ranges from such problems as thread stripping in weak materials to wear and corrosion. Inserts may be

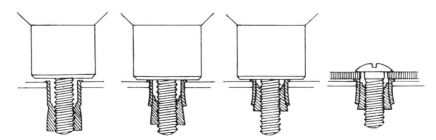

Fig. 14.55 Installation of tubular shear nut rivet. Courtesy of PSM International Ltd, Willenhall

FASTENERS

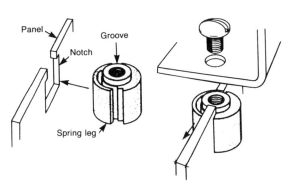

Fig. 14.56 Sheet edge fastener. Courtesy of Southco Inc., Concordville, USA

designed in to the initial assembly or used as a salvage activity. Benefits are as follows:

1. Reduced thread friction;
2. Greater loading for given torque;
3. Minimum torsional bolt strain;
4. Greater load capacity;
5. Maintained bolt preload stretch;
6. Cancels out thread pitch and angle errors;
7. Saves space over threaded bushes;
8. Screw engaged throughout length of insert.

Disadvantages are the use of special taps and assembly tooling. The thread insert has inherent axial and radial elasticity, enabling each thread to adapt itself independently and compensate for the pitch and angle errors that exist and develop between the threaded hole and the bolt as it is torqued up. With the inserted coil, the load and resultant stress is more evenly distributed over all the threads in the assembly; there is no concentration of stress at any specific point. Once preloaded, assembly tightness remains constant even under shock and vibration.

A range of wire thread inserts are also available that provide a screw-locking function (Fig. 14.57). One or more of the intermediate coils being polygonal in design. When the bolt is screwed into the insert, this feature serves to create a strong elastic pressure on the thread flanks, thereby preventing loosening of the assembly. This feature does not inhibit screw removal and reassembly.

14.7.3 Plastic moulding inserts

The increasing use of plastic mouldings brings problems in assembly and servicing, not least is the requirement for adequate strength threads. If directly tapping a hole

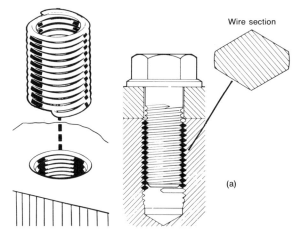

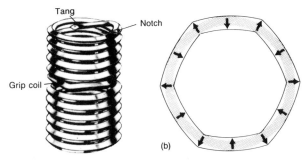

Fig. 14.57 Wire thread inserts: (a) wire thread insert and application; (b) locking thread insert type. Courtesy of Armstrong Fastening Systems, Hull

is not adequate, there are numerous types of metal inserts available designed to cater for a range of plastic materials (Fig. 14.58). Placing inserts after moulding is considered not only to increase tool productivity but also minimize the risk of tool/mould damage from displaced inserts. For post moulding insert placing, there are a few guidelines:

1. Hole depth less than insert length plus two pitches;
2. Inserted screw must not bottom in hole;
3. No counterbore or countersink in mould;
4. Wall thickness must be adequate;
5. Attached components should sit flush on top of insert.

A list of installation methods, plastic types and suitability is given in Table 14.14, some of these applicable to other soft materials apart from plastics. Various types are as follows:

1. *Vaned expansion type.* In some types, inserting the fastener causes the slots to close, while in other types they are formed closed. Insertion of the fixing screw

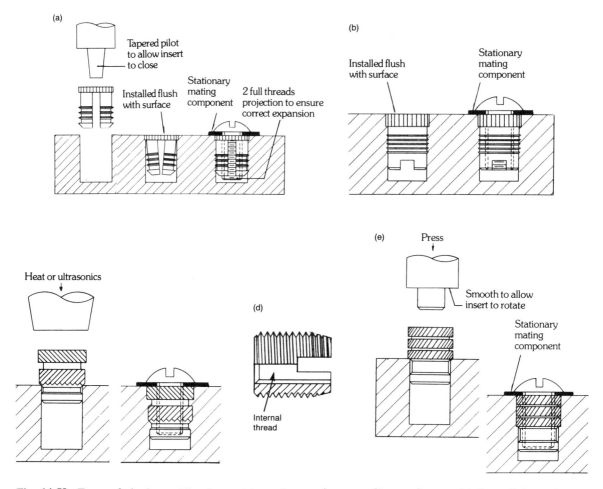

Fig. 14.58 Types of plastic moulding insert: (a) vaned expansion type; (b) press-in type; (c) thermal/ultrasonic insert; (d) self-tap type; (e) self-broaching type. Courtesy of PSM International Ltd, Willenhall

reopens the slots, forcing the vanes on the outer curved surfaces into the wall of the hole to form the grip.

2. *Knurled expansion type*. Similar to the vaned type except that the insert outers are knurled. To assist the resulting grip some types contain a captive eared spreader which may be forced down before inserting the screw in order to expand the shield, without contracting the threads. In thin sections, special flanged inserts can be inserted from the opposite face in order to achieve load-carrying capacity.

3. *Press-in type*. The inserts are a one-piece design requiring no special installation tooling. The grip pattern on the shields ensure wall penetration, while the top knurled section resists rotation under torque.

4. *Self-tapping*. Three equidistant flutes running the full length of the truncated external form provide the tapping function, securely embedding in the plastic parent material.

5. *Self-broaching*. These provide a high torque thread in harder thermosets. The smooth loading punch allows the insert to rotate as it enters the plastic, the tooth shape and helix angle persuade the insert to broach its way in with a spiral motion, thus ensuring minimum bursting stress.

6. *Thermal or ultrasonic types*. Installed by gentle pressure and softening of the plastic, ensuring a flow of the plastic melt into and around the knurled areas, a complete filling of the grooves for torque resistance equivalent to moulding-in. No displaced material is

FASTENERS

Table 14.14 *Application and performance of various insert types*

Insert type	Thermoplastics					Thermosets			Pull-out force (kN)	Thread size	Minimum wall (mm)
	N	AB	PP	PC	A	PU	PE	PH			
Vaned expn.	*	*	*		*				—	M4	2.8
Knurled expn.								*	—	M4	4.0
Push-in	*	*	*		*				0.7(PP)	M3	2.0
Self-tap	*	*	*		*	*	*	*	10(DMC)	M8	11.0
Self-broach								*	0.62(PH)	M4	2.5
Thermal/US	*	*	*		*				6.2(N)	M6	3.3
Thermal				*					4.8(PPO)	M5	2.7
Ultrasonic				*					2.4(PC)	M3	1.9
Anchor pin	*	*	*		*			*	—	—	—

Notes: N = nylon; PP = polypropylene; PC = polycarbonate or PPO; A = acetal; PU = polyurethane; PE = polyester; PH = phenolic/ureas; AB = ABS.

extruded out of the hole. Technique ideal for notch-sensitive plastics.

14.8 Self-tapping screws

Self-tapping screws originally owed their existence to the substantial production cost savings from elimination of the separate hole-tapping operation in assembly fastening operations. Thread-cutting and thread-forming types of self-tapper are used where a location hole is already drilled or punched. The next development stage was the self-piercing, self-tapping screw which could drill, tap and clamp a component assembly in a single operation. Various designs of thread and point form exist depending on the application, the material being threaded, the clamp load required and the production situation. A summary of the common types of self-tapping screw is presented in Fig. 14.59.

14.8.1 Screw and thread form

There is no complete standard for these screws; some are based closely on normal ISO metric machine screws while other types are based loosely on the unified screw thread:[15,16]

1. Machine screw forms — either American or ISO coarse, sometimes fine thread series.
2. Spaced American forms — basically the same diameter for a given screw size number but twice the thread pitch.

For machine screw forms, the metric thread is preferred in Europe (section 14.2.2) but for the spaced thread forms, the specification still remains based on American Standard threads, (Fig. 14.60). This latter type of self-tapper is largely aimed at the plastics fastening sector, thread dimensions and helix angles varied to suit (Table 14.15).

Except for the high helix angle spaced thread, all thread forms can be obtained in thread-forming or thread-cutting options. Head styles are typically those outlined in section 14.2.3. Establishing the correct torque-up figure for self-tapper fastening applications tends to be empirically determined since by virtue of their attraction, usage is non-standard. Some comparisons are given in Fig. 14.61. The important parameters are as follows:

1. *Driving torque*. The tapping torque to form a thread should ideally be as low as possible, but dependent on length of engagement, hole size, material and joint tension required.
2. *Stripping torque*. This is the torque necessary to start thread shearing during screw tightening. Naturally this needs to be as high as possible. Control of screw design to produce a high stripping torque often implies a high drive torque. Optimum stripping torque is only achieved with the correct pilot hole diameter.
3. *Application torque (tightening torque)*. The torque-up value used in practice. Determined from testwork, this represents a safety factor applied to the minimum stripping torque.
4. *Pull-out strength*. The axial load necessary to cause thread failure. Clamping loads should exceed this by a safety margin.
5. *Stripping/drive torque ratio*. In any given application, this should be as high as possible, but it is important to realize that the coarser the screw pitch and the thinner the materials, then the smaller the ratio becomes, such that the application torque setting becomes more critical.

No.	Type	Form	Description and use
AB	Thread forming		Spaced thread and gimlet point. Use intended for light sheet metal and soft plastics. Used in pierced or punched holes where sharp point for starting is required
B	Thread forming		Spaced thread and blunt end, single or twin. Start with 1–2 pitch taper. Used in sheet metal, non-ferrous castings and soft plastics
BP	Thread forming		Same as B, but has a 45° included angle unthreaded cone point. Used for locating and aligning holes or piercing soft materials (self-piercing)
TT	Thread forming		Metric coarse thread, trilobular. Used in any ductile metal. Can be obtained with tapered lead (illustrated), a sharp crested reduced diameter taper lead for 2–3 pitches or a similar stabilizing lead for 4 pitches. Can be used for metals and plastics
2/50	Thread forming		Twin start, high helix angle spaced thread. Mainly intended for plastics. Can be obtained with or without a 1–2 pitch taper lead
Hi-Lo	Forming or cutting		Twin start, high helix spaced thread. Two different thread heights. Thread form designed for difficult applications in plastics, etc., where thread crumbling or stripping is a problem. Blunt, cone and nail points available
O	Thread forming		Metric coarse thread with wedge shaped indents in thread. Tapered point 2–3 pitches. Used for mild steel and non-ferrous materials
Y	Thread cutting		Spaced thread, blunt points with slight taper and cutting flutes extending length of screw. Used with plastics and non-ferrous metals
BT	Thread cutting		Similar to type B, but with one cutting flute extending a short distance. Used for plastic and sheet metal fastening
T	Thread cutting		Metric coarse thread with blunt slightly tapered points and one or more cutting flutes. Used in soft metals and plastics where a standard thread form is required
D	Thread cutting		As type T, but with a single narrow slot for the cutting edge, radial to the screw centre. Used for low strength metals and plastics, high strength brittle metals and for rethreading clogged pre-tapped holes

Fig. 14.59 End-forms summary

With soft plastics in particular, it used to be generally found that the stripping/drive torque ratio was inherently low. Thread-cutting screws with low drive torques do offer an improvement, the spaced thread form being best in this regard (Figs 14.60 and 14.61). General application recommendations for use of self-tapping screws may be summarized as follows:

1. Material and thickness characteristics determine screw type to be used. Choice and type of screw determines the pilot hole diameter.
2. Production hole size tolerance variations can significantly modify the fastener performance.
3. Stripping torque should be at least equal to the torsional strength of the screw.
4. In practice, tapping torque and tightening torque should be approximately one-third and two-thirds of the stripping torque respectively.
5. Plunged (extruded), rather than drilled holes provide a stronger fastening in thin sheet steel.

14.8.2 Thread-forming screws

As the name implies, this category of self-tapping screw, rolls or swages a screw thread in a mating material, but at the risk of creating a high burst load if the correct locating hole tolerance are not observed.[16]

Such screws are normally made of case-hardened or carburized steel in order to resist the very high local pressures caused by plastically deformed material in rubbing contact with the threads. By displacing or forming

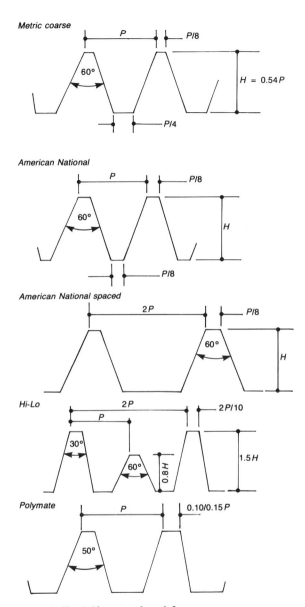

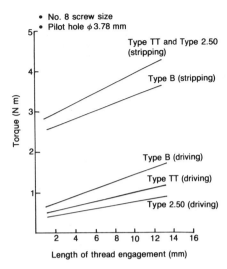

Fig. 14.61 Torque comparison of three thread-forming screws in acetal

stresses are experienced compared to thread forming, which may be an important factor with some materials. All are intended for use with ductile metals and plastics and all types are surface hardened, basically to ensure a keen thread-cutting edge. As such, therefore, these screws have a high tensile strength which does imply that if over-torqued, it is possible to strip the thread easily in the workpiece. With thread-forming screws into metal this aspect is not so obvious since work hardening occurs.

14.9 Pins

Pin fasteners represent an economical and reliable fastening and attachment method, generally enabling simplified component design. Pins provide one of the basic methods of joining parts and may be used as pivots, shafts, hinges, retainers, stops, etc. in numerous locations where the loading is primarily shear. Traditional forms such as tapered, dowel and cotter pins are among the oldest fastening elements in use and are still finding valid applications today. However, the greatest potential for fastening usage is now offered by the elastically distorting pins such as groove and spring grip pins, where hole tolerances are more relaxed (Fig. 14.62).

Fig. 14.60 Self-tapper thread forms

the material in the wall of the pilot hole, a closely mating thread is formed, a much better fit than that achieved by a normally tapped hole and mating screw.

14.8.3 Thread-cutting screws

This type of screw is provided with cutting edges and chip flutes to produce a mating thread via material removal, and as such enables a lower drive torque. Lower radial

14.9.1 Cotter pin

The split cotter pin is assembled into a hole and secured in place by spreading the split ends. As such the cotter

Table 14.15 *Non-metric self tap thread form summary*

Screw size	Basic major diameter		Threads/in. for form type		
	(in)	(mm)	American National spaced	Twin start 2/50	Twin start Hi–Lo
0	0.060	1.524	40		
1	0.073	1.854	36		
2	0.086	2.184	32	24	
3	0.099	2.515	28		
4	0.122	2.845	24	20	
5	0.125	3.175	22		21
6	0.138	3.505	20	18	
7	0.151	3.835			19
8	0.164	4.166	18	15	18
10	0.190	4.826	16	13	
12	0.216	5.486	14		
14	0.242	6.147	14		
16	0.268	6.807	12		
Helix angle range, deg.:			7/8	16/18	12/17
Screw thread action:			Forming cutting	Forming rolling	Forming cutting
Materials:			Plastics, sh. metal, soft non-ferrous	Nylon, acetal, polycarbonate, polypropylene, struct. foam	thermoplastics, thermosets

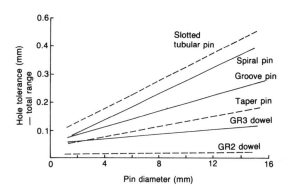

Fig. 14.62 Permissible hole tolerances for various pin types

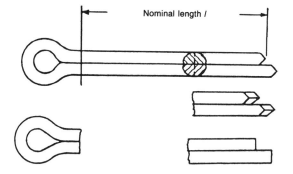

Fig. 14.63 Split cotter pin alternatives

pin is really a locking device for other fasteners, but it is simple, popular and reliable. The split is formed from a single length of substantially half-round wire, the head loop either symmetrical or asymmetrical. The legs usually are of unequal length, ends optionally pointed. An alternative for square ends may have the top leg turned down. Various pin forms are illustrated in Fig. 14.63.[17] Recommended hole sizes vary from 125 per cent of shank diameter to 105 per cent for small to large size pins respectively.

14.9.2 Clevis pin

Generally used as a clearance fit connector device for couplings, linkages, etc. not requiring arduous duty or exacting motion. Pins may be supplied with or without a head and with or without split-pin holes (Fig. 14.64).

14.9.3 Dowel pin

These are used either to retain parts in a known fixed position or to preserve alignment. Steel parallel dowel

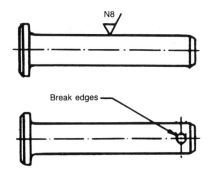

Fig. 14.64 Clevis pin

pins are divided into three grades[18] which provide different degrees of pin accuracy:

1. Grade 1 or precision ground pin, above 4 mm diameter hardened to 700 kg mm^{-2}, $m5$ tolerance on diameter. When hardened cylindrical pins are inserted into soft items, the hole should be reamed 0.025 smaller than the dowel. If inserted into a hardened item, the hole should be lapped 6 μm undersize.
2. Grade 2 is a normally ground pin, generally unhardened to an $h7$ tolerance on diameter.
3. Grade 3, machined, bright rolled or drawn finished low-carbon steel, normally unhardened to an $h11$ tolerance on diameter.

One end of a dowel pin is chamfered to provide a lead (Fig. 14.65), the other end may be similarly chamfered or domed. For use in a blind hole, a flat is provided the length of the pin to allow entrapped air to escape.

For parts which may need to be taken apart frequently, and where continual driving out of the pin would tend to wear the hole(s), the taper dowel is preferable. The wedging action of this type of pin is obtained by a force fit into a tapered hole that often requires three drilling operations plus a final ream. Constraint is only possible for a limited amount of axial travel, the taper for metric series pins being 1:50.

14.9.4 Groove pins

This is a cylindrical pin with three equispaced longitudinal grooves which are rolled or pressed into the body to deform the pin stock outward within controlled limits. Some types of groove pin available with typical use summaries are given in Fig. 14.66. When this type of pin is forced into a drilled hole of the proper size, a locking fit is obtained. When pressed into a hole, the side bulges partially elastically deform back into the grooves, the elasticity of the pin material against the hole wall produces very high pressures locally, hence an exceptionally tight fit, both shock and vibration proof. Since the pin behaves elastically, it can be reused. Best results with this type of pin are obtained with the hole size close to the pin diameter. Undersize hole specification should be avoided.

14.9.5 Spring pins

Also termed radial locking pins, the most common are the tubular or roll pin and the spirally wrapped coil pin (Fig. 14.67). Radial forces induced from pin compression during assembly exert side pressures on the walls of the hole to ensure a good frictional grip, enhanced by the deliberate pin construction. Normal tolerance drilled holes are acceptable and the pins are reusable if need be.

(a) Spiral pin These incorporate a two and a quarter to a two and a half turn spiral construction (Fig. 14.67), offering cross-sectional characteristics that withstand high shock and vibration loadings. the multiple coil configuration permits the use of thin strip which, while reducing material stresses, allows good pin flexibility and fatigue life when properly installed. Choice of strip thickness enables pins to be produced in light, standard and heavy-duty ratings, i.e. offering a choice of radial tension and flex strengths. Ends have a uniform swaged chamfer to facilitate fitting.

(b) Roll pin The spring steel roll pin is a hollow (tubular) straight pin incorporating a full-length longitudinal split. When driven into a matched size hole, $H12$ tolerance, the pin is compressed diametrically with slot width, pin/hole diameter and material elasticity designed to ensure positive locking action. The stress distribution around the hole will be greatest opposite the gap hence

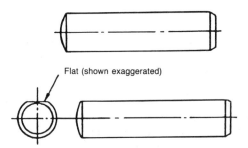

Fig. 14.65 Dowel pins

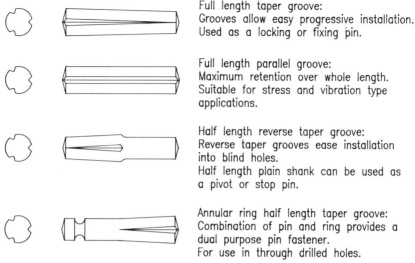

Fig. 14.66 Examples of groove pin. Courtesy of PJM International Inc., Willenhall

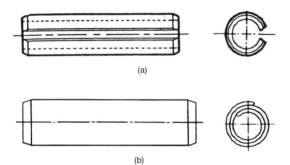

Fig. 14.67 Types of spring pin: (a) roll pin; (b) spiral pin

where maximum strength is required, the slot gap should be aligned with the direction of loading. A higher shear strength may be obtained with these pins if two are used in conjunction. Shear loads are then additive.

While roll pins, or hollow cylindrical spring tension pins, offer a very low-cost pin fastening, they do, nevertheless, lack the strength and flexibility of the groove pin or the spiral-type pin. However, in many applications the roll pin provides an adequate lightweight solution.

14.10 Quick-release fasteners

Quick-release fastener is a generic term to describe just that, a type of fastening which can be separated quickly. Typical applications might be the removal of machine covers, service panels, internal guides, etc. for cleaning and maintenance purposes, the quick-release items normally behaving as load-bearing fasteners. As such therefore, latches, catches and sliding bolts are not included in this category. There are no applicable design standards for these devices, they are proprietary manufactured products which have been accepted in the industry over a period of time and are known as mechanical components in their own right.

14.10.1 Expanded collar

Typically used for fastening material in sheet form, this type produces an axial compression of a collar or grommet which then expands or deflects radially to grip the sheet or section to which a removable sheet is being fastened. It is unusual as a fastening form in that all components are attached to the removable sheet and located through a clearance hole in the fixed sheet.

In its simplest form, it is illustrated in Fig. 14.68. Tightening the screw causes rubber expansion. This is not a rapid-release fastener *per se*, although types with multistart thread or camming mechanism are termed 'fractional turn fasteners'. One commercial form of this concept is the Dzus Arrow which employs a four-section, flexible plastic grommet with a diametrically stepped axial plunger (Fig. 14.69). Depressing the plunger flexes the grommet legs, moving these radially to produce interference with and clamping to the fixed sheet.

By virtue of the longer leg length the Dzus Arrow allows a large thickness tolerance of the sheets to be

FASTENERS

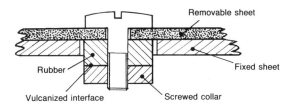

Fig. 14.68 General principle of expanded collar fastener

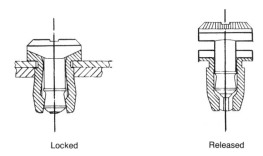

Fig. 14.69 Dzus Arrow quick-release fastener. Courtesy of DZUS Fasteners Europe Ltd

fastened. To release, rotation of the pin causes it to cam-out.

14.10.2 The quarter turn

Rotary stud or quarter-turn fasteners comprise a solid fastener, stud or pin which passes through and is loosely held captive by the removable panel plus an anchor member or receptacle, secured to the inner face of the fixed panel. After initial engagement, a quarter turn locks the fastener. A spring or spring compensation element is present to take up tolerances, remove backlash and provide some clamp tension. This may be part of the anchor member design or controlled beneath the head of the stud.

Usually the fixed-length rotary stud fasteners are based on some form of helical cam or bayonet principle. The cams engage with mating parts of the anchor member (Fig. 14.70). Designs differ inasmuch that the cam ramp is formed in the anchor member itself, the stud provided with projecting pins to provide the engaging element (Fig. 14.71). In some others, initial stud engagement tends to be of self-leading with axial pressure only required (Figs 14.72 and 14.73). Various forms of friction and depitching methods are used to lock the unit against accidental release under load.

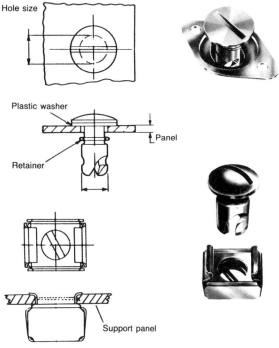

Fig. 14.70 1/4 turn spiral cam fastener into wire spring receptacle. Courtesy of DZUS Fasteners Europe Ltd

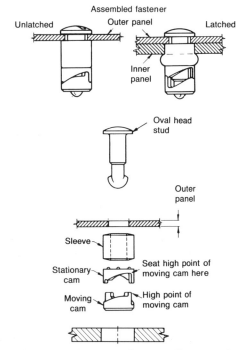

Fig. 14.71 Fractional turn fastener. Courtesy of Southco Fasteners Ltd, Worcester

439

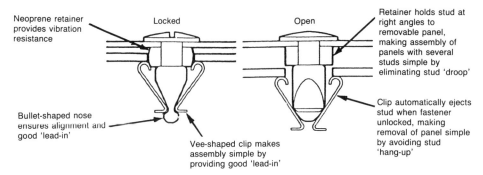

Fig. 14.72 The Oddie quick release fastener. Oddie Quarter Turn Quick Release Fastener. Manufactured by Ross Courtney & Co.

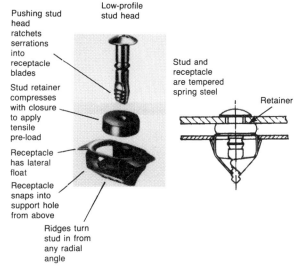

Fig. 14.73 The DZUS pilot fastener. Courtesy of DZUS Fasteners Europe Ltd

References

1. BS 4933:1973 *Specification for ISO cup and countersunk bolts and screws with hexagon nuts*
2. BS 6104:1981 *Mechanical properties of fasteners*
3. BS 4183:1967 *Specification for metric series machine screws and machine screw nuts*
4. BS 3643:1981 *ISO metric screw threads*
5. BS 6293:1982 *MJ threads for aerospace construction*
6. BS 3692:1967 *Specification for ISO metric precision hexagon bolts, screws and nuts*
7. BS 4168:1981 *Hexagon socket screws and wrench keys*
8. BS 3580:1985 *Guide to design considerations on the strength of screw threads*
9. Nut design distributes stress. *Engineers Digest* **47**(6): 28 (1986)
10. BS 4320:1968 *Specification for metal washers for general engineering purposes*
11. Mechanical tightening. *Fastening* Sept 1986: 6
12. Bickford J H Designing reliable bolted joints. *Machine Design* April 1989: 109
13. Sensing bolt re-writes the rules on load. *Eureka* **7**(6): 38 (1987)
14. BS 3673:1968 *Specification for spring retaining rings*
15. BS 1981:1972 *Specification for unified machine screws and machine screw nuts*
16. BS 4174:1972 *Specification for self-tapping screws and metallic drive screws*
17. BS 1574:1972 *Specification for split cotter pins*
18. BS 1804:1968 *Specification for parallel steel dowel pins*

Notes

TORX is a registered trade mark of the Camcar Division of Textron Inc.; Posidriv is a registered trade mark of Phillips Screw Co.; Supadriv is a registered trade mark of European Industrial Services Ltd.

CHAPTER 15

THERMAL MANAGEMENT

15.1 The nature of thermal design

Thermal phenomena are a factor in the design of nearly every product, process or machine that converts energy in some form or alternatively is exposed to a dynamic thermal environment. However, the importance of thermal design may range from minor to major; in the latter case, for example, component reliability may be dependent on actual temperature and the effective temperature difference with the environment. One or all of the modes of heat transfer, conduction, convection and radiation, may come into play in a particular thermal design problem, thermal design being the logical application of materials, physical principles and devices for the control and management of heat flow and temperature. Many practical cases will involve all three modes of heat transfer, e.g. as schematically shown in Fig. 15.1.

Applied to the solution of thermal problems, heat transfer techniques may be divided into two areas:

1. *Internal heat transfer*. This deals with those factors that need to be considered in the removal of heat from individual power-dissipating components, devices or subassemblies, i.e. the first stage of heat removal from the source of its production.
2. *External heat transfer*. This takes into account the transfer of heat into the ultimate heat sink, i.e. the second stage of heat removal, dumping the heat into a body of large thermal capacity. In the majority of cases the final heat sink will be an air mass.

Thermal management techniques gives careful consideration to the transport and transfer of heat. Thermal design must often be integrated with mechanical, structural and electrical design and often the interrelationship with other objectives and performance requirements may constrain the achievable thermal behaviour. A systems approach, rather than a component approach, should always be considered in problems of heat removal, defining flow paths for the entire equipment at an early stage of the design process. An example of thermal design planning is shown in Fig. 15.2, with air-flow paths channelled through the 'hot areas' and at the same time integrated with other machine functions. The thermal design should be put on a mathematical foundation as the development process progresses. Where control of heat flow is required, the designer may work by governing temperature difference or thermal resistance. Conversely, when temperature control is required, heat flow or thermal resistance are possible design variables.

There are two types of cooling that can be contemplated:

1. *Passive thermal control*. Utilizes the intrinsic thermal properties of materials in a largely static manner, i.e. the system is engineered so that sufficient heat transport can be achieved by the naturally occurring temperature difference between source and sink, e.g. conduction, heat pipes.
2. *Active thermal control*. Utilizes applied thermal power, a directed fluid or mechanical movement to effect heat transfer to the eventual sink via an intermediate heat-accepting medium, e.g. forced air cooling, heat exchanger, vortex tube.

Devices and systems for the excitation and transport of thermal energy exist in many forms. These are discussed later in this chapter. Apart from the total machine, a major area where thermal management is involved in its own right is electronics packaging. Increase of board population density and component power densities has brought thermal analysis and thermal design of PCBs to the front. Aspects of electronic component

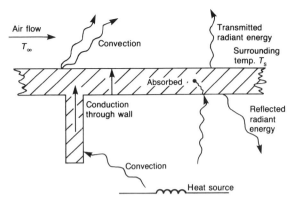

Fig. 15.1 Modes of heat transfer

packaging design likely to be associated with electromechanical design include methods of mounting components on to boards, wiring and interconnection techniques, cooling and protection.

Because of the diverse nature, by definition, of thermal management problems, the design application of different techniques and devices will be dealt with in the following sections as appropriate. However, as a general statement, the following steps normally characterize any thermal design process:

1. Define the thermal control objective;
2. Define any other machine functional or process areas that are affected or may relate to the thermal control investment;
3. Formulate an analytical model based on the known design constraints;
4. Evaluate the heat transfer rates and temperature distributions;
5. Determine material and control hardware requirements;
6. Formulate and specify the control strategy and its operational latitude.

15.2 Principles of heat flow

All thermal energy transfer occurs along a temperature gradient, requiring a temperature difference $\Delta T°$ to drive it. Thus in the case of a piece of equipment, the sum of all the ΔT's associated with cooling individual subsystems or components is the total T required to drive heat from the source to the eventual sink (surrounding environment). Treating a product, subassembly or component as a 'black box', the flow of heat to the environment depends on the overall temperature rise of the 'box' above ambient. In a practical sense, any product assembly is a collection of a number of distinct components, some of which dissipate power fed in externally, some of which transfer heat to other components internally by either conduction, convection or radiation or a combination of these. The heat fluxes at any particular instant will depend on the temperature of all components.

15.2.1 Conduction heat transfer (one dimension)

When a temperature gradient exists in a body, energy is transferred by conduction from the high-temperature

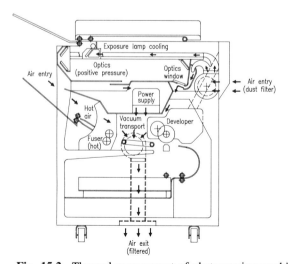
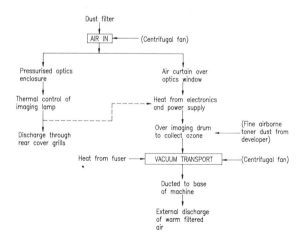

Fig. 15.2 Thermal management of photocopying machine

THERMAL MANAGEMENT

Table 15.1 *Thermal properties of some materials*

Material	Properties at 20 °C				Thermal conductivity k (W m^{-1} °C) (temperature variation where appropriate)				
	ρ (kg m^{-3})	C_p (kJ kg^{-1} °C)	k (W m^{-1} °C)	α (m^2 s^{-1} × 10^5)	−100 °C	0 °C	100 °C	200 °C	300 °C
Pure aluminium	2,700	0.89	204	8.42	215	202	206	215	228
Duraluminium, 95% Al	2,780	0.88	164	6.67	126	159	182	194	—
0.5%C steel	7,830	0.46	54	1.47	—	55	52	48	45
1.0%C steel	7,800	0.47	43	1.17	—	43	43	42	40
5.0%Cr steel	7,830	0.46	40	1.11	—	40	38	36	36
Pure copper	8,950	0.38	386	11.24	407	386	379	374	369
Aluminum bronze	8,660	0.41	83	2.33	—	—	—	—	—
70/30 brass	9,520	0.39	111	3.41	89	—	128	144	147
Silver	10,520	0.23	419	17.0	419	417	415	412	—
Window glass	2,700	0.84	0.78	3.4	—	—	—	—	—
Polycarbonate	1,200	1.17/1.25	0.19	6.6	—	—	—	—	—
LDPE	920	—	0.33	10/20	—	—	—	—	—
HDPE	950	—	0.49	10/20	—	—	—	—	—
PMMA	1,190	1.5	0.21	5/9	—	—	—	—	—
PTFE	2,170	1.0	0.25	10	—	—	—	—	—
Nylon 6 : 6	1,140	1.7	0.25	8	—	—	—	—	—
Polyurethane	1,200	1.8	0.21	10/20	—	—	—	—	—

region to the low-temperature region, the heat transfer rate proportional to the temperature gradient. The defining equation for thermal conductivity, Fourier's law, is as follows:

$$q = -kA \frac{dT}{dx}$$

where q = heat transfer rate in the x-direction (W), k = thermal conductivity (W m^{-1} K^{-1}), dT/dx = temperature gradient and A = area normal to temperature gradient (m^2).

The higher is k, the faster heat will flow in a given material. In general k is strongly temperature dependent, typically $k = k_0(1 + \beta T)$. Values of k are given in Table 15.1. The − sign in the above equation indicates that heat must flow downhill on a temperature scale. Heat flow and temperature effects are a common concern in thermal design. An associated parameter, thermal diffusivity, is defined by

$$\alpha = \frac{k}{\rho C}$$

where α = thermal diffusivity, ρ = density (usually kg m^{-3}), C = specific heat (usually kJ kg^{-1} K^{-1}), ρC = thermal heat capacity (volume basis) (J m^{-3} K^{-1}) and β = temperature coefficient of thermal conductivity.

The larger the value of α the faster heat will dissipate through the material. The lower the thermal heat capacity the less of the energy moving through the material is absorbed in raising the material temperature. Values of α are given in Table 15.1.

Considering the heat transfer rate, q, as a flow, the temperature as a potential or driving function, then a relationship can be produced analogous to Ohm's law, i.e.

$$q = \frac{\Delta T}{\Delta x / kA}$$

where Δx = material thickness and q = quantity of heat entering element per second at equilibrium, i.e.

$$\text{Heat flow (W)} = \frac{\text{thermal potential difference (K)}}{\text{thermal resistance (K W}^{-1}\text{)}}$$

Using the electrical analogue for heat flow through a multi-layer material (Fig. 15.3)

$$q = -k_A A \frac{T_2 - T_1}{\Delta x_A} = -k_B A \frac{T_3 - T_2}{\Delta x_B} = -k_C A \frac{T_4 - T_3}{\Delta x_C}$$

$$= \frac{T_1 - T_4}{\Delta x_A / k_A A + \Delta x_B / k_B A + \Delta x_C / k_C A}$$

Since thermal resistances in series and parallel can be evaluated in the same way as electrical resistances, then for Fig. 15.3

$$q = R_\theta A \Delta T$$

where R_θ = overall heat transfer coefficient

$$\frac{1}{\Delta x_A / k_A + \Delta x_B / k_B + \Delta x_C / k_C}$$

(i) Thermal contact resistance
Practical engineering surfaces are not smooth, they have

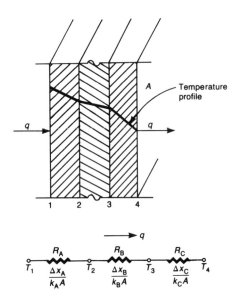

Fig. 15.3 One-dimensional heat transfer through a composite structure plus electrical analogy

inherent roughness. When two unbonded engineering surfaces are held together therefore, the areas of true surface contact depend both on the actual surface texture and the interfacial load applied. For most materials, this can be defined as

$$A_r = \frac{W}{3H}$$

where A_r = real area of contact, W = applied load (kg) and H = hardness (kg mm^{-2}).

For two conducting surfaces in contact, the mechanical constriction at the interface results in a thermal contact resistance (Fig. 15.4), i.e.

$$q = \frac{T_1 - T_3}{\Delta x_A/k_A A + 1/cA + \Delta x_B/k_B A}$$

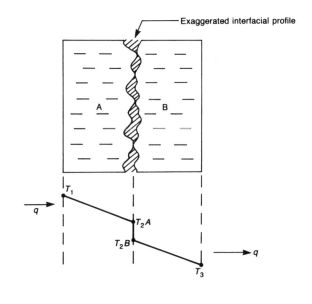

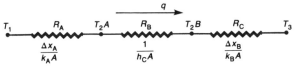

Fig. 15.4 Thermal contact resistance and electrical analogy

where $1/cA$ = thermal contact resistance and c = contact coefficient.

The contact coefficient is an important factor in a number of applications because of the likelihood of fastened surfaces involved in the heat transfer process, especially where the thicknesses of contacting material in the direction of heat flow are small. The two principal contributions to heat flow at the joint will be solid—solid contact points and conduction through trapped gases in

Table 15.2 *Contact conductances of typical surfaces in air*

Surface type	Roughness (μm)	Temperature (°C)	$1/h_c$ (m² °C W^{-1} × 10⁴)
416 stainless, ground, air	2.54	90–200	2.64
304 stainless, ground, air	1.14	20	5.28
416 stainless, ground, with 0.001 in brass shim, air	2.54	30–200	3.52
Aluminium, ground, air	2.54	150	0.88
	0.25	150	0.18
with 0.001 in brass shim, air	2.54	150	1.23
Copper, ground, air	1.27	20	0.07
milled, air	3.81	20	0.18

the void spaces. Some values of contact resistance are given in Table 15.2. The thermal resistance can be reduced considerably, up to 75 per cent, by using 'thermal grease'.

(ii) The conduction shape factor

In a two-dimensional system where only two temperatures are involved, a conduction shape factor, S, such that

$q = kS \Delta T$ overall

The shape factor analysis can be used typically for walls, pipes, etc.[1] This can also be extended to a three-dimensional wall where separate shape factors are used to calculate the heat flow through the edge and corner section, i.e.

$$S_{\text{wall}} = \frac{A}{L}$$
$$S_{\text{edge}} = 0.54L$$
$$S_{\text{corner}} = 0.15X$$

where A = wall area, X = wall thickness and L = length of edge, so that the total shape factor then becomes

$$S = S_{\text{wall}} + S_{\text{edge}} + S_{\text{corner}}$$

(iii) Unsteady-state conduction

If a solid body is suddenly subjected to a change in environment, some time must elapse before an equilibrium condition will prevail in the body. In the transient heating or cooling period that takes place in the interim before equilibrium is established, the analysis should be modified to take into account the change in internal energy of the body with time. In industrial application a large number of heating and cooling processes require calculation. The analysis may be referred to as the lumped-heat-capacity method and assumes that the internal thermal resistance of the body is negligible compared to the external resistance. The general form of calculation of heat transfer for a change of system temperature is

$$Q = \int_{T_1}^{T_2} mC \, dT$$

where Q = total heat transferred between a system and its surroundings or amount of heat energy required to increase temperature by $T°$ (J), m = mass (kg), C = specific heat and mC = specific heat capacity (Jkg^{-1}K^{-1}). For temperature differences where the change in specific heat capacity is not large

$$Q = mC \Delta T \quad \text{or} \quad q = \frac{mC \Delta T}{t}$$

where t = time (s), i.e.

$$\frac{\text{Watts supplied for}}{\text{heating}} = \frac{\text{thermal mass}}{\text{time to reach temperature}}$$

where thermal mass = $mC \Delta T$

15.2.2 Convection heat transfer

Convection is the transport of energy to or from a surface by the combined action of conductive heat transfer and 'fluid' motion, convective heat transfer at solid/fluid surfaces being of most practical interest. In the majority of cases, the 'fluid' will be the air surrounding the item held at elevated temperature, and for the purposes of this chapter will be assumed to be the case. There are two types of convection process recognized:

Free convection. When the airflow results from the influence of gravity on air density differences induced by the heat transfer process itself.

Forced convection. Where the heat transfer involves air movement generated by some external mechanical means.

Newton's law of cooling expresses the overall effect of convection, i.e.

$$q = h_c A = (T_1 - T_\infty)$$

where q = rate of heat transfer (W), T_1 = surface (wall) temperature, T_∞ = surrounding air temperature and h_c = convective heat transfer coefficient (W m^{-2} K^{-1}). In terms of the electrical analogy:

$$q_{\text{(conv)}} = \frac{T_1 - T_\infty}{1/h_c A}$$

where $1/h_c A$ = convective resistance.

The coefficient h_c will be a function of a particular 'fluid' and its flow rate. It is sometimes called the film conductance because of its relation to a thin layer conduction process. Except for certain well-defined situations, exact analytical solutions of h_c do not exist; correlations are by and large empirical, so for design purposes a tolerance factor of 15 per cent should be applied. For a given convection problem, h_c can vary with many system parameters, i.e.

$$h_c = f(\text{shape}, L_c, T_1, T_\infty, \rho, \mu, k, c)$$

where L_c = the characteristic dimension, which depends on the geometry of the situation. For a vertical plate it is the height of the plate, for a horizontal cylinder it is the diameter and so forth, ρ = 'fluid' density and μ = dynamic viscosity, etc.

The principles of dimensional analysis are used to

reduce this list to a manageable set of parameters. In its dimensionless form, h_c may be expressed as

$$Nu = \frac{h_c L_c}{k} = \text{Nusselt no.}$$

$$= \text{ratio of } \frac{\text{area averaged convective conductance } h_c}{\text{thermal conductance } k/L_c}$$

The larger Nu the more effective the convection, e.g. turbulent flow over a long vertical plate. The ratio of fluid buoyancy forces to viscous forces in a free convective flow system may be expressed by

$$Gr = \frac{g\beta(T_1 - T_\infty) L_c^3}{\nu^2} = \text{Grashof no.}$$

where β = coefficient of thermal expansion, ν = kinematic viscosity, g = acceleration due to gravity and $\beta \equiv 1/T$ for an ideal gas where T = absolute temperature (K) (thermodynamic temperature).

At large Gr values, the flow tends to be turbulent. At low Gr values, viscous effects predominate and the flow stays laminar. Using the Reynolds no.

$$Re = \frac{vL_c}{\nu}$$

where v = 'fluid' velocity. Then if $Gr \gg Re^2$, natural convection is dominant; if $Gr \approx Re$, then mixed convection occurs; if $Gr \ll Re^2$, forced convection dominates.

In situations where a constant heat transfer rate exists but the temperature difference is unknown, a modified Grashof no. Gr^* can be introduced where

$$Gr^* = GrNu = \frac{g\beta q'' L_c^4}{k\nu^2}$$

where q'' = constant wall heat flux = q/A.

A further dimensionless 'fluid' property may be defined by

$$Pr = \frac{\nu}{\alpha} = \text{Prandtl no.}$$

$$= \text{ratio of } \frac{\text{momentum diffusion}}{\text{thermal diffusivity}}$$

$$= \frac{\mu/\rho}{k/\rho c} = \frac{c\mu}{k}$$

where α = thermal diffusivity.

For free convection, empirical heat transfer coefficients can be found from the following form for a variety of situations:

$$Nu_f = A(Gr_f Pr_f)^p$$

Table 15.3 *Constants for use in determination of convective heat transfer coefficients*

Geometry	$Gr_f Pr_f$	A	p
Vertical planes and cylinders	10^4–10^9	0.59	$\frac{1}{4}$
	10^9–10^{13}	0.021	$\frac{2}{5}$
	10^9–10^{13}	0.10	$\frac{1}{3}$
Horizontal cylinders	0–10^{-5}	0.4	0
	10^4–10^9	0.53	$\frac{1}{4}$
	10^9–10^{12}	0.13	$\frac{1}{3}$
Upper surface of heated plates or lower surface of cooled plates	2×10^4–8×10^6	0.54	$\frac{1}{4}$
Upper surface of heated plates or lower surface of cooled plates	8×10^6–10^{11}	0.15	$\frac{1}{3}$
Lower surface of heated plates or upper surface of cooled plates	10^5–10^{11}	0.58	$\frac{1}{5}$

or

$$h = \frac{kA}{L_c} (Gr_f Pr_f)^p$$

where A and p are constants (Table 15.3).

Where there is an appreciable variation between wall and stream conditions, the film temperature T_f can be used, defined by

$$T_f = \frac{T_1 + T_\infty}{2}$$

where $10^4 < Gr_f Pr_f < 10^9$, flow is laminar; where $Gr_f Pr_f > 10^9$, flow is turbulent.

for *forced* convection, in a general empirical sense,

$$Nu \propto Re^s Pr^r$$

where r and s are constants such that for heat transfer at the surface of a plate

$$Nu = 0.332 Pr^{0.33} Re^{0.5}$$

or

$$h_c = \frac{k}{L_c} 0.332 Pr^{0.33} Re^{0.5}$$

(i) Conduction–convection system

For a partition or enclosure wall with hot air on one side and cold air on the other (Fig. 15.5), then the heat transfer rate under steady-state conditions is given by

$$q = h_{c1} A (T_A - T_1) = \frac{kA}{\Delta x}(T_1 - T_2) = h_{c2}(T_2 - T_B)$$

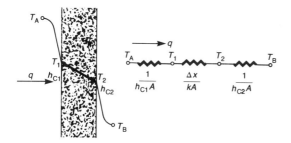

Fig. 15.5 Overall heat transfer through a plane

and representing the heat transfer process by a resistance network, then

$$q = \frac{T_A - T_B}{1/h_{c1}A + x/kA + 1/h_{c2}A}$$

and

$$q = R_\theta A T_{\text{overall}}$$

where R_θ = overall heat transfer coefficient

$$\frac{1}{1/h_{c1} + \Delta x/k + 1/h_{c2}}$$

(ii) Unsteady state conditions

In unsteady-state conditions the convection loss from a body is evidenced as a decrease in the internal energy of a body, i.e.

$$q = h_c (T_1 - T_\infty) = -\frac{mC \, \Delta T}{t}$$

Now $T_1 = T_0$ at $t = 0$, hence

$$\frac{T_1 - T_\infty}{T_0 - T_\infty} = \exp[-(h_c A/mC)^t]$$

and in the electrical analogy

$$R_{\text{th}} = \frac{1}{h_c A} = \text{thermal resistance}$$

$$C_{\text{th}} = mC = \text{thermal capacitance}$$

and

$$R_{\text{th}} C_{\text{th}} = \frac{mC}{h_c A} = \text{time constant of the system (since equation has units of time)}$$

15.2.3 Radiation heat transfer

Thermal radiation is one of the basic mechanisms for energy transfer between bodies at different temperature. It is a form of electromagnetic energy, travels in straight lines and does not depend on the presence of intermediate media. It is characterized by a frequency range 0.1–100 m (Fig. 15.6), in relation to wavelength bands of other parts of the electromagnetic spectrum. The analysis of thermal radiation is more complex than conduction or convection, but while it may be a dominant mode of heat transfer at high temperatures, it has only limited significance at moderate and low temperatures. Often thermal radiation acts in parallel with conduction and convection.

The amount of heat radiated from a surface depends on the emittance ϵ where

$$\epsilon = \frac{\text{amount of radiant energy emitted by a body}}{\text{amount of radiant energy emitted by an ideal radiator (black body)}}$$

The value of ϵ depends on the surface condition of the radiator, i.e. its temperature, surface roughness, finish and coating and, if a metal, the degree of surface oxidation. A black body is an idealized surface, a perfect emitter (and absorber) such that

$$E_b = \sigma T^4$$

where E_b = emissive power of a black body, σ = Stefan–Boltzmann constant = 5.67×10^{-8} W m^{-2} K^{-1}.

However, real (non-black) surfaces emit (and absorb) less readily, hence total emissive power

$$E = \epsilon \sigma T^4$$

where $\epsilon = E/E_b$ and E = emissive power of a body.

In general, dull, dark surfaces are good absorbers and

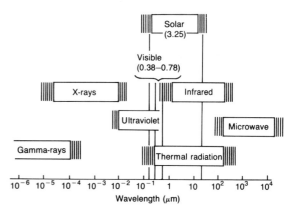

Fig. 15.6 Thermal radiation in the electromagnetic spectrum

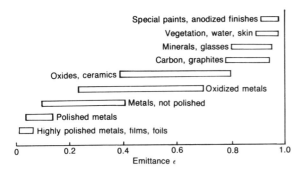

Material	Emissivity, ϵ
Aluminium	
Polished	0.06
Commercial sheet	0.09
Rough polish	0.07
Iridited	0.10
Gold, highly polished	0.018–0.035
Steel, polished	0.06
Iron, polished	0.14–0.38
Cast iron, machine cut	0.44
Brass, polished	0.06
Copper, polished	0.023–0.052
Polished steel casting	0.52–0.56
Glass, smooth	0.85–0.95
Aluminium oxide	0.33
Anodized aluminium	0.81
White enamel on rough iron	0.9
Black shiny lacquer on iron	0.8
Black or white lacquer	0.80–0.95
Aluminium paint and lacquer	0.52
Rubber, hard or soft	0.86–0.94
Water (0–100 °C)	0.95–0.96

Fig. 15.7 Typical emissivity values at 100 °C

emitters of heat, hence have high ϵ values, while the emissivity of metallic and polished metallic surfaces is generally small (Fig. 15.7). Oxide layers significantly increase the emissivity of metallic surface, and for non-conductors ϵ is relatively large, > 0.6 (Fig. 15.7). Radiation really involves an exchange of energy since part of the energy received is always reflected back, but for the vast majority of common engineering problems, reflected energy can be taken care of by the emissivity and absorptivity of each body involved in the exchange. The general radiation equation for the exchange of radiant heat energy between two non-black bodies is

$$q = \sigma F \epsilon A (T_1^4 - T_2^4)$$

where F = a view factor modifier and A = surface area.

(a) Surface area A Transmitted heat is directly related to surface area; if the surface is irregular, the projected area should be used. A finned surface increases the effective surface area for convective heat transfer, but has little effect on the effective area for radiation heat transfer. Radiating surfaces must be able to 'see' each other in order to transfer heat. Fins block the line of sight so only the projected area is effective for the transfer of heat (Fig. 15.8).

(b) Radiation view factor The view factor is defined as the fraction of radiation leaving surface 1 in all directions which is intercepted by surface 2 where

$$A_1 F_{1-2} = A_2 F_{2-1}$$

Since radiation heat transfer depends on the line of sight, the view factor between surfaces is a function of the geometry. Some view factor formulae are given in Fig. 15.9 for some selected surface configurations. The view factor is not easily determined for radiation involving irregular shapes unless the radiating surface is totally enclosed, in which case $F_{1-2} = 1.0$. When partial surfaces are involved, the view factor needs to be calculated or estimated. Approximations may be made as to what fraction of the radiation leaving one surface in all directions (180°) is intercepted by another.

Thus in the more specific case of two surfaces exchanging heat with each other and nothing else, then the net heat transfer is given by

$$q_{\text{net}\,(1 \to 2)} = \frac{\sigma(T_1^4 - T_2^4)}{(1-\epsilon_1)/\epsilon_1 A_1 + 1/A_1 F_{12} + (1-\epsilon_2)/\epsilon_2 A_2}$$

$$= \frac{E_{b1} - E_{b2}}{(1-\epsilon_1)/\epsilon_1 A_1 + 1/A_1 F_{12} + (1-\epsilon_2)/\epsilon_2 A_2}$$

which in terms of the electrical analogy, the elements may be described by Fig. 15.10, as follows:

$E_{b1} - E_{b2}$ = potential difference term.

$\dfrac{1 - \epsilon_i}{\epsilon_i A_i}$ = 'surface resistance' term, the surface resistance to radiation heat transfer

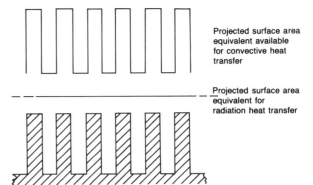

Fig. 15.8 Heat transfer characteristics of a finned surface

THERMAL MANAGEMENT

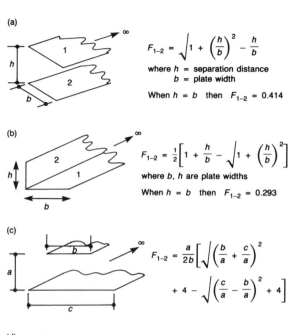

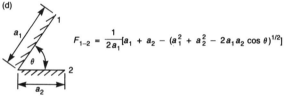

Fig. 15.9 Summary of view factors for some common surface configurations: (a) two infinite rectangular parallel plates; (b) two infinite rectangular plates intersecting each other at 90°; (c) two infinite long parallel plates of different width; (d) two intersecting planes at an angle

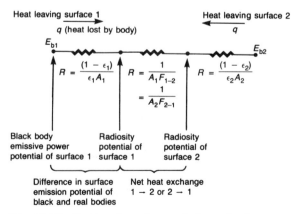

Fig. 15.10 Construction of network for a simple radiation heat transfer situation between two surfaces

and

$$\frac{1}{A_i F_{i-j}} = \text{'space resistance' across the media gap between the two surfaces.}$$

The effects of reflectance and emittance are in fact taken into account by connecting a black body potential node E_b to each nodal point in a network by means of a finite resistance $(1 - \epsilon)/EA$. In the case of a black body, this resistance is zero since $\epsilon = 1$.

When infinite parallel planes are considered for simplicity, A_1 and A_2 are equal, the radiation shape factor = 1.0 since all radiation leaving one plane is intercepted by the other, thus the net heat transfer is now given by

$$q = \frac{\sigma A(T_1^4 - T_2^4)}{1/\epsilon_1 + 1/\epsilon_2 - 1}$$

Similarly, when two long concentric cylinders exchange heat,

$$q = \frac{\sigma A_1(T_1^4 - T_2^4)}{1/\epsilon_1 + (A_1/A_2)(1/\epsilon_2 - 1)}$$

where A_1 = surface area of inner cylinder and A_2 = surface area of outer cylinder.

(i) Radiation heat transfer coefficient

Since radiation heat transfer problems are often closely associated with convection problems and the total convective and radiation heat transfer is often the objective of an analysis, they may be put on a common basis by defining a radiation heat transfer coefficient h_r. As long as the surface temperature difference ΔT is small compared to the absolute values of T, then

$$q_{rad} = \epsilon_1 h_r A_1 (T_1 - T_2)$$

where

$$h_r = \sigma(T_1 + T_2)(T_1^2 + T_2^2) \quad \text{and} \quad h_r \equiv 4\sigma T_m^3$$

where T_m = mean temperature at which radiative heat transfer takes place. Hence the total convective and radiation heat transfer is given by

$$q = (h_c + h_r \epsilon_1) A_1 (T_1 - T_\infty)$$

At 300 °K, the value of $4\sigma T_m^3$ is of the same order as the natural convection coefficient for heat transfer in air.

15.3 Forced airflow

Moving air is the main functional element in many heating and cooling applications, involving a change to its ambient velocity, temperature and pressure condition. As a heat transfer medium, air is propelled continuously by a fan against pressure loss in a system, the fan overcoming the resistance the system imposes against the flow. In the present context a 'fan' designates a combination of impeller and casing necessary to its performance and includes centrifugal, mixed flow and axial types. The rotating blades increase the total energy content of the air handled. The design of a forced airflow system, typically for heat removal, is a continuous compromise between the amount of cooling actually required, the space available, the acceptable power consumption and the allowable noise levels.

15.3.1 Terminology

In terms of performance, it is customary for fan manufacturers to quote inlet volume flow and fan pressure for some 'standard' air condition, agreed to be density 1.2 kg m^{-3}, atmospheric pressure 1 bar (100 kPa), temperature 16 °C, humidity 65 per cent RH.

An air-moving device (AMD) is a power-driven fan moving a continuous flow of air. There are three main components in a fan, the impeller (wheel, rotor), a means of driving it and a pressure conversion housing (scroll). The housing plays an important part in the aerodynamic performance of the fan and is often the major element in converting velocity energy (kinetic energy) into useful static pressure (potential energy). For convenience, and because the magnitude of pressures developed by fans is fairly small, units of measurement are in millimetres of water column, where

1 mm water = 9.8 Pa = 9.8 N m^{-2} pressure

Fan performance curves will show the static pressures that the fan will develop as a function of flow rate (e.g. Fig. 15.11). The graph of fan pressure versus volume flow is known as the 'fan characteristic'. The term 'fan duty' is given to inlet volume flow at a rated fan pressure. As air flows through a system, the total energy of the system may be expressed in terms of an equivalent total pressure P_t such that

Total pressure, P_t = mean static pressure P_s + velocity pressure P_v

These air pressures are gauge pressures measured by a manometer connected to a pitot tube or to pressure taps. The principle is illustrated by Fig. 15.12. Velocity pressure is pressure due to air speed, the pressure exerted by air against obstructions and may be expressed by

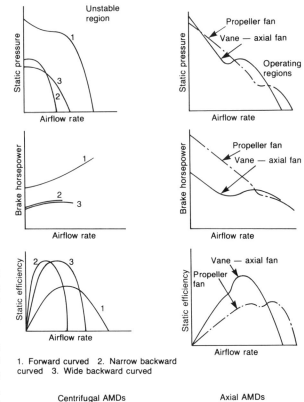

1. Forward curved 2. Narrow backward curved 3. Wide backward curved

Centrifugal AMDs   Axial AMDs

Fig. 15.11 Summary of fan performance characteristics

$$P_v = \left(\frac{V}{4.04}\right)^2 \text{ mm water}$$

where V = air velocity (m s^{-1}) = $\sqrt{(2gH_p\rho_w/\rho_a)}$ and g = acceleration due to gravity = 9.8 m s^{-2}, H_p = pressure head (mm water), ρ_w = density of water and ρ_a = density of air.

Static pressure is the pressure that maintains air velocity against a resistance. For example, if outlet flow is reduced by an obstruction or decrease in duct size, static pressure will build up to move the air against the resistance, i.e. static pressure is the fan's potential to do work.

The power delivered by a fan may be expressed in terms of P_s or P_t, i.e.

Output power = $2.72 \times 10^{-3} Q_v P_s$ Static air (watts)

or

Output power = $2.72 \times 10^{-3} Q_v P_t$ Total air (watts)

where Q_v = flow rate (m^3 per hour) = $3.6 \times 10^3 VA$ and where A = cross-sectional area of the duct (m^2). Fan efficiency, the rate at which energy must be added

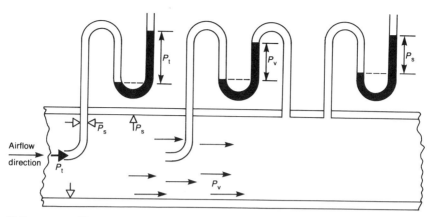

Static pressure: The pressure measured in the air in such a manner that no effect on the measurement is produced by the velocity of the air
Velocity pressure: Pressure that exists at a point in an airstream by virtue of the air density and its rate of motion
Total pressure: At a point in an airstream is the algebraic sum of the static pressure and the velocity pressure

Fig. 15.12 Schematic representation of air flow pressures

to the airstream to maintain motion, may be likewise defined in two ways:

$$\text{Static efficiency} = \frac{\text{air power (static)} \times 100\%}{\text{measured fan input power}}$$

$$\text{Total efficiency} = \frac{\text{air power (total)} \times 100\%}{\text{measured fan input power}}$$

The total (mechanical) efficiency may be regarded as a true fan efficiency while the static efficiency is probably nearer to a usable parameter in practice. Fan efficiencies are illustrated in Fig. 15.11. For best overall performance, a fan should be operated near the region of maximum efficiency. A parameter sometimes used here is the specific speed parameter, N_s, which is the speed a fan would run to give unit volume flow and unit fan pressure,

$$N_s = \frac{nQ_v^{0.5}}{P_s^{0.75}} \left(\frac{\rho_a}{\rho_0}\right)^{0.75}$$

where n = speed (r.p.m.), Q_v = volume flow (m³ per hour), ρ_a = ambient air density and ρ_0 = standard air density.

The significance of specific speed is that for any particular fan design there is a unique attainable value of efficiency for a given value of specific speed (i.e. Fig. 15.13). The needs of the fan application determine the specific speed; optimum fan selection clearly results if a fan is chosen with a high efficiency at the specific speed dictated.

The pressure–volume relationship of a fan is not generally capable of being expressed as a simple mathematical one, hence performance curves need to be determined empirically. However, by considering a single point on the fan characteristic, e.g. at the point of rating, it is possible to derive simple relationships, generally known as the 'fan laws'. These are a family of analytical relationships that permit new values of flow, pressure and power to be determined for changes in operating variables at the same point of rating, but can only be applied to changes of fan involving exact geometric similarity. The performance of a fan in terms of pressure, volume flow and power depends on the following factors:

Volume flow $Q_v \propto nd^3$
Fan pressure $P \propto n^2d^2$
Input power $W \propto n^3d^5$

where d = impeller diameter.

The use of these relationships is summarized in Table 15.4. Since fan volume flow varies with impeller speed, then fan pressure varies with air density and (volume flow)².

15.3.2 Types of fan

Fan types are usually categorized into two basic configurations, axial and radial flow. However, variants are likely to be encountered in engineering design such

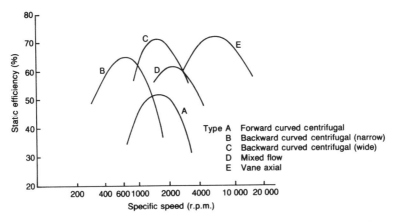

Fig. 15.13 Specific speed and aerodynamic efficiency of air-moving devices

Table 15.4 *Summary of fan law relationships*

Fixed variables	Variable parameter	Ratio	Flow Q_v	Pressure P	Power output	Air noise	Efficiency
AMD type system air density size	Speed	$\dfrac{n_2}{n_1}$	Increases directly with ratio	$(\text{ratio})^2$	$(\text{ratio})^3$	Changes $50 \log\left(\dfrac{n_2}{n_1}\right)$	Approx. same
AMD type system speed size	Air density	$\dfrac{\rho_2}{\rho_1}$	No change — changes directly			Minimal change	Minimal change
AMD type system air density speed	Size (impeller diameter and width change)	$\dfrac{d_2}{d_1}$	Increases directly with $(\text{ratio})^3$	$(\text{ratio})^2$	$(\text{ratio})^5$	Changes $70 \log\left(\dfrac{d_2}{d_1}\right)$ [a]	Minimal change

Notes: 1 = rating point; 2 = new condition; n = impeller speed; ρ = air density; d = impeller diameter.
[a] If $d_2 > d_1$, then add noise to baseline noise; if $d_1 < d_2$, then subtract noise from baseline noise.

as the mixed flow that ingests air axially and discharges at an angle between 0 and 90° to the axis of entry and the tangential or cross-flow fan that receives and discharges air tangentially. Various types of fan are illustrated and compared in Fig. 15.14.

(i) Axial fans

An axial fan propels air in an axial direction with a swirling tangential motion created by the rotating impeller blades. The impeller consists of a number of contoured blades mounted on a spider or a hub at a specified pitch angle (Fig. 15.14). Where the blade tip angles approach ~ 20°, the fan performance characteristic may show a discontinuity corresponding to the aerodynamic 'stall' of part or all of the aerofoils. Normal operation should be at airflows above the discontinuity or instability. Pressure–flow performance is similar for all types of axial fan (Fig. 15.11), except that the usable condition appears further down the performance curve for propeller fans. Static efficiency for both propeller fan and axial flow fan is highest just to the right of the stall range.

(ii) Centrifugal fans

These are the most common type of ventilating fan. Air enters the centre of the rotating impeller axially, is accelerated radially by the fan blades, collected in a volute casing and propelled out of the discharge opening at an angle 90° to the direction of entry (Fig. 15.15). Velocity pressure is largely converted to static pressure by the time air is discharged from the housing. Three types of blade form are available, backward, radial and forward curved blade shapes at the impeller periphery (Fig. 15.14). In addition to the type (and size) of impeller, fan performance depends on the housing design, the inlet ring

THERMAL MANAGEMENT

Centrifugal fans

Forward curved fan

Sometimes known as 'volume', 'squirrel cage' or 'sirocco' blowers
Blades at the discharge are included in the direction of rotation
Air is accelerated in housing to velocities higher than rotational velocities
Unstable without a scroll or housing
Can handle large air volumes at low running speed
Compact installation
Peak η lower than backward-curved fans
Blades inclined in direction of rotation
Deliver highest airflow rates in a given system resistance for centrifugal fans
Smallest package at often the lowest cost

Backward curved fan

Blades inclined away from direction of rotation
Less air leaving backward curved blower wheel is in the form of P_v compared to the forward curved wheel
Since more energy is in the form of P_s, less energy is lost in converting from P_v to P_s in the housing
Can operate well with no housing
Non-overloading power curve, i.e. max. power occurs at a volume near to that at max. efficiency
Largest fan for given duty

Radial-bladed fan

Radially oriented blades
Not widely used type
Performance capability in between forward and backward curved impellers
Overloading power characteristic
Medium efficiency, 50–60%
Self-cleaning impeller
Blade shape allows a higher running speed in general than other types of centrifugal blade
Typically 6–16 blades

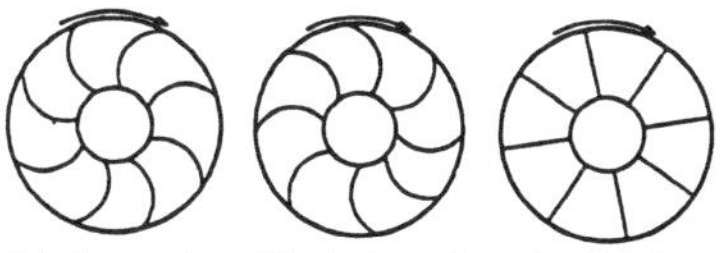

(a) Forward   (b) Backward   (c) Radial

Types of centrifugal fan

Axial fans

Propeller fan

Lowest-cost fan for low-pressure duties
Low efficiency
Lowest power requirement for free delivery applications
Unsuitable for high pressure
Very large axials have no venturi and are just large air stirrers, usually limited performance
Designed to move air from one enclosed space to another at relatively low static pressures
Static efficiency = 40–45%

Axial-flow fan

Usual fan for cooling electronics
Normally designed with an optimized venturi to ensure maximum airflow
Non-overloading power characteristic
High efficiency
Capable of high pressures by multi-staging
Fan is ducted; duct contains air on inlet and discharge sides of the impeller
Low noise
Wide usable range of performance
Airfoil sections run with a fine tip clearance in the duct

Propeller fan

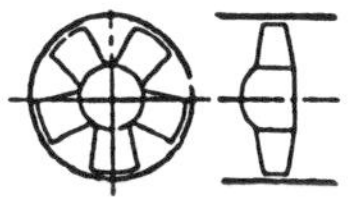

Axial-flow fan

Other fan types

Mixed-flow fan

Can be designed to produce the desired pressure : volume relationship
Capable of handling large volumes associated with axial fans at the high pressures of centrifugal fans
Performance similar to backward curved centrifugal fans

Cross-flow (tangential) fan

Unpredictable in the design sense
Use in low-pressure domestic-type applications
Can be as wide as necessary; discharge is in effect a long slot, i.e. a curtain of air
A long thin airflow path may result in a lower total static pressure drop
Static efficiency only 30–40% compared to up to 75% with centrifugal blowers
Poor performance against system pressure; do not use for cooling electronics in enclosures

Mixed-flow fan

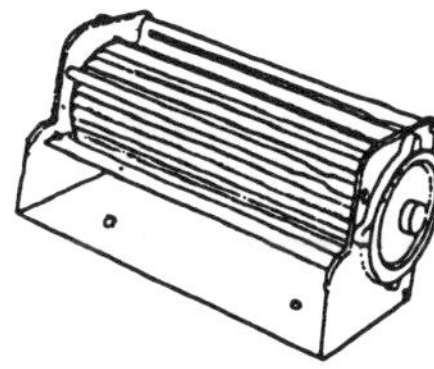

Tangential fan

Fig. 15.14 Summary of fan performance characteristics

diameter/impeller diameter ratio, the location and shape of the cut-off (which prevents recirculation of air around the blower wheel) and the shape of the discharge opening. In Fig. 15.15, the housing height, depth and width are commonly called its *A, B* and *C* dimensions respectively. A summary of the different blade types is given in Fig. 15.14, and a comparison of performance, efficiency and input power in Fig. 15.11.

(iii) Mixed-flow fans

The impeller consists of contoured blades mounted between a curved inlet ring and backplate, the blades axially shaped on the intake side and centrifugally shaped on the discharge periphery (Fig. 15.14). The action of the blower is both axial and centrifugal. The impeller is normally installed in a housing similar to the standard centrifugal wheels, but because a substantial amount of

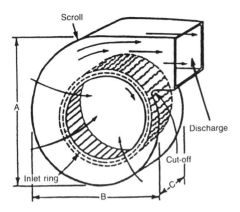

Fig. 15.15 Design aspects of the centrifugal fan

the pressure conversion takes place in the blade passages of a mixed-flow type, its performance is similar to the backward curved type in depending far less on housing configuration than the forward curved type.

(iv) Cross-flow, transverse or tangential fan

This is a long cylindrical impeller with a large number of forward curved shallow blades (Fig. 15.14). Due to the shape of the case, air enters along one side of the impeller making in effect two passes through each blade and leaving the impeller with a high-velocity component. Because of its low efficiency, this type of fan is usually found in low-pressure domestic applications only.

15.3.3 System characteristics

All elements in an airflow system restrict the flow of air. The resistance imposed against the flowing air is expressed as a pressure drop through the system, i.e. the pressure given to the airflow by the fan is progressively lost by cross-sectional changes in ductwork, turbulence at bends, losses through filters and grilles. The loss of pressure due to all these sources is, for most practical purposes, proportional to the square of the airflow at the point of loss, ie.

$$P \propto (Q_v)^2$$

This is termed the 'system characteristic' and is independent of the source of the flow (see section 15.3.1). If the system resistance curve is superimposed on a fan pressure–flow curve, the intersection represents the 'operating point', the point at which

Fan pressure rise (energy added) = resistance pressure drop (energy consumed)

Whatever fan is used, the operating point must fall on the system resistance curve. This point satisfies both fan and system conditions and shows the airflow rate that the fan will deliver in a particular system for that particular speed. However, when considering a particular fan performance, it should be remembered that catalogue data are concerned with the maximum possible airflow a fan can deliver, blowing freely without any external resistance to flow. The system is therefore idealized, approximating only to long straight ducts with no cross-sectional area changes, an operating system in practice that does not exist. For example, Fig. 15.16 illustrates the actual performance curve for the same fan with different air inlet configurations. In practice, therefore, the following concepts need to be introduced:

Internal pressure drop. The losses the fan must overcome within the fan unit itself; excludes ductwork.

External pressure drop. The losses imposed on an installed unit by general airflow obstructions, ducts, grills, filters, etc.

By reducing the fan static pressure at each flow rate by an amount a indicated by the system resistance curve at the same flow rate (Fig. 15.17), the residual characteristic in curve 2 is obtained which represents the performance curve of the actual airflow system. Thus curve 2 is simply fan pressure (F) minus system pressure (R) at each volume flow and so represents the external characteristic of a fan/system assembly. The true operating point is therefore point X.

If hardware is available, the easiest way to establish

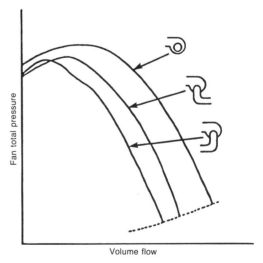

Fig. 15.16 Effect of bends at the inlet on the performance of a centrifugal fan

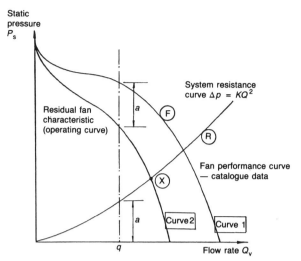

Fig. 15.17 Residual characteristics of a fan and system

the system curve is by experimentation. However, at the design stage where no prototype is available, dynamic losses may be estimated by the use of empirically determined loss factors or local loss coefficients. It is customary to calculate pressure loss by multiplying average velocity pressure in a section by its loss factor K. Hence for each item in the system,

$$\Delta p = KP_v = \frac{K\rho_a V^2}{2} = \frac{K\rho_a}{2}\left(\frac{Q_v}{A}\right)^2$$

where ρ_a = density of air, A = cross-sectional area of air passage, V = mean flow velocity and Δp = local pressure drop.

Hence the pressure loss is expressed as a proportion of the velocity head at some loss point, i.e. that proportion of the velocity pressure at the section indicated which has to be supplied by the fan in order to ensure that the required quantity of air will pass through the obstruction. In a system, the overall pressure loss is given by

$$\Delta p_{tot} = \Sigma \Delta p$$

Within the system flow path, pressure loss can be applied to static or velocity situations, P_v may be converted to P_s and vice versa, just as other forms of kinetic and potential energy may be interchanged. From the above, the overall system pressure loss varies with the density of the air and the square of the volume flow rate. In a typical systems resistance curve (Fig. 15.17),

$$\Delta p_{tot} = BQ^y$$

where B = system characteristics constant, $y = 1$ for laminar flow and $y = 2$ for turbulent flow.

A summary of some typical loss factors is given in Fig. 15.18. In any thermal management system, a certain air volume throughput will be required. This then enables air velocity, hence P_v estimates for each system resistance point. Using estimates of K (Fig. 15.18), the

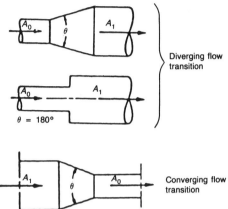

Conical transitions, round and rectangular:

	A_0/A_1	30°	40°	60°	180°
Diverging transition:	0.1	0.52	0.75	0.93	0.83
	0.2	0.41	0.59	0.74	0.65
	0.4	0.23	0.33	0.41	0.37
	0.6	0.10	0.15	0.18	0.16
Converging transition:	0.1	0.02	0.03	0.07	0.45
	0.2	0.02	0.03	0.07	0.40
	0.4	0.02	0.03	0.06	0.30
	0.6	0.01	0.02	0.04	0.20

Duct entry:	Plain duct	= 0.9
	Flanged end	= 0.5
	60° into cone	= 0.2
	Well-rounded entry	= 0.04

Duct exit:	Square exit into pipe	= 1.0
(discharges)	20° cone exit	= 0.4

Bends:

radius/diameter ratio	Rectangular elbow aspect ratio					Circular elbow
	0.25	0.5	1.0	2.0	4.0	
Mitre	1.3	1.3	1.2	1.1	0.9	1.2
0.5	1.3	1.0	0.93	0.99	1.2	0.71
1.0	0.36	0.26	0.21	0.20	0.21	0.22
1.5	0.18	0.12	0.09	0.08	0.07	0.15
2.0	0.11	0.07	0.05	0.04	0.04	0.13
3.0	0.05	0.03	0.02	0.02	0.01	0.11

Screen: (rectangular or circular duct)

A_f/A_0	0.3	0.4	0.6	0.8	1.0
	6.2	3.0	0.97	0.32	0

A_f = free area
A_0 = duct area

Fig. 15.18 Summary of some loss factors K

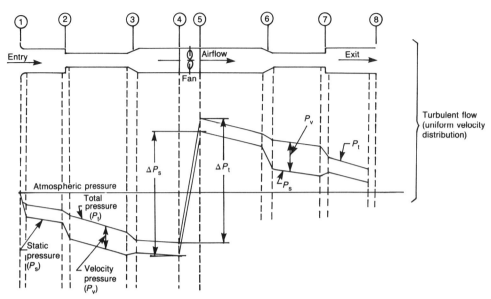

Fig. 15.19 Pressure changes during air flow in ductwork

total pressure drop Δp can be found for that point; P_s follows from knowledge of Δp and P_v.

A typical example of pressure drops through a series of ductwork resistances is shown in Fig. 15.19. It follows that a constriction will result in a decrease in static pressure, but an increases in velocity pressure, whereas the reverse is true for an expansion. Because the fan inlet and outlet performance is outlined in Fig. 15.19, fan pressures ΔP_s and ΔP_t are represented as pressure at the fan outlet minus pressure at the fan inlet.

15.3.4 Filtration

Airflow filters are often used both to prevent ingress of dirt and dust into a machine or electronics area and also to prevent discharge of any machine emissions of process materials into the surrounding environment. Thus for example in Fig. 15.2, the thermal management and process system based on a forced air throughput also deals with control of contamination. Air filters are relevant to the design of an airflow system in respect of pressure drop. The pressure drop across the filter can be high if the air velocity through the filter is too high. A lower air velocity will reduce the pressure drop through the filter even more, but will require a large filter surface area. Some filter manufacturers will recommend a minimum airflow velocity to obtain acceptable filtration efficiency.

Filter media may be manufactured from non-woven fabrics, synthetic fibres, glass fibre, porous solids, perforated solids, paper, particulate solids, open-cell foams, etc. all of which will have their own area of suitability. The characteristics of a filter that are most important are its efficiency, its dust retention capability and its initial pressure drop. The dust or contamination loading that a filter will accept for a given increase in pressure drop is as important as its efficiency. The mechanical performance of a filter relates mainly to the strength it possesses to withstand the filtration process, i.e. burst, tear and tensile strength. It is thus very important to select a suitable material and to choose the correct specification for the performance required. A range of typical contaminants is illustrated in Fig. 15.20.

The effect of placing a filter at the exit of an airflow system, e.g. as in Fig. 15.19, is shown in Fig. 15.21. By plotting system resistance curves on both sides of the fan (not necessarily equal), and since the airflow through the system is fixed, inlet vacuum and pre-filter air outlet pressures are the same line. The effect of contamination loading can also be shown in Fig. 15.21, as the filter becomes clogged, airflow decreases, the pre-filter outlet pressure increases and inlet vacuum decreases.

15.3.5 Performance application of fans

When considering forced airflow for heat removal or thermal management applications, design of such a system is basically one of system sizing as follows:

1. Firstly establish the airflow requirement, i.e.

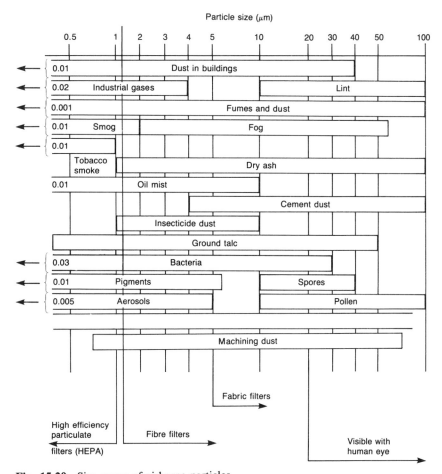

Fig. 15.20 Size range of airborne particles

$$V = \frac{Q}{C_p \rho \Delta T} \text{ cubic metres per second}$$

where Q = heat dissipated (W), C_p = specific thermal capacity of air (J kg^{-1} K^{-1}), ρ = density of air (kg m^{-3}) and ΔT = mean temperature difference between air inlet and outlet.

Assume that exiting air ~70 °C (typical for small electromechanical devices), then knowing the maximum input ambient temperature, once the heat dissipation of individual components is summed, the airflow requirement may be calculated.

2. If an air tunnel is available, establish the external system resistance curves for fan inlet and outlet experimentally. With actual products, this is best done by using a physical mock-up of internal items and introducing leakage paths, variable heat dissipation, etc.

3. If an experimental test bench is not available, establish the total pressure drops by calculation assuming a constant airflow rate (section 15.3.3). Assume turbulent flow and plot a system resistance curve.

4. Having identified a likely fan and plotted its characteristic from the manufacturer's data, the residual fan characteristic can be determined and a true operating point found. The flow rate at the true operating point should have ~25 per cent safety factor above the airflow requirement initially calculated in (1).

5. Aim to test the system configuration as early in the design cycle as possible, evaluating temperatures at critical components over a range of operating conditions and ambient temperature range. This checks that the exit temperature chosen, e.g. 70 °C,

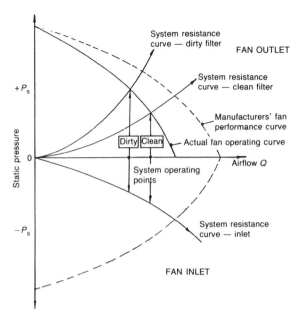

Fig. 15.21 Effect of exit filter contamination (blocking) on system air flow characteristics

and airflow patterns are adequate for individual components, i.e. electronic or any other thermally sensitive item.

Forced air system performance design is quite straightforward once basic fan and flow concepts are understood. The requirement for air 'volume' to remove heat or ventilate is fairly obvious; what is often ignored is that since no air can move without the application of force, the pressure must also be determined and specified before a fan can be selected. The following summary applies to airflow system design:

1. It is misleading to consider air volume alone without calculating the system resistance.
2. Determine the largest possible performance window required to maintain acceptable machine functions.
3. The fan power requirement is directly governed by system resistance for a given flow rate and fan efficiency. Aim should be to keep system resistance to an absolute minimum by maintaining the lowest and most even velocity attainable.
4. If flow paths and airways are cramped with tight bends, sharp changes in cross-section and rough surfaces, then pressure losses will be high and the fan will be bigger, noisier and more expensive.
5. Losses in always vary with $1/(\text{diameter})^5$, thus any chance to increase the diameter should be taken since substituting a 63 mm diameter duct for a 50 mm duct reduces pressure loss and fan power by ~70 per cent.
6. Since p varies with Q_v^2, in a fixed system design, in order to double the volume flow, four times the initial pressure must be applied.
7. The physical size of a fan is determined by its output power. Sufficient space must be provided to accommodate an efficient unit with low noise level.
8. Airflow noise varies as the fifth power of the air speed, hence the lower the air velocity the better. The wider the fan the quieter it is at a given airflow. For optimum noise and efficiency, an impeller aspect ratio of 1:2 is often used (impeller blade width:diameter).
9. There is very little reduction in inlet flow from a side wall or panel even when mounted close to the fan. As long as a clearance of one-quarter of the inlet diameter is maintained, full flow will occur.
10. Locate fans at the bottom or back of equipment so that sound is dissipated through natural obstructions.
11. Consider the direction of airflow relative to the user; all discharges should be away from the operator.

15.4 Other heat transfer devices

When standard cooling/heating techniques are no longer adequate for a particular application, it is usual to consider heat exchangers or heat transport devices as alternatives. Heat exchangers involve the transfer of heat between two separate and different 'fluid' streams while heat transport devices involve moving thermal energy between two different locations. A separate division is sometimes made on active or passive grounds (section 15.1).

15.4.1 Heat exchangers

Where the enclosure or assembly must remain sealed, forced air cooled heat exchangers are the most straightforward option. A simple, reliable and low maintenance design based on the use of heat sinks is shown in Fig. 15.22. In this case artificial air movement may be used as effectively inside the enclosure as well as outside. Performance calculation with such a heat exchanger are dealt with in section 15.5.6. Many different types and style of heat exchanger exist in industrial applications, e.g. plate, fin, tube or pipe. A hollow core fin-type arrangement for cooling two PCBs is shown in Fig. 15.23 and a direct cold wall heat exchanger in Fig. 15.24. The latter will have fluid-filled pipes running along

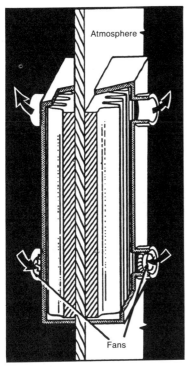

Fig. 15.22 A simple air-to-air heat exchanger constructed from commercial heat sinks. Courtesy of Redpoint Ltd, Swindon

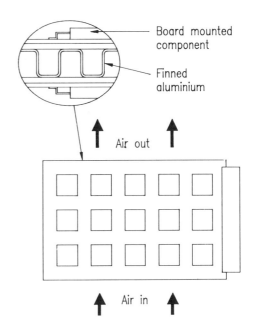

Fig. 15.23 Example of PCB heat exchanger

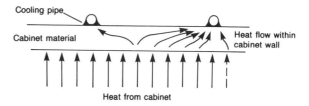

Fig. 15.24 Heat flow in the plane of a cabinet wall when cooling pipes are widely spaced

the wall of the equipment or enclosure to be cooled (or heated). If the cooling fins are close enough together and the liquid flow is high enough, the outer surface may be considered as a constant temperature plane. If the cooling pipes are not close together, there may be an obvious temperature differential between the backplate and the pipes.

15.4.2 Heat pipes

Heat pipes are a special class of passive heat transport device, available in a number of geometries. They are used to transport heat from the source to the sink directly without any external power. They can be used to eliminate hot spots and may accept heat that has a high power density. A typical application may involve a component, subassembly or PCB that has a high power dissipation and a relatively long heat flow path to the heat sink. They resemble solid conductors in many ways except that their effective conductivity can greatly exceed that of metals, ~300 times that of copper for example.

A typical heat pipe consists of a partially evacuated, hermetically sealed tube, containing a fluid that transfers heat when it evaporates, and a wick (capillary pump) that brings the fluid back to its starting-point when it condenses (Fig. 15.25). The liquid and vapour phases are initially in equilibrium. When heat is applied at one end of the tube, the fluid within the pipe vaporizes or boils, raising the pressure locally in the vapour phase and causing the vapour to flow to the unheated, cooler sections of the pipe where heat is removed, the excess vapour condenses and the wick draws the fluid back to the starting-point where the process is repeated. Heat is thus transported in the form of the latent heat of vaporization of the working fluid. Suitable fluids are water, methanol, ammonia, freon, sulphur dioxide, etc. The pumping action is a

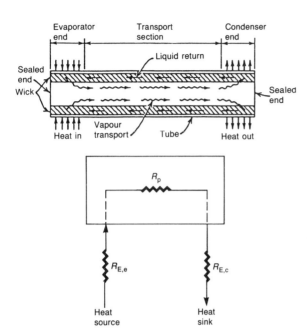

Fig. 15.25 Section through the length of a typical heat pipe together with the thermal model equivalent

surface capillary effect requiring the presence of a structure with small pores, typically 0.025–0.25 mm diameter, i.e. a wick. Wicks are usually made or porous ceramic, porous metal or woven stainless mesh. The evaporation and condensation cycles take place over a very narrow temperature band and as a result there is only a small temperature difference between both ends.

Heat transport capability can be defined in terms of the maximum axial heat flow rate q_{max} where

$$q_{max} = \frac{2K_w A_w N_1 F_1}{r_c L_{eff}}$$

and

$$N_1 = \frac{\gamma \sigma \rho_1 \lambda}{\mu_1}$$

$$L_{eff} = 1/2 L_e + L_a + 1/2 L_c$$
$$= \text{effective length of the heat pipe}$$

where K_w = permeability of the wick, A_w = cross-section area of the wick, r_c = pore radius of the wick, F_1 = ratio of viscous pressure loss in liquid to total pressure loss, γ = surface tension of liquid, λ = latent heat of liquid, ρ_1 = density of liquid, μ_1 = dynamic viscosity of liquid, L_e = length of evaporator section, L_a = length of adiabatic section and L_d = length of condenser section.

However, heat pipes are frequently specified in terms of QL capability, since for a given heat pipe design the product $q_{max}L$ is effectively fixed. Doubling the effective working length L results in halving the maximum load the heat pipe can carry. In practice, corrections are also made to q_{max} values specified by manufacturers to allow for working temperature and orientation, in addition to length, i.e.

$$q_{max}L = \text{design} \times \text{temperature} \times \text{orientation}$$
$$\text{point} \quad \text{correction} \quad \text{correction}$$

These corrections are typically based on a 'design point' of pipe length 100 mm, horizontal orientation and a 50 °C mean pipe temperature. The maximum axial flux (q_{max}/cross-sectional area) allowable at a given operating temperature for both methanol and water filled heat pipes is given in Fig. 15.26.

1. *Temperature correction.* A heat pipe cannot operate below the freezing-point of its working fluid or beyond its maximum heat transport capability. Correction factors are given by Fig. 15.27 for typical heat pipes.
2. *Orientation correction.* The wick is probably the most critical part of the design in that it determines the capillary wicking action available for drawing

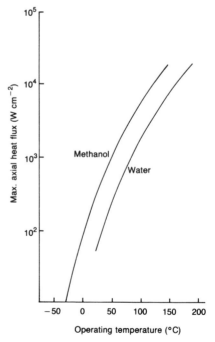

Fig. 15.26 Maximum axial heat flux allowable for a given heat pipe operating temperature

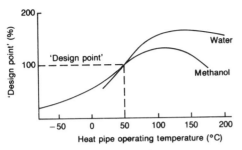

Fig. 15.27 Heat pipe correction factor as a percentage of the design point

the condensed liquid back to the evaporator end. A wick must be capable of raising the fluid against gravity and at the same time have minimum internal resistance to flow. Pipe orientation has virtually no effect on vapour flow, but gravity can act either to assist or impede the flow of the liquid. The performance of the heat pipe is degraded when it is forced to work against gravity. If the evaporator is located above the condenser, heat transport capability is multiplied by a factor $(1 - \eta)$ to allow for pipe orientation where

$$\eta = \frac{r_c(H_t - d_i \cos \alpha)}{2H_w}$$

where H_t = height of condenser above evaporator (negative if vice versa), d_i = interior diameter of pipe, α = tilt angle (always > 0), H_w = wicking height factor = $\sigma/\rho_l g$ and g = acceleration due to gravity. The factor H is temperature sensitive. Values of H for different wicking fluids are given in Table 15.5.

Table 15.5 Values of wicking height factors for various fluids (units of H_w in $m^2 \times 10^{-6}$)

Temperature (°C)	Ammonia	Freon	Methanol	Water
−73	5.97	—	—	—
−53	5.62	—	—	—
−33	5.06	—	—	—
−13	4.49	—	—	—
7	3.94	—	3.03	7.61
27	3.39	1.25	2.87	7.36
47	2.79	1.10	2.70	7.08
67	2.14	0.95	2.54	6.77
87	1.48	0.80	2.37	6.46
107	0.85	0.66	2.17	6.14
127	0.21	0.52	1.95	5.82
147	—	—	1.68	5.48
167	—	—	1.38	5.12
187	—	—	1.06	4.74
207	—	—	0.71	4.33

Heat transport capability defines the limit of operation, whereas thermal resistance is the important design parameter. A thermal model of a heat pipe is given in Fig. 15.25. Simplifying,

Overall thermal resistance = $R_p + R_{E,e} + R_{E,c}$

where R_p = heat pipe thermal resistance and $R_{E,e}$, $R_{E,c}$ are external resistances determined by the installation and not the heat pipe.

Once thermal resistances are known, together with transported watts, the temperature differentials between input and output regions may be calculated (see section 15.2.1). The total temperature drop between device and cold wall consists of

Temperature drop at evaporator end + axial temperature drop + temperature drop at condenser end

If required, the axial temperature drop can be calculated as for a conventional length of material bar (see section 15.2.1). Because of the high effective conductivity values, there will only be very small temperature differences between evaporation and condensation zones. For example, a 100 mm long pipe, 6.0 mm diameter can pump 110 W of heat at 50 °C with a 4.5 °C temperature difference along the transport section of the pipe.

Two developments of the heat pipe are variable conduction heat pipes (VCHPs) and electrohydrodynamically enhanced condensation (EHD). The primary application of VCHPs is to control the temperature of a heat source when the heat load varies or the temperature of the heat sink is not constant. As such, 'intelligent' temperature stabilization can either be achieved electronically or by adding a non-condensing gas to the working fluid. The theory is complex; EHD-enhanced condensation is based on the Senftleben effect, by subjecting a dielectric fluid to a strong electric field, the rate of heat transfer can be increased significantly over conventional systems. Typically voltages of 10–25 kV are involved, but the energy required to maintain the effect is small compared to the thermal transfer achieved.

15.4.3 Thermoelectric heat pump

Thermoelectric heat pump (THP) cells or thermoelectric coolers are solid-state devices driven by low-power d.c. supplies and used to cool loads down to −40 °C. For lower temperatures, multi-stage cells may be used. These devices are based on the Peltier effect, i.e. if a current source is connected by conductors of similar material to the ends of a third conductor of a different material, then one end of the latter will cool while the other will warm up. The rates of heating and cooling depend on the junction temperature and current.

In its simplest form the THP consists of two pillars of solid-state semiconductor material, one p-type, the other n-type sandwiched between metal plates (Fig. 15.28). When current flows in the direction shown, junctions XX absorb heat (current passes from the n-type to the p-type material) cooling the plate while junction YY discharges heat (current passes from the p-type to the n-type material). The heat is thus 'pumped' across the 'cold' face to the 'hot' face. If the current direction is reversed, heat flow is reversed, thus THPs can be used for cooling or for maintaining a temperature above ambient. The heating junctions are connected to a heat sink (Fig. 15.28). In practice a number of pillars are used in the cooler, the semiconductor elements connected electrically in series but thermally in parallel.

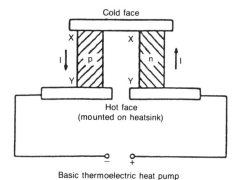

Basic thermoelectric heat pump

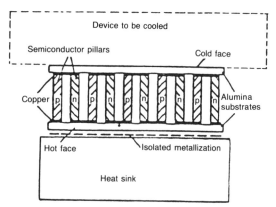

Section through 10-element single-stage thermoelectric heat pump

Fig. 15.28 Construction and application of single and multi-element heat pump. Courtesy of Philips Infrared Defence Components

The heat absorbed by the cold junction as a result of the Peltier effect is reduced by the heating effect of the current and the heat transmitted by thermal conduction from the hot junction, i.e.

$$Q_n = Q_p - Q_j - Q_c$$

where Q_n = heat extracted from cold junction, Q_p = the Peltier heat = $T_c I(\delta_p + \delta_n)$, Q_j = heat generated by passage of current = $0.5I^2L(\rho_n + \rho_p) = 0.5I^2R$, Q_c = conducted heat = $(A/L)(k_n + k_p)\Delta T$ and L = length of n and p pillars, A = cross-sectional areas, ρ_n, ρ_p = resistivities of semiconductor materials, R = total resistance, k_p, k_n = thermal conductivities of n and p materials where typically $k_n \equiv k_p$, ΔT = temperature differential between hot and cold junctions, T_c = cold junction temperature, I = current and δ = Seebeck coefficient.

When a temperature differential is established between the ends of a semiconductor material a voltage is generated, the Seebeck voltage. This voltage is proportional to the temperature differential; the constant of proportionality may be derived for THPs assuming that the properties of n and p materials are essentially equivalent, i.e.

$$I(\text{max}) = \frac{2\delta T_c}{R}$$

$$Q_n(\text{max}) = \frac{4\delta^2 T_c^2}{2R} - k\Delta T$$

= maximum net heat pumping capacity
= amount of heat required to suppress the temperature differential to zero

and

$$\Delta T(\text{max}) = \frac{4\delta^2 T_c^2}{2Rk}$$
$$= 2ZT_c^2$$

where

Z = figure of merit for a semiconductor material
$$= \frac{(\text{Seebeck coefficient})^2}{\text{electrical resistance} \times \text{thermal conductivity}}$$

and

COP = coefficient of performance
$$= \frac{Q_n(\text{max})}{VI} = \frac{\text{net heat absorbed at the cold end}}{\text{applied power input (W)}}$$

The coefficient of performance is used to define cooling efficiency, whereas in the selection of thermoelectric materials, the main objective is to maximize the figure of merit.

There are three usual objectives in the application of thermoelectric modules:

1. *Maximum heat pumping.* Operation at the voltage and current that pumps the maximum amount of heat over the specified temperature differential. As the load increases, the temperature differential that can be achieved decreases. Figure 15.29(a) gives cold face temperature as a function of heat pumping rate at maximum current.
2. *Maximum COP.* The current for operating at the maximum COP is generally less than the current for maximum heat pumping (Fig. 15.29(b)). As the current increases, the cold face temperature drops, reaching a minimum at the peak of the curve.
3. *Maximum efficiency.* For heat pumps to operate at maximum efficiency, the waste heat at the hot face should be removed as quickly as possible. Thermal design of the system and the heat sinking of the hot face should ensure that the temperature of the hot face is maintained as close as possible to ambient.

The practical load for a THP ranges from a few milliwatts to tens of watts, the application limitation being related to the competitiveness of other cooling techniques. To achieve very low temperatures down to $-100\,°C$, multi-stage pumps must be used (Fig. 15.30). These are single-stage cells mounted vertically in pyramid configuration. Because the lower stages (closest to the hot face) need to pump heat dissipated by the upper stages as well as heat from the load, more semiconductor elements are required on the lower stages.

15.5 Thermal management of electronics

Heat flows will be caused by electrical losses. The difference between the power taken from the power supply and the output power of the equipment is taken as the heat loss.

Most electronics are heat sensitive. Excessive heat or cold can impair equipment reliability. Ideally electronics should be kept between -5 and $+65\,°C$ for proper functioning. In sparsely populated PCBs, natural convection may be the simple approach, but as board densities increase and package sizes decrease, so the requirement for planned thermal management becomes unavoidable. Higher power densities of components reflect the demand for faster, more powerful yet smaller PCBs and at the same time require better heat dissipation. However, excessive cooling capability or cooling devices can escalate manufacturing costs, hence heat transfer and thermal management need to be closely planned as the design progresses. In a typical PCB there will be any

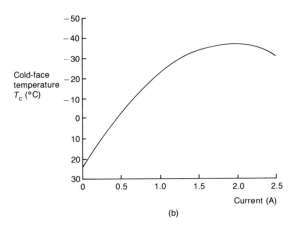

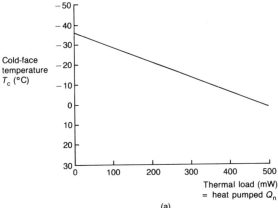

Fig. 15.29 Characteristics of a single-stage cooler. Courtesy of Philips Infrared Defence Components

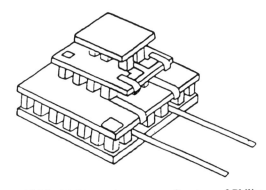

Fig. 15.30 Multistage heat pump. Courtesy of Philips Infrared Defence Components

number of components, some of which will generate heat, some will not and some will be more temperature sensitive than others. Consequently 'hot spot' locations and thermal paths need to be monitored and heat-dissipating devices developed based on a proper mathematical understanding.

Surface-mounted components will result in the average power dissipation density on PCBs and other substrates increasing considerably. This may well change thermal management strategy towards board-level cooling rather than device-level cooling in the future.

15.5.1 Thermal planes

A passive technique often used with PCBs is to remove heat by conduction via a thermal overlay (Fig. 15.31). This can take the form of heat sink strips running beneath uniformly distributed components such as ICs (Fig. 15.31(a)), a conductive plate of aluminium or copper bonded to the PCB (Fig. 15.31(b)) or an overlay placed on top of the PCB and the heat-dissipating electronics placed in contact with it, sitting on the thermal plane (Figs 15.31(c) and 15.32). However, with PCB thermal planes, heat from hot components can raise the temperature of the more sensitive cooler components; so to avoid this situation, path routes need to be carefully constructed to a cold wall. In practice the conducting metal network or plane will conduct heat from a device to the edges of the PCB where a clamping arrangement or guide will facilitate the passage of heat down the temperature gradient into the material of the enclosure or frame.

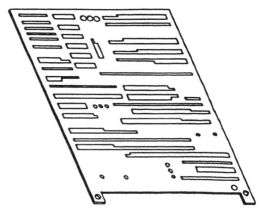

Fig. 15.32 Thermal plane. Courtesy of ENCO Industries Ltd

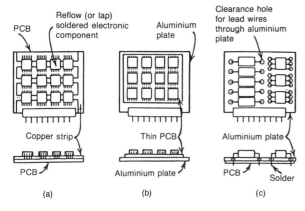

Fig. 15.31 Various methods for conductively cooling PCB mounted components: (a) copper strips under components; (b) aluminium plate bonded to the PCB; (c) clearance holes in aluminium plate with PCB on back. From Steinberg G., *Cooling Techniques for Electronic Equipment*, copyright J. Wiley 1980. Reprinted by permission of John Wiley and Sons

The thermal overlay technique can yield a reproducible thermal profile on the PCB. It may not necessarily be a separate component but could be built into the PCB, as a laminated core or as one of the layers of a multi-layer board, but in this case allowing for some heat conduction through the PCB material itself. If a bonded plate is used (e.g. Fig. 15.31(c)), components would be lap soldered, e.g. surface mount device (SMD) level cooling. Aluminium or copper heat sink planes and cores are capable of conducting away large amounts of heat if the proper construction techniques are used. The body of a component with a high heat dissipation, $>0.6\,\mathrm{W\,cm^{-1}}$, should be fastened to the PCB so that there is a low thermal resistance path from the body of the component to the heat sink plate.

Also used in conjunction with thermal planes are the thermal bridge, thermal spring and thermal module. The thermal bridge (Fig. 15.33) can be used either to protect a sensitive component or to remove heat from a component with a high power consumption directly to a cold wall. The thermal module performs the same function as a thermal bridge, but extended across the whole of the PCB (Fig. 15.34). The module is configured to mount into card guides, the whole assembly accessing the circuit guide and hence the cold wall. Thermal contact, module to component, is via beryllium copper thermal springs (Fig. 15.34). Sensitive component protection is achieved by removal of the module extrusion locally.

15.5.2 Heat sinks

The main function of a heat sink is to protect an electrical device from the heat it produces as a by-product of normal operation. Some electrical components dissipate more heat

THERMAL MANAGEMENT

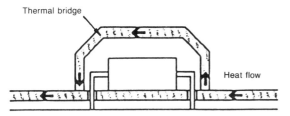

Fig. 15.33 The thermal bridge. Courtesy of ENCO Industries Ltd

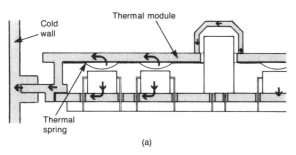

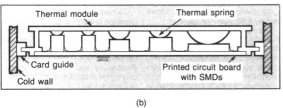

Fig. 15.34 The thermal module (a) for conventionally populated PCBs and (b) for surface mount PCBs. Courtesy of ENCO Industries Ltd

than they can dissipate unassisted, and if not removed will cause the device to exceed its safe operating temperature. The ultimate heat sink of course is the air mass surrounding the electronics package, but initially, heat is usually dissipated at board level by the use of mounted heat sinks (e.g. Fig. 15.35). A finned aluminium heat sink is no more than a pathway through which heat travels by conduction from the source to the air mass. For the most part, cooling is achieved by convection, often forced to some degree.

The junction in transistors and semiconductors where current control takes place is where power dissipation is greatest, hence the junction temperature becomes a critical control requirement. It is an accepted fact that for every 10 °C reduction of device operating temperature, operating life doubles. Most air-cooled heat sinks are manufactured either as stampings, dissipating up to 10 W or as extrusions dissipating up to 100 W, depending on size and type.

Fig. 15.35 Example of a PCB mounted heat sink. Courtesy of IMI Marston Ltd, Wolverhampton

(i) Heat sink theory

Junction temperature is a function of the sum of the thermal resistances between the junction and ambient, the amount of heat being dissipated and the ambient air temperature, i.e.

$$q = \frac{T_j - T_a}{R_{tot}}$$

where T_j = junction temperature of component, T_a = ambient temperature and

$$R_{tot} = R_{jc} + R_{cs} + R_{sa}$$

where R_{jc} = thermal resistance junction–case, R_{cs} = thermal resistance, case to mounting surface and R_{sa} = thermal resistance, heat sink to ambient.

The thermal parameters are represented pictorially in Fig. 15.36. Heat sink selection is normally represented in terms of R_{sa} with values for case dissipation (q), junction–case thermal resistance (R_{jc}) and maximum junction operating temperature (T_j) obtainable from manufacturers' data sheets, i.e.

$$R_{sa} = \frac{T_j - T_a}{q} - (R_{jc} + R_{cs})$$

and

$$R_{sa} = \frac{L_1}{kA_1} + \frac{1}{h_c A_c + h_r A_p}$$

where L_1/kA_1 is the conduction component, $1/h_c A_c$ is the convection component and $1/h_r A_p$ the radiation component (Fig. 15.37), and A_1 = average area available for heat flow conduction across the heat sink, L_1 = average path length across the heat sink section,

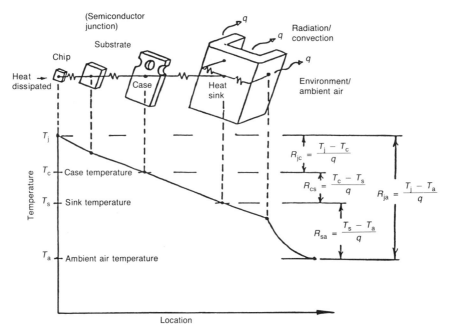

Fig. 15.36 Schematic heat flow representation of semiconductor component and its heat sink

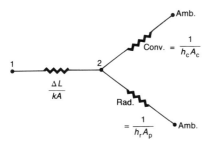

Fig. 15.37 Thermal resistance components of R_{sa}

h_c = convective heat transfer coefficient, A_c = exposed heat sink surface area, h_r = radiation heat transfer coefficient and A_p = projected surface area for radiation heat transfer. For natural convection conditions, the above equation can be simplified by assuming that the conduction loss is small and can be ignored, i.e. the temperature gradient through the heat sink is negligible, hence

$$R_{sa} = \frac{1}{h_c A_c + h_r A_p} \text{ degrees Centigrade per watt}$$

A general guide characterizing heat sink size and performance is given in Fig. 15.38. However, for individual determination of the transfer coefficients, the

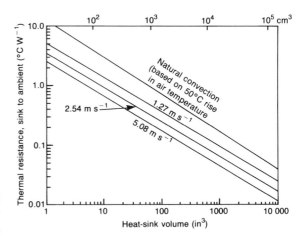

Fig. 15.38 Relationship between thermal resistance and heat-sink volume. From Guger E.C., *Handbook of Applied Thermal Design*, McGraw-Hill Inc., 1989. Reproduced with permission of McGraw-Hill Inc.

convective heat transfer coefficient for an isolated vertical surface in a natural convective laminar flow is given by

$$h_c = 1.41 \left(\frac{\Delta T}{L_v} \right)^{0.25} \text{ watts per square metre per degree Kelvin}$$

where ΔT = assumed surface–ambient temperature difference and L_v = vertical height of surface.

A quick determination of h_r is graphically presented in Fig. 15.39 based on $\epsilon = 0.9$. For ϵ values other than 0.9, the coefficient may be modified:

$$h_r^* = \frac{\epsilon_s h_r}{0.9}$$

where h_r = surface radiation coefficient from Fig. 15.39, ϵ_s = emissivity in question and h_r^* = corrected radiation coefficient.

In natural convection conditions, convective heat transfer is high in comparison with radiation, hence it is usual to ignore radiative effects. But with some heat sinks, especially the larger asymmetrical types, the internal temperature gradient needs to be considered, hence

$$R_{sa} = \frac{L_1}{kA_1} + \frac{1}{h_c A_c}$$

In this case L_1 is a mean conduction path length, and for forced convection

$$h_c = 3.86 \sqrt{\frac{V}{L_f}}$$

where V = linear air velocity (m s^{-1}) and L_f = length of fin parallel to the airflow (m).

The result of the R_{sa} calculation taken for maximum values and worst case operation provides a thermal resistance value that must be equalled or bettered by the heat sink selected for the duty, i.e. the heat sink must have a published thermal resistance rating that does not exceed the calculated R_{sa}.

(ii) Surface finish

Heat sinks are available with a range of surface treatments, from a black anodized finish approximating to a black body to untreated aluminium. A typical comparison for an extruded heat sink in natural convective conditions is given in Fig. 15.40. In forced convection however, thermal performance is essentially the same for all types of finish.

(iii) Thermal interface material

The interface thermal resistance R_{cs} between the device and its heat sink is controllable and can be established with respect to the electronics design being considered. It is usual practice to employ a thermally conductive interface material, most of which also provide electrical isolation. If isolation is not required, then the optimum resistance may be obtained by using a thermal compound (thermal grease) on its own. A comparison between a number of materials, both with and without thermal grease, is given in Table 15.6 for a TO3 style case. The results apply to interface pressures of 0.5 N mm^{-2} and above (screw torque > 0.11 N m). Above these levels, the interface thermal resistance is not greatly affected by surface pressure or surface roughness.

Although thermal grease has a major effect as a thermal interface material (e.g. Table 15.6), it is unreliable as a production tool and is liable to more damage than a

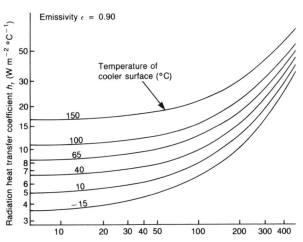

Fig. 15.39 Relationship between radiative coefficient and surface temperature difference. From Guger E.C., *Handbook of Applied Thermal Design*, McGraw-Hill Inc., 1989. Reproduced with permission of McGraw-Hill Inc.

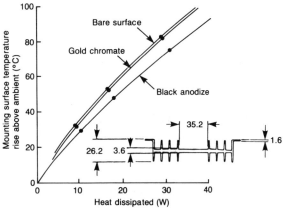

Fig. 15.40 Natural convective performance of extruded heat sink as a function of surface finish. Courtesy Thermalloy International Ltd, Corby

Table 15.6 *Thermal interface materials*

	R_{cs} °C W^{-1}		Max. service temperature (°C)	Comments
	With thermal grease	No thermal grease		
Polyimide film (0.05 mm)	0.55	1.45	400+	Tough and flexible; low-cost mica alternative
Mica (0.05–0.0075 mm)	0.35	1.2	550+	Good electrical properties
Silicone rubber (0.20 mm)	—	1.10	150	Good thermal performance without grease; glass fibre impregnated to improve strength. Low k but high deformity gives large contact area
Anodized aluminium (0.5 mm)	0.28	1.15	500	Abrasion resistant; durable. High thermal conductance
Al$_2$O$_3$ (1.5 mm)	0.32	1.0	1700+	Stable oxide; high strength
Beryllium oxide (1.5 mm)	0.15	0.56	2150+	Optimum thermal performance choice; use with grease to prevent cracking. Expensive. Rigid and brittle but equivalent thermal conductivity to aluminium
Bare joint	0.14	0.5	—	

Table 15.7 *Summary of thermal pad materials*

Binder	Silicone	Silicone	Silicone	Silicone	Polyurethane	Polyurethane	Silicone
Reinforcement	Glassfibre	Glassfibre	Polyimide film	Polyimide film	Polyimide film	Polyester film	Polyimide film
Filler	BN	Al$_2$O$_3$	BN	MgO	BN	MgO	BN
Thermal impedance (°C in^2 W^{-1})	0.15/0.18	0.30/0.34	0.25	0.35	0.25	0.40	0.40
Thickness (mm)	0.25	0.25	0.13	0.13	0.13	0.13	0.18
Usage temp. (°C)	−60/200	−60/200	−60/200	−60/200	−60/150	−60/150	−60/200

thermal pad should the device require replacement. For this reason there are a number of one-component interface materials available for greaseless application based on reinforced silicone or urethane elastomers with ceramic fillers. The ceramic fillers provide stiffness to avoid creepage of the elastomer under pressure. A summary of the thermal interface performance of these materials is given in Table 15.7.

(iv) The heat sink selection process

The selection process can be defined as a number of stages:

1. Once the semiconductor device is known and the requirement for heat sinking established, the following data can be established from semiconductor manufacturers' catalogues: (a) the heat dissipation of the case, q; (b) the junction–case thermal resistance, R_{jc}; (c) the maximum operating junction temperature, T_j.
2. Decide whether natural or forced convection conditions are to be applied to the calculation.
3. The design engineer then needs to establish R_{sa} for the following set of conditions; (a) likely ambient temperature maximum T_a; (b) a value for R_{cs} depending on whether an insulator is used, with or without thermal grease and the torque figure used for the device fasteners — such data are usually available from catalogues (see also Tables 15.6 and 15.7).

4. An alternative route is to determine R_{sa} based upon the type of convection involved and the heat transfer coefficients deduced.
5. Once R_{sa} is known, the heat sink volume can be roughly established from Fig. 15.41. Heat sink manufacturers normally quote a R_{sa} value against individual heat sinks for a fixed temperature rise. An initial determination of the heat sink type can thus be made and compared against the space allocation available. At this point the length of an extruded heat sink will be nominal length specified by the manufacturer.
6. In the case of a stamped heat sink profile, manufacturers either make available graphs of R_{sa} versus q or T versus q for natural convection conditions, and air velocity versus R_{sa} for forced convective conditions. In any case the value of R_{sa} derived for the original value of q should not exceed the value calculated in (3) above.
7. In the case of extruded heat sinks, the calculated value of R_{sa} from (3) needs to be expressed in terms of the length of heat sink required. Manufacturers will either make available data of R_{sa} versus length or quote R_{sa} values based on a standard 75 mm length. With the latter, a ratio of

$$\frac{R_{sa}\text{ (calculated)}}{R_{sa}\text{ (75 mm)}}$$

can be applied to the normalized curve in Fig. 15.42 to determine the length of extrusion required. However, there is no reason why the evaluation in (6) above should not also apply to a fixed 75 mm extrusion as a standard heat sink. In this case there are convective advantages in having the fins vertical (e.g. Fig. 15.43, also Fig. 15.40).
8. If the results do not compare or are marginal, clearly further iterations need to be performed. Finally, the heat sink should be tested out in the application.

(v) Heat sink selection summary

The following points should prove useful:

1. Select a standard stamping if possible since it is cheaper.
2. If more heat is to be removed, an extrusion may be required.
3. Under conditions where a thermal gradient is likely in the heat sink, die-cast heat sinks have 20 per cent less conductivity than extruded heat sinks.
4. Select a less expensive finish where possible to save cost.
5. Black anodizing is not necessary in forced convection; a chromate conversion coat provides the same performance at 40 per cent the cost of anodizing.
6. For natural convective situations, a chromate coating has a 10–20 per cent performance reduction over black anodizing.
7. To improve thermal transfer, use a thermal joint compound, a thermal interface material, employ flat surfaces and use high torques for medium- and high-power devices.

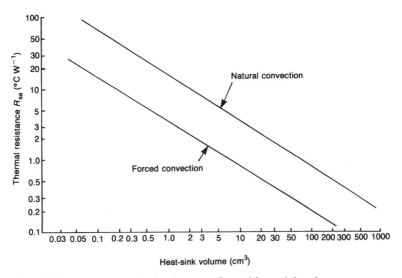

Fig. 15.41 General correlation between R_{sa} and heat sink volume

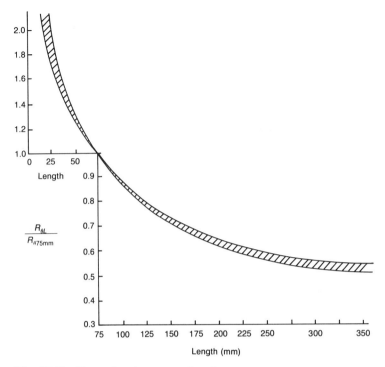

Fig. 15.42 Thermal resistance vs. length

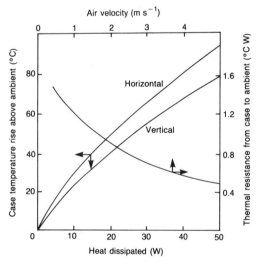

Fig. 15.43 Effect of fin orientation on heat-sink performance (same heat-sink as Fig. 15.39). Courtesy Thermalloy International Ltd, Corby

8. Expend effort on lowering junction temperatures to increase system reliability.
9. It is expensive to integrate cooling problems with packaging problems.
10. Stacking of PCBs should be vertical and heat sink fins mounted in the vertical plane.
11. With forced air convection, let the flow work in conjunction with the natural convection path.
12. Forced convection requires an airflow calculation based on anticipated air temperature rise to determine airflow. Choice of fan/blower is then based on the known system resistances.

15.5.3 High-power stud-mounted components

High-power components require special care for mounting, since temperatures can rise very rapidly and failures can occur if the mounting interface conditions change. In high-power components, e.g. thyristors, diodes and rectifiers, junction temperatures need to be controlled. For high reliability this is typically 125 °C. The mounting position of a high-power device within a chassis can often control the hot spot temperature, e.g. when mounted in the centre of a PCB it will be hotter than when mounted at the edge of the same PCB.

Stud-mounted components should be assembled with care, typically as in Fig. 15.44. The insulating washer should not cold flow under pressure. The nut should be

THERMAL MANAGEMENT

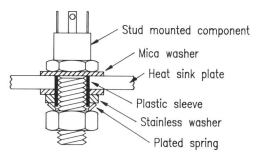

Fig. 15.44 Cross-section through a stud-mounted component. From Steinberg G., *Cooling Techniques for Electronic Equipment*, copyright J. Wiley 1980. Reprinted by permission of John Wiley and Sons

torqued to a consistent value and the spring washer maintains interface pressure. Thermal grease can increase conduction values by up to 30 per cent.

15.5.4 General aspects of PCB cooling

Often PCBs are mounted vertically in chassis open at the top and bottom to allow free circulation of cooling air. To avoid pinching or choking the natural convective flow, space boards are positioned a minimum of 2 cm apart. In such circumstances radiation cooling effects can be ignored as the boards 'see' each other. Apart from the power dissipation on the component side, conduction via leads to solder pads and trackwork on the rear of the PCB provide heat spreaders capable of providing additional surface area (30 per cent say) for convective heat transfer.

For initial design evaluation, PCBs can be modelled in an airflow enclosure using light bulbs, resistors, etc. as heat sources to approximate internal component dissipation. By resiting components experimentally, thermal improvements may be gained before commitment to final PCB layouts.

Plug-in PCBs are often used with edge guides. Since these guides are usually fastened to the walls of a chassis or enclosure, they can therefore provide a heat conduction path. Many different types of edge guides are used in industry, and various types and their thermal resistances are shown in Fig. 15.45.

Some general considerations of PCB cooling can be summarized as follows:

1. Those PCBs with air gaps below components should not be used for vacuum applications since there will be no heat transfer across the gap.
2. In cases where a PCB is placed so close to a wall that natural convection is inhibited, heat transfer may still take place from 'conduction' across the air gap.
3. With lap-soldered joints, heat will not flow to the rear face of the board so easily.
4. Thermal paths should be carefully planned to carry the maximum quantity of heat over the shortest possible distance.
5. Locate high heat-generating components near the top of the board and if possible mount directly in contact with the external structure.
6. If vertical stacking, stagger components to provide clear convective paths.
7. Mount temperature-sensitive components near the bottom of the board.
8. If temperature-sensitive components must be located near hot components, use a thin sheet of polished aluminium as a reflecting shield.
9. Rather than increasing airflow, attempt to resite components or locally deflect the flow.
10. If transferring heat by conduction, a direct metallic path with thermal joints must be provided.
11. Conductive members should be short with a large cross-section and high thermal conductivity.
12. To decrease thermal resistance, limit the number of thermal interfaces and increase the contact surface area.

15.5.5 Encapsulation

Electronic components with high densities are often encapsulated in various types of epoxy or silicone

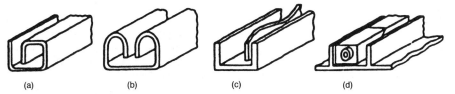

Fig. 15.45 Board edge guides with typical thermal resistivity: (a) G guide 4.7 °C cm W^{-1}; (b) B guide 3.4 °C cm W^{-1}; (c) U guide 2.4 °C cm W^{-1}; (d) wedge clamp 0.8 °C cm W^{-1}

compounds. Discrete electronic components are often potted to improve the vibration and shock resistance of the PCB. However, with discrete components, heat removal can be a problem, even with relatively low power dissipation unless proper attention is paid to the packaging of these components. In this case lead wire lengths within the module should be minimized, and internal heat sinks or heat sink strips used to transfer heat to the surface of the potted module where it can then be effectively removed. Since many potting compounds have a high coefficient of thermal expansion, a resilient conformal coating (e.g. silicone, acrylic, polyurethane) should be applied to the PCB before it is potted in order to avoid cracked components.

15.5.6 Cooling aspects of enclosures

In this section the optimum cooling of electronic equipment in enclosures and the design parameters that influence the situation are considered.

(i) Cooling with total enclosure

In certain situations the electronics enclosure must remain sealed. This has the advantage of course that in the inside of the equipment is protected against the ingress of dirt, but the disadvantage that heat is largely transferred by convection via the case. The heat flow from the inside of the equipment through the case to the outside can be given by

$$q = R_\theta A(T_a - T_i)$$

where q = heat flow (W), T_a = external temperature, T_i = internal temperature and R_θ = overall heat transfer coefficient (W m^{-2} K^{-1}).

Now the performance of an enclosure is determined by the unit heat dissipation, also referred to as heat flow density, heat flux density or heat flow rate intensity, W, where

W = an approximate measure of the amount of heat that can be effectively dissipated from the surface of an enclosure

$$= \frac{\text{estimated power dissipation of unit}}{\text{total area of dissipating surfaces}}$$

hence

$$R_\theta = \frac{W}{\Delta T} = \frac{1}{1/h_c + x/k + 1/h_r}$$

but since $x/k \ll 1/h$ it is neglected and cooling then becomes independent of case material. The heat flow density achieved for varying cases of heat transfer as a function of temperature difference is given in Fig. 15.46. Here one needs to distinguish between vertical and horizontal walls. The heat flow per unit of surface area is modest, but surface areas can be expanded by corrugation or, if necessary, by a finned surface. With forced flow both on the inside and the outside of the case, this then produces a heat exchanger. A further parameter used is the heat concentration ϕ, an approximate measure of the amount of heat generated by the components packaged inside the enclosure, i.e.

$$\phi = \frac{\text{estimated power dissipation}}{\text{internal volume of packaging}} \text{ W m}^{-3}$$

The maximum capability of internal cooling techniques in this context can be summarized as (W m^{-3}): natural cooling 0.15, potted components 0.25, metallic conduction 5.0, forced air 7.0.

(ii) Cooling with open case

If only a small proportion of the case area is opened by cooling slots then the heat flow density will greatly exceed that for a closed case (Fig. 15.46). Consequently this simple and effective cooling method is well known and widely used. Cooling slots or holes should be provided on the top and bottom of the enclosure to guarantee flow through the equipment. The area of intake and exhaust holes required to create a proper convective draught can be estimated by

$$A \approx \frac{0.08q}{\rho_i (T_i - T_a)^{1.5}}$$

where ρ_i = density of internal air (kg m^{-3}) and where T_i in this case is the maximum allowable internal temperature.

The equation does not apply where $T_i - T_a > 35$ °C. The total area of the exhaust holes should be greater than the intake area to allow for air expansion with temperature. However, the cooling requirement of most commercial electronic equipment is such as to need a single-stage forced circulation system. With a 10 per cent open area, the increase in heat flow density achieved by forced flow is significant (Fig. 15.47), providing heat transfer rates typically 10× those available from natural convection and radiation. The penalty for increased cooling is higher cost, power, noise and complexity, but on the plus side are the reduced size of the system and a higher component density potential. Usually PCBs are arranged in a 'toast rack' fashion and air forced between the boards. A blowing fan system will raise the internal air pressure, help to keep out dirt and dust and will

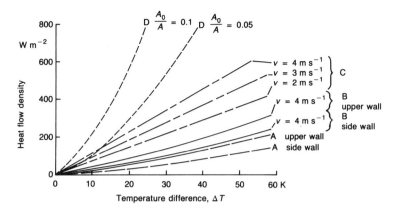

Case A: Natural convection/radiation head transfer only
Case B: With fan placed inside case so that air is circulated with a better heat transfer to the case and more even temperature distribution achieved internally by forced convection. Increasing flow velocity, $v > 4\ \text{m s}^{-1}$ has little effect
Case C: Heat transfer improved by forced flow on the outside of the case as well as the inside, hence greater flow densities
[Case D: With cooling slots, 10% and 5% open area]

Fig. 15.46 Characteristics of different types of enclosed cooling. Courtesy of S. Harmsen, Papst Motoren GmbH and Co. KG

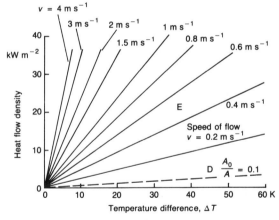

Fig. 15.47 Comparison of fan cooling (E) with free convective cooling (D); 10% open area. Courtesy of S. Harmsen, Papst Motoren GmbH and Co. KG

produce a useful degree of turbulence in the box to improve internal heat transfer characteristics. However, air passing over the fan motor may itself be heated on entry to the enclosure, and air filters are required to prevent contamination of the electronics. In an exhaust system, air entering the box is cooler.

Reference

1. Holman J P *Heat transfer*. McGraw-Hill 1986

CHAPTER 16

NOISE REDUCTION

16.1 Introduction

For practical purposes the term 'sound' is taken to mean vibratory motion perceptible through the hearing organ. Vibratory energy (or sound) travels in waves. It is the frequency of the waves that determines the speed at which the ear-drum and other parts of the hearing organ vibrate while the pressure of the sound affects the magnitude of the oscillation. The brain registers these movements as pitch and loudness respectively.

Loud and other unwanted acoustic stimuli are called noise. The level of annoyance we feel depends not only on the quality of the sound but also on our attitude towards it. Noise is a form of air pollution, and like other forms of pollution it not only affects the quality of life but where the noise source is sufficiently concentrated and the exposure to it sufficiently long, can cause physiological effects leading to deafness.

The recognition of noise as a hazard has been identified within EC legislation[1-3] and there is a responsibility on manufacturers of certain machines to define the sound power level of the product and thereby provide the user with the information he/she needs to assess his/her own local sound protection measures. In addition there is increased attention focusing on acoustic noise as an unwanted by-product of office equipment, domestic appliances, commercial machinery, etc. Consequently there is a continuing trend towards noise control in product design. In Europe, noise labelling is complementary to legislation, but there is a growing recognition that interplay of such a complementary strategy is essential to successful noise management programmes.[4]

16.2 Fundamentals

16.2.1 Nature of sound

If the pressure disturbance as a function of time is sinusoidal, this is a single frequency — a pure tone. However, real sounds are rarely pure sinusoidal tones; they tend to be either a combination of discrete sinusoidal frequencies or a disturbance varying erratically with time but with a frequency band spectrum. Some typical sounds are shown in Fig. 16.1. Complex periodic steady sounds have a pattern which repeats itself at a constant rate. Speech displays a waveform whose character varies with time, is of finite length but possesses a clear form. Industrial noise typically displays complex patterns, a very disordered and random waveform of pressure versus time. Such a wave has no periodic component, but it may be represented by a collection of waves of all frequencies and is known as broadband noise.

(i) Frequency
The number of pressure variations per second is referred to as the frequency of the sound and is measured in cycles per second (Hz). The frequency of sound produces its distinctive tone. The normal range of hearing for a healthy young person extends from approximately 20 Hz up to 20 kHz (the range of notes on a piano is from 27.5 Hz to 4.2 kHz). The wavelength of sound (λ) is related to frequency as follows:

$$\lambda = \frac{c}{f} \text{ metres}$$

where c = speed of sound (m s^{-1}) = $\sqrt{(E/\rho)}$, f = frequency (s^{-1}), E = Young's modulus for solids and bulk modulus for fluids and ρ = density.

In air the speed of sound is a function of absolute temperature K, i.e. $c = 20.05\sqrt{K}$ m s^{-1} = 344 m s^{-1} at 21 °C.

It is worth noting that at 20 Hz, one wavelength is just over 17 m while at 20 kHz it is only 1.7 cm. In materials, sound travels faster when elasticity is high and density is low. Material acoustical properties are summarized in Table 16.1.

(ii) Amplitude
Another quantity used to describe a sound is the size or

NOISE REDUCTION

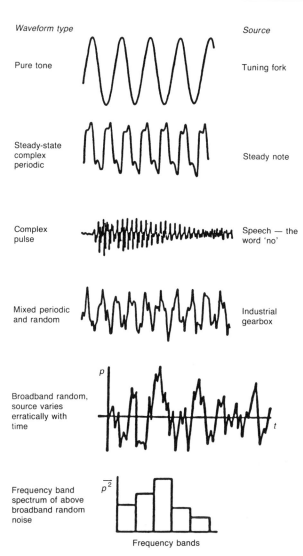

Fig. 16.1 Examples of typical sounds

Table 16.1 *Properties of materials*

Material	Young's modulus $E(10^9 \text{ Nm}^{-2})$	Density $\rho(\text{kg m}^{-3})$	Speed of sound $c(\text{m s}^{-1})$
Air (20 °C)	—	1.21	344
Fresh water	—	998	1481
Aluminium	71.6	2700	5150
Steel	210	7800	5200
Lead	16.5	11300	1210
Concrete	19.6	1700	3400
Glass	67.6	2500	5200
Teak	17.0	900	4350
Pine	8.0	550	3800
PVC	2.4	1400	1310
Polyurethane	1.6	900	1330
Polyethylene	0.2	930	460
Nylon 6:6	2.0	1140	1320
Brass	105	8500	3200

amplitude of the pressure fluctuations. The minimum sound pressure amplitude that a healthy ear can detect is 20 μPa, some 5×10^9 less than normal atmospheric pressure, whereas the greatest sound pressure before pain is 60 Pa. Since the acoustic intensity is proportional to (sound pressure)^2, a linear scale would require 10^{13} unit divisions to cover the range of human experience. However, as a best match to the considerable dynamic range of the ear, a logarithmic scale, rather than linear, equates to subjective response. The logarithmic scale provides a convenient way of comparing the sound pressure of one sound with another. to avoid a scale which is too compressed, a factor of 10 is introduced, giving rise to the decibel (dB).

The decibel is not an absolute unit of measurement, therefore it is a ratio between a measured quantity and an agreed reference level. The reference level is the hearing theshold 20 μPa. This is defined as 0 dB, hence the decibel scale compresses a range of 10^6 into a range of 120 dB. Since the ear reacts to a logarithmic change in level, 1 dB is the same relative change anywhere on the scale.

16.2.2 Definitions

(i) Sound pressure level
This is the disturbance of the static atmospheric pressure caused by the presence of sound. It is measured with a sound level meter and at any point depends upon the acoustic power output of the noise source, the distance from the source and the environment in which the noise source and receiver are placed. It is common practice to refer to a sound pressure level (SPL or *Lp*) in decibels as the square of the sound pressure with reference to 20 μPa. The frequency band or weighting must be specified (section 16.3):

$$Lp \text{ (or SPL)} = 10 \log_{10} \left(\frac{P}{P_{\text{ref}}}\right)^2$$

$$= 20 \log_{10} \left(\frac{P}{P_{\text{ref}}}\right) \text{ dB}$$

where P = r.m.s. value of sound pressure (Pa), $P_{\text{ref}} = 20 \times 10^{-6}$ Pa or

$$Lp = 20 \log P + 94 \ldots \text{ dB}$$

(ii) Sound power level

Sound power is the rate at which acoustic energy is emitted from a sound source, and together with wave intensity, is a true power-related quantity in acoustics. The unit of sound power is the watt and the normal range of interest is 10^{-9}–10^{-2} W. It is common to express sound power as the sound power level (PWL or Lw) in decibels with reference to 10^{-12} W. Again the adoption of a standard or reference value for the denominator allows the decibel scale to be used as an absolute scale of magnitude. The frequency band or weighting must be specified (section 16.3):

$$Lw \text{ (or PWL)} = 10 \log_{10} \frac{\text{(sound power)}}{\text{(reference power)}} \text{ dB}$$

$$= 10 \log_{10}\left(\frac{W}{W_{\text{ref}}}\right)$$

$$= 10 \log W + 120 \text{ dB}$$

where W = acoustic power of source (W) and W_{ref} = reference power (10^{-12} W).

Sound power level cannot be measured directly and has to be deduced from SLP measurements corrected for measurement conditions. It is a useful means for quantifying the acoustic power of a source in a way, which for most practical purposes, is independent of the source environment.

(iii) Sound intensity level

The sound intensity level L_i of the intensity I may be defined as follows:

$$L_i = 10 \log_{10} \frac{\text{(sound intensity)}}{\text{(reference sound intensity)}} \text{ dB}$$

hence

$$L_i = 10 \log_{10} I + 120 \text{ dB}$$

where reference sound intensity = 10^{-12} W m^{-2}.

Sound intensity at a given point in a sound field in a specified direction is defined as the average sound power passing through a unit area perpendicular to the specified direction at that point, and given by

$$I = \frac{P^2}{\rho c}$$

where the quantity ρc is known as the 'characteristic impedance' and

$$L_i = Lp = 26 - 10 \log_{10}(\rho c) \text{ dB}$$

then given ρc for air = 416 kg m^{-2} s^{-1} (Table 16.1)

$$L_i = Lp - 0.2 \text{ (dB)}$$

16.2.3 Frequency analysis

To diagnose machine noise, it is convenient to analyse using a series of frequency bands, normally in the audible frequency range 20 Hz to 20 kHz as a constant percentage bandwidth spectrum, i.e. the bandwidth or frequency range over which the sound energy is integrated is a constant percentage of the centre frequency of the band. There are internationally agreed 'preferred' frequency bands for sound measurement and analysis.

The widest band used for frequency analysis is the octave band, i.e. the upper frequency of the band is exactly twice the lower limit, extending from $1/\sqrt{2}$ to $\sqrt{2}$ times the centre frequency (Table 16.2). Occasionally, more detailed information on the noise structure is required, e.g. with one-third octave narrower bands. These are bands approximately one-third the width of an octave band (Table 16.2), with their centre band frequencies adjusted slightly so that they repeat on a logarithmic scale. For example, the sequence 31.5, 40, 50 and 63 have logarithms 1.5, 1.6, 1.7 and 1.8. The corresponding frequency bands are sometimes referred to as 15th, 16th, 17th, etc. (Table 16.2).

A typical example of the frequency analysis of a complex sound is given in Fig. 16.2. The gain of a real filter outside its bandwidth is not zero, but should reduce rapidly as the frequency moves away from the band edges. Filter bandwidth is defined at a point 3 dB below the peak.

The simplest analysers produce only octave or one-third octave spectra, which may have limited analytical use. A narrow bandwidth analysis, however, allows consideration of the frequency and repetition rates of the sound (Fig. 16.3). This information may then be used to link the sounds to the frequencies of the machine elements. However, narrow-band frequency analysis usually gives little information about noise processes that are essentially broad band in nature, e.g. flow noise.

16.2.4 Decibel arithmetic

Because of their logarithmic nature, noise levels expressed in decibels cannot be added or subtracted directly in the normal arithmetical manner. Instead the values must be converted from decibels to powers or power ratios which can be added or subtracted as illustrated in Fig. 16.4. Alternatively, a simple chart may be used (Fig. 16.4). To add, or subtract, a series of levels, group them in pairs from lowest to highest, combine each pair then combine the results of pairs and so on. Note that fractions are of no significance in the final answer as 1 dB is defined as

Table 16.2 *Preferred centre frequencies and corresponding wavelength*

Band number	Octave band centre frequency	One-third octave band centre frequency	Band limits		Wavelength in air (m)
			Lower	Upper	
14		25	22	28	13.76
15	31.5	31.5	28	35	10.92
16		40	35	44	8.60
17		50	44	57	6.88
18	63	63	57	71	5.46
19		80	71	88	4.30
20		100	88	113	3.45
21	125	125	113	141	2.75
22		160	141	176	2.15
23		200	176	225	1.72
24	250	250	225	283	1.38
25		315	283	353	1.09
26		400	353	440	0.86
27	500	500	440	565	0.69
28		630	565	707	0.55
29		800	707	880	0.43
30	1 000	1 000	880	1 130	0.34
31		1 250	1 130	1 414	0.28
32		1 600	1 414	1 760	0.22
33	2 000	2 000	1 760	2 250	0.17
34		2 500	2 250	2 825	0.14
35		3 150	2 825	3 530	0.11
36	4 000	4 000	3 530	4 400	0.09
37		5 000	4 400	5 650	0.07
38		6 300	5 650	7 070	0.05
39	8 000	8 000	7 070	8 800	0.04
40		10 000	8 800	11 300	0.03
41		12 500	11 300	14 140	0.028
42	10 000	16 000	14 140	17 600	0.022
43		20 000	17 600	22 500	0.017

the smallest discernible difference between two sounds of different levels heard under ideal conditions.

The procedure described only applies to unrelated sounds. To add or subtract related signals such as two sine waves of the same frequency, it is necessary to take into account the phase difference between the two.

16.2.5 Loudness levels

Although the healthy ear can detect sound at all frequencies within the 20 Hz to 20 kHz range, it does not allocate the same importance to each frequency. In other words, the ear is frequency sensitive. By using observers to compare two sounds of different frequency content, a 1 kHz reference tone and a pure tone of some other frequency adjusted by the observer to the judged equal in loudness to the reference tone, a mean result map of 'equal loudness curves' is produced (Fig. 16.5).

The units used to label the equal loudness contours are called phons. The phon is a unit of loudness such that at 1 kHz, the number of phons = sound pressure level. Lines are constructed so that all tones of the same number of phons sound equally loud. For sounds of low amplitude, the ear has a marked frequency sensitivity; a sound at 30 Hz would have to be approximately 50 dB higher than a sound at 1 kHz to be judged equally as loud (Fig. 16.5). This level of insensitivity or 'deafness' does, however, clearly depend on amplitude. This non-linearity of subjective judgement means that sound soes not appear 'twice as loud' by doubling the intensity. In the mid-frequency ranges at sound pressures ≥ 40 dB, the

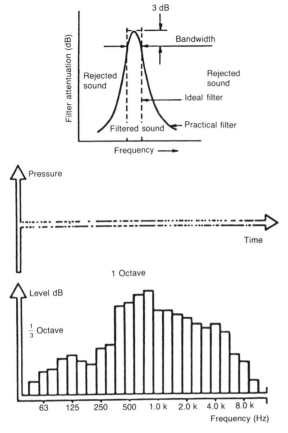

Fig. 16.2 Frequency analysis of a complex sound plus definition of filter bandwidth

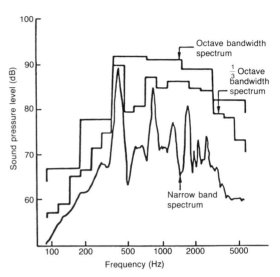

Fig. 16.3 Comparison of octave, one-third octave and narrow-band spectra of the same sound. From Middleton A.H., *Machinery Noise*, Engineering Guide 22. Reproduced by permission of Oxford University Press

subjective effects of sound level changes are given in Table 16.3.

16.2.6 Pitch

Pitch is the subjective response to frequency; low frequencies are identified as 'low-pitched' while high frequencies are identified as 'high-pitched'. When presented with a single 'pure' tone or 'clean' timbre, the perceived pitch is related to its frequency and the loudness mainly to its intensity. Presented with two tones simultaneously, the ear can behave as a frequency analyser and split the complex sounds into its component tones if the frequencies are widely spaced. If closely spaced, the perceived sound is a mixture of the two.

For tones with separation not exceeding 20 Hz, they sound like a single tone fluctuating in loudness, the fluctuations are also known as 'beats'.

16.3 Measurement of sound

16.3.1 Weighting networks

It has been previously shown that the apparent loudness of a sound varies with frequency as well as sound pressure (Fig. 16.5). Sound-measuring instruments are designed to make allowances for this behaviour of the ear by the use of electronic 'weighting networks'. The inverse of the equal loudness curve in Fig. 16.5 gives a fairly good representation of the amount of weighting needed for the various frequency components of sound, but because the curves in Fig. 16.5 flatten out as the SPL increases, more than one weighting is required to cover the whole range of sound amplitude. The various international standards bodies recommend the use of three weighting networks, A, B and C, as well as a linear (unweighted) network for use in sound-level meters. The weighting curves are illustrated in Fig. 16.6, the 'A' weighting for SPLs up to 55 dB, 'B' for levels between 55 and 85 dB and 'C' for levels above 85 dB respectively.

The 'A' weighting was originally designed to approximate the response of the human ear at low sound levels, but the general trend today would appear to be towards its exclusive use. Table 16.4 shows the correction which must be added to a linear reading to obtain the weighted reading for a particular frequency. A noise measurement of machinery given as dB(A) without further

NOISE REDUCTION

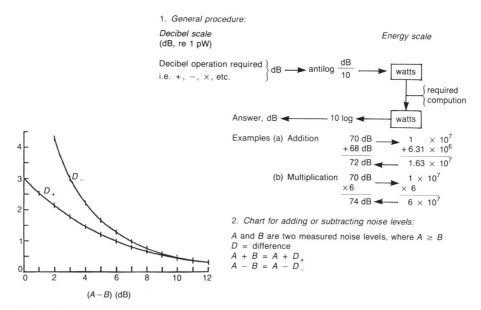

Fig. 16.4 Examples of decibel manipulation: (1) general procedure; (2) chart for adding or subtracting noise levels

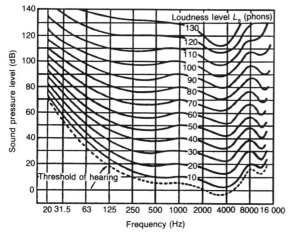

Fig. 16.5 Equal loudness contours

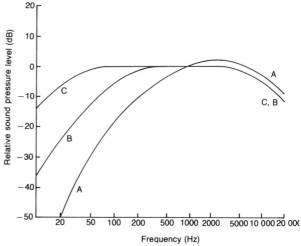

Fig. 16.6 International Standard 'A', 'B' and 'C' weighting curves for sound level meters

Table 16.3 *Subjective effect of changes in sound pressure level*

Change in sound level (dB)	Change in power		Change in apparent loudness
	Decrease	Increase	
3	1/2	2	Just perceptible
5	1/3	3	Clearly noticeable
10	1/10	10	Half or twice as loud
20	1/100	100	Much quieter or louder

qualifications is normally taken to be referencing a continuous steady noise.

However, measuring noise in terms of a single number tells us virtually nothing about the frequency content, however representative the weighting shape may be. As long as the weighted overall level is acceptable this perhaps does not matter, but as soon as the noise is found

479

Table 16.4 A-weighting network corrections (dB)

Frequency (Hz)	10	12.5	16	20	25	31.5	40	50
A-weighting correction	−70.4	−63.4	−56.7	−50.5	−44.7	−39.4	−34.6	−30.2
Frequency (Hz)	63	80	100	125	160	200	250	315
A-weighting correction	−26.2	−22.5	−19.1	−16.1	−13.4	−10.9	−8.6	−6.6
Frequency (Hz)	400	500	630	800	1 000	1 250	1 600	2 000
A-weighting correction	−4.2	−3.2	−1.9	−0.8	0.0	0.6	1.0	1.2
Frequency (Hz)	2 500	3 150	4 000	5 000	6 300	8 000	10 000	12 500
A-weighting correction	1.3	1.2	1.0	0.5	−0.1	−1.1	−2.5	−4.3
Frequency (Hz)	16 000	20 000						
A-weighting correction	−6.6	−9.3						

Table 16.5 Frequency analysis and overall sound pressure level

	Frequency band (Hz)									Result[a]
	31.5	63	125	250	500	1k	2k	4k	8k	
Sound pressure level (dB) re − 20 μ Pa	95	95	90	85	80	81	75	70	65	99
'A' weighting (dB)	−39	−26	−16	−9	−3	0	+1	+1	−1	—
'A' weighting sound pressure level in dB(A)	56	69	74	76	77	81	76	71	64	85

[a]The overall sound pressure level (L_{Pt}) can be calculated from the individual band levels using the relationship:

$$L_{Pt} = 10 \log_{10} \sum_{i=1}^{n} 10^{L_{Pi}/10} \text{ as outlined in Fig. 16.4.}$$

to be over a given limit, noise control techniques will require a frequency distribution spectrum of levels. The manipulation of frequency levels is an extension of the principles outlined in section 16.2.4 and as illustrated by Table 16.5.

16.3.2 Measurement instrumentation

(i) Sound-level meter

This is the basic instrument for experimental noise evaluation. In its simplest form, a sound-level meter may indicate only dB(A) levels, but at its most complex it may include A, B and C weightings, octave band levels, peak levels, instantaneous peaks, impulse values, integrated averages (linear and weighted) and sample times. The various components of a sound-level meter are illustrated by Fig. 16.7.

The microphone signal is first amplified. An input attenuator at the first stage of amplification is provided to ensure that it is not overloaded. Next the various filter circuits are introduced. The all-pass (linear) mode accepts and weights equally all frequencies within the range of the instrument. Sounds encountered in practice are seldom ever steady in level, fluctuations in level are always encountered, sometimes significantly so. To accommodate this phenomenon, the sound-level meter is usually provided with three responses:

Fast. This has a time constant of 125 ms and provides a fast reacting display designed to approximate the response of the ear. Follows rapid fluctuations, but difficult to read the meter accurately.

Slow. This has a time constant of 1 s. While not simulating the response of the ear, it is useful for determining mean levels when the sound fluctuates continuously during the course of measurement. Normally used for continuous noise. Enables the scale of an analogue meter to be read which would otherwise be impossible using the fast time constant.

Impulse. A time constant of 35 ms used for the measurement of impulsive or impact-type noises.

Measurement specifications and standards usually specify the meter response to be used. If the background noise is less than 10 dB below the total level when the source is turned on, a correction must be made to each

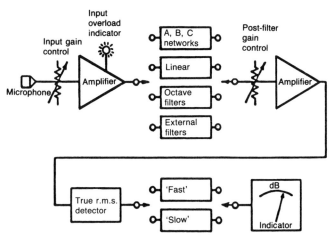

Fig. 16.7 Block diagram of a typical sound level meter. From Jones D et al *Noise and Society*. Copyright © John Wiley 1984. Reprinted by permission of John Wiley & Sons Ltd

reading as per Fig. 16.4. If the difference between the total noise level and the background noise in any octave band is <3 dB, then meaningful acoustic measurements in that band probably cannot be made. The grades of sound-level meter used for engineering measurements are as follows:[5]

Type 1. Precision sound-level meter, for laboratory or field use where the acoustical environment may be closely controlled.
Type 2. General-purpose sound-level meter for general field use and for recording noise level data for subsequent analysis.
Type 3. Survey sound-level meter, intended for preliminary investigations only.

(ii) Microphones

Sound measurement primarily depends on the microphone and its associated pre-amplifier which together generate an electrical signal from a sound pressure. The microphone is a crucial part of any measuring system, a sound-measuring microphone being the most accurate and reliable class of microphone available. The most common type is the condenser microphone and to a lesser extent the piezo-electric microphone. Both have very uniform frequency response and long-term sensitivity stability. However, the piezo-electric microphone tends to be microphonic, i.e. it may respond equally well to vibration and sound, whereas a condenser microphone is relatively insensitive to vibration.

The sensitivity of a microphone is expressed in decibels relative to a reference level:

$$\text{Sensitivity } s \text{ (dB re 1 V Pa}^{-1}) = 20 \log_{10}\left[\frac{\text{sensitivity (mV Pa}^{-1})}{1000 \text{ mV Pa}^{-1}}\right]$$

or, in terms of SPL,

$$s = 20 \log_{10} E_0 - L_p + 94 \text{ dB}$$

where E_0 = output voltage for input L_p.

Typical values of sensitivity would be in the range 10–50 mV Pa^{-1} at 250 Hz. Since the output voltage of a microphone is proportional to the area of the diaphragm, the smaller the microphone of a given type, the smaller will be its sensitivity. Thus a microphone with large diaphragm diameter will produce a large output voltage, but the requirement for high-frequency response is against a large diameter. The problem is that at high frequencies, sound wavelengths are small so for any microphone there is a frequency at which the wavelength of sound and diaphragm diameter are equivalent. This is termed a diffraction effect. In terms of general-purpose characteristics though, 12.5 mm microphones tend to be a popular choice. However, since for practical reasons of sensitivity the problem of diffraction cannot be avoided, it is necessary to consider the orientation of the diaphragm with respect to the sound field being measured and calibrate the microphone response accordingly. Three types of microphone response are recognized:

1. Free-field-response microphones are used for measuring sound mainly coming from one direction, i.e. flat response for normal incidence.
2. Random-response microphones have a flat frequency response in diffuse sound fields where sound arrives

from all angles, i.e. a reverberant sound field characterized by sound intensity incident from all directions.
3. Pressure-response microphones measure the actual sound pressure level at the diaphragm; uses include measuring surface sound pressure levels where the microphone is flush mounted. They can be used as free-field microphones if orientated 90° to the sound direction, but frequency response may then be reduced.

16.3.3 Time varying sound

(i) Equivalent sound level

Noise is not necessarily steady and continuous but tends to fluctuate in level over a period of time. Historically several single number measures for time-varying sound levels have been developed, e.g.

1. Noise exposure forecast (NEF);
2. Composite noise rating (CNR);
3. Community equivalent noise level (CENL, sometimes L_{dn});
4. Level exceeded 90, 50 or 10 per cent of the time (L_{90}, L_{50}, L_{10} ... L_x);
5. Equivalent sound level L_{eq}.

In the case of the L_x value, L_{10} the level exceeded 10 per cent of the time is a measure of the higher-level (more intrusive) components of a noise, whereas L_{90} is a measure of the background, residual or ambient level. With a sound meter set to fast response, $L_{10} \equiv$ average of the maximum deflections and $L_{90} \equiv$ average of the minima. However, to foster uniformity and simplicity of measurement and monitoring, the equivalent sound level concept, L_{eq}, has generally been adopted in preference. Both L_{eq} and L_x are expressed in dB(A) units.

If the A-weighted sound level L varies with time as in Fig. 16.8, the equivalent sound level for the time T is defined as

$$L_{eq} = 10 \log_{10} \left[\frac{1}{T} \int_{t_1}^{t_2} 10^{L/10} \, dt \right]$$

where $T = t_2 - t_1$. The value in brackets gives the mean value of P_A^2/P_{ref}^2 over the period concerned where $P_A =$ the 'A'-weighted r.m.s. sound pressure and $P_{ref} =$ reference sound pressure, 20 μPa.

Since sound intensity is approximately proportional to the square of the sound pressure, the equivalent continuous sound level is the constant sound level which would expose the ear to the same amount of A-weighted sound energy or noise dose as does the actual time-varying

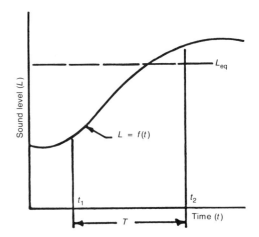

Fig. 16.8 A continuous variation of sound level with time

sound over the same period. Thus L_{eq} is also called the energy equivalent sound level. Here, $L_{eq,T}$, can be measured directly using an integrating sound-level meter, sometimes called a noise average meter, over a preset period of time T. This would normally be between 60 s and 2.8 hours for a single instrumentation read-out. Over longer periods of time, or where integrating meters are not available, the variation of sound level as a function of time is given as a histogram (Fig. 16.9). Discrete levels $L_1, L_2, L_3 \ldots L_n$ are plotted for the given time x, that they occur. Hence in this case

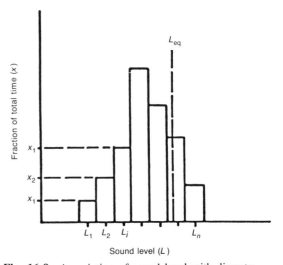

Fig. 16.9 A variation of sound level with discrete intervals

NOISE REDUCTION

$$L_{eq} = 10 \log_{10} \left[\sum_{i=1}^{i=n} x_i \, 10^{L_i/10} \right]$$

In the case of a single constant sound level L_i which occurs for a fraction x_i of the total time and for which the background sound level is negligible then

$$L_{eq} = L_i + 10 \log_{10} x_i$$

If the background level $\neq 0$, then the L_{eq} is a sum of the two partial L_{eq} (section 16.2.4). In its usual context L_{eq} equates the measured fluctuating noise level to a steady noise level considered to be received over an 8-hour period, written $L_{Aeq,8h}$. An integrated noise level measured for a period less than 8 hours is not necessarily the same value.

(ii) Sound exposure level (SEL)

A feature often found on the more elaborate integrating sound-level meters is referred to as SEL. Measured in dB(A) it is defined as that level which, lasting for 1 s, has the same acoustic energy as a given transient noise event lasting for a time T (e.g. Fig. 16.10). It is a measure of acoustic energy and thus can be used to compare unrelated noise events because the time element in its definition is always normalized to 1 s. It is a useful parameter for comparing single events of different levels and duration, e.g. vehicle pass-by, workshop processes. It is defined by

$$L_{EA,T} = 10 \log \frac{1}{T_{ref}} \int_0^T \left[\frac{P_{A(t)}}{P_{ref}} \right]^2 dt$$

where $T_{ref} = 1$ s, and T = time interval.

Sound exposure levels of unrelated transient noise events can be combined on an energy basis and be converted to an L_{eq} acting over a specified time-scale by subtracting, say, $(10 \log_{10} 60)$ dB to give $L_{Aeq,60s}$.

(iii) Impulse sound

Impulse noise is defined as a short-duration sound characterized by a shock front pressure waveform (Fig. 16.11). The major characteristics of impulse noise are (a) the extremely fast rise time and (b) the high peak levels attained for brief moments (<0.5 s). Typical inpulse sounds are metal stamping operations, mechanical impacts, etc. The duration of impulsive noises may vary from microseconds to 50 ms. If the peak levels of 140–170 dB were maintained for a full 1 s, it is certain that ear damage would occur. Most damage risk criteria for impulse noise are based on exposure to repetition.

The difficulty with impulse sound is measuring it. The normal fast response available on sound-level meters cannot follow short-duration impulse noises. Some meters

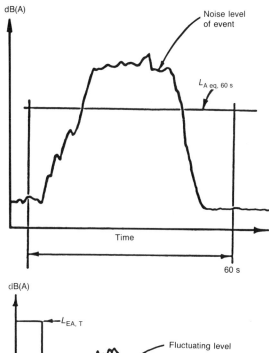
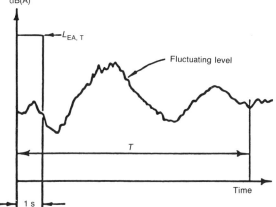

Fig. 16.10 Sound exposure for both a single event and continuous sound

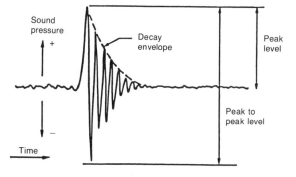

Fig. 16.11 The impulsive noise characteristic

are fitted with impulse or impulse-hold settings which are intended to display a reading corresponding to the subjective impression of an impulsive sound, referred to as the 'I' characteristic. Although this is short enough to enable some detection and display of transient noise, the true maximum sound pressure level is not measured because a 35 ms response time is relatively long (Fig. 16.11). Meters with a peak hold facility can measure true instantaneous sound levels but cannot measure the total sound received during the measurement period, i.e. the peak hold figure is unrelated to the length of time the signal is produced. A fast impulsive setting gives a reading which is the approximate average of the peak hold and slow readings and although it will not record sharp pulses, nor is it a good measure of steady noise, it does record all the fluctuations. The measurement of sound energy. L_{eq}, compensates for all the shortcomings of the above methods.

Figure 16.12 is a schematic representation of measured sound-level formats as determined for a free-field single impulsive noise. The implications of these varying methods of measurement are important. All are correct but their relevance to the situation being evaluated has to be decided.

For continuous noise sources, if the difference between the A-weighted impulse sound pressure level and the A-weighted level, $[Lp(AI) - Lp(A)]$ is >3 dB, then the noise is considered to be impulsive, and the greater the correction the larger the impulse content.

16.3.4 Positions of noise measurement

The noise field around a machine will vary with location of measurement, time, operating conditions, the machine mounting and the acoustic environment in which the measurement is made. Different noise origins, i.e. aerodynamic, electromagnetic or mechanical, will emit noise in different directions depending on the machine shape. Non-steady-state processes may well emit impulsive noise. The noise field is even more complex if there are reflecting surfaces near the machine, reflected sound waves interacting with direct waves at the point of measurement to increase or decrease the sound pressure level depending on whether they are in or out of phase. When accurate machine sound levels are required, the machine should be placed in the centre of an anechoic chamber or suspended in a large open area and multiple sound pressure levels recorded.

(i) The measurement surface

Sound pressure levels around a machine vary irregularly from point to point. If SPL measurements are made at a number of points on an imaginary surface enclosing the machine, it is standard practice to calculate a unique value for the sound in each frequency band emitted from the machine. Operating, location and mounting conditions would be specified. Although test codes exist for different types of equipment, the use of different measurement surfaces may yield different noise estimates for the source.

According to BS 4196,[6] it is first necessary to establish a reference box, the smallest rectangular box or parallelepiped that just encloses the source, i.e. length l_1, width l_2 and height l_3 (Fig. 16.13). For general machinery and business equipment the preferred measurement distance d is 1 m, but in no case less than 0.25 m.[7] The types of measurement surface are as follows:

1. For floor-mounted equipment, the measurement surface is a rectangular parallelepiped (or 'shoebox'), all faces coplanar with the reference box and located at a distance d, typically 1 m (Fig. 16.13), terminating in a reflecting plane (floor). Nine microphone positions are located on this (imaginary) measurement surface. This is the most popular and commonly used method. The measurement area, S, is equal to the sum of the areas of these faces.

2. For floor-mounted equipment and an alternative to (1) is the conformal surface, defined as being everywhere located a distance d from the nearest point on the envelope of the reference box.[6] In this case there are eight key microphone positions and

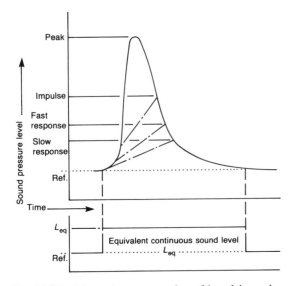

Fig. 16.12 Schematic representation of impulsive noise evaluations

Coordinates of key measurement points

No.	x	y	z
1	a	0	h
2	0	b	h
3	-a	0	h
4	0	-b	h
5	a	b	c
6	-a	b	c
7	-a	-b	c
8	a	-b	c
9	0	0	c

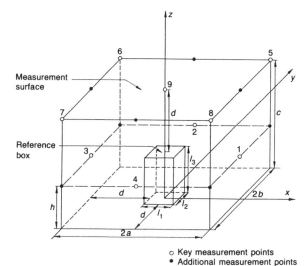

Fig. 16.13 Nine-microphone position measurement surface

the surface area of the conformal measurement surface is defined as

$$S_{con} \equiv 4(ab+bc+ca)\left[\frac{a+b+c}{a+b+c+2d}\right]$$

where $a = 0.5l_1 + d$
$b = 0.5l_2 + d$
$c = l_3 + d$

where l_1, l_2, l_3 are the length, depth and height respectively of the equipment.

3. In the case of wall-mounted equipment or equipment exclusively installed against a wall, the centre of the measurement surface co-ordinates in (1) and (2) above are the centre of the side plane and the base centre of the side plane respectively which are the reference side walls of the equipment.
4. For very small or hand-held equipment, a sphere, hemisphere or quarter sphere of radius r may be

chosen as the measurement surface provided that the condition $r \geq 3d$ is met where d is the distance of one of the upper corners of the reference box from the origin of the co-ordinates. This requires 10 microphone measurement positions.[6]

For the 'A' weighted SPL and the level in each frequency band of interest, a mean value of Lp is calculated from the measured sound pressure levels Lp_i, after correction for background noise if necessary from Table 16.6, by use of the following:

$$L\bar{p} = 10 \log_{10} \frac{1}{N} \sum_{i=1}^{N} 10^{0.1 Lp_i}$$

where $L\bar{p}$ = surface SPL averaged over the measuring surface (dB re 20 Pa), Lp_i = 'A' weighted or band pressure level resulting from the ith measurement (dB re 20 Pa) and N = total no. of measurements.

When the range of Lp_i values does not exceed 5 dB, a simple arithmetic average will not differ by more than 0.7 dB from the value calculated with the above equation.

Table 16.6 Corrections for background noise

Difference between sound pressure level measured with sound source operating and background sound pressure level alone	Corrections to be subtracted from sound pressure level measured with sound source operating to obtain sound pressure level due to sound source alone
dB	dB
<6	Measurements invalid
6	1.0
7	1.0
8	1.0
9	0.5
10	0.5
>10	0.0

When the level of background noise is more than 6 dB below the sound pressure level at each measurement point and in each frequency band, the measured sound pressure levels shall be corrected for the influence of background noise using the formula

$$B = L_c - 10 \lg (10^{0.1 L_c} - 10^{0.1 L_b})$$

B is the correction, in decibels, to be subtracted from the sound pressure level measured with the sound source operating to obtain the sound pressure level due to the sound source alone; L_c is the measured sound pressure level, in decibels, with the sound source operating and L_b is the level of background noise alone, in decibels.

When the level of the background noise is less than 6 dB below the sound pressure level at each measurement point and in each frequency band, the accuracy of the measurements is reduced and no corrections shall be applied for that band. The results may, however, be reported and may be useful in determining an upper limit to the sound power level of the equipment being tested.

(ii) Operator position

For operator-attended machines, the sound power level may be given or supplemented by the noise level at the operator position(s). It may be referred to also as the work station noise emission level or 'ISO Position' noise level. Microphone location relative to the reference box is given for seating and standing operation in Fig. 16.14.[8]

(iii) Bystander position

For equipment that does not require operator attention while in the operating mode, at least four bystander positions are selected instead of the operator position. A bystander is defined as an individual who is not the operator of the equipment but whose position lies within the sound field produced, either occasionally or continuously. The bystander positions are typically 1 m from the reference box in the horizontal plane and 1.5 m above the floor at the front, rear, right- and left-hand sides of the equipment.[8]

16.3.5 Noise measurement environment

Ideally the test environment should be free from reflecting objects providing a free field, or more usually with engineering measurements, a free field over a reflecting plane such as on the ground in an open area or in a semi-anechoic chamber (an anechoic chamber with a hard reflecting floor). Free-field conditions are not usually encountered in typical rooms and laboratory areas where sources are normally installed, hence for measurements in these locations, corrections may be required to account for background noise or undesirable reflections. When a source produces sound in a room, the sound field is only partially composed of sound radiating directly from the source (direct sound field). Reflections from room surfaces contribute to the overall sound field (reverberant or semi-reverberant sound field).

Sound pressure level is the quantity most directly related to the response of people to airborne sound, but it is not the most convenient descriptor of a noise source since it depends on the distance from the source and the environment in which the source and measurement position are located. A better descriptor is the sound power level of the source, and in many cases this is sufficient for the following reasons:

1. Allows a noise comparison between different machines;
2. Allows verification that the noise produced by a particular machine meets the noise level control specification;
3. Provides a means of predicting noise levels in free-field and reverberant spaces.

A source may radiate sound more effectively in some directions than in others. The directivity index DI may be calculated from measurements in a free field over a reflecting plane using the following:

$$DI = Lp_i - L\bar{p} + 3 \text{ dB}$$

and the directivity factor, D, calculated from

$$D = \text{antilog}_{10} \frac{DI}{10}$$

When a source radiates into free space, the interest is in directivity as well as sound power level, but if a source radiates into a reverberant field, directivity information is lost (i.e. average SPL essentially independent of distance from the source), hence sound power level is all that needs to be defined regarding the source. In terms of sound pressure level at a point and its directivity factor, providing the measuring point is not close up to the machine, then

$$Lw = Lp - 10 \log_{10} D + 10 \log_{10} S_a$$

where Lw = sound power level and S_a = area of measurement surface (m^2).

The most widely accepted methods of determining sound power are based on solid pressure measurements made in the vicinity of the source. A number of methods are in common use.

(i) Noise measurement in a free field

This requires negligible reverberant sound, usually only realized in anechoic rooms[9] although similar but less precise measurements may be made in the open or in large

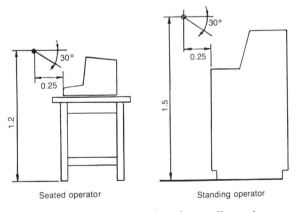

Fig. 16.14 Microphone positions for standing and seated operators (dimensions in metres)

rooms which are near anechoic.[6] The principle is to determine the sound power of a machine by integrating the sound intensities at a number of points on the surface (section 16.3.4(i)). In terms of SPL

$Lw = Lp + 10 \log_{10} S_a$

If the SPL at one distance (e.g. d_0) from the source is known, the SPL at a greater distance (e.g. d_1) may be calculated using the 'two-surface method' without having to evaluate the sound power level of the source, e.g.

$SPL_0 - SPL_1 = 20 \log d_1 - 20 \log d_0$

where SPL_0 = SPL at distance d_0 and SPL_1 = SPL at distance d_1. Or

$SPL_0 - SPL_1 = 20 \log \dfrac{d_1}{d_0}$ dB

which represents the well-known inverse square law, i.e. sound pressure level decreases by 6 dB per doubling of distance from the source.

It also follows that

$Lw = SPL_0 + 10 \log S_0 - C$

where S_0 = area of control surface (m²) where SPL_0 measured, S_1 = area of control surface (m²) where SPL_1 measured and C = correction factor (Fig. 16.15).

(ii) Noise measurement in a reverberant room

A reverberant room has highly sound reflecting, non-parallel walls, ceiling and floor. Ideally, as the sound from a source is reflected back and forth continuously between the surfaces of the room, the multiple reflections create a completely 'diffuse' sound field, i.e. having equal energy density at all points. Such a room is considerably less expensive to build than an anechoic chamber, but cannot measure directivity nor give accurate sound power results where there are predominant tones.

In a reverberant room, the sound power level of a source is given by

$Lw = Lp_{ss} + 10 \log_{10} V - 10 \log_{10} T - 14$ dB

where Lp_{ss} = steady-state sound pressure level (dB), V = room volume (m³), T = reverberation time (s) = time taken for a sound in the room to decrease by 60 dB when the sound is suddenly stopped, i.e.

$\dfrac{0.161V}{S\bar{\alpha}}$

and S = surface area of the room (m²), α = absorption coefficient defined as the fraction of incident energy that is absorbed at a surface — Table 16.7 gives values of common constructional materials, and $\bar{\alpha}$ = average absorption coefficient of the sides of the room, i.e.

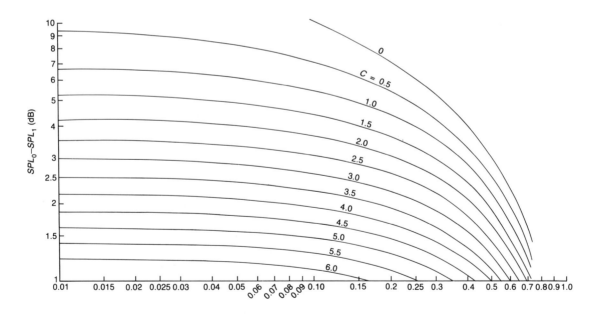

Fig. 16.15 Sound power level correction factor

Table 16.7 Absorption coefficients for some common materials

Material	Thickness including any air space (mm)	63	125	250	500	1 k	2 k	4 k	8 k
Brickwork	—	0.05	0.05	0.04	0.02	0.04	0.05	0.05	0.05
Concrete	—	0.01	0.01	0.01	0.02	0.02	0.02	0.03	0.03
Up to 4 mm thick glass	—	0.25	0.35	0.25	0.20	0.10	0.05	0.05	0.05
6 mm plate glass	—	0.08	0.15	0.06	0.04	0.03	0.02	0.02	0.02
Mineral fibre blanket	25	0.05	0.10	0.35	0.60	0.70	0.75	0.80	0.75
Mineral fibre blanket faced with 10% open area	25	—	0.10	0.25	0.50	0.75	0.75	0.55	—
Open-cell polyurethane foam	25	0.10	0.15	0.30	0.60	0.75	0.85	0.90	0.90
Rigid polyurethane foam	—	—	0.20	0.40	0.65	0.55	0.70	0.70	—
Acoustic plaster	12	0.05	0.10	0.15	0.20	0.25	0.30	0.35	0.35
Fibreboard	12	0.05	0.05	0.10	0.15	0.25	0.30	0.30	0.25
Plywood	9.5	—	0.28	0.22	0.17	0.09	0.10	0.11	—
Medium pile carpet, foam underlay	10	0.05	0.05	0.10	0.30	0.50	0.65	0.70	0.65

$$\frac{S_1\alpha_1 + S_2\alpha_2 + S_3\alpha_3 + \ldots}{S}$$

where S_1 = area of room having an absorption coefficient, etc.

However, since SPLs in an actual reverberation room will vary from point to point, a number of microphone positions and locations[10] can be used to give $L\bar{p}_{ss}$.

(iii) Noise measurement in a semi-reverberant room
Most rooms in which machines are installed are neither well damped (anechoic) nor highly reverberant (diffuse), but it is nevertheless still possible to obtain a sound power level/SPL sufficiently accurate for most engineering purposes. Ordinary rooms, laboratories, offices, etc. are semi-reverberant. It is therefore necessary to include a correction factor for room absorption. The sound power level of the machine can be determined as

$$Lw = L\bar{p} - 10 \log_{10}\left[\frac{D}{4\pi r^2} + \frac{4}{R_c}\right] \text{ dB}$$

where R_c = room constant for a given frequency band (m²) = $S\bar{\alpha}/(1 - \bar{\alpha})$ and r = distance from source.

In the absence of a detailed knowledge of the particular source characteristics, it is usually sufficient to assume a unidirectional source and use the values in Table 16.8. Considering the above equation, there is clearly a distance r from the source where the direct energy $D/4\pi r^2$ starts to dominate the reverberant energy quantity $4/R_c$, i.e. a 'critical' distance. Below this distance, because the direct field is dominant, the sound level is free field. Above this point, the sound field is entirely reverberant, the distance r from the source does not enter the equation and the level is everywhere constant. In other words, up to the 'critical' point the SPL follows the inverse square law, but beyond this point the SPL evens out to remain virtually constant (Fig. 16.16). For acoustically 'hard' rooms, the reverberant field will dominate up to relatively short distances from the source. The reverse occurs when the room is highly absorptive.

If it is not possible to compute the room constant easily, the sound power level of a source can be determined by taking sound pressure measurements on two separate surfaces having different radii in the far (reverberant) field, i.e.

$$Lw = L\bar{p}_2 - 10 \log_{10}[S_1^{-1} - S_2^{-1}]$$
$$+ 10 \log_{10}[10^{(Lp_1 - Lp_2)/10} - 1]$$

where Lp_1 = average SPL measured at the smaller test

Table 16.8 Values of directivity factor and directivity index for a unidirectional source located in a large room

Position of source	Directivity factor D	Directivity index DI (dB)
Near centre of room	1	0
At centre of wall, floor or ceiling	2	3
Centre of edge formed by junction of two adjacent surfaces	4	6
Corner formed by junction of three adjacent surfaces	8	9

NOISE REDUCTION

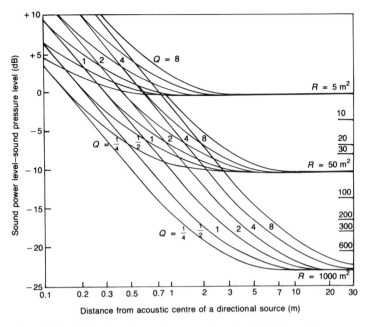

Fig. 16.16 Variation of sound pressure level with distance from source

surface area S_1 and Lp_2 = average SPL measured at the smaller test surface area S_2.

Once Lw has been determined, R_c can be found. However, the method assumes that the machine noise is at least 10 dB above the background level. If this is not the case, the SPL has to be corrected for background noise (section 16.2.4).

16.4 The generation of noise

Machine elements and processes generate noise either as the result of vibration of machine components, structure, drives, panels, etc. or the result of air turbulence created by fans and blowers. If airborne or structure-borne mechanical noise is generated at low levels, high levels may be apparent through resonance of the machine case (covers) or of resonant surfaces on which the machine is mounted. Structure-borne vibrations may be persistent at considerable distances from the source as there is little loss of vibrational energy though structural transmission. Because a machine structure consists of a large number of mechanical elements each with their own resonant frequency, the noise spectrum contains numerous peaks and troughs (Fig. 16.17).

Machinery noise can be basically divided into two groups:

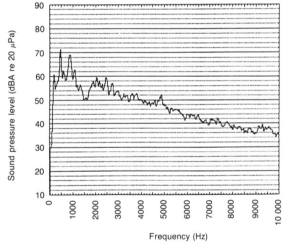

Fig. 16.17 Tonal assessment of banknote counting machine

1. *Aerodynamic.* Machines which interact with the surrounding air by adding energy to it or subtracting energy from it. All aerodynamic processes produce some degree of turbulence by virtue of the work done in moving air, typical sources being fan blades, airflow around obstacles, grills, etc.

2. *Structural.* Machines or mechanisms containing a

vibrating source from which energy is transmitted by structural paths to the outer boundaries of the machine. Causes of vibration are unbalanced rotating or reciprocating parts, fluctuating loads and forces, misalignment, loose mountings or loose fastenings. The exciting force in most cases is a repetitive series of impulses, and resultant vibrations generated at frequencies within the audio range are heard as noise. A summary of the frequency content of different vibration types is given in Table 16.9. In real machines there are many transmission paths and many sections of the structure that radiate noise, every vibrating surface being coupled directly or indirectly to a source of mechanical vibration.

Noise generation can be visualized in terms of three functional aspects, the noise source, the noise path and the receiver. A schematic of the noise process on this basis is illustrated by Fig. 16.18.

1. *Source*. Where the sound originates. The exact definition of the source depends on the scope of the problem.
2. *Path*. The route the sound travels between source and receiver. The path can exist in solid, liquid or gaseous media and often involves multiple paths in one or more (e.g. Fig. 16.18).
3. *Receiver*. The person or object whose response to the sound is of interest. The response of people is highly variable and subjective, depending on many factors. Although noise is a recognized unwanted sound, it is difficult to quantify in a simple manner for design purposes.

In applying the acoustical system, there are multiple sources and paths to be considered. Each source will usually have distinct sound characteristics and each path will have a characteristic response to the sound. A typical example of a business machine is given in Fig. 16.19.

16.4.1 Noise generated by machine elements

(i) Rolling element bearings
Together with residual imbalance in shaft or rotor, rolling bearings are a common source of mechanical noise (section 7.5.1). The two items are interrelated since imbalance will increase vibration and thus increase bearing wear, hence bearing noise. The noise level will tend to increase with speed. The causes of vibration are departures from the ideal circular profiles of the inner ring, the outer ring and the rolling elements. Full hydrodynamic lubrication is seldom present, so that any skidding that occurs will affect the amount of vibration generated.

The following guidelines apply:

1. High standards of machining accuracy of housings and shafts are required to ensure bearing mounting and running concentricity.
2. Rings and rolling elements damaged as a result of poor fitting are a common cause of noise.
3. In preference, bearings should be grease, rather than oil lubricated.
4. Resonance effects can be reduced or eliminated by employing a housing material with good damping characteristics (Fig. 16.20).

The fundamental frequency of rolling bearing noise in terms of different bearing aspects is given in Table 16.10. Not only can these frequencies give rise to second, third and higher harmonics, but the number of possible frequencies generated can be further increased by the different sources of excitation which may be present.

Table 16.9 *Vibration from machine elements*

	Frequency of vibration				
Element	Order of shaft r.p.m.	High	Low	Impact	Comments
Rotating parts	*				Depends on degree of imbalance
Misalignment	*				Can produce marked axial vibration
Mountings	*				Can be resonant if loose
Belts and pulleys	*				
Bearings		*			Broadband frequency generated
Gears		*		*	Depends largely on gear type and accuracy of manufacture
Solenoids				*	
Clutches		*		*	
Covers			*		

NOISE REDUCTION

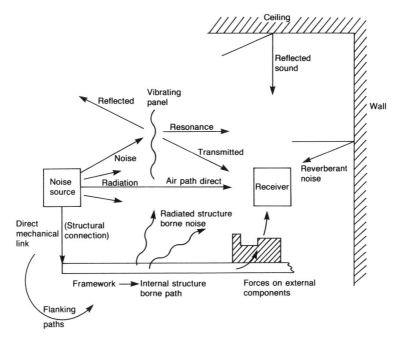

Fig. 16.18 Schematic of the noise process

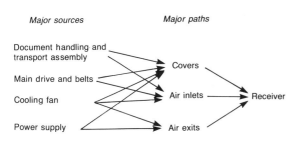

Fig. 16.19 Example of noise sources and paths in typical business machine

Hence, in practice, the noise may appear as wide band. The amplitude of all excited frequencies tends to be small, but vibration monitoring of the bearing frequency spectrum is a standard technique for detecting the early onset of bearing failure.[11,12]

(ii) Plain bearings

All plain bearings are subject to friction which in itself can be a source of noise if an adequate oil film thickness is not maintained. Instability and noise can result from shaft vibration. Noise levels in all cases increase with bearing wear, the process accelerated by vibration.

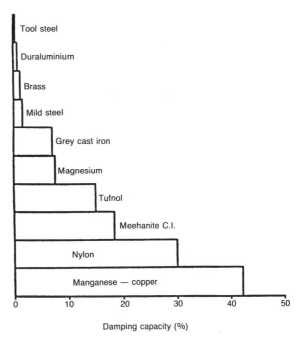

Fig. 16.20 Relative damping capacities of some common materials

Table 16.10 *Fundamental noise frequencies associated with rolling element bearings. Courtesy of Elsevier Advanced Technology, Oxford*

Source	Fundamental Frequency	
	In terms of element speed	In terms of shaft speed[a]
Imbalance	$\frac{n_s}{60}$ Hz	$\frac{n_s}{60}$ Hz
Eccentricity	$\frac{n_s}{60}$ Hz	$\frac{n_s}{60}$ Hz
Irregularity of rolling elements	$\frac{n_t}{60}$ Hz	$\frac{am\, n_s}{120(a+b)}$
Irregularity of cage	$\frac{n_t}{60}$ Hz	$\frac{am\, n_s}{120(a+b)}$
Irregularities on inner race	$\frac{n_s - n_t}{60}$	$\frac{n_s}{60}\left(1 - \frac{am}{2(a+b)}\right)$
Irregularities on outer race	$\frac{n_t}{60}$	$\frac{amn_s}{120(a+b)}$

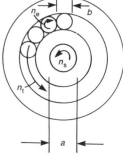

n = speed in rpm

[a] m = number of spots of irregularity.

Machine and shaft balance is thus important in controlling noise and wear levels.

If 'stick-slip' is present, this tends to excite the rotating system at its natural frequency (section 5.2.6), tending towards an audible 'squeal' with some materials in dry sliding conditions. 'Chatter' will result from excess clearance in the bearing, i.e. if worn.

(iii) Gears

Gears are a common source of vibration excitation that results in noise. Resonant modes of the gears, connected structures and/or panels are readily excited by the vibration frequencies arising out of the tooth-meshing process (section 10.5.6). Gears invariably form part of a mechanical transmission system and may be associated with other inherently noisy items in the drives, thus noise reduction in respect of gears tends to be a systems activity. The noise from enclosed gearboxes is usually dominated by noise from gearbox structure resonance effects.

(iv) Timing belts

Synchronous timing belt systems may be a major source of noise problems in some machines, especially office machines where the user environment may be inherently quiet (or semi-acoustic) and noise regulations may be stringent. Timing belt noise is generated by the meshing and unmeshing of the teeth (section 9.2.3(ii)). The following general guidelines[13] apply to timing belt noise:

1. The finer the tooth pitch and the larger the pulley diameter, the lower the noise emitted since the load carried by each tooth is reduced.

2. An increase in belt tension affects the load carried by each tooth and so may lead to higher belt noise. However, in most drive systems there is an optimum tension for reasons of belt wear, power rating, creep, etc.

3. An increase of belt speed will increase the tooth-pulley meshing (impact) rate and so generate increased noise. Doubling of belt speed can increase noise levels by up to 10 dB(A). Belt speed is the most important parameter in timing belt noise.

4. The preferred pulley material is in the order steel—aluminium—polycarbonate. The difference in noise performance is due to material stiffness and as such, solid pulleys are generally quieter. In practice, however, moulded pulleys are normally adopted because of low inertia and because they are more cost-effective.

5. Flanges on pulleys in undesirable locations will increase noise levels compared to a drive system without flanges. It should be the case to provide flanges only where necessary, preferably only on the drive pulley.

6. Misaligned shafts increase the noise level.

7. As long as belts do not sag, increase in belt width has no effect on belt noise.

(v) Transformers

Magnetic noise or 'hum' is generated in iron-cored transformers and chokes. The most probable effect is the induced vibration of the mounting structure at power line frequency harmonics (i.e. $2f$), although harmonics will also be present over the complete noise spectrum. In general, the smaller the transformer the higher the frequency of the loudest harmonic and vice versa. The most effective method of reducing transformer noise is either enclosure or isolation, or a combination of the two.

(vi) Fans
Fan noise is somewhat difficult to define when it relates to a piece of equipment in a quiet office where the primary source is a small cooling fan. Such noise is of a distinctly noticeable annoyance level. Noise sources in fans can be divided into aerodynamic and mechanical in origin. Aerodynamic noise may be divided into:

1. *Broadband noise (unpitched)*. Created by air turbulence and vortex action;
2. *Discrete tone noise*. Generated by the rotational pulsing of the air as the blades pass a given point.

The mechanical sources of noise in a fan are bearings, imbalance, magnetic forces, etc. Some fans are quieter than others of the same air-handling capacity, but there is a minimum amount of noise that will always result from the air-handling process for a required capacity. Some examples of fan noise levels are given in Table 16.11. For axial fans

$$Lp \propto \frac{x(\Delta pQ)^2}{\phi^2}$$

where p = static pressure, x = no. of blades, Q = airflow and ϕ = fan diameter.

For fans of similar design producing constant airflow, fan noise is inversely proportional to fan diameter (i.e. large fans will be quieter) and directly proportional to fan work (i.e. high-efficiency fans will be quieter) (section 15.3.5). The several mechanisms that combine to produce the noise depend mostly on the generation of turbulence, an area that is not so well defined as mechanical noise-generating mechanisms. Fan tip radius, tip gap and speed are controlling factors.[14,15] While steady loading of the fan blade (lift, drag) results in the forces that cause thrust (and torque), unsteady loading on the blade results in forces that cause noise.

Tonal noise results from the interaction of the regular blade wake with fixed obstructions. Typical non-uniformities are struts, local change of housing shape, casting variations, lugs, counterbored holes, vanes, heads of fasteners, etc. Tones are normally generated at the blade passing frequency:

$$f = \frac{nx}{60} \text{ Hz}$$

where n = fan speed (r.p.m.) and x = no. of fan blades.

The noise may or may not include several harmonics; the harmonics may have sound pressure levels as high or higher than the fundamental frequency f. These tones are more annoying to the ear than broadband noise and are generally the reason why a particular fan is described as noisy. In the case of a centrifugal fan, if there is insufficient clearance between the fan wheel and the scroll cut-off, a very pronounced frequency spike will be introduced.[16]

Discrete tones can be minimized by the fan designer by using smooth aerodynamic surfaces and avoiding sharp edges and corners. The axial clearance between blades and motor or housing support legs should be at least equal to the width of the legs. Irregular blade spacing will break up the 'whine' from regular blade interactions. This does not greatly reduce sound power levels but it does redistribute the harmonics to make an effective reduction in the A-weighted level.[14] Randomly spaced blades have a similar effect in improving the subjective quality of the radiated fan noise.

(vii) Electric motors
The noise generated by rotating electrical components originates from the noise sources given under the headings below.

(a) Windage The noise arising from aerodynamic action in motors is commonly referred to as windage. This includes the noise produced by the motion of the rotor as well as any integral cooling fan. Ideally the design itself should endeavour to provide as smooth a passage as is possible for the airflow with streamlining of edges, support arms and any other obstacles in the airstream, avoiding abrupt changes in direction or section. If windage noise is high, one method of treatment is by attaching sound-attenuating ducts to the motor inlet and

Table 16.11 *Axial fans; comparison of sound power levels with performance rating*

Air flow (m³h⁻¹)	Static pressure (mm)	Sound power level re — 1 pW at given frequency						
		125	250	500	1k	2k	4k	8k Hz
34	0.5	—	46	46	36	—	—	—
41	0.5	—	49	54	47	35	—	—
86	0.64	—	51	55	43	38	36	—
122	2.03	60	58	59	56	50	50	43
139	2.29	62	63	60	58	52	50	42
269	7.6	61	64	63	60	59	55	48
345	15.8	68	62	66	65	60	56	52
442	20.8	64	62	67	69	61	58	56

outlet openings, provided that the noise is not resonated from the motor frame.

(b) Mechanical noise This is essentially due to bearing noise, residual imbalance in the rotor and, where fitted, the brush gear. Balancing of electrical motors is usually carried out in two balancing planes on the assumption that the rotor is rigid. With small fractional horsepower (FHP) motors, imbalance is generally corrected by removing material from the laminations (grinding or machining), while with larger motors it is more satisfactory to add weights.

Brush gear noise is generated by brush−commutator segment impact or chatter, which may be amplified by brush holder resonance. It is likely to be low compared to the overall noise level of the motor, but may become significant at high operating speeds. Friction-excited brush squeal can become objectionable due to its high frequency, 4−6.5 kHz. The elements causing chatter and vibration are difficult to analyse, but some solutions to brush noise problems are as follows:

1. Use a softer brush material. Disadvantage, however, is increased brush wear hence decreased brush life.
2. Improve surface finish of commutator. Also reduces brush friction. Motor operation should be such as to minimize the brush−commutator arcing as this will cause the commutator to deteriorate in use.
3. Adjust brush pressure to an acceptable minimum. Must be consistent with electrical requirements of motor.
4. Modify brush shape. For example by sloping brush in direction of motion.

(c) Electromagnetic noise Electromagnetic noise is a result of variations in the circumferential magnetic field producing pulsating forces in the air gap between rotor and stator and thus consequent vibration. The effect is enhanced by the presence of any stator−rotor eccentricity together with any resonance that may occur in the motor frame components. Compensation may be applied, e.g. skewing of the rotor winding slot.[17] Since an electric motor may operate over a very wide speed−torque range, the frequency of the electromagnetic noise can vary widely and so be difficult to avoid totally. Magnetic noise may be reduced by increasing the air gap, but this affects the degree of saturation; usually there is an optimum air gap consistent with minimum noise.

The vibration spectrum of the case of an induction motor is shown in Fig. 16.21, with vibration peaks for imbalance (shaft speed and harmonics), misalignment at higher frequencies than imbalance, and slot passing frequency (shaft speed × number of armature slots).

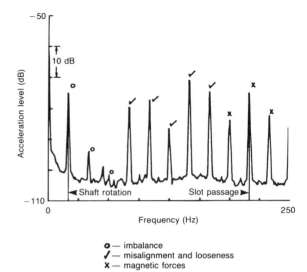

Fig. 16.21 Acceleration spectrum of motor with imbalance and misalignment. From Lyon R.H., *Machinery Noise and Diagnostics*, Butterworth Heinemann Ltd

16.4.2 Noise radiation from covers

The radiation of noise from machinery and equipment is usually dominated by the radiation from panels and covers. Panels and covers provide an interruption to the airborne sound path, and when properly designed can be an effective means of noise control, but an incorrect design may actually result in an amplification of sound power levels at some frequencies.

When sound energy is incident on a panel or cover, some is reflected and the remainder passes into the cover. Of the sound energy passing into the cover, some is absorbed (transformed into heat inside the structure as a result of overcoming internal damping) and some travels through to the outer surface of the cover and so reappears as airborne sound. The chain of events is thus airborne sound → mechanical vibration of the cover → back to airborne sound. Any flanking energy or structure-borne noise may add to the radiated sound level. Any acoustic insulation will increase the absorption.

The acoustical performance of covers is often described in one, or more, of the following ways:

Noise reduction. The difference in SPL on each side of the cover or panel.

Insertion loss. The difference in decibels between two SPLs which are measured at the same point both with and without the cover(s) attached to the machine. It is a useful parameter for noise control since it is easy to measure and straightforward to relate to.

Transmission loss (sound reduction index). This is a property of the cover material. It is more difficult to measure, but it is useful for the analysis of alternative sound transmission paths. When sound is incident on a panel or cover, the fraction of incident energy that is transmitted is termed transmission coefficient τ, i.e.

$$\tau = \frac{W_\text{t}}{W_\text{i}} = \frac{\text{sound power transmitted through panel}}{\text{sound power incident on panel}}$$

from which the transmission loss is defined by

$$R = 10 \log_{10}\left(\frac{1}{\tau}\right) \text{ dB}$$

In general R and τ depend on the angle of incidence of the incident sound. Normal incidence, random incidence and field incidence transmission loss are commonly used terms. Published values of R, as a function of frequency, will relate to infinitely large panels, hence should be treated only as an approximation to field conditions. Small material physical changes, i.e. weight, thickness or mounting can alter the R values significantly. Values of R are based on the assumption that the cover is 'forced' by acoustic pressures. If the cover is mechanically excited, i.e. not totally structure isolated, then the expected sound reduction index will be reduced.

(i) Panel transmission loss

The general pattern of the field incidence panel sound reduction index is illustrated in Fig. 16.22 by four design regions:

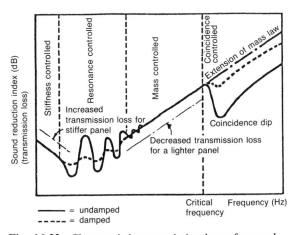

= undamped
----- = damped

Fig. 16.22 Characteristic transmission loss of a panel

1. Stiffness controlled;
2. Resonance controlled (damping);
3. Mass controlled (6 dB per octave);
4. Coincidence controlled (damping).

The design of panels and covers must therefore consider the region of interest for the noise that is causing the problem.

(a) Stiffness-controlled region At low frequencies, transmission loss is controlled by the stiffness of the cover. This region is generally of reduced interest because (a) the resonance frequency is generally quite low, and (b) sound sources do not emit much power at low frequencies. When the enclosure of a low-frequency sound is necessary, maximum transmission loss can be obtained by using stiff, lightweight panel materials such as structural foam or honeycomb materials, i.e. a stiff but not a massive cover with a high resonance frequency. The resonant frequency of a panel may be increased using stiffening ribs, but the achievable increase in transmission loss is generally quite limited.

(b) Resonance-controlled region In the region of the fundamental resonance and up to four times this frequency (harmonics), transmission loss is determined by resonance and damping. An undamped panel may have zero or negative transmission loss in this region which is not a problem unless the sound source emits significant power in this region. In the case of a simply supported uniform rectangular panel, the lowest resonance frequency is given by

$$f_1 = 0.45 C_\text{L} t \left(\frac{1}{a^2} + \frac{1}{b^2}\right) \text{ Hz}$$

where t = panel thickness (m), a = width of panel (m), b = length of panel (m) and C_L = longitudinal wave velocity of sound in the panel material = $E/\rho(1-\nu^2)$ m s^{-1} = Poisson's ratio, where E = Young's modulus (N m^{-2}) and ρ = density (kg m^{-3}).

Values of fundamental resonance frequencies and some relative comparisons for a number of common engineering materials are given in Table 16.12. This illustrates how the resonant frequency reduces with increase of panel area as the effective panel stiffness decreases. Table 16.12 also shows that there is little advantage in changing from steel or cast iron to other materials if the panel thickness remains unchanged. However, if the same panel weights are maintained by increasing thickness when going to a lower density material, then there will be a substantial increase in the panel resonant frequency. Plastics are poor compared to metals although they compare well at equal panel weights.

Table 16.12 *Fundamental resonances and critical frequencies of typical panels*

	Material				
	Steel	Aluminium	Glass	PMMA	Lead
Density (kg m^{-3})	7 700	2 700	2 500	1 150	11 300
Longitudinal wave velocity (m s^{-1})	5 050	5 150	5 200	1 800	1 210
Surface density (kg m^{-2})	97 500	32 200	38 000	35 400	600 000
Panel thickness (mm)	1.5	2.3	4.6	4.6	2.3
Fundamental resonant frequency (Hz) for panel size					
0.6 × 0.75 m	16	24	49	17	5.7
0.3 × 0.5 m	52	81	163	56	19
0.15 × 0.3 m	191	296	598	207	70
Critical frequency of coincidence (Hz)	—	8 440	5 588	3 300	6 690	21 770

These comparisons do not take into account relative strength and stability. Clearly, covers should be so designed that the resonant frequencies of the separate units are not in the range where appreciable sound attenuation is required.

(c) Mass-controlled region At frequencies above the resonance point, the transmission loss is controlled by the surface density of the cover, the loss increasing at a rate of 6 dB per octave. This is the region of primary interest for most noise-control situations, most panels behaving according to the mass law, i.e.

$$R = 20 \log f + 20 \log \rho_s - 48 \text{ dB}$$

where f = frequency (Hz), ρ_s = surface density (kg m^{-2}) = mass per unit area of the panel (Table 16.12).

From the above, the higher the cover mass the greater the inertial force resisting movement and the higher the sound reduction index. When normal cover materials cannot provide adequate transmission loss, it may be necessary to use two panels. Special materials such as lead/foam or vinyl/foam laminates are designed for this purpose. For complex structures, the mass law may well not apply.

(d) Coincidence-controlled region Coincidence is an unfortunate effect of panel stiffness. The critical coincidence frequency, f_c, occurs when the sound wavelength in air matches the wavelength of bending waves in the panel. The effect is similar to that of low-frequency resonance and the sound reduction index falls below the mass law prediction (Fig. 16.22). Some values of critical frequency are given in Table 16.12, depending on materials and thickness. Critical frequency is given by

$$f_c = \frac{c^2 \rho^{0.5}}{1.8 t E^{0.5}}$$

where c = speed of sound in air = 330 m s^{-1}.

Coincidence often leads to loss of acoustic insulation in the mid- to high-frequency range; the extent of the mass law region depends on the stiffness of the panel (e.g. Fig. 16.23). For a high sound reduction index over a wide frequency range, the general requirement is for high mass and low stiffness. For example, lead exhibits coincidence well into the ultrasonic frequency range and its large internal damping greatly suppresses the first resonance so that its behaviour is essentially mass law controlled over the entire audio frequency range (Table 16.12).

(ii) Radiated sound power
The sound power radiated by a machine surface, panel or cover can be estimated by using an accelerometer and sound meter or vibration analyser. Measurements should not be too close to the edges of the panel or cover. The noise radiated can be estimated from the acceleration level on the panel:

$$P_{rad} = \left[\frac{u}{2\pi f}\right] ab\rho c \eta$$

where u = r.m.s. acceleration (m s^{-2}) over octave or one-third the octave bandwidth, P_{rad} = radiated acoustic power (W), f = band centre frequency (Hz), ρc = specific acoustic impedance of air = 430 kg m^{-2} s^{-1} and η = radiation efficiency.

The quantity η is a measure of the efficiency of

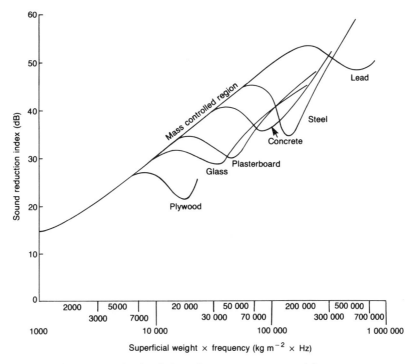

Fig. 16.23 Sound reduction index of some common materials

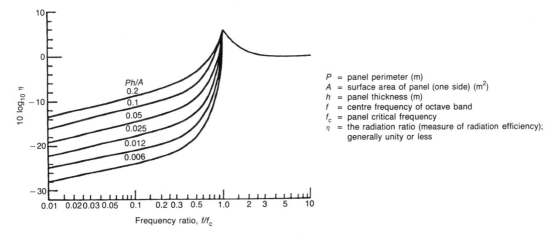

Fig. 16.24 The radiation efficiency of steel and aluminium panels

radiation, its value usually unity or less. A means for estimating η for flat plates is given in Fig. 16.24. Corrections to the above may need to be made to take into account the noise measurement environment (section 16.3.5). The following assumptions apply to the above formula:

1. The vibrating panel is attached at its edges to a rigid frame;
2. The panel edges are simply supported;
3. Sound is radiated into a rectangular space;
4. Ribbed panels are considered as a series of sub-panels, vibrating independently of one another.

16.5 Noise reduction methods

There are two basic methods of noise reduction, (1) reduce the noise at source; or (2) noise control in the path. Great efforts generally go into attenuating the noise once it has left the fundamental source; a whole industry has evolved and is still developoing to meet this need.[18] Far less emphasis tends to be placed on reducing the actual noise source, but it is at the noise origin where the potential returns are greatest in terms of smoother operation, less stress fluctuation and hence greater reliability and life. Bringing in considerations of noise and vibration on a near 'frozen' design as an afterthought results in a second-best solution at an increased cost. Noise control requires an input into the design team at the early stages of development in the same way as other performance and function parameters are specified.

16.5.1 Noise control at the source

The fundamental aim of noise control at the source is to reduce such vibration amplitudes as may be present and so reduce the radiated acoustic power. This in effect means reducing the energy available for driving the vibratory system. Direct treatment in such cases involves a change in structural mass or stiffness to shift the resonance frequency outside of the vibration range. Indirect treatment involves containing the vibrational energy within the original system, decoupled from adjacent structures.

(i) Defining a noise source

To investigate noise sources, it is normally necessary to use a combination of vibration measurements and one-third octave sound power levels. There are some general investigation principles as follows:

1. The sound power levels of the radiated noise should be analysed to determine the predominant frequencies.
2. Those parts of the configuration that are the most efficient noise radiators should be identified from testing various source-path combinations and comparing their mechanical vibration analysed for frequency content.
3. Vibration measurement techniques can be used to track down the real source of the noise.
4. Very often a structural resonance will amplify an original insignificant vibration so that it becomes a major noise source.
5. Sound power levels of sources can be added on an energy basis to determine the total sound power of the device. A-weighted levels are then possible by calculation.

For acoustic analysis it is necessary to use sound power levels for the following reasons:

1. Data for different sources, sizes and mounting arrangement can be compared.
2. Data are a characteristic only of the source and are essentially independent of the measurement environment.
3. Data from all sources can be added to determine the total sound power level of the machine.

When noise sources have been located, the question is how much noise reduction can be gained. The effective contribution of each separate noise source should be listed and ranked. Effective noise control must always address the dominant source(s) to produce measurable results. For example, in Fig. 16.25, source A contributes most to the total power at high frequency, source B contributes most at low frequencies, while source C is not a significant noise power source. Both A and B contribute equally to the A-weighted level so both must be controlled to achieve significant noise reduction.

(ii) Basic sources of vibration

The dominant sources involving a transformation of energy into noise are as follows: (a) imbalance; (b) impact

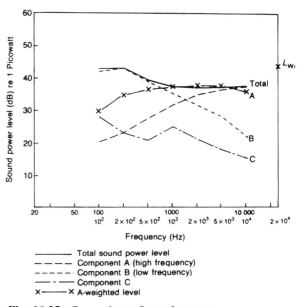

Fig. 16.25 Comparison of sound sources

forces; (c) aerodynamic forces (fans and blowers); (d) electromagnetic forces (section 16.4.1).

(a) Imbalance The presence of imbalance in rotating machines is a primary cause of vibration. The basic treatment is therefore the adoption of suitable levels of balance.[19] Specific recommendations for levels of balance on rotating machinery will depend on a number of factors such as rotational speed, the form of the rotating assembly, the bearings, etc. It is usual to consider the residual imbalance as a measure of the degree of balance, specified in terms of vibration amplitude. Figure 16.26 is a general guide illustrating the relationship between amplitude of vibration (mass eccentricity) and rotational speed relative to subjective running behaviour.

Specific balance requirements are empirically defined internationally in terms of the quality grade G (Fig. 16.26), where

$$G \text{ (or } Q) = ew$$

where e = eccentricity, dynamic imbalance or permissible residual mass centre displacement (μm), i.e.

$$= \frac{\text{Force couple imbalance}}{\text{Wt of rotating unit}} = \frac{U}{m} \text{ g mm kg}^{-1}$$

and U = permissible residual imbalance value per unit of rotor mass, m = mass (kg) and w = angular velocity (rad s^{-1}).

The implication of Fig. 16.26 should clearly be taken into account in machine design, especially if the operation involves a speed change. The data in Fig. 16.26 refer only to a rigid rotating assembly where operating speed is $<0.8 \times$ critical speed. (Critical speed is defined as that rotational speed that corresponds to a resonant frequency of the system, i.e. the excitation of a bending mode.) The following comments apply to shaft design and operation:

1. The shaft should be stiff enough so that the operating speed is always below the critical speed.
2. Avoid simple fractions of the critical speed, i.e. $\frac{1}{2}n$, $\frac{1}{3}n$ etc. for continuous operation.
3. Optimum speed range is 0.5–0.7 of the critical speed.
4. Stiffening the shaft for higher running speeds minimizes vibration problems and makes balancing easier.

(b) Impact forces Many machines rely on impulsive-type mechanisms for their operation, e.g. cams, clutches, solenoids. Like imbalance, the forces tend to be periodic at the operating rate of the machine, but since they are very abrupt the impact frequency spectrum contains a much larger range of harmonics. The impacted surface will vibrate at one or more of its resonant frequencies, the noise dying away as the amplitude of vibration decays. There are three broad methods of dealing with impact noise:

1. Modify the impulsive characteristics of the source;
2. Reduce the amount of impulsive energy reaching the radiating surface;
3. Inhibit the response of the radiating surface.

In practical terms, the following design suggestions should apply:

1. Maximize timing to reduce peak velocity changes in the driven load, i.e. reduce accelerations.
2. Area and mass of the driven mechanism should be reduced as much as possible without affecting structural strength or function.
3. Resilient stopping devices should be used at the ends

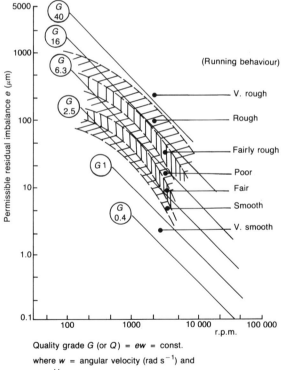

Quality grade G (or Q) = ew = const.
where w = angular velocity (rad s^{-1}) and
$e = \frac{U}{m}$ g mm kg^{-1}
where U = permissible residual imbalance value per unit of rotor mass, m = mass and e = permissible residual mass centre displacement

Fig. 16.26 Relationship between vibration amplitude and rotational speed

of mechanism travel where possible, e.g. solenoids. Where impacts are incidental to the main machine operation, e.g. conveyors or collection chutes, resilient liners should be used.

4. Inhibiting the radiation from a surface can be achieved with an unconstrained layer of damping compound (see section 16.5.2(iv)).
5. If the radiating surface is solid, such as a structural component item, consider replacement with a high damping alloy or material to improve the dissipation of energy more rapidly (Fig. 16.20).
6. In a one-piece structure, most energy loss occurs at the fixing points. Dissipation is increased in riveted or bolted constructions due to relative movement of components; increased friction may also be deliberately introduced at the joints.

(c) Aerodynamic forces — fans and blowers Causes of fan noise have been discussed in section 16.4.1(vi). Designing to minimize the sound power of any air-moving system requires careful attention to detail of all parameters that can affect performance at the earliest possible stage of equipment design. Systems designed into left-over space are typically inefficient and hard to quieten. As a general rule for low noise, use as large a unit as possible with low impeller/blade speed. To minimize fan noise as a source, the following guidelines should apply:

1. Determine the minimum air volume and velocity requirements necessary for adequate performance.
2. Minimize flow resistance by providing short, streamlined airflow paths for orderly airflow and large cross-sectional paths for low velocities and minimum static pressure.
3. Avoid unnecessary obstructions and sharp transverse edges in the flow path, and ensure any flow path cross-sectional changes are gradual and changes in flow direction minimal.
4. Inlet and outlet airflow is critical to quiet operation; grills and guards create turbulence and noise. Guards usually have less effect on the outlet than the inlet. Vertical or horizontal baffles are worst. Surface area reduction should not exceed 20 per cent.
5. If a filter is necessary on the inlet, it should be large enough to avoid large-scale turbulence.
6. Fan or impeller selection should be to maximize static efficiency at desired flow and presssure conditions (Fig. 16.27).

(iii) Minimizing source noise

A number of techniques and methods for reducing the source noise have been referred to already. These can be supplemented by the general comments detailed below.

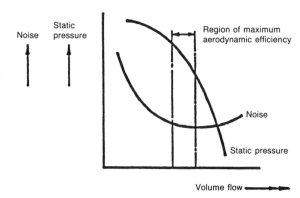

Fig. 16.27 Effect of fan duty on noise

(a) Sound source specification

1. Calculate the allowable sound power level for a source (from the end-user criterion) based on all the sound sources in the machine and the probable amplification/attenuation paths.
2. Advise suppliers of the low noise level criteria at the earliest possible stage in the design process.
3. Preliminary design specifications should include acoustic requirements and in a form that is related to the operating parameters of the equipment.

(b) Correct design

1. A detailed frequency analysis of the sound source will often reveal areas where a small design change can have a significant effect on the noise output.
2. Separate operating speeds and resonant frequencies.
3. Use shock-absorbing techniques to absorb impact energies.
4. Reduce sound-radiating areas.
5. Replacing the source with another having different sound characteristics can be an effective means of noise control in some circumstances.
6. Equipment with components designed to run at maximum efficiency will produce the quietest result.

(c) Correct operation

1. Sound output is likely to change when the functional operating parameters are changed.
2. All equipment should be operated at the design conditions.
3. If guards, covers, etc. are fitted, make sure that they are in place and that openings can be acoustically seated.
4. Maintain good dynamic balance in the equipment and ensure bearings are replaced correctly.

5. Ensure lubrication is correct.
6. Control the mechanical run-out of shafts — this improves balance. By reducing vibration sources, many components become less noisy and achieve longer lives.

16.5.2 Noise control in the path

The sound transmission path often provides most opportunities for noise control, the intent being to make the transmission path as inefficient as possible, causing the sound energy to be dissipated before reaching the receiver. The control methods generally used are as follows:

1. *Sound absorption*. Acts to reduce reflected sound energy. Sound-absorbing materials are porous, the effectiveness increasing with density, number of pores per linear centimetre and its thickness.
2. *Enclosures*. Can be totally enclosed or partial barriers to sound transmission. Usually materials are rigid or flexible sheets of impervious material. To be fully effective, a barrier must enclose a noise source almost completely.
3. *Muffling*. General technique to control openings in an enclosure. Air passages and outlets lined with sound-absorbing material, i.e. acoustical lining.
4. *Vibration damping*. Application of deadening or damping material to a vibrating surface in order to reduce the amplitude of vibration and resonance response by absorbing the vibrational energy.
5. *Vibration isolation*. A general technique to separate a vibrating source from the adjacent structure. Decoupling is accomplished with resilient mounts, springs, dampers, etc. In some cases, isolation treatment may be combined with damping treatment.

Each of the above will have their own advantages and disadvantages. Most noise-control problems in practice will require a combination of techniques. A summary of the materials used for noise reduction is given in Table 16.13.

(i) Sound absorption

Sound absorption materials are the most familiar acoustic materials. They reduce the acoustical energy reflected from surfaces by allowing the sound waves to enter and dissipate within the material. They may be used for sound reduction or reverberation control or both, the important applications including the reduction of reverberant sound in an enclosure and the reduction of sound propagated through ductwork. Sound absorption has no effect on direct sound (Fig. 16.28). A sound absorbent is a material or composite structure which can provide sound energy absorption by one or more of the following:

1. Friction between the fibres of a porous fibrous material;
2. Absorption in the voids of a porous non-fibrous material;
3. Absorption within narrow entries to an air space.

The performance of a sound absorption material is

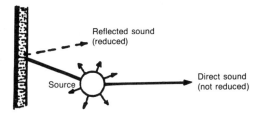

Fig. 16.28 Effect of sound-absorbing materials

Table 16.13 *Materials for sound reduction*

Material type	Panel deadening (vibration damping)	Sound barrier	Sound absorbing	Vibration isolating
Function	Decreases amplitude of vibration of the panel at resonant frequencies	Stops transmission of sound through the air	Reduces amplification of sound by reflection and can improve the performance of a sound barrier	Reduces transmission of conducted sound
Materials	Asphalted felt Adhesives Damping tapes Fibrous blankets Mastics Sponge rubber	Plaster board Hardboard Plywood Plastic sheet Leaded vinyl sheet Heavy foil sheet	Mineral fibre Felts Foamed plastics with interconnecting cells Acoustic tiles Perforated panels with absorbent backing	Springs Elastomers Cork Felt Compressed mineral fibre

defined by the proportion of incident sound absorbed, α (section 16.3.5(ii)), as a decimal rather than as a percentage. It may also be termed the noise reduction coefficient. Normally random incidence absorption coefficients are referred to in product literature, but in some cases the normal absorption coefficient may be quoted (incidence angle 90°). There is no direct relationship between the two. When comparing data from different sources, test conditions may vary between laboratories, so expect some variance. Strictly speaking, therefore, the quantity is defined in terms of the experimental conditions by which it is measured. The benefit of using sound absorption will depend on the relative importance of the direct and reflected sound at a particular frequency, the total sound energy at a point = sum of direct + reflected sound energy (Fig. 16.28). When considering the use of acoustically absorbing material, first determine whether the reverberant sound field dominates the direct sound field at the point where the improvement is required. If this is the case, then the decrease in the sound pressure level, ΔLp, may be calculated as follows:

$$\Delta Lp = 10 \log_{10} \left[1 + \frac{(S\bar{\alpha})_i}{(S\bar{\alpha})_T} \right]$$

where $(S\bar{\alpha})_T$ = original amount of absorption and $(S\bar{\alpha})_i$ = increased sound absorption (see also section 16.3.5(ii)). If $(S\bar{\alpha})_T$ is large, then the amount of absorption to be added must be very large to make ΔLp significant.

Sound absorption materials take the form of porous blankets of glass fibre/mineral wool or open-cell foam. These materials have fine pores through which air has difficulty in travelling. A blanket of sound absorption material must usually be associated with a hard (reflecting) surface to produce the absorption effect. Common thermal insulation materials are generally good acoustic absorbers, being most effective at the higher frequencies. The performance of a number of materials is illustrated by Table 16.14. Most porous absorbents have an optimum density and flow resistance at which maximum absorption is achieved. Too small a pore structure will restrict the passage of sound waves while too large a pore structure will offer low frictional resistance, both extreme cases resulting in low absorption.

The performance of an absorption material can be improved by mounting it so that an air gap exists between the material and the hard backing wall. The optimum air gap depth is equal to the thickness of the porous liner. In practice, it is often necessary to protect the absorption material from contamination by dust, moisture, etc. which would render it useless. For this reason, many commercial materials have thin surface membranes or coatings. Some facings are also perforated, partly for decorative purposes, partly for protection, but also as a means of controlling the absorption behaviour of the panel. Typically the facing effect is minimized with open areas of 10–20 per cent. A further reduction in open area alters performance by increasing the low-frequency absorption characteristic and decreasing high-frequency absorption efficiency. Unperforated panels thus act as low-frequency absorbers. Therefore, the manner in which sound absorption materials are constructed has a profound effect on performance. Application parameters are described in Table 16.15.

Sound absorption practice may be summarized by the following points:

1. The absorption coefficient depends on frequency, tending generally, but not universally, to increase with frequency.
2. The absorption coefficient is not an absolute quantity for any material. As a general rule, the increase in thickness of fibrous or porous absorbents will improve sound absorption mainly over the low and middle frequency ranges.
3. The general classification for sound absorption materials is as follows: low, $\alpha = 0.25$ or less; moderate, $\alpha = 0.25-0.4$; absorbent, $\alpha = 0.4-0.7$; highly absorbent, $\alpha > 0.7$.
4. It is false economy to use expanded polystyrene or closed-cell polyurethane foam since these are of little use in absorbing sound.
5. Carpeted floors help to reduce reverberant noise.
6. Each doubling of total sound absorption, $\Sigma\, S\alpha$, will reduce the sound pressure level by 3 dB in the frequency band of interest.
7. 'Miracle' materials do not exist in acoustical control. The best solution is always to use a conventional material in a cost-effective manner.
8. To obtain good absorption at a particular frequency, the material thickness should be one-tenth the wavelength or more (Table 16.2), i.e. 25 mm foam has good properties above 1 kHz, 12 mm foam above 2 kHz, etc.
9. Special materials such as coated foam, or membrane/panel construction are very useful for low-frequency sound absorption.

(ii) Acoustic barrier

An acoustic barrier or acoustic insulation placed between a radiating surface and the outside will retard the transmission of airborne sound. Acoustic materials which may be good sound absorbers are generally poor as

NOISE REDUCTION

Table 16.14 Summary of the sound absorption and sound barrier performance of typical acoustic products

Material	Density (kg m^{-3})	Composite thickness (mm)	Random absorption coefficient[a]								Sound reduction index (dB)[b]							
			63	125	250	500	1k	2k	4k	8k(Hz)	63	125	250	500	1k	2k	4k	8k(Hz)
Mineral fibre/lead/mineral fibre	—	45	0.05	0.17	0.28	0.4	0.75	0.8	0.85	0.9	22	16	20	24	35	46	53	46
		80									22	21	28	35	43	48	50	50
		125									28	27	34	41	49	50	50	50
Mineral fibre + perforated sheet steel outer panel	—	50	0.1	0.2	0.45	0.65	0.76	0.8	0.8	—	—	20	28	35	43	50	—	—
Fire-retardant acoustic foams																		
Polyether	30	25	—	0.12	0.22	0.48	0.64	0.7	0.82	—	Not recommended							
Polyester	30	25	—	0.1	0.27	0.48	0.9	0.95	0.95	—	Not recommended							
Polyurethane + aluminized polyester surface film	340	25	—	0.22	0.45	0.79	0.2	0.3	0.75	0.85	Not recommended							
Polyurethane + perforated PVC surface film coat	30	25	—	0.22	0.44	0.82	0.34	0.34	0.54	0.58	Not recommended							
Polyurethane + internal lead barrier (laminate)	—	31	0.04	0.14	0.24	0.34	0.44	0.37	0.78	0.84	23	22	21	20	31	46	46	44
Cork board	90	—	—	0.1	0.1	0.32	0.6	0.32	0.48	—	Not recommended							
Flexible heavy layer mineral filled polymer monolayer sound barrier mat	10	2.5	Not recommended								—	23	26	30	35	38	42	48
Flexible spaced layer sound barrier mat; mat + low-density foam + 1.2 mm steel	8	8	Not recommended								—	17	20	25	38	48	53	58

[a] Measure of reflected absorption, section 16.5.2 (i).
[b] Measure of reduced transmission, sections 16.5.2 (ii) and 16.4.2.

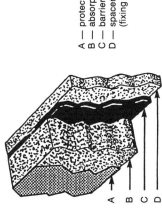

Typical construction of an acoustic laminate

A — protective facing (optional)
B — absorption layer, e.g. foam, mineral wool
C — barrier layer (lead or heavy polymeric)
D — spacer layer (deadzone isolation) (fixing side)

regards sound insulation, i.e. they may be good at reducing the reflection of sound waves incident on their surface, but are usually poor at reducing sound energy transmitted through the material. Both reverberant and direct fields will contribute to the sound radiated by covers or enclosures, hence for best results the internal surfaces should be coated with a sound-absorption material. The effect of inadequate absorption in machine covers may be quite noticeable, since when sound travels in an enclosed space, many reflection paths will repeat themselves to form resonance modes. If an enclosed sound source emits sound corresponding to a resonant frequency, amplification may occur.

Barrier materials are summarized in Table 16.13 and performance figures are given in Table 16.14. The performance of a sound-insulating material is expressed in terms of the sound reduction index (section 16.4.2). The sound reduction index is frequency dependent, the lower frequencies much more readily transmitted. For normal working, however, an averaged value is commonly used (Table 16.14). As with absorption materials, the sound reduction index is derived empirically. Application parameters are given in Table 16.15.

To be fully effective, a barrier must almost totally enclose the noise source since sound can easily flow around edges and through openings and air gaps. This is, however, less important for higher frequencies. Air gaps usually occur around removable panels, doors, cover bases, etc. A reduction in transmission loss occurs because the opening has a lower transmission loss than the panel. The effective sound reduction index, Re, that would exist with a small opening, or different type of panel construction, may be estimated from

$$Re = 10 \log_{10} \left[\frac{S_1 + S_2 \ldots}{\tau S_1 + \tau S_2} \right]$$

where τ_n = transmission coefficient of surface n; for an air gap $\tau = 1$ and S_n = area of surface n.

As an example of use, if suffix 1 is taken to be the transmission coefficient for the main metal panel itself, suffix 2 for an opening, suffix 3 for a plastic access door, etc. Fig. 16.29 shows the importance of openings in the covers or machine panels. For example, if a cover has a transmission loss of 30 dB at a particular frequency, an area opening of 1 per cent of the total enclosure panel area will reduce the transmission loss to 20 dB. If the opening was 10 per cent of that total, the transmission loss is only 10 dB (Fig. 16.29). Typically the gaps around covers result in ~ 1 per cent open area and ventilation grill openings ~ 10 per cent.

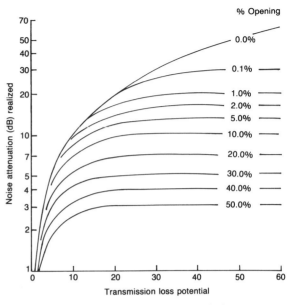

Fig. 16.29 Effect of openings on transmission loss

(iii) Muffling devices

Mufflers or silencers provide a means of sound control so that openings do not degrade the acoustical performance of an enclosure. In other words, mufflers impede the transmission of sound with minimum resistance to the airflow. There are two general types of muffler: (1) dissipative (or absorbtive), where much of the performance is due to flow-resistive material (sound absorbent) and (2) reactive, where the acoustical performance is due to shape rather than flow-resistive material.

(a) Dissipative These are typically ducts, bends and chambers lined with sound-absorbing materials and used to cover medium and high frequency bands. They provide attenuation over a broad range of frequencies and are simple to design. The acoustical effectiveness achieved will depend upon the following:

1. The properties (flow resistance) of the sound-absorbing lining material;
2. The thickness of the sound-absorbing material relative to wavelength;
3. The size of the duct/air passage/diameter/width relative to wavelength;
4. The length of the muffler.

Table 16.15 *Summary of application parameters for acoustic materials*

Acoustic foam	Acoustic barrier mats	Cork
Open-cell foams are best for sound absorption but they do not impede sound transmission	Gives a combination of sound absorbing and sound insulating properties	Insulation corkboard is produced in a wide density range
Use closed-cell foam if an effective seal for an opening is required	Are dense, flexible, heavyweight materials primarily designed for the reduction of transmitted noise through a panel	Densities $80-100$ kg m^{-3} have good sound absorption properties in the middle frequency range
Open-cell foams collect or shed dust particles so may need to use a coated foam if clean air is important	Effectiveness as a barrier directly related to superficial weight (mass per unit area)	Densities >175 kg m^{-3} are used for thermal insulation and anti-vibration treatments
The rigidity of the foam has little effect on acoustic performance	Monolayer type is used for frequencies below 400 Hz	**Mineral or glass fibre**
The higher the permeability the greater the sound absorption	Spaced layer type has the barrier layer laminated to a suitably spaced material such as P/U foam and used to improve insulation at frequencies >400 Hz	Excellent thermal resistance and sound absorber
The thinner the foam the higher the flow resistivity (i.e. the lower the permeability) needs to be for optimum performance	It is desirable to use a sound barrier material with a surface weight equivalent to panel to which it is applied	Widely used for manufacture of acoustic tiles, boards and panels
The smaller the pore size, the greater the absorption capability, other factors being equal	Best results with spaced layer barrier, the filler of sound absorbent material should be resilient enough not to conduct vibration directly to the outer sheet, the inner sheet being air impervious	
Particularly suitable for medium and high frequency sound treatment		
Low-frequency performance is improved with an air gap between the foam and a hard surface		
Common type is open-cell P/U, inert and fire retardant		
12 mm foam is recommended as a guide for frequencies down to 2 kHz; 25 mm down to 1 kHz and 50 mm at lower frequencies		
25 mm foam is the minimum recommended thickness for most purposes		
Low-frequency absorption of film faced foam is greatly improved when the film is not totally adherent to the foam surface		
Foam–lead–foam laminates will improve transmission loss but they will not correct deficiencies in enclosure design		
The acoustic laminate principle is that the absorption layer reduces reverberant noise, the barrier layer blocks transmitted noise and any spacing layer further improves performance		

Typical varieties of dissipative mufflers are shown in Fig. 16.30. The maximum attenuation for a lined duct per length equivalent to duct diameter or width is about 4 dB and this will occur at a frequency where the wavelength is one to two times the duct width. For maximum attenuation at this frequency, a lining thickness of at least one-tenth wavelength is required. Assuming that the whole inner surface of the duct is lined,

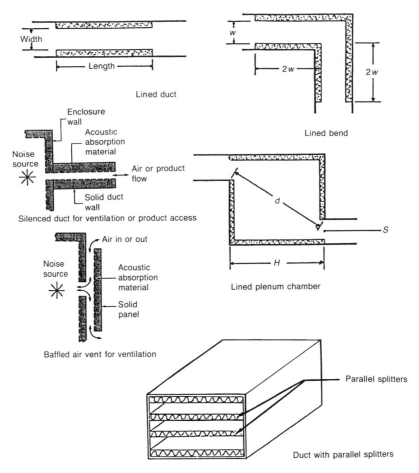

Fig. 16.30 Examples of various types of muffler device

$$\frac{\text{Attenuation per}}{\text{unit length}} \approx 1.03\alpha^{1.04}\left[\frac{\text{perimeter of duct}}{\text{cross-sectional area of duct}}\right]$$

At higher frequencies where wavelength is less than duct width, the attenuation will decrease due to direct sound transmission. In this case splitters can be used to reduce effective duct width, or the use of lined bends or other methods to block the line of sight will improve high-frequency performance (Fig. 16.30). The lined length on each side of a bend should exceed the duct width. Bends which are not associated with a lined duct will produce little attenuation.

Lined ducts must often be very long to provide the required insertion loss, hence a lined plenum chamber is frequently a more practical solution (Fig. 16.30). The plenum chamber combines the effect of an abrupt change in cross-section with the effect of an absorptive lining to provide good attenuation at all frequencies. The attenuation from a plenum chamber may be estimated as

$$IL = -10\log_{10}\left[A\left(\frac{H}{2\pi d^3} + \frac{1-\bar{\alpha}}{S\bar{\alpha}}\right)\right] \text{ dB}$$

where IL = insertion loss = the difference in decibels between two SPLs measured at the same point before and after the plenum is installed, A = outlet area, H = plenum height and d = slant distance, entry to exit.

To be effective, the plenum cross-section must be approximately five times that of the duct. The greater this ratio and the greater the thickness of the absorption layer the greater the insertion loss. Guide vanes should not be installed in a plenum or around a bend as they reflect sound waves around the bend and so reduce the effectiveness of the muffler.

(b) Reactive Reactive silencers are tuned devices which need to be specifically designed for the application. They

are normally used for low-frequency noise, such as low-speed reciprocating machinery, the muffler dimensions being small relative to the wavelength of interest. They are designed to cause an impedance mismatch at the source so that the sound power level is reduced.[20] With large or multi-stage blowers, the use of a reactive expansion chamber or resonator may be necessary to attenuate discrete frequencies.

(iv) Vibration damping

The sound transmission characteristics of a structure are determined primarily by three physical properties — mass, stiffness and damping. While mass and stiffness are predictable features, damping is more difficult. The primary effects of panel, surface or structural damping are as follows:

1. A reduction in the amplitude of resonant vibrations to give a reduction in the radiated sound level.
2. A more rapid decay of free vibrations and corresponding reduction of noise generated by repetitive impacts.
3. Attenuation of structure-borne waves propagating along the surface.
4. Increased transmission loss above the coincidence frequency of the surface.

The only way to reduce vibration-excited sound is to dissipate as large a proportion as possible of the vibrational energy in the form of heat in a damping layer directly attached to the radiating surface. It is more convenient to use applied damping materials rather than metallic covers or panels with high internal damping. The use of stiffeners on their own does not provide damping, but rather shifts the resonant frequency. Damping materials are usually viscoelastic layers, either bituminous, PVC or PVA high-hysteresis materials. They may be unconstrained (simple surface layer applied) or constrained (with surface layer covered) (Fig. 16.31).

The performance of a dampened panel is defined in terms of β, the loss factor or damping factor, as follows:

$$f_d = f_0 \sqrt{\left(1 - \frac{\beta^2}{4}\right)} \text{ Hz}$$

where f_d = damped resonance frequency and f_0 = undamped resonance frequency.

For laminated material, published manufacturers' data may refer to 'β comb'. Some loss factor values for various damping materials are given in Table 16.16. The maximum loss factor that can theoretically occur in practice is 1. For conventional sheet metal components, β lies between 0.001 and 0.01.

(a) Unconstrained layer damping An unconstrained or free-layer damping treatment consists of a simple two-part construction (Fig. 16.31), a highly damped material being applied directly to the structure to be controlled. The effectiveness of the damping treatment depends upon forcing the damping material to move and dissipate energy as it contracts and extends longitudinally on the base. The material may be sprayed or trowelled on to a structure or available in self-adhesive sheet form for bonding. To achieve a reasonable degree of damping, an applied material weight ≡ base panel weight is desirable.

Because of the viscoelastic nature of the damping materials, performance is temperature dependent (Fig. 16.32). Optimum performance is obtained with the damping layer two to three times the panel thickness.

(b) Constrained layer damping A constrained layer of damping material consists of a composite laminate made up as a damping core material sandwiched between skins of structural material in three-part construction (Fig. 16.31). The damping action is a direct result of shearing the viscoelastic core as the laminate is flexed.

The loss factors produced with a sheet metal sandwich are usually high (Table 16.16). Ideally it is preferable to

INTEGRAL DAMPING
applying damping material between the skins of plates or beams (referred to as constrained layer damping material)

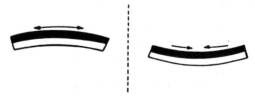

ADDITIVE DAMPING
applying damping material to the surface of a beam or plate (referred to as unconstrained layer damping material)

Fig. 16.31 Construction of damping materials

Table 16.16 *Sheet metal laminate loss factors*

Material	Thickness(es) (mm)	Loss factor, β, at given frequency				
		100	200	500	1k	2k(Hz)
4 : 1 asymmetric viscoelastic composite on sheet steel	1.5	0.14	0.14	0.12	0.12	0.10
Laminated sheet metal	0.7 steel + 0.7 steel	0.52	0.7	0.7	0.65	0.55
Steel–bitumen–steel laminate	4 + 1.8 + 2	—	0.4	—	0.3	—
Steel sheet (reference)	1.6	—	0.001	—	—	—

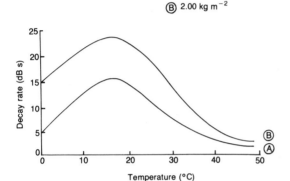

Fig. 16.32 Temperature dependence of damping material performance

have unequal metal thicknesses in the sandwich (asymmetric construction), so ensuring that the damping layer is subjected to shear deflection as it does not lie on the neutral axis. For steel the ideal ratio is 4:1. Having equal sheet metal thicknesses is a less efficient way of using the damping material. Fabrication of sound-deadening sheet material requires some adjustment of practice, i.e. sharp bends should be avoided to avoid metal tearing and damping reduction. Unequal metal lengths will also occur off the bend. All fastening methods tend to reduce the effectiveness of the damping layer by preventing relative motion.

(c) Structual damping In the case of structure-borne noise, the actual design of the structure should be investigated first. If the structure is driven mechanically by attachment to some other vibrating structure or by turbulent air impact, then the response will be dominated by resonant modes and damping will be effective. Where a structure is vibrating in forced (or non-resonant) mode, damping will be ineffective, i.e. excitation is at a frequency away from the resonant frequency. Fastening methods will influence damping in a practical application, e.g. damping is increased in order of welded joint → bolted joint → riveted joint → adhesive joint.

(d) Summary Vibration damping practice may be summarized with the following points:

1. The simplest form of treatment for sound damping involves the addition of a high-hysteresis material as a surface layer. Choice of material must also take into account the resonant frequency and temperature.
2. As with sound absorption materials, many misleading claims are made about the damping characteristics of various materials.
3. A number of acoustical materials such as foam–lead–foam composites are incorrectly called damping materials. They do not damp a panel, they add mass to increase transmission loss. If damping is also required, a layer of vibration-damping material should also be added.
4. Commercial damping materials are generally expensive and may be totally ineffective if used incorrectly.
5. Damping materials become progressively less effective as structure becomes heavier.
6. The addition of damping material to a panel will generally only have a small effect on the transmission of sound through the panel.
7. For optimum results the weight of the damping material should be at least equal to that of the base panel.

(v) Vibration isolation

Vibration isolation provides a discontinuity in the sound structure-borne path just as an enclosure provides a discontinuity in an airborne path. In a noise problem, the propagation path may not always be obvious, but the best enclosure can be made ineffective if the structure-borne vibration is not properly isolated.

In many cases, isolation is necessary to solve low-frequency noise problems. Isolation must permit the relative motion between two parts of a device, and by allowing this movement to occur, most of the energy in the vibrating part is then not transferred to the other parts of the structure. All isolators involve the use of a resilient element or spring. Design for vibration isolation involves two constructional aspects (Fig. 16.33):

1. Design of a resilient mounting system;
2. Design of all other interfaces to avoid flanking transmission around the mounting system.

In a simple undamped single degree-of-freedom system, the resonant frequency of the system is given by

$$f_0 = \frac{1}{2\pi} \sqrt{\frac{k}{m}}$$

where f_0 = natural frequency of the resilient (isolator–load) system (Hz), k = spring stiffness (N m^{-1}) and m = vibrating mass (kg), and the resonant frequency is related to the loaded static deflection of the spring by

$$f_0 = \frac{1}{2\pi} \sqrt{\frac{g}{d}}$$

where g = acceleration due to gravity (9.8 m s^{-2}) and d = static displacement of spring when loaded with stationary mass m.

The effectiveness of isolation is expressed in terms of transmissibility, T_r, as follows:

$$T_r = \frac{1}{1 - [f_f/f_0]}$$

where f_f = forcing frequency of force excited or motion excited vibration present, from which

$$\text{Isolation efficiency per cent} = 100 \left(\frac{1}{1 - [f_f/f_0]} \right)$$

The principle involved in isolation treatment is to mount the source in such a manner that the natural frequency of the isolation system is substantially lower than the forcing frequency of the vibration to be isolated. A minimum design goal should be 80 per cent isolation (Fig. 16.34), i.e. 0.2 transmissibility. Vibration at any frequency $<\sqrt{2}$ times the natural frequency will be amplified with a maximum response near the national frequency. With undamped 'springs' (Fig. 16.35), the transmissibility can be determined from the amount of deflection due to its supported load and a knowledge of the exciting frequency, and so limited to the desired percentage. Isolation performance may be summarized on this basis by Fig. 16.34.

Numerous types of isolator are available and they should always be selected according to the load they are to carry, i.e. the natural frequency of the spring–mass system determined from the performance data published by manufacturers. A manufacturer of vibration isolators should provide data relating load to performance characteristics and include spring rate, natural frequency and static deflection.

The preferred location for vibration isolators is in the same plane as the centre of gravity of the isolated body (Fig. 16.36). When this is not practical, the distance from the mounting plane to the centre of gravity should be much smaller than the distance between mounts, i.e. $A \gg B$ (Fig. 16.36). Such configurations are dynamically stable. Mounting points on the isolated structure should be rigid in order to allow the mount to function properly. Alternate structural paths should be as least as flexible as the mounting system (Fig. 16.33).

The most commonly used types of isolation mounting/materials are as follows:

Metal springs — as springs or mesh pads;
Cork/felt — compression pads;
Rubber — compression or shear pads;
Air springs — shock absorbers.

(a) Metal springs One of the most practical types of vibration or shock mounting, having high strength, low internal damping, a constant spring rate (helical normal) and relative temperature independence (Ch. 12). Can be designed to provide isolation at virtually any frequency from 1.3 Hz upwards. However, where springs are used to isolate a major low-frequency disturbance, high-frequency noise is transmitted along the spring wire. In such situations it may be necessary to use a rubber pad between the spring and the machine frame. Some examples of isolation spring mounts are given in Fig. 16.37. Since metal springs have very little useful internal damping (~0.1 per cent), damping can be introduced in the form of viscous fluid, air or friction damping.

(b) Rubber mounts Rubber is one of the most effective

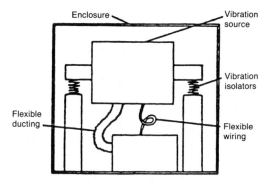

Fig. 16.33 Use of vibration isolation with an enclosure

Forcing frequency / Mount frequency	Transmissibility	Isolation efficiency	Result
1	Amplified	—	Worse than rigid mount
$\sqrt{2}$	1.0	0	Same as rigid mount
1.5	0.80	20%	Very poor
2	0.33	67%	Fair to good
2.5	0.19	81%	Good
3	0.125	87.5%	Very good
4	0.06	94%	Excellent
5	0.04	96%	—
6	0.02	98%	—
10	0.01	99%	Virtually vibration free mounting

(a)

(b)

Fig. 16.34 Typical vibration isolator performance

materials because of its viscoelastic damping properties (section 12.8). For a given volume of rubber, deflection is least in compression and greatest in shear (Fig. 16.38). Loading in compression thus provides the greatest load capacity and loading in shear the greatest flexibility (deflection). Combined loading in compression and shear is provided by an angled mount (Fig. 16.38). Load capacity and deflection then depend on the angle selected for the mount. The most common usage is the isolation of medium to lightweight machinery where the rubber mount acts in shear. The resonant frequency varies from 5 Hz upwards, making them useful for isolation in the mid-frequency range.

(c) Compression pads Principally cork or reinforced rubber materials, designed for a load-bearing function, but at the same time providing isolation and vibration damping from the support structure. See also vibration-damping materials in section 16.5.2(iv). Unlike rubber, cork is a compressible material and may accommodate deflection up to 30 per cent. Natural frequency depends on density and loading, but load limits are applied to achieve low-frequency isolation since large material thicknesses are then required.

(d) Air springs These consist of an enclosed volume of air compressed behind a piston or diaphragm (Fig. 16.39). An air spring can be so designed to give practically any spring rate, the lower limit to the resonant frequency being ~1 Hz. Frequency is controlled by the air volume under compression plus the area of the piston (bellows), the natural frequency being independent of the applied loading. Damping can be controlled by installing an orifice or throttling valve.

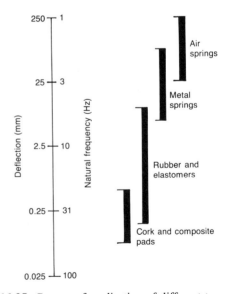

Fig. 16.35 Ranges of application of different types of isolator

NOISE REDUCTION

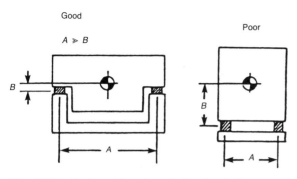

Fig. 16.36 Preferred location of vibration isolators

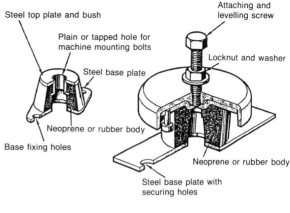

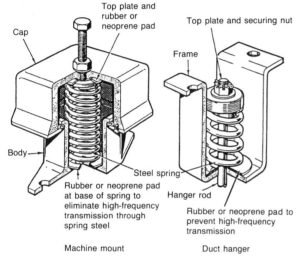

Fig. 16.37 Examples of isolation spring mounts. Courtesy of Woods of Colchester Ltd

Fig. 16.38 Examples of rubber isolation mounts. Courtesy of Woods of Colchester Ltd

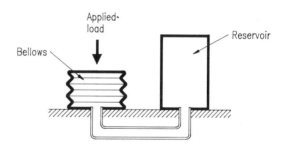

Fig. 16.39 Air spring

16.5.3 Active noise cancellation

This is a technique involving the cancelling of one sound by the transmission of a secondary sound in precise antiphase.[21] The principle has been recognized for many years, but it has required powerful computing techniques (digital electronics) to enable it to become a viable noise attenuation technology. The concept is illustrated simplistically in Fig. 16.40, a sound wave being blocked by an inverse wave of equal amplitude and frequency, the resultant cancellation energy dissipated as low-level heat.

There are two types of situation where it has proved possible to apply active noise cancellation:

1. Repetitive noise, i.e. cyclical, predictable or noise highly correlated with some measurable source activity. These systems would use a pulse to lock the electronics into the timing of the noise source. The signature of the noise source can then be accurately predicted and antiphase sound produced quickly with each cycle.

2. Randomly generated noise or vibration. In these systems, rather than predict the unwanted noise signature, a source monitor would be so positioned to perform predictions of the antiphase waves after a delay from monitor position to the selected sound point. Very high-speed signal processing is required.

511

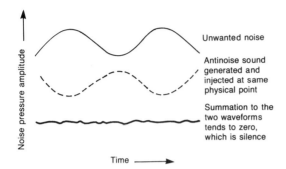

Fig. 16.40 Active noise cancellation concept

Although the basic theory is straightforward, the skill lies in implementing the process at the required instant. Typical areas where the technique is being applied are engine silencers, in-vehicle quieting systems, continuous process machinery, etc. Active noise cancellation has proven most effective at low frequencies, <600 Hz. Active control of local repetitive sound fields, i.e. from process machines, will allow normal conversation unimpeded.

16.6 Practical noise diagnosis and treatments

Before redesigning components or attenuating path noise, there has to be an intelligent identification of the noise source. The various identification methods can be summarized as follows:

1. The human ear is an exceptionally good sensor for machinery noise analysis. It has good directional and frequency discrimination.
2. If possible turn off various parts of the machine in turn and measure the noise level at each step to recognize the parts of the machine contributing most to the general noise level. By isolating machine elements in turn, the load may alter on another part so affecting its noise performance.
3. If a machine is producing an impulsive noise (section 16.3.3(iii)), it is desirable to examine the time history of the noise profile when related to mechanical events. Once the relationship is established, vibration measurements can relate the machine structure vibration to the noise level.
4. Frequency analysis is the simplest process that can be applied to a noise level to aid diagnosis and discriminate various frequencies (section 16.2.3).
5. The rate of change of noise level with speed can indicate the noise source in some cases by running a machine through its speed range. This would indicate that the noise source mechanisms are changing in order of importance and also whether resonances within the machine are causing high noise levels.
6. If the machine is placed in an anechoic chamber, directional patterns of noise radiation can be determined from equal noise contours plotted around the machine. Contours can be plotted for any frequency bandwidth. Alternatively, noise sources at some particular position (e.g. operators) can be investigated with a directional microphone.
7. With machine covers or panels, an accelerometer or vibration transducer can be used to estimate vibration level and dimension, hence noise contribution, of each panel.

Sections 16.4 and 16.5 deal in detail with the fundamentals of machine element noise generation, causes and avoidance, plus practical methods and techniques of reducing the effects. Because machine configurations vary so significantly, it would be difficult to define a precise investigative route for dealing with a noise problem; rather apply basic noise reduction principles to the source and the path. The rules are as follows:

1. Always minimize the excitation forces.
2. The degree of transference should be evaluated and minimized. The total noise level may be increased above that of the primary source through energy transference interactions between parts of the machine, structural or air path.
3. Maximize the absorption of radiated sound, direct source as well as aerodynamically generated.
4. Reduce noise dissipated from vibrating surfaces by damping them, resonant panels and resonant structures.

Not all procedures are necessary for all machines, though some experience is necessary to judge relative effectiveness. Noise problems are better dealt with during the early stages of the design process; it is not always an easy or cheap process to carry out retrospective noise reduction of an existing machine.

(i) Exciting forces
Some guidance for minimizing exciting forces is as follows:

1. Balance rotating machinery (section 16.5.1(ii)).
2. Run at the slowest possible speed. Reduce inertia of moving parts.

3. Form gears accurately, fit plastic gears where possible (section 16.4.1(iii)).
4. Reduce impacts by fitting resilient stops or reduce impact velocity by modifying mechanism or mechanism actuation.
5. Modify the action of the machine to reduce the rate of change of forces, i.e. the acceleration and deceleration rates, drive take-up time, etc.
6. Use materials having better damping properties, i.e. use of machine castings, plastic cams, etc.
7. If transmission noise is essentially tonal, examine drive profiles, tension, the need for synchronization, etc.
8. Avoid obstructions and abrupt section changes close to fan blades (sections 16.5.1(ii) and 16.4.1(v)). Check correct fitting of grills, vanes and dampers.
9. Ensure fastenings are secure and cannot work loose.
10. Use only minimum forces and drive sources required for operational efficiency. Where possible replace a cyclic force with a constant one.

(ii) Noise radiation

Both direct sound from the source and transferred exciting forces cause vibration and subsequent noise radiation. Some guidance for minimizing noise radiation is as follows:

1. Minimize the area of the noise radiating surface, i.e. panels and covers (section 16.4.2).
2. Provide a resilient link in the structural path between the source and the vibration radiator (section 16.5.2(v)).
3. Use damped sheet metal in preference to undamped sheet (section 16.5.2(iv)).
4. The outer surface of a machine should consist of highly damped panels attached to the structural framework at positions where vibration of the framework is minimal. Normally these would be points of mass/stiffness concentration remote from the vibrating source. This minimizes vibration input as well as the resonant amplification of the outer surface.
5. Stiffen covers in order to raise the resonant frequencies above the machine excitation range. Maximum excitation is at the panel centre.
6. Avoid attaching vibrating parts to covers and panels.
7. Ensure that the noise transmitted through a panel or cover is not greater than that emitted as a result of vibration excitation of the panel (sections 16.5.2 and 16.4.2).
8. Do not mount vibration sources on a lightly damped fabricated structure.
9. Ensure panels and covers are not subjected to any other direct vibration excitation such as air discharge jets, etc.
10. Reduce noise within the machine by adding acoustic absorption to the inside of panels (section 16.5.2(i)).

16.7 Noise labelling

In its simplest form, noise labelling requires that manufacturers and retailers attach to particular categories of machinery and equipment, a label which informs potential customers of the noise emitted during a prescribed test.[4] In this context, it is to be hoped that market power itself can be harnessed to progress towards a quieter environment. 'Less noise' thus becomes positive information alongside machine performance. However, although this subject has been considered within various standards-making bodies for many years, nothing much happened until the emergence of two EEC directives, the Noise Directive[1] effective 1990 and the Machinery Directive[2] effective 1994. These are applicable both to machines and working environments, laying down for the first time a noise control obligation on machine manufacturers, as follows:

1. The Noise Directive, EEC 86/188, states that 'the risks resulting from exposure to noise must be reduced to the lowest level practicable, taking account of technical progress and the availability of measures to control the noise, particularly at source'. The following goals are laid down:
   (a) measurement of noise exposure at the workplace;
   (b) a general obligation to reduce noise;
   (c) adequate information on the noise emission from machines;
   (d) limit to the noise exposure in the working environment.
2. The Machinery Directive, EEC 89/392, puts the noise reduction obligation into a firm requirement, stating 'machinery must be so designed and constructed so that risks resulting from the emission of airborne noise are reduced to the lowest level taking account of technical progress and the availability of means of reducing noise, particularly at source'. The directive provides for:
   (a) determination of machine noise emission;
   (b) declaration of noise emission level in the technical documentation;
   (c) a general obligation to reduce noise;
   (d) installation instructions must indicate requirements for reducing noise and vibration.

Directive 86/188 requires a noise declaration, but gives only a general statement. Directive 89/392 goes far deeper into the subject and specifies which noise emission quantities have to be declared, expressed in the following terms:

1. An A-weighted continuous sound pressure level at the workplace, L_{Aeq};
2. A C-weighted sound peak pressure level at the workplace, Lp_C;
3. The A-weighted acoustic power level of the machine.

The noise emission data declared by the machine manufacturer may be 1+2 or 1+2+3. The applicable noise threshold levels are given in Table 16.17. There is a further EEC Directive, 86/594,[3] that deals with airborne noise emitted by domestic appliances, but compliance is optional on member states. France and Germany in particular[4,22,23] have made progress with framing noise emission regulations which influence consumer choice through 'less noise'.

In the noise-labelling requirement, noisy as well as quiet machines will be declared. However, there is a degree of discussion within various countries as to how the noise statement should actually be made. It has been suggested[24] that noise labelling should be reserved for the few machines on which a label must be stuck to comply with a national requirement. Outside this restriction a 'noise declaration' is acceptable in commercial and technical documents and need not appear on the machine. A unified noise declaration procedure is expected to be in force throughout the EEC and the European Free Trade Association (EFTA) by 1993.[25]

The reasons advanced for noise labelling are many, but the criteria can be summarized into four basic groups[24] as follows:

1. *Economical and marketing.* (a) To increase the quality of products; (b) to make possible a more loyal dialogue between manufacturer and consumer; (c) to make 'less noise' a sale factor; (d) to make 'less noise' a purchase factor; (e) to identify the quietest machines.
2. *Statutory.* (a) To help legislators fix reasonable noise emission limit values if required; (b) to fulfil a national regulation.
3. *Prevention during the design process.* (a) To encourage manufacturers to reduce noise at the source; (b) to predict better the noise impact of a new machine in the workplace.
4. *European.* (a) To suppress barriers to trade in EEC and EFTA countries; (b) to contribute to a quieter Europe/world.

Before noise labelling can function effectively, a number of conditions need to be fulfilled:[4]

1. There should be a clearly defined and universally acceptable noise descriptor that easily communicates to the buyer the noise emission level of the machine.
2. The noise descriptor should be linked to a clearly defined measuring method.
3. There needs to be a test house methodology for certification and verification of labelled values.
4. The subject of noise labelling will require public education.
5. Consumer rights against false labelling should be accessible in law.
6. Noise abatement legislation should advantage manufacturers and favour consumers of low-noise products.

Although EEC directives have been published[1-3] outlining the general requirement, European standards are required to supplement these and so create test and performance uniformity (see Ch. 17). These must be based on international standards and measurement codes wherever possible so that the declarations may be compared at an international level. The following standards family will apply to noise labelling within the EEC:[25]

1. The general standards for the determination of emission characteristics of sound sources are ISO 3740–3747 + 6081 (Table 16.18).
2. 'Noise test codes' specific to a family of machines and stating values for machine parameters, i.e. operating conditions, mounting conditions, measurement positions etc., during noise measurement, e.g. EN704, ISO 1680, 9296 (Table 16.18).
3. Specific noise declaration standards, ISO 4871 + 7574 (EN 27574) (Table 16.18).

Table 16.17 *Noise labelling thresholds and labelled quantities according to EEC directive 89/392*

Noise emission at the workplace(s). Thresholds for noise labelling	Noise emission information to be yielded
if L_{Aeq} < 70 dB(A)	'L_{Aeq} < 70 dB(A)'
if 70 dB(A) ≤ L_{Aeq} < 85 dB(A)	'L_{Aeq} = ... dB(A)'
if L_{Aeq} ≥ 85 dB(A)	'L_{Aeq} = ... dB(A)' L_{WA} = ... dB(A)'
if L_{pC} < 130 dB(C)	'L_{pC} < 130 dB(C)'
if L_{pC} ≥ 130 dB(C)	'L_{pC} = ... dB(C)'

Table 16.18 *International acoustics standards*

Standard	IEC/EN	ISO	BS	EEC Directive
Emission and labelling Acoustics — determination of sound power level of noise sources — guidelines for the use of basic standards and for the preparation of noise test codes		3740 (1980)	4196 Pt 0	
Acoustics — determination of sound power sources — precision methods for broad-band sources in reverberation rooms		3741 (1975)	4196 Pt 1	
Acoustics — determination of sound power level of noise sources — precision methods for discrete-frequency and narrow-band sources in reverberation rooms		3742 (1975)	4196 Pt 2	
Determination of sound power levels of noise sources — engineering methods for special reverberation test rooms		3743 (1976)	4196 Pt 3	
Acoustics — determination of sound-power levels of noise sources — engineering methods for free-field conditions over a reflecting plane		3744 (1981)	4196 Pt 4	
Acoustics — determination of sound-power levels of noise sources — precision methods for anechoic and semi-anechoic rooms		3745 (1977)	4196 Pt 5	
Acoustics — determination of sound-power levels of noise sources — survey method		3746 (1979)	4196 Pt 6	
Acoustics — determination of sound-power levels of noise sources — survey method using a reference and sound source		3747	4196 Pt 7	
Acoustics — determination of sound-power levels of noise sources — engineering method for small, omnidirectional sources under free-field conditions over a reflecting plane		3748		
Acoustics — noise emitted by machinery and equipment — guidelines for the preparation of test codes of engineering grade requiring noise measurements at the operators position		6081 (1986)	7025 (1988)	
Acoustics — noise labelling of machinery and equipment		4871 (1984)	—	
Statistical methods for determining and verifying the noise emission values of machinery and equipment. General	27574 Pt 1	7574/1	6805 Pt 1 (1987)	
Statistical methods for determining and verifying the noise emission values of machinery and equipment. Method for determining and verifying stated values for individual machines	27574 Pt 2	7574/2	6805 Pt 2 (1987)	
Statistical methods for determining and verifying the noise emission values of machinery and equipment. Method for determining and verifying stated values for batches of machines using a simple transition method	27574 Pt 3	7574/3	6805 Pt 3 (1987)	86/594
Statistical methods for determining and verifying the noise emission values of machinery and equipment. Method for determining and verifying stated values for batches of machines	27574 Pt 4	7574/4	6805 Pt 4 (1987)	86/594

Table 16.18 (*Cont.*)

Standard	IEC/EN	ISO	BS	EEC Directive
Acoustics. Determination of emmision — relevant sound power levels of multi-source industrial plants. Engineering method	—	8297	—	
Acoustic power measurement — intensity method	—	9614	—	
Determination of sound power levels of noise sources; characteristics and calibration of reference sound sources	—	6926	—	
Emission specifications				
Test code for the determination of airborne acoustic noise emitted by household and similar appliances — general requirements	704-1 (HD 423.1)	—	6686 Pt 1 (1986)	
Test code for the determination of airborne acoustic noise emitted by household and similar appliances — particular requirements	704-2-1 (1984) 704-2-2	—	6686 Pt 2.1 (1990) 6686 Pt 2.2 (1990)	Vacuum cleaners Forced draught convection heaters
Acoustics. Measurements of airborne noise emitted by computer and business equipment. Methods of measurement	—	7779	7132 Pt 1 (1989)	
Noise emitted by computer and business equipment. Method of measurement of high-frequency noise.	—	9295	7135 Pt 2 (1989)	
Noise emitted by computer and business equipment. Method of determining and verifying declared emission values	—	9296	7135 Pt 3 (1989)	
Airborne noise emitted by computer and business equipment. Method for determination of sound pressure levels at the operators and other specified positions around the machine tool	—	7960 Pt 1	≡4813 (1972)	
Airborne noise emitted by machine tools. Operating conditions for metal-working machines	—	7960 Pt 2	—	
Airborne noise emitted by machine tools. Operating conditions for wood-working machines	—	7960 Pt 2	—	
Fans		5136 10302		
Textile machines		9902		
Test code for the measurement of airborne noise emitted by rotating electrical machinery; Pt 1: Engineering method for free-field conditions over a reflecting plane.	[34 Pt 9] HD 53.9	[4999 Pt 109 (1987)] 1681 Pt 1		
Test code for the measurement of airborne noise emitted by rotating electrical machinery; Pt 2 Survey method		1680 Pt 2		
Fans for general purposes		[5136 (1990)]	848 Pt 2 (1985)	

Whatever means of measuring the noise level is chosen, the result can be affected by the acoustic environment, the measuring equipment, test laboratory operator, etc. in other words a noise declaration needs to be statistically based to take experimental tolerances into account. Within international standards (Table 16.18), care is taken to introduce clear statements regarding the measurement uncertainties involved. A-weighted sound pressure levels are measured at the operator position (section 16.3.4(ii)). If no operator position is specified, a noise level is determined by averaging the measured values at the four bystander positions, (section 16.3.4(iii)). Because the purchasers and authorities will wish to make their own evaluation, a statistical reproducibility of noise testing is introduced.[26] For a 95 per cent probability of acceptance:

Individual machines $L_d \geq L_m + 1.645\,\sigma_R$ dB

where L_d = declared noise emission value (sound power or pressure level), L_m = measured value of individual machine and σ_R = standard deviation of measured values obtained under reproducible conditions, i.e. different test labs, different times, different operators. If σ_R is not known, or not specified, an estimated standard deviation can be used as follows (see also section 16.3.2(i)):

Type 1 accuracy class, $\sigma_R = 1.5$ dB
Type 2 accuracy class, $\sigma_R = 2.0$ dB
Type 3 accuracy class, $\sigma_R = 5.0$ dB

Batches of machines

$$L_d \geq L_m + 2\sigma_t$$

where L_m = average measured value of machines in the batch, σ_t = total standard deviation of measured values for the batch = $(\sigma_R^2 + \sigma_p^2)^{1/2}$ and σ_p = the standard deviation of measured values obtained on different batches, same measuring equipment under repeatable conditions.

If σ_p is not known for the batch, use 1.5 dB.

References

1. EEC Directive 86/188 on the protection of workers from the risks related to noise exposure at work. OJ of the European Communities L137 24 May 1986
2. EEC Directive 89/392 on the approximation of laws of member states relating to machinery. OJ of the European Communities L183 29 June 1989
3. EEC Directive 86/594 dealing with airborne noise emitted by household appliances. OJ of the European Communities 6 Dec. 1986
4. Eddington N Noise labelling and legislation. *Noise & Vibration Control Worldwide* April 1989: 118
5. BS 5969:1981 (IEC 651) 'Sound level meters'
6. BS 41196:Pt 4:1981 (ISO 3744) *Determination of sound power levels of noise sources; engineering methods for free field conditions over a reflecting plane*
7. ECMA 74 *Measurement of airborne noise emitted by computers and business equipment*
8. BS 7135:Pt 1:1989 *Noise emitted by computer and business equipment* — Pt 1 *Method of measurement of airborne noise*
9. BS 4196:Pt 5:1981 *Precision methods for determination of sound power levels for sources in anechoic and semi-anechoic rooms*
10. BS 4196:Pt 2:1991 *Precision methods for determination of sound power levels for discrete frequency and narrow band sources in reverberation rooms*
11. Detecting faulty rolling element bearings by vibration analysis. *Des. Eng.* **19**: 16
12. Angelo M Vibration monitoring of machines. *B & K Technical Review* **1**: 1 (1987)
13. Wong C J Timing belt noise of office machines. *Noise Control Engineering Journal* **35**(3): 103 (1990)
14. Russell M F et al Noise from cooling fans for small rotating machines. *Proceedings of the Institution of Mechanical Engineers Conference. 'Engineering a Quieter Europe'*. Birmingham 1990
15. Gray L M A Review of the physics of axial fan acoustics and aerodynamics with a view towards noise control. *Proceedings Noise Control 'Quieting the Noise Source'*. INCE 1983
16. Wilson E A Air movers and audio noise in electronic equipment. *Des. Eng.* Feb. 1983: 55
17. Lyon R H *Machinery Noise and Diagnostics*. Butterworths 1987
18. Bull D G The control of noise at machinery source, a practical systematic teaching scheme. *Proceedings of the Institution of Mechanical Engineers Conference. 'Engineering a Quieter Europe'*. Birmingham 1990
19. BS 6861:Pt 1:1987 (ISO 1940) *Balance quality requirements of rigid rotors*
20. Bies D A, Hansen C H *Engineering noise control*. Hyman 1988
21. Williams J E Anti-Noise. *Proceedings Noise Control 85. 'Computers for noise control'*. Institute of Noise Control Engineers 1985
22. Mola S Overview of the improvement in the acoustic quality of electrical household appliances Inter-noise 90. *Proceedings of the International Conference on Noise Control Engineering*. Gothenburg Aug. 1990
23. Maue J H Noise labelling practice in the case of industrial vacuum cleaners. Inter-noise 90. *Proceedings of the International Conference on Noise Control Engineering*. Gothenburg, Aug. 1990
24. Jaques J Machine noise labelling — requirements for an effective practical implementation *Proceedings of the Conference 'Engineering a Quieter Europe'*. Institution of Mechanical Engineers 1990
25. Jaques J Noise labelling. An overview of the concept and its practical implementation. Inter-noise 90. *Proceedings of the International Conference on Noise Control Engineering*. Gothenburg, Aug. 1990
26. Higginson R F Measurement uncertainites in determination of noise emission. Inter-noise 90. *Proceedings of the International Conference on Noise Control Engineering*. Gothenburg, Aug. 1990

CHAPTER 17
DESIGN CERTIFICATION

17.1 Introduction

This chapter concerns the subject of product or design certification which, for the vast majority of electrotechnical products and equipment, means compliance with a defined safety standard* and the recognition of that compliance by third-party assessment. The setting of standards of safety implies the definition of what is acceptable as a reasonable level of risk in the foreseeable use or even misuse of a product or process. Risk management in the context of safety hazards and product liability as part of the design process is dealt with in Chapter 2. Here we consider the principle of reference to standards to demonstrate the conformity or fulfilment by a product or equipment to laid-down specification requirements. Such approvals are also often a legal requirement before a product may be placed in a particular market-place. Conformity is assessed by notified bodies or test houses. An overall view of product certification is represented by Fig. 17.1, the certification system itself contained by particular standards and the standards in turn embraced by regulations.

However, the standards scene has been a rapidly changing one since the creation of the European Single Market in 1993 and will continue to develop further for some time. Thus the theme of this chapter will concentrate on the implications of the European economic area in regard to design certification rather than our own base of national British Standards, although cross-references will be given as appropriate.

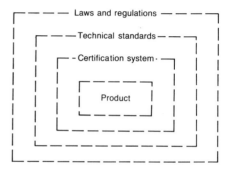

Fig. 17.1 Overview of certification

Within the European economic area (EEA), there are 18 countries, including members of the European Economic Community (EEC) and the European Free Trade Association (EFTA) with some 360 million people affected: but the existence of national regulations for the protection of safety, and so on, creates an obvious form of barriers to trade.† Each national system may reflect different traditions and approach. The result can make trading within Europe of goods subject to regulation expensive in money and time and can be one of the major cost penalties associated with not yet having a true Single European Market. National technical barriers will be progressively eliminated as the Single Market is completed such that any product that can be marketed in any member state of the EEC can be freely sold in all other parts. By providing for the implementation of harmonized technical criteria, one-stop testing will

---

*A standard may be defined as a 'document established by consensus and approved by a recognized body that provides for common and repeated rules, guidelines or characteristics aimed at achieving the optimum degree of order in a given context, consolidating the results of science, technology and experience aimed at the promotion of optimum community benefit'.[1]

†An EEA agreement establishing the world's largest free trade area was negotiated between the EEC and EFTA in May 1992. It is reasonable to suppose that most EFTA countries will eventually become an integral part of the Single Market.

become the European quality goal for marketing in Europe.

Although guidelines regarding the requirements of standards and directives are given in this chapter, it should be emphasized that this information should not be regarded as a substitute for reading the specific documents. Similar standards do in fact differ in certain details hence the standard itself should always be checked for precision of requirements.

17.2 Approaches to certification

17.2.1 Testing and standards

Standards for electromechanical equipment started to take shape in the early 1950s. It has been an evolving situation with new standards being produced as new products come on to the market. At first there were national standards with specific requirements made by individual authorities. However, since the early days there has developed a firm commitment to harmonizing of standards in Europe and to assist manufacturers in production for broader markets. Manufacturers today have to design their products to comply with international standards and preferably demonstrate compliance with a third-party type testing and certification. They have the responsibility for product liability and often use certification as an important sales and marketing argument.

The testing of single item (product) is known as type testing and the results obtained relate only to the product which has been subjected to examination. The certification body and test laboratory perform a type test to demonstrate the product is in conformity with a specific standard. The development of mass production has led to the need for type testing being integrated into product approval systems which ensure continued quality for the whole of production. To follow up that production continues to comply with the standard, the certification body may operate factory inspections, market surveillance and/or repeated type tests. The approvals or certification bodies themselves are, in turn, subjected to a process of certification.

Today more and more products are covered by European harmonized standards developed within the European Committee for Electromechanical Standardization (CENELEC), making efficient European co-operation in product testing and certification possible. Test results from one country can be accepted in the others, saving time and money. Likewise, harmonized International Electrotechnical Commission (IEC) standards are necessary for international certification. A summary of international and national certification activity is given in Table 17.1. The types of standard involved are summarized under the headings below.

(i) ISO/IEC standards

These are likely to form the basis of European standards, especially so in the electrotechnical requirements area. These express, as nearly as possible, an international consensus of opinion on the subjects dealt with and so constitute recommendations rather than mandatory requirements and will contain specific national differences where agreement is not found.

(ii) Harmonized document (HD)

This is a European standard issued by the European Committee for Standardization (CEN) or CENELEC on the basis of general agreement over the majority of requirements for a new standard. It will contain specific national deviations which cannot be resolved on a European basis. Failure to reach complete agreement may hinge on the fact that there are historical differences of approach that cannot be changed quickly.

(iii) The European norm (EN)

This indicates a standard that is fully harmonized and accepted by all CEN/CENELEC member countries. The only national deviations allowed are due to legal requirements. The process of harmonization has in the main been a process of each member state arguing for the inclusion of its own requirements. This naturally brings about a good deal of compromise, but at the end of the process, essential safety requirements are nevertheless embodied in this emerging European standard. Fully harmonized national standards carry an EN reference number and, except for the language, are identical in all member countries. Equipment design based on such standards is therefore acceptable throughout the Community.

17.2.2 International certification

In order to facilitate international trade in electrotechnical equipment, there exists convenient processes for mutual use and recognition of test results from other countries. These are known as international certification schemes. As a minimum basis for certification, each participating country must have its own notified and accredited test house, referred to as an NCB (national certification body). A list of NCBs along with their approval marks is given in Fig. 17.2. The general procedure for international

Table 17.1 *Summary of global, European and national certification activities*

		International		European (EEC+EFTA)		National (British)
Standards-making bodies	ISO:	International Organization for Standardization. Comprises national standards bodies of some 88 countries. Wide field of responsibility covering all areas of standardization outside electrotechnology. Promotes international acceptance of particular standards	CEN:	European Committee for Standardization. European equivalent of the ISO	BSI:	British Standards Institution National member body of ISO and CEN
	IEC:	International Electrotechnical Commission. The electrotechnical arm of the ISO. Comprises national standards bodies of some 43 countries.	CENELEC:	European Committee for Electrotechnical Standardization (formerly CENELCOM). Many European standards ratified by CENELEC are based on IEC documents	BEC:	British Electrotechnical Committee. National member body of IEC and CENELEC
			ETSI:	European Telecommunications Institute. Standardization body in telecommunications policy		
Certification schemes	IECQ:	IEC Quality Assessment system for Electronic Components. International equivalent of the CECC.	CECC:	Cenelec Electronics Components Committee. Covers certification and quality assessment of electronic components; some 15 participating countries. Also publishes standards.	BSI safety marks:	A product certification scheme for products that conform to a BS specifically concerned with safety or to the safety requirements of standards that cover other product characteristics as well
	IECEE:	IEC system for Conformity Testing to Standards of Safety of Electrical Equipment	ECITC:	Covers different certification aspects of information technology products according to CEN and CENELEC standards.	BSI Kitemark:	Indicates on a product that BSI has satisfactorily and independently tested samples of the product for compliance against the appropriate British Standard together with an assessment of the manufacturer's quality system.
	CB:	Certification Body: IECEE's certification system, originally developed for use with CEE standards. Integrated into the IEC in 1985 as the IECEE. Comprises some 29 countries with three further applicants.	HAR:	European marking of low-voltage cables and cords tested and approved in accordance with standards HARmonized in line with CENELEC HD standards. HAR agreement is a multilateral recognition agreement		

DESIGN CERTIFICATION

Standards used		BASEC:	British Approvals Service for Cables. The organization nominated by the UK Government under the Low Voltage Directive for certification of cables, cords and wires.
	CCA:	CENELEC Certification Agreement. A multilateral recognition agreement for low voltage equipment, mainly electrotechnical appliances, components and equipment	BEAB: British Electrotechnical Approvals Board. UK notified member for participation in CCA and CB schemes
	LOVAG:	Low Voltage Agreement group for low-voltage industrial equipment	HS: Harmonized Standard. This means a published British Standard that is aligned with a CENELEC Harmonized Document.
	EMC:	EMC testing and certification co-ordinating committee under consideration within CENELEC	NS: National Standard. Not aligned with a Harmonized Document. Types of British Standard are specifications, codes of practice, test methods or reference glossaries.
	HD:	Harmonized Document. This is a standard which has been ratified by the CENELEC technical board	Normalized standard: A British Standard aligned with a European Norm.
	EN:	The European Norm. is the standard which will eventually become the common document used throughout CENELEC	Quality system: Usually against ISO 9002, EN 29000 or BS 5750–2
	IEC:	International Standard produced by an IEC Technical Committee. Represents an international consensus of opinion on requirements as accepted by the National Committees. Any divergence between IEC recommendations and national rules will be indicated.	Note: BEAB Approvals would be based upon the hierarchy of standards set out in the Low Voltage Electrical Equipment (Safety) Regulations, i.e. (a) An HD, HS, EN or NS. (b) An appropriate IEC LVD Standard if any. (c) A British Standard or standard from another CENELEC country accepted under mutual recognition rules.
	ISO:	International Standard produced by an ISO Technical Committee	

AUSTRALIA	Standards Association of Australia, Sydney CB		AUSTRIA	Osterreichisches Verband für Elektrotechnik, Vienna OVE CCA, HAR, CB		BELGIUM	Comité Electrotechnique Belge, Linkbeek CEBEC CCA, HAR, CB
CHINA	China Commission for Conformity Certification of Electrical Equipment CCEE CB		CZECHOSLOVAKIA	Electrotechniky zkusebni ustav, Prague CB		DENMARK	DEMKO, Herlev, Copenhagen DEMKO NCA, CCA, HAR, CB
FINLAND	Finnish Electrical Inspectorate, Helsinki SETI NCA, CCA, HAR, CB		FRANCE	Union Technique de l'Electricité, Paris UTE CCA, HAR, CB		GERMANY	VDE- Prufstelle, Offenbach VDE CCA, HAR, CB
GREECE	Hellenic Organization for Standardization, Athens ELOT CCA, HAR, CB		HUNGARY	Hungarian Institute for Electrical Equipment Testing, Budapest MEEI CB		IRELAND	National Standards Authority of Ireland, Dublin NSAI CCA, HAR, CB
ISRAEL	Standards Institution of Israel, Tel-Aviv SII CB		ITALY	Istituto Italiano del Marchio de Qualita, Milan IMQ CCA, HAR, CB		JAPAN	IECEE Council of Japan, Tokyo JET CB
NETHERLANDS	N.V. KEMA, Arnhem KEMA CCA, HAR, CB		NORWAY	NEMKO, Oslo NEMKO NCA, CCA, HAR, CB		POLAND	Central Office for Quality of Products, Warsaw CBJW CB
PORTUGAL	Instituto Portugues da Qualidada, Lisbon IPQ HAR		RUSSIA	GOOSTANDART, Moscow GOST HAR, CB		SPAIN	Asociacion Electrotecnica y Electronica Espanola, Madrid AEE CCA, HAR, CB
SWEDEN	SEMKO AB, Stockholm SEMKO NCA, CCA, HAR, CB		SWITZERLAND	Schweizerischer Elektrotechnischer Verein, Zurich SEV CCA, HAR, CB		UNITED KINGDOM	British Electrotechnical Approvals Board, Walton-on-Thames BEAB CCA, HAR(BASEC), CB

Fig. 17.2 Summary chart of national test bodies and international certification

certification schemes is that the applicant submits the product to an NCB, the primary NCB, generally in the country of his choice. Once certified by a primary NCB, then testing laboratories in other countries, secondary NCBs, within the certification scheme would undertake to accept the test results from the primary NCB. Depending on the degree to which the testing standards are harmonized, and national deviations considered by

DESIGN CERTIFICATION

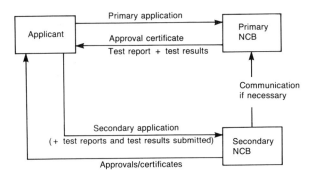

Fig. 17.3 Certification scheme process

the primary NCB, the secondary NCB may insist on limited testing. The certification process is depicted schematically in Fig. 17.3.

(i) Certification body (CB) scheme

This is the IEC global scheme (Table 17.1), a mutual recognition scheme covering a substantial part of the world market. Members are identified in Fig. 17.2. The principle is that a manufacturer may obtain a CB test certificate for a defined product from a primary NCB which may than be presented to the NCBs in other countries whose certification marks are requested. Test results are recognized according to IEC standards. In practice the scheme is used mostly by controls manufacturers. However, the different NCBs may choose to participate in the CB scheme just for the product standards of their own choice, thus causing a certain limitation in flexibility. In the CB scheme there is no co-operation in the field of factory inspection or market surveillance procedures.

(ii) CENELEC certification agreement (CCA) scheme

This is a scheme for international recognition of products basically covered by the Low Voltage Directive,[2] and based on HDs or ENs. A type test may be performed by any of the CCA NCBs within the EC and European Free Trade Association (EFTA). A test report, NTR, is then recognized by the others as a base for their certification or approval. It is a choice of the applicant to ask the primary NCB also to test for possible national deviations of other countries. In most cases the secondary NCB will require a sample of the product. The CCA NCBs use factory inspection and/or market surveillance for follow-up of certified products.

(iii) Nordic certification scheme

The five Nordic countries run the Nordic Certificate Service under the auspices of EMKO, the Joint Body of the Electrical Testing Institutes of the Nordic countries (Fig. 17.2). In this case, when one of the NCBs conducts complete Nordic testing, all deviations of any of the individual countries are taken into account. If the applicant informs the primary NCB about representatives in the Nordic countries marketing the product, these agents receive approval certificates from secondary NCBs after information is given direct from the primary NCB. Complete Nordic testing can be based on a CCA NTR or on a CB certificate.

17.3 Conformity in the Single Market

One of the principal aims of the Single Market is to break down the barriers between member states and remove the burdens for firms trading in more than one country. Nowhere have the obstacles been more apparent than in the field of national standards. As the regulatory barriers to free circulation of industrial products within the Community are removed, the Single Market will only become a reality for European industry as common technical standards are developed at the European rather than national level. To this end a 'new approach' was adopted by the Council of Ministers in 1985 to provide technical regulations at the Community level. Instead of setting detailed technical requirements, EC directives made under this approach are confined to a statement of the essential safety requirements or any other requirements in the general interest. EC directives are addressed to member states. They do not instantly create Community law, but they do state an objective which must be realized by member states within an allowed time-frame and do allow member states to take into account any special domestic circumstances.

Thus in the 'new approach', EC legislation confines itself to laying down the 'essential requirements' (ERs) of the directives. In respect of each directive, harmonized European standards are then developed in order to provide manufacturers with a set of technical specifications giving a presumption of conformity to the ERs. Both CEN and CENELEC are responsible in this regard for the harmonization of divergent national standards or the preparation of new standards if none exist. However, such standards remain voluntary. Manufacturers are still able to put products into the market-place that meet other standards, or no standards at all, subject to the fulfilling of conformity assessment procedures laid down in the directive. Demonstration of conformity is known as 'attestation'. All products must demonstrate conformity with the relevant ERs. Manufacturing in conformity with harmonized standards is one way of so doing.

Those 'new approach' directives relevant to electromechanical products are as follows:

1. *Construction products* (89/106/EEC). Implemented in the UK by the Construction Products Regulations 1991 (SI 1991/1620) (no date published yet for compliance);
2. *Machinery* (89/392/EEC and 91/368/EEC). Implemented in the UK by the Supply of Machinery (Safety) Regulations 1992 (SI 1992/3073) (compliance period to 31.12.94);
3. *EMC (electro-magnetic compatibility)* (89/336/EEC as amended by 92/31/EEC). Implemented in the UK by the Electromagnetic Compatibility Regulations 1992 (SI 1991/2372) (transition period for compliance to 31.12.95);
4. *Product safety.* Under consideration (9284/91) (compliance period to June 1994);
5. *Product labelling* (noise, efficiency, etc.). Under consideration, see section 16.7.

In addition to these there is the Low Voltage Directive (LVD) 1973 (Council Directive 73/23/EEC), not a 'new approach' directive but nevertheless an important directive on which all current electrical equipment safety regulations are now based. Where a simple manufacturers' declaration of conformity is not considered adequate on the grounds of a combination of scale of risk and likelihood of failure, the intervention of a third party is required. Such third parties are required to be 'notified' as competent bodies by governments of member states. A 'notified body' will conform to the requirements of EN 45000 (BS 7500).

17.3.1 The CE mark

When a manufacturer makes a product in conformity with one or more of the Community's 'new approach' technical directives, he is required to put the CE mark on the product (Fig. 17.4). The mark represents a presumption of conformity with all or any of the directives that apply to it including any conformity assessment requirements (section 17.3.2). Products not covered by the 'new approach' directives are not entitled to carry the mark. The specific requirements for conformity are set out in the directives themselves. Some products are covered by more than one 'new approach' directive. The CE mark is not a consumer mark nor is it a quality symbol; it reflects the fact that legal safety provisions apply to that particular product placed in the market-place, i.e. it indicates that the ERs are met and therefore is not related to a specific standard. If, however, a product standard incorporates the ERs, the product will be deemed to comply with an EN and so must be given the CE designation. An exception is made in the case of the LVD, applying probably to some 75 per cent of electrical equipment in the Single Market, in that no marking is required. However, in the future this position will change to bring the LVD in line.

Strictly speaking, the CE mark is for the authorities, i.e. when a product carries a CE mark, the authorities in the EC member states must assume that the EC directives are being adhered to and the product is thereby allowed to be freely offered as in the Single Market, no member state may refuse access on technical grounds.

Needless to say, there is a degree of confusion regarding the placing of the mark, what it actually means and the question of its coexistence with national marks. However, the Commission has attempted to clarify the situation with a proposed regulation*.[3] The proposed rules for the use of the CE mark are as follows:

1. The person affixing the CE mark to a product assumes responsibility for its conformity to all Community directives and regulations, i.e. that person makes the declaration of conformity. In the event of a challenge, the declaration must be kept available to enforcement authorities for 10 years following the placement of the equipment in the market-place.
2. The mark should be fixed to the product or on the rating plate. However, where the particular product type does not allow for this, the mark can be fixed to the packaging or accompanying documents if the directives allow. There must be enough information on the product to enable traceability back to the

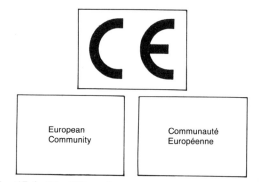

Fig. 17.4 The CE mark

*Regulations in Community legislation are directly applicable, they do not have to be confirmed by national parliaments to have a binding legal status.

manufacturer or agent if there are doubts over conformity.
3. Information concerning the compliance directives, etc. should not appear with the CE mark, but either on a declaration of conformity, certificate of conformity or product documents.
4. Where a notified body is involved at the production phase, the CE mark should be accompanied by the identification number of that body.
5. The product may bear other marks, i.e. national marks, but these should not be capable of being confused with the CE mark.

17.3.2 Declaration of conformity

The common factor relating all the 'new approach' directives is that in addition to applying the CE mark, the manufacturer or his authorized representative in the Community is required to make a declaration of conformity, also referred to as the EC declaration of conformity. This would typically contain the following particulars:

1. Name and address of person, manufacturer or representative in the Community affixing the CE mark;
2. Description of the equipment, product or machinery;
3. Relevant provisions complied with;
4. Details of the 'attestation' process. Attestation of conformity describes the processes and procedures that were necessary to demonstrate the conformity of the equipment with the technical specifications defined by the ERs. These procedures may range from a manufacturer's statement to third-party testing and certification.

In making a declaration of conformity, the attestation requirements will depend on the directive itself. The EC Commission has developed a set of modules describing elements of conformity assessment procedures relating to the design and production phases of manufacture,[4] elements of which will apply to a directive as appropriate. This has also been termed the 'modular approach' policy. For the first time, quality assurance is recognized as a technique for establishing conformity. The modules range from a simple manufacturer's declaration with only the requirement of a technical file, to varying degrees of third-party intervention. The normal level applied to the declaration of conformity will most likely be compliance to a relevant harmonized specification coupled with compliance with the requirements for a factory production control system (dual module), i.e. the process typically applied today by a number of national marks bodies. A simplified summary of the modular approach is given in Fig. 17.5. In practice the Commission would decide what variants are applicable to a particular product group and the choices available within those variants.

For CE mark purposes, the Commission may decide that the attestation process should include one of the following:

1. An EC-type examination carried out by a notified body to ascertain and certify that an example of the equipment satisfies the relevant provisions of a directive;
2. Certification of conformity with specification carried out by a notified body against a harmonized technical standard; and possibly
3. Certification of conformity of the factory production control system by a third-party certification body (inspection body).

It should be noted that bodies for the above assessment and certification activities have to be designated by member states for the purpose and notified to the Commission as such. The terms 'approved', 'designated' and 'notified' tend to be used interchangeably in EC documentation.

17.3.3 European Organization for Testing and Certification (EOTC)

It is difficult to imagine that European industry can continue to live with a system of national marks within the context of Community legislation providing for the use of the CE mark. The various certification schemes in section 17.2 have proved their usefulness, but have not led to the development of a true European culture in matters relating to testing and certification. To this end the Council of Ministers agreed in 1990 to the setting up of the EOTC, to operate alongside CEN and CENELEC as a new organization. The purpose of EOTC is to act as a catalyst to bring conformity assessment bodies together to establish mutual recognition arrangements or certification schemes. The ultimate aim is 'one-stop testing', inspection or certification, so that a product examined by one body in one European country is accepted by others without the need for further examination or tests. This is the aim of the Single Market and beyond. It requires mutual confidence achieved via the use of agreed criteria and procedures 'agreement' groups set up within EOTC. These will be formed by testing, inspection or certification bodies active in the market prepared to sign and manage flexible mutual recognition agreements or certification schemes at the European level.

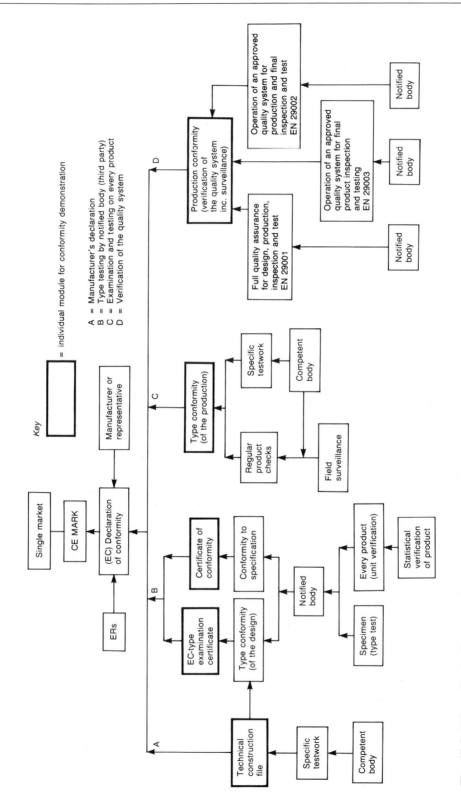

Fig. 17.5 EC conformity demonstration routes

Of all the European standards in need of harmonized application, the EN 29000 standard for quality systems is probably the most obvious example. Assessment of company quality systems is a well-known source of multiple activity and a barrier to trade. Through EOTC the application of this standard will be harmonized so that confidence may be placed in third-party assessments.

17.3.4 The European standard

The true European standard does not yet exist in its own right. Present standards harmonized within CEN and CENELEC have no formal status until the national standardization bodies transpose the contents as one or more national standards and withdraw any conflicting provisions. It is questionable whether two-stage standardization is in the interests of customers since it implies different national marks need to be acquired in different parts of the Community — this is not what the Single Market programme is all about. It is considered that future European standards should exist in their own right and should not need to be transposed at national level before they can be used. A single conformity mark (ES) would contribute to a clearer public conception of European standardization.[5]

17.4 Safety standards

Safety standards applicable to small machines of an electromechanical nature will recognize the internationally accepted level of protection required against electrical, mechanical and thermal hazards. Application of such standards will encompass the principles of safety intended to prevent injury or damage when products and machinery are normally operated in accord with the manufacturer's instruction. Abnormal situations expected in practice, i.e. misuse, would also be covered. In standards documentation the term 'hazard' is universally used to foresee a risk situation (see Ch. 2). Such hazards may be summarized in general terms as follows:

1. *Electric shock hazard.* Due to passage of current through the human body. Parts that have to be touched or handled should be at earth potential or properly insulated. Internal wiring and connections need to comply with specified standards and be adequately secured. It is normal to provide two levels of protection for the equipment operator to prevent electric shock in the case of a fault. It is assumed that service personnel are competent in dealing with obvious hazards, but the design should protect against unexpected hazards and provide warning (Ch. 2). Voltages up to 40 V peak or ~ 60 V d.c. are not generally regarded as dangerous.
2. *Mechanical hazard.* The general requirement is to prevent injury to the operator from parts of equipment exposed in normal use and to prevent access to moving parts or electrical connections. Also to ensure that the equipment is mechanically stable and structurally sound and avoid the presence of sharp edges and points.
3. *Heat hazard.* The general requirement is to prevent injury due to the high temperatures of parts that may be accessible to the operator.
4. *Fire hazard.* Temperatures that would cause a fire risk may result from overloads, component failure, insulation breakdown, high resistance or loose connections. Fires originating from within the equipment should be contained. Such design objectives may be met by taking all reasonable steps to avoid high temperatures that might cause ignition, controlling the quantity of combustible materials, ensuring that combustible materials have low flammability and using suitable equipment enclosures.
5. *Radiation hazard.* This can cover a wide field, such as acoustic, UV radiation, laser, radio-frequency (RF) emission, high-intensity visible light sources, etc. Unsafe levels are not allowed.
6. *Chemical hazard.* Regulations covering the use and application of chemicals are generally defined and understood. Reliance is mainly placed on equipment warnings and usage instructions.

The guiding principle in protection from any of the above hazards is the 'principle of double improbabilities'. In other words, if one level of protection fails, then there is another level that will provide protection. Most of the requirements of safety standards are just common sense and good engineering practice. Within Europe (EEC and EFTA) the small machinery product range is covered by the following standards:

Domestic machinery:	EN 60335-1	BS 3456-201[6]
Office machinery:	EN 60950	BS 7002[7]
Industrial machines:	EN 60204-1	BS 2771-1[8]
		BS 5304
Electrical equipment:	IEC 348	BS 4743[9]
	IEC 1010	
	IEC 601-1	BS 5724-1

The above are typically Part 1 or parent documents specifying the general requirements. Within the EN 60335 series, specific standards also exist for different types of appliance. While domestic and office machinery may be certified by an approved body to comply with a safety standard, the industrial machine standard only lays down the basics required for design liability — it is a code of practice to be observed by the manufacturer. Clearly a formal independent approval is not possible with large, custom-built or specialized machinery where, in addition, the build quantities may be low.

All of the above safety standards will call up other specialized standards and test procedures as required, many of which will be common to all the above standards. These are cross-referenced in Table 17.2 against the IEC, BS and HD equivalents.

In those cases where a manufacturer's declaration of conformity to a standard is not considered adequate on the grounds of scale of risk and failure probability or simply on the basis of a legal requirement, then the intervention of a third party is required to demonstrate conformity, i.e. the involvement of a competent certification body or test house for type testing. Tests are made on a single representative sample, as delivered. The objective is to determine whether the equipment, as designed, is capable of meeting the requirements of the standard by withstanding all the relevant tests. Except where specific test conditions are stated elsewhere in the particular standard, tests are usually carried out with the equipment or any movable part of it placed in the most unfavourable position that could occur in normal use and under the most unfavourable combination within the manufacturers' operating specification of the following:

1. Supply voltage and tolerance (+6 to −10 per cent);
2. Supply frequency;
3. Operating mode;
4. Adjustment of operator-accessible regulating and control devices.

Where the safety of individual components is relevant to the overall equipment safety, then components would either comply with the requirements of the equipment standard or with individual component standards. Typical components would be relays, transformers, motors, etc. The component would be checked for its correct application in accordance with its rating to determine compliance.

17.4.1 Definitions

Before describing the nature of product safety features expected from compliant equipment and products, a few definitions are necessary; these are given under the headings below.

(i) Electrical insulation

Basic insulation. This is the insulation applied to live parts to provide basic protection against electrical shock. It is commonly the first line of protection from electrical hazards.

Supplementary insulation. This denotes independent insulation (secondary protection) applied in addition to basic insulation to ensure protection against electric shock in an event of failure of the basic insulation.

Double insulation. Denotes insulation comprising both basic insulation and supplementary insulation.

Reinforced insulation. A single insulation system applied to live parts which provides a degree of protection against electric shock that is equivalent to double insulation.

(ii) Class of equipment

Class 0 (not legal in UK). In this case, protection against electric shock relies on basic insulation, i.e. there are no means for the connection of conductive accessible parts, if any, to the protective conductor in the fixed wiring of the installation. Reliance in the event of failure of the basic insulation is placed upon the environment. A class 0 appliance either has an enclosure of insulating material or a metal enclosure separated from live parts by appropriate insulation.

Class 01. Equipment has at least basic insulation throughout and is provided with an earthing terminal, but has no earth connection via the supply cord or plug/socket.

Class I. Equipment having at least functional insulation throughout and provided with a terminal or contact for protective earthing. Accessible metal parts are connected to the protective earthing conductor in the fixed wiring of the installation in such a way that conductive accessible parts cannot become live in the event of basic insulation failure.

Class II. Protection against electric shock relies on double or reinforced insulation throughout. There is no provision for protective earthing or reliance on installation conditions. If the outer enclosure is an insulating material, any screws, nameplates, etc. must be isolated from live parts by at least reinforced insulation.

(iii) Degree of enclosure protection

Certain environments and conditions of use require that equipment and machinery are adequately protected against

DESIGN CERTIFICATION

Table 17.2 Cross-reference of standards called into product and machinery safety standards

Subject of standard	Household and commercial products to 250 V a.c. BS 3456-201 EN 60335-1	Business equipment to 600 V a.c. BS 7002 EN 60950	Industrial machines to 1000 V a.c. BS 2771 EN 60204-1	Specification	Test method	Code of practice	BS	HD	IEC	Other
Screw lampholders	*			*			6776		238	EN 60238
Lamp caps and holders	*			*			5101-1	65	61-1	
Appliance couplers	*			*			4491		320-2	EN 60230
Isolating transformers	*	*	*	*			3535-1	21	742	EN 60742; CEE 15
PVC insulated cables	*	*	*	*			6500	22	245	
Rubber insulated cables	*	*	*	*			6500		245	
Miniature cartridge fuses	*			*			4265	109	127	
Conduits and thread sizes			*		*		6053	393	423	
Impact test apparatus					*		7003		817	
Thermal classification of electrical insulation	*	*			*		2757	495	85	
AC motor capacitors	*			*			5267		(252)	
Metal oxide thickness measurement					*		5411-5			ISO 1463
Coating thickness measurement					*		5411-11			ISO 2178
Proof tracking indices	*	*			*		5901	214	112	
Insulating material flammability	*	*			*		6334	441	707	
Flammability tests, various					*		6458-2	444-2	695-2	
Industrial plugs and sockets		*	*	*			4343	196	309	CEE 17
Graphical symbols		*	*			*	6217	243	417	
Degrees of protection		*	*	*			5490	365	(529)	
Cable sheath insulation		*			*		6469	385	540	
Low-voltage systems		*	*	*			PD 6469		664	
Metric threads			*	*			3643			
Mains electrical equipment		*	*	*			415	195	65	
Controls	*	*		*			3955		730	
Environmental testing procedures			*		*	*	2011-2	323	68-2	
Colours of indicator lights			*			*	4099		73	
Contactor control gear (to 1000 V)			*	*			5424	419: 420	408 158	
Control switches (to 1000 V)		*	*	*			4794	420	337	
Air break switches (to 1000 V)			*	*			5419	422	408	
Switchgear and control gear				*			5486		439	EN 60439
Terminal marker system			*			*	5559	241	445	
Actuator movement indication			*			*	6013	331	447	
Controls interface specification			*			*	5782	373	550	
Electrical conduits		*	*	*			6099	394-1	614-1	
RFI measurements					*		800			CISPR 14; EN 55014
Measurement methods for RFI					*		4999: 5000	231	72	CISPR 11
Dimensions and ratings, rotating M/c				*			5070	246	113	
Diagrams, charts and tables			*			*	88		269	
Fuses				*			2754	366	536	
Protection against electric shock			*			*				

access of solids and liquids. In the case of user finger access to live parts, the converse applies inasmuch that the user or operator requires the protection. The end result, however, as far as the design aspect is concerned, is the same. The IP or the international protection rating is the means used to grade the protection level afforded by the equipment covers or enclosure. A summarized pictorial layout of the IP ratings is shown in Fig. 17.6. Typically two numbers are used, i.e. IPn_1n_2. The first number determines the protection offered against the intrusion of solid objects such as levers, screwdrivers, fingers, etc. and the second figure indicates the degree of protection against water intrusion. The first number will probably not be a requirement of European standards, hence the degree of moisture protection will be described by IPXn, where n is from the range 0 to 8 (Fig. 17.6). This supersedes the drip-proof, splash-proof and watertight categorization.[6,10] In Fig. 17.6 a third number is included, protection against mechanical shock. It is used by Continental manufacturers to define the impact or shock resistance of an enclosure. The grade defines the energy content allowable of a mechanical shock.

17.4.2 Identification and instructions

In order that electrical equipment and products may be properly identified, installed and maintained, a certain amount of technical information must be made available, some of which needs to be clearly marked on the product and the other detail in the form of installation instructions.

(i) Marking
Products, or any detachable part of the item in question, should be marked with the following information:

1. Rated voltage(s) or voltage range(s);
2. A symbol for the nature of the supply (if applicable);
3. Rated frequency;
4. Rated input in watts or kilowatts or rated current in amperes;
5. Rated current of appropriate fuse link (if applicable);
6. Maker's or vendor's name, trade mark or identification;
7. Maker's model or reference number;
8. Rated operating time (if applicable);
9. Symbol for class II construction (if applicable);
10. Symbol for degree of moisture protection, IPn_1n_2.

Such marking information should be so placed as to be unambiguous and clearly visible after installation, e.g. either on the outside or directly visible when opening a door or cover. Normal practice would be to use a rating label or rating plate. In all cases the information displayed must be legible, durable and capable of withstanding a rub test with petroleum spirit. In addition, the following information may need to be clearly displayed:

1. A visible warning that before gaining access to live terminals, supply circuits must be isolated.
2. Protective earth terminals must be identified with an earthing symbol placed on a permanent surface.
3. If, to preserve the equipment from damage, it is necessary to refer the user to the instruction manual, the equipment should be marked with a caution symbol.[11]

Where symbols are used to convey the above information, they should be in an accepted format (Table 17.3).

(ii) Instructions
The information necessary for the installation, operation and maintenance of equipment should be supplied in the form of instruction sheets, diagrams, charts, etc. as appropriate depending on the complexity of the product and the nature of any safety risks. See also Chapter 2. Such information would be in an agreed language, but typically in the language of the country into which the product is being sold. Such information may consist of the following:

1. A wiring diagram fixed to the appliance, unless the

Table 17.3 *Recognized format for the display of information on a product*

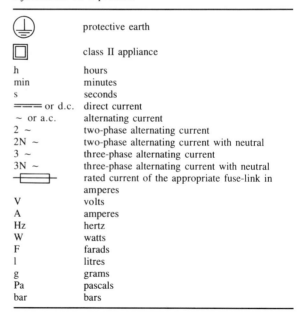

⏚	protective earth
☐	class II appliance
h	hours
min	minutes
s	seconds
⎓ or d.c.	direct current
∼ or a.c.	alternating current
2 ∼	two-phase alternating current
2N ∼	two-phase alternating current with neutral
3 ∼	three-phase alternating current
3N ∼	three-phase alternating current with neutral
⊸⊏⊐⊸	rated current of the appropriate fuse-link in amperes
V	volts
A	amperes
Hz	hertz
W	watts
F	farads
l	litres
g	grams
Pa	pascals
bar	bars

DESIGN CERTIFICATION

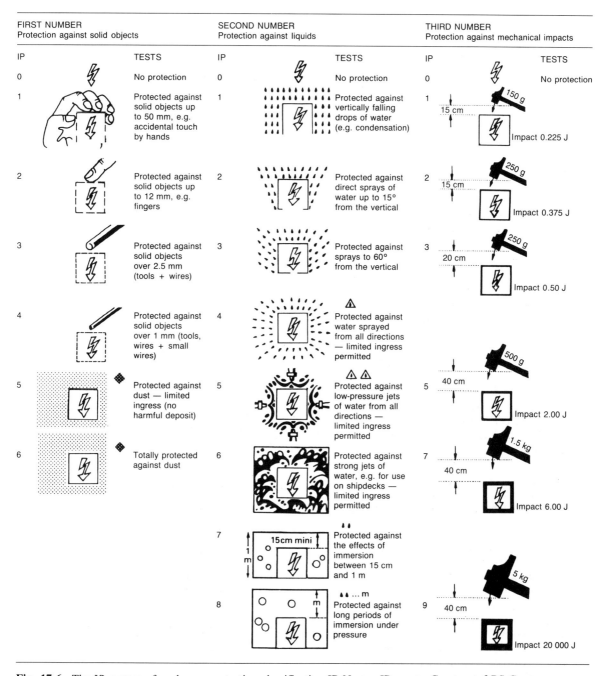

Fig. 17.6 The IP system of enclosure protection classification IP No. = IP$n_1 n_2 n_3$. Courtesy of RS Components Ltd, Corby

    mode of connection is obvious;
2. A circuit block diagram plus interconnection diagrams where appropriate;
3. An installation drawing and instructions;
4. A description of operation;
5. Maintenance instructions and adjustment procedures;
6. A parts list where appropriate;
7. A recommended spares list.

17.4.3 Electrical safety requirements

In general terms, electrical equipment should provide protection to the user against electric shock, both in normal operation and in the case of a fault (abnormal operation). The following comments will apply:

1. Products should be so constructed and enclosed that there is adequate protection against accidental contact with live parts. Compliance is checked with a standard articulating test finger, a test probe or test pin as appropriate.
2. Appliances intended to be connected to the supply by means of a plug shall be so designed that there is no risk of electrical shock from charged capacitors when touching the pins of the plug.
3. Where live parts are located in an enclosure, the use of a tool is necessary to open the enclosure or to disconnect the supply before the enclosure can be opened.
4. Live parts must be covered by durable insulation.
5. Proper functioning of the equipment should not be impaired by momentary voltage interrupts, dips or spikes.
6. Where appropriate, safety interlocks should be provided to stop the operations concerned.
7. All electrical components, e.g. automatic controls, switches, terminal blocks, transformers, should comply with the safety requirements specified in the relevant standards (Table 17.2), and used in compliance with their marked operating characteristics. If not marked, or where no relevant standard may be called up, the component is tested under the conditions occurring in the equipment. However, wherever possible, safety critical components should be certified as complying with a relevant standard by a recognized competent authority.

The following sections will deal with the more specific electrical safety design requirements.

(i) Electrical insulation and leakage current

At operating temperature, insulation should be adequate and leakage current in normal use not excessive.

An insulation test or electric strength test comprises a test voltage, 50 or 60 Hz, applied between live parts and accessible parts for 1 min. No flashover or breakdown should occur during the test. In the case of class II equipment, the test voltage is applied between live parts and parts separated from live parts and basic insulation only. The test voltages applied are as follows:

500 V for basic insulation subject to an extra low voltage;
1000 V for other basic insulation;
2750 V for supplementary insulation;
3750 V for reinforced insulation.

Where humid conditions occur in normal use, an additional insulation test will be called for following a high humidity soak.[6,12]

Leakage current is measured at approximately $1.1 \times$ supply voltage between any pole of the supply and accessible metal parts or metal foil in contact with accessible insulated surfaces. In normal operation and after acclimatization at high humidity, the leakage current should generally not exceed the following:

Class 0, 01 and III equipment	0.5 mA
Portable class I equipment (including hand held)	0.75 mA
Stationary class I equipment	3.5 mA (0.75 mA if subject to high humidity)
Class I stationary heating appliances	0.75 mA per rated kW input to 5.0 mA maximum
Class II equipment	0.25 mA

(ii) Creepage and clearance

Creepage distances are the distances measured over the surface of insulation between a live part and, for instance, a live part or different potential or any other metal part. Clearances are measured over the shortest distance between these parts over the surface, through air or both. Basic insulation is often just fresh air if the clearance distance is reliably maintained. Some examples of creepage are given in Fig. 17.7. Allowable minimum creepage and clearance distances are given in Table 17.4.

In addition to these limits, the distances through insulation for working voltages up to 250 V between metal parts should not be less than 1.0 mm if separated by supplementary insulation and not less than 2.0 mm if separated by reinforced insulation. Standards usually show preference for basic and supplementary insulation systems since they automatically incorporate the 'double improbability' approach. A reinforced or single layer is only really advised where it is 'impracticable' to achieve two separate layers.

(iii) Internal wiring

In the placement and routeing of internal wiring, a few design rules should be followed:

DESIGN CERTIFICATION

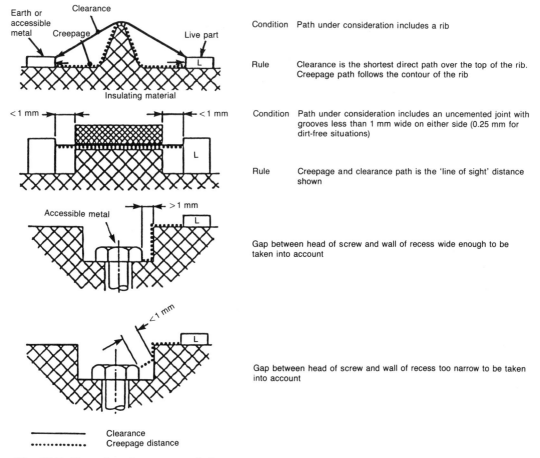

Fig. 17.7 Examples of creepage and clearance

1. Wire routeings should be smooth and free of sharp edges. Electrical wiring should be so protected to avoid contact with burrs, cooling fins, moving parts, etc. that may cause damage to the insulation of conductors. Holes in metal through which insulated wires pass shall have smooth, well-rounded surfaces or shall be provided with bushings.
2. Internal wiring should be routed, supported, clamped or secured in a manner that prevents:
   (a) excessive strain on the wire and on the terminal connections;
   (b) loosening of terminal connections;
   (c) damage to conductor insulation.
3. For uninsulated conductors, it should not be possible to reduce creepage and clearance distances below specified values in normal use.
4. Stranded conductors should not be consolidated by lead–tin soldering where they are subject to contact pressure unless the clamping means is so designed that there is no risk of bad contact due to cold flow of the solder.
5. Where sleeving is used as supplementary insulation, it should be retained in position by positive means.

(iv) Supply connection

Both for internal wires and external cables, the cross-sectional area should be adequate for the current they are intended to carry when the equipment is working under normal load such that the maximum permitted temperature of the insulation is not exceeded. Terminals for the connection of fixed wiring should allow for conductor sizes as given in Table 17.5. Such terminals should be so designed that they clamp the conductor between metal surfaces with sufficient contact pressure and without damage to the conductor. Terminals should not be accessible without the aid of a tool and be so located,

Table 17.4 *Minimum creepage distances and clearances*

Distances (mm)	Class III appliances and constructions		Other constructions					
			Working voltage ≤ 130 V[a]		Working voltage > 130 V and ≤ 250 V		Working voltage > 250 V and ≤ 480 V	
	Creepage distance	Clearance	Creepage distance	Clearance	Creepage distance	Clearance	Creepage distance	Clearance
Between live parts of different potential:[b]								
— if protected against deposition of dirt[c]	1.0	1.0	1.0	1.0	2.0	2.0	2.0	2.0
— if not protected against deposition of dirt	2.0	1.5	2.0	1.5	3.0	2.5	4.0	3.0
— if lacquered or enamelled windings	1.0	1.0	1.5	1.5	2.0	2.0	2.0	3.0
Between live parts and other metal parts over basic insulation:								
— if protected against deposition of dirt[c]								
(i) if ceramic or mica	1.0	1.0	1.0	1.0	2.5[d]	2.5[d]	—	—
(ii) if of other material	1.5	1.0	1.5	1.0	3.0	2.5[d]	—	—
— if not protected against deposition of dirt	2.0	1.5	2.0	1.5	4.0	3.0	—	—
— if the live parts are lacquered or enamelled windings	1.0	1.0	1.5	1.5	2.0	2.0	—	—
Between live parts and other metal parts over reinforced insulation:								
— if the live parts are lacquered or enamelled windings	—	—	6.0	6.0	6.0	6.0	—	—
— for other live parts	—	—	8.0	8.0	8.0	8.0	—	—
Between metal parts separated by supplementary insulation:	—	—	4.0	4.0	4.0	4.0	—	—

[a] For printed circuit boards, these values may be reduced between tracks to 150 V per mm (min. 0.2 mm) if protected against the deposition of dirt and 100 V per mm (min. 0.5 mm) if not protected against deposition of dirt.
[b] The clearances specified do not apply to the contact gaps of thermostats, switches, overload protection devices, relays etc.
[c] The interior of an appliance having a reasonably dustproof enclosure is considered to be protected against deposition of dirt provided the appliance does not generate dust within itself. Hermetic sealing is not required.
[d] May be reduced to 2.0 mm if the parts are rigidly held.

guarded or insulated that should a strand of flexible conductor escape when fitted, there is no risk of accidental contact between such a strand and accessible or unearthed conductive parts.

Where a non-detachable power cord is provided, this should be restrained via a cord anchorage. Supply cords should not be in contact with sharp points or edges.

For primary power isolation, a disconnection device should form part of the installation — a hand-operated device for disconnection during machine cleaning, maintenance and long service interruptions. Such a device should have a contact separation of at least 3 mm and for single-phase equipment both poles should disconnect simultaneously and in the case of three-phase, all phase conductors simultaneously. Where applicable, an emergency stop may be provided to stop the machine in case of danger.

(v) Earthing provision

Accessible metal parts of class 01 and I products which may become live in the event of an insulation fault should be permanently connected to an earthing contact. Class

Table 17.5 *Conductor sizes*

Rated current of appliance (A)	Nominal cross-sectional area (mm²)	
	Flexible cables and cords	Cables for fixed wiring
Up to and including 3	0.5–0.75	1–2.5
Over 3 up to and including 6	0.75–1	1–2.5
Over 6 up to and including 10	1–1.5	1–2.5
Over 10 up to and including 16	1.5–2.5	1.5–4
Over 16 up to and including 25	2.5–4	2.5–6
Over 25 up to and including 32	4–6	4–10
Over 32 up to and including 40	6–10	6–16
Over 40 up to and including 63	10–16	10–25

0, II and III products should have no earthing provision. An earthing or protective circuit should effectively connect all exposed conductive parts of the equipment. Continuity is assured by the following points:

1. Connections should not have any additional fastening functions.
2. The clamping means of earth terminals should be adequately locked against accidental loosening and it is not possible to loosen them without the aid of a tool.
3. A connection between earthing terminal or earthing contact and parts required to be connected should be of low resistance, i.e. $\not> 0.1\,\Omega$.
4. Flexible metal conduits should not be used as earth conductors.
5. When parts are removed for maintenance, the protective circuit should not be interrupted.
6. If a sub-assembly can be withdrawn while energized, it should still be connected to earth.
7. A protective circuit must not comprise switching devices, overcurrent protection devices or plugs.
8. Self-tapping screws may be used to provide earth continuity provided the connection is not disturbed in normal use and that at least two screws are used for each connection.
9. Screws and nuts that make a mechanical connection and provide earth continuity must be secured against loosening, e.g. with lock washers.
10. Conductive parts in contact at protective earth connections and terminals should not be subject to corrosion in any working, storage or transportation environment conditions.
11. Green/yellow conductors should only be used for earthing.

17.4.4 Abnormal operation

The equipment should be so constructed such that as a result of abnormal or careless operation resulting in a fire risk or mechanical damage, operator safety or protection against electric shock is not impaired. Electronic circuits should be so designed that a fault condition occurring would not render the equipment unsafe. After abnormal operation or a fault, the equipment should remain safe for an operator but typically need not be in full working order. Fused links, thermal cut-outs, overcurrent protection devices and the like may provide adequate protection. Components and circuitry are tested for compliance as described under the headings below.

(a) Motors Under locked rotor and other abnormal conditions, motors should not cause hazards because of excess temperatures. With the motor in the stalled condition, at the end of the test period or when motor protection devices operate (cut-outs, etc.) the temperature of the windings should not exceed those in Table 17.6. A normal running overload test is also applied in most cases, the torque load increased such that the current through the windings increases by 10 per cent. Allowable steady-state maximum winding temperatures are given in Table 17.6.

(b) Transformers In the event of short circuits which are likely in normal use, excess temperatures should not occur in the transformer or in the circuits associated with the transformer. Compliance would be checked with the most unfavourable short circuit or overload likely. Winding temperatures should not exceed those in Table 17.6.

(c) Other electromechanical components These would be tested by locking the mechanical movement in the most disadvantageous position and then continuously energizing for a defined period. The fault condition should not result in a hazard situation.

(d) Electronic circuits Except where there is a low power circuit or where protection against electric shock does not rely on the correct functioning of the electronic circuit, then fault conditions are considered and if necessary applied one at a time. In general, examination of the appliance and its circuit diagram will reveal the fault conditions expected to give the most unfavourable response. Compliance would be checked by simulating the following types of condition:

1. Short circuit of creepage and clearance distances between live parts;

Table 17.6 *Allowable temperature rise during abnormal operation*

Type of equipment	Insulation Class; Limiting Temperature (°C)				
	Class A	Class E	Class B	Class F	Class H
1. Stalled rotor test					
Equipment other than those operated until steady conditions are established	200	215	225	240	260
Equipment operated until steady conditions are established.					
— if impedance protected	150	165	175	190	210
— if protected by protection devices that operate during the first hour, max, value.	200	215	225	240	260
— after the first hour, max. value.	175	190	200	215	235
— after the first hour, arithmetic average	150	165	175	190	210
2. Running overload test					
Normal operation at max. rated voltage until steady-state conditions established, then torque load increased to increase current in windings in steps of 10%. Max. winding temperature value	140	155	165	180	200

2. Open circuit at component terminals;
3. Short circuit of capacitors and other components;
4. Failure simulation of integrated circuits (ICs) and microprocessors.

17.4.5 Mechanical safety requirements

The mechanical design and construction of the equipment must be adequate to withstand such handling as may be expected in normal storage, transportation and use such that no hazard may be created. In particular:

1. Units and equipment in normal use should not become physically unstable to a degree that they become a hazard to operators and service personnel. Stability tilt tests would be applied with doors, drawers, slides, etc. in the most unfavourable position.
2. Moving parts of motor-operated appliances should, as far as possible, be so arranged to provide adequate protection against personal injury. Protective guards, covers and enclosures should not be removable without the aid of a tool.
3. Edges and corners should be rounded and smoothed when otherwise they could be hazardous.
4. Equipment should have adequate mechanical strength, constructed against rough handling in normal use (impact loading) and in transportation (drop test, bump test and vibration testing). The product should not show any damage to impair compliance with other safety requirements.

Specific safety requirements of the construction are as follows:

1. Adequate degree of protection against moisture or liquids.
2. Minimum risk of a hazard due to entry of foreign objects from above or below.
3. Equipment producing dust or using powders, liquids or gases should be so constructed that no concentration hazard can exist to affect compliance of other safety requirements.
4. Handles, knobs, levers, etc. must not work loose in normal use and create a hazard.

DESIGN CERTIFICATION

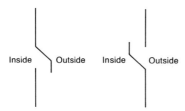

Fig. 17.8 Examples of safe louvre design

5. Louvres located on the side of an electrical enclosure should not exceed 5 mm, allow entry of foreign objects or allow contact with live parts (Fig. 17.8).
6. Accidental changing of control or reset devices should not result in a hazard.
7. Parts that affect the degree of moisture protection should not be removable without the aid of a tool.
8. If supplementary or reinforced insulation is removed during routine servicing, the appliance or product should be made inoperable.
9. Creepage and clearance distances should not be reduced as a result of wear, deposition of dust or contamination during use.
10. Direct contact between live parts and thermal insulation should be prevented.
11. Bare heating elements should be supported in case of rupture or sagging.
12. Where a heating appliance is located against a wall, spacers must be provided to prevent wall overheating, and not be removable from the outside.
13. No pointed ends of self-tapping screws or other fasteners should be touchable by the user during normal use or user maintenance.
14. No asbestos or PCB-containing oils should be used.
15. When a class II applicance is installed and connected to fixed wiring, the required degree of protection should be maintained.
16. Capacitors of class II equipment should not be connected to accessible metal parts.
17. Handles held in normal use must not become live in the event of an insulation fault.

Certain equipment, e.g. that used for cutting and grinding, and that where parts are exposed to hot gases and fluids and any other situation where a clear hazard exists when the equipment is used as intended, is permissible provided that there are sufficient warnings marked on the equipment, instructions for use are supplied and that, wherever possible, guarding is provided. Special regulations may apply to machinery when used in factories.[9]

17.4.6 Thermal safety requirements

The thermal risk is basically contained in two ways, firstly by ensuring that in normal use no part of the equipment attains an unsafe temperature, and secondly by ensuring that non-metallic parts are adequately resistant to ignition and to the spread of fire. Where fire may be caused by overload or insulation breakdown, then suitable fusing or thermal cut-outs may be used to assist in providing protection. In modern safety standards, some part of the equipment catching fire in a fault condition is allowed as long as the design and materials chosen are able to contain the source of the fire.

In normal use, in achieving a steady thermal state, the allowable temperature rises of various parts of the equipment are given in Table 17.7. Equipment would be tested in accord with manufacturers' installation and usage instructions.

The heat resistance of external parts of non-metallic material, insulating material supporting live parts, thermoplastic materials providing supplementary insulation, etc. is examined by various test methods to ensure that they are adequately resistant to ignition and spread of fire in case of a fault condition, i.e. the double improbability principle. Fiercely burning materials must not be used in any construction unless suitably impregnated. The test procedure sequence typically applied is given in Fig. 17.9. The tests are summarized under the headings below.

(i) Ball pressure test

This is a 5 mm diameter ball loaded against the test surface with a 20 N force, the material thickness $\not< 2.5$ mm. The temperature of the test specimen is at least 75 °C for external parts and 125 °C for surfaces suppporting live parts.[5,11] After 1 hour the diameter of the impression should not exceed 2 mm.

(ii) Burning test

This defines the flammability of a material, i.e. its ability to burn with a flame under specified test conditions. By arrangement of the test specimen in either a horizontal or vertical position, it is possible to distinguish between the different degrees of flammability of materials.[7,13] The horizontal position of the test specimen is particularly suitable for evaluating the extent of burning and/or velocity of flame propagation, i.e. the burning rate. The vertical position of the test specimen is particularly suitable for evaluating the extent of burning after extinction of the flame. A summary of test conditions and results is given in Table 17.8. Results from the two methods are not equivalent. Most design engineers are

Table 17.7 *Allowable temperature rises during steady-state operation*

Component parts of equipment under test	Temperature rise (°C)
Windings	
Class A insulation	75[a]
Class E insulation	90[a]
Class B insulation	95[a]
Class F insulation	115[a]
Class H insulation	140[a]
Connection terminals (including earth) of fixed appliances	60
Without temperature marking	30
With temperature marking	T-25
Rubber or PVC insulation of internal and external wiring including power supply cables	
Without temperature marking	50
With temperature marking	T-25
Materials used for insulation (other than that specified for wires and windings)	
Impregnated or varnished paper or press board	70
Epoxy bonded PCBs	120
Glass-filled polyester	110
Urea-formaldehyde mouldings	65
Silicone rubber	145
PTFE	265
Mica and dense ceramics	400
Wood	65
External enclosure of motor-operated equipment held in normal use	60
Handles, knobs, grips and similar parts which are continuously held in normal use:	
Of metal	30
Of porcelain or vitreous material	40
Of moulded plastic, rubber or wood	50
Handles, knobs, grips and similar parts which are held for short periods only in normal use:	
Of metal	35
Of porcelain or vitreous material	45
Of moulded plastic, rubber or wood	60

[a] Temperatures obtained using the resistance method, the temperature rise calculated from the following formula:

$$\Delta t = \frac{R_2 - R_1 (k + t_1) - (t_2 - t_1)}{R_1}$$

where Δt is the temperature rise of the winding R_1 is the resistance at the beginning of the test, R_2 is the resistance at the end of the test, k is equal to 234.5 for copper windings, t_1 is the room temperature at the beginning of the test and t_2 is the room temperature at the end of the test.

more familiar with the UL94 flammability categories.[14] The equivalence of UL ratings is apparent in Table 17.8. The minimum requirement in the IEC is for FH3(HB) classification (Fig. 17.9). Separate standards apply to foamed material.[7]

(iii) Glow-wire test

If separately moulded samples are not available for (ii), the glow-wire test is carried out (Fig. 17.9).[15] This test simulates thermal stresses which may be produced by heat or ignition sources, e.g. glowing elements, overloaded resistors. The glow wire is electrically heated to a given temperature, applied at a force of 1 N for 30 s. The specimen is considered to have passed if either (1) there is no flame or glowing, or (2) any flames or glowing extinguishes within 30 s of the removal of the glow wire, assuming the surrounding parts have not completely burnt. Wire temperature guidance is given in Table 17.9. If insulated material supports connection carrying >0.5 A, higher temperature glow-wire tests are also done

DESIGN CERTIFICATION

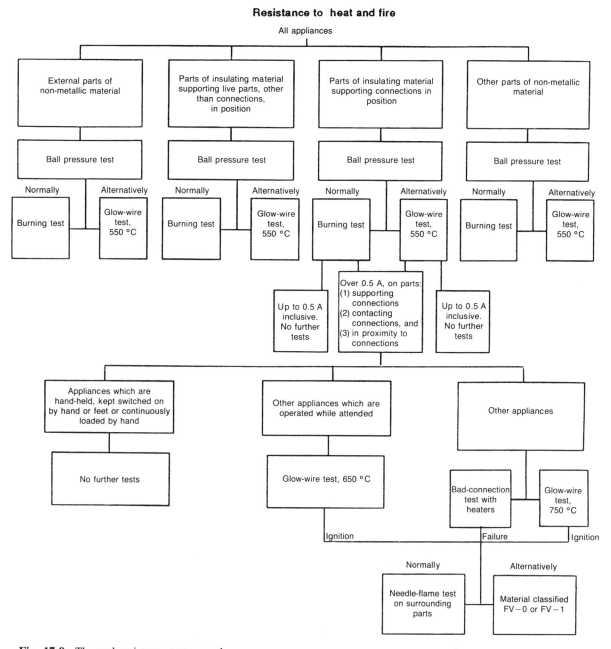

Fig. 17.9 Thermal resistance test procedure sequence

in close proximity (< 3 mm) to these connections (Fig. 17.9). If such parts flame during the application of the glow wire an additional needle-flame test may be applied.[15] There is no requirement to carry out a needle-flame test on materials rated FVO and above.

17.5 EC directives

In completing this chapter on design certification and standards, summaries of present and future possible directives in the electromechanical area are included in

Table 17.8 *Summary of burning tests*

1. *Horizontal burning test*	
Test:	Samples 130 × 13 × 3 mm; a 25 mm blue flame applied to one end for 30 s.
Results categories	
Category FH1:	No visible flame during the test.
Category FH2:	The flame ceases to burn before the 100 mm mark is reached by the front of the flame. The length of the burnt area is noted.
Category FH3:	The front of the flame reaches the 100 mm mark. The burning rate in mm/min should be noted.
Category HB:	Either the front of the flame does not reach the 100 mm reference mark or the burning rate does not exceed 40 mm/min. (75 mm/min for sample thicknesses < 3 mm)
2. *Vertical burning test*	
Test:	Specimen size 130 × 13 × 3 mm, held vertically. A 20 mm blue flame is applied to the lower end for 10 s, withdrawn and the duration of flaming noted. After flaming ceases, the burner is replaced for another 10 s and so on
Results categories	
Category V0:	This classification applies if the average duration of flaming does not exceed 5 s.
Category V1:	This classification applies if the average duration of flaming does not exceed 25 s.
Category V2:	As for V1 but flaming dipping that ignites cotton may occur.
Category 5V:	Using a 130 mm flame with a 40 mm inner cone, the test flame is applied for 5 s and removed for 5 s, the operation repeated for 5 cycles. The material should not continue to burn longer than 60 s after the fifth ignition. There should neither be any dripping nor significant destruction of the test specimen.

Table 17.9 *Guidance temperatures for glow-wire test*

	Parts of insulating material	
Temperature	In contact with current-carrying parts or retaining them in position	For enclosures and covers not retaining current-carrying parts in position
550°C	To ensure a minimum level of ignition of, and spread of fire by, parts liable to contribute to a fire hazard, and which are not subjected to other tests in this respect (to avoid fiercely burning material)	
650°C	Equipment for attended use	Fixed accessories in installations
750°C	Equipment for attended use but under more stringent conditions	
	Fixed accessories in installations	Equipment intended to be used near the central supply point of a building
	Equipment for unattended use but under less stringent conditions	
850°C	Equipment for unattended use continuously loaded	
960°C	Equipment for unattended use, continuously loaded but under more stringent conditions	
	Equipment intended to be used near the central supply point of a building	—

this section. At the time of writing, while a number of these have been adopted, none of the 'new approach' have been strictly implemented, mainly as a result of the harmonized standards not yet in place or the attestation procedures not properly formalized by the EC. It is therefore likely that transition periods will be allowed in most cases before the CE mark becomes mandatory. However, while manufacturing design should be moving closely towards the ERs, it should be borne in mind that compliance with standards (especially national ones) does not necessarily mean compliance with directives.

17.5.1 Low-Voltage Directive

This directive results from early EC legislation and

harmonizes electrical equipment safety standards for all products designed for use within the voltage range 75–1500 V d.c. and 50–1000 V a.c. The directive has been implemented in the UK by the Low Voltage Electrical Equipment (Safety) Regulations 1989.[2] This is not a 'new approach' directive so does not include the requirement for the CE mark. To counter any possible commercial disadvantages on that score the CE mark regulations will be amended in the future to make them applicable to the LVD while otherwise leaving that directive alone.[17]

The principal elements[2] may be summarized as follows:

1. The product description and characteristics to ensure safe use of the product in the application for which it is made should be marked on the product.
2. The manufacturer's brand name or trade mark should be clearly printed on the electrical equipment.
3. The electrical equipment should be made in such a way to ensure that it can be safely and properly assembled and connected.
4. The electrical equipment should be so designed and manufactured to ensure proper protection against such hazards as physical injury, electrical shock, high temperatures, dangers revealed by experience, environmental conditions, overload, etc. provided that the equipment is used in applications for which it is made and is adequately maintained.
5. The electrical insulation must be suitable for foreseeable conditions.

17.5.2 Machinery Safety Directive

This directive applies to 'functioning machines', i.e. powered assemblies of mechanically linked parts, at least one of which moves. The range of application covers anything from an electrical motor to a power station and is likely to have a major impact on manufacturers of machinery in the EEC. It does not cover manually powered machines, road vehicles, lifts or second-hand machinery (unless imported into the EEC). The ERs are expressed in terms of potential dangers to operators, i.e. hazards from the materials and products used in the construction, adequate lighting, noise and dust emissions, adequate design for handleability, protection from moving parts, adequate instructions for use, ease and safety of maintenance operations, typically as summarized by section 17.4. Additional essential requirements will apply to the more dangerous categories of machine i.e. agricultural foodstuffs machinery, portable or hand-held machinery, woodworking machinery, saws, presses and mobile machinery. This directive also introduces the noise reduction obligation (section 16.7).

Three levels of enabling machinery standards are in preparation within CEN to enable demonstration of conformity with the ERs. These are as follows:

First level A standards — comprising the general principle of machinery design and safeguarding, equivalent to BS 5304 (e.g. BS EN292 and BS EN414);

Second level B standards — covering specific safety devices and ergonomic aspects, i.e. visual displays, control layouts, guards, interlocks, stop switches;

Third level C standards — specific types of machinery calling up the appropriate items from A and B.

For most machinery, the manufacturer (or his representative) must draw up, and keep available, a technical construction file for at least 10 years from the date of manufacture. The technical construction file should comprise the following:

1. An assembly drawing of the machine plus control circuit layouts;
2. Detailed drawings, calculations and test results, etc. required to check conformity with the ERs;
3. A list of applicable ERs, standards and technical specifications defining the design;
4. Details of any hazard reduction methods developed;
5. Technical report(s) from a competent body;
6. Machinery instructions;
7. Confirmatory tests on safe installation.

In all cases an EC declaration of conformity must be drawn up for the machinery (section 17.3.2), and issued to the purchaser.

17.5.3 Construction Products Directive

The products covered by this directive are those 'produced for incorporation in a permanent manner into construction works, including both buildings and civil engineering works'. However, two other conditions must also be met before a manufacturer or supplier must comply with the requirements of the directive, namely (a) at least one ER must apply to the product in use; and (b) the 'works' where the product is in use must be subject to Building Regulations or other national legal provisions.

The ERs are as follows:

1. Mechanical resistance and stability;
2. Safety in case of fire;
3. Hygiene, health and the environment;

4. Safety in use;
5. Protection against noise;
6. Energy economy and heat retention.

Depending on the intended use of the product and the particular regulatory requirements, then some or all or none of the ERs may apply. The connection between ERs and specific products will in most cases be made through harmonized European standards, interpretive documents and European technical approvals (ETAs). Modules for conformity assessment will be allocated product by product. Acceptable methods of test will be devised by agreement groups working under the umbrella of the EOTC.

17.5.4 The Electromagnetic Compatibility Directive

The definition of electromagnetic compatibility (EMC) is the ability of a device, equipment or system to function satisfactorily in its electromagnetic environment without introducing intolerable electromagnetic disturbances to anything in that environment. The electromagnetic environment includes a range of phenomena such as transients and noise on mains supplies, electromagnetic fields from power cables, etc. plus electrostatic discharges from personnel, i.e. man-made disturbances. It is usual to exclude natural phenomena, such as lightning from the environment. All the above types of disturbance can be propagated by conduction, induction and radiation into equipment along power and interconnecting cables and on to sensitive circuits. The achievement of electromagnetic compatibility involves the limitation of noise generation and unwanted signals, a reduction of the extent of coupling between noise-generating and sensitive systems and the provision of adequate immunity of sensitive circuits to disturbances. Thus the EMC Directive requires that all electrical and electronic equipment sold in Europe (with some specific exceptions) will need to comply with the following requirements:

1. *The radio emissions produced*. The electromagnetic disturbance generated by an equipment should not exceed a level that does not allow radio and telecommunications equipment to operate as intended.
2. *Immunity to incoming electromagnetic signals*. The apparatus should have an adequate degree of intrinsic immunity to electromagnetic disturbances to enable it to operate as intended.

In addition, there are some further classifications:

1. *Noise sources*. May be divided into broad and narrow band (Table 17.10). To some extent the distinction is artificial in that it depends on the characteristics of the measuring instrument or system being disturbed. Nevertheless it is often a convenient distinction to make in assessing interference effects.
2. *Conduction*. Propagation of electromagnetic disturbance into equipment and systems via power supply cables, signal cables, earth connections, etc.
3. *Induction*. An important coupling mechanism between cables and also other components such as relays, transformers, inductors, etc. Current flowing in one conductor induces currents in adjacent conductors by magnetic induction.
4. *Radiation*. Electromagnetic radiation is produced by conductors in which currents or voltages are changing. Energy is propagated regardless of the presence of other conductors.
5. *Emissions*. Investigation may well involve the measurement of complex waveforms varying considerably in amplitude and time. Several methods of measurement have been devised to provide a realistic laboratory simulation giving consistent results in relation to the annoyance caused by the interfering signal (Fig. 17.10). In the commercial world, it is usually possible to divide test methods into conducted emission measurement at low frequency and radiated emission measurement at high frequency (Table 17.11).
6. *Immunity*. Similar principles are applied as in (5) but to external disturbances. These fall into four categories (Table 17.11):
   (a) disturbances conducted via power lines;
   (b) disturbances induced in external input/output cables;
   (c) radiated fields from RFI 'transmitters';
   (d) electrostatic discharges due to human body charging.

For non-telecommunications equipment, compliance with the directive may be demonstrated by either of the following:

1. *Self-certification*. This is expected to be the main route to compliance for the majority of products and is where the manufacturer himself declares the product compliant by using a specified European standard. Here the manufacturer is required to perform testing and satisfy himself that the product meets the technical requirements of the relevant standard before making a declaration.
2. *A technical construction file*. In those cases where there are no harmonized standards available, only

DESIGN CERTIFICATION

Table 17.10 *EMC noise sources*

	Continuous	Discontinuous	Notes
Broad band	Electrical machinery	Electrical contact devices	Most of these devices are subect to limitations of noise emission by existing standards and regulations, e.g. BS 800
	Computator motors	Thermostats	
	Flourescent lighting	Relays	
	Tyristor phase control circuits	Automatic control circuits	Transients due to switching operations which may be very fast and hence covering a wide frequency spectrum, are discontinuous in nature but not subject to any form of control
	Ignition systems		
Narrow band	Harmonics	Oscillators	
		RF heating equipment and cooking apparatus	
		Pulse generators	
		Digital equipment (all types)	

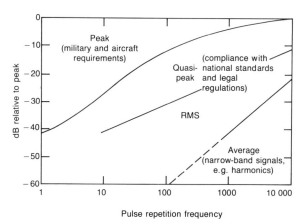

Quasi-peak detector: A detector which delivers an output which is a defined fraction of the peak value of an impulsive input signal, the fraction tending towards unity as the repetition frequency increases

Fig. 17.10 Response of various EMC emissions detectors

national standards have been used or where the manufacturer chooses not to use standards, the manufacturer may decide on a technical construction file. However, this must include an EC-type examination certificate from a competent body who will have verified the performance and results of the EMC testing.[18]

In order that manufacturers are able to declare conformity to the ERs, suitable technical standards are being generated (Table 17.11). However, it is not possible to produce dedicated emissions and immunity standards for all the electrical and electronic product types that currently exist, therefore CENELEC have developed both generic and product-related standards. Generic standards may be applied across a broad range of products, but where a product-orientated standard exists, it will take precedence.

17.5.5 Product Safety Directive

At present there is a draft directive designed essentially as a 'catch-all' to introduce a general duty on manufacturers, importers and the like to supply only safe products. When implemented, it will apply to all products, howsoever used, and therefore would cover consumer products, goods used at work, food, medicines, transport, etc. There would be no exclusions except where provisions of existing directives already cover particular safety features. A safe product (not presenting an 'unacceptable risk') is taken to mean a product that does not present in respect of its design, composition, execution, functioning, wrapping, conditions of assembly, maintenance or disposal, instructions for handling and use, or any other of its properties, an unacceptable risk for the safety and health of persons, either directly or indirectly, through its effect on other products or in

Table 17.11 Summary of EMC standards for emmissions and immunity

Emissions (EN 50081-1)

Harmonized standard applicable	Based on	Equipment covered	Type of interference	Frequency range	Limits
EN 55022	CISPR 22 BS 6527	Information technology equipment	Mains conducted Radiated	0.15–30 MHz 30–1000 MHz	46–56dB(μV) av. 56–66 dB(μV) quasi-peak 30dB(μV/m) up to 230 MHz; 37dB(μV/m) above 230 MHz at 10 m
EN 55014	CISPR 14 BS 800	Domestic, commercial and light industrial. (Toy trains to vending machines, e.g. equipment incorporating electric motors or generating impulsive 'click-type' interference)	Mains conducted Radiated (from mains load) Radiated (for battery operated equipment only) Discontinuous (clicks per min.)	1.15–30 MHz (LM/MF/HF) 30–300 MHz (VHF) 30–300 MHz	56–66 dB(μV) quasi-peak depending on frequency 45–55 dB(pW) quasi-peak increasing linearly with frequency under consideration upper quartile permitted level of clicks by test
EN55011	CISPR 11 BS 4809	Industrial, scientific and medical equipment. (Intentional generators of RF)	Mains conducted Radiated	0.15–30 MHz 0.15–1000 MHz	0.003–5 V quasi-peak 34–110 dB(μV/m) at 100 m in frequency range 0.15–30 MHz and 30–130 dB(μV/m) at 30 m in frequency range 300–1000 MHz.
EN 60555-2 (BS EN 61000)	IEC 555-2 BS 5406-2	Electric power systems. (limitation of disturbances to supply network caused by products equipped with electronic devices)	Harmonics	0–2 kHz	2.3–0.15 A for odd harmonic orders 3–39; 1.08–0.23 A for even harmonic orders 2–40. (presently limited to domestic equipment but under review)
EN 60555-3	IEC 555-3 BS 5406-3	Domestic, commercial and light industrial electrical and electronic equipment connected to the mains supply.	Voltage fluctuations (mains flicker)	—	RMS voltage variation $\Delta V/V(\%)$ as a function of no. of voltage variations per min. Perceptibility Unit (PU) = 1 max.

Immunity (EN 50082-1)

Type of interference	Basic standard	Test specification	Purpose of simulation	Performance criterion
1. Radiated interference to enclosure:				
(a) Electromagnetic field	IEC 801-3 (HD 481.3)	27–500MHz 3 Vm^{-1} (unmodulated)	Susceptibility of instrumentation to radiated electromagnetic energy.	A

DESIGN CERTIFICATION

(b) Electrostatic Discharge	IEC 801-2 EN 60801-2 (HD 481.2)	8 kV discharge	B	Electrostatic discharge from persons, e.g. operator
2. Interference on signal and control lines (I/O).	IEC 801-4	0.5 kV (peak) 5 kHz 5/50 ns Tr/Th	B	Applicable only where cables exceed 3 m. Disturbances induced in external output/input cables often due to noise on parallel power cables
3. Interference on d.c. I/O power lines. Fast transients, common mode	IEC 801-4	0.5 kV (peak) 5 kHz 5/50 ns Tr/Th	B	Disturbances conducted via power lines
4. Interference on a.c. I/O power lines. Fast transients, common mode	IEC 801-4	1 kV (peak) 5 kHz 5/50 ns Tr/Th	B	Disturbances conducted via power lines

Summary of proposed Immunity Standards for inclusion at some future date:

1. Enclosure:
   (a) Magnetic Field; 50Hz, 3A/m; criterion A.
   (b) RF Electromagnetic Field, amplitude modulated; IEC 801-3: 80–1000MHz 3V/m, 80%AM(1kHz); criterion A.
   (c) RF Electromagnetic Field, pulse modulated; 1.89GHz, 3V/m; 50% duty cycle, criterion A.
   (d) Electrostatic Discharge; IEC 801-2; 4kV contact, 8kV air discharge; criterion B.

2. Signal and Control Lines:
   (a) Power Frequency, common mode; 50Hz, 10V rms; criterion A.
   (b) RF, common mode; 150kHz–10MHz, 3V/m, 150Ω source impedance; criterion A.

3. DC Power Lines:
   (a) Supply Voltage Deviation (short time); criterion C.
   (b) Supply Voltage Variation (long time); criterion A.
   (c) Transients on Power Line; 0.5kV, 1.2/50(8/20)μs Tr/Th; (IEC 801-5) criterion B.
   (d) Conducted RF Immunity; 150 kHz to 80 MHz, disturbing signal 80% AM (1 kHz), IEC 801-6, criterion A (may supersede IEC 801-3 up to 80 MHz a preferred test procedure)

4. AC Power Lines:
   (a) Voltage Dips (brown outs); criterion B/C.
   (b) Voltage Interruptions (drop outs); criterion C.
   (c) Voltage Fluctuations (oscillation bursts at regular intervals); criterion A.
   (d) Transients on Power Line (simulation of switching spikes); 2kV(c.m.), 1kV(d.m.), 1.2/50(8/20)μs Tr/Th, IEC 801-5 criterion B.
   (e) Conducted RF Immunity; 150 kHz to 80 MHz, disturbing signal 80% AM (1 kHz); IEC 801-6, criterion A (may supersede IEC 801-3 up to 80 MHz as preferred test procedure).

Note: Immunity Performance Criteria Definitions:
   A: No user noticeable loss of function or performance when used as intended and specified by the manufacturer.
   B: No user noticeable loss of function. Degradation of performance only when the test is being made. No change of operator state. No change in stored data.
   C: No faulty operation. Temporary loss of function allowed but system must recover itself or be able to be restored by the operator.

Tr = rise time 0.1–0.9V
Th = impulse duration, 50% value
c.m. = common mode
d.m. = differential mode

combination with them. The Directive exempts second-hand goods.

The liability of the design function is described by article 4 in ref. 16. A product or equipment that has been manufactured in accordance with specific Community or national standards that lay down the relevant requirements in terms of health and safety shall be presumed to comply with the obligation to place only safe products on the market. In the absence of specific rules for a particular product or product family, compliance with general safety requirements would be examined having regard for the state of the art, the state of technical knowledge, national regulations, practical feasibility and good business practice.

References

1. Christen A Role of standards in product certification. *Proceedings ERA Conference on Product Safety and Design for Certification* Sept 1991. ERA Report 91-0349
2. The Low Voltage Electrical Equipment (Safety) Regulations 1989; Statutory Instrument 728. HMSO 1989
3. Proposal for a council regulation concerning the affixing and use of the CE mark of conformity on industrial products. *Official Journal of the E.C.* C160/14: 20 June 1991
4. Modules for various phases of the conformity assessment procedures intended for use in the technical harmonisation directives. *Official Journal of the E.C.* L380/13 Dec. 1990: 90/683/EEC
5. Commission Green Paper on the Development of European Standardisation COM(90) 456: Oct. 1990
6. BS 3456-201/EN 60335-1 *Safety of household and similar electrical appliances*
7. BS 7002/EN 60950 *Specification for safety of information technology and electrical business equipment*
8. BS 2771-1/EN 60204-1 *Specification of electrical equipment of industrial machines*
9. BS 5304 *Safety of machinery*
10. BS 5490/IEC 529 *Classification of degrees of protection provided by enclosures*
11. BS 5378 *Safety signs and colours*
12. IEC 335-1 (3rd edn) *Safety of household and similar electrical appliances*
13. BS 6334/IEC 707 *Determination of the flammability of solid electrical insulating materials when exposed to an igniting source*
14. UL 94. Underwriters Laboratories Inc.
15. BS 6458/IEC 695 *Fire hazard testing for electrotechnical products*
16. Proposal for a Council Directive of General Product Safety. EEC 9284/91 Nov. 1991
17. Council Proposal 93/C28/02. *Official Journal of the E.C.* C28/16: February 1993
18. Guidance document on the preparation of a technical construction file. Department of Trade and Industry 1992

INDEX

A-motor torque spring 343
abnormal operation 535, 536
   circuits 535
   motors 535
   transformers 535
abrasivity of fillers 107
abrasive wear 372
absorption coefficient 502
absorption materials 502
acceleration torque 197, 198
a.c. brushless motor 205
a.c. commutator motors 212
   repulsion type 214
   universal type 212
a.c. induction motor 200
a.c. motors 187
a.c. motor family 191
a.c. servomotor 205
acoustic analysis 498
acoustic barrier 502
acoustic barrier materials 502
acoustic insulation 502
acoustic power 475
acoustic properties of materials 475
acoustic standards 515
acquisition phase 12
   build cycle 15
   costs 15
   design evaluation 15
   design review 16
   design team 13
   drawing definition 14
   pre-production build 13
   prototype build 12
   specification factors 14
   value engineering 14
acrylic-based adhesive 90
active noise cancellation 511
adaptive design application 25
addendum 275, 289

addendum circle 280
adherend
   surface preparation 78
adhesion process
   adherend requirements 77
   adhesive requirements 77
   thermodynamic basis 77
   wetting and spreading 76
adhesives 75
adhesive bond
   adhesive failure 81
   cohesive failure 81
   elements 80
   starved joint 80
   strength in shear 81, 82
adhesive bonding, advantages and disadvantages 76
adhesive bonding contacting process 75
adhesive bonding of plastics 79
adhesive bonding rules 75
adhesive compatibility 79
adhesive joint
   butt joint 81
   flanged joint 81
   lap joint 81
   scarf joint 81
adhesive joint design 80, 102
adhesive layer 79
adhesive nut locking 411
adhesive requirements 77
adhesive selection chart 94
adhesive selection procedure 102, 103
adhesive setting
   polymerization 87
   solvent release 87
   thermoplastic cooling 86
adhesive system definition
   design optimization 103
   performance required 102
   production process 103

adhesive types
   anaerobic adhesive 90
   anaerobic structural adhesive 92
   cyanoacrylate adhesives 93
   epoxy 98
   hot melt 96
   pressure sensitive adhesive 99
   reactive acrylic adhesive 95
   toughened adhesive 90
adhesive wear 372
adjusted rating life
   material factor 169
   operating condition factor 170
   reliability factor 169
adjusted rating life (L_{10}) 169
aerobic adhesive 95
aerodynamic forces; fans and blowers 500
AGMA 284
airborne particle sizes 457
airflow system characteristics 454
air filtration 456
air flow 449
air flow pressures 451
air movement
   fan pressure 454
   loss factors 455
   pressure–flow characteristics 454
   pressure drop 454
   pressure loss 455
air moving device 450
air pressures 450
alochrom process 368
aluminium alloy finishing guide 364
American Gear Manufacturers Association 268
amorphous thermoplastic 106
ampere-turns 303, 308, 322
anaerobic adhesive types
   fast cure 90

INDEX

anaerobic adhesive types (cont'd)
  joint design parameter 91
  joint strength 91
  non-curing 90
  performance limitations 92
  retaining 91
  slow cure 90
  structural 91, 92
  threadlocking 90
anaerobic structural adhesive 92
anechoic chamber 484, 512
anechoic room 486
angular backlash 282
annular contact ball bearing 152
annulus gear 299
anodizing 362
anthropometrics 10
anthropometric design details 11
approach action 280
arc contact factor 245
armature current 214
armature inductance 219
armature resistance 214
armature resistance speed control 221
armature shunt resistance speed control 221
armature voltage speed control 222
asynchronous induction motor 200
attestation 523, 525
austenitic nitrocarburizing 375
autocatalytic reduction 360
availability 55, 58
axial clearance 162
axial fan 452
axial fan noise 493

B-motor torque spring 343
B–H curve 304
backlash 281
back emf 214, 225
balance quality grade 499
ball bushing 184
ball nut 183
base circle 278
base pitch 279
basic dynamic load rating 163
basic insulation 528
basic rating life (L_{10}) 168
basic static load rating 164
bathtub curve 45
bathtub curve characteristics 45
bearing cages 155
bearing clearance 143
bearing design allowances 126
bearing fatigue life 168
bearing greases 175
bearing noise frequencies 492

bearing oil viscosity 172
bearing preload 162, 163
bearing seals 155
bearing thrust factor 165, 168
bearing tolerances 161
bearing torque 173
bell locking system 412
belt and pulley forces 252
belt centre distance 243
belt creep 243, 248
belt hysteresis 242
belt stress patterns 251
belt stretch 243
belt tension
  bending 242
  centrifugal 242, 248, 251
  peak 242
belt tensioning 244, 248
belt tension ratio 241
belt width 245
belt wrap 241
belville washer 416
bevel gear
  bending stress 290
  contact stress 290
bevel gears 289
bevel gear terminology 290
bias spring 351
biometrics 12
blackodizing 369
blade passing frequency 493
board edge guide 471
bonderizing 368
boriding 376
boundary lubrication 141
brainstorming 28
brainstorming subjects 29
brake and clutch type comparisons 312
brake design procedure 320
breakdown torque 197
brushless dc motor
  hybrid stepper 234
  permanent magnet 224
  permanent magnet stepper 234
  stepper 226
  switched reluctance 236
  variable reluctance stepper 234
build process 15
burning tests 540
bystander noise measurement position 486

cable drive chain 255
caged nut 413
cantilever spring 333
capacitor shorting 211
capacitor start motor 201
carbonitriding 376

carburizing 375
cardan 260
case hardening 372
catenary pull 257
CE mark 524
CENELEC 519
CENELEC certification agreement 523
centrifugal fan 452
certification body 525
certification body scheme 523
certification of conformity 525
certification process 523
certification scheme process 523
certification standards 519
certification summary 520
certification testing 519
chain articulation 256
chain chordal action 257
chain components 256
chain drive 255
chain drive arrangement 258
chain drive centre distance 258
chain drive design practise 257
chain fatigue life 256
chain life 256
chain tension 258
chain types 255
change control 18
change notice 22
change request 19
characteristic life 47
characteristic sound impedance 476
chemical conversion coating 367
chemical vapour deposition 369
chordal action of chain 257
chromating 368
chromating processes
  chromate 368
  chromate–phosphate 368
  Dacrotizing 368
chromium plating 360
circlips 425
circlips load capacity 426
circlip types 426
class of equipment 528
clevis pin 436
climatic testing
  humidity 69
  temperature 69
close mesh centre distance 278, 282
closed loop control system
  phase locked 220
  position 220
  torque 220
  velocity 220
clutch/brake disengagement 309
clutch/brake heat dissipation 306
clutch/brake holding power 315

INDEX

clutch/brake response time 308
clutch/brake rotor 314
clutch/brake slip power 314
clutch/brake stator 314
clutches and brakes 304, 309
clutch and brake type
   eddy current 311
   electromagnetic friction 314
   hysteresis 311
   magnetic particle 310
   tooth clutch 315
   wrap spring 316
clutch coil switching 307
clutch design procedure 318
clutch slip 311
clutch tooth styles 315
clutch torque 305
clutch torque–speed behaviour 306
coefficient of friction 111, 115
cogged belt 246
colour schemes 10
community noise level 482
commutation limit 219
competitive products 6, 7
compliance demonstration 528
composite noise rating 482
composite tolerance factor 279
compression spring 334
compression spring formulae 336
concession 20
concurrent engineering 39
condenser microphone 481
conduction–convection heat transfer 446
conduction angle 237
conduction heat transfer 442
conduction shape factor 444
conductor sizes 535
configuration control 26
conformal measurement surface 485
conformity declaration 525
conformity demonstration 523
conformity testing 528
constant force tensator spring 342
constant step rate control 233
constrained layer damping 507
Construction Products Directives 524, 541
Consumer Protection Act 31, 32
consumers risk 73
contact angle 162, 165
contact arc correction factors 246
contact conductance 444
contact ratio 279, 284, 289
contact stress modification factors 285
contacting process 75
contractual liability 31
convection heat transfer 445

convective conductance 445
convective heat transfer coefficient 445
convective resistance 445
conversion coating 383
copyright 36
Copyright, Designs and Patents Act 36
coreless motor 217
corner radius 179
cost tracking 15
cost of ownership pattern 57
cotter pin 435
counterface material 119
countersunk head 401
coupler engagement 260
coupling torsional stiffness 260
creativity
   brainstorming 28
   lateral thinking 28
   morphological charts 29
   synectics 29
creep 344
creepage and clearance 532
creepage and clearance distances 534
crinkle washer 415
crinkle washer load 415
critical frequency 496
cross-flow fan 454
crystalline thermoplastic 106
current decay in inductive winding 308
current time lag in inductive winding 308
curved washer 415
curved washer load 415
CVD process
   plasma 369
   thermal 369
   laser 369
cyanoacrylate adhesive
   chemical type 93
   setting mechanism 94
   viscosity grade 93
cycloidal gear profile 271

Dacrotizing process 368
damping 495
damping loss factor 507
damping material construction 507
d.c. motor braking
   reverse current 223
   rheostatic 224
d.c. motor family 191
d.c. motor speed control
   armature resistance 221
   armature shunt 221
   armature voltage 222
   field weakening 221
d.c. motor type
   compound wound 215

disc armature 218
   permanent magnet 217
   series wound 215
   servos 218
   shell armature 217
   shunt wound 215
d.c. motors 187, 214
De Saint Venant coefficient 336
decibel 474
decibel arithmetic 476
declared noise emission value 517
deep groove ball-bearing 151
deformed thread nut 410
design aesthetics 9
design assessment 59
design certification 518
design creativity 27
design engineer functions
   configuration control 26
   configuration responsibility 25
   development input 25
   drawing issue 26
   performance monitoring 26
design engineer role 24
Design for Assembly (DFA) 40
Design for Logistics (DFL) 40
design for maintainability 52
Design for Manufacture (DFM) 40
design for reliability 52
design functions 2, 24
design hazard 32
design innovation 27, 30
design liability 31, 528
design life cycle 18
design of hazard warning label 32
design phases 2
design process
   departmental activity 2
   phases of activity 3
design protection 36
design review 16, 31
design review team 16
design right 36
design risk 31
design standards 34
design team 26
detonation spray deposit 379
development cost 5
diametral pitch 275
differential nut thread pitching 408
diffuse sound field 487
dip coating 386
direct sound field 486
disk clutch and brake 305
disk spring 416
   coned height 417
   load–deflection characteristics 417
   stacking 417

549

disk washer 416
dispersion spray coating 386
dissipated power 217
dissipative muffler 504
double insulation 528
dowel pin 436
drawing documentation 19
drawing release 26
drive belts 240
drive belt type
   flat 240
   toothed belts 249
   V-belts 246
drive couplings 258
drive motors 187
drive shaft 264
drop-out times 309
drop point of grease 175
dry bearings 105
dry bearing bore closure 126
dry bearing design
   aspect ratio 124
   clearance 125
   counterface material 119
   wall thickness 125
dry bearing design procedure
   defining the application 128
   life estimate 132
   material selection 128
   temperature rise estimation 128
   wear rate estimation 131
dry bearing friction 115
dry bearing heat removal 113
dry bearing lubrication 119
dry bearing materials 105
   fillers 114
   fluorocarbons 120
   high temperature polymers 121
   lubricant additives 114
   polyacetals 121
   polyamides 120
   polyethylene 121
dry bearing noise 118
dry bearing running clearance 127
dry bearing swell allowance 126
dry bearing temperature 111
dry bearing thermal allowance 127
dry bearing type
   effect of fillers 107
   integrally lubricated polymers 124
   split bearing liners 123
   thin layer 121
dry bearing wear
   PV factor 108
   running-in 106
   static load carrying capacity 109
   surface roughness 106
   temperature rise 110, 111

test methods 110
transfer film 107
Duane model, reliability growth 62
ductwork air flow 456
durability 58
dynamic equivalent radial load 165
dynamic imbalance 499
dynamic torque 229
Dzus fastener 438

earthing provision 534
eccentricity 499
eccentricity ratio 141
EC conformity demonstration 526
eddy current clutch/brake 311
EEC Directives
   product liability 31
effective dynamic load rating 163
effective pull 241, 245
effective sound reduction index 504
effectiveness factor 170
efficiency of clutch/brake 314
elastomer spring 344
   design applications 344, 346
   series/parallel stack 344
   stiffness 346
elastomer spring frequency ratio 344
elastomer spring materials 344
electric braking 223
electric motor deficiency 199
electric motor noise 493
electric motors 187
electrical insulation 528, 532
electrical safety requirements 532
   creepage and clearance 532
   earthing 534
   insulation 532
   leakage current 532
   supply 533
electrical time constant 216, 308
electrodeposited composite coating 361
electroless composite coating 362
electroless plating 360
electrolytic composite coating 362
electromagnetic actuator 303
electromagnetic brake 303
electromagnetic clutch 303
electromagnetic friction clutch/brake 314
electromagnetic noise emissions 542
electromagnetic noise immunity 542
electromechanical conversion coefficient 214
electromechanical torque 196
electron beam hardening 373
electronic component encapsulation 471
electropainting 385
electrophoretic coating 385

electroplating 354
electrostatic powder coating 386
EMC directive 524, 542
EMC noise sources 542, 543
EMC standards 544
emissive power 447
emissivity values 448
emittance 447
EMKO 523
enamelling 377
enamelling of
   aluminium 379
   cast iron 378
   sheet steel 377
enclosure protection 528
enclosure protection classification 531
engineering change request 19
engineering project plan 8
engineering responsibility 2
entrepreneurial function 30
environmental testing
   climatic tests 67
   transportation tests 68
EOTC 525
epicyclic gears 298
epicyclic gear outputs 299
epoxy adhesives
   performance factors 98
   setting mechanism and cure 98
   types 98
equipment repair time 53
equivalent bearing load 167
equivalent load calculation 167
equivalent sound level 482
equivalent static load 168
ergonomics 10
essential requirements 523
Euro-standard 527
European economic area 518
European Economic Community 518
European Norm standard 519
European Single Market 518
exponential distribution 47
extended life equation 168
external rotor motor 206

failure analysis 60
failure distribution
   exponential 47
   log-normal 46
   normal 45
   Pareto 47
   Weibull 47
failure modes 52
failure mode and effect analysis 32, 60
failure mode testing 65
failure patterns 43
failure rate 43

INDEX

fan aerodynamic noise 493
fan application
   fan power requirement 458
   performance suitability 456
   system design 458
fan characteristic 450
fan components 450
fan duty 450
fan efficiency 451
fan laws 451
fan mechanical noise 493
fan noise 493
fan output power 450
fan performance 450, 453
fan pressure 454
fan type
   axial 452
   centrifugal 452
   mixed flow 453
   tangential 454
fasteners 395
fastener drive methods
   cruciform recess 401
   hexagon recess 402
   high torque 402
   security 403
   slot 401
fastener drive style 399
fastener head forms
   cheese head 401
   countersunk 399
   hexagon head 401
   pan head 401
fastener head style 399, 400
fastener inserts 428
fastener insert type
   plasting moulding 431
   sheet metal 428
   wire thread 431
fastener point style 403
fastener thread run-out 424
fastener torque/angle preload control 424
fastener types
   inserted 428
   pins 435
   plastic 407
   quick release 438
   screws 399, 401
   self-tapping 433
   SEMS 418
   set screw 405
fault tree analysis 61
feasibility phase 7
   human factors 9
   industrial design 9
   product architecture 9
   product introduction plan 7

technology tasks 8
ferritic nitrocarburizing 374
field performance feedback 73
field service
   access 53
   facilities 54
   latches and fasteners 54
   maintenance skills 54
field service function 3
field weakening speed control 221
film thickness/roughness ratio 170, 283
filters 456
filter bandwidth 476
filter contamination 458
finned surface heat transfer 448
fire retardancy rating 537
flame spray deposit 379
flanking transmission 509
flash temperature 111
flat belts 240
   elastic belt 241
   non-stretch belt (inelastic) 241
   power transmission 241
   semi-elastic belt 241
flat belt design method 245
flat belt drive geometry 242
flat belt operation 244
flat spring 333
flat spring equations 333
flexible coupling 258
   misalignment 259, 263
flexible coupling type
   bellows 262
   gear 262
   helical beam 262
   multi-jaw 262
   Oldham 263
   rubber sleeve 260
   Schmidt 264
   spider 260
   universal joint 263
   universal lateral 264
flexible drive 265
flexible shaft 265
fluidized bed coating 386
flux density 196, 304
FMEA worksheet format 61
foamed hot melt adhesive 98
forced air flow 450
forced convection heat transfer 445
Fourier's law 443
fractional turn fastener 438
free convection heat transfer 445
free field 486
freewheeling diode 222
frequency analysis 512
frictional heating 111
friction coefficient 111

friction factor 173
friction–velocity effect 118
full-load torque 197

galvanizing 392
gas spring strut 348
Gaussian distribution 45
gears
   advantages 268
gear centre distance 278
gear contact stress factors 284
gear fatigue 282
gear features 269
gear formulae 276
gear interference 280
gear lubrication 283
gear materials 271, 272
gear meshing 281
gear noise 283, 492
gear pinion 268
gear pitting resistance 282
gear power transmission 285
gear rating 284
gear scuffing 282
gear selection
   gear size 301
   operating conditions 300
gear terminology 271
gear tolerances 279
gear tooth correction 281
gear tooth distress 283
gear tooth engagement 271
gear tooth errors 278, 283
gear tooth thickness 281
gear train
   gear ratio 298
   velocity ratio 298
gear transmissions 297
gear type
   bevel 270
   helical 270
   spur 268
   worm 270
gear undercutting 280, 281
gear wear 282
gear wheel 268
glow-wire test 540
glue line integrity 79
Grashof no. 446
gravity idler 244
grease channelling 175
grease consistency 175
grease lubrication 174
grease lubrication mechanism 177
grease packing 177
grease stability 175
green window 424
groove nut 409

551

INDEX

groove pin 437
grooveless retaining ring 427
grooveless shaft fixings 427
grub screw 405

hair spring 341
hammer paint finish 385
hank nut 428
harmonic drive 299
harmonized document 519
harmonized standards 519
heat dissipation 216, 219
heat flow 443
heat flow principles 442
heat pipe 459
  correction factors 460
  heat transport capability 460
  QL capability 460
  thermal resistance 461
heat pipe heat flow rate 460
heat resistance tests
  ball pressure 537
  burning test 537
  glow wire test 538
heat sink
  case dissipation 465
  convective heat transfer 466
  radiation heat transfer 467
  surface finish 467
  thermal interface material 467
  thermal resistance 465
heat sink selection 468
heat sink theory 465
heat transfer 441
heat transfer coefficient 443, 446
heat transfer devices
  heat exchanger 458
  heat pipe 459
  thermoelectric heat pump 461
heat transfer modes 442
heat transfer rate 443
helical gears 287
  contact ratio factor 288
  contact stress 287
  geometry factor 289
  helix factor 289
  overlap ratio 288
  root bending stress 289
helical gear axial thrust 287
helical gear forces 288
helical spring 334, 337
helical spring washer 415
helical torsion spring 340
helix angle 278, 287, 291
Hertz contact stress 284
hexagon head style 401
hexagonal socket drive 403
high torque drive 404

holding torque 227
Hooke's joint 263
hot melt adhesive 86
  composition 96
  performance parameters 97
  types 96
human error 59
human factors 9
hybrid control system 220
hybrid stepper 234
hydrodynamic lubrication 140
hysteresis clutch/brake 311
hysteresis synchronous motor 205

IEC standards 519
imbalance 499
impact forces 499
impact plating 392
impulse noise 483
induced emf 214
induction motors 200
induction motor braking 211
induction motor noise spectrum 494
induction motor slip 200
induction motor speed control 207
induction motor speed factors
  number of poles 208
  slip control 208
  supply frequency 210
induction motor type
  capacitor start 201
  external rotor 206
  hysteresis synchronous 205
  permanent magnet synchronous 205
  permanent split capacitor 202
  reluctance synchronous 204
  shaded pole 202
  torque motor 207
  two capacitor 202
industrial design 9
innovation 30
insertion loss 494, 506
instructions 12, 530
instruction manual 33
integrated noise level 482
integration time constant 233
intelligent manufacturing systems (IMS) 40
interfacial free energy 76
interim maintenance 56
internal wiring 532
international certification 519
International Standards Organisation 268
inverter 210
involute gear profile 271
involute gear system 275
ion implantation 371

ion plating 371
ion vapour deposition 371
IP rating 530
ironless rotor 217
ISO 284
ISO metric screw thread 396
ISO position noise level 486
ISO standards 519
isolation efficiency 509

Japanese design process 37, 38
  business environment 37
  design cycle 38
  product acquisition 38
  product initiation 38
  vendor utilization 39
  work regime 37
joint shear 422
joint stiffness 421
joint tightening
  preload 420
  stiffness constant 421
  theory 419
  torque coefficient 421
  torque–tension relationship 419
just in time (JIT) 40

key microphone positions 484
knowledge based engineering (KBE) 40

laminated strip coating 387
lap shear adhesive joint
  adherend stresses 82
  adhesive stresses 83
  adhesive type 84, 85
  cleavage 83
  strength 84, 85
  types 85
laser surface hardening 372
LAT's 327
lateral thinking 28
launch problems
  design error 58
  handling error 58
  manufacturing error 58
  operator error 58
lead angle 278
lead of worm gear 291
leakage current 532
Lewis bending stress 286
Lewis equation 297
life adjustment factor 169
life of linear bearing 186
limited angle torquers 327
limiting PV value 109–111
limiting speed factors 173
linear backlash 282
linear ball bearings 184

INDEX

linear motion equations 196
linear ramped control 233
linear solenoid
　plunger and core design 322
line of action 271, 279
Lloyds K-factor 285
load carrying capacity 163
load inertia 230
load line displacement 140
load safety factor 168
loading roughness 44
lock washer 199
locked rotor torque 197
locking nut system 410
locking screw
　encapsulated adhesive 404
　polymer insert 404
　torque definitions 403
locking screws 403
losses in electric motors 199
loudness contours 479
loudness levels of noise 477
louvre design 537
Low Voltage Directive 523, 540
lubrication regimes 137
lumped heat capacity 445

Machinery Directive 513, 524
machinery noise
　aerodynamic 489
　structural 489
Machinery Safety Directive 541
magnetic circuit 303
magnetic drive 305
magnetic flux 214, 303
magnetic forces 196
magnetic noise 492
magnetic particle clutch/brake 310
magnetic pull 303
magnetizing curve 304
magnetizing force 303
magnetomotive force 196, 200, 214, 303
maintainability
　access 52
　adjustments 53
　identification 53
　manual 53
　test features 53
maintainability definitions 54, 55
maintenance action rate 55
maintenance categories 55
maintenance distributions
　log–normal 55
　negative exponential 55
man–machine interface 9, 12
manufacturing responsibility 2
marketing function 3

maturity phase 17
mean component reliability 45
mean down time 55
mean time between failures (MTBF) 43
mean time in between repairs 55
mean time to failure 54
mean time to repair 55
mechanical properties of fasteners 396
mechanical safety requirements 536
mechanical time constant 216, 309
metallic paint finish 385
metric gears 275
metric screw thread
　profile standards 396
　thread designation 396
　tolerance class 398
　tolerance grade 398
　tolerance position 398
microphones 481
microphone positions 486
microphone sensitivity 481
microprocessor speed control 233
microstepping 230
misalignment 259
mitre gears 270
mixed flow fan 453
models 10
modification classification guide 20
modular approach policy 525
module 275, 291
morphological charts 29
motor acceleration 198
motor bandwidth 219
motor constant 215
motor damping constant 216
motor duty 198
motor duty cycle 197
motor efficiency 199, 215
motor electromechanical noise 494
motor imbalance 494
motor inertia 197
motor load curves 197
motor loading 197
motor mechanical noise 494
motor noise 493
motor noise spectrum 494
motor power output 215
motor regulation constant 216
motor selection criteria 237, 238
motor speed control
　power transistor control 208
　thyristor control 207
　triac control 207
motor torque definitions 197
motor torque production 191
motor voltage constant 214
motor windage 493
mufflers 504

muffler sound attenuation 506
muffler types 506
multistage composite coating 362
multi-waved washer 415

national certification body 519
national test bodies 522
natural frequency 509
needle roller bearing 154
neoprene belt characteristics 250
net pull 241
New Approach Directives 523, 524
　construction products 524
　machinery 524
new product team 6
Newton's law of cooling 445
nitriding 373
noise 474
noise average meter 482
noise control at source 498
noise control in the path 501
noise control methods
　acoustic enclosure 502
　muffling device 504
　sound absorption 501
　vibration damping 507
　vibration isolation 508
noise declaration standards 517
noise diagnosis 512
Noise Directive 513
noise emission declaration 513
noise exposure forecast 482
noise field 484
noise frequency analysis 476
noise from machine elements
　fans 493
　gears 492
　motors 493
　plain bearings 491
　rolling element bearings 490
　timing belts 492
　transformers 492
noise generation 489
noise labelling 513, 514
noise loudness 477
noise measurement environment 486
noise measurement surface 484
noise path 490, 491
noise performance of covers 494
noise radiation 494
noise radiation reduction 513
noise receiver 490
noise reduction 474, 500
noise reduction coefficient 502
noise reduction methods 498
noise reference box 484
noise source 490
noise source definition 498

553

INDEX

noise test codes 514
noise treatment practice 512
non-conversion coating 383
non-metallic bearings 105
non-stick coating 382
Nordic certification scheme 523
normal distribution 45
notified body 518, 524
Nusselt no. 445
nuts 407
   basic design 407
   strength grade 408
   stress distribution 408
nut clamp load 421
nut tightening theory 419
nut type
   captive 412
   locking/torque 409
   spring steel 411
Nyloc nut 409

octave band 476
oil film thickness 170
oil leakage factor 140
oil retaining bearing 135
oil viscosity 172
on-condition maintenance 56
one-way clutch 183
open loop control 233
operation instructions 33
operator maintenance 56
operator position noise level 486
operator reliability 59
organic coating
   application factors 382
   paint 383
   plastic substrate 391
   polymers 385
   pre-coated strip 386
overhung nut 409
oxide coatings 368
oxide coating process
   blackodizing 369
   chemical anodizing 368
   chemical oxidation 368

packaging 35, 59
paint 383
   application methods 385
   constitution 383
   curing methods 383
   decorative options 385
paint spray methods 385
paint types 384
painted plastics 391
panel damping 507
panel resonance 496
panel transmission loss 495
Pareto distribution 47, 60

Pareto 'trival many' 50
Pareto 'vital few' 47, 58, 60
parkerizing 368
patents 36
pcb cooling 471
peak motor torque 197
Peltier effect 461
performance evaluation 16
performance stress testing 66
permanent deformation 164
permanent magnet brushless motor 224
permanent magnet motors 215
permanent magnet stepper motor 234
permanent magnet synchronous motor 205
permanent split capacitor motor 202
permeability factor 140
perspectives 10
phase angle 201
phase control of motor 209
phase locked control system 220
phon 477
phosphating 367
physical vapour deposition 371
physiological factors 10
piezo-electric microphone 481
pin fasteners 435
pin type
   clevis 436
   cotter 435
   dowel 436
   groove 437
   spring 437
pitch 291
pitch angle 289
pitch circle 271
pitch point 271
plain bearing noise 491
plain washer 413
plane of engagement 260
planetary gears 299
planned maintenance 56
plasma carburizing 375
plasma nitriding 373
plasma spray deposit 379
plasma spray gun 381
plastic chain 259
plastic fasteners 407
plastic gears 293
   backlash 296
   centre distance 296
   design allowance 296
   fillet radius 294
   lubrication 296
   metal−plastic difference 294
   temperature rise 295
   tip modification 294
   tooth form 294

plastic gear rating
   bending stress 297
   contact stress 297
plastic moulding insert 431
plastic moulding insert type
   broaching 432
   knurled expansion 432
   press-in 432
   self-tapping 432
   ultrasonic 432
   vaned expansion 431
plastic powder coatings 388
plastic rolling element bearings 181
plastic screw torque 407
plastic spring type
   disk 348
   helical compression 348
   snap-fit hook 347
   sulcated 348
plated plastics 365, 367
   electroless plating 367
   electroplating 365
   guidelines 365
plenum chamber 506
plugging 211, 223
Poka-Yoke 40
polar moment of inertia 197
polymer coating
   design rules 385
   methods 386
porous bearing oil circulation 139
porous metal bearing
   clearance 143
   density 137
   installation 146
   lubrication methods 139, 146
   permeability 137
   porosity 137
   running-in mechanism 144
   shaft material 145
   shaft roughness 145
   types and composition 136
porous metal bearing design procedure
   application definition 147
   bearing material 147
   life estimate 149
   running temperature 149
porous metal bearing lubrication regime
   boundary 141, 149
   hydrodynamic 139, 149
porous metal bearing temperature 142
porous metal bearings, advantages and disadvantages 135
position control system 220
power balance equation 214
power components 470
power factor 200
Prandtl no. 446

INDEX

pre-assembled nut/washer 418
pre-coated strip 386
pre-finished steel 390
preload indicating washer 417
preload measurement methods
   bolt extension 423
   bolt yield 424
   nut rotation 424
   torque 423
   torque/angle control 424
pre-production change request 19
pre-production model 13
press nut 428
pressure angle 271, 278, 281, 287
pressure drop 454
pressure sensitive adhesives
   application 101
   mechanism 100
   strain rate performance 99
   tape construction 101
prevailing torque nut type
   deformed thread 410
   integral spring 410
   material insert 409
primary NCB 519
printed circuit motor 218
probability density function 43
probability of failure 42
probability of survival 42
problem set 67
process maintenance 56
producers risk 73
product acquisition documentation 3
product acquisition hardware 3
product architecture 9
product definition 31
product delivery process (PDP) 40
product durability 58
product family strategy 6
product features 4
product identification 530
product labelling 32
Product Labelling Directive 524
product liability 31
product maintenance 54
product manual 33
product presentation 4
product reliability 42
product safety 31
Product Safety Directive 524, 543
profit return 5
project plan 418
projection type preload washer 418
proposal phase 4
   engineering input 7
   field service input 6
   marketing input 6
protection levels 527

prototype model 12
psychological factors 10
pull-in time 308
pull-in torque 204, 230
pull-out torque 204, 230
pulley crowning 243
pulse width modulation 210
PV factor 108
PV value
   limiting 110
   normal 110
PVD processes
   evaporation 371
   ionization 371
   sputtering 371

quadrature 227
quadrature component 196
quality department function 2
quality engineering (QE) 40
quality of design 2
quick release fasteners 438
quick release fastener type
   ¼-turn 439
   expanded collar 438

radial clearance 162
radial load factor 165, 168
radiated sound power 496
radiation heat transfer 447
radiation heat transfer coefficient 449
radiation potential difference 448
radiation shape factor 449
radiation space resistance 449
radiation surface area 448
radiation surface resistance 448
radiation view factor 448
radius of gyration 197
random incidence absorption coefficient 502
rated belt power 245, 247
rated drive power 306
rating life 168
reactive acrylic adhesive 95
reactive muffler 506
recess action 280
recessed head style 402
recirculating ball bushing 184
recirculating roller bearing 186
recognised body 518
referred inertia 197
referred load 197
reflecting plane 486
registered design 36
reinforced insulation 528
reliability basics 42
reliability evaluation 65
reliability factor 169

reliability growth 62
reliability manual 51
reliability prediction
   Duane growth 62
   failure mode and effect analysis 60
   fault tree analysis 61
   reliability growth 62
reliability programme 50
reliability stressing 46
reliability testing
   compliance testing 65
   determination testing 65
   performance testing 65
   production testing 65
   stress testing 65
reliability–maintainability balance 57, 58
reluctance 303
reluctance synchronous motor 205
repulsion motor 214
residual characteristic of fan system 455
residual mass centre displacement 499
resilient mounts 508
resonance frequency 260, 495, 507
resource requirement 5
retaining ring type
   circlip 425
   grooveless 427
   tolerance ring 427
return on investment 5
reverberant sound field 486
reverberation control 501
reverse current braking 211, 223
Reynold's no. 446
rheostatic braking 211, 224
ring compression preload washer 418
risk 5
risk management 32
rivet bush 428
rivet bush assembly 428
rivet nut 429
roll pin 437
roller bearing 153
roller chain 255
roller clutch 182
rolling element bearings 151
rolling element bearing cages 155
rolling element bearing coding
   boundary dimensions 158
   clearance 162
   precision class 161
rolling element bearing equivalent load 165
rolling element bearing fits 178
rolling element bearing friction 173
rolling element bearing housing 179
rolling element bearing load ratings 163

INDEX

rolling element bearing lubrication 173
rolling element bearing materials 154
rolling element bearing noise 177, 490
rolling element bearing preload 162
rolling element bearing rating life 168
rolling element bearing seals 155
rolling element bearing selection
   procedure
  application definition 179
  installation 181
  life calculation 181
  load capacity 180
  requirement specification 179
  speed limitation 180
rolling element bearing speed limits 171
rolling element bearing steel 156
rolling element bearing type
  angular contact bearing 152
  deep groove ball bearing 151
  needle roller bearing 154
  roller bearing 153
  self-aligning ball bearing 153
  tapered roller bearing 154
  thrust bearing 154
rolling element code structure 160
rolling element frictional losses
  lubricant 173
  rolling 173
  sliding 173
room constant 488
rotary motion equations 196
rotary solenoid 324
rotary stud fastener 439
rotating system inertia 198
rotor field strength 311
rotor inertia 216, 226
rotor pole pitch 237
running temperature 142, 149
running-in process 146

safe operating area 216
safety 31
safety devices 32
safety features 32
safety margin 44
Safety of Goods Act 31
safety specification requirement 518
safety standard reference 529
safety standards 518, 527
safety standards criteria
  chemical hazard 527
  electric shock hazard 527
  fire hazard 527
  heat hazard 527
  mechanical hazard 527
  radiation hazard 527
salient pole 226, 236

SCR control 222
second generation adhesive 90
Seebeck effect 462
self-aligning ball bearing 153
self-certification 542
self-clinch fastener 428
self-pierce and clinch fastener 428
self-tapper threads 435
self-tapping screw
  application guidelines 434
  thread forms 433
  torque definitions 433
  types 434, 435
SEM screws 419
semi-anechoic chamber 486
sequential testing 72
service factors 245
service organisation
  see field service
servomotor performance 218
set screw
  axial holding 406
  performance 406
  point penetration 406
  types 405
set screw point style 406
shaded pole induction motor 202
shading ring 203
shaft angle 289
shaft attachment methods
  concentric fasteners 267
  keying 266
  pins 266
  set screws 266
  tapered collar 266
shakeproof washer 416
shape factor analysis 445
shape memory alloy spring 350
shear nut tubular rivet 429
sheet edge locking insert 429
sheet metal insert type
  rivet bush 428
  self-clinch 428
  tubular locking 429
sheradizing 392
silencers 504
simultaneous engineering 39
Single European Market 518
single market 523
single wave washer 415
sintered metal bearing 135
slide : roll ratio 270
SMA actuator design 351
solenoids 303, 320, 322
solenoid alignment 326
solenoid application 325
solenoid closure force 322
solenoid coil temperature rise 322

solenoid contamination 325
solenoid design procedure 325, 326
solenoid disengagement 320
solenoid duty cycle 322
solenoid failure modes 326
solenoid force/stroke characteristics 322
solenoid friction 325
solenoid mounting 326
solenoid noise 326
solenoid power 322
solenoid pull 322
solenoid rating 321
solenoid response time 320
solenoid type
  latching (bistable) 320
  rotary 320, 324
solid height 334
solubility parameter 78
solvent-based adhesive 87
Sommerfeld reciprocal 140
sound
  amplitude 474
  frequency 474
  intensity level 476, 482
  loudness 477
  pitch 478
  power level 476
  pressure level 475, 477, 482
sound absorption 501
sound absorption coefficient 487
sound absorption materials 502
sound absorption practice 502
sound directivity factor 486
sound directivity index 486
sound exposure level 483
sound field type
  direct 486
  free 486
  reverberant 486
sound frequency analysis 480, 500
sound level meter
  grade 480
  microphone types 481
  types of response 480
sound measurement 478
  equivalent sound level 482
  impulse sound 483
  sound exposure level 483
sound muffling 504
sound peak pressure level 514
sound power level 486, 498, 500, 514
sound power level correction 487
sound power measurement
  in a free field 486
  in a reverberant room 487
  in a semi-reverberant room 488
sound pressure level 484, 486, 514
sound reduction index 495, 504

INDEX

sound reduction index (cont'd)
   coincidence controlled 496
   mass controlled 496
   resonance controlled 495
   stiffness controlled 495
sound reduction materials 501, 503, 505
sound source specification 500
sound transmission coefficient 495
sound weighting curves 479
sound weighting network
   'A' weighting 478
   'B' weighting 478
   'C' weighting 478
spattered paint finish 385
specific speed parameter 451
specific wear rate 108, 112
spiral pin 437
spiral torsion spring 341
   angular deflection 342
split bearing liners 123
split phase induction motor 201
spray and fuse deposits 381
spray deposit
   coating characteristics 380
   materials 380
   plastic substrate 382
   polymer impregnation 382
spray methods
   detonation 379
   flame 379
   plasma 379
sprayed surface coatings 379
springs 329
spring active height 338
spring deflection 333, 334, 338
spring design method 337, 340, 341
spring diameter 336
spring end stresses 339
spring ends 335, 339
spring index 334, 335
spring materials 329
spring mounts 511
spring natural frequency 336
spring rate 334
spring solid height 335, 338
spring steel nut 411
spring stress guidelines 331
spring tension 338
spring type
   elastomer 344
   flat (cantilever) 333
   gas 348
   helical (compression) 334
   plastic 347
   shape memory alloy 350
   tapered helical 337
   tensator 342

   tension 338
   torsion 340, 341
spring washer 413
spring wire section 336
spring working deflection 335
spur gear
   bending stress 286
   contact stress 284
   contact stress endurance limit 284
   contact stress safety factor 284
   endurance limit 287
   geometry factor 286
   Lewis form factor 286
   material quality factor 285, 287
   permissible contact stress 284
   root bending stress 287
spur gear contact stress 284
squirrel cage 200
SR drive 236
stack side tension 241
staircase plot 73
stall current 215
static pressure 450
staytite nut 410
steady state conduction 444
step angle 229
stepper motor 226
stepper motor control mode 229
stepper motor slew mode 229
stepping motor drive system
   bipolar drive 232
   chopper drive 232
   constant step rate 233
   dual voltage 232
   linear ramped 233
   microprocessor 233
   R/L drives 230
   unipolar drive 232
stick-slip 118
stiffnut 409
stress relaxation 344
stress testing
   controllable parameter 66
   performance latitude 66
   sensitivities 66
   uncontrollable parameter 66
Stribeck curve 143, 149
structural damping 508
sulcated spring 348
sun gear 299
superglue 93
supplementary insulation 528
supply connection 533
surface damping 507
surface energy 76
surface engineering 353
surface engineering basics 353
surface hardening 372

surface hardening methods
   austenitic nitrocarburizing 375
   boriding 376
   carbonitriding 376
   carburizing 375
   electron beam 373
   ferritic nitrocarburizing 374
   laser transformation 372
   nitriding 373
   particle injection 373
surface hardness profile 374
surface modification processes 354
surface sound pressure level 485
surface tension 76
surface thermal treatments 372
surface transmission coefficient 504
surge protection circuitry 309
switched reluctance motor 236
symbols 12
synchronous belt 249
synchronous induction motor 200, 204
synchronous motor speed 204
synectics 29
system resistance 454

T-bend test 387
tack 97, 99
Taguchi 40
tangential fan 454
tapered nut thread form 408
tapered roller bearing 154
technical construction file 542
teeth in mesh 251, 253
temperature gradient 443
temperature rise allowable 538
tensator constant torque spring
   A-motor 343
   B-motor 343
tensator spring 342
tensator spring deflection 343
tensator spring force 342
tensator spring storage drum 343
tensator spring stress 343
tensator spring type
   constant force 342
   constant torque 342
tension ratio 241
tension spring 338
tension spring design method 340
test house 518
thermal bridge 464
thermal capacitance 447
thermal conductivity 443
thermal contact coefficient 444
thermal contact resistance 443
thermal control 441
thermal design 441
thermal diffusivity 443, 446

INDEX

thermal enclosures
  cooling slots/openings 472
  heat concentration 472
  heat dissipation 472
  open case cooling 472
  totally enclosed cooling 472
thermal grease 444, 467
thermal heat capacity 443
thermal interface material 467, 468
thermal management 441
thermal management of electronics
  encapsulation 471
  enclosures 472
  heat sink 464
  pcb cooling 471
  power components 470
  thermal plane 464
thermal module 464
thermal pad materials 468
thermal planes 464
thermal potential difference 443
thermal properties 443
thermal radiation spectrum 447
thermal resistance 443, 447
thermal resistance coefficient 111, 142
thermal resistance test procedures 539
thermal safety requirements 537
thermal spring 464
thermal time constant 217
thermochemical surface treatment 372
thermoelectric heat pump
  coefficient of performance 462
  efficiency 462
  electrical resistance 462
  heat pumping capacity 462
thin-layer bearing materials
  porous bronze 122
  PTFE dispersion 122
  PTFE weave 121
  reinforced PTFE 122
thread–shank stress relief 425
thread crest diameter 398
thread cutting screw 435
thread end-forms 434
thread forming screw 434
thread plating thickness 355
thread profile 397
thread tolerance limits 400
threaded fasteners
  bolt 395
  material parameters 395
  screw 395
threadlocking adhesive 91
thrust ball bearing 154
tight side tension 241
timing belt 249
timing belt noise 492
tolerance ring 427

tooth clutch 315
tooth gears 268
tooth meshing 492
toothed belt 249
  high torque drive profile 249
  modified curve linear 250
  recessed parabolic profile 253
toothed belt construction 249
toothed belt design method 254
toothed belt noise 252
toothed belt power rating 253
toothed belt power transmission 251, 253
toothed belt service factors 254
toothed belt tension factors 251, 254
toothed belt tracking 252
torque–ampere ratio 237
torque–angle monitoring 424
torque–speed curves 199
torque angle 196
torque capability 306
torque constant 214
torque control system 220
torque motor 207
torque nut 409
torque nut type
  free spinning 410
  prevailing torque 409
torque ripple 225
torque sensitivity 214
torque to inertia ratio 219
torsion spring
  angular deflection 340
  diameter 341
torsion spring design method 341
torsion spring ends 341
torsional spring rate 341
tortuous liability 31, 32
total pressure 450
total sound energy 502
toughened adhesive 90
transformer noise 492
transient noise 483
transit damage 59
transmissibility 509
transmission loss 495, 504
transportation testing
  bump test 72
  shock testing 69
  vibration tests 69, 72
tubular locking insert 429
tubular rivets 430
two-capacitor motor 202
type test 519, 523, 528

unconstrained layer damping 507
universal motor 212
unregistered design right 36

unscheduled maintenance 56
unsteady state conduction 445, 446
utilization factor 55

V-belt, light duty
  premium construction 246
  standard wrapped 246
V-belt correction factors 249
V-belt design method 249
V-belt forces 248
V-belt power transmission 247
V-belt size 247
value engineering 14
vapour coating 369
variable reluctance stepper 234
velocity control system 220
velocity pressure 450
vibration amplitude 499
vibration damping 507
vibration damping practice 508
vibration excitation forces 512
vibration isolation 508
vibration isolator
  air spring 510
  compression pad 510
  rubber mounts 509
  springs 509
vibration measurement 498
vibration sources
  aerodynamic forces 500
  electromagnetic forces 499
  imbalance 499
  impact forces 499
viscosity ratio 170
vitreous enamelling 377

Wahl correction factor 334, 337
walterizing 368
warning label 31, 32
washers 413
washer type
  crinkle 415
  curved 415
  disc 416
  helical spring 415
  plain 413
  preload indicating 417
  spring 413
  toothed lock 416
washer types, assembled 418
wave generator 300
wavelength of sound 474
wear coefficient 108
wear equation 108
wear factor 108, 112
wear rate estimation 131
wear rates of coating 360
wedge belt 247

INDEX

wedging factor 247
Weibull distribution 47, 168, 169
wettability 77
wetting 76
winding current 201
wire thread insert 430
Wood's correction factor 335
work of adhesion 77
work of cohesion 77
worked penetration of grease 175
worm gear
   bending stress 292
   contact stress 291

output power 292
permissible torque 293
sliding velocity 292
tangential load capacity 292
worm gear efficiency 291
worm gear terminology 292
worm pitch diameter 291
wound field motors 214
wrap-spring clutch/brake type
   brakes 317
   electromagnetic 318
   on—off clutch 317
   overrunning clutch 317

single revolution clutch 317
wrap-spring clutch/torque 316
wrap-spring drive 305

zero-rated spring 342
zinc coating
   hot dip galvanizing 392
   impact plating 392
   life 391
   protection mechanism 391
   sheradizing 392